ECOLOGICAL
ENTOMOLOGY

ECOLOGICAL ENTOMOLOGY

Edited by

CARL B. HUFFAKER

Division of Biological Control
Department of Entomological
Sciences
University of California, Berkeley
and the International Center
for Integrated and Biological
Control University of California,
Berkeley and Riverside

ROBERT L. RABB

Department of Entomology
North Carolina State University
at Raleigh

Sponsored by

The International Center for
Integrated and Biological Control,
The University of California and
North Carolina State University,
Raleigh

A WILEY-INTERSCIENCE
PUBLICATION

JOHN WILEY & SONS

New York   Chichester   Brisbane
Toronto   Singapore

Library of Congress Cataloging in Publication Data:

Main entry under title:

Ecological entomology.

"A Wiley-Interscience publication."
Includes index.
1. Insects—Ecology.   2. Insect control.   I. Huffaker,
Carl B., 1914–   II. Rabb, R. L.   III. International
Center for Integrated and Biological Control.   IV. North
Carolina State University.

QL463.E18   1984   595.7'05   84-2177

ISBN 0-471-06493-9

CONTRIBUTORS

P. L. ADKISSON, Office of Agriculture and Renewable Resources, Texas A&M University, College Station, Texas

HERBERT G. BAKER, Department of Botany, University of California, Berkeley, California

ALAN A. BERRYMAN, Department of Entomology, Washington State University, Pullman, Washington

GUY L. BUSH, Department of Zoology, Michigan State University, East Lansing, Michigan

L. E. CALTAGIRONE, Department of Entomological Sciences, University of California, Berkeley, California

B. A. CROFT, Department of Entomology, Oregon State University, Corvallis, Oregon

R. H. DADD, Department of Entomological Sciences, University of California, Berkeley, California

D. L. DAHLSTEN, Department of Entomological Sciences, University of California, Berkeley, California

DIANE W. DAVIDSON, Department of Biology, University of Utah, Salt Lake City, Utah

G. R. DeFOLIART, Department of Entomology, North Carolina State University, Raleigh, North Carolina

FRANZ ENGELMANN, Department of Biology, University of California, Los Angeles, California

H. T. GORDON, Department of Entomological Sciences, University of California, Berkeley, California

FRED GOULD, Department of Entomology, North Carolina State University, Raleigh, North Carolina

A. P. GUTIERREZ, Department of Entomological Sciences, University of California, Berkeley, California

K. S. HAGEN, Department of Entomological Sciences, University of California, Berkeley, California

M. P. HASSELL,   Department of Pure and Applied Biology, Imperial College, London, England

MARJORIE A. HOY,   Department of Entomological Sciences, University of California, Berkeley, California

C. B. HUFFAKER,   Department of Entomological Sciences, University of California, Berkeley, California

R. D. HUGHES,   Division of Entomology, Commonwealth Scientific & Industrial Research Organization, Canberra, Australia

D. H. JANZEN,   Department of Biology, University of Pennsylvania, Philadelphia, Pennsylvania

ROBERT L. JEANNE,   Department of Entomology, University of Wisconsin, Madison, Wisconsin

R. E. JONES,   Department of Zoology, James Cook University, Townsville, Australia

G. G. KENNEDY,   Department of Entomology, North Carolina State University, Raleigh, North Carolina

PETER G. KEVAN,   Department of Environmental Biology, University of Guelph, Guelph, Ontario, Canada

J. E. LAING,   Department of Environmental Biology, University of Guelph, Guelph, Ontario, Canada

J. H. LAWTON,   Department of Biology, University of York, York, England

R. F. LUCK,   Division of Biological Control, University of California, Riverside, California

S. MASAKI,   Laboratory of Entomology, Hirosaki University, Hirosaki, Japan

EUGENE MUNROE,   Lyman Entomological Museum, Macdonald College, McGill University, Ste. Anne de Bellevue, Canada

PETER W. PRICE,   Department of Biological Sciences, Northern Arizona University, Flagstaff, Arizona

R. L. RABB,   Department of Entomology, North Carolina State University, Raleigh, North Carolina

MARY ANN RANKIN,   Department of Zoology, University of Texas, Austin, Texas

JOHN REESE,   Department of Entomology, Kansas State University, Manhattan, Kansas

G. A. SIMMONS,   Department of Entomology, Michigan State University, East Lansing, Michigan

MICHAEL C. SINGER,   Department of Zoology, University of Texas, Austin, Texas

R. E. STINNER,   Department of Entomology, North Carolina State University, Raleigh, North Carolina

R. W. SUTHERST, Division of Entomology, Commonwealth Scientific & Industrial Research Organization, Long Pocket Laboratories, Indooroopilly, Australia

EDWARD S. SYLVESTER, Department of Entomological Sciences, University of California, Berkeley, California

C. A. TAUBER, Department of Entomology, Cornell University, Ithaca, New York.

M. J. TAUBER, Department of Entomology, Cornell University, Ithaca, New York

JOHN N. THOMPSON, Departments of Botany and Zoology, Washington State University, Pullman, Washington

R. M. TRIMBLE, Research Station, Agriculture Canada, Vineland Station, Ontario, Canada

THOMAS J. WALKER, Department of Entomology and Nematology, University of Florida, Gainesville, Florida

Y. H. WANG, Formerly with the Division of Biological Control, University of California, Berkeley, California (now in private industry)

W. G. WELLINGTON, Institute of Animal Resource Ecology, University of British Columbia, Vancouver, British Columbia, Canada

PREFACE

This book is concerned with the very broad and sometimes ill-defined area of the ecological relationships of insects. We have sought to bring together the known facts and reasonable concepts and theories relating insects to the factors of their environment, both biotic and abiotic, over both the short term (population dynamics) and longer (biogeography) spans of time. Ecology is not really separable into plant and animal ecology, or the latter into vertebrate and invertebrate ecology, because the concept embraces a holistic view of the interrelationships of all the organisms and all their environmental parameters interacting in a given physical setting of adequate dimensions in space and time. Nevertheless, there is much to be learned by looking at insects from an ecological perspective. The entomologist certainly needs to know the ecology of the insects he/she studies in order to understand them and to orient efforts to manage populations of either harmful or beneficial (to humans) species. The general ecologist cannot evaluate the potential significance of the insects present in an ecosystem under study without a rather detailed knowledge of their basic ecology. This book attempts to bring together in one volume and in considerable, if still inadequate, scope and detail the ecological information about insects that is needed in each instance. Its plan of organization is described in Chapter 1. The book is intended as both a reference source for researchers and for use as a text at the advanced student level, by selection of chapters or parts of chapters relevant to the intent of specific courses.

The above explains why the title is *Ecological Entomology* instead of *Insect Ecology*. Although the book emphasizes insects—and most of the research on which it is based was designed to learn how environmental factors affect growth, behavior, survival, reproduction, and thus the population fluctuations and economic impact of many species of both basic and economic interest—we envision a broader goal. That is, to understand the significance of insects within the structure of the many ecosystems of which they are a part: as selective, stabilizing, and in the broad sense symbiotic factors as well as destructive ones. This is a long-range holistic viewpoint that gives equal attention to all factors, including evolutionary processes, instead of focusing solely on temporary immediate dynamics or "solutions" to insect "problems."

Since many kinds of knowledge may in some way be useful, it has been our aim to present this very broad and rapidly developing field in unusual generality, breadth, and detail. No book exists that approaches it in this respect. At its inception in the early 1960s, the senior editor hoped to write it with the help of one or two colleagues, but after roughing in some five or six chapters himself and obtaining no collaboration, it soon became clear that two or three authors could not acquire a mastery of all the relevant research in the many fields of specialization. The work would require many authors, each closely familiar with the literature in one area. We started this in 1977, well aware that it would present some problems of integration and unevenness in style and detail, even though we urged authors to emphasize ecological relevance. Extensive editing and cross-referencing among chapters have been used to reduce these problems. The original intent was to have most chapters of approximately equal length, with concise presentation of principles and a few illustrative examples, and citing many references for the use of readers needing more detail. It proved impossible to achieve this in many instances since some authors were unable to be so concise and others, despite our pleadings, delayed submission of manuscripts for several years. The editors accept responsibility for any demerits, as they claim some credit for the merits, of a pioneering work whose completion proved to be a much more difficult task than they had anticipated.

The book thus has two specific goals: (1) to look at the significance of insects to the structure of the ecosystems of which they are a part and (2) to look at the significance of the other ecosystem components to the insects present there. The second goal has been dealt with much more intensively than the first, a reflection perhaps of the much greater research efforts to understand population fluctuations and economic impact of insects than to their roles in natural community structure and energy flow.

For a book of this sort it is never possible to acknowledge appropriately all the many individuals and institutions that have helped make it possible. We wish, however, to acknowledge first of all the patient and persistent efforts of every author, those who collaborated with them in any way to produce the work, and all the secretarial and technical assistants who have been of such great help. We appreciate the opportunity afforded the editors by the Rockefeller Foundation to meet and work on the book in its planning stage at the Foundation's study center at Bellagio, Italy. We express our deep-felt apology to those many authors who produced their chapters quickly and whose work was in a sense penalized by the great delay consequent upon the completion of the final chapters. To Mrs. Nettie Mackey we acknowledge the carrying of a tremendous coordinating responsibility as secretary and editorial assistant.

CARL B. HUFFAKER
ROBERT L. RABB

Berkeley, California
Raleigh, North Carolina

April 1984

CONTENTS

SECTION III NATURAL CONTROL OF INSECT POPULATIONS

SECTION IV ROLES OR EFFECTS OF INSECTS IN ECOSYSTEMS

SECTION V   APPLICATION OF ECOLOGY TO INSECT POPULATION MANAGEMENT

ECOLOGICAL ENTOMOLOGY

Section I

INTRODUCTION

Our interest in insects stems largely from their numbers and diversity, as well as from the pest status of some species and the beneficial activity of others. Why are there so many insects? What roles do they play in ecological systems? What ecological factors determine their spatial and temporal patterns? How might we respond more appropriately to their presence? Such questions embody much of the rationale for ecological entomology, as discussed in Chapter 1.

In Chapter 2 the ecosystem concept is suggested as an appropriate frame in which to view insects as structural and functional elements of nature and as a useful point of departure in ecological and evolutionary studies. The general notion of nature as a complex of interacting systems is widely accepted, but many questions remain as to how these systems are arranged and to what degree they are interdependent. To what extent are structure and function deterministic or probabilistic? To what degree should we treat elements of the systems as relatively fixed or subject to evolutionary change? In our models how can elements and systems with vastly different spatial and temporal scales of reference be integrated?

We suggest that studies of insects within the ecosystem context may provide more useful answers to such important and challenging questions.

Chapter 1

Meaning of Ecological Entomology— The Ecosystem

C. B. HUFFAKER, H. T. GORDON, and R. L. RABB

1.1 INTRODUCTION—MEANING OF ECOLOGICAL ENTOMOLOGY; THE ECOSYSTEM

This book seeks to explain the roles of insects in the economy of nature, and the interplay between all environmental factors, biotic and abiotic, that may affect these roles. We deal with both the diversity and numbers of insects, as demes, as populations, and as species. This is the scope and meaning of ecological entomology. Two basic goals of the book are

1. to describe the positions of insects in food chains and their functions in the economy of nature and
2. to describe and explain the spatial and temporal patterns of insect populations.

Ecology itself is often defined as the study of the total relationships (interactions implied) of organisms with their environments, at the level of the individual (autoecology) or that of variously constituted groups (synecology—a community ecology). This is a very broad and diffuse scope. Krebs (1972) illustrated diagrammatically the dependence of ecology on four other closely related biological disciplines: genetics, evolution, physiology, and behavior, as shown in Fig. 1.1. No simple diagram can reveal the intricacy of interconnections among these major fields. Genetics explores the storage and usage of information on life processes, from the molecular level to its expression at the cellular and higher organizational levels (where it merges with physiology), its transmission and its modification by mutation, recombination, and selection, and its diversity within demes, species, and higher taxa. Physiology elucidates the mechanisms of cell growth, multiplication and differentiation, and the integration of cells into the specialized tissues and organs of individual living organisms. Behavior is the facet of physiology (or ecophysiology) that emphasizes those processes that directly interact with the external world and allow the individual to evaluate, respond, and adapt to, modify, or even select its environment. Biological evolution and systematics deal with the present state and also the long history of the innumerable expressions of the torrent of genetic information, worldwide, over eons of time. Ecology cannot encompass the infinite detail and complexity of its sister sciences, however much it may be influenced by them. Ecology's special domain has become the analysis of a great variety of *ecosystems:* subdivisions of the global environment containing a restricted number of living species adapted to survive within well-defined abiotic and biotic conditions. Although such partitioning simplifies these "arenas" of interaction, they are still so large and complex that individual organisms tend to become mere units in population counts (often only estimates from samplings). Mathematical descriptions of population numbers and distribution over time and space tend to be

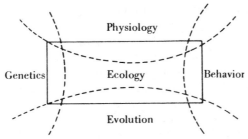

Fig. 1.1   Schematic diagram of relation of ecology to other biological sciences. (From Krebs 1972.)

abstract and probabilistic but sometimes promise sufficient prediction of the future state of the system to be useful in rationalizing major economic decisions.

The concept of an ecosystem in which the organic and inorganic world interact was foreshadowed by Edward Forbes in 1843, wherein "the multiplication of individuals dependent on the rapid reproduction of successive generations of Mollusca [in the Aegean Sea] will of itself change the ground and render it unfit for the continuation of life in that locality until a new layer of sedimentary matter is deposited." A more intricate web of relationships between coexisting organisms and between organisms and their environment—seen as a struggle for existence—was the basis for Wallace's (1876) and Darwin's (1859) concept of natural selection in the origin of species. The idea of order in biotic assemblages evolved further with Semper's (1881) recognition of the role of specificity in predators, the concept of "key-industry" animals, and the principle now known as the pyramid of numbers or biomass (cf. Elton 1927).

The term *ecosystem*, originated by Tansley (1935), came into vogue to emphasize that the biotic world (organisms, populations, communities) and the physical or abiotic world *act together* at a given place to produce the observed phenomena of natural history. Organisms depend entirely upon the physical environment for life-sustaining energy, water, and mineral nutrients, and their distributions are limited by their tolerances of physical conditions, while they eventually return or cycle materials back to the abiotic sphere. Study of the interactions and processes involved and their consequences for the organisms and the environments constitute the discipline of ecology—with the emphasis on the distribution and abundance of the organisms (Andrewartha & Birch 1954, Krebs 1972).

This book deals extensively with ecosystems of various levels of complexity and kind. In Chapter 2 Price elaborates on the concept of ecosystem, its organization, structure, and trophic dynamics.

1.2 ORGANIZATION OF THIS BOOK

Each chapter surveys a different topic from a somewhat independent, specialist point of view. Some overlap was unavoidable. Space limitations account for the brevity of some presentations and for minor omissions or oversimplifications. The bibliography is extensive but could not include all possible relevant primary references (though many of them are retrievable from cited references). The sequence of chapters can be roughly subdivided into sections. Section I, Chapters 1 and 2, is introductory and might be said to be looking at events from the "top-down" (as also in Section III), while in Section II the concentration is at the lower, organismal level of integration, the "bottom-up" approach.

Section II, Chapters 3–5, thus emphasizes largely intrinsic physiological processes involved in growth, development, selection and utilization of foods, and reproduction that are best studied in simple, controlled, artificial environments and are best seen in individual responses. Chapters 6 and 7 move further afield and survey the physiological–behavioral adaptive responses that allow many insects to endure (through dormancy) or escape from (by migration) harsh natural environmental conditions. Chapter 8 introduces the extraordinary diversity of special ecological niches—ways of life—to which various insects have adapted. Genetics, as related to variability within populations, to speciation, and to evolutionary adaptations over wide geographic areas and long eras of geological history is given less detailed treatment in Chapters 9 and 10. However, genetic variation and natural selection are so basic to all of biology that they are implicit everywhere in this book.

The last 15 chapters are primarily ecological. Section III, Chapters 11 to 18, deals with the major theories of population dynamics, with emphasis on the factors that regulate or control the diversity, numbers, and distribution of insect populations in natural environments. Chapter 11 provides a methodology by which one can investigate these questions. Chapter 12 concentrates on the theoretical and mathematical concepts that have been advanced to explain the "natural control" of populations, a subject that has long been controversial. Chapters 13 and 14 elaborate on the heterogeneity of natural environments, particularly as related to weather and food supplies. Chapter 15 deals with interspecific competition and Chapter 16 with the principles of predation. Chapter 17 examines the question of "self-regulation" in populations, while Chapter 18 analyzes the evidence for population regulation in social insects.

Section IV, Chapters 19–22, deals with some special ecological relationships of insects that are so significant that they have been extensively studied [that is, trophic levels (Chapter 19), pollination (Chapter 20), pathogen transmission (Chapter 21), and the impact of phytophagous insects on plant populations (Chapter 22)] but could not be adequately presented in other chapters.

Section V consists of the final three chapters (23–25), which deal with applications of ecological principles in the management of insect pests by humans. In the last few millennia human activity (which, in retrospect, can be seen as imperfectly planned) has profoundly altered our planet. Hundreds of insect species have successfully adapted to these changes and have become serious enemies of man; the need to redress the balance has stimulated the growth of many sciences, including entomology and ecology. The next section of this chapter will explore some examples of this theme.

A brief introduction to Sections I, II, III, IV, and V precedes each section.

The multiauthor nature of this book has caused it to be more wide-ranging and heterogeneous than earlier texts on the ecology of insects, and certain topics recur in several chapters with, at times, somewhat different points of view. Many prior works have been drawn on by the authors, and some major ones deserve to be acknowledged here: in general ecology those of Clements and Shelford (1939), Odum (1959, 1971), Collier et al. (1973), Krebs (1972), and Ricklefs (1973); in animal ecology those of Elton (1927, 1966), Hesse et al. (1937), Allee et al. (1949), Andrewartha & Birch (1954), Lack (1954), Slobodkin (1961), MacFadyen (1963), and Dempster (1975); in insect ecology those of Chapman (1931), Bodenheimer (1938, 1958), Clark et al. (1967), Varley et al. (1974), and Price (1975). The emphasis in Clements and Shelford on the inseparability of plant and animal communities has now been adopted by most ecologists. Noteworthy also are the intriguing works of Nicholson on mimicry (1927) and animal population regulation through intraspecific competition (1933, 1954), of Uvarov (1931) on climatic influences on insects, of Smith (1929, 1935) and Thompson (1929, 1939) on effects of insect parasitoids on their host populations, of Pearl on logistic growth of populations (Pearl & Reed 1920), and of Lotka (1925), Volterra (1926, 1927), Gause (1934), and Kostitzin (1937) on predator–prey interactions (cf. Scudo & Ziegler 1978), as well as the major theoretical contributions of G. Evelyn Hutchinson (cf. Slobodkin 1961, acknowledgments), Robert MacArthur (cf. Cody et al. 1975), and many others.

1.3 HUMAN INFLUENCES AND SOME ECOLOGICAL INTERACTIONS INVOLVING INSECTS

The importance of understanding ecological interactions was discussed briefly above, and examples where such knowledge has been used in pest control are given in Chapters 23–25. However, in this introductory chapter we should like to give added emphasis to the importance of a broad appreciation of the diverse roles of arthropods as a basis for developing and maintaining appropriate relations with them. Here we discuss a few field and laboratory studies illustrating some subtle and unexpected conse-

quences that can ensue from an inadequate data base or erroneous interpretation of population events. The need for simplifying methods of analysis seems obvious in confronting ecological complexity; however, our current
examples also will stress the limitations of such methods (cf. Chapter 11).

1.3.1 The Clear Lake Gnat, Chemicals, and Predatory Fish in California

The Clear Lake gnat in California historically was a major problem in that
resort area because the lake produced so many adult midges as to be a
great nuisance. In 1949 TDE, one of the "miracle" insecticides, was applied
to the lake water at a low dosage (1:70,000,000). This largely eliminated
the problem until about 1953 when the midges began to increase again.
In 1954 a second introduction of this persistent chemical was made at an
increased dosage (1:50,000,000), and while this treatment seemed successful, its effects were less prolonged than the first, and a third treatment
followed in 1957 which was a failure. The insects were becoming resistant
to the chemical, a phenomenon now known to have occurred in several
hundred well-established examples (e.g., Georghiou 1972, Plapp 1976). An
unexpected side effect was that the chemical was found to be killing some
water fowl inhabiting the lake, particularly grebes. Tests of dead bird tissues
showed very high concentrations of chlorinated hydrocarbons, assumed to
be TDE (Prine et al. 1975). Later it was shown that lake fish had also
concentrated chlorinated hydrocarbons (DDT run-off into the lake from
the pear treatments, as well as the TDE) to the point of being unsafe to
eat. Thus, a main attraction of the resort (sport fishing) was threatened.
Gnat populations were exploding despite the TDE treatments and TDE
was not used after 1958 (Hunt and Bischoff 1960). But the mind-set on
chemical controls did not end there, and from 1962 to 1975 Clear Lake
was again treated chemically, this time with the less persistent methylpar-
athion, also less toxic to fish (Hazeltine 1963).

Meanwhile, Cook (1968), using an ecological systems approach to the
problem, was investigating the potential of various fish species as gnat
predators. After 7 years of study into the complexities of food-web relationships, Cook introduced the Mississippi silverside, a small fish from the
midwest, into Clear Lake in 1967 for gnat control. With the cessation of
chemical treatments in 1975, which again were proving less effective than
initially, the impact of the silversides on gnat populations and on chironomid midges became apparent, although controversy persists regarding
the specific role of the fish (Prine et al. 1975). According to Cook (1981 &
pers. comm.) the Clear Lake gnat now appears to be constrained well within
human tolerance limits through this example of biological control (through
its direct predation and indirect negative effects on the food resources of
midge larvae).

1.3.2   The Cyclamen Mite and Predatory Mites in Strawberries

Studies by Huffaker and Kennett (e.g., 1956) illustrate how fundamental interactions of various natural factors, acting jointly or in sequence, account for variations in the populations of the cyclamen mite (*Steneotarsonemus pallidus*) and its impact on strawberries in California. At that time strawberries were grown as a 3- or 4-yr crop, on which cyclamen mites became a plague during the second year unless controlled by chemicals, which were improperly and excessively used in most instances. In contrast, these phytophagous mites seldom damaged older plantings that had not had acaricide treatment and where predatory mites (*Typhlodromus* spp.) were quite active. By introducing both pest and predatory mites into new plantings, damaging mite populations could be forestalled.

The basic interactions were best illustrated through the use of a greenhouse ecosystem (Fig. 1.2). The physicochemical environment sets the stage for strawberry growth in that the plants require much water and nutrients, and having such, they grow profusely, providing a favorable physical environment and food for cyclamen mites, which are extremely sensitive to dry, hot air. In the absence of predatory mites (Fig. 1.2A), heavy cyclamen mite densities develop, and their feeding then causes delayed but extreme

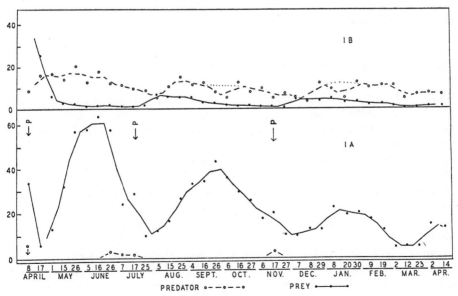

Fig. 1.2   Changes in density of *Tarsonemus pallidus* (= prey) in predator-present and predator-inhibited (free) plots and *Typhlodromus* (= predator) frequencies. One plot (1B) with full predator presence; the other (1A) with the predators largely removed with parathion (P) treatments. See Huffaker and Kennett (1956) for details and further data.

losses of plant nutrients and a distorted, open, plant-growth form with sparse foliage. Consequent to this cyclamen mite-induced deterioration in its own habitat (desiccating stresses of the physical environment and lack of nutrients), a crash in the mite population then occurs. However, when the mite populations have been at low levels for a time, the plants recover and regain a profuse condition of growth. This is followed by a second increase in mites. The second mite peak, however, is not as high as the first because of the residual damage to the plants by the first high-density peak. This sequence of interactions between plants and the phytophagous mites in the absence of predators was repeated with declining amplitude until the plants died or reached a state with no potential for plant regeneration through runners.

In the presence of *Typhlodromus* (Fig. 1.2B), however, the strawberry plants remained vigorous and set many runners because the predators regulated the cyclamen mites at a very low density. Thus, the predatory mites prevented the engagement of high-density, intraspecific regulation that would have been imposed on the cyclamen mite from its self-induced food and habitat limitations.

Though proven effective, inoculating strawberry plantings with pest and predatory mites was adopted by very few growers, perhaps because of a mind-set towards chemicals. Nevertheless, this system well illustrates how both density-related and density-independent factors may interact to produce natural control (cf. Chapter 12).

1.4 WHY ARE INSECTS SO SUCCESSFUL?

There are various views on this question. Our own emphasizes body-plan. Of the 1,067,000 known species of animals, nearly 72% are insects (Daly et al. 1978). Although all hexapod apterygotes were formerly included in the Insecta, studies of their morphology (Manton 1977) indicate that hexapody evolved independently in the Collembola, Diplura, Protura, and Thysanura by loss of all but the three anterior pairs of legs from similar myriapod ancestors. These wingless forms comprise less than 0.5% of all hexapods. Within the Insecta (Pterygota) 12–14% of the species are exopterygotes (cf. Chapter 3), more than half of them Hemiptera, and 84–88% are endopterygotes. The largest and possibly oldest endopterygote order (Coleoptera) radiated extensively, its 280,000 species occupying a great variety of niches. The other major endopterygote branches diversified into many orders, each primarily adapted to more restricted niches (e.g., most Lepidoptera are leaf-feeders, most Hymenoptera are predators or parasites of other insects).

The evolution of the Insecta is only sketchily known. Fossils of many exopterygote orders occur in Carboniferous strata some 300 million years old. It is probable that these had evolved from apterygote ancestors (akin to the Thysanura) existing in the Devonian, more than 350 million years ago, but no indisputable fossils have yet been discovered. Acquisition of thoracic wings enabled fast, directed dispersal of reproductive adults over a large area, certainly advantageous in competition with less mobile rivals. This advance was facilitated by the earlier concentration of locomotor muscles in the thorax, driving three pairs of legs, with a pleuron-coxa articulation permitting greater stability and power than that developed in other hexapod groups (where design favors horizontal rather than vertical-climbing locomotion).

Manton's anatomical work suggests a hypothetical evolutionary sequence from primitive myriapods adapted to life on land to hexapods adapted for climbing, then also to gliding, and finally to powered flight. This would require synchronous specialization of the anterior and posterior body regions, already initiated by hexapody. The head is equipped with a great variety of tactile, chemosensory, and optical organs. The mouthparts have, in various species, become modified for efficient consumption of diverse foods. The brain, though minute, has a complex repertoire of instinctive behavior (Milne & Milne 1978). Trophic-metabolic and reproductive systems are concentrated in the abdominal region. The persistence of this efficient body-plan during eons of time suggests that it is an optimal one, although innumerable variants of it have evolved in adaptation to special niches.

The most spectacular and successful of these adaptations was the drastic modification of the larval body-plan in endopterygotes, perhaps initiated as a reduction and eventual loss of the long jointed legs to facilitate burrowing in confined spaces (soil or the interior of living or dead plants and animals). Involved also were other adaptive structural changes that in time converted the larval form into an efficient specialized feeding machine utterly different from the adult, so that an intermediate pupal stage was required for the *de novo* construction of the adult form (usually though not always alate). Since food reserves of the pupa allow survival for long periods when the environment is unfavorable, the adults can emerge, disperse, and reproduce when conditions favorable for the next generation have been restored.

Insofar as taxonomic diversity is a criterion of success in the insects, it is clearly related to the exploitation of nearly every possible food source, usually with a high degree of host specificity (Southwood 1978). Most of the terrestrial plant species are primary hosts for at least one exopterygote (usually hemipteran) and for several endopterygotes (of various orders, often at least one coleopteran) that specifically feed on their roots, stems, leaves, phloem, flowers, seeds, or fruits. Then, too, as Huffaker (1974) observed, every insect species that evolves to exploit a new niche becomes a resource for parasite or predator species that may coevolve to exploit *it*. The more successful and abundant the phytophagous species becomes, the greater the opportunity for exploitation of its eggs, larvae, or pupae by specific parasitoids. About 25% of all insect species are predators or parasites, mostly on other insects, and the great majority of these are Hymenoptera. Insects are also the primary resource for most arachnids, many kinds of microorganisms, certain vertebrates, and even a few insectivorous plants (cf. Chapter 19).

The multiplicity of predatory and parasitoid attacks on the insect fauna exerts selective forces that increase its diversity even beyond that compelled by competition for special niches, since there may be many distinct ways of avoiding or resisting any attack. Similarly, adverse selection pressure by phytophagous insects (Chapter 22) and the beneficial coselection pressure from insect pollinator specialization (Chapter 20) may have played a major role in the rapid diversification of the angiosperm flora and the pollinator fauna since the Cretaceous. These flowering plants, mostly of relatively small size, comprise 89% of the extant species of higher land plants. Families or genera have evolved diverse chemical and other defenses that discourage attack by most insects and are overcome only by relatively few insect species that coevolved countermechanisms (such as chemical resistance or symbiotic microorganisms) to a particular array of defense mechanisms—thus coevolved specialists arose. The mutualistic interactions, as in the replacement of wind pollination by insect pollination (Chapter 20), may have been as important in both plant and insect diversification as the antagonistic ones [not only direct attack (Chapter 22) but transmission of pathogens (Chapter 21)].

An often-stated reason for the evolutionary success of the Insecta is their relatively small body size, which allows them to exploit many small special niches and to have short life cycles, high birth rates, and large rapidly increasing populations even in restricted ecosystems. Immense numbers of individuals allow great genetic diversity and fast adaptive response (extending to speciation) to changing environment. However, there is a 100,000-fold range of adult body size in the insects (from tens of micrograms to many grams); the duration of the life cycle is more a function of synchronization with external environmental cycles than of body size, and fecundity (not always high) leads to explosive population growth only when resources are ample and mortality factors (cf. Section 1.5) are inactivated. Any species (including man) can "explode" if such conditions persist. Body size does not differentiate insects from the apterygotes, myriapods, or arachnids—of no less ancient terrestrial lineage but more restricted radiation, much of the restriction perhaps imposed by rapid preemption of most new niches by the pterygote radiation. Nevertheless, pterygotes have not yet driven wingless forms out of the many niches to which they were perfectly adapted. The many traits to which we attribute insect success have not made them irresistible. Nor has large body size notably hindered evolutionary advance in the vertebrate line. In the words of Paul Errington, "nature's way is any way that works," and many ways do.

1.5 *k*-FACTOR ANALYSES OF POPULATION CHANGE

Populations tend to fluctuate with time in localities where they are well established. The extent of these fluctuations varies in different years and different places and with different species. Dempster (1975) noted that the aim of population studies is to identify and measure the factors that limit such temporal fluctuations and that restrict the spatial distribution by leading to extinction in some localities. Long-term age-specific life table data, gathered from a locality in which the impact of various factors that affect mortality or natality (Chapter 11) can be assessed, may be analyzed by the *k*-factor method expanded by Varley and Gradwell (1963). Population loss during each of a sequence of time intervals ($i = 0, 1, 2, \ldots$) is measured as k_i, the difference between the logs of the initial number and the surviving number. The total population loss, K, during the entire life or annual cycle is the sum of the component k values. A component k with a high positive correlation to the total K during many cycles identifies a "key factor" time period, during which a fluctuation-controlling loss recurs. If a comparable sequence of biotic or abiotic factors that independently cause population loss can be constructed, another set of k values can be calculated within which key factors may be detectable. If one is found, the correlation between its k value and the corresponding log population density can be calculated to test density dependence; if density dependence is nonlinear, more complex regression analysis may be required. The method has the merit of being empirical and inflexibly rooted in the real world; it is an opposite pole to theoretical ecology that can add or adjust constants or even modify fundamental equations to fit any data.

No *k*-factor analysis is "typical." Figure 1.3 is therefore only one example,

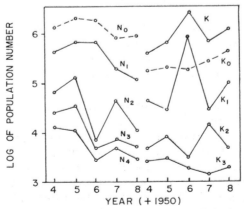

Fig. 1.3 A *k*-factor analysis of the population fluctuations of the broom beetle (*Phytodecta olivacae*). (Data from Richards & Waloff 1961.)

based on five consecutive annual studies of populations of the broom beetle, *Phytodecta olivacea*, in an 0.8-ha field in England (data of Richards & Waloff 1961). Each point in the left half of the figure is the log of a numerical estimate of one of five stages during the years 1954–1958. Points for the same stage are joined by lines to make recognition easier. The values for N_0, the initial population of eggs, are quite unreliable since they are calculated by multiplying the number of adult female survivors from the preceding winter by 250, the average fecundity in laboratory studies. Values of N_1 (eggs + larvae), N_2 (pupae), N_3 (emerged adults in the late autumn), and N_4 (adult survivors in early spring) are derived from sampling. Each point on the right half of the figure represents a difference between pairs of values in the left half, but only the fluctuations are correct since each series is offset to maintain the sequence. The k line therefore shows the total population loss in each annual cycle, k_0 the natality loss and k_1, k_2, and k_3 the mortality losses during the egg and larval stage, the pupal stage, and the adult overwintering stage. Losses in each stage are attributable to different biotic and abiotic factors.

It is important to distinguish between a "key factor" (highly correlated with total K) and a regulating factor, key or contributory, which is one that exerts a stabilizing influence (cf. Varley et al. 1974, pp. 120–125). Moreover, a regulating or stabilizing influence can derive from a simple direct density-dependent action or as a delayed, density-dependent response, of a parasitoid, for example, where the parasitoid population's efficiency is limited by its own density (mutual interference) and the host and parasitoid populations are reciprocally linked generation to generation (Hassell 1978, cf. Chapter 16). If a plot of a k value against the population density on which it acts suggests direct density dependence, this must be validated by a two-way regression since log N and the k-value calculation are not independent measurements (Varley et al. 1974). If the b_{xy} and b_{yx} lines fall on the same side of the line $b = 1$, the suggestion is said to be validated, but if otherwise not so (cf. Huffaker & Stinner 1971). Delayed density dependence is tested by plotting the k values of the suspected factor against the N values on which it acts, with the points joined in a time series. A spiralling form suggests a delayed density-dependent component, a stabilizing one if it spirals inward.

There is no solid key factor for change in Fig. 1.3, although k_1 alone dominates K at the single high peak in 1956 despite low values for k_0, k_2, and k_3. The form of k_2, pupal stage mortality seemingly caused by carabid beetles and other soil predators, is such as to suggest compensation relative to the action of k_1 since in every case the trend lines are inverse; yet Dempster found no statistical significance in this inverseness. Also, the pattern of k_2 alone is like that of its N value. Yet, Dempster did not find k_2 to be density-dependent.

Regarding k_3, Dempster tested the action of the fungus *Beauveria bassiana* on overwintering adult *Phytodecta* for density dependence. The test showed density dependence, but Dempster concluded that "any stablizing action of *Beauveria* . . . is likely to be lost [masked?], however, because of disturbing action of other factors making up k_3, especially *Periletus.*"

The k_2 mortality was greatest in 1957, although pupal density was then relatively low, so that simple tests of density dependence fail. However, mortality by predation is not a function of prey density alone, but also of the prey–predator ratio. This is complicated by searching efficiency, the presence of alternative prey, and mortality factors acting on the predator populations. One cannot conclude that a mortality factor is density-independent, nor that it is failing to contribute to a density-dependent complex of factors merely because the population loss it causes is not statistically correlated with the initial density. The action of any one factor is influenced by many other factors that previously or concurrently alter the structure of the ecosystem (cf. Chapter 12).

This special topic was included here because no other chapter deals with it and we think the methodology and ideas represented should be encouraged in the next generation of ecologists. A great many factors within an ecosystem exert some influence on each of its component populations. Some influences are compensated by other factors and some are masked by others. Yet we believe that improved methods of analysis may show that control is often effected by a few dominant factors interacting in diverse ways. Elucidating these will be a major advance in the science of ecology.

REFERENCES

Allee, W. C., A. E. Emerson, O. Park, T. Park, and K. P. Schmidt. 1949. *Principles of Animal Ecology.* W. B. Saunders, Philadelphia.

Andrewartha, H. G. and L. C. Birch. 1954. *The Distribution and Abundance of Animals.* University of Chicago Press, Chicago.

Bodenheimer, F. S. 1938. *Problems of Animal Ecology.* Oxford University Press, Oxford.

Bodenheimer, F. S. 1958. *Animal Ecology Today.* W. Junk, The Hague.

Chapman, R. N. 1931. *Animal Ecology with Special Reference to Insects.* McGraw-Hill, New York.

Clark, L. R., P. W. Geier, R. D. Hughes, and R. F. Morris. 1967. *The Ecology of Insect Populations in Theory and Practice.* Methuen, London.

Clements, F. E. and V. E. Shelford. 1939. *Bio-ecology.* Wiley, New York.

Cody, M. L. and J. M. Diamond (Eds.). 1975. *Ecology and Evolution of Communities.* Belknap Press and Harvard University Press, Cambridge, Mass.

Collier, B. D., G. W. Cox, A. W. Johnson, and P. C. Miller. 1973. *Dynamic Ecology.* Prentice-Hall, Englewood Cliffs, New Jersey.

Cook, S. F. Jr. 1968. *Calif. Vector Views* **15:** 63–70.

Cook, S. F. Jr. 1981. *Environment* **23:** 25–30.

Daly, H. V., J. T. Doyen, and P. R. Ehrlich. 1978. *Introduction to Insect Biology and Diversity.* McGraw-Hill, New York.

Darwin, C. 1859. *The Origin of Species by Means of Natural Selection or the Preservation of Favored Races in the Struggle for Life.* Murray, London.

Dempster, J. P. 1975. *Animal Population Ecology*. Academic, London.

Elton, C. S. 1927. *Animal Ecology*. Macmillan, New York.

Elton, C. S. 1966. *The Pattern of Animal Communities*. Methuen, London.

Forbes, E. 1843. *Rept. Br. Assoc. Adv. Sci.* **13:** 130–193.

Gause, G. F. 1934. *The Struggle for Existence*. Williams & Wilkins, Baltimore.

Georghiou, G. P. 1972. *Annu. Rev. Ecol. & Syst.* **3:** 133–168.

Hassell, M. P. 1978. *The Dynamics of Arthropod Predator–Prey Systems*. Princeton University Press, Princeton, N.J.

Hazeltine, W. E. 1963. *J. Econ. Entomol.* **56:** 621–626.

Hesse, R., W. C. Alee, and K. P. Schmidt. 1937. *Ecological Animal Geography*. Wiley, New York.

Hunt, E. G., and A. I. Bischoff. 1960. *Calif. Fish and Game* **46:** 91–106.

Huffaker, C. B. 1974. *Environ. Entomol.* **3:** 1–9.

Huffaker, C. B. and C. E. Kennett. 1956. *Hilgardia* **26:** 191–222.

Huffaker, C. B. and R. E. Stinner. 1971. Pp. 333–350 in Hidaka, Asahina, Gressitt, Nishida, and Nomura (Eds.), *Entomological Essays to Commemorate the Retirement of Professor K. Yasumatsu*. Kokuryukan Publishing, Tokyo.

Kostitzin, V. A. 1937. *Biologie Mathematique*. Librarie Armand Colin, Paris.

Krebs, C. J. 1972. *Ecology. The Experimental Analysis of Distribution and Abundance*. Harper & Row, New York.

Lack, D. 1954. *The Natural Regulation of Animal Numbers*. Clarendon, Oxford.

Lotka, A. J. 1925. *Elements of Physical Biology*. Williams & Wilkins, Baltimore.

MacFadyen, A. 1963. *Animal Ecology*. Pitman, London.

Manton, S. M. 1977. *The Arthropoda: Habits, Functional Morphology and Evolution*. Clarendon, Oxford.

Milne, L. J. and M. Milne. 1978. *Insect Worlds*. Scribner's, New York.

Nicholson, A. J. 1927. *Australian Zool.* **5:** 1–104.

Nicholson, A. J. 1933. *J. Anim. Ecol.* (Suppl.) **2**(1): 132–178.

Nicholson, A. J. 1954. *Australian J. Zool.* **2:** 9–65.

Odum, E. P. 1959. *Fundamentals of Ecology*. W. B. Saunders, Philadelphia.

Odum, E. P. 1971. *Fundamentals of Ecology* (3rd ed.). Saunders, Philadelphia.

Pearl, R. and L. J. Reed. 1920. *Proc. Natl. Acad. Sci. USA* **6:** 275–288.

Plapp, F. W. 1976. *Annu. Rev. Entomol.* **21:** 179–197.

Price, P. W. 1975. *Insect Ecology*. Wiley, New York.

Prine, J. E., G. G. Lawley, and P. B. Moyle. 1975. *Bull Soc. Vector Ecologists* **2:** 21?–31.

Richards, O. W. and N. Waloff. 1961. *Phil. Trans. Roy. Soc. Lond. (B)* **244:** 205–257.

Ricklefs, R. E. 1973. *Ecology*. Chiron, New York.

Scudo, F. M. and J. R. Ziegler. 1978. In *The Golden Age in Theoretical Ecology: 1923–1940. Lecture Notes in Biomathematics*, S. Levin (Ed.). Springer-Verlag, Berlin.

Semper, K. G. 1881. *Animal Life as Affected by the Natural Conditions of Existence*. Appleton, New York.

Slobodkin, L. B. 1961. *Growth and Regulation of Animal Populations*. Holt, Rinehart & Winston, New York.

Smith, H. S. 1929. *Bull. Entomol. Res.* **20:** 141–149.

Smith, H. S. 1935. *J. Econ. Entomol.* **28:** 873–879.

Southwood, T. R. E. 1978. Pp. 19–40 in C. A. Mound and N. Waloff (Eds.), *Diversity of Insect Faunas*, Symposium of the Royal Society of London. Blackwell, London.

Tansley, A. G. 1935. *Ecology* **16:** 284–307.

Thompson, W. R. 1929. *Parasitology* **21:** 269–281.

Thompson, W. R. 1939. *Parasitology* **31:** 299–338.

Uvarov, B. P. 1931. *Trans. Entomol. Soc. Lond.* **79:** 1–247.

Varley, G. C. and G. R. Gradwell. 1963. *Proc. Ceylon Assoc. Advan. Sci.* **18:** 142–156.

Varley, G. C., G. R. Gradwell, and M. P. Hassell. 1974. *Insect Population Ecology.* University of California Press, Berkeley.

Volterra, V. 1926. *Mem. Accad. Lincei* **2:** 31–113.

Volterra, V. 1927. *R. Comitato Talassografico Italiano, Memoria* **131:** 1–142.

Wallace, A. R. 1876. *The Geographical Distribution of Animals.* Macmillan, New York.

Chapter 2

The Concept of the Ecosystem

PETER W. PRICE

The ecosystem concept is one of the oldest and best known concepts in ecology. Stephen Forbes had a clear picture of it in 1887, when in his celebrated paper, "The Lake as a Microcosm," he wrote that whatever affects any species in the lake "must have an influence of some sort upon the whole assemblage," emphasizing "the necessity for taking a comprehensive survey of the whole as a condition to a satisfactory understanding of any part." (Forbes 1887). Much later the phytosociologist Arthur Tansley (1935, 1939) coined the term *ecosystem* for this "whole system" that included the organisms in their physical environment. He stressed that such systems are the basic units of nature because it is impossible to study realistically any natural phenomenon in isolation. "There is constant interchange of the most various kinds within each system, not only between the organisms but between the organic and the inorganic" (Tansley 1935). The term *ecosystem* was soon adopted by ecologists, and the concept stimulated a major thrust in ecological research toward the understanding of movement of materials in whole systems, an early influential study being Lindeman's (1942) paper, "The Trophic-Dynamic Aspect of Ecology." The first major text on general ecology by Odum (1953) gave the concept a prominent place, and the hope expressed by Evans (1956) that the term "ecosystem" would be adopted universally has been satisfied.

Thus, the ecosystem concept has remained intact for almost a century. Lindeman (1942) defined the ecosystem as "the biotic community *plus* its abiotic environment," and Odum (1953) and Evans (1956) continued to stress the dynamic state in ecological systems.

> In its fundamental aspects, an ecosystem involves the circulation, transformation, and accumulation of energy and matter through the medium of living things and their activities. Photosynthesis, decomposition, herbivory, predation, parasitism and other symbiotic activities are among the principal biological processes responsible for the transport and storage of materials and energy, and the interaction of the organisms engaged in these activities provide the pathways of distribution. The food-chain is an example of such a pathway. In the nonliving part of the ecosystem, circulation of energy and matter is completed by such physical processes as evaporation and precipitation, erosion and deposition. The ecologist, then, is primarily concerned with the quantities of matter and energy that pass through a given ecosystem and with the rates at which they do so. Of almost equal importance, however, are the kinds of organisms that are present in any particular ecosystem and the roles that they occupy in its structure and organization. Thus, both the quantitative and qualitative aspects need to be considered in the description and comparison of ecosystems (Evans 1956).

2.1 COMPONENTS

Although most authors stress the basic nature of the ecosystem as a natural unit, it contains many components that may be viewed as a hierarchy of increasing complexity. *Individual organisms* exist in *populations* defined by

Mayr (1963) as a group of potentially interbreeding individuals at a given locality. Populations with members that actually or potentially interbreed with individuals of other populations are grouped into *species*. Such species are more or less discrete from others because they are reproductively isolated. Species may be classed into *guilds* of species that exploit the same resource in a similar manner (Root 1967, e.g., Fig. 2.1, this volume). Several

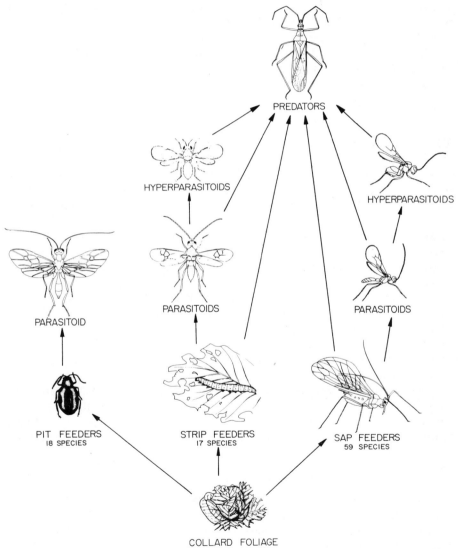

Fig. 2.1   The food web on collards. Note that three distinct guilds are involved in this component community: the pit feeders, strip feeders, and sap feeders (after Root 1973). Most of the herbivores are specialized to feed on members of the Cruciferae because plants in this family contain toxic mustard oils.

guilds may form a *component community*, "an assemblage of species associated with some microenvironment or resource" such as tree holes, leaf litter, or a particular plant taxon as in Fig. 2.1 (Root 1973). Component communities existing in the same area, such as a patch of vegetation or a pond, "merge and interact to varying degrees" to form *compound communities* (Root 1973). It is these compound communities of organisms and their physical environment that form ecosystems.

Ecosystems are also a part of grander organizational entities and cannot be regarded in isolation (as illustrated later). Ecosystems of a similar vegetation type are classed into *biomes* such as tropical rain forest, desert, tundra, and steppe, and ultimately all biomes form the biosphere composed of all parts of the earth in which life exists. Eventually, perhaps, the biocosmos may have to be recognized, as Wald (1954) and Sagan (1977) are persuasive that life must exist in other planetary systems.

2.2   INTERACTIONS

The core of the ecosystem concept is that components of any system interact. Indeed, our ecological language classifies organisms according to the mode by which they relate to other parts of the system. *Autotrophs* or *primary producers* are self-nourishing. They fix light energy and utilize inorganic chemicals to form complex organic molecules on which all life depends. Green plants are the major autotrophs in most ecosystems. *Heterotrophs* or *consumers* utilize autotrophic organisms as food, either directly or by feeding on other heterotrophs. *Decomposers* or *saprophages* are heterotrophic but feed on dead organic matter, playing an important part in the recycling of nutrients, a major process in ecosystems, as described later. Heterotrophic organisms are subdivided according to what they consume and how they consume it. *Herbivores* eat plants; *carnivores* eat animals. *Predators* first kill and then eat their food. *Parasites* sap energy and nutrients from a living source of food and reside in or on their food. *Grazers* and *browsers* feed on plants by plucking parts. Many organisms live in close association with others, forming mutualistic relations in which exchange of nutrients and/or energy is more or less essential to each (e.g., mycorrhizal fungi and plants, microbes in the digestive tract or ruminants and herbivorous insects, and pollinators and plants in which the plant offers energy-rich nectar in exchange for the transport of pollen from one flower to another).

Such extensive linkage between organisms has a profound impact on the organization in ecosystems through what Elton (1927) called *food chains*, or linear-feeding associations, and *food cycles*, which we now call *food webs*, formed of interlocking food chains (as in Figs. 2.2 and 2.3). Not only are there direct feeding links, but frequently communities are held together by chemical links related to discovery of food and mates (Fig. 2.4). As Evans

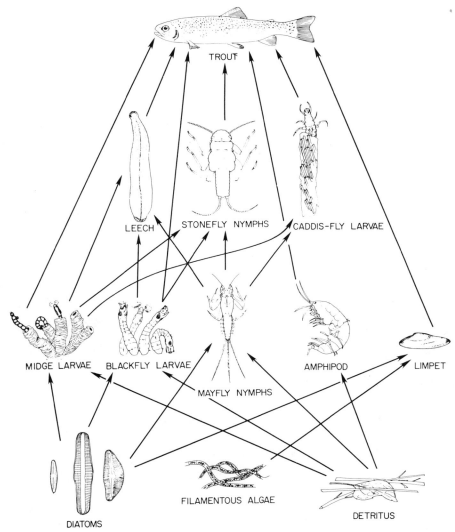

TROUT

LEECH   STONEFLY NYMPHS   CADDIS-FLY LARVAE

MIDGE LARVAE   BLACKFLY LARVAE   AMPHIPOD   LIMPET

MAYFLY NYMPHS

FILAMENTOUS ALGAE

DIATOMS

DETRITUS

Fig. 2.2 An example of some of the major links in a food web in an aquatic community. Note the importance of insects in this stream ecosystem (after Badcock 1949).

(1956) pointed out in the quotation above, these feeding relationships greatly influence the pathways along which energy and matter pass, the rates of passage, and ultimate destinations.

There exist important interactions between ecosystems, and it is now clear that the whole biosphere is linked together in many ways. Products of man, such as pesticides and polychlorinated biphenyls, occur around the globe. The burning of fossil fuels has a global impact by increasing

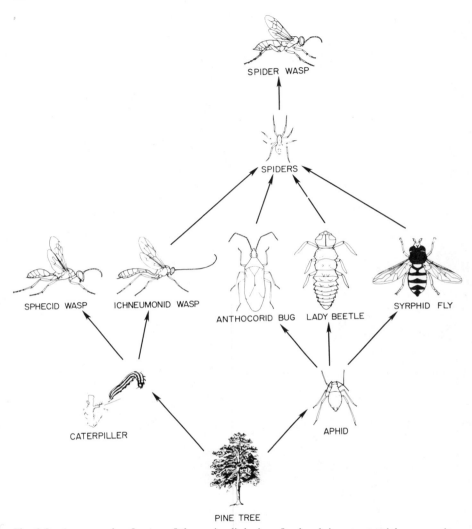

Fig. 2.3   An example of some of the major links in a food web in a terrestrial community.
Note again the importance of insects in this woodland ecosystem (after Richards 1926).

concentrations of carbon dioxide in the atmosphere (Broecker et al. 1979,
Williams 1979). Urban ecosystems impact on forest ecosystems through
development of acid rain which reduces primary productivity of forests
(Bormann & Likens 1979, Likens 1976).

2.3   CONSTRAINTS

A global view of ecosystems and biomes such as that developed for vegetation by Holdridge (Fig. 2.5) illustrates clearly that every species is con-

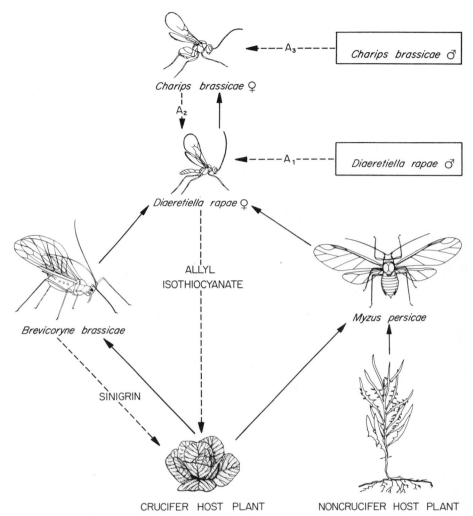

Fig. 2.4   Linkage in part of the food web based on collards in which chemical cues play a vital role in the discovery of food. Broken lines indicate chemical attraction and solid lines indicate passage of food through the food web. A_1 and A_3 are sex pheromones of *D. rapae* and *C. brassicae* and A_2 is a chemical attractant in the primary parasite used by the hyperparasite to locate its host (after Read et al. 1970).

strained in its distribution to a relatively small part of the globe. This is seen on a local scale by the zonation of vegetation on an altitudinal gradient (Fig. 2.6). Merriam (1894, 1898) called these bands of vegetation life zones and formulated laws on how they were determined by temperature, but today it is clear that not just temperature but also precipitation and the interaction between them (evapotranspiration) are principal determinants of plant species distribution, as in the Holdridge system.

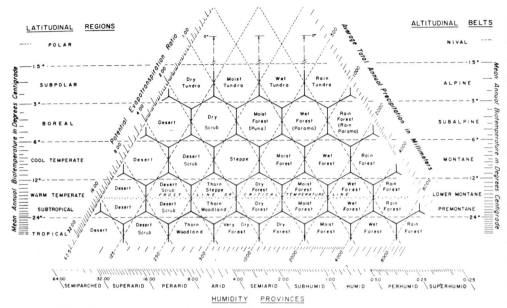

Fig. 2.5 Life zones on a global scale. By using temperature, rainfall, and potential evapo-transpiration, each zone is delimited in Holdridge's system (from Holdridge et al. 1971). (With permission from L. R. Holdridge and Pergamon Press.)

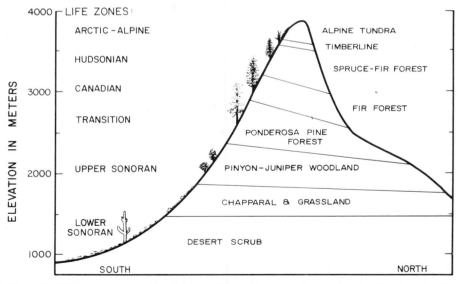

Fig. 2.6 Life zones on an altitudinal gradient with vegetation as seen by Merriam on the San Francisco Peaks in Northern Arizona and south to the Sonoran Desert (see Lowe 1964 for details).

26

In general, every species is limited in its distribution by physical factors, food availability, presence or absence of nesting sites, and so on. Each essential factor can be arranged as a gradient, or ordered, and the population distribution on each axis can be plotted. Distribution on one gradient may frequently be normal or bell shaped with an optimal location, marginal sites, and uninhabitable sites (Fig. 2.7). By plotting distribution on another gradient at right angles to the first, and the population size along this gradient, the area of suitable conditions for the species can be defined (Fig. 2.7). A third dimension would yield a volume of habitable space. Eventually, enough essential resources could be defined to describe the *ecological niche* of the species, which delimits the "space" in which a species can reproduce indefinitely (for other definitions see Elton 1927, Macfadyen 1957, Hutchinson 1957, Whittaker et al. 1973, and a compendium by Whittaker & Levin 1975).

All places in which a species can reproduce indefinitely may not be available because resources have been preempted by another species (competition becomes important) or impact of enemies (predators, parasites) may be too severe. Thus, biotic constraints on distribution in ecosystems may be as important as abiotic factors. We must, therefore, differentiate between the ecological niche that could be utilized in the absence of competition and other enemies (the *fundamental niche* or *potential niche* of a species) from the space really utilized by a species (the *realized* or *actual niche*).

Thus, species with similar requirements living in the same community frequently show significant differences in morphological characters which

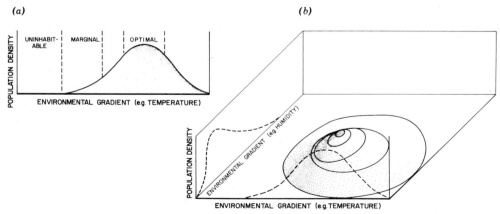

Fig. 2.7 The ecological niche of a species defined on (a) one gradient and the population response on this gradient and (b) two gradients and the population response on each. The gradients might be temperature and moisture or any other essential resources.

presumably result in exploitation of significantly different parts of resource gradients. Coexisting tiger beetles have different-sized mandibles (Pearson & Mury 1979); proboscis lengths of bumblebees living together are significantly different (Inouye 1977); and parasitoids utilizing the same host in the same place show differences in ovipositor length (Price 1972). Therefore, coexistence of species is possible because competition between them is minimized.

Where species remain very similar in their utilization of resources, competition may become too severe, and one species may be limited to a small part of its fundamental niche. Such *competitive exclusion* or *competitive displacement* from part of an organism's niche has been documented for insects several times. The most widely quoted case concerns parasitoids of the California red scale, *Aonidiella aurantii*, studied by DeBach and Sundby (1963). *Aphytis chrysomphali* was displaced from large parts of California by *Aphytis lingnanensis*, and both were subsequently displaced from much of their range by *Aphytis melinus*. Parasitoids of the oriental fruit fly, *Dacus dorsalis*, have interacted in a similar manner (Bess & Haramoto 1958, and Chapter 15). The Oriental fruit fly itself displaced the Mediterranean fruit fly, *Ceratitis capitata*, at lower elevations in Hawaii (Keiser et al. 1974).

Of course, species can show complementary distributions in time or space for many different reasons, competition being just one, although frequently invoked as the mechanism. Physical conditions may change in a fluctuating fashion, favoring one species and then another. For example, during periods of subnormal precipitation, the migratory grasshopper, *Melanoplus bilituratus*, becomes abundant; but when precipitation is supernormal, it is replaced in abundance by the differential grasshopper, *Melanoplus differentialis* (Wakeland 1961). During ecological succession, vegetation and other conditions change, favoring one species and then another. The spider *Arctosa perita* is found in very early stages of vegetational succession on sand dunes, while *Pardosa lugubris* occurs mostly in wooded sites found in late succession (van der Aart 1974; Fig. 2.8). Great care is always needed in order to establish the real cause and effect relationships in species distributions and abundances and the constraints imposed by abiotic and biotic factors.

2.4 MAJOR PROCESSES

Although a large ecosystem may be exceedingly complex, with hundreds of species that interact, the major processes may be categorized into four types representing fundamentally different kinds of processes: energy flow, biogeochemical cycling, ecological succession, and the evolution of species. These will be discussed in turn.

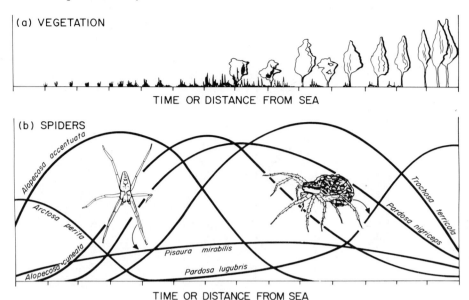

(a) VEGETATION

TIME OR DISTANCE FROM SEA

(b) SPIDERS

Alopecosa accentuata

Arctosa perita

Alopecosa cuneata

Pisaura mirabilis

Pardosa lugubris

Trochosa terricola

Pardosa nigriceps

TIME OR DISTANCE FROM SEA

Fig. 2.8   Ecological succession on sand dunes on the west coast of Holland with (a) change in vegetation from bare sand (left) to woodland (right) and (b) the change in spider species abundance during succession (after van der Aart 1974).

2.4.1   Energy Flow

All energy utilized by organisms is fixed in a utilizable form by the autotrophs. The major source of energy is from the sun, and it is converted by green plants through photosynthesis into carbohydrates. Energy available for photosynthesis differs over the globe, increasing towards the equator: at 52° N it is about 2.5×10^8 cal/m²·yr, at 42° N about 4.7×10^8 cal/m²·yr, and at 32° N about 6.0×10^8 cal/m²·yr (Phillipson 1966). The vast majority of this energy is lost as heat, with only 1–5% being used in photosynthesis to convert carbon dioxide and water into carbohydrates, the chemical energy that supports the whole food web in any community. The matter produced or energy fixed by autotrophs is called *primary production*.

Primary production differs enormously around the earth as solar energy input and availability of water varies. For example, when solar energy and water are abundant, as in tropical rain forests, primary production may reach 5000 dry g/m²·yr, whereas temperate forests produce only about half that. In extreme conditions of low solar energy input primary production may reach only 10–400 dry g/m²·yr as in tundra and alpine ecosystems, or under very low precipitation, 10–250 dry g/m²·yr as in desert scrub ecosystems (Whittaker 1970).

Herbivores utilize the autotrophs as food, and the matter produced, or

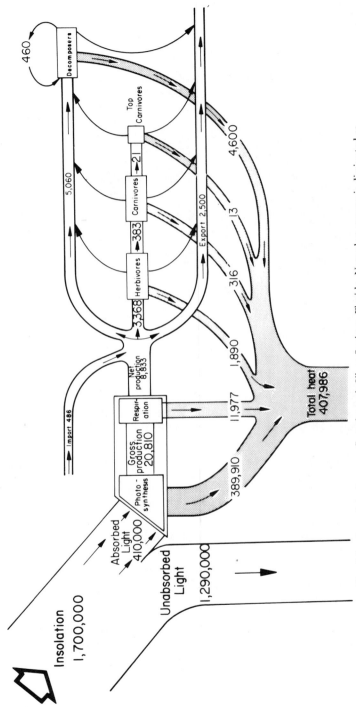

Fig. 2.9 A model of energy flow through Silver Springs, Florida. Note that energy is dissipated as heat in the ecosystem and is totally utilized except for a small proportion exported to adjacent systems (after Odum 1957 and Phillipson 1966). Numbers indicate the rate of energy flow in kcal/m² · yr. (With permission from H. T. Odum. 1957. Trophic Structure and Productivity of Silver Springs, Florida. *Ecological Monographs* **27**: 55–112. Copyright 1957 by the Ecological Society of America.)

energy fixed, at the second trophic level is called *secondary production*. This energy source is then utilized by carnivores, and so energy flows through the ecosystem. At each step most *gross production* is utilized for respiration, and less than half becomes available to the next trophic level as *net production* (e.g., Fig. 2.9). Net production is utilized by an organism for growth and reproduction; some is lost as elimination of dead material or when secretions are released, and some may be consumed by organisms of the next trophic level. Thus, all inputs and outputs of energy may be represented as in Fig. 2.10. This model may be applied with little modification to populations of organisms or to ecosystems. Indeed, any unit in an ecosystem may be usefully modeled in terms of energy flow, as for a bumblebee colony illustrated by Heinrich (1979) and Fig. 2.11.

As the model of energy flow for Silver Springs (Fig. 2.9) and that for individual organisms (Fig. 2.10) indicate, much of the net production at each trophic level passes as dead organic matter to the decomposers in the ecosystem. In an oak-pine forest, 2650 dry g/m²·yr is fixed and 1200 dry g/m²·yr is invested in net primary production. Of this, 360 dry g/m²·yr results in litter available to decomposers or the detritus food web, whereas only 30 dry g/m²·yr is utilized by herbivores (Woodwell 1970). This illustrates the relative importance of decomposers and herbivores in the movement of energy in temperate forest ecosystems. In grassland and aquatic ecosystems, however, a much greater percentage of net primary production is normally utilized by herbivores, leaving much less for the detritus food web (cf. Petrusewicz & Grodzinski 1975, and Fig. 2.12).

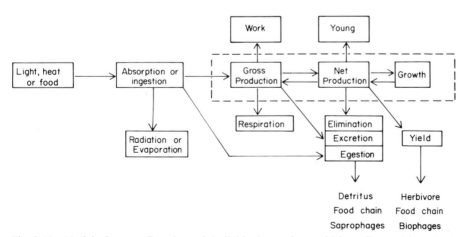

Fig. 2.10   Model of energy flow through individual organisms, which applies to either plants (autotrophs) or animals (heterotrophs). The dashed line represents the boundary of the organism. [With permission from G. O. Batzli. 1974. Production, Assimilation and Accumulation of Organic Matter in Ecosystems. *Journal of Theoretical Biology* **45:** 205–217. Copyright by Academic Press Inc. (London) Ltd.]

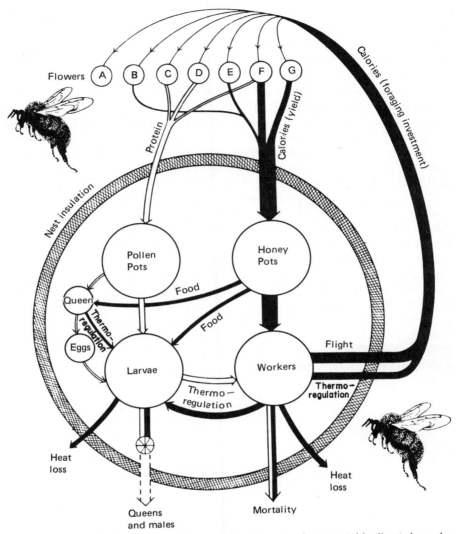

Fig. 2.11 A model of movement of energy (black lines) and matter (white lines) through a bumblebee colony. (With permission from B. Heinrich. 1979. Bumblebee Economics. Copyright by Harvard University Press.)

Ultimately, all solar energy fixed in a closed ecosystem is dissipated as heat by autotrophs, heterotrophs, or decomposers. Energy *flows* through the system, and once it has passed through the autotrophs, none is available again to the first trophic level. No recycling of energy can occur. In open ecosystems some energy may be lost to the system, being exported as dead organic matter, or through the emigration of organisms (Fig. 2.9), but eventually this energy will be utilized and dissipated.

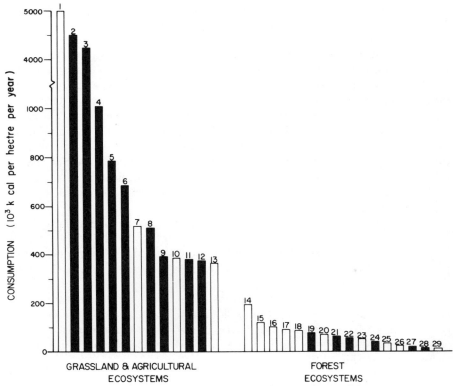

Fig. 2.12 Comparison of herbivore consumption in temperate latitude grassland and agricultural ecosystems compared with forest ecosystems. Insect examples are given as solid bars, mammals as open bars (after Petrusewicz and Grodzinski 1975).

Energy, of course, is vital to all life, and therefore the understanding of energy flow in ecosystems is a significant component of knowledge on any organism in the system. Energy may be an important limiting factor in the life of organisms (cf. Chapter 4). Since Lindeman's (1942) classic paper on community energetics, a major thrust in ecology has been the description of energy production and utilization for many areas of the earth (overview by Wiegert 1976). Insects play important roles in the flow of energy in ecosystems as herbivores, carnivores, and detritivores, as illustrated in food webs (Figs. 2.1–2.3) and the histogram of energy consumed by herbivores (Fig. 2.12) (cf. Mattson 1977).

2.4.2 Biogeochemical Cycling

The cycling of chemicals is a very different kind of process from the flow of energy because, unlike energy, chemicals are not lost to the system when a transfer takes place, as when chemicals in plants are converted to animal tissues or when organic matter is decomposed to inorganic compounds. Energy flow is like a one-way street, whereas chemicals move as if on a roundabout.

Many biogeochemical cycles are of interest to the ecologist concerning the movement in ecosystems of such elements and compounds as water, carbon, nitrogen, oxygen, phosphorus, sulfur, potassium, and so on (general ecology texts such as Smith 1966, Odum 1971, Collier et al. 1973, and Ricklefs 1973 cover a selection of these cycles). Many chemicals manufactured by man become involved in these cycles, although recycling through decomposition is frequently exceedingly slow because of the supernormal quantities produced or because many unnatural chemicals are not readily decomposed as are those in a natural cycle. Thus, application of fertilizers and pesticides to land or water disrupt significantly the natural biogeochemical cycles.

The nitrogen cycle provides an example of a biogeochemical cycle that is almost complete in undisturbed ecosystems (cf. Delwiche 1970). It is an important cycle because the element is a constituent of amino acids and proteins required by all organisms (Chapter 4), and it may be a major limiting factor, particularly for plants and herbivores.

Free nitrogen gas (N_2) in the atmosphere is converted to compounds containing N by lightning and by biological fixation, fixation meaning the conversion of N_2 to compounds, particularly ammonia and nitrates, that can be utilized by plants, and then animals. Biological fixation involves a series of steps (Fig. 2.13):

1. *Nitrogen fixation.* N_2 is converted to ammonia largely by free-living bacteria in the soil such as species of *Azotobacter* and *Clostridium*, free-living blue-green algae in aquatic systems (e.g., *Anabaena* and *Nostoc*), and symbiotic bacteria associated with many plants, particularly *Rhizobium* species in the root nodules of legumes.

2. *Nitrification.*   Ammonia, or the ammonium ion, is oxidized to (a) nitrous acid (which then forms nitrite salts) by bacteria and (b) then to nitric acid and nitrates.

The bacterium *Nitrosomonas* is an autotroph deriving chemical energy by oxidation of the ammonium ion to nitrites, a reaction that yields 65 kcal/mole of ammonium. *Nitrobacter* species obtain energy by oxidizing nitrites to nitrates. Nitrates are generally the most readily available source of N to plants because the nitrate ion is not absorbed onto soil particles, whereas ammonium ions tend to be adsorbed onto clay particles because of their complementary charge. Plants absorb ammonium ions or nitrate ions and use energy derived from photosynthesis to synthesize amino acids and then proteins. Animals consume plants and the N continues the cycle along the various routes in the ecosystem's food web.

Plants and plant parts die, animals excrete waste products and die, and

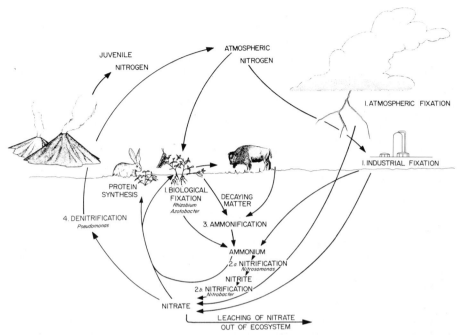

Fig. 2.13   The nitrogen cycle (see Delwiche 1970 for details). The major processes in the cycle are numbered 1–4 as in the text.

all dead organic matter is decomposed. The N cycle is completed in at least two further steps.

3. **Ammonification.**   N in organic compounds is liberated as ammonium by microorganisms in decaying matter. This ammonium is then utilized locally in the nitrification process to form nitrates.

4. **Denitrification.**   Any nitrates or nitrites in the soil, whether fresh from the nitrogen fixation–nitrification route, or from the living organism–ammonification route, may be broken down to N_2 by denitrifying bacteria such as *Pseudomonas denitrificans*, under anaerobic conditions deep in the soil or in stagnant water.

Additional natural sources of N are from volcanic activity and the weathering of rocks. N is lost from a specific ecosystem or short-term cycle through sedimentation of N-containing particles or the leaching of nitrates into ground water and its passage into rivers, lakes, and perhaps the sea.

The natural N cycle has been disrupted by a massive input of the element to agricultural ecosystems as fertilizers derived from industrial fixation of atmospheric N. Delwiche (1970) estimated that industrial fixation contributed 30 million metric tons/yr to the biosphere, and by the year 2000 such fixation may exceed 100 million tons. Compared to this figure, biological fixation is likely to remain stable at about 44 million metric tons in the terrestrial biosphere and 10 million tons in the marine environment. The results of this massive input of N fertilizers is, of course, a spectacular increase of net primary production in agricultural systems, great increases in secondary production through improved nutrition of domestic animals and man himself, but also an excess fixation and nitrification compared to denitrification, with resultant pollution problems in ecosystems receiving water passing through agricultural ecosystems. Nitrates not only contribute to blooms of algae in lakes, but they become toxic in drinking water and even fatal to cattle and young children at concentrations greater than 38 mg/ liter (Edmondson 1971).

The great increases in crop yields achieved by application of N graphically illustrates how N is a major limiting factor in plant growth, and in most ecosystems. For the second trophic level N is also limiting because animals must concentrate it from their plant food. Insects feeding on leaves and stems typically have two to four times as much N/unit of tissue as their food (Slansky 1974), and much evidence suggests that N deficiency for herbivores is a major regulating factor in herbivore populations (e.g., T. C. R. White 1978, and cf. Chapter 4).

It has taken many years of study by many investigators to work out the N cycle, and much more study is needed before the complete linkages between the chemicals, biotic process, and geological processes are understood. But the rate at which humans are altering the environment and adding new chemicals to it demands rapid answers to the now inevitable questions on their environmental impact; their toxicity to all members of the food webs they enter, their biodegradeability, and their movement within and between ecosystems. As ecologists and entomologists we have a particular responsibility to ensure that environmental damage is minimized wherever an insecticide is used, and that such use is fully justified in terms of all costs and benefits and all alternative actions possible. How can we do a responsible job without first liberating toxic substances and following their movements and effects in real ecosystems? The use of model (laboratory) ecosystems has become an integral part of studying the fate and effects of new insecticides and other novel chemicals produced by man. This laboratory study of new compounds in biogeochemical cycles, or their resistance to natural mechanisms resulting in cycles, is discussed below.

The model ecosystem first developed (Metcalf et al. 1971) consisted of a 10-gallon aquarium, with white sand forming the terrestrial part of the system and 7 liters of water into which aquatic organisms were placed (Fig. 2.14). Radio-labeled insecticide was sprayed onto plants growing in the sand and these were consumed by caterpillars. The insecticide passed into the aquatic part of the system when frass and dead insects fell onto the moist sand and into the water. It was then absorbed by algae and diatoms, which in turn were fed upon by snails and water fleas and other planktonic

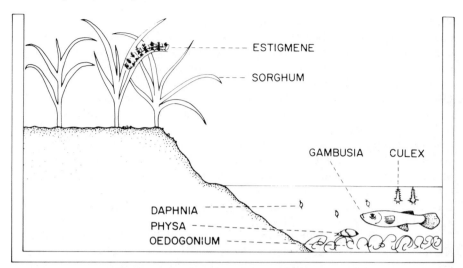

Fig. 2.14   Cross section of the model ecosystem.

species, and these in turn were eaten by mosquito larvae. Mosquito fish were the *top predators* in the system (Fig. 2.15). Each experiment lasted only 33 days when each component and the water were analyzed by thin-layer chromatography, radioautography, and scintillation counting to determine the qualitative and quantitative nature of the radio-labeled products and their location in the ecosystem.

Thus, in 33 days, it was possible to demonstrate the dangerous properties of DDT and its degradation products in an ecosystem, with the rapid biological concentration of toxic chemicals as they pass up the food chain.

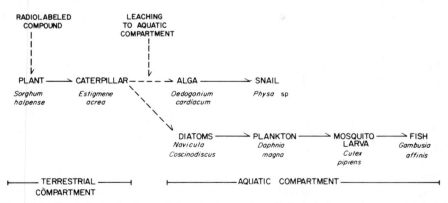

Fig. 2.15   Food chains in the model ecosystem and the linkage between the terrestrial and aquatic compartments. Solid lines indicate trophic linkage. Dashed lines indicate movement of chemicals independently from the food chains.

When ^{14}C-labeled DDT was applied at the equivalent of 1 lb/acre, it accumulated in the higher trophic levels as DDE, DDD, and DDT and was concentrated 10,000–100,000 times compared to concentrations in the water (Table 2.1). ^{14}C-DDE was concentrated to 30,000–50,000 times that in the water. If these kinds of experiments could have been done when DDT was first used in the mid-1940s, the results would have supported in a very significant manner the early pleas that the insecticide should be utilized with care because of its threat to fish, birds, and mammals (cf. Cottam & Higgins 1946, Woodwell 1967). In the absence of model ecosystem studies, it was not until the early 1960s when the forecast danger of DDT was fully realized, with its highly toxic effects on the top predators (Hunt & Bischoff 1960, Carson 1962, Rudd 1964, Woodwell 1967), and it was used for 28 years in the United States before it was legally banned for agricultural uses in 1973 (Metcalf 1977).

The danger with DDT is that it does not readily enter natural biogeochemical cycles. It is degraded to DDE which is extremely persistent and resistant to biodegradation. Added to this is the greater solubility of DDT and its breakdown products in fats than in water, so it persists in fatty tissues of animals and becomes concentrated there. Indeed, the less water soluble a pesticide is, the more it is subject to biological concentration and thus the more dangerous it is. An extensive study of 49 pesticides by Metcalf & Sanborn (1975) illustrates this relationship vividly (Fig. 2.16).

The model ecosystem has many potential uses (Metcalf 1977). The biodegradability of new pesticides can be examined, and those that enter biogeochemical cycles can be identified. For example, methoxychlor has many of the good properties of DDT, but its biodegradability was much greater than that of DDT and it did not reach high concentrations in mosquitoes and fish, although it may pose a threat to snails (Table 2.1). Many other pesticides, polychlorinated biphenyls, and other pollutants, carcinogens (Lu et al. 1977), drugs (Coats et al. 1976), products from coal (Lu et al. 1978), food additives, and heavy metals, have also been run

Table 2.1 Biological concentration of radio-labeled DDT, DDE, DDD, and methoxychlor in a model ecosystem[a]

	Concentration in ppm			
Pesticide	Water	Snail	Mosquito	Fish
DDT	0.004	22.9	8.9	54.2
DDE	0.0053	103.5	159.5	145.0
DDD	0.006	5.6	5.8	39.1
Methoxychlor	0.0016	15.7	0.48	0.33

[a]Data from Metcalf et al. (1971).

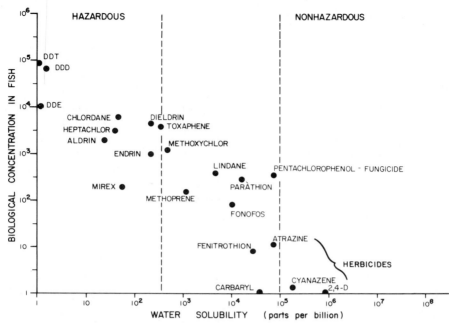

Fig. 2.16   Solubility in water of pesticides and the biological concentration in mosquito fish in the model ecosystem (after Metcalf and Sanborn 1975).

through model ecosystems, and impact on plant growth has been examined (Cole et al. 1976). Thus, the use of model ecosystems holds great promise for providing essential information on how chemicals resulting from man's technology will behave in real ecosystems.

The main properties of biogeochemical cycling are well illustrated by the N cycle and the movement of DDT in model and real ecosystems. In natural systems chemicals essential to life are usually involved in complete cycles involving geophysical and biotic components of ecosystems. The processes involved with synthesis of chemicals operate at similar rates to those of degradation. A balanced system prevails. The addition of massive quantities of chemicals, even though these are found in nature, is disruptive within the ecosystem treated, and to others into which excess materials flow. The manufacture and application of novel chemicals has even more drastic effects; since natural decomposition pathways frequently do not exist in the biosphere, compounds accumulate indefinitely and may tend to be concentrated in natural food webs, including those in which man is an important omnivore. Other examples of biogeochemical cycles will not be given here but are treated in the ecology texts cited above, in the September 1970 issue of *Scientific American*, and by Likens et al. (1977).

2.4.3 Ecological Succession

Ecological succession is the process and pattern of change in the biotic community through time. Change occurs at any site because organisms arrive at different times and, once established, they alter the living conditions for themselves and other organisms, benefiting some and forcing others into local decline or extinction. Again, this is a very different kind of ecosystem process from the major processes already considered. Of course, as communities change with time, so do patterns of energy flow and biogeochemical cycling so the major processes are closely linked. This linkage and the trends with time in ecological succession have been nicely summarized by Odum (1969), and Table 2.2. As noted by Odum, the fourth major process in ecosystems—evolution—is also linked to ecological succession because selective pressures naturally change as living conditions change.

The change in vegetation through time was particularly evident to early ecologists where sand dunes were formed on lake shores and seashores (Cowles 1899). Blown sand builds up on windward shores forming new areas for colonization by plants and animals, resulting in a display of ecological succession in space. Studies of plant and animal succession on sand dunes have been particularly revealing on the processes involved, starting with Cowles's work on the dunes of Lake Michigan and spreading to studies in many parts of the world (e.g., Tansley 1939, Shelford 1963, Chevin 1966, Bakker et al. 1974). It was soon recognized that as vegetation changes, the animals respond to this change, with notable studies on change in arthropod communities on dunes [Shelford (1907) on tiger beetles, Chevin (1966) on whole communities, van der Aart (1974) on spiders, and Weygoldt (1969) on pseudoscorpions] (Figs. 2.8 and 2.17).

During ecological succession from blown sand on a seashore to the establishment of trees, many factors are constantly changing. Sand mobility decreases with time as plants colonize, root in the sand, and so stabilize it. Salt concentration decreases as it is leached out of the sand, and winds become less salt laden with distance from the shore. Organic matter and nutrients first decrease as debris from the sea decays, and then they increase as plant matter dies and decomposes. The number of species able to colonize increases as conditions become less extreme. These trends fit into the general picture summarized in Table 2.2

But ecological succession is not only limited to change of vegetation and associated animals. There is a succession of species observed on practically any resource, some proceeding slowly and others rapidly. Even though plant species may be very persistent, a succession of arthropods is often observed on these plants [e.g., on broom (Waloff 1968) and pine (Martin 1966)]. Decaying organic matter frequently becomes available to a sequence of arthropods, each adapted to a different stage in the process, as seen in rotting logs (Blackman & Stage 1924) and stored grain (Coombs & Woodroffe 1973). Even on living animals a succession of parasite species may

Table 2.2   A tabular model of ecological succession: Trends to be expected in the development of ecosystems[a]

Ecosystem attributes	Developmental stages	Mature stages
Community energetics		
1. Gross production/community respiration (P/R ratio)	Greater or less than 1	Approaches 1
2. Gross production/standing crop biomass (P/B ratio)	High	Low
3. Biomass supported/unit energy flow (B/E ratio)	Low	High
4. Net community production (yield)	High	Low
5. Food chains	Linear, predominantly grazing	Weblike, predominantly detritus
Community structure		
6. Total organic matter	Small	Large
7. Inorganic nutrients	Extrabiotic	Intrabiotic
8. Species diversity—variety component	Low	High
9. Species diversity—equitability component	Low	High
10. Biochemical diversity	Low	High
11. Stratification and spatial heterogeneity (pattern diversity)	Poorly organized	Well organized
Life history		
12. Niche specialization	Broad	Narrow
13. Size of organism	Small	Large
14. Life cycles	Short, simple	Long, complex
Nutrient cycling		
15. Mineral cycles	Open	Closed
16. Nutrient exchange rate, between organisms and environment	Rapid	Slow
17. Role of detritus in nutrient regeneration	Unimportant	Important
Selection pressure		
18. Growth form	For rapid growth ("r selection")	For feedback control ("K selection")
19. Production	Quantity	Quality

Table 2.2 (*Continued*)

Ecosystem attributes	Developmental stages	Mature stages
Overall homeostasis		
20. Internal symbiosis	Undeveloped	Developed
21. Nutrient conservation	Poor	Good
22. Stability (resistance to external perturbations)	Poor	Good
23. Entropy	High	Low
24. Information	Low	High

*a*With permission from E. P. Odum, in *Science* **164:** 262–270, April 18, 1969. Copyright 1969 by the American Association for the Advancement of Science.

occur as with parasitoids on a population of the spotted alfalfa aphid (van den Bosch et al. 1964).

Early ecologists studying succession noticed that there is a radical change in the characters of species from early to late seral stages. Newly disturbed ground is rapidly colonized by short-lived species with very effective means of dispersal: good colonizers such as weedy species of plants. They have evolved with a pattern of life characterized by small size, short generation

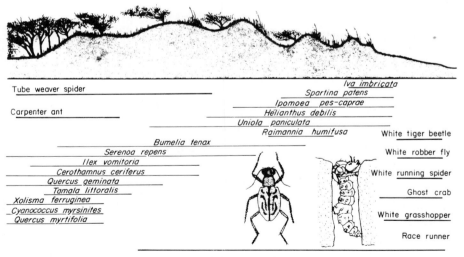

Fig. 2.17 Ecological succession on sand dunes on the east coast of Florida showing the distribution of plants, anthropods, and a lizard in space and time (after Shelford 1963). The sea is to the right and succession progresses from right to left. A tiger beetle adult and larva are illustrated which are found on open sand early in succession.

time (perhaps only a few weeks), and rapid production of a large number of small seeds adapted for colonizing distant, new open patches of ground. Such pioneering species are often poor competitors, being adapted to reproduce before a site becomes overcrowded. They are opportunists but they also can for a time become overcrowded. They do not exist just at the low end of a logistic curve until the site conditions are changed or better K species come in. They seem to be adapted to maximize instantaneous rate of population increase, symbolized as r, and were termed r-selected species by MacArthur & Wilson (1967; and Table 2.2, "Selection Pressure") (cf. Chapter 12).

These characters change as succession proceeds so that at the other extreme the plant species are large, long lived and highly competitive, producing relatively large seeds with low dispersal potential. Such species are adapted to persist in a site for a long time once they have colonized. One generation may be succeeded by another of the same species, giving the community an aspect of persistence, stability, and maturity known as the climax community. MacArthur & Wilson (1967) termed such species K-selected because they are adapted to persist under highly competitive conditions, close to the carrying capacity K of the environment (Table 2.2).

However, ecological succession is frequently set back to an earlier stage before maturity is reached. In agricultural ecosystems land is usually kept in an early stage of succession in which weedy species thrive. In natural ecosystems repeated fires may keep the vegetation at an intermediate stage (Loucks 1970), and recurrent epidemics of forest insects (e.g., spruce budworm, pine sawflies, and bark beetles) modify succession considerably (Mattson & Addy 1975, Amman 1977, and cf. Chapter 22), as do leaf-cutting ants (Jonkman 1977).

Ecological succession is a major process, producing many predictable patterns. Although one of the oldest concepts in ecology, there is still much to be learned about the mechanisms involved with succession: how one plant species replaces another, how animals modify plant species assemblages by affecting location, reproduction, and survival (cf. Chapter 22), the mechanisms whereby mature communities persist, and the role of other major ecosystem processes (cf. Harper 1977, and an overview by Golley 1977).

2.4.4 Evolution of Species

It has not been possible to dicuss one major ecosystem process without the inclusion of others; all are so intricately entwined. Thus, the subject of organic evolution has already entered into discussions. In the previous section it was inferred that strongly competing species tend to evolve characters that minimize the probability of competitive displacement. The evolution of life history patterns was discussed in relation to ecological succession. It is also implicit in the discussions on energy flow and biogeochemical cycles that where a particular resource becomes adundant, then species are likely to evolve to exploit it; hence, the movement of energy and matter in ecosystems. Dobzhansky et al. (1977) defined evolution as follows:

> Organic evolution is a series of partial or complete and irreversible transformations of the genetic composition of populations, based principally upon altered interactions with their environment. It consists chiefly of adaptive radiations into new environments, adjustments to environmental changes that take place in a particular habitat, and the origin of new ways for exploiting existing habitats. These adaptive changes occasionally give rise to greater complexity of developmental pattern, of physiological reactions, and of interactions between populations and their environment.

Evolution results from "the differential perpetuation of genotypes" (Mayr 1963) caused by natural selection. It is evident how tightly the evolution of a population or species is linked to its ecology. The two subjects cannot be considered realistically in isolation as Darwin (1859) clearly realized in *The Origin of Species* and as recent texts on evolutionary biology illustrate (e.g., Dobzhansky et al. 1977, Futuyma 1979). Indeed, the subjects of evolution, ecology, and behavior form the basis for understanding any species (e.g., cf. Mayr 1963, Krebs & Davies 1978, Ehrman & Parsons 1976), and all are synthesized in the modern discipline of evolutionary biology. Evolution involves genetic changes in populations and species and emergence of new species. The processes involved are discussed in the references cited above and, for example, in Lewontin (1974), M. J. D. White (1978), and Ayala (1976) (cf. Chapter 9).

Examples of rapid evolutionary responses by insects to environmental change have been well documented for the development of industrial melanism and insecticide resistance. It is interesting that the melanic morph of the peppered moth, *Biston betularia*, was becoming more abundant in Britain in response to effects of industrial pollution while Darwin was alive. Unfortunately he was not aware of this change, for as Kettlewell (1959) pointed out, it provided "Darwin's missing evidence" for evolutionary change in direct response to environmental change. The now extensive literature on industrial melanism is summarized by Kettlewell (1973), and research on complexities in the selective factors involved is continuing to provide a fascinating literature (e.g., Bishop et al. 1975, Steward 1977a,b).

Evolution of insecticide resistance in arthropod populations provides good examples of the processes involved (reviewed by Georghiou 1972, Brown 1978, Plapp 1976, Georghiou & Taylor 1977, cf. Chapter 9). First of all, the importance of evolution in survival of populations is illustrated by the hundreds of cases of insecticide resistance known by 1977: 139 cases in arthropod species of medical importance and 225 cases in agricultural systems, for a total of 364 species that have shown this evolutionary response to a change in environmental resistance (Georghiou & Taylor 1977). No doubt, in the next decade, many other species will be added since in the period 1967–1975, the number of known resistant species increased by 62.5%.

Of course, many pest species were routinely exposed to natural toxic compounds in the environment—plant toxins, products from bacterial decay, and so on—and enzymes involved with their detoxification were probably common before pesticides were used (cf. Chapter 4). So some individuals in populations were preadapted to cope with pesticides by having genes coding for enzymes that detoxified both natural and man-made toxic chemicals [see Gould (1979) for evidence for preadaptation in mites for host transfer to toxic host plants]. When insecticide was applied, those individuals with appropriate enzyme systems for detoxification survived, reproduced, and fostered resurgence of the pest population largely composed of resistant individuals. Application of new insecticides selected for individuals with additional resistance factors until populations, such as those of the housefly, typically contained an arsenal of genetically coded defenses against insecticides.

Genetic studies on houseflies have shown that every chromosome has genes involved with resistance (Plapp 1970). The location of some are illustrated by Brown (1978). These genes are responsible for a remarkable variety of adaptive traits (reviewed by Plapp 1976). Three genes are known to confer resistance to DDT: DDT-dehydrochlorinase (*Deh*), knockdown resistance (kdr), and DDT-microsomal detoxification (DDTmd). For each gene there may be several isozymes that confer different levels of resistance. For example, reduced sensitivity of acetylcholinesterase (AChE) to organophosphates and carbamates is controlled by at least four isozymes. Plapp (1970) recognized seven major genes for insecticide resistance, with major mechanisms involving reduced sensitivity of AChE, decreased absorption of insecticide, and increased rates of detoxification. Georghiou (1972) discusses additional mechanisms.

Thus, the housefly and most other insect populations have a most versatile battery of potential defenses against insecticides, with rapid evolutionary responses in populations under insecticide stress. Naturally, any chemical man uses in an attempt to control insects acts as an artificial selective agent with a consequent evolutionary response by the insects. Thus, insects developed resistance to unusual levels of insect growth-regulating chemicals such as juvenile hormone analogs as rapidly as to conventional insecticides, and with a similar array of defenses (Plapp 1976). These so-called third-generation insecticides produce the same evolutionary responses in insects as did the first generation insecticides used at the turn of the century.

Much more will be said about evolutionary processes in Chapter 9. It is important here to emphasize that evolution is a major process in ecosystems and that insects have supplied some of the most dramatic examples of rapid evolutionary change. This should not be overlooked whenever we study natural ecosystems, agricultural or urban ecosystems, or whenever we keep insects in cultures.

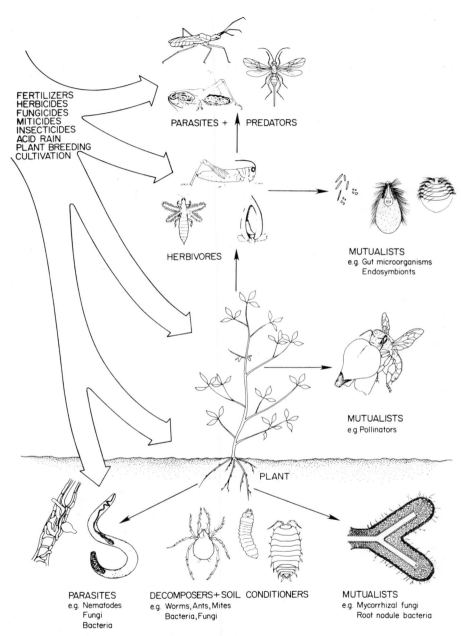

FERTILIZERS
HERBICIDES
FUNGICIDES
MITICIDES
INSECTICIDES
ACID RAIN
PLANT BREEDING
CULTIVATION

PARASITES + PREDATORS

HERBIVORES

MUTUALISTS
e.g. Gut microorganisms
Endosymbionts

MUTUALISTS
e.g. Pollinators

PLANT

PARASITES
e.g. Nematodes
Fungi
Bacteria

DECOMPOSERS + SOIL CONDITIONERS
e.g. Worms, Ants, Mites
Bacteria, Fungi

MUTUALISTS
e.g. Mycorrhizal fungi
Root nodule bacteria

Fig. 2.18 The components of a typical terrestrial ecosystem and some influences imposed by man either within that ecosystem or from other ecosystems (e.g., acid rain). Whenever any one of these factors is manipulated, the impact is felt by all components of the system.

2.5 CONCLUSIONS

Although the ecosystem concept has been with us a long time, and although its value is appreciated by most ecologists, in practice it is seldom applied. We tend to simplify interactions down to a one-to-one relationship: prey and predator, one competitor versus another, one pesticide on one crop species, one plant species and its resistance to one herbivore pest species. Seldom are other components in the system observed simultaneously, and it is probably safe to say that never are all components studied.

Of course, there is always a need for compromise, and one must distinguish between what is necessary and what is merely desirable. My contention is that more components in ecosystems need to be studied simultaneously than is usual today. For example, the interaction between plants and herbivores cannot be fully understood without considering the enemies of herbivores: predators, parasites, and diseases (Price 1981, Price et al. 1980). No fertilizer or pesticide should be applied without knowing its impact on symbiotic mycorrhizal fungi and other soil-dwelling organisms (Ruehle & Marx 1979). We cannot describe ecosystems realistically as static states, for they are very dynamic, all major processes resulting in change. All components in an ecosystem are linked by the major processes. We must acknowledge this fact more in our studies of them (cf. Chapters 24 and 25).

Perhaps it is fitting to end this chapter with an illustration of significant components of an ecosystem and the many ways in which we attempt to manipulate them, usually without knowing how more than one or two organisms will be affected by such manipulation (Fig. 2.18).

ACKNOWLEDGMENTS

I am very grateful to Pamela Lungé for all the artistic work in this chapter. I thank the National Science Foundation for financial support through grant no. DEB-7816152.

REFERENCES

Amman, G. D. 1977. Pp. 3–18 in Mattson (Ed.), 1977, referenced here.

Ayala, F. J. (Ed.). 1976. *Molecular Evolution*. Sinauer, Sunderland, Mass.

Badcock, R. M. 1949. *J. Anim. Ecol.* **18:** 193–208.

Bakker, K., I. H. Bosch, R. van Crevel, N. C. Michielsen, G. Drost, E. A. J. Wanders, V. Westhoff, and M. van Wijngaarden (Eds.). 1974. *Meijendel: Duin-water-leven.* van Hoeve, The Hague.

Batzli, G. E. 1974. *J. Theor. Biol.* **45:** 205–217.

Bess, H. A. and F. H. Haramoto. 1958. *Proc. Tenth Int. Congr. Entomol., Montreal, 1956.* **4:** 835–840.

Bishop, J. A., L. M. Cook, L. Muggleton, and M. R. D. Seaward. 1975. *J. Appl. Ecol.* **12:** 83–98.

Blackman, M. W. and H. H. Stage. 1924. *New York State Col. Forest. Syracuse Univ. Tech. Publ.* **17:** 1–269.

Bormann, F. H. and G. E. Likens. 1979. *Pattern and Process in a Forested Ecosystem.* Springer, New York.

Broecker, W. S., T. Takahashi, M. J. Simpson, and T.-H. Peng. 1979. *Science* **206:** 409–418.

Brown, A. W. A. 1978. *Ecology of Pesticides.* Wiley, New York.

Carson, R. 1962. *Silent Spring.* Houghton-Mifflin, Boston.

Chevin, H. 1966. *Mem. Soc. Nat. Sci. Natur. Math. Cherbourg.* **52:** 8–137.

Coats, J. R., R. L. Metcalf, P-Y. Lu, D. D. Brown, J. F. Williams, and L. G. Hansen. 1976. *Environ. Health. Persp.* **18:** 167–179.

Cole, L. K., J. R. Sanborn, and R. L. Metcalf. 1976. *Environ. Entomol.* **5:** 583–589.

Collier, B. D., G. W. Cox, A. W. Johnson, and P. C. Miller. 1973. *Dynamic Ecology.* Prentice-Hall, Englewood Cliffs, N.J.

Coombs, C. W. and G. E. Woodroffe. 1973. *J. Anim. Ecol.* **42:** 305–322.

Cottam, C. and E. Higgins. 1946. *J. Econ. Entomol.* **39:** 44–52.

Cowles, H. C. 1899. *Bot. Gaz.* **27:** 95–117, 167–202, 281–308, 361–391.

Darwin, C. 1859. *The Origin of Species by Means of Natural Selection.* Murray, London.

DeBach, P. and R. A. Sundby. 1963. *Hilgardia* **34:** 105–166.

Delwiche, C. C. 1970. *Sci. Am.* **223**(3): 136–146.

Dobzhansky, T., F. J. Ayala, G. L. Stebbins, and J. W. Valentine. 1977. *Evolution.* Freeman, San Francisco.

Edmondson, W. T. 1971. Pp. 213–229 in W. W. Murdoch (Ed.), *Environment, Resources, Pollution and Society,* Sinauer, Stamford.

Ehrman, L. and P. A. Parsons. 1976. *The Genetics of Behavior.* Sinauer, Sunderland, Mass.

Elton, C. 1927. *Animal Ecology.* Macmillan, New York.

Evans, F. C. 1956. *Science* **123:** 1127–1128.

Forbes, S. A. 1887. *Bull. Illinois Natur. Hist. Survey* **15:** 537–550.

Futuyma, D. J. 1979. *Evolutionary Biology.* Sinauer, Sunderland, Mass.

Georghiou, G. P. 1972. *Annu. Rev. Ecol. Syst.* **3:** 133–168.

Georghiou, G. P. and C. E. Taylor. 1977. *Proc. 15th Int. Congr. Entomol., Washington, D.C.,* pp. 759–785.

Golley, F. B. (Ed.). 1977. *Ecological Succession.* Dowden, Hutchinson and Ross, Stroudsburg, Penn.

Gould, F. 1979. *Evolution* **33:** 791–802.

Harper, J. L. 1977. *Population Biology of Plants.* Academic, New York.

Heinrich, B. 1979. *Bumblebee Economics.* Harvard University Press, Cambridge, Mass.

Holdridge, L. R., W. C. Grenke, W. H. Hatheway, T. Liang, and J. A. Tosi, Jr. 1971. *Forest Environments in Tropical Life Zones: A Pilot Study.* Pergamon, Oxford.

Hunt, E. G. and A. I. Bischoff. 1960. *Calif. Fish. Game* **46**(1): 91–106.

Hutchinson, G. E. 1957. *Cold Spring Harbor Symp. Quant. Biol.* **22:** 415–427.

Inouye, D. 1977. Pp. 35–40 in Mattson (Ed.), 1977, referenced here.

Jonkman, J. C. M. 1977. Pp. 85–87, in Mattson (Ed.), 1977, referenced here.

Keiser, I., R. M. Kobayashi, D. H. Miyashita, E. J. Harris, E. L. Schneider, and D. L. Chambers. 1974. *J. Econ. Entomol.* **67**: 355–360.

Kettlewell, H. B. D. 1959. *Sci. Am.* **200**(3): 48–53.

Kettlewell, H. B. D. 1973. *The Evolution of Melanism: The Study of a Recurring Necessity with Special Reference to Industrial Melanism in the Lepidoptera.* Oxford University Press, Oxford.

Krebs, J. R. and N. B. Davies. 1978. *Behavioral Ecology, an Evolutionary Approach.* Sinauer, Sunderland, Mass.

Lewontin, R. C. 1974. *The Genetic Basis of Evolutionary Change.* Columbia University Press, New York.

Likens, G. E. 1976. *Chem. Eng. News* **54**: 29–44.

Likens, G. E., F. H. Bormann, R. S. Pierce, J. S. Eaton, and N. M. Johnson. 1977. *Biogeochemistry of a Forested Ecosystem.* Springer, New York.

Lindeman, R. L. 1942. *Ecology* **23**: 399–418.

Loucks, O. L. 1970. *Am. Zool.* **10**: 17–25.

Lowe, C. H. 1964. *Arizona's Natural Environment: Landscapes and Habitats.* University of Arizona Press, Tucson.

Lu, P-Y., R. L. Metcalf, and E. M. Carlson. 1978. *Environ. Health Persp.* **24**: 201–208.

Lu, P-Y., R. L. Metcalf, N. Plummer, and D. Mandel. 1977. *Arch. Environ. Contam. Toxicol.* **6**: 129–142.

MacArthur, R. H. and E. O. Wilson. 1967. *The Theory of Island Biogeography.* Princeton University Press, Princeton, N.J.

MacFadyen, A. 1957. *Animal Ecology: Aims and Methods.* Pitman, London.

Martin, J. L. 1966. *Can. Entomol.* **98**: 10–27.

Mattson, W. J. (Ed.). 1977. *The Role of Arthropods in Forest Ecosystems.* Springer, New York.

Mattson, W. J. and N. D. Addy. 1975. *Science* **190**: 515–522.

Mayr, E. 1963. *Animal Species and Evolution.* Belknap Press of Harvard University Press, Cambridge, Mass.

Merriam, C. H. 1894. *Natl. Geogr. Mag.* **6**: 229–238.

Merriam, C. H. 1898. *U.S. Dept. Agric. Div. Biol. Surv. Bull.* **10**: 9–79.

Metcalf, R. L. 1977. *Annu. Rev. Entomol.* **22**: 241–261.

Metcalf, R. L. and L. R. Sanborn. 1975. *Illinois Natur. Hist. Survey Bull.* **31**: 381–436.

Metcalf, R. L., G. K. Sangha, and I. P. Kapoor. 1971. *Environ. Sci. Technol.* **5**: 709–713.

Odum, E. P. 1953. *Fundamentals of Ecology.* Saunders, Philadelphia.

Odum, E. P. 1969. *Science* **164**: 262–270.

Odum, E. P. 1971. *Fundamentals of Ecology.* 3rd ed. Saunders, Philadelphia.

Odum, H. T. 1957. *Ecol. Monogr.* **27**: 55–112.

Pearson, D. L. and E. J. Mury, 1979. *Ecology* **60**: 557–566.

Petrusewicz, K. and W. L. Grodzinski. 1975. Pp. 64–70, in D. E. Reichle, J. F. Franklin, and D. W. Goodall (Eds.), *Productivity of World Ecosystems.* Nat. Acad. Sci., Washington, D.C.

Phillipson, J. 1966. *Ecological Energetics.* Arnold, London.

Plapp, F. W. 1970. Pp. 179–192 in R. D. O'Brien and I. Yamamoto (Eds.), *Biochemical Toxicology of Insecticides.* Academic, New York.

Plapp, F. W. 1976. *Annu. Rev. Entomol.* **21**: 179–197.

Price, P. W. 1972. *Ecology* **53**: 190–195.

Price, P. W. 1981. Pp. 251–279, in D. A. Nordlund, R. L. Jones, and W. J. Lewis (Eds.), *Semiochemicals: Their Role in Pest Control.* Wiley, New York.

Price, P. W., C. E. Bouton, P. Gross, B. A. McPheron, J. N. Thompson, and A. E. Weis. 1980. *Annu. Rev. Ecol. Syst.* **11:** 41–65.

Read, D. P., P. P. Feeny, and R. B. Root. 1970. *Can. Entomol.* **102:** 1567–1578.

Richards, O. W. 1926. *J. Ecol.* **14:** 224–281.

Ricklefs, R. E. 1973. *Ecology.* Chiron, Newton, Mass.

Root, R. B. 1967. *Ecol. Monogr.* **37:** 317–350.

Root, R. B. 1973. *Ecol. Monogr.* **43:** 95–124.

Rudd, R. L. 1964. *Pesticides and the Living Landscape.* University of Wisconsin Press, Madison, Wis.

Ruehle, J. L. and D. H. Marx. 1979. *Science* **206:** 419–422.

Sagan, C. 1977. *The Dragons of Eden: Speculations on the Evolution of Human Intelligence.* Random House, New York.

Shelford, V. E. 1907. *Biol. Bull.* **14:** 9–14

Shelford, V. E. 1963. *The Ecology of North America.* University of Illinois Press, Urbana.

Slansky, F. Jr. 1974. *Energetic and Nutritional Interactions between Larvae of the Imported Cabbage Butterfly, Pieris rapae L., and Cruciferous Food Plants.* Ph.D. dissertation, Cornell University, University Microfilms, Ann Arbor, Mich.

Smith, R. L. 1966. *Ecology and Field Biology.* Harper & Row, New York.

Steward R. C. 1977a. *Ecol. Entomol.* **2:** 231–243.

Steward, R. C. 1977b. *J. Anim. Ecol.* **46:** 483–496.

Tansley, A. G. 1935. *Ecology* **16:** 284–304.

Tansley, A. G. 1939. *The British Islands and their Vegetation.* Cambridge University Press, London.

van den Bosch, R., E. I. Schlinger, E. J. Dietrick, J. C. Hall, and B. Puttler. 1964. *Ecology* **45:** 602–621.

Van der Aart, P. J. M. 1974. Pp. 178–182 in K. Bakker et al. (Eds.), *Meijendel: Duin-water-leven.* van Hoeve, The Hague.

Wakeland, C. 1961. *U.S.D.A. Prod. Res. Rep.* **42:** 1–9.

Wald, G. 1954. *Sci. Am.* **191**(2): 44–53

Waloff, N. 1968. *Adv. Ecol. Res.* **5:** 87–208.

Weygoldt, P. 1969. *The Biology of Pseudoscorpions.* Harvard University Press, Cambridge, Mass.

White, M. J. D. 1978. *Modes of Speciation.* Freeman, San Francisco.

White, T. C. R. 1978. *Oecologia* **33:** 71–86.

Whittaker, R. H. 1970. *Communities and Ecosystems.* Macmillan, New York.

Whittaker, R. H. and S. A. Levin (Eds.). 1975. *Niche Theory and Application.* Dowden, Hutchinson and Ross, Stroudsburg, Pa.

Whittaker, R. H., S. A. Levin, and R. B. Root. 1973. *Am. Natur.* **107:** 321–338.

Wiegert, R. G. (Ed.). 1976. *Ecological Energetics.* Dowden, Hutchinson and Ross, Stroudsburg, Pa.

Williams, J. (Ed.). 1979. *Carbon Dioxide, Climate and Society.* Pergamon, Oxford.

Woodwell, G. M. 1967. *Sci. Am.* **216:**(3): 24–31.

Woodwell, G. M. 1970. *Sci. Am.* **223:**(3): 64–74.

Section II
BASIC
BIOLOGICAL
AND
ECOLOGICAL
ADAPTATIONS,
PROPERTIES,
BEHAVIORS,
AND PROCESSES

The purpose of this section is to focus attention on morphological, physiological, behavioral, and genetic attributes and processes fundamental to an understanding of spatiotemporal patterns of particular insect species (Section III), each of which is adapted somewhat differently to particular resources and hazards.

The size range and flight propensities represented among insect species underlie their success in utilizing such a wide diversity of resources. However, successful exploitation requires that development, growth, and longevity be synchronized with resource patterns (Chapter 3). And beyond temporal synchrony each potential food source presents many challenges to candidate consumers (Chapter 4). Many food resources of insects occur in small quantities, widely dispersed in an otherwise hazardous environment. Consequently, food finding evokes specialized sensory and behavioral adaptations. Additional physiological and behavioral mechanisms are required before found food can be utilized (e.g., defensive mechanisms of hosts must be countermanded in extracting adequate nutrition).

Much of the energy value extracted from food fuels the reproductive process, the "bridge" between the individual and the population—a process with seemingly endless variations among insects (Chapter 5). Although insect populations grow rapidly under favorable conditions due to reproductive propensities, they enter various dormancy states (Chapter 6) or seek more benign environments through dispersal and migration (Chapter 7) when favorability of a site deteriorates.

Of course there are many individuals and many local populations eliminated by hazardous physical conditions and/or biotic agents, such as predators, parasites, pathogens, and competitors; however, insects have evolved a rich array of defensive mechanisms in countering enemies and in recruiting allies (Chapter 8).

The sum total of phenotypic and genetic adjustments to environmental heterogeneity comprises nature's fitting of lineages for different niches. Since both lineages and environments are open-ended, the fitting is a continuing selective process resulting in extinction for some lineages and evolutionary change in others (Chapter 9). A consideration of the wide-scale and long-term patterns that result (Chapter 10) serves to remind us that our population studies (Section III) are but snapshots, or at most, brief episodes, of a continuum with biospheric spatial and infinite temporal dimensions.

Chapter 3

Growth and Development of Insects

H. T. GORDON

3.1 INTRODUCTION

Growth is change in size of an individual (or any part of it); development is change in form. Both phenomena are complex, variable, interrelated, and hard to quantitate. Some numerical data (such as biomass, food consumption, or energy production) may prove to be useful parameters in ecological models. Other data, such as changing hormone levels during the life cycle or the intricate mechanisms involved in food selection and utilization or growth regulation, may never be explicitly useful even in the most elaborate descriptions of population dynamics and interactions. However, a rudimentary understanding of how external and internal forces control growth, development, and survival rates and a knowledge of the variability and range of these rates may lead to more realistic ecological hypotheses.

The available information is massive but fragmentary in many areas and superabundant in others. It is dealt with here selectively (brief glimpses and examples of some major processes and their quantitation and environmental interactions). More comprehensive presentations exist in Wigglesworth (1972), Schwerdtfeger (1963), and the multivolume treatise edited by Rockstein (1973). Awareness of the diversity of information on one intensively studied genus (*Drosophila*) can be had from the treatise edited by Ashburner and Wright (1978). Seldom do the authors try to interpret their results in an ecological or evolutionary perspective. Data are derived from small samples of a population, assumed to be representative of the insect species and homogeneous (variability caused by random factors). Often only one parameter is measured and reported as a mean and variance, although when more than one is measured for each individual (as in numerical taxonomy) the presence of subgroups can be detected. Figure 3.1 plots the pupal weight of *Tribolium* females versus the larval development rate, revealing two biotypes in the population (Howe 1966). The major subgroup (above the broken line) tends to develop faster into smaller pupae, and the positive correlation between weight and rate suggests that the parameters are not independent. The minor subgroup consists of heavier but slower developing individuals, which in a smaller sample might be considered rare aberrations. Any kind of cluster analysis is unusual in studies of growth (Kogan 1972). Natural populations are heterogeneous (often with balanced polymorphism) and undergo continuous selection by local environmental fluctuations (cyclic or random) and by geographic clines (Dobzhansky 1951, Masaki 1978, Chapters 6, 9, 12, herein). Experimental populations may be unrepresentative, consisting of more than one subgroup in a ratio influenced by the artificial conditions of laboratory rearing. Data and generalizations based on them may be imperfectly extrapolable to ecosystems. Heterogeneity is often revealed, as in selection of fast- and slow-developing strains of *Drosophila* from a seemingly homogeneous parent strain (Bakker 1969); critical analysis of this phenomenon by Lints and Gruwez (1972) showed that (as in *Tribolium*) development time

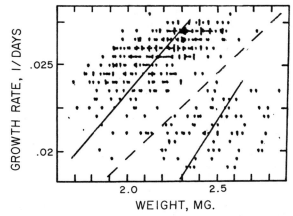

Fig. 3.1   Scatter diagram of larval growth rate (1/days) vs. pupal weight (mg) for several hundred individual female *Tribolium castaneum*. Modified from Howe (1966).

is not an independent parameter, being inextricably correlated with biomass and the development time of the female parent generation.

3.2   THE EGG STAGE

Reviews by Anderson (1972), Sander (1976), and Hinton (1981) have much detail on egg development. Egg size commonly ranges from $\frac{1}{10}$ to $\frac{1}{1000}$ of adult size, although a single overwintering egg of an aphid is nearly equal to the adult size. Biomass ranges from less than 0.1 μg (minute endoparasitic Hymenoptera) to hundreds of milligrams (giant tropical Coleoptera), mostly microscopic yolk particles (phospholipoproteins, glycoproteins, glycogen, nucleic acid, and a triglyceride energy reserve that may comprise as much as 50% of the dry weight) and an often complex protective shell (chorion). Both yolk and chorion are greatly reduced in most endoparasitic species, as well as in ovoviviparous and viviparous ones, since embryos draw on nutrients from host or maternal tissues. Embryogenesis normally follows fertilization, although parthenogenesis occurs in many Orders.

Development rate is usually measured by the time between egg laying and emergence of the first-instar larva. The common range at optimal temperature is 4–40 days for species in the Exopterygota, in which the cells of the embryonic primordium form a minute embryo that gradually increases in size to a larva morphologically similar to the adult, but from 1 to 14 days in the Endopterygota, in which the primordium is larger and develops into a structurally simpler larva. In some species development may be halted by a long period of diapause (Chapter 6). Temperature is the dominant environmental factor; the rate may decline to less than $\frac{1}{10}$ of the maximum near the lower limit. Brief exposure to supraoptimal temperatures may give rise to abnormalities observable in a later stage, especially in endopterygotes.

3.3 THE LARVAL STAGE OF EXOPTERYGOTA

First-instar larvae are structurally similar to adults, but the proportions of the three major body regions differ (head relatively large, abdomen small) and the wings and reproductive system are undeveloped. Growth is therefore allometric (heterogonic), with different regions and organs growing at different (not always constant) rates. It is also discontinuous, since the cuticular exoskeleton must be shed and a larger one constructed when biomass has increased from two- to fourfold. Each molt (ecdysis) terminates an instar, and there are commonly from five to seven, the final one being of longer duration and terminated by a conspicuous metamorphosis to the (usually) winged adult.

Each molt is initiated by release of a low-molecular-weight protein hormone from neurosecretory cells in the brain, which causes a thoracic gland to secrete the steroid hormone ecdysone. This induces the epidermal cells to digest most of the cuticle protein and its reinforcing chitin microfibrils for reuse in synthesis of new cuticle, leaving only a thin outer layer. By swallowing air (or water), the internal body pressure is raised to expand the body volume, split the residual cuticle, and allow emergence of the next-stage insect. The outer layer of the new cuticle is made rigid by cross-linking the peptide chains of the protein matrix by quinones, formed by phenoloxidase-catalyzed reaction of molecular oxygen with ortho-diphenols derived from tyrosine (Cottrell 1964). The relatively thicker endocuticle is not "tanned," and the interscleritic regions are kept flexible by the absence of chitin fibrils and by a high content of a highly elastic protein, resilin. To reduce evaporative water loss, the outer surface of the cuticle is coated by a layer of wax less than 1 micron thick, secreted and continuously renewed by special epidermal gland cells whose major components are hydrocarbons and/or free and esterified long-chain alcohols and fatty acids. Wax composition varies greatly in different species; for each there is a critical transition temperature (ranging from 30 to as high as 58°C) above which the rate of water loss increases sharply.

The juvenile hormone (JH) plays an as-yet-undefined role in larval growth, adult oogenesis, and metamorphic change (Gilbert 1976). Adult metamorphosis is determined by a temporary inactivation of the corpora allata (CA, that secrete JH) at the time of the penultimate (last larval) molt; this somehow "programs" the epithelial cells to develop adult structures at the next molt. The balance between JH and ecdysial hormones also determines the polymorphism of termite castes and of aphids and can be somehow influenced by factors such as diet or day length.

Internal growth processes are no less complex, involving either an increase in cell size, or in number of cells, or both. Neurones may increase several hundredfold in size, with little or no increase in number; the trophic glial cells of the nervous system tend to increase in number more than 20-fold, showing (like epidermal cells) high mitotic activity at molting. Muscle fibers are multinucleate, with very diverse growth patterns. During the early instars, some only grow in size, multiplying (by longitudinal cleavage) in later ones; other fibers multiply early and then only increase in size. The thoracic flight musculature undergoes great change in size and form in late instars and the early adult. The gonads increase many thousandfold in cell number, especially in later instars, and undergo differentiation in the last instar and adult (Smith, 1968). All these qualitative changes associated with larval growth make significant quantitation difficult. As general references for this and the next section, see Etkin and Gilbert (1968), Jungreis (1979), Chen (1971), and Reynolds (1980).

3.4   THE LARVAL AND PUPAL STAGES OF ENDOPTERYGOTA

Although some exopterygote larvae have a few structures adaptive to special habitats (e.g., for respiration in aquatic or endoparasitic forms) that disappear in the adult, the larvae of endopterygotes differ greatly from their adult form, not only in outer structure but in musculature, nervous system, tracheation, and many internal organs. Clusters of embryonic cells (imaginal buds or discs) are present, which during the pupal stage will develop into the typical insectan adult structures (compound eyes, legs, wings, genitalia). The larval form is adapted to invade diverse habitats and utilize food sources of many kinds and can feed and grow at phenomenal rates. A *Drosophila* larva can increase nearly a thousandfold in size in 4 days (primarily by increase in cell size, not number), a necessary adaptation to evanescent decaying food sources. Even in non-dipteran Orders, 200-fold increase in size in less than 2 weeks is not uncommon, and the growth rate verges on exponential. The expansibility of the body wall allows more than a six-fold increase in each instar, and there is no large decline (as in most exopterygotes) of the rate of food intake, growth, and oxygen consumption that would lengthen duration of the larval stage. However, when environmental conditions are suboptimal, larvae can survive for many months or even years at much slower growth rates, though in many species unfavorable conditions do not greatly prolong the larval stage but cause development to viable adults 20–30% of normal size.

Control of the molting cycle is similar to that in exopterygote (and other arthropod) larvae, except that the presence of JH causes retention of the unique larval form. A fall of the JH level during the final larval instar causes metamorphosis to the pupa at the next molt. The pupa is a quiescent, nonfeeding stage (often adapted to survival in unfavorable environmental conditions). Histolysis of larval tissues predominates during the early period, histogenesis of adult structures in the later period. The pupal stage duration ranges commonly from 4 to 14 days, but may be greatly prolonged by lower temperatures or by a diapause state.

3.5 ENERGY REQUIREMENT FOR MAINTENANCE, ACTIVITY, AND GROWTH

As general references, see Rockstein (1973, Vol. 6) and Kammer and Heinrich (1978). The most common measure of energy production is the rate of oxygen consumption, often expressed in the classic units of Q_{O_2} (mm³ O_2/mg live biomass/hr). Whether carbohydrate or fat is being oxidized, each mm³ (at NTP equal to 44.2 nanomole) represents 5 mcal (millicalories) of energy, of which only 2 mcal is as high-energy phosphate while 3 mcal are liberated as heat. Q_{O_2} values for insects range from less than 0.01 (in diapause) to 100 (in active flight), and for "normal activity" commonly from 0.5 to 2.5 at temperatures near 25°C; normal values show both erratic and rhythmic fluctuations, with peaks at the time of molting.

The minimal levels of the diapause state, sustainable for many months or even years, involve drastic changes in metabolism (such as primarily nonmitochondrial respiration) and may be surprisingly constant over a wide temperature range (Bodine & Evans 1932). In nondiapause states, maintenance energy production declines nonlinearly as temperature falls to a value of the order of a 1 mcal/mg · hr below which survival time (that normally lengthens at lower temperature) quickly falls to zero.

Growth processes require a higher level of energy production, but it is very difficult to isolate the energy cost of growth. Complete development of the egg of *Oncopeltus* at 25°C requires 300 mcal/mg live biomass (Richards & Suankrasa 1962), so the cost of embryonic growth must be somewhat lower than this value. For larval growth a tentative estimate based on analysis of biomass gain versus food intake of *Blattella* at 30°C is of the order of 1000–2000 mcal/mg live biomass (Gordon 1972). The simpler, fast-growing larvae of endopterygotes may be more efficient. Such estimates are quite different from the biomass energy *content* (measured by total combustion calorimetry) used in some ecological studies; a large fraction of the biomass plays a purely structural role, and building and maintaining these structures demands energy. Most of this is not for biosynthesis but for the activity of muscle and other tissues involved in the intake and utilization of food, and supplied with oxygen by the tracheal system.

This system consists of from 1 to 10 primary tracheal tubes on each side, invaginations of the cuticle from spiracular openings that can be closed by valves and sphincters. The tracheae interconnect and branch extensively to finer tubulatures extending deep into various organs, ultimately connecting to the tracheoles, tubes less than 1 μm in diameter and 150–300 μm long that are formed by single cells (tracheoblasts) in close proximity to active tissues. At each molt the tracheal lining is shed and larger diameter tubes form; old tracheoles persist and many new ones form and establish connections with the fine tracheae. The density of tracheolation parallels the energy demand of each tissue, since respiratory O_2–CO_2 exchange rate is a function of tracheolar surface area, and can supply oxygen at rates above 400 mm^3/mg · hr to tissues such as flight muscle.

3.6 MEASURING THE GROWTH RESPONSE

The most easily measurable growth parameter is the time interval between two recognizable (though under different environmental conditions, or for different individuals, not necessarily identical) states, for example, between egg laying and hatching or between two successive larval molts. Its reciprocal is the pure *time rate*, in units of 1/day. The rate of gain in live biomass (mg/day) is more informative but tends to increase as body size becomes larger. Comparison of biomass growth rates of an individual in successive instars, or of different insect species, requires correction for size difference by calculating the relative rate (mg gain/per g biomass/per day). Although biomass has a complex, variable composition and structure, it can be viewed broadly as a two-component system, of active and reserve-storage tissues that can "grow" independently. Storage tissues (primarily the fat body, cf. Kilby 1963) have a low water content since triglyceride droplets contain very little while glycogen and protein granules contain less than 70%. Active tissues usually contain 90% water and make up more than half of live biomass (and most of the metabolic activity underlying growth) even when they make up less than half of the dry biomass. Live biomass is a better measure of active tissue than dry biomass, although probably more variable since water is volatile. Parameters such as potassium content (so far unused) or carbon monoxide binding might be even better indices of active biomass, but compositional analysis during growth is rarely done. It is relevant to meaningful calculation of relative rates since some major biomass components, such as reserve (or cuticular–structural) proteins and triglycerides, play no active role in growth. Relative rates measured for insect species range from 1 to 1000 mg dry wt. gain/per g dry wt./per day (Slansky & Scriber 1982).

The problem of calculating the mean biomass value needed for a relative rate has not been resolved. It is usual to measure biomass at the beginning and end of an instar, but the arithmetic mean is unsuitable since biomass gain is rarely linear with time. Waldbauer (1968) determined an "integral mean" by a series of weighings during the instar and measuring the area under the weight–time curve. Gordon (1968) preferred an "exponential mean" (the mean of the natural logs of the initial and final biomass), and this in fact yields the mathematically correct value for the mean of the instantaneous relative rate during the entire time interval (Fisher 1921, Radford 1967).

Table 3.1 presents different numerical estimates of the growth rate of *Pieris* larvae on eight species of cruciferous food plants. The first two rates (1/days instar duration, and mg dry biomass gain/day) are taken directly from the work of Slansky and Feeny (1977). However, values for integral mean live biomass are recalculated from their values of the integral mean dry biomass and mean water content. The two values of relative growth rate are even less reliable, being ratios of two mean values instead of means of individual values. For each column the coefficient of variation is calculated, not as a valid random statistic but as an index of the growth-regulating power of the larvae when utilizing different food sources. By this criterion, live biomass is relatively well regulated (the CV for the corresponding dry biomass is 14%), lending some support to the view (Gordon 1972) that regulatory mechanisms tend to stabilize active tissues (with a high water content) more than the relatively inert storage ones. The pure time rate is also well regulated, suggesting that instar termination is not determined by attaining a fixed increment in biomass (in fact, starved larvae may eventually molt to the next stage despite a loss of biomass). The high CV for the absolute rate of dry biomass gain (also true for body nitrogen gain) suggests that this is the most sensitive index of plant–insect interaction. The two relative rates show an interesting difference. The one based on the well-regulated live biomass closely parallels the absolute rate of biomass gain, though slightly less sensitive. The one based on dry biomass is more highly correlated with the pure time rate and is likewise well regulated. The major conclusion of this work was that larvae can regulate nitrogen intake and retention even when the nitrogen content of the plant varies over a wide range. Much more work will be required before general rules about regulation of instar duration and biomass gain can be derived.

All numerical indices of growth imperfectly measure the interactions between larva and environment even for one instar, and values may be strongly influenced by earlier conditions. If these were different, the response may be a "transition effect" valid only for that specific transition state. Even if conditions are uniform, interaction during the entire larval stage is not well defined by any single numerical value (cf. next section). The time (τ) required for larval growth to the pupal or adult stage is an often-used simple measure, but it is poorly correlated with the percent survival (S). Howe (1971) proposed that $(\log_e S)/\tau$ be adopted as a more descriptive index of interaction, since this is a key factor in his simplified estimate of the population increase parameter r.

3.7 GROWTH CURVES

The incredibly complex phenomena underlying larval growth (egg and pupal growth still lack numerical data) generate deceptively simple curves of live biomass w versus time t. These are sigmoid, delimited by the initial

Table 3.1 Growth rate values for fifth-instar *Pieris rapae* larvae reared on leaves of various crucifers[a]

Food Plant[b]	Exp.#[c]	1/days	mg/day	mean live wt, mg	mg/g · day (live)	mg/g · day (dry)
Lunaria annua	1	0.39	8.1	95	86	500
Lepidium virginicum	2	0.40	6.3	97	65	440
B. o. (acephala)	1	0.45	11.1	120	93	587
B. o. (acephala)	2	0.47	9.5	112	85	562
B. o. (acephala)	2	0.48	9.7	118	82	560
B. pekinensis	2	0.49	9.5	116	82	524
B. o. (capitata)	1	0.49	12.5	120	104	581
Barbarea vulgaris	1	0.50	10.9	117	93	573
B. o. (acephala)	3	0.51	9.9	119	83	607
Dentaria diphylla	1	0.52	14.7	120	122	622
B. juncea	2	0.55	11.2	105	106	662
CV		10%	21%	8%	17%	11%

[a]Adapted from Slansky and Feeny (1977).
[b]*B.* = *Brassica* and *o.* the species *oleracea*, with variety in parentheses.
[c]Exp.# indicates one of three series of experiments. Listed in order of increasing time rates.

and final values (W_i and W_f), but are usually transformed to a quasilinear form by plotting biomass on a logarithmic scale. As a further simplification, molting discontinuities are ignored. Actual growth data can be roughly approximated by a differential equation that assumes constant, exponential growth:

$$\frac{dw/w}{dt} = \frac{d(\ln w)}{dt} = \frac{\ln(W_f/W_i)}{\tau} = k$$

where τ is the total development time.

The nonlinearity of growth data is caused by a decrease in k as w increases. Three simple equations have been used to fit actual data. All cause the growth rate to become zero when $w = W_f$. The monomolecular (decaying exponential) is

$$\frac{dw/w}{dt} = k(W_f/w - 1)$$

The logistic (autocatalytic, or Verhulst) is defined by

$$\frac{dw/w}{dt} = \frac{k(W_f - w)}{W_f} = k(1 - w/W_f)$$

The Gompertz equation is

$$\frac{dw/w}{dt} = k \ln\left(\frac{W_f}{w}\right)$$

All three are special cases of a more general equation

$$\frac{dw/w}{dt} = \frac{k[(W_f/w)^m - 1]}{m}$$

since this becomes the monomolecular if $m = 1$ and the logistic if $m = -1$. Although indeterminate at $m = 0$, its limit as m approaches zero is in fact the Gompertz. This was recognized by Richards (1959) in his analysis of the mathematically equivalent form of the theoretical equation of von Bertalanffy (Fabens 1965), whose hypothesis required that m be greater than zero, restricting the family of curves to intermediates between the monomolecular and the Gompertz. Richards found that the "infra-logistic" range (m less than -1) often fits actual data better. The value of m controls the shape of the sigmoid curve so that its point of inflection (maximum dw/dt) can be at any value of w instead of at $0.5W_f$ (logistic) or $0.37W_f$ (Gompertz).

Much theorizing underlies attempts to fit growth data (Gray 1929). Equations can be derived in which change is an exponential function of t instead of w (Backman 1938). The concept of relative (nonlinear) "biological time" (Chapters 11, 24) recurs in the work of Laird et al. (1965); this also recognizes that the terminal phase of growth has a major linear (accretionary instead of self-multiplicative) component. The exponential component is described by

$$\frac{dw/w}{dt} = ke^{-\alpha t}$$

and its integral resembles that of the Gompertz. The exponential decay constant α, relatively constant during active growth, is called a *coefficient of biological time*. Since the Laird equation has one less constant than the Richards equation, it is less flexible and good fit requires prior correction for the linear growth component.

Although different parts of the body may grow at different rates, Huxley's allometry rule often holds. This states that the ratio of the logarithms of the size of any two parts is relatively constant during growth (Huxley 1932). Laird (1965) applied her equation to data for various body parts; with rare exceptions, values of k (the relative growth rate at zero time) and α were closely similar to those for the whole body. All this suggests that unifying, integrative forces are at work, but mere mathematical analysis cannot elucidate their nature or complexity since the information content of the data is so meager.

3.8 EFFECT OF TEMPERATURE ON GROWTH AND DEVELOPMENT

Sacher (1967) devised a novel analysis of temperature effects (Fig. 3.2) based on fairly complete data for *Tribolium* larvae. The growth rate G ($1/\tau$, where τ is duration of the whole larval stage) curve is bell shaped with an optimum near 34°C. The metabolic rate curve M has the rough linearity of most Arrhenius plots of rate versus reciprocal of the absolute temperature. Sacher expressed the energy cost of growth by an "organizational entropy function":

$$S_{org} = R \ln (M/G) + C$$

where R is the universal gas constant (1.986 cal/mole $\cdot$ deg.) and C is an empirical constant. Entropy curve S is an inverted bell with a minimum at 26°C, which may be a more realistic biological temperature optimum than the maximum G value. An independent measure of "organizational disorder" is the percent larval mortality (curve D) that has a broad minimum in the 26–32°C range. Other data (David & Clavel 1967a,b) suggest that the coefficient of variation of G likewise increases rapidly outside the mid-

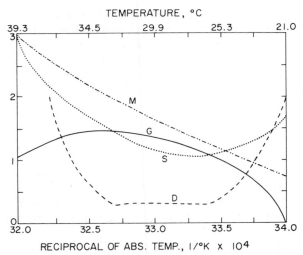

Fig. 3.2   Effect of temperature on four parameters of larval growth of *Tribolium confusum*. Curves *M* and *G* are Arrhenius plots of the mean metabolic rate and the mean growth rate (left ordinate, units are natural logarithm of reciprocal time). Curve *D* is mortality (units of 25%). Curve *S* is "organizational entropy" (units of 0.5 cal. mole^{-1} deg^{-1}). Redrawn from Sacher (1967).

range of temperature, as do developmental anomalies. The inability of any individual to grow optimally over a very wide temperature range presumably underlies selection of genetic heterogeneity, balanced polymorphisms, geographic clines, and (for prolonged temperature extremes) diapause states (Chapter 6).

Many of the simpler mathematical "models" devised to fit experimental data were critically discussed by Wigglesworth (1972). Their equations deal only with pure time rates (ignoring the complexities of biomass, survival, metabolism, etc.) and with the midrange of temperature where the relation is roughly linear. The most commonly used equation is

$$\frac{1}{\tau} = \frac{T - T_z}{K}$$

where τ is the mean (or median) development time, T is the temperature in °C, T_z is a hypothetical "biological zero" at which no development is possible, and K is the "thermal constant" in units of day-degrees. When this equation is used to predict the development time for 50% of a population at the fluctuating temperatures of normal environments, it is assumed that the distribution of individual times is unimodal normal and that "thermal summation" occurs. This means that development time will be the same if the mean temperature is the same, no matter how the temperature varies with time. This is usually true when both the maximum and minimum temperature lie within the range where the equation holds. Since the curve over a wider temperature range is sigmoidal, K actually decreases at lower temperatures and increases at higher ones. Development rate tends to be faster than predicted when one temperature extreme is below the midrange, and slower when it is above it, and the "biological zero" is also not truly constant (Messenger & Flitters 1959). Huffaker (1944) found that a higher frequency of temperature alternation (6 hr at high and 6 hr at low) gave a rate for *Anopheles* larvae 20% faster than that predicted by summation when mean temperature was low (20–24°C) and the temperature differential 9°C or more. Acceleration was less with a diurnal alternation of 6–12 hr at high and 18–12 hr at low, and even less with 15–18 hr at high and 9–6 hr at low. He concluded that the longer exposure at low temperature has a "stimulating" effect; this may be a hysteresis or transition effect. Richards and Suankrasa (1962) found that the Q_{O_2} of developing *Oncopeltus* eggs transferred from 15 to 25°C for a 5-hr period rose to the 25°C equilibrium rate in 1.5 hr, but on return to 15°C required more than 5 hr to fall to the 15°C equilibrium level; development was faster and used less metabolic energy than at the equivalent mean constant temperature. Transition to a new dynamic equilibrium is not instantaneous. The living system may be better adapted to normal environmental fluctuations than to an artificial constant state.

Development at different constant temperatures may affect not only the rate but the final state attained. Mean pupal weight of female *Lymantria monacha* was 300 mg for populations reared in the 12–20°C range, 500 mg at 25°C, and 400 mg at 28°C (the optimum for larval growth rate, Zwölfer 1935). Larvae of another moth, *Scrobipalpa ocellatella*, developed eventually into large adults in the 17–28°C range, where the primary effect was a reduction of the larval τ from 28 to 13 days; in the 30–33°C range larval development was only slightly accelerated (10 days), the major effect being a striking reduction in body size (Weismann 1959). The temperature response also *changes during growth*, reflecting the often-forgotten fact that the organism is becoming *different*. Huffaker (1944) found that the temperature optimum (for minimum development time) of *Anopheles* declined from 33.3°C in the egg stage to 30.3°C in both the last larval and pupal stages. Although the absolute time rates of different stages cannot be compared, one can calculate the *ratio* of the rates at two different temperatures for each stage. Relative to the rate at the midrange temperature of 25.4°C, the rate at 10°C is 0.09 for the egg stage and near 0.2 for all later stages, but at 14.5°C it is less than 0.3 from egg to third instar and near 0.4 for the two final stages. Not only the optimum but the shape of the curve is changing during development; the relatively fast growth of the early stages is more depressed by low temperature than the slower growth of the later stages. Variation of temperature response also exists for different stages of the life cycle in *Lymantria* (Zwölfer 1935); this univoltine species lays eggs in the spring, and the high-survival temperature range is 10–23°C (optimum 22°C) for the egg stage, but 14–31°C (optimum 28°C) for the larval stage. Neither the breadth nor the limits of the survival temperature range are fixed, since "conditioning" can occur (Mellanby 1954). Interspecific differences in range can be very great, 12–50°C for *Thermobia*, and −3–12°C for *Grylloblatta*; however, temperature optima for most species lie in the range from 22 to 38°C, with the lower limit of growth lying from 12 to 22°C below the optimum (Taylor 1981). Values are adaptive to the variety of climates encountered (since the temperature response of biotypes of one species differs) and also influenced by the food supply (Maercks 1935, Sharpe & Hu 1980). Laudien (1973) reviewed over 1000 publications in this general area (cf. Clarke 1967).

The foregoing discussion of the complexity and variability of the temperature response suggests that mathematical models must be in reality empirical and descriptive, even when (as in growth curves) the equations arise from theorizing about the underlying phenomena. In fact, equations used to describe growth as a function of time can be fitted to the S-shaped curve for rate–temperature in the region below the optimum temperature by redefining the parameters. However, two equations allow fitting of rate data even above the optimum temperature since they generate bell-shaped curves. The older is the generalized inverted catenary of Janisch (1932):

$$\frac{1}{\tau} = \frac{2a}{b^{T-T_m} + b^{T_m-T}}$$

where τ is the development time at temperature T (°C), T_m is the temperature at which τ is a minimum (maximal rate, and in that sense the optimum), a is the measured rate at T_m, and b expresses the rate of decline in growth as T diverges from T_m. Good fit is obtained for larval growth of *Anopheles* (Huffaker 1944) and of several species of muscids with widely different temperature optima and ranges (Larsen & Thomsen 1940). A newer equation (likewise frankly empirical) is based on the normal curve (Taylor 1981)

$$\frac{1}{\tau} = ae^{-b(T-T_m)^2}$$

For many chemical reactions the Arrhenius equation, which in logarithmic form is

$$\ln r = \ln a + \frac{\ln b}{T}$$

fits data for reaction rate (r) as a function of the absolute temperature (°C + 273.2) well, and it can also be fitted to biological rate data over a fairly wide range. It was assumed that the constant b had a simple thermodynamic significance, allowing calculation of the activation energy of a hypothetical "master reaction" that determined the measured rate. The literature was critically analyzed by Johnson et al. (1954), who concluded that control of a complex system by a single reaction is improbable. The concept has been extended (by adding two hypothetical temperature-inactivation reactions) and a complex equation derived that can be fitted to biological rate data over the entire temperature range (Schoolfield et al. 1981). Unless their thermodynamic constants can be shown to correspond with actual enzyme reactions that truly control the rate, simpler empirical equations (Stinner et al. 1974, Logan et al. 1976) may be preferable. It is also an open question whether all data points should be given equal weight in such exercises in least-squares curve fitting.

3.9 WATER BALANCE

Water is both a major (50–90%) and a volatile component of biomass, as well as a minor but important constituent of the atmosphere. Pure water has a vapor pressure (p_0, mm Hg) that is approximately calculable from

$$\log p_0 = 9.1 - \frac{L}{2.3\ RT} = 9.1 - \frac{2300}{T}$$

where L is the latent heat of vaporization (near 10,440 cal/mole in the biological temperature range), R the gas constant (1.986 cal/mole $\cdot$ °C), and T the absolute temperature (°K). In the 0–40°C range the numerical value of p_0 is also roughly equal to the saturation concentration (mg water/liter of air). The relative humidity (RH) is the percentage of the actual water concentration to the saturation value.

In living organisms a fraction of the water content is bound to tissue constituents, and the equilibrium vapor pressure (p) is lowered according to the approximate equation

$$\log p = \log p_0 - \frac{PV}{2.3RT} = \log p_0 - \frac{0.095P}{T}$$

where P is the osmotic pressure (in atmospheres), V the molar volume of water (0.018 liter/mole), and R the gas constant (0.082 liter-atmospheres/°C.). Osmotic pressure in insects is commonly measured as freezing-point depression (°C) of minute samples of hemolymph, convertible to atmospheres by multiplying by 41.7 or to milliosmoles by multiplying by 537.6. The P value of insects is commonly near 450 milliosmoles, in equilibrium with RH near 99%, and seldom exceeds 700 (RH 96%). Few environments are so humid, and many mechanisms have been evolved to lower the rate of water loss (Gilby 1980), especially in species adapted to very dry environments (Chapter 6).

In feeding stages the inevitable water loss to the air is replaced by ingestion of free water and/or water content of the food source. However, in eggs or pupae of many species, and in larvae adapted to dry habitats, metabolic water production can compensate for low rates of water loss, since each mm^3 of O_2 produces 0.56 µg of water from oxidation of 0.5 µg of fat. In some insect and mite species metabolic energy can be utilized to extract water even at RH as low as 45% (Wharton & Arlian 1972, Machlin 1979). The free-energy cost (e, mcal/µg water extracted) is approximately 0.033 $\ln(a_v/a_w)$, where a_v is the atmospheric RH and a_w the equilibrium RH for the internal P, not far from 99%. Values of e are roughly 0.026 at RH 45% and 0.009 at 75%. Metabolic efficiency is not known, but assuming it is 10% the oxidation of 1 µg of fat would extract 40 µg of water at RH 45%, far more than the metabolic water production of 1.13 µg.

The effect of humidity on insect development cannot be described by any general rules like those that seem to govern temperature effects, and is often influenced by temperature (Schwerdtfeger 1963). Some species show RH optima, from as low as 30% to as high as 90%, occasionally with a narrow range permitting high survival. Others show little effect until RH falls to a critical level that is temperature-dependent. Others show a gradual lengthening of development time as RH declines. The egg and pupal stages are often unaffected by humidity over a very wide range, except at temperature extremes. The interaction between temperature and humidity is often presented as a thermohygrogram, these two parameters being the axes, with data curves either for equal percentage mortality or equal development time, the mortality effect usually being more sensitive. An example is Fig. 3.3, that shows the 5% and 50% mortality curves for three successive stages of *Lymantria monacha*, a species adapted to a relatively cool and moist climate. First-instar larvae are less tolerant to low humidity than either eggs or second-instar larvae. Humidity is more difficult to control than temperature, and the effect of diurnal fluctuations is not known, though in natural environments it tends to be inversely related to temperature.

3.10 EFFECTS OF QUANTITY AND QUALITY OF FOOD SUPPLY

Food is by far the most complex environmental factor, and the nature of the food determines much of the structure and functioning of insects (Chapters 4, 5, 6, 7). The basic quantitative measure is the intake in grams (dry weight) that is often determined indirectly as the area of leaves consumed or the weight of frass excreted (Waldbauer 1968); if a nonabsorbable marker (such as chromic oxide or alumina) can be added at a known percentage to the food, the quantity of marker excreted is a good index of intake (McGinnis & Kasting 1964, Van Herrewege & David 1969). The

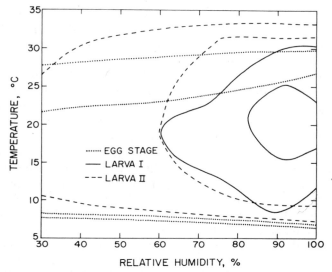

Fig. 3.3   Thermohygrograms for the developing egg and the first two larval instars of *Ly-mantria monacha* (adapted from Zwölfer 1933). Curves represent equal percentage mortality, 5% for the innermost and 50% for the outermost of each pair.

techniques are laborious, especially for minute insects or mites (Stepien & Rodriguez 1972) and not applicable to many species that require an aquatic or gel medium.

Food quality involves many parameters other than chemical composition (that is known, to a limited extent, only for very few natural food sources— cf. Chapters 4, 20). Physical factors such as hirsute leaves or tough seed-coats may make feeding difficult or impossible. A more subtle barrier is palatability (willingness to feed), determined by the action of a variety of volatile or water-soluble constituents on insect chemoreceptors, with either a phagostimulant or a phagodeterrent effect (Dethier 1980, Bernays & Chapman 1977, Chapters 4,5). The rate of intake of an acceptable food is determined by a complex sequence of processes. Food that is not liquid or particulate must be comminuted by sclerotized mouthparts and mixed with saliva into a brei. Most adult insects (and exoterygote larvae) ingest discrete meals at fairly long intervals, storing the brei in a saclike crop. Crop capacity sets an upper limit to meal size, usually a small fraction of body weight (hematophages sometimes ingest many times their weight). Crop contents are released gradually into the midgut, a long tube lined by a single layer of endothelial cells, enveloped in muscle fibers that control mixing and rearward flow, and usually protected from direct contact with the food by a thin microporous peritrophic membrane. Most cells are differentiated for enzyme secretion and nutrient absorption, but embryonic cells are present that generate new trophic cells (especially at each molt).

"Batch feeders" tend to have relatively low mean food intake rates since transit time (from ingestion to excretion of indigestible material) may be 24 hr or longer, correlated with relatively efficient digestion of food biopolymers (e.g., proteins, polysaccharides, nucleic acids). Food intake rate is to some extent determined by digestibility since dilution with an equal weight of an indigestible material (e.g., as cellulose) may nearly double the rate. Such a feeding strategy is advantageous if there is no strong selection pressure for very fast growth and if acceptable food sources are limited in number and/or scattered in space. However, sap-feeding Homoptera and most endopterygote larvae are "continuous feeders," with transit times of a few hours. Their relatively long midguts have various functional adaptations to optimize efficiency of digestion and absorption, but these are almost never complete since their rates undergo severely diminishing returns as the substrate concentration within the gut falls. Incomplete digestion of food consumed at a high rate will maximize growth rate if transit time is adjusted to maximize the quantity of products absorbed per unit time. This strategy is enhanced if the digestive and metabolic machinery is adapted to a specific food source, as in oligophages; for example, *Manduca sexta* larvae deprived of maxillary chemoreceptors consume leaves of a nonhost plant at the same high rate as those of a host plant, but digest the nonhost only half as efficiently (Waldbauer 1968). Food intake rate of *Pieris* larvae is higher when the N content is lower, and protein digestion is more efficient, suggesting regulatory mechanisms to optimize growth rate (Slansky & Feeny 1977). Compensatory mechanisms may cause similar growth response to food sources quite different in composition (Gordon 1972, Barton Browne 1975). This is evident from the fundamental dry-mass-balance equation

$$G = F - S - O = I - O$$

in which each term is in relative rate units (mg/g live biomass/day), summing up many sequential and simultaneous activities with complex interactions and controls that undergo both gradual and abrupt changes during growth and maturation and that all have upper and lower limits. I is the rate of inflow of nutrients to the living system, equal to the difference between the food intake rate F and the rate of excretion of unutilized solids S. The value O is the rate of loss of such nutrients (that may in part have been previously incorporated in the biomass) by oxidative reactions, complexly related to the Q_{O_2} and the associated production of CO_2, H_2O, and energy. G is the rate of incorporation of the remaining nutrients into both active and storage biomass, primarily controlled by the composition of I (current nutritional status) but also influenced by the composition of the live biomass (reflecting prior nutritional status) that may contain reserves of certain nutrients sufficient to compensate temporarily for deficiencies in the current food supply. Egg reserves of indispensable polyunsaturated fatty acids may be large enough to permit normal growth of larvae on diets lacking them (Dadd 1973).

Studies of the adequacy of food sources for growth can be misleading if their duration is restricted to a few instars or to a single generation. Larvae previously reared on high-quality food may temporarily grow well when transferred to an inadequate food; for example, early fifth-instar larvae of *Prodenia eridanea* from a culture on leaves of *Chenopodium album* grow faster (with lower mortality) on *Malus floribunda* than on *Ulmus pumila*, though only *Ulmus* leaves continue to support growth during the sixth instar (Soo Hoo & Fraenkel 1966). The *Prodenia* study showed that this polyphagous species can grow (for the last two instars, after transfer from *Chenopodium*) on a great variety of plant species. However, another polyphage (*Hyphantria cunea*) could not be maintained for several generations on many host plants that allowed good growth and reproduction of a single generation (Morris 1967), though continuous rearing on alder was possible. Reliance on "primary hosts" is one possible polyphage strategy, others being host alternation or biotypes of varying host specificity (Hovanitz & Chang 1963, Bush 1969, Lachaise 1974). The distinction between polyphagy and oligophagy is lessened by the observation that many first-instar larvae of *Manduca sexta* will feed on a wide spectrum of plants and remain "polyphagous" if reared on nonsolanaceous foods but become progressively "conditioned" to feed exclusively on a solanaceous host if this is the sole food source during the early instars (Yamamoto 1974). Transition phenomena and long-term selection of new biotypes in nature or in laboratory rearing imply that generalizations about food range may need revision.

Since G for complete development is $(1/\tau) \ln (W_f/W_i)$, any factor that lowers growth rate may entail either longer development time, lower final biomass, or both. Larval female grasshoppers, fed only 25% of normal daily food consumption, prolong the early instars to maintain body weight above 50% of normal, but shorten the final instar by molting to adults of only 30% normal size (Muthukrishnan & Delvi 1974). Larvae of the parasite *Pimpla turionellae* can develop from single eggs laid in lepidopteran pupae exhibiting over 50-fold difference in size (Fig. 3.4); mean time to adult emergence is well regulated ($\pm 16\%$) at about 19 days for males and 22 days for females, but body size varies by a factor of 9 for males and 5 for females. In small hosts G is much lower, but food conversion efficiency is higher since the ratio of adult parasite weight to host weight varies from 0.4 in the smallest hosts to less than 0.1 in the largest.

Even when food supply is not limited, G is lowered by qualitative differences of many kinds, such as presence of growth inhibitors (Waiss et al. 1981) or deficiency or excess of nutrients. Efficient incorporation is possible only if all essential nutrients are present in the proper balance. The concept of balance is illustrated in Fig. 3.5, plotting pure time rate, as $\log(1/\tau)$, versus relative concentration of nutrients in agar gel. The curve for a well-balanced diet (COM) shows that below a critical dilution the rate declines

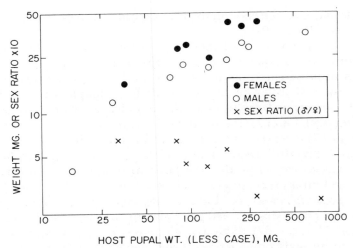

Fig. 3.4 Mean adult body weight of *Pimpla turionellae* reared on pupae of nine different species of Lepidoptera (mean pupal weight range from 16 to 850 mg, excluding weight of pupal case). The mother lays only unfertilized (male-determined) eggs in the smallest pupae. As host pupal weight increases, male/female ratio declines from 0.66 to 0.25 (the stable value at host weights above 200 mg). Plotted from data of Arthur & Wylie (1959).

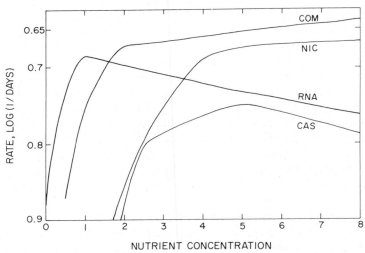

Fig. 3.5 Growth rate, log (1/τ days), of *Drosophila* larvae on artificial diets of varying nutrient concentration. Logs are actually negative values, so a smaller number means a faster rate. COM is dilution of a complete diet (units 0.5 g/250 ml) in agar gel (David et al. 1970); ordinate incremented by 0.3 to facilitate comparison. Other curves redrawn from Sang (1959) vary only single components of a basal diet, NIC units μg/5 ml, RNA units g/250 ml, CAS units g/100 ml.

sharply, presumably because feeding at the maximum rate cannot provide enough nutrients to sustain fast growth; above that level feeding rate can be lowered to maintain growth rate at the highest possible value. The other curves are for diets in which concentration of only one component is varied (all others being at the same optimal level). Nicotinamide (NIC) is a vitamin required only in minute amount; below a critical level it limits growth, while above it imbalance has no effect since metabolism can easily store or degrade minute excesses. Ribonucleic acid (RNA) can be biosynthesized (since slow growth is possible if it is omitted), but growth is accelerated as its concentration is increased to an optimum level where all absorbed RNA can be metabolized quickly; above that level the rate of degradation and excretion of the excess RNA becomes growth limiting. The curve for a major dietary component (CAS, bovine milk protein, casein, providing all required amino acids) needs more complex interpretation since several different imbalances are involved. The composition of casein (adequate for a slow-growing mammal) is severely imbalanced for fast-growing insects. Some of its amino acids (not always the same ones) are growth limiting at various points of the CAS curve; below the optimum a deficiency controls growth, above it an excess, but metabolic interactions are so complex that the limiting factors are unidentifiable.

The intricacy of the problems is revealed in the work of Rock (1972) on growth of *Argyrotaenia velutinana* on amino acid mixtures. One might expect that the optimum diet composition would resemble the body composition, especially for those amino acids that the larva cannot biosynthesize. However, body composition changes during growth since different parts have different composition, and growth is allometric. Also, biochemical adaptations are to extraspecific foods usually differing in composition, not to cannibalism. Casein has a lower content of arginine (and several other essential amino acids) and a much higher content of serine than those of *Argyrotaenia*, but growth ($1/\tau$) on a casein diet is accelerated by supplementation either with arginine or with serine. Arginine corrects a major deficiency, so that a given intake of casein can be utilized faster and more efficiently. Serine is a metabolic precursor of the formate and glycine required for purine synthesis (not only for nucleic acids but for uric acid synthesis, the major pathway for excretion of excess ammonia), and so allows faster degradation of amino acids (such as leucine) of which casein has a large excess, making possible a faster rate of intake. Since only the pure time rate was measured, interpretation is speculative.

Figure 3.6 shows how growth-regulating mechanisms stabilize the major parameters of G when either total nutrient concentration (COM) or concentration of only one component (two vitamins, thiamine and pyridoxine, and one structural phospholipid component, choline) are progressively lowered. The response to imbalance initially involves a longer development time, allowing maintenance of normal final biomass (because prolonged feeding time, possibly aided by a higher rate of intake, enables a sufficient

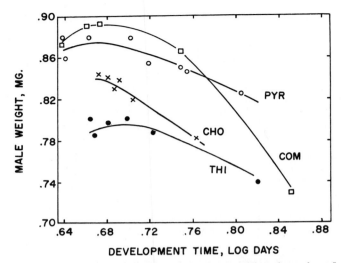

Fig. 3.6 Relation of final adult weight to larval development time, log τ days, for *Drosophila* males reared inmedia in which all nutrients (COM) or only single components are present in decresing concentration (left to right). COM data taken from David et al. (1970), other data from Sang (1959). COM and CHO curves slightly relocated to facilitate comparison. Since basal diet compositions differed in the experiments, actual weight and rate maxima varied.

total food intake). When τ exceeds the normal by roughly 30%, the larvae sacrifice normalcy of body size to avoid excessive prolongation of development. Lowering G even by a factor of 2 causes no great increase in larval mortality, but the variance of both τ and W_f increases. It is also worth noting that τ is nearly the same for males and females, though W_f of males is 33% less.

3.11 OVERVIEW

Development is a sequence of distinct stages: egg, larval-feeding instars, a pupal instar (in endopterygotes), and imago (adult). Duration of each stage, changes in size and/or form, and probability of survival are determined by complex interactions of internal growth-regulating mechanisms, the current environment, and many cumulative residua (e.g., nutrient reserves) of earlier stages, including the parental generation. Measurable parameters include time, biomass, energy, food intake, and percent survival, from which various indices can be derived. Every environmental factor, when varied in an otherwise constant environment, has a range that allows high survival, although other developmental parameters may show large quantitative variations. The range (and various optima within it) may differ for different stages and even more for different biotypes of one species that may coexist in one population. Some environmental interactions can be described by useful empirical equations, sometimes derived from simplifying theoretical assumptions.

REFERENCES

Anderson, D. T. 1972. Pp. 95–242 in Counce and Waddington (Eds.), 1972, referenced here.

Arthur, A. P. and H. G. Wylie. 1959. *Entomophaga* **4:** 297–301.

Ashburner, M. and T. R. F. Wright. 1978–1980. *The Genetics and Biology of Drosophila*, Vols. 2a, b, c, d. Academic, New York.

Backman, G. 1938. *Archiv. Entwicklungsmech. Org.* **138:** 37–58.

Bakker, K. 1969. *Neth. J. Zool.* **19:** 541–595.

Barton Browne, L. 1975. *Adv. Ins. Physiol.* **11:** 1–116.

Bernays, E. A. and R. F. Chapman. 1977. *Ecol. Entomol.* **2:** 1–18.

Bodine, J. H. and T. C. Evans. 1932. *Biol. Bull.* **63:** 235–245.

Bush, G. L. 1969. *Evolution* **23:** 237–251.

Chen, P. S. 1971. *Monogr. Develop. Biol.* **3:** 1–230.

Clarke, K. U. 1967. Pp. 293–352, in Rose (Ed.), 1967, referenced here.

Cottrell, C. B. 1964. *Adv. Ins. Physiol.* **2:** 175–218.

Counce, S. J. and C. H. Waddington (Eds.). 1972. *Developmental Systems: Insects*, Vol. I. Academic, New York.

Dadd, R. H. 1973. *Annu. Rev. Ent.* **18:** 381–420.

David, J., and M.-F. Clavel. 1967a. *Naturaliste Can.* **94:** 209–219.

David, J. and M.-F. Clavel. 1967b. *J. Insect Physiol.* **13:** 717–729.

David, J., P. Fouillet, and J. van Herrewege. 1970. *Ann. Soc. Ent. Fr. (N.S.)* **6:** 367–378.

Dethier, V. G. 1980. *Am. Natur.* **115:** 45–66.

Dingle, H. (Ed.). 1978. *Evolution of Insect Migration and Diapause.* Springer-Verlag, New York.

Dobzhansky, T. 1951. *Genetics and the Origin of Species.* Columbia University Press, New York.

Etkin, W. and L. I. Gilbert (Eds.). 1968. *Metamorphosis.* Appleton-Century-Crofts, New York.

Fabens, A. J. 1965. *Growth* **29:** 265–289.

Fisher, R. A. 1921. *Ann. Appl. Biol.* **7:** 367–372.

Gilbert, L. I. (Ed.). 1976. *The Juvenile Hormones.* Plenum, New York.

Gilby, A. R. 1980. *Adv. Ins. Physiol.* **15:** 1–33.

Gordon, H. T. 1968. *Am. Zool.* **8:** 131–138.

Gordon, H. T. 1972. Pp. 73–105 in Rodriguez (Ed.), 1972, referenced here.

Gray, J. 1929. *Br. J. Exp. Biol.* **6:** 248–274.

Hinton, H. E. 1981. *Biology of Insect Eggs*, Vols. I, II, and III. Pergamon, Oxford.

Hovanitz, W. and V. C. S. Chang. 1963. *J. Res. Lepid.* **2:** 185–200.

Howe, R. W. 1966. *J. Stored Prod. Res.* **2:** 117–134.

Howe, R. W. 1971. *J. Stored Prod. Res.* **7:** 63–65.

Huffaker, C. B. 1944. *Ann. Entomol. Soc. Am.* **37:** 1–27.

Huxley, J. S. 1932. *Problems of Relative Growth.* Methuen, London.

Janisch, E. 1932. *Trans. Entomol. Soc. Lond.* **80:** 137–168.

Johnson, F. H., H. Eyring, and M. J. Polissar. 1954. *The Kinetic Basis of Molecular Biology.* Wiley, New York.

Jungreis, A. M. 1979. *Adv. Insect Physiol.* **14:** 109–183.

Kammer, A. E. and B. Heinrich 1978. *Adv. Ins. Physiol.* **13:** 133–228.

Kilby, B. A. 1963. *Adv. Ins. Physiol.* **1:** 111–174.

Kogan, M. 1972. Pp. 107–126 in Rodriguez (Ed.), 1972, referenced here.

Lachaise, D. 1974. *Ann. Soc. Entomol. Fr.* **10:** 3–50.

Laird, A. K. 1965. *Growth* **29:** 249–263.

Laird, A. K., S. A. Tyler, and A. D. Barton. 1965. *Growth* **29:** 233–248.

Larsen, E. B. and M. Thomsen. 1940. *Vidensk. Med. dansk. naturh. Foren.* **104:** 1–75.

Laudien, H. 1973. Pp. 355–399 in Precht et al. (Eds.), 1973, referenced here.

Lints, F. A. and G. Gruwez. 1972. *Mech. Age. Dev.* **1:** 285–297.

Logan, J. A., D. J. Woolkind, S. C. Hoyt, and L. K. Tanigoshi. 1976. *Environ. Entomol.* **5:** 133–1140.

Machlin, J. 1979. *Adv. Ins. Physiol.* **14:** 1–48.

Maercks, H. 1935. *Ar. Physiol. angew. Entomol.* **2:** 175–195.

Masaki, S. 1978. Pp. 72–100 in Dingle (Ed.), 1978, referenced here.

McGinnis, A. J. and R. Kasting. 1964. *Science* **144:** 1464, 1465.

Mellanby, K. 1954. *Nature* **173:** 582, 583.

Messenger, P. S. and M. E. Flitters. 1959. *Ann. Entomol. Soc. Am.* **52:** 191–204.

Morris, R. F. 1967. *Can. Entomol.* **99:** 24–33.

Muthukrishnan, J. and M. R. Delvi. 1974. *Oecologia* **16:** 227–236.

Precht, H., J. Christophersen, H. Hensel, and W. Larcher. 1973. *Temperature and Life.* Springer-Verlag, New York.

Radford, P. J. 1967. *Crop Sci.* **7:** 171–175.

Reynolds, S. E. 1980. *Adv. Ins. Physiol.* **15:** 475–595.

Richards, A. G. and S. Suankrasa. 1962. *Entomol. Exp. Appl.* **5:** 167–178.

Richards, F. J. 1959. *J. Exp. Bot.* **10:** 290–300.

Rock, G. C. 1972. Pp. 183–197 in Rodriguez (Ed.), 1972, referenced here.

Rockstein, M. (Ed.). 1973. *The Physiology of Insects* (2nd ed., 6 vols.). Academic, New York.

Rodriguez, J. G. (Ed.). 1972. *Insect and Mite Nutrition.* North-Holland, Amsterdam.

Rose, A. H. 1967. *Thermobiology.* Academic, New York.

Sacher, G. A. 1967. *Ann. N.Y. Acad. Sci.* **138:** 680–712.

Sander, K. 1976. *Adv. Ins. Physiol.* **12:** 125–238.

Sang, J. H. 1959. *Ann. N.Y. Acad. Sci.* **77:** 352–365.

Schoolfield, R. M., P. J. H. Sharpe, and C. E. Magnuson. 1981. *J. Theor. Biol.* **88:** 719–731.

Schwerdtfeger, F. 1963. *Autokologie. Die Beziehungen Zwischen Tier und Umwelt.* Verlag Paul Parey, Berlin.

Sharpe, P. J. H. and L. C. Hu. 1980. *J. Theor. Biol.* **82:** 317–333.

Slansky, F. and P. Feeny, 1977. *Ecol. Monogr.* **47:** 209–228.

Slansky, F. and J. M. Scriber. 1982. *Bull. Entomol. Soc. Am.* **28:** 43–55.

Smith, D. S. 1968. *Insect Cells, their Structure and Function.* Oliver & Boyd, Edinburgh.

Soo Hoo, C. F. and G. Fraenkel. 1966. *J. Ins. Physiol.* **12:** 711–730.

Stepien, Z. A. and J. G. Rodriguez. 1972. Pp. 127–151 in Rodriguez (Ed.), 1972, referenced here.

Stinner, R. E., A. P. Gutierrez, and G. D. Butler. 1974. *Can. Entomol.* **106:** 519–524.

Taylor, F. 1981. *Am. Natur.* **117:** 1–23.

Van Herrewege, C. and J. David. 1969. *Ann. Nutr. Alim.* **23:** 253–268.

Waiss, A. C. Jr., B. G. Chan, C. A. Elliger, D. L. Dreyer, R. G. Binder, and R. C. Gueldner. 1981. *Bull. Entomol. Soc. Am.* **27:** 217–221.

Waldbauer, G. P. 1964. *Ent. Exp. Appl.* **7:** 253–259.

Waldbauer, G. P. 1968. *Adv. Ins. Physiol.* **5:** 229–288.

Weismann, L. 1960. Pp. 384–387 in Symposium on the Ontogeny of Insects, Prague, 1959. Academic, New York.

Wharton, G. W. and L. G. Arlian. 1972. Pp. 153–165 in Rodriguez (Ed.), 1972, referenced here.

Wigglesworth, V. B. 1972. *The Principles of Insect Physiology* (7th ed.). Chapman & Hall, London.

Yamamoto, R. T. 1974. *J. Insect Physiol.* **20:** 641–650.

Zwölfer, W. 1933. *Z. angew. Entomol.* **20:** 1–50.

Zwölfer, W. 1935. *Z. angew. Entomol.* **21:** 333–384.

Chapter 4

The Food of Insects

K. S. HAGEN, R. H. DADD,
and JOHN REESE

4.1. INTRODUCTION

Insects occur in every habitat save the depths of the seas. They owe much of their diversity and abundance to the evolution of a variety of feeding habits that facilitate utilization of nearly every natural organic substance. In achieving this, members exhibit diverse adaptations which include: specializations relating to digestion; detoxifying enzymes for combating plant poisons; interdependencies with mutualistic or symbiotic microorganisms; extreme changes in life-style associated with metamorphosis and polymorphism; exploitation of all conceivable modes of dispersal; and entry into various dormant conditions that tide them over the worst seasonal and environmental exigencies. Many of these are directly or indirectly involved with food.

Nutritional ecology attempts to relate what the insect or insect population requires of its food to what its environment can provide throughout changing seasonal and other circumstances. An ideal study (not yet produced) would involve both detailed knowledge of the insect's nutrititional and other dietetic and alimentary needs and a complementary knowledge of the potential of natural foods to assuage these needs, with consideration of the flux of both food chemistry and the insect's nutrient requirements at different seasons and life stages. While much is now known of the nutrient needs of many insects, and recent years have seen an explosion of information on their requirements for chemicals related to their feeding, digestion, and other behaviors, a complementary detail in our knowledge of the ability of potential foods to provide these chemicals is lacking, and this clouds all attempts to integrate the two sorts of knowledge. Hence, for all their potential importance to ecology and pest control, the ideas and hypotheses put forward in the area of nutritional ecology tend to be speculative, tentative, and often contentious.

Growth, development, and reproduction of insects are directly dependent on the quantity and quality of food ingested. Moreover, ingestion of food depends upon its being found and accepted. And besides being available, acceptable, digestible, assimilable, and able to provide all nutrients required for energy production and biomass increase, food must also provide the many chemicals other than nutrients that influence the necessary behavior of insects, whether directly involved in alimentation or as effectors of functions distinct from alimentation. Semiochemicals, or effectors of behavior, include attractants for orientation to food, stimulants to feeding, and some, conversely, have roles in deterring and repelling insects from much that might otherwise be eaten; they also include many chemicals that are precursors or sources of the hormones, pheromones, kairomones, and allomones involved in mating, oviposition, defense, and other intra- and interspecific phenomena in the ecology, psychology, and sociology of insects. Social insects order their caste heirarchies in large part via the mediation of such behavioral chemicals, and semiochemicals indirectly contribute to the organization of biotic communities, since they affect food and hence spatial partitioning that must enter the calculus of possible food webs in ecosystems. Thus insects have evolved requirements for, or have adapted to, a variety of dietetic chemicals, with consequences which ramify the interrelations that constitute the structure of ecology through the cellular, organismal, intraspecific, interspecific, and community levels.

Before examining particular cases of alimentary specializations, we review what insects *require* of food for growth, development, and reproduction—the province of dietetics and nutrition. There is some confusion in the use of these terms, so we impose our distinction on them at the start. Nutrition in the strict sense deals with the delineation of the minimal complement of nutrients required for successful growth, development, reproduction, and indefinite continuation of the species. It also covers the quantitation of required nutrients: their balance in relation to total food intake, absolute amounts of them required in relation to production of biomass, and, when alternatives supporting the same functions are available, their relative utilizability and efficiency. As is commonly used, "dietetics" comprises two types of information, both more vague than "nutrition," but potentially broader in scope. First, dietetics sometimes embraces all natural foods that can be used for successful functioning, viz., the sense employed in human studies and in those observational determinations of the natural foods, the *dietaries,* of many insects that provided the basis of Brues's (1946) classic compendium. Alternatively, dietetics may be viewed as the delineation of all necessary attributes of an organism's food, not only of the nutrients but also such features as feeding stimulants, structural or other characteristics that ensure feeding, amenability to digestion, and passage of the food through the gut. An artificial mixture comprising all, and only, the simple chemical nutrients required by an insect will generally not support good growth unless formulated to include also various nutritionally inert components such as agar or cellulose powder to provide a texture akin to that of the natural food, or perhaps a bulkiness in the gut to favor gut mobility. Additionally, specific chemical feeding stimulants may be necessary. All such nonnutrient components of food, artificial or natural, are extraneous to nutrition in the strict sense but are of equal importance together with nutrients in terms of dietetics, for without them feeding is impaired, and any possibility of adequate nutrition is abrogated, however nominally adequate the nutrients. In practice, nutrition usually cannot be studied without some concern for dietetics in both its connotations: A knowledge of natural dietaries suggests what potential nutrients to study in nutritional studies; and an appreciation of the necessary nonnutrient dietetic attributes (above) precedes the possibility of concocting the chemically defined diets upon which the study of specific nutritional requirements depends.

Nutrition in the narrow sense defined above encompasses two sorts of information: qualitative determination of all types of nutrient required, with the ultimate aim of defining all needed nutrients in the simplest chemical terms in which they can function, and quantitative determinations of how much of any nutrient is required and how well it is utilized in terms of energy and biomass formation. Acquisition of the first type of information requires, first, some knowledge of the chemical composition of foods in the insect's natural dietary, then proceeds to formulation of artificial diets from food-related chemicals, and a determination of the components necessary for growth or other indices of performance. As knowledge of food composition sharpens, the definition of synthetic diets increases, and hence the knowledge from studies using such diets becomes more refined. Thus, this line of enquiry usually departs from experimentation with natural foodstuffs to use of evermore chemically defined synthetic mixtures. Study of quantitative nutrition, by contrast, can proceed from any level of knowledge of food composition. At its simplest, it would determine how much of a natural food is needed to produce so much growth, and the general utilizability could be determined from the proportion wasted as excrement. Such quantitative data can be gathered for any level of chemical definition of a foodstuff, natural or artificial. Indeed, with vertebrates this type of study, at various levels of precision, is mandated by the economic importance of knowing, quantitatively, how well expensive foodstuffs are transformed to meat with highest overall efficiency. With insects it is otherwise; very detailed qualitative studies of the specific nutrient requirements of many species have been made, with little attention to quantitative considerations beyond empirical determinations of optimal proportions of nutrients in synthetic diets.

With a few exceptions, quantitative insect nutrition has dealt mainly with natural foodstuffs, sometimes with artificial foods composed mainly of crude fractions of natural foods, and only occasionally with synthetic diets. This is a serious weakness since lack of quantitation makes it difficult to precisely relate nutritional findings to other aspects of insect growth and behavior. This is unfortunate for this account of nutritional ecology, the essence of which should be to show quantitative patterns governing the sorts and magnitudes of insect populations that can be supported by the food potentials of various habitats. We regretfully note that with a near absence of such quantitative information, nutritional ecology has to be largely conjectural.

4.2 NUTRITIONAL REQUIREMENTS

About 30 reviews of insect nutrition have appeared over the past 40 years, recent ones being by Rodriguez (1972), House (1972, 1974), Dadd (1970 a,b, 1973, 1977), and Friend & Dadd (1982). From these all previous reviews can be accessed, both the general ones and those dealing with particular groups of insects, and via these the complete primary literature of insect nutrition. Here we present a synopsis, for the most part with minimal further citation, since the core information accessible via these reviews now forms a massively validated canon. Our account is intentionally unbalanced. Since its purpose is to provide a framework against which the bearing of nutrition on insect ecology may be discussed, most attention is focused on those classes of nutrients most frequently invoked in an ecological context.

It is now recognized that the overall qualitative nutrient requirements of all insects are basically similar and are, in turn, similar to those of animals in general as exemplified by the thoroughly investigated domestic vertebrates (Maynard & Loosli 1978). It was previously thought that insects might have nutrient requirements of unique and special types because so many insect species fed and flourished with foodstuffs which seemed utterly different from those of the generality of animals, while some fed only on a single plant species. This is now known to be the result either of very specific requirements for inciting feeding, discussed later, or, with many insects whose foodstuffs superficially seem totally unnutritious, utilization depends on assistance of microorganisms, commensal or symbiotic. In our synopsis here we include both the overt foodstuff and any microorganismal products associated with it as the extrinsic nutrient source.

Being heterotrophs, all animals require exogenous protein or some other source of amino acids for tissue construction, usually as a major fraction of their food intake. A further major fraction of intake is required for the metabolism that provides energy, and this fraction may be carbohydrate, fat, some portion of the amino acids of protein, or various combinations of these classes of nutrient. Different types of insects have evolved diverse facilities whereby their energy needs may be satisfied via extremely different combinations of these three possible energy-producing classes of nutrient, and this, together with their differing facilities for using diverse balances between tissue-building amino N nutrients and energy-providing nutrients, to a large extent governs the type of natural foodstuffs constituting their dietaries and hence their place in the ecological web. We later return to this, but first we consider the physiologically very important but quantitatively minor classes of nutrient.

4.2.1 Mineral Nutrients

All insects require various inorganic chemicals as a small fraction of their nutrient intake, and although insect mineral nutrition has been neglected, one can say that all insects require appreciable amounts of potassium, magnesium, and phosphate. Since many basic biological functions depend on sodium, calcium, and chloride ions, one can assume that these too are always essential. However, in the few attempts made, it has usually been difficult, often impossible, to demonstrate in insects a requirement for calcium or sodium, which appear to be "trace" requirements, ubiquitously present in adequate amounts as impurities in other dietary components.

This is in marked contrast to vertebrates, which require much calcium for bone construction and much sodium and chloride for special homeostatic functions of blood and lymph, functions which have no exact parallel in insects. Nevertheless, though the needs of insects for calcium, sodium, and chloride are low, successful synthetic diets usually provide for these needs by including a mammallian salt mixture having very high proportions of these ions, arguing great versatility in the ability of most insects to deal with extremely disproportionate mineral intakes. This being so, we would generally expect that the common essential minerals would not greatly affect the adequacy or optimality of natural foodstuffs for insects, notwithstanding several early reports showing different growth rates for insects consuming foodplants grown with different mineral treatments that altered the plant mineral content. In these studies it could have been equally possible, indeed likely, that the treatments altered protein and carbohydrate, as well as levels of other nutrients in the plants. Nonetheless, it is of interest that phytophagous insects in general have especially high potassium levels in their hemolymph and low sodium, a balance that characterizes the mineral composition of leafy plant tissues. Though many phytophagous insects have been reared apparently normally on synthetic diets high in sodium and with excessive calcium, a small improvement in fitness might still attach to some optimal mineral balance within a range characteristic of the natural food.

Besides the major mineral ions, vertebrates require an array of trace metals known to function as necessary cofactors for specific enzyme reactions: of prime importance are iron, zinc, manganese, and copper, but recently the essentiality of ultra-trace amounts of cobalt, molybdenum, tin, vanadium, silicon, selenium, fluoride, and other nutrients has been shown for various vertebrates (Miller & Neathery 1977). Since cofactor functions of, for example, iron and zinc are considered of almost universal importance to organisms, such trace metals can be assumed, tentatively, to be essential for insects. Recently the essentiality of iron, zinc, manganese, and copper has indeed been demonstrated for several insects. To our knowledge, there are no examples suggesting that availability of trace metal nutrients affect the success of insect populations in nature, but it is worth bearing in mind that vertebrate populations having otherwise apparently good nutrition can be stressed by trace mineral deficiency in forage as exemplified by cobalt deficiency malaise of sheep. Aphids are sensitive to lack of trace metals in synthetic diets, possibly via adverse effects on their symbiotes, since many microorganisms are very sensitive to these deficiencies (Mittler 1971). Iron deficiencies are frequent in plants growing on certain soils, again a possible indirect effect of soil microorganisms (Schroth & Hancock 1982). In such a situation it could be hypothesized that subtle differences in available iron from the host plant could delicately regulate the success of aphids on otherwise adequate host plants, though again, as for variations of major inorganics in plants, we would warn that many other nutrients would probably also fluctuate with the plant metal deficiency.

4.2.2. Vitamins

A quantitatively minor but vitally important fraction of the nutrient intake for any animal comprises a variety of vitamins and other growth factors. Following vertebrate usage, these may be categorized as water-soluble vitamins (B complex vitamins and ascorbic acid, or vitamin C), fat-soluble vitamins A, D, E, and K, and the essential fatty acids; in addition, for insects, and probably arthropods in general, a dietary sterol is essential, in contradistinction to vertebrates, which can synthesize all the sterol they require.

4.2.2.1 B Vitamins. B vitamins comprise a heterogeneous assortment of chemicals having in common two functional features: first, all act, variously, as necessary cofactors of enzymes basic to the universal metabolic pathways of all organisms; secondly, whereas higher plants and many microorganisms can biosynthesize all or most of them, animals lack this ability and must obtain them exogenously. The B vitamins essential to probably all insects, as to vertebrates, are thiamin, riboflavin, pyridoxine, nicotinamide, folic (pteroylglutamic) acid, biotin, and the cobalt-containing vitamin B_{12}, cyanocobalamin. Excepting B_{12}, all are generally found essential for all insects studied, excluding species dependent upon symbiotes, which in several cases are known to supply vitamins to their host insect. There is also evidence that some insects are able to synthesize folic acid, although at a low and inadequate rate (Sang 1978).

Distinct in function from B vitamins, though originally included with them, is choline, essential to all insects in amounts much above those characteristic of typical vitamins. Choline and the cyclic alcohol, meso-inositol (which is essential to some but not all insects), are sometimes called *lipogenic factors* because as subcomponents of the phosphatidylcholine (lecithin) and phosphatidyl-inositol types of phospholipid, they are involved in lipid membrane structure and in transport lipoproteins. Choline is thought not to be synthesized by transmethylation of ethanolamine in insects (as can occur in mammals), and hence its essentiality, which is additionally dictated by the necessity for choline as a precursor of the neurotransmitter, acetylcholine. A requirement for carnitine (vitamin B_7), chemically related to choline and able to partially spare it for some insects, is a unique nutritional characteristic of tenebrionids. In fatty acid oxidation carnitine has a necessary physiological function in the transport of acyl coenzyme A from cytosol to mitochrondria in insects, as in other animals, all of which have evidently retained the ability to synthesize it, excepting tenebrionids.

4.2.2.2 Ascorbic Acid. A pending review (Kramer & Seib 1982) bibliographs the insect literature on ascorbic acid and is the source material on which this summary is based. Ascorbic acid, the vitamin C of human nutrition, has been found dietarily necessary for nearly all insects studied whose natural dietaries necessarily include substantial amounts of fresh plant tissues; thus, ascorbic acid is a vitamin for sap-sucking Homoptera and leaf-eating Orthoptera, Coleoptera, and Lepidoptera. A few phytophagous Lepidoptera do not require it; no Diptera have been shown to require it, though few of those studied are typical fresh plant tissue feeders. On the other hand, and without exception so far, vitamin C has proved dispensable from synthetic diets for all insects whose natural dietaries do not necessarily include fresh plant material: fleshflies, filthflies, stored products insects (both carnivorous and graminivorous), wood borers, saprivores, detrivores, other omnivores such as roaches and crickets, and hymenopterous parasitoids.

There is thus a clear ecological significance to the ascorbic acid requirement, in that its ready availability in large amounts to true phytophages appears to have led to an adaptational loss of ability to synthesize it, assuming a synthesis ability was indeed the primitive insectan condition. Because of evidence for the presence and biosynthesis of ascorbic acid in tissues of many insects that do not require it in the diet, it was formerly assumed that insects generally must synthesize it, but this is now in question. Doubt has been cast on the ability of any insect to accomplish the synthesis since insects lack a terminal enzyme of the biosynthetic pathway from hexose to ascorbic acid (Kramer & Seib 1982). However, it is difficult to bring this into justification with the massive evidence for dietary dispensability for nonphytophagous insects, unless it is supposed that these latter utilize quite different metabolites for the same physiological purposes that are ascorbate-dependent in phytophages. Little is known of the biochemical functions of ascorbic acid in either vertebrates or insects. It is evident that, nutritionally, the majority of true phytophages depend critically on a substantial intake of vitamin C; however this dependence arose, it contributes to the restriction of such insects to ecosystems centered on living plant tissues.

4.2.2.3 Fat-Soluble Vitamins.
With respect to the fat-soluble vitamins of vertebrate nutrition, the steroidal vitamins D are not known to be required or have any function for any insect, nor are they able to satisfy or spare the sterol requirement.

However, vitamin A (retinol) or its provitamin, β-carotene, is dietarily essential for formation of insect visual pigments, and studies have shown behavioral, neurological, and retino-morphological defects in various species deprived of it (Carlson et al. 1967, Harris et al. 1977). The silkworm provides good evidence for a separate growth-promoting effect of vitamin A (Shimizu et al. 1981). Similar but unconvincing evidence has been obtained with other insects. As carotene is always present in plants and at least traces of vitamin A would be present in the visual apparatus of all animals, there are likely to be few ecospheres in which this vitamin would be limitingly deficient. Carotenoids are utilized by many insects in their pigmentation, which in turn can influence their aposematic and heterochromy (background constructing) functions (Rothschild 1973).

Vitamin E (*a*-tocopherol), the antisterility factor of vertebrate nutrition, is widely used in artificial insect diets as a protective antioxidant to delay breakdown of essential polyunsaturated fatty acids; such a generalized function can be equally well served by quite unrelated antioxidants such as ascorbic acid and gallates (Dadd & Kleinjan 1979). Beyond this, a specific vitamin E function is known for a few insects in which dietary deficiency manifests as reproductive failure (House 1966, Hagen 1952, McFarlane 1966). Because signs of deficiency imposed throughout larval growth only appear at adult maturity in the few known cases of a vitamin E requirement, it is probable that the need is widespread but overlooked because most nutritional studies terminate with adult emergence. A small growth-promoting function of vitamin E distinct from its essentiality for sperm viability was detected in both sexes of the house cricket; this function could apparently be supported also by vitamin K_1 (but not other K vitamins), and this was ascribed to the fact that like tocopherol, vitamin K_1 has a phytyl side chain to the molecule. This is the only positive evidence for any role of the K vitamins among insects (McFarlane 1978).

4.2.3 Nucleic Acids and Nucleotides

Nucleic acids, or their component nucleotides, nucleosides, and bases, form another category of water-soluble growth factors. They are needed to build DNA and RNA. However, with few doubtful exceptions, they are exogenously required only by Diptera among the insects. Other insects, like higher animals, can biosynthesize all the component nucleotides they require. Many Diptera also have such synthetic abilities, but so feebly as to limit growth. Such flies complete a normal larval development on nucleic acid-deficient diets, but do so slowly, requiring dietary RNA (or more probably, constituents of it, such as adenine compounds for *Drosophila melanogaster* and the housefly) to optimize performance. However, certain *Drosophila* strains and mutants have an absolute dietary need for nucleic acid constituents, as does the screwworm, *Cochliomyia hominivorax,* and several mosquitoes. Completely adequate synthetic mixtures of nucleic acid constituents have been devised for *Culex pipiens,* and its complex minimal requirements determined and compared with the known nucleotide needs of other Diptera (Dadd 1979). There appears to be no overall pattern to the nucleic acid constituents known to be required by these various flies, and indeed recent work (Dadd unpubl.) indicates quite different patterns of requirements for two other mosquitoes, *Culex tarsalis* and *Anopheles stephensi.* Since the vast majority of insects have no dietary requirement for nucleic acids, this class of nutrient is probably of little interest here. The best studied Diptera exemplifying the requirement are mostly fruitflies, filthflies and mosquitoes, whose natural larval dietaries comprise a high proportion of microorganisms, or fleshflies that consume animal tissues rich in nucleic acids and unlikely to offer significant natural deficiencies.

However, little is known of the nucleic acid nutrition of truly phytophagous Diptera; those borers that have been studied, such as the onion maggot, are probably largely saprophytic on decay microorganisms. If true phytophagous flies also depend critically on dietary nucleic acids, the low nucleic acid content of much plant tissue might prevent optimal growth and development. The report that dietary RNA affects adult fecundity of some fruitflies (Tsiropoulos 1980) indicates this possibility.

4.2.4   Sterols

So far as is known a sterol is essential in the diet of all insects because sterols, which insects cannot synthesize, play essential structural roles in cellular membranes and transport lipoproteins and are required as precursors of steroid hormones (the ecdysteroid molting hormones) for insects. Until recently the "animal sterol," cholesterol, seemed totally adequate in synthetic diets for all insects studied. Certain zoophagous insects such as dermestids and screwworms require cholesterol specifically, being unable to develop with only phytosterols, a situation readily understood for insects that obtain cholesterol directly in their normal food. Phytophagous insects, however, normally ingest and utilize mainly phytosterols, not cholesterol, though many grow well on synthetic diets with either cholesterol or a phytosterol, the latter being different for different species. On the assumption that phytophagous insects, too, would physiologically require intrinsic cholesterol, it was early supposed that they must convert phytosterols to cholesterol, and this was demonstrated for many phytophages. Much recent work has been devoted to elucidation of phytosterol dealkylation and metabolism to cholesterol; from the cholesterol so formed, or, in the case of zoophages, ingested already formed, further metabolism via 7-dehydrocholesterol leads to formation of ecdysones (cf. reviews of Clayton 1964, Robbins et al. 1971, Svoboda et al. 1975).

An important concept for understanding sterol nutrition of insects is the distinction between *essential* and *sparing* sterols and, associated with this, the postulation of distinct metabolic and structural sterol functions. Insects such as *Dermestes vulpinus* and *Drosophila melanogaster* fail to develop with only the saturated sterol, cholestanol, in the diet, but develop normally if a minute amount of cholesterol is added that, alone, would be totally inadequate (Clark & Bloch 1959, Kircher & Gray 1978). Thus, the cholestanol is said to "spare" a major part of the normal cholesterol requirement, this major requirement thought to be physiologically necessary for relatively nonspecific structural functions, while the minute essential residual requirement for cholesterol, as such, was perceived to support a micronutrient metabolic function subsequently equated with the provision of ecdysone precursor. This sparing phenomenon allows many insects, particularly phytophages, to utilize a wide array of sterols without further modification for structural purposes as long as a minute amount of cholesterol or other ecdysone precursor is also available. This has important implications for the ability of insects to utilize foods such as plants containing a range of phytosterols that may not be efficiently metabolizable by particular insects; once it is realized, contrary to initial beliefs, that many plants have traces of cholesterol, it becomes clear that they may present perfectly adequate mixtures of sparing phytosterols and trace essential cholesterol.

A reassessment of insect sterol nutrition was inaugurated with the discovery of insects for which cholesterol is inadequate. *Drosophila pachea*, monophagous on rotting senita cactus, requires the sterol unique to this cactus, schottenol (Δ^7-stigmastadienol) or related Δ^7 sterols such as 7-dehydrocholesterol or lathosterol (Δ^7-cholestenol) (Heed & Kircher 1965). The ambrosia beetle *Xyleborus ferrugineus*, while able to develop through the larval stages on a diet with only cholesterol as a sterol component, fails to pupate unless it eats a suitable Δ^7 sterol such as ergosterol, the sole sterol of its normal ectosymbiotic fungal food, or the related 7-dehydrocholesterol (Chu et al. 1970); in this case the primary required sterol is ergosterol, with cholesterol acting as a sparing sterol. Recently it appears that the tea tortrix *Homona coffearia* may also require primarily a Δ^7 sterol, since on artificial diet the pupal–adult molt aborts with only cholesterol but is perfect with ergosterol (Sivapalan & Gnanapragasam 1979). These cases, fascinating in purely nutritional terms, take on particular interest in the context of nutritional ecology since such specific sterol requirements sharply demarcate the range of possible foodplants on which such insects could develop, thereby playing a major role in defining their ecological niches.

The currently most active studies of insect sterol nutrition are revealing an increasing diversity of ways in which phytophagous insects metabolize and utilize particular phytosterols in various plants (cf. reviews by Dadd 1977, Svoboda et al. 1975, 1978). A few examples will suffice. Although most phytophagous and omnivorous insects studied convert dietary phytosterols such as sitosterol, campesterol, and stigmasterol to cholesterol, several utilize them unchanged in their tissues for structural functions. The female housefly must, additionally, obtain cholesterol during either larval or adult feeding to produce eggs. This may be so also for *Trogoderma granarius* since it accumulates in its tissues the traces of cholesterol in its cereal food. In contrast, the milkweed bug *Oncopeltus fasciatis* may dispense with cholesterol entirely since it can form an unusual ecdysteroid molting hormone, makisterone A, more potent for this bug than ecdysone and probably biosynthesized from the related phytosterol, campesterol, of its food seeds. The honeybee also utilizes, without metabolism, a variety of worker-acquired phytosterols, but the workers preferentially incorporate 24-methylenecholesterol, common in pollens, *into* the food given to brood larvae, in whose tissues this sterol is always predominant by the time they pupate regardless of workers' dietary sterols. This suggests that 24-methylenecholesterol is required for some structural function in this insect.

Other phytophagous insects metabolize food sterols to tissue sterols other than cholesterol. The Mexican bean beetle converts soybean sterols to saturated stanols and Δ^7 stenols to form its main tissue structural sterols; and the beetle *Tribolium confusum* converts dietary phytosterols (or cholesterol if provided) into a tissue mixture that is predominantly 7-dehydrocholesterol and secondarily cholesterol. Finally, it was found (Ritter et al. 1982) that certain fungus-cultivating attine ants have tissue sterols devoid of cholesterol and consisting solely of $\Delta^{7,24}$-methylenecholesterol with two other related Δ^7 sterols, all conjecturally derived from the ergosterol that would predominate in fungal food.

These specialized sterol utilization patterns recently revealed among phytophagous insects presumably represent evolutionary adaptations to the available plant sterols characteristic of particular plant eco-niches, and once the adaptation has been accomplished, the distribution of such an insect is to that extent determined. These sterol nutritional specializations are somewhat analogous in effect to the adaptations of oligophagous insects to particular plant allelochemics. Interestingly, the two sorts of adaptations sometimes go hand in hand. Thus, *Oncopeltus* normally requires specific chemical phagostimulants from its food seeds to promote ready feeding, and *Drosophila pachea* has evolved an imperviousness to alkaloids of the senita cactus that are toxic to many other insects. Such double specializations are doubtless reinforcing in delimiting the ecological distributions of such insects.

4.2.5 Essential Fatty Acids

Though many, perhaps most, insects can utilize dietary fat, none, with the possible exception of certain parasitoids (Bracken & Barlow 1967, Thompson 1981), is known to require dietary fats or oils as a bulk energy-producing nutrient in synthetic diets, not even *Dermestes* beetles or waxmoth larvae whose normal dietaries are high in fats. However, certain polyunsaturated fatty acids are dietarily essential as accessory growth factors for the majority of insects studied critically, often required for good growth, and usually necessary for a normal molt to the adult. The subject of essential fatty acids for insects was reviewed by Dadd (1981), and this synopsis draws mainly from his exhaustively bibliographed discussion.

The essential fatty acid requirement of all Orthoptera, Lepidoptera, Coleoptera, and Hymenoptera so far studied can be satisfied by dietary linoleic (C18 : 2ω6) or linolenic (C18 : 3ω3) acids, the parent members of the ω6 and ω3 families of polyunsaturated fatty acids, which are primarily essential, respectively, for warm-blooded vertebrates and fish. For mammals the physiologically essential fatty acid is arachidonic acid (C20 : 4ω6), utilized directly if present in the diet or metabolically derived from lower members of the ω6 family (such as linoleic acid) if only these are dietarily available, as for herbivores. In contrast to the ability of vertebrates to satisfy their needs with dietary arachidonic acid or other long polyunsaturates, arachidonic acid was useless as a replacement for linoleic or linolenic acids in the diet of those Orthoptera and Lepidoptera that have been studied. Taken in conjunction with the almost complete absence, until recently, of records of arachidonic acid or other long-chain polyunsaturates from the hundreds of fatty acid analyses performed on insect tissues, it might be concluded that the physiological functions of essential fatty acids for vertebrates and insects differ in some fundamental way.

Although dietary linoleic and linolenic acids were initially thought to be interchangeable for insects having an essential fatty acid requirement, it has emerged that several Lepidoptera require linolenic acid specifically for a normal adult molt and some require both linoleic and linolenic acid for optimal growth. Recently, the status of essential fatty acids for insects was further complicated by discovery that several mosquitoes require vitamin level quantities of arachidonic or certain structurally related fatty acids of both the $\omega6$ and $\omega3$ families, a requirement that cannot be satisfied by linoleic or linolenic acids. This suggests a commonality in essential fatty acid physiological functions between insects and vertebrates and is supported by recent detection of low levels of many long-chain polyunsaturates in insects of all sorts (usually prominent only in certain tissues or specific lipid classes such as phospholipids) and the detection in a few insects of prostaglandins (Loher et al. 1981), metabolites of eicosapolyenoic acids having localized hormonelike functions and present in mammalian tissues.

The physiological ramifications of these recent findings can be pursued via the speculations in Dadd's reviews (1981, 1982). Here we comment primarily on the possible ecological significance of these diverse essential fatty acid requirements of insects. First, the majority of insects (all those studied, excluding mosquitoes) for which the C18 polyunsaturates suffice fully will generally obtain ample amounts from most sorts of foods: both linoleic and linolenic acids are richly provided by most living plant materials, are present universally in seed oils, occur substantially in animal tissues, and are maintained even in dead organic tissues. In contrast, arachidonic acid and other long-chain polyunsaturates are absent from plant tissues but are characteristic of vertebrate and many other animal tissues; thus, mosquitoes and any other insects for which long-chain polyunsaturates may prove to be dietarily essential must include some animal-derived material in their dietary. For mosquitoes, the adult female blood-sucking habit would provide substantial amounts that can be carried over via the egg to provide for larval development (Stanley-Samuelson & Dadd 1981). Also, though filter-feeding larvae are thought in nature to feed primarily on bacteria, which lack such fatty acids, to the extent that animal-derived detritus and protozoa may be ingested, this would augment egg-derived and ultimately blood-derived polyunsaturates. Nevertheless, for mosquitoes, habitat suitability might be affected by the availability of animal-derived detritus and animal-like microbiota; indeed this would be the only source of long-chain polyunsaturates for the progeny of autogenous females.

4.2.6 Energy Sources

The major part of food ingested is used to provide energy. Carbohydrates, fats, and proteins are all involved in cycles of energy-producing reactions. Carbohydrates are the main energy source for most insects, but not for all. Many carnivorous species, for example, larval flesh flies grow normally without any carbohydrates. Detritivores such as *Culex pipiens* and *Aedes aegypti* complete development without carbohydrates (at a reduced rate) (Sneller & Dadd 1981). But carbohydrates of some sort (i.e., various sugars or polysaccharides, depending on the insect's digestive enzymes) are needed by most larval phytophagous insects. Seed and cereal feeders require carbohydrate at levels of 20 to 70% of their dietary solids and phloem sap-sucking aphids at about 80% of their nutrient solids (by wt.). Finally, many insects, such as mosquitoes, that require no carbohydrate as larvae require only sugar as adults. Thus, dietary carbohydrate needs vary greatly between species and sometimes between juveniles and adults of the same species.

None of the 100 or so insect species reared on defined diets requires bulk lipid, but if it is present, many would use it efficiently for energy. Even the waxmoth, whose natural diet is beeswax, grows normally on an artificial diet with carbohydrate completely replacing wax as the major energy source. Insects thus seem not to need lipid beyond its contribution to the quantitatively minor, yet very important, growth factors discussed above.

Since many dipterous larvae grow best with diets lacking both carbohydrate and bulk lipid, they evidently satisfy their energy needs entirely by oxidation of protein-derived amino acids. Like the mosquito, several phytophagous larval insects require substantial carbohydrate for optimal growth but can complete development at a reduced rate without it (Harvey 1974), and thus facultatively they can operate an amino acid-based energy metabolism. The hide beetle, *Dermestes maculatus*, grows well with triglyceride completely replacing the carbohydrate otherwise required (Applebaum et al. 1971) and, thus, like the waxworm, utilizes lipid and carbohydrate interchangeably. Probably most insects have some ability to induct energy from protein and fat, but mostly as an adjunct to primary dependence on carbohydrate.

4.2.7 Amino Acids and Protein

All insects studied require the same 10 amino acids found necessary for the rat, if we exclude species that use symbiotes to supply essential amino acids lacking in the diet. The 10 essential amino acids are arginine, histidine, isoleucine, leucine, lysine, methionine, phenylalanine, threonine, tryptophan, and valine. For several insects additional amino acids are essential. Larvae of the flesh fly, *Phormia regina,* and the silkworm require proline and glutamic or aspartic acid. Occasional earlier data suggested that some insects required a sulfur-containing amino acid, cystine, and a phenolic amino acid, tyrosine. Important in schlerotization and melanization of insect cuticle, tyrosine is derived metabolically from phenylalanine, which is essential; hence tyrosine might *appear* necessary if phenylalanine, which it spares, is inadequate. Cystine (cysteine) can sometimes spare the essential sulfur-containing amino acid, methionine, and may *appear* essential since it sometimes acts as a supplementary sulfur source in sulfate-short diets.

C. pipiens (larvae) not only requires dietary proline but is the first insect shown to require asparagine (Dadd 1978) at much the same dietary concentration as other amino acids; asparagine is thus a major structural element rather than a vitaminlike growth factor. Another mosquito, *Aedes aegypti* (larvae), requires neither asparagine nor aspartic acid (Singh & Brown 1957, Dadd 1978).

The 10 rat-essential amino acids are also required by adult insects for egg production. Many species, however, can carry them over from larval feeding; such adults do not have to ingest them. However, for optimum egg production, many species must ingest them as adults (e.g., anautogenous mosquitoes, many cyclorrhaphous Diptera, predaceous insects, parasitic Hymenoptera that host feed as adults, and some butterflies).

Because nitrogen (N) plays a central role in all metabolic processes and in genetic coding, of all the food constituents perhaps it is the quantity and quality of N (protein and/or amino acids) available that generally limits growth and fecundity of insects. The importance of N to phytophagous insects has been reviewed recently (Mattson 1980, McNeill & Southwood 1978, Scriber & Slansky 1981, White 1978). Mycophagous and saprophagous insects rely mainly on microorganisms as a source of N, and the carnivorous ones, at least in their larval stages, obtain N from either eating whole animal tissues or from the hemolymph of their prey. Hemotophagous insects, of course, obtain their required N from blood.

4.2.8 Water

Like all organisms, insects require water. Most terrestrial insects contain at least 70% water by weight; the extremes range from 46 to 92% (Rapoport & Tschapek 1967). Usually, insects ingest water with their food; the water in the food can have profound effects—the amount can vary from 1 to over 90%. Some insects also drink water, some absorb it. Terrestrial mites and insects often exchange significant amounts of water with ambient air (Wharton & Arlian 1972). Larvae of stored products insects can utilize very dry foods. The Mediterranean flour moth can survive with only 1% water in its food. It has been claimed that this insect produces water metabolically from oxidation of fat in the food (Frankel & Blewett 1944). Andrewartha and Birch (1973) did not believe it had been definitely shown that insects actually use metabolic water to maintain their water balance. However, recent studies tend to support the claim for metabolic water. Certain desert tenebrionids produce sufficient metabolic water to maintain a constant water content at the expense of lipid conversion; however, the hemolymph volume decreases, but upon drinking water it is restored to normal (Nicolson 1980). The willow aphid taps the plant nutrient sap stream in the phloem sieve tubes. The sap contains only 8% dry matter and is taken up by aphids weighing 2–5 mg at a rate of 1–2 µl/hr; the excretion rate is very similar to the intake rate (Mittler 1958). Silkworms feeding on mulberry leaves containing at least 70% water excreted about 50%, utilizing 60% for growth (Legay 1958). Scriber and Slansky (1981) calculated nutritional performance of last-instar foliage-chewing larvae to be significantly greater on leaves of 75–95% water. They also found that the generally low performance of tree feeders, contrasted to forb feeders, is correlated with the lower water content of tree foliage (but also a more rapid decline in N); Scriber (1979) suggests that low leaf water may be more suppressing than low N content.

Adults of many holometabolous insects drink water, but so do some larvae and nymphs. Eggs and larvae of many insects absorb water. Thus, *Tenebrio* has one generation a year on dry food, but if allowed to drink water, it can have six generations (Mellanby & French 1958). *Syrphus ribessi* larvae estivate when desiccated, and when moistened, take up water through the anal papilli and resume growth (Schneider 1948). Various immature stages of species that diapause during dry seasons take in water when the rainy season begins; this is often associated with termination of diapause (Chapter 6). Locust eggs take in water through their micropyles and some lepidopterous larvae or prepupae take in "contact water" through the cuticle (Lees 1955).

Many phytophagous insects require a high moisture intake (Waldbauer 1962, 1964, 1968), yet, water dilutes the nutrients and *Celerio euphorbiae* larvae ate correspondingly more (House 1965). Dilution increased efficiency of conversion in *Prodenia eridania* (e.g., Soo Hoo & Fraenkel 1966). However, others have found a decrease with decreasing water in the food, involving various Lepidoptera (Feeny 1975), *Hyalophora cecropia* (Scriber 1977), and black cutworm (Reese & Beck 1978). Optimal water level for growth, however, is quite different from that for conversion efficiency, due to the interaction between efficiency and the amount of dry material ingested. *Pieris rapae* larvae on low N plants consumed food faster and utilized N more efficiently than larvae on high N plants (Slansky & Feeny 1977).

4.3 FOOD UTILIZATION AND ITS MEASUREMENT

Once an insect locates its food, plant, prey, or saprolite, this food substrate must support growth, development, and reproduction, in which case it is considered suitable. Thus, the food ingested must not only contain the basic nutrients but also be amenable to assimilation and conversion into energy and structural substances required for normal activity and development (Beck & Reese 1976, Beck 1972, and Chapter 3).

4.3.1 Nutritional Indices

Nutritional indices are employed to assess food suitability and bioassay compounds used in artificial diets. Such techniques have shown, for example, that a phytophagous insect may grow unequally on different plants even if no apparent behavioral feeding barriers exist. Waldbauer (1968) reviewed the methods of assessing food intake and utilization by insects, stimulating research on quantitative insect nutrition. In the last 10 years over 300 papers involving quantitative work on 330 species have appeared. Slanskey and Scriber (1982) summarized this later work in terms of the following five different nutritional indices relative to growth. Two of these indices involve food consumption and growth, namely, relative consumption rate (RCR) and relative growth rate (RGR):

$$RCR = \frac{\text{amount of food ingested (fresh or dry wt. of food eaten)}}{\text{duration of feeding period (days)} \times \text{mean wt. of insect during feeding period}}$$

$$RGR = \frac{\text{wt. gain of insect during feeding period (fresh or dry wt.)}}{\text{duration of feeding period (days)} \times \text{mean wt. of insect during feeding period}}$$

Three indices involve digestibility and efficiency of conversion, that is, assimilation efficiency or approximate digestibility (AD), efficiency of conversion of assimilated or digested food (ECD), and efficiency of conversion of ingested food (ECI):

$$AD = \frac{\text{amount ingested (mg)} - \text{feces (mg)}}{\text{amount ingested (mg)}} \times 100$$

$$ECD = \frac{\text{weight gained (mg)}}{\text{amount ingested (mg)} - \text{feces (mg)}} \times 100$$

$$ECI = \frac{\text{weight gained (mg)}}{\text{amount ingested (mg)}} \times 100$$

These five indices were used and the data presented under different consumer types. Slanskey and Scriber found that across all feeding categories of immature stages, the ranges in mean rate and efficiency values were: RGR, 0.003–3.39 mg/day · mg; RCT, 0.04–2.3 mg/day · mg; AD, 9–88%; ECD, 18–89%; and ECI, 0.6–68%. The highest RGR values were for herbivores, especially foliage-chewing Lepidoptera and sucking insects on forbs. Carnivores had higher AD values compared to herbivores; however, Slansky and Scriber cautioned on comparing such performance values, since considerable variation exists within the same order (e.g., Lepidoptera) feeding on the same food in the same way. The wide variation in performance values noted stems from differences in food quality involving variations in any nutrient components (including seasonal trends) and is often coupled with changes in allelochemics (Mattson 1980, Scriber & Slansky 1981). Chapters 7, 18, and 20 provide data on insect energetics and flight.

4.3.2 Energy Flow Through Trophic Levels

Various formulas deal with energy flow and the efficiency of its utilization. Kozlovsky (1968) defined ecological efficiency as the ratio (for any of the various parameters) of energy flow within or between trophic levels of a natural community, within or between populations, or within or between individual organisms. He defined and equated the terminology for 20 different ecological efficiency ratios. The formulas involve the energy (in food) ingested and assimilated, tissue or biomass produced, and energy lost through respiration and excretion. Determining energy budgets involves these same parameters, and the biomass consumed or produced is usually expressed in calories.

Energy budget data are used in assessing the energy efficiency of transfer processes. Patterns of energy processing related to growth and reproduction of the pea aphid, for example, indicate that this aphid is very efficient in biomass production ($x = 49\%$) and assimilation ($x = 83\%$) (Randolph & Randolph 1975). For syrphid larvae preying on pea aphids, 64% of the energy consumed was assimilated, 54% going to tissue and 10% to maintenance (Barlow 1979).

Chaplin and Chaplin (1981) compared growth energetics of a migratory and nonmigratory milkweed bug under different temperatures. In each case the migratory species grew more rapidly than the nonmigratory one because it maintains a higher assimilation efficiency (i.e., reared on its nearly monophagous diet) and its growth rate is higher over a broad temperature range. The authors compared these results with other data on energetics and expressed support for the hypothesis that monophagous herbivores are more efficient in converting food into tissue than polyphagous species, as suggested by Waldbauer (1968) and Whittaker and Feeny (1971). An *apparent* exception, the monarch butterfly larva, a specialist on milkweed, has a net production efficiency similar to or lower than polyphagous Lepidoptera. To explain this, Schroeder (1976) suggested that food is not limiting for monarch larvae, and hence, there is selection for rapid growth at low food conversion, rather than slow growth at high efficiency.

4.3.3 The Role of Nitrogen

Slansky and Feeny (1977) reported that growth of cabbage butterfly larvae is limited by the N in their food, N budgets for such larvae are ecologically more significant than energy budgets, and natural selection favors the *rate* (power output) rather than *efficiency* of a biological process and proposed that herbivores should adjust consumption to the lowest rate at which rate of N accumulation is maximal. Mattson (1980) suggests, however, that the above hypothesis may be valid in the context of evolutionary time, but not necessarily valid in ecological time, since insects often feed in discrete periods, or feeding may be constrained (e.g., by dietary moisture, fiber, allelochemical levels, temperature, and light rhythms). Mattson (1980) plotted the efficiency of conversion of ingested food values (body growth divided by amount of food consumed) against food N concentration for many invertebrate herbivores and found in general that efficiency of conversion values vary from nearly 0.3% to a high near 58%, with low values of about 1% being associated with aquatic and terrestrial arthropods that feed on N-poor wood, litter, and detritus. The highest values, 40–50%, were for insects that feed on certain seeds, phloem saps, and pollen nectar stores. The high conversion values of about 50% were associated with predaceous and parasitic insects. Mattson (1980) also concluded that organisms on low N diets must consume substantially more food than those on high N diets. Again, the form and quantity of the available N source (proteins and/or amino acids) is the central limiting feature for growth, development, and fecundity of insects.

4.3.4 Sources and Patterns of Nitrogen

The N in insect food from plants varies in quantity and quality, depending on the source (e.g., leaf, fruit, sap, nectar, pollen, wood, detritus). The N content can vary from 0.08 to 7% dry wt., depending on the plant part and the plant's seasonal growth cycle. Highest concentrations (3–7%) occur in young, actively growing tissue and in propagules such as seeds and bulbs. Leaf N concentrations decline with age to 0.5–1.5% at leaf abscission (Mattson 1980). The lowest N levels occur in the sap. Xylem saps vary from 0.0002 to 0.10% N w/v, and phloem sap from 0.004 to 0.60% N w/v. Variation in N content corresponds with seasonal growth cycles. Since N is highest in young, growing tissue (above) it declines through the season. Also, as widely proven, N level varies between plant life forms and species (e.g., Mattson 1980). C_4 grasses usually have lower N contents than C_3 grasses, and evergreen gymnosperms have about half the foliar N of comparable deciduous angiosperm tissues (1–2% vs. 2–4%). Both N-fixing legumes and nonlegumes have N contents ranging between 2 and 5% dry wt.

Pollens and nectars are rich sources of N. Pollens contain 7–35% protein and, with exceptions, all known protein amino acids; free amino acids occur in large amounts, with the exception of tryptophan and phenylalanine in a few pollens (Barbier 1970) (cf. Chapter 20). Floral and extrafloral nectars also contain a diversity of free amino acids. Baker et al. (1978, cf. Chapter 20) compared amino acid complements of extrafloral nectars of 33 species of tropical and temperate-zone flowering plants with the floral nectars of 248 species, and also the extrafloral and floral nectars from the same species. The extrafloral nectar and floral nectar always differed in their complement of amino acids, and certain ones of only moderate frequency in the floral nectars had a higher frequency in extrafloral nectars, particularly those in the cysteine group. Nonprotein amino acids occurred more frequently in extrafloral nectar. The extrafloral nectars of only 3 of 21 angiosperm species contained all 10 of the essential amino acids, and the floral nectar of only 1 of the same 21 species contained all 10. None of the 17 species that did not contain all 10 in either their floral or extrafloral nectars had all 10 even with their nectars combined.

4.3.5 Sources and Patterns of Carbohydrates

Many holometabolous insects have completely different diets as adults than as larvae, and some, as adults, can survive on sugar solutions alone. Floral nectar, extrafloral nectar, honeydew, and pollens are natural foods for many adult Diptera and Hymenoptera and some Heteroptera, Coleoptera, and Lepidoptera. Floral nectars vary in their sugar composition and have different influences on their insect pollinators (Chapter 20). Sugars such as sucrose are more common than hexoses in flowers with unprotected nectaries. Well-protected nectars contain mostly glucose and fructose and little sucrose (cf. Chapter 20, Baker et al. 1978, Baker & Baker 1982, Percival 1961, Waller 1972). Nectar sugars vary in caloric content, and perception and ingestion of them by nectarivores depend upon the concentration of each sugar or mixture of sugars (Mitchell 1981). Nectars of extrafloral nectaries contain mainly sucrose, glucose, and fructose, but seven other sugars have also been reported (Bentley 1977). Honeydews excreted by Homoptera commonly contain sucrose, glucose and fructose. Aphids and coccids with filter chambers excrete honeydew of 25–40% sucrose (dry wt.) compared to less than 1% for aphids lacking filter chambers (Kunkel & Kloft 1977). The trisaccharides excreted vary, depending upon the homopterous species: fructomaltose has been found in honeydews of citrus mealybugs and certain scales and aphids (Gray & Fraenkel 1953), and melezitose and glucosucrose in honeydews of a variety of scales and mealybugs (Ewart & Metcalf 1956), while trehalose, common in insect hemolymph, seems to occur in every honeydew reviewed by Kunkel and Kloft (1977).

Up to 14 different carbohydrates, including the common sugars, are found in certain pollens, and many pollens also contain starch (cf. Chapter 20 & Barbier 1970).

Total carbohydrate content in leaves varies from about 3.5 to 10%, depending on the plant species; in fruits, 4–20%, and in nuts, 11–20% (Albritton 1953). Plants can be divided into two main groups, depending on their carbon-fixation pathways during photosynthesis, and are referred to as C_3 and C_4 plants. This distinction may have played a role in selective herbivory. The C_4 (Krantz syndrome) plants initially fix C into 4-carbon acids in the mesophyll cells and then, via the Calvin–Benson cycle, into 3-carbon molecules. Most plants utilize only the Calvin–Benson cycle and are C_3 plants (Smith & Brown 1973). The C_3 species have lower net rates of photosynthesis than C_4 species, starch accumulation in C_4 species occurs primarily in chloroplasts of the bundle sheath cells rather than throughout the mesophyll, as in C_3 species. Caswell et al. (1973) hypothesize that C_4 plants are generally a poorer food for herbivores, particularly insects, than C_3 plants, and herbivores tend to avoid C_4 species.

4.4 ROLES OF SEMIOCHEMICALS ASSOCIATED WITH THE FOOD OF INSECTS

Semiochemicals are involved in behavioral or physiological interactions between organisms. Intraspecific semiochemicals are termed *pheromones*. Interspecific ones are termed *allelochemics*, which includes kairomones, allomones, and synomones. Kairomones are chemical signals that benefit the receiver; allomones benefit the emitter; and a synomone is a substance produced or acquired by an organism such that, when it contacts another species, it evokes a behavioral or physiological reaction from the receiver that is adaptively favorable to both emitter and receiver (Nordland et al. 1981).

Among the myriad of semiochemicals that insects respond to, many are associated with plants. There has been an "exponential" increase in interest in semiochemicals since Fraenkel's paper (1959), "The raison d'être of secondary plant substances." Much of the research involves the biochemical interactions between host and herbivore (cf., e.g., Beck & Reese 1976, Chapman & Bernays 1978, Dethier 1970, Harborne 1982, Jermy 1976, Király & Szalay-Marzsó 1971, Nordlund et al. 1981, Rosenthal & Janzen 1979, Sondheimer & Simeone 1970, Van Emden 1973, Wallace & Mansell 1976) and the use of semiochemicals in arthropod control (Dethier 1947, Lewis et al. 1980, Harris 1980, Nordlund et al. 1981, Ritter 1979, Shorey 1976, Wood et al. 1970).

Various secondary plant substances protect plants from herbivores and plant pathogens—from insect attack by repelling oviposition, deterring feeding, reducing digestive processes, and/or by modifying assimilation of foods. Such chemicals are allomones since they benefit the plants containing

them. However, some of these same substances can be considered kairomones since they also benefit the organisms perceiving them. Some insects
are attracted by certain secondary plant substances, stimulated to mate,
oviposit, feed, and utilize such chemicals as precursors of hormones, pheromones, and allomones. Thus, secondary plant substances initially selected
in plants for defense became kairomones to insects that evolved mechanisms
to negate these toxins, thus leading to host plant specificity (Dethier 1970,
1980, Ehrlich & Raven 1965, Feeny 1975, 1976, Fraenkel 1969, Harborne
1982, Reese 1979, Schoonhoven 1968, 1981).

4.4.1   Roles in Host and Food Finding and Acceptance

The main chemoreceptors responsible for accepting or rejecting food by
insect larvae are found on the maxillary palpi (Dethier 1970, Schoonhoven
1973, 1981). Secondary plant substances so detected usually have a deterring role, but some are phagostimulatory (Kogan 1977, Schoonhoven 1981,
Rosenthal & Janzen 1981); a substance that deters a generalist feeder may
be a phagostimulant for a specialist.

Some of the ways that plant-associated semiochemicals influence behavior and physiology in insects are shown in Figs. 4.1 and 4.2. Coevolution
of plants and pollinating insects is one of nature's great wonders. Not only
are colors involved in attracting the insects, but there are also chemicals
that attract by odor and chemicals that entice feeding. Figure 4.1A depicts
a hawk moth feeding on floral nectar. Sugars and amino acids in the nectar
stimulate the moth to feed. In Fig. 4.1B a bee is landing on a flower that
has attracted it by volatile chemicals; it is then guided into the flower where
it collects nectar and pollen for use as food for the colony. Since the flowers
in either of these cases are pollinated by these insects (Chapter 20, Harborne
1982), the semiochemicals involved in this mutualistic attraction and acceptance are called synomones. Another coevolutionary mutualistic relation
between insects and plants is suggested by the presence and function of
extrafloral nectaries (Fig. 4.1C), which produce solutions containing sugars
and amino acids that attract and reward predaceous insects that protect
the plants from phytophagous insect attack (Bentley 1977). The monarch
butterfly (Fig. 4.1D) is attracted to its milkweed host and the larvae se-
quester cardenolides from these plants and carry these distasteful chemicals
(allomones) into their subsequent pupal and adult stages; this protects all
stages from predatory birds. Birds quickly associate the monarch's vivid
colors (aposematic) with a vomiting experience (Roeske et al. 1976, Rothschild 1972). Cardenolide toxicity to vertebrates is associated with high sodium ion concentrations; insects tolerate the chemicals since they have
potassium as a major hemolymph cation, with sodium having a minor role
(Brattsten 1979). Figure 4.1E suggests a boll weevil attracted to a cotton
plant by volatile essential oils. A complex of 14 compounds in cotton stimulates the weevil to feed. The male weevil must feed to attract females; this

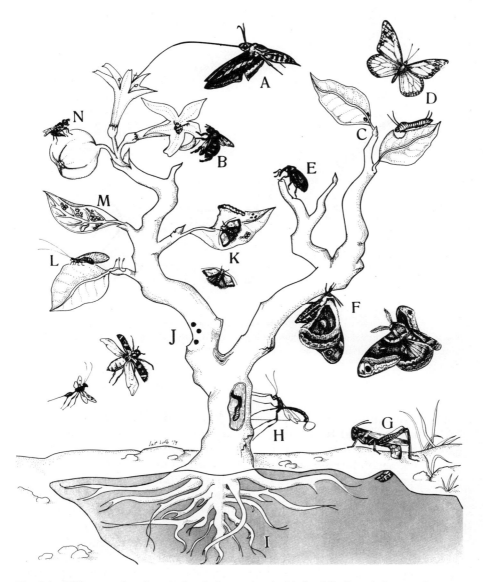

Fig. 4.1   Different roles of semiochemicals associated with food finding, defense and reproduction of insects. See text for explanation of associations.

suggests that cotton contains a constituent that can be converted to sex
pheromones. About 80 volatile compounds were identified in male and
female boll weevils and their frass by Hedin et al. (1976). Figure 4.1F
suggests a male saturniid moth attracted to a female via a sex pheromone.
Triggering pheromone release by the female depends on a volatile hydro-

NEGATIVE INFLUENCES OF SECONDARY PLANT METABOLITES IN NUTRITION OF INSECTS

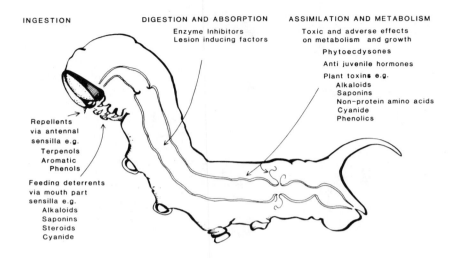

INGESTION   DIGESTION AND ABSORPTION   ASSIMILATION AND METABOLISM

                     Enzyme Inhibitors   Toxic and adverse effects
                     Lesion inducing factors   on metabolism and growth

                                                   Phytoecdysones

                                                   Anti juvenile hormones

                                                   Plant toxins e.g.
                                                      Alkaloids
                                                      Saponins
                                                      Non-protein amino acids
                                                      Cyanide
                                                      Phenolics

Repellents
via antennal
sensilla e.g.
   Terpenols
   Aromatic
   Phenols

Feeding deterrents
via mouth part
sensilla e.g.
   Alkaloids
   Saponins
   Steroids
   Cyanide

References: Harborne(1982); Rosenthal & Janzen(1979).

Fig. 4.2   Different roles of semiochemicals associated with food that restrain feeding, digestion, assimilation and metabolism of insects.

carbon emanating from its host plant (Riddiford & Williams 1967). Similarly, male tephritid fruit flies are attracted to volatile chemicals from host
fruit trees that bear fruits acceptable for oviposition; thus, the plant odors
may act as secondary sex attractants or rendezvous stimulants to bring the
sexes together at a place where suitable host fruits are available (Bush 1966,
Metcalf et al. 1979). Figure 4.1G suggests conditions that determine grasshopper reproduction. Given lush quality grass, grasshoppers initiated and
completed egg development, whereas on poor quality grass, they laid no
eggs (McCaffery 1976). Phytophagous Orthoptera and Lepidoptera, especially, benefit from high N in their food, particularly if in the form of
easily digested amino acids (White 1978, Slansky & Feeny 1977). Grasshoppers detect and preferentially feed on grasses with higher proline and
valine (Haglund 1980), which increase in concentration in drought-stressed
plants (Hsiao 1973). Figure 4.1H shows a large ichneumonid, *Pseudorhyssa*,
ovipositing through 2–3 in. of wood to reach a *Sirex* larva parasitized by
another ichneumonid *Rhyssa*. The cleptoparasitic wasp locates the siricid-
infested trees by detecting chemicals produced by a symbiotic fungus associated with *Sirex* feeding and feces and utilizes the drilled shaft made by
Rhyssa (Spradbery 1969). Figure 4.1I reminds us that roots also have sec-

ondary plant substances. These chemicals (allomones) offer protection against soil-inhabiting pathogens (McKey 1979) and insects, the latter through chemicals like rotenoids (Harborne 1982). But some plant allomones may serve as kairomones to certain insects; volatile organic sulfur compounds produced by onion both attract adult female onion flies and stimulate oviposition (Matsumoto 1970). Increased oviposition occurs in the presence of certain soil microbes, which Hough et al. (1981) suggest may metabolize precursors in onions and convert them to volatile, onion-fly stimulatory compounds. A braconid parasitoid, *Aphareta pallipes,* finds the onion by odor and then locates the fly maggot by touch (Salkeld 1959). Figure 4.1J represents three bark beetle entrance holes on the trunk of a pine tree. Initially, the female beetles were attracted to volatile terpenes coming from the tree. Upon boring in the bark, female *Dendroctonus brevicomis* release a sex pheromone that attracts males. This pheromone is a complex made up of myrcene sequestered from a bark leoresin, *exo*-brevicomin, and frontalin; the latter two chemicals are synthesized by the female (Wood 1982). After a certain density of beetles have colonized a tree, either verbenone or *trans*-verbenol, or both, are released by the beetles and this repels further attacks at that pheromone source. Bedard et al. (1980) hypothesize that at some distance from a tree under attack, *trans*-verbenol *increases* response to attractants, but as the beetles approach the source of attractants, verbenone plus *trans*-verbenol increasingly *interrupts* the attractance (cf. Renwick & Vité 1970). Some of these same pheromones act as kairomones for natural enemies of bark beetles. Thus clerid beetles are attracted also to terpenes and dolichopodids to bark beetle pheromones (Bakke & Kramme 1981, Wood et al. 1968, Williamson 1971). A pteromalid parasitoid was more attracted by a host tree terpene than to bark beetle pheromones (Camors & Payne 1972). Figure 4.1K depicts a moth ovipositing. Scales from her body fall on the eggs and leaf surface. The egg chorion and scales bear the hydrocarbon tricosane, a kairomone for *Trichogramma*, which are parasitoids of lepidopterous eggs (Lewis et al. 1972). Kairomones utilized by parasitoids in finding caterpillar hosts also come from lepidopterous larval feces and chewed leaf areas (Weseloh 1981, Vinson 1981). Figure 4.1L depicts a *Chyrsopa carnea* female feeding on honeydew excreted by aphids. She has deposited her stalked eggs on the leaf. She is attracted to the honeydew by a kairomone produced when the honeydew amino acid tryptophan is oxidized (Hagen et al. 1976, van Emden & Hagen 1976). Figure 4.1M shows a colony of aphids on a leaf, stimulated to feed by the presence of certain sugars and the amino acid methionine (Mittler 1970). Lastly, Figure 4.1N shows a tephritid female ovipositing into a fruit. On completion of ovipositing, she will drag her ovipositor over the fruit surface, marking the fruit with a pheromone that deters other females of her species from ovipositing in that fruit; she then continues her search for unstung fruits. This marking pheromone can also serve as a signaling kairomone for braconid parasites, inducing more intensified searching (Prokopy 1981).

4.4.2 Host Food Suitability

The behavioral aspects of host plant finding and acceptance are certainly primary. If, however, such behavior is not reinforced in some way, it would probably break down eventually. Thus, if ovipositional "errors" are not "punished" by some adverse physiological or metabolic effects, behavioral patterns conveying host fidelity would probably quickly break down and the insect would increasingly utilize other hosts. Interestingly, there are cases of larval growth rates that do not closely follow host plant choice (Smiley 1978). So we must also look at the whole question of host *suitability*.

According to some researchers (Feeny 1975, 1976, Rhoades & Cates 1976), some plants have compounds that reduce digestibility after being ingested. A field of theory has developed regarding this area of food acceptance and suitability and the relative feeding pressures by hervibores. In this, the large, less-selective grazing vertebrates may have had different influences than the small, more selective feeders among the insects. Not able to select so specifically in taking large mouthfuls from a mixture of grasses and forbs, the large grazers, including the ancient reptiles, could not develop the close species to species (and plant part) fidelities that close food selection can lead to and is exhibited by many specific small-niche utilizers among the insects. This area is in its infancy, and we will not dwell on it (cf., e.g., Feeny 1975, 1976, Rhoades & Cates 1976). We deal here with the chemicals in plants that are associated with host suitability once ingested. Rather well-known, different hosts vary in their suitability after ingestion, even for rather polyphagous species (Busching & Turpin 1977, Hough & Pimentel 1978).

Various interactions between allelochemics and nutrients may affect the suitability of insect food. Thus, not only presence of nutrients, but also their "bioavailability," may be significant. Many examples of such interactions that interfere with bioavailability exist in the vertebrate literature, but little work has been done with insects (Reese 1977, 1978, 1979). Allelochemics may resemble essential nutrients closely enough to compete metabolically. L-Canavanine seems sufficiently similar to L-arginine to be incorporated (in an insect) into proteins in place of L-arginine, but not similar enough to yield functional proteins (e.g., Rosenthal & Bell 1979). Tannins can block availability of proteins by forming complexes, but perhaps not in all cases. Fox and Macauley (1977) found high levels of condensed tannins in some species of *Eucalyptus,* but much lower levels in others. Yet ECI values for the eucalypt insect *Paropsis atomaria,* feeding on the high and low tannin eucalypts were essentially the same. They conclude that the tannins and other phenols did not affect the nutritional physiology. In certain grasshopper species, hydrolysable tannin has a deleterious effect because it passes through the peritrophic membrane, but in other species, it does not, causing no bad effect (Bernays 1978).

To pass across the gut wall, nutrients must be in a proper form. Proteins must be broken down to their component amino acids; this is accomplished by proteases. Thus, protease inhibitors decrease availability of proteins. The effects of protease inhibitors on insects have been reviewed (Ryan 1973, Ryan & Green 1974). Birk and Applebaum (1960) reported adverse effects of soybean trypsin inhibitors on development and protease activity in *Tribolium castaneum*. In *Sitophilus oryzae*, high doses of soybean trypsin inhibitor caused increased adult mortality (Su et al. 1974).

Possibly a plant can produce much higher levels of an enzyme inhibitor after being attacked, and thus the plant would not expend energy for synthesis of material until the need arises. Green and Ryan (1972) found that wounding leaves of potato or tomato by adult Colorado potato beetles induced rapid increase of protease inhibitor, and this response spread from the injured tissue to other parts of the plant. Though not preventing further attack or development of this specialized insect, such a plant reaction could protect it from nonspecialists unable to handle the protease inhibition, but this situation relieves this beetle of the competition from the nonspecialists.

Some adverse effects may be due to allelochemic interactions with particular nutrients. Some, especially those with ortho-dihydroxy arrangements on an aromatic ring, may chelate essential minerals. Gossypol decreases assimilation by *Heliothis zea* larvae, but has no measureable effect on *H. virescens* (Shaver et al. 1977). Sinigrin reduces assimilation by *Papilio polyxenes* (Erickson & Feeny 1974). Many plant allelochemics reduce insect growth (e.g., Beck & Reese 1976, Chan et al. 1978, Dahlman 1977, Berenbaum 1978, Elliger et al. 1976, Waiss et al. 1979), but in only a few cases is anything known about growth inhibition. "Digestibility-reducing" factors in creosote resins apparently block nutrient digestibility in certain Orthoptera (Rhoades & Cates 1976). Interestingly, as higher levels of cholesterol were added to an artificial diet, adverse effects on growth of the sunflower moth to certain diterpene acids diminished (Elliger et al. 1976).

In recent years the hormonal effects of plant compounds have been examined. While juvenile hormone (JH) analogs are most effective during the later stages of metamorphosis, antijuvenile hormones should be most effective during earlier stages and in adults (Bowers 1977). Many phytoecdysones have been isolated from plants (Russell 1977), but conclusive proof that they are effective against phytophagous insects has been elusive; however, Harborne (1982) lists five points that could be argued in favor of an ecological function for plant ecdysones.

We note here that some insect-resistant varieties retain their resistance for long periods. This suggests a complex of resistance mechanisms. Within an insect population, there may be individuals that can detoxify a chemical resistance factor (below). If this factor were the only resistance mechanism of the plant population, selection pressure would favor insects that can detoxify it, and resistance would break down. A better chance of relative permanence exists if the cultivar contains allelochemics adverse to both

behavior and metabolism. Two mutations, one behavioral and one metabolic, would have to arise simultaneously before a resistant insect biotype
could arise (Beck & Schoonhoven 1979, Erickson & Feeny 1974).

4.4.3 Detoxification of Adverse Chemicals by Consumer Insects

As noted above, in some insects L-canavanine competes with L-arginine
and thus forms defective proteins. The beetle *Caryedes brasiliensis*, however,
lives on seeds high in L-canavanine. The larvae have an arginyl *t*-RNA
synthetase that discriminates between L-arginine and L-canavanine (Rosenthal et al. 1976), and L-canavanine is apparently even utilized as a N
source (Rosenthal et al. 1978). The mixed-function oxidases are enzymes
that can detoxify a range of toxic substances, including, probably, many
plant compounds. Highly polyphagous insects are exposed to a larger number of such compounds, and indeed their mixed-function oxidase levels
tend to be high (Krieger et al. 1971). These enzymes are located mainly in
the fat body, Malpighian tubules and midgut; the midgut is especially
important in Lepidoptera (Dauterman & Hodgson 1978).

4.5 OVERVIEW

Figure 4.2 sketches major body parts of an insect larva that are involved
in finding, accepting, and rejecting food, and if ingested the various functions that become involved to handle secondary plant compounds. This
figure, in combination with Fig. 4.1A–N, presents a pictorial summary of
many features in the nutritional ecology of insects.

The same chemicals required by a protozoan or cells of a vertebrate for
growth are also required by the cells of an insect. If any of the required
organic chemicals are lacking in the food and the insect still develops, it is
because various symbiotic microorganisms are providing them. Such mutualistic relationships occur in insects of many orders (Brooks 1963, Koch
1967). The nutritional requirements of insects at the organismal level are
similar to those of vertebrates except that insects have no requirement for
the vitamin D of vertebrates but do have a requirement for an extrinsic
sterol.

The specificity of phytophagous insects to their plant hosts is largely
determined by secondary plant metabolites, and these chemicals often serve
directly or as precursors of hormones, and the behavior chemicals known
as pheromones, allomones, and kairomones (Duffey 1980, Harborne 1982,
Nordlund et al. 1981, Rosenthal and Janzen 1979). These secondary plant
substances can be considered ecological chemical requirements since they
serve to tie insects to their hosts and protect plants from other insects,
pathogens, and general herbivores that have not broken the chemical defenses of these plants.

The quality and quantity of food determine rates of development and reproduction, but it seems that the food does not commonly determine average population densities of phytophagous insects (cf. Chapter 15). The dominant species of major vegetation types seem to be protected by strong general chemical deterrents and their growth characteristics (Feeny 1976, in part), while certain of the associated species (various herbs) that are not protected from their host-specific phytophagous insects by natural enemies may in fact be regulated at low densities by these specialists (cf. Chapter 22). Partial host plant resistance can slow the rate of an insect's population increase but even here the insect's natural enemies commonly regulate its average population density (van Emden & Wearing 1965). The average density of entomophagous insect populations is largely governed by the quantity of prey available, and there are interactions among the three trophic levels, with the plants influencing the basic interactions between the insect herbivores and their natural enemies (Price et al. 1980, Rhoades & Cates 1976).

However, even when a phytophagous insect species has broken the chemical defenses of a plant, there may still not be a strong selective pressure for development of still other chemicals to deter it due to the density-dependent regulation on the population of the insect by its natural enemies. Consequently, such density-dependent relationships between the second and third trophic levels provide for a measure of long-term ecological stability at the primary trophic level—the plant community level.

REFERENCES

Albritton, E. C. 1953. *Standard Values in Nutrition and Metabolism.* Wright Air Develop. Cent. Tech. Rep. 52-301.

Andrewartha, H. G. and L. C. Birch. 1973. *The Distribution and Abundance of Animals.* University of Chicago Press, Chicago.

Applebaum, S. W., A. M. Konijn, and B. Menco. 1971. *Insect Biochem.* **1:** 1–13.

Baker, H. G., P. A. Opler, and I. Baker, 1978. *Bot. Gaz.* **139:** 322–332.

Bakke, A. and I. Kramme. 1981. *J. Chem. Ecol.* **7:** 305–312.

Barbier, M. 1970. *Progress in Phytochem.* **2:** 1–34.

Barlow, C. A. 1979. *Can. Entomol.* **111:** 897–904.

Beck, S. D. 1972. Pp. 1–6 in Rodriguez (Ed.), 1972, referenced here.

Beck, S. D. and J. C. Reese. 1976. *Recent Adv. Phytochem.* **10:** 41–92.

Beck, S. D. and L. M. Schoonhoven. 1979. Pp. 115–135 in F. G. Maxwell and P. R. Jennings (Eds.), *Breeding Plants Resistant to Insects.* Wiley, New York.

Bedard, W. D., P. E. Tilden, K. Q. Lindahl, Jr., D. L. Wood, and P. A. Rauch. 1980. *Chem. Ecol.* **6:** 997–1013.

Bentley, B. L. 1977 *Annu. Rev. Ecol. Syst.* **8:** 407–427.

Berenbaum, M. 1978. *Science* **201:** 532–534.

Bernays, E. A. 1978. *Entomol. Exp. & Appl.* **24:** 244–253.

Birk, Y. and W. Applebaum. 1960. *Enzymologia* **22:** 318–326.

Bowers, W. S. 1977. Pp. 129–142 in G. B. Marini-Bettolo (Ed.), *Natural Products and the Protection of Plants.* Pontifica Academia Scientiarum, Vatican City.

Bracken, G. K. and C. A. Barlow. 1967. *Can. J. Zool.* **45:** 57–61.

Brattsten, L. B. 1979. Pp. 199–270 in Rosenthal and Janzen (Eds.), 1979, referenced here.

Brooks, M. A. 1963. Pp. 200–231 in *Symbiotic Associations. 13th Symp. Soc. Gen. Microbiol., Roy. Inst., London.* Cambridge University Press, London.

Brues, C. T. 1946. *Insect Dietary.* Harvard University Press, Cambridge, Mass.

Busching, M. K., and F. T. Turpin. 1977. *Environ. Entomol.* **6:** 63–65.

Bush, G. L. 1966. *Bull. Mus. Comp. Zool., Harvard Univ.* **134:** 431–562.

Camors, F. B. Jr. and T. L. Payne. 1972. *Ann. Ent. Soc. Am.* **65:** 31–37.

Campbell, B. C. and S. S. Duffey. 1979. *Science* **205:** 700–702.

Carlson, S. D., H. R. Steeves III, J. S. Van de Berg, and W. E. Robbins. 1967. *Science* **158:** 268–270.

Caswell, H., F. Reed, S. N. Stephenson, and P. A. Werner. 1973. *Am. Natur.* **107:** 465–480.

Chan, B. G., A. C. Waiss, Jr., R. G. Binder, and C. A. Elliger. 1978. *Entomol. Exp. & Appl.* **24:** 294–300.

Chaplin, S. B. and S. J. Chaplin. 1981. *J. Anim. Ecol.* **50:** 407–420.

Chapman, P. J. and E. A. Bernays (Eds.). 1978. *Insect and Host Plant.* Proc. 4th Int. Symp. Nederlandse Entomologische Vereniging.

Chu, H.-M., D. M. Norris, and L. T. Kok. 1970. *J. Insect Physiol.* **16:** 1379–1387.

Clark, A. J. and K. Bloch. 1959. *J. Biol. Chem.* **234:** 2583–2588.

Clayton, R. B. 1964. *J. Lipid Res.* **5:** 3–19.

Dadd, R. H. 1970a. Pp. 35–95 in M. Flarkin and B. T. Scheer (Eds.), *Chemical Zoology,* Vol. 5, Arthropoda, Part A. Academic, New York.

Dadd, R. H. 1970b. *Chem. Zool.* **5:** 117–145.

Dadd, R. H. 1973. *Annu. Rev. Entomol.* **18:** 381–420.

Dadd, R. H. 1977. Pp. 305–346 in *CRC Handbook Series in Nutrition and Food.* CRC Press, Cleveland, Ohio.

Dadd, R. H. 1978. *J. Insect Physiol.* **24:** 25–30.

Dadd, R. H. 1979. *J. Insect Physiol.* **25:** 353–359.

Dadd, R. H. 1981. Pp. 189–214 in G. Bhaskaran, S. Friedman, and J. G. Rodriguez (Eds.), *Current Topics in Insect Endocrinology and Nutrition.* Plenum, New York.

Dadd, R. H. 1982. Pp. 107–147 in T. E. Mittler and R. H. Dadd (Eds.), *Metabolic Aspects of Lipid Nutrition in Insects.* Westview Press, Colorado.

Dadd, R. H. and J. E. Kleinjan. 1979. *Entomol. Exp. & Appl.* **26:** 222–226.

Dahlman, D. L. 1977. *Entomol. Exp. & Appl.* **22:** 123–131.

Dauterman, W. C. and E. Hodgson. 1978. Pp. 541–577 in M. Rockstein (Ed.), *Biochemistry of Insects.* Academic, New York.

Dethier, V. G. 1947. *Chemical Insect Attractants and Repellents.* Blakiston, Philadelphia.

Dethier, V. G. 1970. Pp. 83–102 in E. Sondheimer and J. B. Simeone (Eds.), *Chemical Ecology.* Academic, New York.

Dethier, V. G. 1980. *Am. Natur.* **115:** 45–66.

Duffey, S. S. 1980. *Annu. Rev. Entomol.* **25:** 447–477.

Ehrlich, P. R. and P. H. Raven. 1965. *Evolution* **18:** 586–608.

Elliger, C. A., D. F. Zinkel, B. G. Chan, and A. C. Waiss, Jr. 1976. *Experientia* **32:** 1364, 1365.

Erickson, J. M. and P. Feeny. 1974. *Ecology* **55:** 103–111.

Ewart, W. H. and R. L. Metcalf. 1956. *Ann. Entomol. Soc. Am.* **49:** 441–447.

Feeny, P. P. 1975. Pp. 3–19 in Gilbert and Raven (Eds.), *Coevolution of Animals and Plants.* University of Texas Press, Austin.

Feeny, P. P. 1976. *Rec. Adv. Phytochem.* **10:** 1–40.

Fox, L. R. and B. J. Macauley. 1977. *Oecologie* (Berl.) **29:** 145–162.

Fraenkel, G. 1959. *Science* **129:** 1466–1470.

Fraenkel, G. and M. Blewett. 1944. *Bull. Entomol. Res.* **35:** 127–139.

Friend, W. G. and R. H. Dadd. 1982. *Adv. Nutritional Res.* **35:** 127–139.

Gray, H. E. and G. Fraenkel. 1953. *Science* **118:** 304, 305.

Green, T. R. and C. A. Ryan. 1972. *Science* **175:** 776, 777.

Hagen, K. S. 1952. *Influence of Adult Nutrition Upon Fecundity, Fertility and Longevity of Three Tephritid Species.* Ph.D. dissertation, University of California, Berkeley.

Hagen, K. S., P. Greany, E. F. Sawall, Jr., and R. L. Tassan. 1976. *Environ. Entomol.* **5:** 458–468.

Haglund, B. M. 1980. *Nature* **288:** 697, 698.

Harborne, J. B. (Ed.). 1972. *Phytochemical Ecology.* Academic, New York.

Harborne, J. B. (Ed.). 1982. *Introduction to Ecological Biochemistry* (2nd ed.). Academic, New York.

Harris, M. R. (Ed.). 1980. *Biology and Breeding for Resistance to Arthropods and Pathogens in Agricultural Plants.* Texas A&M University, College Station.

Harris, W. A., D. F. Ready, E. D. Lipson, and A. J. Hudspefh. 1977. *Nature* **266:** 648–650.

Harvey, G. T. 1974. *Can. Entomol.* **106:** 353–365.

Hedin, P. A., A. C. Thompson, and R. C. Gueldner. 1976. *Rec. Adv. Phytochem.* **10:** 271–350.

Heed, W. B. and H. W. Kircher. 1965. *Science* **149:** 758–761.

Hough, J. A., G. E. Harmon, and C. J. Eckenrode. 1981. *Environ. Entomol.* **10:** 206–210.

Hough, J. A. and D. Pimentel. 1978. *Environ. Entomol.* **7:** 97–102.

House, H. L. 1965. *Can. Entomol.* **97:** 62–68.

House, H. L. 1966. *Can. Entomol.* **98**(11): 1121–1134.

House, H. L. 1972. Pp. 513–573 in R. N. T-W-Fiennes (Ed.), *International Encyclopedia of Food and Nutrition.* Pergamon, Oxford.

House, H. L. 1974. Pp. 1–62 in M. Rockstein (Ed.), *The Physiology of Insecta.* 5. Academic, London.

Hsiao, T. C. A. I. 1973. *Rev. Plant Physiol.* **24:** 519–570.

Jermy, T. (Ed.). 1976. *The Host-Plant in Relation to Insect Behaviour and Reproduction.* Plenum, New York.

Király, Z. and L. Szalay-Marzsó. 1971. *Symposium on Biochemical and Ecological Aspects of Plant-Parasite Relations.* Akadémiai Kiadó, Budapest.

Kircher, H. W. and M. A. Gray. 1978. *J. Insect Physiol.* **24:** 555–559.

Koch, A. 1967. Pp. 1–106 in S. M. Henry (Ed.), *Symbiosis,* Vol. 2. Academic, New York.

Kogan, M. 1977. *Proc. 15th Inter. Congr. Entomol., Washington, D.C.,* pp. 211–248.

Kozlovsky, D. G. 1968. *Ecology* **49:** 48–60.

Kramer, K. J. and P. A. Seib. 1982. Pp. 275–291 in Seib and Tolbert (Eds.), *Ascorbic Acid Chemistry, Metabolism and Uses.* Amer. Chem. Soc., Washington, D.C.

Krieger, R. I., P. P. Feeny, and C. F. Wilkinson. 1971. *Science* **172:** 579–581.

Kunkel, H. and W. Kloft. 1977. *Apidologie* **8:** 369–391.

Lees, A. D. 1955. *The Physiology of Diapause in Arthropods.* Cambridge University Press, London.

Legay, J. M. 1958. *Annu. Rev. Entomol.* **3:** 75–86.

Lewis, W. J., K. S. Hagen, W. L. Roelofs, and L. M. Schoonhoven. 1980. *FAO Plant Protection Bull.* **28:** 121–128.

Lewis, W. J., R. L. Jones, and A. N. Sparks. 1972. *Ann. Entomol. Soc. Am.* **65:** 1087–1089.

Loher, W., I. Ganjian, I. Kubo, D. Stanley-Samuelson, and S. S. Tobe. 1981. *Proc. Natl. Acad. Sci.* **78:** 7835–7838.

Matsumoto, Y. 1970. Pp. 133–160 in D. L. Wood, K. M. Silverstein, and M. Nakajima (Eds.), *Control of Insect Behavior by Natural Products.* Academic, London.

Mattson, W. J. 1980. *Annu. Rev. Ecol. Syst.* **11:** 119–161.

Maynard, L. A. and J. K. Loosli. 1978. *Animal Nutrition.* McGraw-Hill, New York.

McCaffery, A. R. 1976. *Symp. Biol. Hunger* **16:** 163–172.

McFarlane, J. E. 1966. *J. Insect Physiol.* **12:** 179–188.

McFarlane, J. E. 1978. *Can. Entomol.* **109:** 329, 330.

McKey, D. 1979. Pp. 55–133 in Rosenthal and Janzen (Eds.), 1979, referenced here.

McNeill, S. and T. R. E. Southwood. 1978. Pp. 77–98 in J. Harborne (Ed.), *Biochemical Aspects of Plant and Animal Coevolution.* Academic, London.

Mellanby, K. and R. A. French. 1958. *Entomol. Exp. & Appl.* **1:** 116–124.

Metcalf, R. L., E. R. Metcalf, W. C. Mitchell, and L. W. Y. Lee. 1979. *Proc. Natl. Acad. Sci.* **76:** 1561–1565.

Miller, W. J. and M. W. Neathery. 1977. *BioScience.* **27:** 674–679.

Mitchell, R. 1981. *Annu. Rev. Entomol.* **26:** 373–396.

Mittler, T. E. 1958. *J. Exp. Biol.* **35:** 626–638.

Mittler, T. E. 1970. *Entomol. Exp. & Appl.* **13:** 432–437.

Mittler, T. E. 1971. *J. Insect Physiol.* **17:** 1333–1347.

Nicholson, S. W. 1980. *J. Insect Physiol.* **26:** 315–320.

Nordlund, D. A., R. L. Jones, and W. J. Lewis. 1981. *Semiochemicals: Their Role in Pest Control.* Wiley, New York.

Percival, M. S. 1961. *New Phytol.* **60:** 235–281.

Price, P. W., C. E. Bouton, P. Gross, B. A. McPheron, J. N. Thompson, and A. E. Weis. 1980. *Annu. Rev. Ecol. Syst.* **11:** 41–65.

Prokopy, R. J. 1981. Pp. 181–213 in Nordlund, Jones, and Lewis (Eds.), 1981, referenced here.

Randolph, P. A. and J. C. Randolph. 1975. *Ecology* **56:** 359–369.

Rapoport, E. H. and M. Tschapek. 1967. *Rev. Ecol. Biol. Soil* **4:** 1–58.

Reese, J. C. 1977. Pp. 129–152 in P. A. Hedin (Ed.), *Host Plant Resistance to Pests.* ACS Symp. Ser. 62, Am. Chem. Soc., Washington, D.C.

Reese, J. C. 1978. *Entomol. Exp. & Appl.* **24:** 625–631.

Reese, J. C. 1979. Pp. 309–330 in Rosenthal and Janzen (Eds.), 1979, referenced here.

Reese, J. C. and S. D. Beck. 1978. *J. Insect Physiol.* **24:** 473–479.

Renwick, J. A. A. and J. P. Vité. 1970. *Contrib. Boyce Thompson Inst.* **24:** 283–292.

Rhoades, D. F. and R. G. Cates. 1976. *Rec. Adv. Phytochem.* **10:** 168–213.

Riddiford, L. M. and C. M Williams. 1967. *Science* **155:** 588–590.

Ritter, F. J. (Ed.). 1979. *Chemical Ecology: Odour Communications in Animals.* Elsevier/North-Holland Biomedical Press, Amsterdam.

Ritter, K. S., B. A. Weiss, A. L. Norrbom, and W. R. Nes. 1982. *Comp. Biochem. Physiol. B.* **71:** 345–349.

Robbins, W. E., J. N. Kaplanis, J. A. Svoboda, and M. J. Thompson. 1971. *Annu. Rev. Entomol.* **16:** 53–72.

Rodriguez, J. G. (Ed.). 1972. *Insect and Mite Nutrition.* North Holland Publishing, Amsterdam.

Roeske, C. N., J. N Seiber, L. P. Brower, and C. M. Moffitt. 1976. *Rec. Adv. Phytochem.* **10:** 93–167.

Rosenthal, G. A. and E. A. Bell. 1979. Pp. 353–385 in Rosenthal and Janzen (Eds.), 1979, referenced here.

Rosenthal, G. A., D. L. Dahlman, and D. H. Janzen. 1978. *Science* **22:** 528, 529.

Rosenthal, G. A. and D. H. Janzen (Eds.). 1979. *Herbivores—Their Interaction with Secondary Plant Metabolites.* Academic, New York.

Rosenthal, G. A., D. J. Janzen, and D. L. Dahlman. 1976. *Science* **192:** 256–258.

Rothschild, M. 1972. Pp. 1–12 in Harborne (Ed.), 1972, referenced here.

Rothschild, M. 1973. Pp. 59–83 in van Emden (Ed.), *Insect-Plant Relationships.* Blackwell, Oxford.

Rothschild, M., R. Aplin, J. Baker, and N. Marsh. 1977. *Nature* **280:** 487, 488.

Russell, G. B. 1977. *New Zealand Entomol.* **6:** 229–234.

Ryan, C. A. 1973. *Annu. Rev. Plant Physiol.* **24:** 173–196.

Ryan, C. A. and T. R. Green. 1974. *Rec. Adv. Phytochem.* **8:** 123–140.

Salkeld, E. H. 1959. *Can. Entomol.* **91:** 93–97.

Sang, J. H. 1978. Pp. 159–192 in M. Ashburner and T. R. F. Wright (Eds.), *The Genetics and Biology of Drosophila,* Vol. 2A. Academic, New York.

Schneider, F. 1948. *Mitt. Schweiz. ent. Ges.* **21:** 249–285.

Schoonhoven, L. M. 1973. *Symp. Roy. Entomol. Soc. Lond.* **6:** 87–99.

Schoonhoven, L. M. 1981. Pp. 31–50 in Nordlund, Jones, and Lewis (Eds.), 1981, referenced here.

Schroeder, L. A. 1976. *Oikos* **28:** 27–31.

Schroth, M. N. and J. G. Hancock. 1982. *Science* **216:** 1376–1381.

Scriber, J. M. 1977. *Oecologia (Berl.)* **28:** 269–287.

Scriber, J. M. 1979. *Entomol. Exp. & Appl.* **25:** 240–252.

Scriber, J. M. and F. Slansky, Jr. 1981. *Annu. Rev. Entomol.* **26:** 183–211.

Shaver, T. N., J. A. Garcia, and R. H. Dilday. 1977. *Environ. Entomol.* **6:** 82–84.

Shimizu, I., S. Kitabatake, and M. Kato. 1981. *J. Insect Physiol.* **27:** 593–599.

Shorey, H. H. 1976. *Animal Communication by Pheromones.* Academic, New York.

Singh, K. R. P. and A. W. A. Brown, 1957. *J. Insect Physiol.* **2:** 199–220.

Sivapalan, P. and N. C. Gnanapragasam. 1979. *J. Insect Physiol.* **25:** 393–398.

Slansky, F. Jr. and P. P. Feeny. 1977. *Ecol. Monogr.* **47:** 209–228.

Slansky, F. Jr. and J. M. Scriber. 1982. *Bull. Entomol. Soc. Am.* **28:** 43–55.

Smiley, J. 1978. *Science* **201:** 745–747.

Smith, B. N. and W. V. Brown. 1973. *Am. J. Bot.* **60:** 505–513.

Snellor, V. P. and R. H. Dadd. 1981. *J. Med. Entomol.* **18:** 235–239.

Sondheimer, E. and J. B. Simeone (Eds.). 1970. *Chemical Ecology.* Academic, New York.

Soo Hoo, C. F. and G. Fraenkel. 1966. *J. Insect Physiol.* **12:** 693–709, 711–730.

Spradbery, J. P. 1969. *Bull. Entomol. Res.* **59:** 291–297.

Stanley-Samuelson, D., and R. H. Dadd. 1981. *J. Insect Physiol.* **27:** 571–578.

Su, H. C., R. D. Speirs, and P. G. Mahany. 1974. *J. Ga. Entomol.* **9:** 86, 87.

Svoboda, J. A., J. N. Koplanis, W. E. Robbins, and M. J. Thompson. 1975. *Annu. Rev. Entomol.* **20:** 205–220.

Svoboda, J. A., M. J. Thompson, W. E. Robbins, and J. N. Kaplanis. 1978. *Lipids* **13:** 742–753.

Thompson, S. N. 1981. *Proc. Symp. 9th Int. Congr. Plant Prot.* **1:** 93–100.

Tsiropoulos, G. L. 1980. *Ann. Entomol. Soc. Am.* **73:** 705–707.

van Emden, H. F., and C. H. Wearing. 1965. *Annu. Appl. Biol.* **56:** 323, 324.

van Emden, H. F., and K. S. Hagen. 1976. *Environ. Entomol.* **5:** 469–473.

van Emden, H. H. (Ed.), 1973. *Insect/Plant Relationships.* Symp. R. Entomol. Soc. Lond., No. 6.

Vinson, S. B. 1981. Pp. 51–77 in Nordland, Jones, and Lewis (Eds.), 1981, referenced here.

Waiss, A. C. Jr., B. G. Chan, C. A. Elliger, B. R. Wiseman, W. W. McMillan, N. W. Widstrom, M. S. Zuber, and A. J. Keaster. 1979. *J. Econ. Entomol.* **72:** 256–258.

Waldbauer, G. P. 1962. *Entomol. Exp. & Appl.* **5:** 147–158.

Waldbauer, G. P. 1964. *Entomol. Exp. & Appl.* **7:** 253–269.

Waldbauer, G. P. 1968. *Adv. Insect Physiol.* **5:** 229–288.

Wallace, J. W. and R. L. Mansell (Eds.). 1976. *Biochemical Interaction between Plants and Insects. Recent Adv. Phytochem,* Vol. 10. Plenum, New York.

Waller, G. D. 1972. *Annu. Entomol. Soc. Am.* **65:** 857–862.

Weseloh, R. M. 1981. Pp. 79–95 in R. J. Jones and W. J. Lewis (Eds.), *Semiochemicals—Their role in Pest Control.* Wiley, New York.

Wharton, G. W. and L. G. Arlian. 1972. Pp. 152–165 in Rodriguez (Ed.), 1972, referenced here.

White, T. C. R. 1978. *Oecologia* **22:** 119–134.

Whittaker, R. H. and P. P. Feeny. 1971. *Science* **272:** 757–770.

Williamson, D. L. 1971. *Ann. Entomol. Soc. Am.* **64:** 586–589.

Wood, D. L. 1982. *Annu. Rev. Entomol.* **27:** 411–446.

Wood, D. L., L. E. Browne, W. D. Bedard, P. E. Tilden, R. M. Silverstein, and J. O. Rodin. 1968. *Science* **159:** 1373, 1374.

Wood, D. L., R. M. Silverstein, and M. Nakajima (Eds.). 1970. *Control of Insect Behavior by Natural Products.* Academic, New York.

Chapter 5

Reproduction

in Insects

FRANZ ENGELMANN

5.1 INTRODUCTION

The rate of propagation of a species depends on a multiplicity of intrinsic and environmental factors that may interact to various degrees. Factors basic to egg output include the inherent capacity of the ovaries to produce a given number of eggs, the acquisition of reserves for making yolk, the hormonal control of vitellogenesis, and the environmental cues that control the timing of hormone synthesis and release. The success of a species depends to a large extent on how it exploits the environment. The great diversity of reproductive patterns found in insects is fascinating in itself (Engelmann 1970). A study of insect reproduction can furthermore be used to decipher basic biological phenomena applicable to other animals. Among these are questions such as how environmental cues are translated by the central nervous system (CNS) into control of hormone release and vitellogenesis. We can approach this question profitably as a physiologist or as a molecular endocrinologist.

5.2 FACTORS AFFECTING FECUNDITY

5.2.1 Intrinsic Capacity

The capacity to produce a given number of offspring resides primarily in the number of ovarioles/ovary, ovariolar structure, and longevity of the species. In the extremes, the combination of these factors allow production of only a single egg in the lifetime of the sexual morph of some aphid species, or laying of several hundred thousand eggs by queens of the social Hymenoptera and Isoptera. An extensive list of observed egg productions in many species from different orders of insects (Engelmann 1970) exemplifies this diversity. Extended longevity as seen in queens of social insects, combined with sociality and community attendance, results in enormous reproduction potential of a single female.

Larval feeding can influence the number of differentiated ovarioles/ovary and the consequent potential for egg production. This is dramatically seen in worker honeybees, where food quality and reduced quantity results in differentiation of only a few ovarioles, whereas queens may have about 160–180 ovarioles/ovary. Poor larval feeding caused by overcrowding in *Sarcophaga bullata* resulted in pupal weights of about one-fifth of normal and differentiation of only 32 ovarioles compared to more than 100 in adequately fed larvae (Pappas & Fraenkel 1977). This illustrates that the number of ovarioles, even though genetically determined, can be modified by quality and quantity of nutrition obtained during differentiation of the ovaries.

114

5.2.2 Nutrition

Insect eggs generally contain all the nutrients necessary for completion of embryonic development. Protein and lipid yolk accumulate in eggs and are derived from larval and adult feeding. Some species, such as many noctuid moths and the Ephemeroptera, never feed on proteins as adults and depend on reserves accumulated as larvae. Other species, particularly among the Diptera, produce a limited number of eggs without feeding as adults, but after protein ingestion, lay many more. The phenomenon of oviposition without prior feeding on protein sources was termed *autogeny* by Roubaud (1929) and has attracted much interest among insect physiologists since its discovery in mosquitoes. Clearly, the quantitative expression of autogeny is based on the quantity of reserves carried over from larval feeding. Many examples, particularly among the mosquitoes, could be cited in support of this (cf. Engelmann 1970). Under constant, most favorable nutritional conditions, some females of the anautogenous *Aedes aegypti* could become autogenous (Lea 1964), a phenomenon that probably does not occur in nature. Other species, such as *Sarcophaga bullata*, may deposit some yolk but never fully mature any egg even under most favorable conditions. Pappas and Fraenkel (1977) termed this *incipient autogeny*. It presumably has no survival value for the population and may just be the expression of an evolutionary transition between anautogeny and autogeny. Autogeny probably has survival value for a species living in the Arctic where mammalian hosts are scarce. Examples are *Aedes impiger* and *A. nigripes* (Corbet 1964), which can propagate to a limited extent without feeding. Interestingly, autogeny is more often found among species living in adverse regions of the earth, which suggests that it may have evolved in these adverse conditions more readily than otherwise.

A well-known observation is that insects that feed on more protein either as larvae (expressed as pupal weight), adults, or both, produce many more eggs than those that take in little. Linear relationships are seen between pupal weights or size of meal and egg output (Engelmann 1970).

An investigation of the requirements for specific food chemicals for reproduction in the many diverse insect species is one of the fascinating aspects. Interpretation of data in numerous investigations is complicated by several components often not considered in these studies. For example, trace amounts of essential minerals or vitamins may be carried over from larval or nymphal feeding or supplied by symbionts in the gut or fat bodies. Thus, it is frequently difficult to determine the actual adult nutritional requirements for reproduction (cf. Engelmann 1970). Relatively few publications report on a systematic analysis of a specie's adult nutritional needs; the overwhelming number of papers are concerned only with effects on larval growth. This, of course, will ultimately affect egg production in many species.

Despite these cautious remarks, some pertinent findings and thoughts may be expressed here. Many insects are monophagous or oligophagous as immature and mature animals. They accumulate reserves subsequently used for egg production from plants that also frequently contain allelochemics, such as toxic alkaloids or glycosides. These secondary plant substances can be tolerated or detoxified by the species that feed on such plants (cf. Chapter 4). Unfortunately, for a majority of species information is only available on immature stages (cf. Beck & Reese 1976), and we are forced to speculate regarding the effects of allelochemics on reproduction and egg output. One interesting case deserves mentioning: identification of Cruciferae as acceptable hosts for crucifer feeders is apparently made by the perception of sinigrin, an otherwise toxic mustard oil glycoside. *Papilio polyxenes*, a species which normally feeds on Umbelliferae such as celery, could be coaxed into receiving sinigrin when the food plants were cultivated in water containing this toxic glycoside (Erickson & Feeny 1974). Concentrations of 0.01% resulted in a pupal weight of about two-thirds of normal and, correspondingly, egg output was significantly reduced in spite of ingestion of normal amounts of food. Concentrations of 0.1%, which are tolerated by crucifer feeders, are toxic to *P. polyxenes*. The mechanism by which sinigrin adversely affects food assimilation is not known.

As illustrated here, the same secondary plant chemical can be detrimental to certain species while beneficial to others. Several species of butterflies, moths, and beetles are attracted to sinigrin-containing plants and oviposit on them preferentially. The same chemical also functions as a phagostimulant for the larvae of these species. An insect–plant interaction has evolved that in these cases is beneficial to the insect and in other cases to the plant when it acts as a feeding deterrent. In a chemically highly heterogeneous environment, only a few signals are important for the success of a given species. Certain key signals are of prime significance, particularly in monophagous or oligophagous species which are at a disadvantage when their food sources are scarce.

While certain chemicals may either induce or deter feeding, it is the quantity of nutrients consumed and utilized that is basic to egg production itself (cf. Engelmann 1970). Some proteins of certain food sources can be more nutritious than others, that is, they may be better assimilated. However, this can only be determined in extensive artificial feeding tests. Moreover, it is extremely difficult to relate laboratory tests on nutritional values of certain proteins to those of natural foods. Greenberg (1951), for example, reported that in *Aedes aegypti*, bovine plasma albumin promoted higher fecundity than casein or gelatin. Similarly, *Erioischia brassicae*, the cabbage root fly, produced about 158 eggs when fed on yeast extract, but only about 104 eggs on fibrin (Finch 1971). In both of these examples the authors compared the effectiveness of proteins unnatural for these species and tested primarily for reproductive potential and not for the components, which are normally important for propagation of these species. The literature is replete with reports of food-related egg production that have no natural basis.

In nature many phytophagous insects feed on plant specimens of variable age. *Leptinotarsa* lays many more eggs when fed on young rather than old potato leaves, but when lecithin was added to old leaves normal egg production was observed (Grison 1948). Also, the beetle *Haltica lythri* laid only about 25% of the normal number of eggs when feeding on senescent leaves (Phillips 1976) (Fig. 5.1). In these cases the actual factors that may allow better egg production when feeding on young leaves are not known. Old leaves may contain less protein or lack certain essential components, be it only in trace amounts.

Minerals and vitamins appear to control the quantitative egg production in an often unknown fashion. This is shown for potassium salts and salt mixtures in *Phormia* and *Sarcophaga*, respectively, which were essential for complete and normal egg output (Pappas & Fraenkel 1977). Vitamin requirements for normal reproduction have been documented repeatedly for many insect species (cf. Engelmann 1970). After feeding on a vitamin mixture, the ichneumonid *Exeristes comstockii* dramatically increased its daily oviposition rate (Bracken 1965). In *Acheta domesticus* vitamin E was not only essential in males for normal spermatogenesis and mating (Meikle & McFarlane 1965) but also improved egg production in females (McFarlane 1976). Yet in *Plusia gamma* this vitamin did not affect egg laying (Macaulay 1973). As these few examples illustrate, for unknown reasons a specific chemical may be very important in some species, but not in others.

Dietary requirements for egg production and spermatogenesis in insects may be as diverse as the species themselves. A most fascinating case of

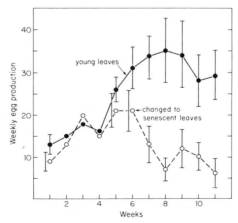

Fig. 5.1   Weekly egg production in two populations of the beetle *Haltica lythri*, one feeding continually on young leaves and the other was transferred to senescent leaves. (Modified after Phillips 1976.)

insect–host adaptation is seen in the rabbit flea *Spilopsyllus cuniculi* (Roths-child & Ford 1966, 1973). This flea remains reproductively inactive when feeding on the anoestrus doe or male rabbit. When feeding on the late pregnant doe, yolk deposition in the oocytes begins, and at parturition the flea transfers to the newborn, copulates, and lays its eggs (Fig. 5.2). The onset of vitellogenesis in this flea coincides with increased corticosteroid and oestrogen levels in the host's blood. An airborne factor (kairomone) from the rabbit nest, presumably emanating from the rabbit urine (Roths-child & Ford 1973), promotes copulation attempts. Within approximately 20 days the fleas transfer back to the doe. Artificially raised corticosteroid levels in the host or spraying the flea with this hormone or oestrogen had the same effect as feeding on late pregnant does. Presently it is unclear

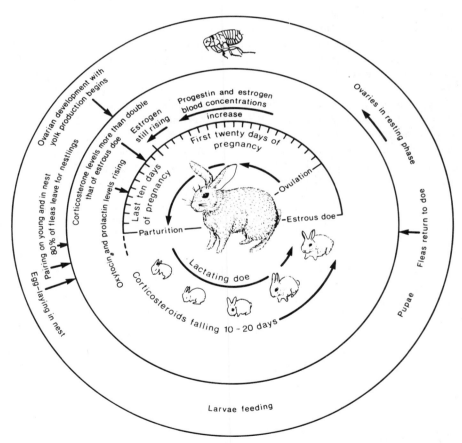

Fig. 5.2 The reproductive interrelationship of the rabbit flea *Spilopsylla* with its host. Various phases of the flea's reproduction are closely linked to the level of several of the host hormones. (Adapted and modified after Rothschild & Ford 1973.)

whether the host's hormones induce vitellogenin production or whether they affect production of JH by the recipient flea. One may expect that similar intimate interrelationships exist between other insect species and their hosts but no detailed analyses have been made in any of these cases.

Host dependencies of insects are readily seen in the blood-sucking and parasitoid species. Insect–plant relationships may have evolved to a similar degree of host dependence by the exploiter insects. One such case is *Drosophila pachea*, which lives exclusively on the Senita cactus of the Mexican Sonoran desert. For growth and reproduction this species depends on a sterol Δ^7-stigmasten-3-ol (Schottenol) contained in this cactus (Heed & Kircher 1965, Kircher et al. 1967). Other species of *Drosophila* cannot live on the Senita cactus because it also contains two toxic alkaloids only tolerated by *D. pachea*.

The above examples represent spectacular adaptations to specific circumstances and are therefore noteworthy. The majority of species are, on the other hand, rather more "ordinary." Specific chemicals consumed by an insect may affect its endocrine system or any of the innumerable biochemical processes required to make an egg. For none of the cases mentioned, nor any others, do we know much beyond the easily observable facts.

5.2.3 The Environment

Photoperiod and temperature are probably the most important environmental factors influencing the process of reproduction; other factors, such as humidity or species density, may be of great significance at times. Optimal temperatures can be as low as 16–20°C, as in *Coccinella septempunctata* (Bodenheimer 1943), or as high as 37–40°C, like in *Piesma quadrata* (Schubert 1927). In nature some species experience rather high temperatures during day times, but cooler conditions at night; that is, the daily average is considerably lower. Laboratory observations are generally made at highly unnatural constant temperature, and studies under fluctuating temperature regimes have been very rare (for development, cf. Chapter 3). When insects experience high temperatures during part of the 24-hr period, mean lower temperature limits for egg production may be rather low. More eggs may be laid under a fluctuating temperature regime of a given arithmetic mean value than at a constant temperature at this value, or at either of the respective constant low, or high, temperatures. For example, *Gryllus bimaculatus* at a constant 20°C laid a lifetime total of 276 eggs and at 30°C about 900 eggs. However, at fluctuating temperatures of 19–31°C in a daily rhythm of 8–16 hr, respectively, a lifetime total of 1433 eggs were laid (Hoffmann 1974). Similarly, females of *Bracon mellitor* at fluctuating temperatures of 21–32°C at a rhythm of 10–14 hr, respectively, laid 310 eggs, but at constant 32°C just 213 eggs, and at constant 21°C only 171 (Barfield et al. 1977). In both of these species fecundity at the constant temperature equal to the arithmetic mean value of the fluctuating regime was far lower than that under the fluctuating regime itself. However, an arithmetic mean is not an appropriate consideration, since most biochemical events are exponentially accelerated with rising temperatures. Obviously, the species mentioned fared better under natural daily temperature changes that were only approximated in the laboratory. It is a matter of speculation why fecundity under daily temperature fluctuations follows neither the calculated mean of the regime of temperatures nor a Q_{10} of approximately 2, as observed for some processes (cf. Chapter 3).

Animals, including insects, have evolved responses to the annually shifting day length; it allows them to anticipate adverse climatic conditions. The cold winters and dry summers of the temperate zones are unfavorable for insect reproduction, and many species go into reproductive diapause (cf. Engelmann 1970, Saunders 1975, Dingle 1978, Beck 1980). A thorough analysis is only available for a few species, notably for several carabid beetles (Thiele 1977) and species of Neuroptera (Tauber & Tauber 1976). This subject is treated extensively in Chapter 6, and consequently its discussion here is restricted.

Short days, often combined with low temperature, induce winter diapause in species such as *Leptinotarsa* (De Wilde et al. 1959), *Chrysopa carnea* (Tauber & Tauber 1973), and many others. Photoperiod is frequently the overriding cue. Aestivation (summer diapause) in the weevil *Hypera postica* (Huggans & Blickenstaff 1964) and the carabid *Orthomus barbarus atlanticus* (Paarmann 1976) is induced and maintained under long-day conditions. In all these cases the induction of reproductive diapause makes "ecological sense," since larval foods for these species would be scarce during winter or dry summers, respectively. Initiation and termination of diapause may have different controls; for example, short photoperiod may induce hibernation, but it may be terminated in midwinter before long photoperiods prevail (Tauber & Tauber 1976, and Chapter 6). Normally, however, low winter temperatures do not allow a "visible" diapause break in these species.

Control of the annual reproductive rhythms among carabid beetles may be complex. Some require a succession of long–short–long photoperiod exposure (Fig. 5.3), others only a long- and short-day succession, and still others reproduce independently of photoperiodic influences (Thiele 1975, 1977). In some species, such as *Pterostichus oblongopunctatus*, a short day has to be combined with temperatures lower than 15°C for subsequent successful initiation of vitellogenesis. Varying interactions of photoperiod and temperature may exist in other, even closely related, species.

In many of the diapausing species studied, perception of either short or long photoperiods is translated into control of endocrine activity [mostly the corpora allata (CA)]. In the Colorado beetle, *Leptinotarsa* (De Wilde & De Boer 1961), for example, short photoperiod (<14 hr of light) induces a diapause syndrome consisting of accumulation of reserves, positive geo-tropic behavior, cessation of mating behavior, and cessation of egg pro-duction. Its diapause is equivalent to pseudoallatectomy. Juvenile hormones (JHs) applied to the diapausing insects broke diapause; that is, they began to exhibit negative geotropic behavior and produced and laid eggs (Schoone-veld et al. 1977). Analogous observations have been made in the butterfly *Danaus plexippus* in which short-day or allatectomy induced reproductive diapause; application of a JH analog reinitiated egg production (Barker & Herman 1973). In both these cases the photoperiod is perceived by the brain, which in turn controls the CA. In species whose egg production is ultimately stimulated by the CA, these glands are obviously inhibited during reproductive diapause. Inhibition could be brought about by lack of an allatotropic hormone, the liberation of an allato-inhibitory hormone, or by nervous controls. Circumstantial evidence is available to support any of these possibilities.

5.2.4 Mating

In many insects the act of mating is an important stimulus for vitellogenesis. In the extreme, such as the cockroach *Diploptera punctata*, virgins do not incorporate yolk into the eggs at all; the mating stimulus is purely me-chanical and is relayed from the genitalia via the ventral nerve cord to the brain which disinhibits the CA. In other cockroach species the degree of stimulation of egg growth by mating varies from almost no role in *Pycnos-celus* to considerable growth acceleration in *Leucophaea* (cf. Engelmann 1970).

The mechanisms by which mating affects vitellogenesis may be quite different for different species. It may be indirect, because virgins often do not lay their matured eggs, and the presence of these in the genitalia blocks further egg growth in the ovarioles, as observed among species of Diptera and Lepidoptera. The actual mode of inhibition of maturation of additional eggs is not known in these cases. In other species factors that are transmitted during mating, either with the spermatozoa or in the seminal fluid, may be responsible for further activation of egg growth. The mechanical filling of the spermatheca may be another trigger for additional egg production. For *Cimex lectularius* it is reported that following hemocoelic insemination, the spermatozoa have to reach the base of the ovarioles before stimulation of vitellogenesis occurs (Davis 1965). As soon as spermatozoa have entered the ovarioles, the CA are further activated and more eggs are then matured.

In reviewing the literature on the effects of mating on egg production, one gains the feeling that a large body of detailed physiological information is waiting to be uncovered. Because of the diversity of mating effects, and because of different requirements in different species, many different modes of action may exist.

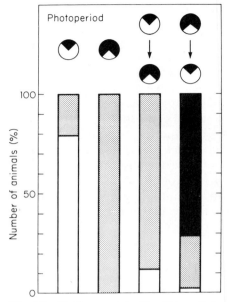

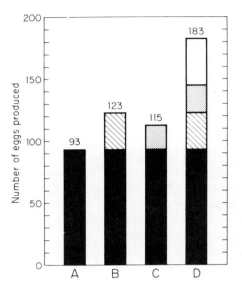

Fig. 5.3 Vitellogenesis in *Pterostichus nigrita* under constant or changed photoperiodic regimes. Only after a transfer from short-day (LD 8:16) to long-day (LD 16:8) was vitellogenesis observed in a substantial number of animals. The experience of short day prior to long day, not the reverse, was essential for subsequent vitellogenesis. White column = poorly differentiated ovaries; lightly hatched column = previtellogenesis; darkly hatched column = complete vitellogenesis. (Modified after Thiele 1977.)

Fig. 5.4 Egg production in *Acrolepiopsis* as affected by the host plant (leek) and mating. (A) Virgin females without the plant. (B) Virgin females with the plant. (C) Mated females without the plant. (D) Mated females with the plant. As is seen in D, the effects of mating and simultaneous presence of the host plant are more than additive. (Modified after Thibout 1974.)

5.2.5 Host Perception

Stimulation of egg production upon perception of a host—either plant or animal—is difficult to analyze. The difficulty resides in the fact that a species may feed on the host while ovipositing, thus egg maturation or oviposition is stimulated, or both. In a few cases it has been shown, however, that the mere perception of the host can indeed stimulate higher egg production (Labeyrie 1978).

The first case to be recognized was that of the ichneumonid *Diadromus pulchellus*, which matured many more eggs when it had access to the host, *Acrolepiopsis assectella*, or its empty cocoons (Labeyrie 1960). Physical contact with the host was not essential; the host is perceived olfactorily. The host plant for the moth *Acrolepiopsis* is leek, *Allium porrum*, and again it was shown that presence of the host plant stimulated egg maturation and laying (Thibout 1974). Egg production in the presence of the host plant was not increased by induced oviposition, however, since virgins that oviposited and those that did not matured nearly equal numbers of eggs. Both the presence of the host and mating stimulated egg production; together, the effects were more than additive (Fig. 5.4). Several other species are known to produce more eggs in the presence of the host. Among these are the bean weevil *Acanthoscelides obtectus* and beet moth *Scrobipalpa ocellatella* (Labeyrie 1978). It remains an enigma how odor or simple contact with the host stimulates vitellogenesis. For certain species one may speculate that the endocrine system is "turned higher" after perception of certain chemicals. In other instances hormones may not be involved at all. We do not know how any of these species translate the perception of the host into higher fecundity. It certainly is not an "all or none" phenomenon since the animals exhibit a graded response to the host chemicals.

5.3 MATING BEHAVIOR

In the vast majority of insect species egg fertilization is essential for initiation and completion of embryonic development. A fascinating array of behavioral patterns can be observed that facilitate finding of the sexes. In some flies and beetles mating behavior can be rather simple or nearly nonexistent, that is, the male approaches anything that moves and is of appropriate size (even an ink dot); only after mounting is the species and sex recognized. On the other hand, courtship may be very complex, involving a sequence of specific stimulus–response interactions between the sexes. This complexity is not related to the degree of advancement on the evolutionary scale. We find, for example, in primitive Apterygota and Blattaria an intricate and sophisticated display of behavioral interactions (cf. Engelmann 1970) as complex as the behavior patterns for the highly evolved drosophilids (Spieth 1974).

The biophysical signals (olfactory, auditory, or visual) involved in mating behavior have been analyzed in many instances. Equally interesting are aspects of circadian or circannual rhythmicity of mating behavior and the chemistry and biochemistry of sex attractants and their associated species specificity. Ultimately we would like to understand how the animals integrate perceived stimuli and filter out from the complex environment only those few which are the important key signals, behaving in a way specific to the species. Implicit is a central integration with, at least in some species, superimposed endocrine modulation of nervous activity. Only a brief survey is possible here, one that cannot do justice to the depth of our knowledge. The reader is referred to several recent books (e.g., Birch 1974, Shorey & McKelvey 1977, Blum & Blum 1979). Older findings and literature are covered by Engelmann (1970).

5.3.1 Pheromones and Sexual Behavior

Chemical communication between animals of the same species, that is, via pheromones, is probably the most common means of insect mating behavior. The very large volume of papers attests to this. Journals such as the *Journal Chemical Ecology* or *Environmental Entomology* contain many papers dealing with various aspects of pheromone research, and an increasing number of symposia and books deal with this topic.

Basically, two different but complementary research tools have been used in behavioral pheromone research: field or laboratory observations of attraction of the sexes and electrophysiology of pheromone perception by insect antennae. The first complete analysis of pheromone-mediated mating behavior was accomplished in *Bombyx mori* and was facilitated by the fortuitous combination of efforts by the laboratories of Butenandt, Schneider, and Schwinck. After more than 20 yr of research, Butenandt et al. (1959) published the identification of the silkmoth pheromone, bombykol. Bombykol is an alcohol, 10-*trans*-12-*cis*-hexadecadienol, which only in the *trans–cis* configuration will elicit a behavioral response in the male comparable to that when exposed to a female. Following this pioneering work and passage of another 20 yr, more than 600 pheromones have been isolated and chemically described.

Upon exposure to the pheromone, a male silkmoth will go through a sequence of activities: increased antennal movements, wing vibrations at small and then large amplitudes, and beginning locomotion that ends in vigorous and erratic runs in an unoriented zigzag fashion. Similar observations have been made for many species. Only if an air current is passed over the "calling" female towards the male will he orient to the female and move upwind until he meets the source of pheromone emission (Schwinck 1954). It is important to realize that male attraction is based on two components: an arousal by the pheromone and an anemotactic orientation to the origin of this scent. Directing a neutral air current towards the male will not cause him to move. Since this early elucidation of how a male moth is led to approach his female, the same two-component principle has been found to apply to many different species (Marsh et al. 1978). Translated to field conditions, it says that a male moth is attracted to a female over many kilometers by orienting anemotactically to her after having been aroused by the female scent.

Receptors for pheromones are found on the antennae. The first neurophysiological verification was obtained by Schneider and Hecker (1956) for *Bombyx* when he successfully recorded summation action potentials from the entire antennae upon exposure to the partially purified pheromone. The slow deflection of the potential, termed *electroantennogram* (EAG), was not evoked by other compounds or air alone. Pheromone-evoked EAGs have since been recorded for many species (Schneider 1966, Seabrook 1978) and are often used in screening during pheromone purifications. At first it was thought that EAGs are evoked only by the species' specific pheromone, but soon it became apparent that on the basis of EAG recording, pheromones of closely related species are frequently perceived as well (Schneider 1966, Schneider et al. 1977). In nature it is rare that a species reacts behaviorally to another species' pheromone. The EAG is an unreliable indicator of normal behavior in *Trichoplusia*, since it can be recorded from male antennae that receive the pheromone and pheromone inhibitor simultaneously (Mayer 1973) or from female antennae (though with smaller amplitude) (Light & Birch 1979); in either case no behavioral response was elicited. Obviously an EAG alone does not represent the response of an intact animal.

One of the most abundant sensory receptors on moth antennae are the sensilla trichodea, long hairlike structures. These sensilla have pores into which dendrites may reach and come in contact with the pheromone molecules (Steinbrecht 1973). Single cell recordings have been made from sensilla in response to the specific pheromone (Albert et al. 1974). One may assume that the pheromone molecule when bound to an acceptor molecule on the dendritic membranes induces a change in membrane permeability. Information leading to the central nervous system (CNS) is coded in the pattern of ensuing action potentials. A first stage of CNS integration probably resides in the olfactory glomeruli where a convergence of information apparently occurs. Phasic tonic trains of action potentials have been recorded in this brain area in *Antheraea polyphemus*, but no recording could be obtained more centrally in response to pheromone stimulation (Boeckh & Boeckh 1979). This first report on CNS processing of pheromone perception is certainly one of the exciting developments that may eventually lead to an understanding of the animal's behavior. One would indeed like to see the convergence of information gathered by a sophisticated technical approach and whole animal observations into a comprehensive model.

Does a pheromone molecule actually function in transmission of species-specific information? The answer is yes. Several species may synthesize and emit the same pheromones yet cross attraction essentially does not occur between two species. Several mechanisms are observed which function in reproductive isolation. One is the synthesis, emission, and perception of different ratios of the geometric isomers of pheromones. To illustrate this, a list of several tortricid species and their pheromone blends of (Z)-11/(E)-11-tetradecenyl acetate is provided (Table 5.1). Field captures are generally

Table 5.1   Examples of blends of geometrical isomers used as pheromones in some tortricid species

Species	Z11-/E11-14:Ac[a]
Choristoneura rosaceana	97:3
Argyrotaenia velutinana	91:9
Archips argyrospilus	60:40
Archips podana	60:40
Archips semiferanus	33:67
Archippus breviplicanus	30:70
Platynota stultana	11:89

[a](Z)-11-/(E)-11-tetradecenyl acetate; ratios found in female moth extracts.

best at these ratios, but anomalies do exist (Roelofs 1978). Additional pheromones blended with these specific isomeric ratios may be important for maximal efficiency. Different blends of geometric isomers of the same pheromone could be extracted from the European corn borer, *Ostrinia nubilalis*, where blends from different locations varied, and each blend was most attractive to the corresponding field population. The blends of the population in New York State was 4:96 for (Z):(E) whereas in populations of Ontario, Georgia, or Iowa it was the opposite, namely 97:3 (Kochansky et al. 1975). The different populations of this species had been introduced to North America from different parts of Europe where they presumably arose in geographic isolation. In Pennsylvania the two strains occur sympatrically, and males of each strain are primarily attracted to females of their own strain.

While specific pheromones and their isomeric blends can be very effective in attracting only the "appropriate" species, in some instances rates of pheromone release and their consequent concentrations may be important. For example, *Trichoplusia ni* and *Autographa california*, which occur sympatrically and use the same pheromone, (Z)-7-dodecenyl acetate, are attracted to high and low concentrations, respectively (Kaae, Shorey & Gaston 1973). Similar situations may also exist for other species.

Pheromone specificity, blends, and concentrations are means by which reproductive isolation is achieved. Additional mechanisms exist, however. For instance, pheromone-producing Lepidoptera generally emit and are attracted to pheromones at certain specific times of the daily light–dark cycles. The functional significance of different circadian mating rhythms of several sympatric noctuid moth species that utilize the same pheromone is illustrated in Fig. 5.5 (Kaae, Shorey, McFarland & Gaston 1973). The activity window is often rather wide, assuring attraction of males to females even at low population density. However, an extremely narrow window is observed in the Queensland fruit fly, *Dacus tryoni*. At dusk, during a period

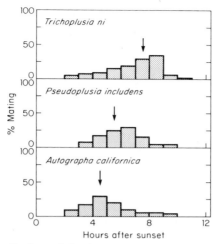

Fig. 5.5 Frequency distributions of the time of mating after sunset in three species of Plusiinae held in outdoor cages. All three species presumably use the same pheromone but emission occurs at different times and consequently the attraction of the wrong species is minimized. (Modified after Kaae, Shorey et al. 1973.)

of only 30 min, the males swarm and then settle on vegetation where they release their pheromone; mating is observed at the same time (Tychsen & Fletcher 1971). The locomotor activity of male and female flies coincided (Fig. 5.6) and could be inhibited by bright light. Dimming the light to 0.8 lumen/ft² induced mating at any time of day, which means the animals are in a state of readiness much of the time; under constant dusk illumination, they exhibited, however, a circadian pattern of mating behavior for a few consecutive days before it dampened out. The adaptive significance of the narrow window in pheromone release, its perception, and following mating behavior lies in the synchronization of both sexes.

Short-range or contact pheromone perception is often equally important in mating behavior as long-distance attraction. Species and sex recognition takes place at contact in several species of cockroaches (Barth 1968) and in many species attracted by visual and auditory cues over long distances. A proper succession of courtship movements may depend on contact chemoreception in the drosophilids (Spieth 1974), as in many other species. The courtship sequences may be terminated if the next proper signal is not received.

5.3.2 Auditory Stimuli in Sex Recognition

The involvement of sound emission and perception in insect courtship has also been studied on two different levels, namely, the behavior of the whole

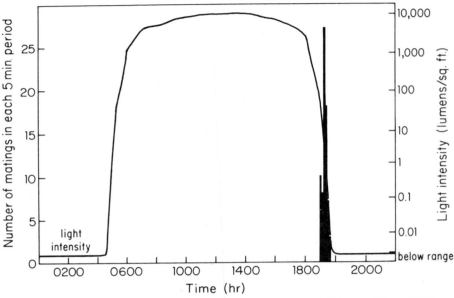

Fig. 5.6 The timing of copulation in *Dacus tryoni* under natural light conditions. Within a narrow range of 30 min during dusk pheromone emission occurs and mating is observed. (Modified after Tychsen & Fletcher 1971.)

animal and the complementary neurobiology. As with chemical communication, sound production and its perception is primarily effective in communication at some distance, perhaps up to approximately 30 m. Sounds and their characteristics have been recorded and analyzed and the associated behavioral aspects described, particularly for Orthoptera (cf. Alexander 1961, Elsner & Popov 1978). Sound communication in crickets and grasshoppers is presumably the "appropriate" means, since they generally live in dense ground vegetation where directional perception of odors is less likely to be successful. In Orthoptera, sounds are produced by scraping the femur against a file on the forewing, scraping the tegmina against each other, or scraping the femur against the abdomen. In the field one hears a multiplicity of sounds emitted by several species simultaneously. How does each species filter the pertinent sounds from the background noise, and which of the sound parameters are most important?

Female crickets may approach the singing male, and as Regen (1913) experimentally demonstrated, this is induced by sound alone: the female walked toward a loudspeaker that emitted the male song (phonotaxis). Phonotaxis as a behavioral response has been extensively used in analysis of sound patterns in several sympatric species. For example, in *Oecanthus turanicus* and *Gryllodinus kerkennensis* (Popov & Shuvalov 1977), which produce a continuous or broken trill, pulse rates or their continuity are the main determining factors. Other species produce species-characteristic chirps, that is, groups of one to five pulses which are spaced differently depending on the species (Fig. 5.7). In *Scapsipedus marginatus* the pulse interval pattern within the chirp is the important parameter, and chirp intervals could be altered extensively without affecting phonotactic behavior (Zaretsky 1972). Other species, such as *Melanogryllus desertus*, modulate pulse amplitude within a chirp (Fig. 5.7), thus determining the accuracy of phonotaxis (Popov & Shuvalov 1977).

The examples given here illustrate how reproductive isolation among species of Orthoptera can be achieved by distant sound communication. Sound perception is, on the other hand, not the sole means by which a species is recognized by the conspecific mate. Upon meeting of male and female, chemical recognition is very important for continuing courtship behavior. This is at least true for certain species (Loher & Rence 1978).

Intrinsic control of sound emission and perception must exist in many species. The males of certain crickets, for example, will not sing and court

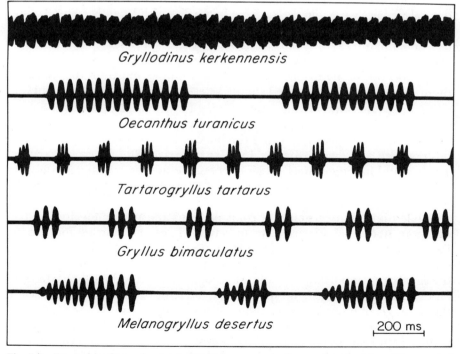

Fig. 5.7   Examples of sound pulses and pulse patterns in five species of crickets. The species' characteristic patterns are readily apparent. (Modified after Popov & Shuvalov 1977.)

unless a spermatophore is ready for transfer. Central nervous control of singing is coordinated with production of a spermatophore; the information is transmitted via the ventral nerve cord. Another example of central control of courtship is seen in the female acridid *Gomphocerus rufus* which will not respond to a courting male unless her CA are active (Loher 1962). In this latter case the innate releasing mechanism is controlled by the superimposed hormonal milieu. No generalization about such superimposed mechanism can be made, since our knowledge for other species is very fragmentary.

5.3.3 Visual Stimuli and Sex Recognition

Visual attraction in insects is primarily found in diurnal species such as butterflies, dragonflies, and some Diptera. However, nocturnal species that produce light themselves (e.g., many Lampyridae) use these visual cues to orient effectively towards their mate. Male butterflies typically recognize the conspecific female by her "style" of flight and color patterns on the wings. Probably the first to realize the importance of moving style and color in butterfly attraction was Tinbergen et al. (1942) for the greyling *Euminis*. Appropriately moved dummies were followed by the male greyling just as natural females were. Similarly, many butterflies with brightly colored wings were shown to exclusively use movement patterns and color for recognition (cf. Engelmann 1970). Characteristically, these species respond better to oversized dummies than to their own normal-sized females. Overoptimal stimuli do not occur in nature, but these findings illustrate the point: in these species visually perceived stimuli release the typical courtship pursuit.

Naturally, the swiftly flying dragonflies, with their large eyes, have adopted similar recognition cues: color patterns, particularly of the wings, and flight movements. In many dragonflies the males establish a territory at the oviposition site and "wait" for the females to arrive. She is recognized and grasped and mating ensues. Males of the same species, which generally are colored differently than the females, are chased away. Color cues are the main signals.

While diurnal insects employ color patterns and movement for sex recognition, nocturnal species have to rely on light emission by their mate. Females of *Lampyris noctiluca* glow for long periods, and only males of the same species "home in" on the glow. For different species the pattern of the lantern is important for species recognition (Schwalb 1961). Males of other Lampyridae flash in species-characteristic fashion (one to several flashes at given intervals) while flying over the vegetation. When the female flashes at given intervals, he will flash again and a dialogue may develop while he approaches the responding light source. Libraries of flash patterns of the many fireflies have been assembled (cf. Lloyd 1971, 1973). Important features for the species-specific response are intervals between flashes (Fig. 5.8) (Papi 1969), number of pulses within each flash, or changing amplitude of the pulses within a flash (Lloyd 1971, 1973). In some species these parameters do not leave much room for variation and thus species recognition is very precise. Obviously, communication by flashes of light is functional over long distances, while short-range communication involves species-specific pheromone perception, be it olfactory or contact chemoreception. For the fireflies the question arises as to the control of precise light emission and perception. Emission is clearly neuronally driven and could be induced by stimulation of the nerves leading to the lantern. Illumination of the animals or only of the eyes inhibited flashing in *Luciola* (Brunnelli et al. 1977).

5.3.4 Control of Mating Behavior and Receptivity

Control of sexual behavior and receptivity in insect species can be both environmental and intrinsic. In most species the controls are not known and one needs to resort to speculation in an attempt to explain the correlations between developmental stages and observed courting behavior. In many short-lived species no control is obvious. This applies to many Lepidoptera and Ephemeroptera that may mate soon after emergence and also mate repeatedly.

Circadian entrainment of sexual activities is frequently implicit (above), and superimposed on this may be gonadal and endocrine maturation processes (cf. Engelmann 1970). For years attempts to show existence of gonadal hormones and their involvement in male and female sexual behavior were unsuccessful. Today evidence is still largely circumstantial. Ovarian maturation depends in the majority of species on juvenile hormone (JH), and in at least a few species it was shown that female receptivity is also influenced by JH. In females of the grasshopper *Gomphocerus* and the cockroach *Leu-*

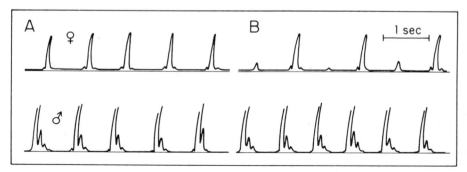

Fig. 5.8 Dialogue of light flashes between male and female fireflies (*Luciola lusitanica*) illustrating the importance of the proper flash intervals for elicitation of a response. (A) The male flashes at regular intervals of 800–900 msec and the female responds normally. (B) The male flashes at 720–820 msec intervals and the response of the female is irregular with alternating strong and weak answers. (Modified after Papi 1969.)

cophaea, mating frequency is reduced after allatectomy, indicating a direct positive effect of JH on the CNS. On the other hand, production of pheromone appears to be dependent on JH in the cockroach *Byrsotria,* that is, the role of JH in mating behavior is indirect in this case. Involvement of neurosecretory hormones in mating behavior is not easy to assess because ablation of neurosecretory cells often involves extensive additional brain damage, and nonspecific side effects may result.

Females of many species do not mate more than once. In some cockroaches, for example, the mechanics of deposition of a spermatophore causes refractoriness to further courtship; transsection of the ventral nerve cord in these species caused them to mate repeatedly. In other species, notably several Diptera, transfer of species-specific compounds with the seminal fluids caused nonacceptance of additional males (Leopold 1976). Secretions of the ejaculatory duct in *Musca domestica* or the accessory glands in *Aedes aegypti* and *Drosophila,* which are part of the seminal fluids, contain substances that induce refractoriness in these species. Either implantations of these organs or injections of semipurified extracts into virgin females rendered them nonreceptive to courting males. Obviously, specific substances act on the CNS of these females, effecting a change in their behavior.

5.4 HORMONES AND VITELLOGENESIS

5.4.1 The Endocrine Glands and Their Activity Control

Many species of insects periodically produce batches of eggs under precisely defined conditions, a phenomenon that may suggest endocrine control of vitellogenesis. *Rhodnius prolixus,* for example, lays a given number of eggs following a blood meal. An insect such as this naturally lent itself for the discovery of the endocrine gland that affects the production of eggs (Wigglesworth 1936). By controlled decapitation at given time intervals after a blood meal, Wigglesworth found that the CA was absolutely essential for making fully grown eggs. Reimplantation of this gland into decapitated females unequivocally showed, for the first time in an insect, hormonally controlled egg growth. Subsequent to this discovery, ablation and reimplantation experiments revealed that the same principle applies to additional species of Hemiptera and those of the Orthoptera, Dermaptera, Blattaria, Diptera, Coleoptera, and some Lepidoptera (cf. Engelmann 1970). Species of other orders presumably follow this principle as well. However, some Diptera, and in particular the noctuid moths, produce large numbers of eggs in the absence of known endocrine glands. Even an isolated abdomen of the *Cecropia* moth can make at least a few complete eggs. To date, we do not know of any endocrine control of vitellogenesis in *Cecropia* (Pan 1977).

Stimulatory and inhibitory controls, or both, for the CA may exist, and these could theoretically be either hormonal or nervous. A common principle cannot be expected, since the many different species have different biological "needs" for the control of the CA. We know, however, that the brain somehow translates environmental cues into control of these glands, either by stimulation or inhibition. Factors such as food intake, mating, photoperiod (long or short day in different species), group effects (pheromones), parental care, and cyclic incubation of embryos in ovoviviparous and viviparous species are known to influence the CA (cf. Engelmann 1970). The fascinating task now is the identification of the mode of control for the different qualities of these cues.

Only a limited discussion of experimental results is possible here, but it may illustrate the essential features. Probably the easiest case to interpret is that of inhibition of the CA in "viviparous" cockroaches (Engelmann 1957, 1970), in which the CA undergo changes from high activity to nearly complete inactivity and thus control the extended cycles of ovarian activity rhythms. The normal biology of these species "requires" cessation of vitellogenesis during gestation periods. Unilateral transsection of the NCCI or the NCA during midpregnancy results in activation of the ipsilateral CA, followed by vitellogenesis at the "wrong" time. The experimental results suggest that the CA are nervously inhibited by the brain during gestation. No inhibitory neurohormone has been identified so far.

Inhibition of the CA by brain neurons may not be unique to viviparous cockroaches. For example, cutting the allatic nerves in virgins of the Hemipteran *Dindymus* activated the CA to a similar degree as mating normally does (Friedel 1974). Normally, mating stimuli cause a lift of this inhibition. In another Hemipteran, *Pyrrhocoris apterus,* short photoperiod causes inactivation of the CA and cessation of egg production. After cutting the nervi allati in short-day animals, this gland became active, as shown by the resumption of vitellogenesis (Hodkova 1977). If, however, brain–CC–CA complexes were transplanted into short-day animals the CA remained inactive. In this case the brain inhibited the CA when it experienced the appropriate short photoperiod. In both of these Hemiptera inhibition by a nervous pathway appeared somewhat incomplete, and one may postulate that additional humoral controls exist.

Stimulatory humoral control of the CA may indeed exist in adults of some species. The evidence for this is, however, still circumstantial, because no neurohormone with such a function has been identified. The idea of a stimulatory humoral control traveling within the NCCI has its foundation in the observation that denervation of the CA in *Schistocerca,* for example, did not allow activation (Pener 1965). Similar observations have been made in *Leptinotarsa* in which, under short-day conditions, isolated CA were ineffective in terminating diapause (De Wilde & De Boer 1961). On the other hand, application of JH to either short-day or allatectomized females induced a temporary break of the diapause and a limited number of eggs were produced (Fig. 5.9) (Schooneveld et al. 1977). *Leptinotarsa* is perhaps

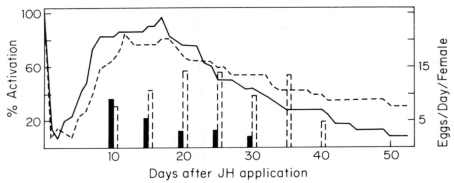

Fig. 5.9   Locomotor activity and rates of oviposition in diapausing (------) or allatectomized (———) *Leptinotarsa* which had received 30 μg of JH. The beetles were continually kept under short-day photoperiods. The higher rate of reproduction of the nonoperated females may indicate that the corpora allata of the recipient females were temporarily turned on by the applied JH. (Modified after Schooneveld et al. 1973.)

a suitable species for further search and identification of such stimulatory neurohormones; in this species the activity pattern of the CA can be manipulated by changing photoperiods.

Of necessity much of the present-day discussion on neurohormonal control of the CA is based on indirect and speculative evidence. We surely would like to be able to identify the stimulatory and inhibitory chemical entities, for only then will we be able to decide on the actual mode of control of these endocrine glands. For a general interpretation, further complications may certainly be anticipated because different species are likely to differ in the mode of control. The question of whether these regulatory hormones travel down the axons that innervate the CA, or whether they reach them via the circulation, needs to be approached as well.

5.4.2   The Yolk Proteins and their Precursor: Vitellogenin

In 1954 Telfer identified in the *Cecropia* moth a blood-borne protein that is immunologically identical to the major yolk protein. This yolk protein precursor, vitellogenin, is recognized by immunological techniques, and quantity and rates of synthesis have been determined in several species (cf. Engelmann 1970, 1979) (Fig. 5.10). In general, 80–90% of the yolk proteins are derived from vitellogenin synthesized by the fat bodies and exported into the hemolymph. Native vitellogenins of the hemolymph are proteins of 400,000–600,000 Daltons in size and are made up of several subunits in most species (Wyatt & Pan 1978). In *Locusta* the primary translation products are two polypeptides of approximately 235,000 and 225,000 Daltons that may later be processed proteolytically to several smaller units.

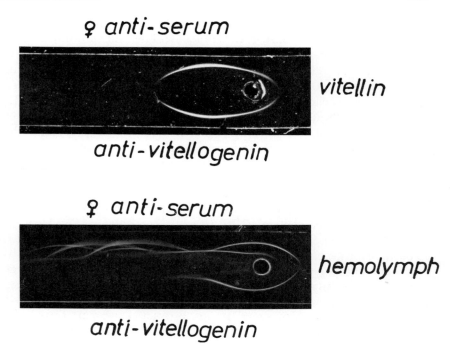

Fig. 5.10   Immunoelectrophoresis and identification of vitellogenin and vitellin in *Leucophaea* by antisera to hemolymph of vitellogenic females and the monospecific antibody to vitellogenin. Vitellogenin and vitellin are immunologically identical.

Also, for *Leucophaea* one can extrapolate from the size of the vitellogenin polysomes that a product of approximately 250,000 Daltons is the primary translation product. In *Drosophila* the precursors for the three yolk proteins of about 47,000 Daltons, recognized on SDS-polyacrylamide gels, are approximately 1000 Daltons larger than the final products. They are products of three distinct genes (Bownes 1979).

   Vitellogenin and vitellin have been the terms used for the hemolymph protein and its derived egg counterpart (cf. Wyatt & Pan 1978, Engelmann 1979). Immunological identity of the two classes of proteins has been shown in all cases studied, yet we prefer the use of different terms to denote the precursor–product relationship. Biosynthesis of the vitellogenins has been shown to be under the control of JH in the majority of insect species of several orders (cf. Engelmann 1979). The nascent vitellogenins (immunologically identified) are produced only when JH has been released by the CA. Under the influence of JH, transcription of specific vitellogenin mRNA occurs, and this messenger is then located in a class of large polysomes of the fat bodies of vitellogenic females. One of the most pertinent questions that needs to be answered is how hormone specificity is determined in the target tissues. Is it the existence of specific JH receptors that mediate the JH action only in these tissues? This will certainly be actively pursued in the near future.

In some species (e.g., *Aedes aegypti* and other Diptera) it has been shown that another developmental hormone, namely Ecdysterone, may direct the appearance of vitellogenin and the following vitellogenesis (Hagedorn 1974). Often, unusually high doses of ecdysone have been needed to induce a modest amount of vitellogenin synthesis. The discovery of induction of vitellogenin by ecdysteroids led to reports that adults of many species synthesize these hormones, and depending on the identification methods used, concentrations of 10^{-7} M were sometimes found; this is a level which may be physiologically meaningful. Maturing ovaries are often shown to be the source of ecdysone and it is assumed that the hormone is "secreted" into the hemolymph (Hagedorn 1974). In order for ecdysone to play a physiological role in vitellogenin biosynthesis, this is precisely what would have to occur. However, there is little evidence that unequivocally demonstrates *in vivo* secretion of ecdysone at physiological levels. Some leakage may occur under experimental conditions. Enormous quantities of this hormone are, on the other hand, sequestered by the growing oocytes. It remains to be documented whether, in some species, ecdysones are part of the vitellogenin induction process and occur in the postulated fashion *in vivo* under normal circumstances.

While it has been made abundantly clear that vitellogenin synthesis is directed by JH in most species, a role of brain hormones in this fundamental event is probably nonexistent. Complete and normal vitellogenesis can be induced in isolated abdomina of several species, provided that adequate nutritional reserves are available and sufficient JH had been applied. JH is also required for vitellogenin uptake by the growing oocytes, at least in a few species tested. JH is thus the key hormone in vitellogenesis. As stated above, vitellogenin synthesis and vitellogenesis occur in some moths without the involvement of any known hormone. Speculation on the evolution of such groups of species remains fruitless until we have the same depth of knowledge in additional species as we have for *Cecropia* (Pan 1977).

5.5 REPRODUCTIVE ADAPTATIONS

Members of the class Insecta have successfully invaded almost every conceivable ecological niche, and this is undoubtedly in part the result of a number of interesting adaptive features. Only a few of these, pertaining to reproduction, can be considered within the limited space available. Some aspects have been mentioned in other contexts within this chapter. Of particular interest to entomologists are aspects of parthenogenesis and viviparity.

5.5.1 Parthenogenesis

Parthenogenesis is the production of offspring from unfertilized eggs, and depending on the sex of the offspring, we distinguish different types of parthenogenesis: thelytoky (only females produced), arrhenotoky (only males produced), and deuterotoky (males and females produced). Each of these modes of propagation entail peculiar reproductive features (cf. Engelmann 1970, Suomalainen et al. 1976). One can envision that thelytokous species, such as many Phasmida, and some Lepidoptera and Coleoptera should be highly successful, perhaps more than the bisexual species, since no reproductive energy is "wasted" in production of male offspring. Indeed, parthenogenetic and bisexual females of the phasmid *Bacillus rossii* laid about the same number of eggs and the successful hatch of parthenogenetic eggs was only slightly lower than those of the bisexual females (cf. Bedford 1978); the parthenogenetic strain produced many more females than the bisexual one. However, the parthenogenetic strain is only abundant in certain regions of Italy. Natural history shows that in the course of evolution of most species the males won the contest, probably because outcrossing allowed a better exchange of genetic material and a consequent better adjustment to changing environmental conditions. When environmental conditions change relatively rapidly, parthenogenetic strains and species lose out, since rates of mutations, the only means to adjust, may be too slow. A combination of parthenogenetic and bisexual reproductive modes may be thought to be ideal and is observed among aphid species (below). Here the advantages of parthenogenesis are readily seen during favorable climatic conditions (large populations develop) and the disadvantages are counteracted by a change to sexual reproduction (cf. Chapter 7).

If thelytokous parthenogenesis entails certain disadvantages for the propagation of the species, arrhenotoky, *per se*, which is seen particularly often among the Hymenoptera, would appear to have only disadvantages. However, females reproducing arrhenotokously when virgin will produce female offspring when mated and, as is seen, populations have survived. If, on the other hand, males do not become available, these species will die out. One may speculate and it has been observed that in some long-lived species, a virgin female may eventually be mated by her own offspring and thus production of females is assured. Those short-lived Hymenoptera do not derive any apparent benefits: energy is "wasted" in production of males which are not used (cf. Chapter 8).

A number of species with exclusive parthenogenetic reproduction are of interest from several biological, cytological, and evolutionary points of view. For instance, a thelytokous species may develop without chromosomal reduction divisions, so-called apomixis. This is observed in many Curculionidae. Development may also be initiated with automixis of egg and polar nuclei. Automixis may be triggered by the entry of the sperm into the egg, a phenomenon known as gynogenesis, and is seen for example in the psychid moth *Luffia*. In this case the spermatozoan nucleus does not contribute to the gene pool of the offspring. In both apomixis and automixis genetic polymorphism is considerably reduced and clones will be established.

Having identified the disadvantages of unisexual reproduction, it is important to note that some parthenogenetic species have successfully established extensive populations. What are the factors that allowed this to occur? Several theories regarding the success of certain parthenogenetic species exist (cf. Cuellar 1977) and are essentially based on certain observed correlations. One of the facts frequently discussed is that parthenogenetic species or parthenogenetic strains of bisexual species are often found in geographic isolation. For example, parthenogenetic populations of the moth *Solenobia* are found in remote Alpine valleys freed of ice relatively recently. Also, many parthenogenetic curculionid weevils are found in isolated valleys of the Alps or climatically unfavorable northern European regions. These observations certainly favor the hypothesis that parthenogenesis evolves in isolation where a single female can establish a clone. Perhaps low population density, that is, unavailability of a mate, was a factor that allowed latent parthenogenesis to be expressed. Establishment of a clone presumably was further facilitated by low vagility (winglessness) of many parthenogenetic species.

As noted, viable parthenogenetic insect populations are frequently found in adverse but stable climates. It is also observed that many of these species or strains are polyploid, a condition which presumably arose by automixis. It is inferred that polyploidy conveys a certain robustness and consequently the species survived under adverse environmental conditions. Whether this inference is valid needs to be proven, but the coincidence of polyploidy and parthenogenesis is just too frequent and cannot be neglected in these considerations. The extensive genetic polymorphism found between curculionid weevil populations is presumably the result of widespread polyploidy among this group of weevils (Suomalainen et al. 1976). Curculionid weevils were highly successful and are widely distributed throughout Scandinavia. Polyploid genotypes are probably balanced and heterozygotic, and this may be the reason for good adaptability of these species. In other words, contrary to what is often believed, under certain circumstances, parthenogenetic species successfully compete with bisexual ones and are able to evolve and adapt to changing habitats.

In some species one to many parthenogenetic generations may alternate with a bisexual generation. This pattern of reproductive biology, termed heterogony, is observed among the Aphidoidea, Cynipidae, and Cecidomyidae. Much of our knowledge regarding these species is descriptive, and with a few exceptions little is known on aspects of controls of the reproductive patterns (cf. Engelmann 1970). In the aphids, depending on the species, one to many parthenogenetic generations may occur, and the number of generations may be obligatorily fixed or controlled by epigenetic factors. Generally, long days that coincide with mostly elevated temperatures and availability of food induces parthenogenetic reproduction in which only females are produced. Some species could be reared indefinitely via parthenogenetic offspring when kept under long-day photoperiod; as soon as short photoperiod prevailed sexuals appeared in the colonies. Besides photoperiod, crowding and nutritional aspects (quality and quantity) may influence a shift from parthenogenetic to sexual reproduction. Sometimes all of these factors operate in conjunction.

While photoperiod is the predominant environmental cue in the mode of aphid reproduction, food quantity appears the exclusive factor that controls heterogony in the Cecidomyidae. Some Cecidomyidae live on fungus hyphae and propagate paedogenetically (cf. Engelmann 1970). Species of this group have been reared only as larvae for several hundred generations, but when crowding was allowed or the food got old, the larvae produced male and female larvae that metamorphosed to winged adults that mated and eggs were laid. The winged sexual forms certainly aid in dispersal of these species; new food sources may be exploited.

As this rather brief narration illustrates, control of heterogony may be different in different species; here we cannot do justice to the complexity and variability known for heterogonic species.

Parthenogenesis as a form of propagation has been successful for a number of species of insects. It is more than just a peculiar mode of production of offspring. As was mentioned above, this mode of reproduction may be an evolutionary dead end for some species because of lack of adaptability to changing environments. For others, particularly in conjunction with polyploidy and associated heterozygocity, it has become a successful mode and these species have outnumbered their bisexual counterparts.

5.5.2 Viviparity

Live bearing species are found in several orders of insects, and for convenience the various forms of viviparity can be classified as follows (cf. Hagan 1951, Engelmann 1970).

Ovoviviparity: A type of viviparity in which the embryos do not receive solid nutrients from the mother; water may be obtained, however. Embryos are generally carried in a brood pouch, an outgrowth of the outer genital ducts. Examples are found among the cockroaches such as *Leucophaea* or *Nauphoeta,* in flies of the genus *Sarcophaga,* or the beetles *Chrysomela* and *Phytodecta.*

Viviparity: Three types of true viviparity are known in which embryos or larvae obtain nutrients from the mother. In *adenotrophic viviparity* the embryos and larvae receive nutrition through modified maternal structures like the accessory sex glands. Examples are Glossinidae and Hippobosci-dae. *Haemocoelic viviparity* is a type in which the embryos live in the hae-mocoel of the mother and obtain nutrition via their epidermis. Species of the Cecidomyidae and the Strepsiptera can be mentioned. In *pseudoplacental viviparity* the embryos reside in the genital tracts of the mother and pre-sumably receive nutrition via placentalike structures developed by the mother or the embryo, or both. *Hesperoctenes* (Hemiptera), *Diploptera* (Blattaria), and *Hemimerus* (Dermaptera) are representatives of this type.

A fairly complete list of species in which viviparity has been reported is found in Engelmann (1970) and not much additional information has be-come available since then. A true test for the requirement of the embryos to be incubated by the mother and successful completion of development is not available for the overwhelming majority of viviparous species. *In vitro* incubations have rarely been attempted. Growth of embryos in haemocoelic viviparous Cecidomyidae and Strepsiptera almost certainly necessitates transfer of nutrients from the female; however, no report is known on attempted artificial extramaternal rearing of larvae. In the adenotrophic tsetse fly *Glossina austeni,* normal and viable development of embryos and larvae is dependent on adequate nutrition from the milk gland, the function of which is impaired after allatectomy (Ejezie & Davey 1976).

Rather little is known for most viviparous species regarding control of egg maturation and the ensuing development of embryos and larvae. Only two groups have been studied in detail, namely, cockroaches and Glossin-idae. In the former, eggs in the ovaries mature periodically, and these periods are separated by intervals of inactivity during gestation of up to $2\frac{1}{2}$ months. Cessation of vitellogenesis during gestation in these ovovivip-arous and viviparous cockroaches results from inhibition of the CA (Engel-mann 1957) (Fig. 5.11). This inhibition is lifted either by removal of the embryos from the brood sac or by cutting the nervi allati during pregnancy. The brain is the controlling center for activation of the CA and production of JH. In these species cyclic inhibition of the CA assures maturation of a new set of eggs at the "proper" time.

Coordination of maturation of ovarian eggs and embryonic development in the uterus exists in other viviparous species as well, but little is known on the mechanisms of control. In the Glossinidae, for example, only one egg becomes vitellogenic while the larva in the uterus develops to comple-tion, and at the time of parturition this egg is ovulated and takes the place of the larva which has just been deposited (Tobe & Langley 1978). There is apparently no need for periodic shut-down of the CA, as in the viviparous cockroaches, since growth of the one egg is slow and maturation is timed for parturition of the previous larva. However, in some unknown fashion a control mechanism for yolk incorporation must exist, because only one of the "waiting" oocytes will begin to grow upon parturition of the larva.

Evolution of viviparity in species of rather primitive as well as those of more advanced orders must have occurred several times. Causes that led

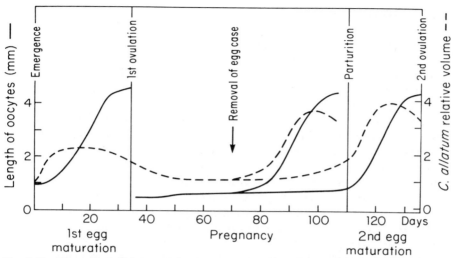

Fig. 5.11 Illustration of the reproductive cycle of the ovoviviparous cockroach *Leucophaea*. The corpora allata are inhibited during pregnancy, but this inhibition is lifted following the removal of the egg case from the brood sac at midpregnancy. (Modified after Engelmann 1970.)

to viviparity remain a topic of speculation. It is noteworthy that *Musca corvina* (Portchinsky 1885) and *Chloeopsis diptera* (Giard 1905) are viviparous in southern but oviparous in northern regions. In warm and dry climates growing embryos and larvae are certainly less vulnerable to desiccation when incubated by the mature female. If these adverse conditions have allowed the evolution of viviparity in these species, this idea certainly does not apply to many aphids or the Cecidomyidae, where we observe just the opposite (cf. Engelmann 1970). These latter species are facultatively viviparous and also produce offspring parthenogenetically as long as conditions are good, but lay eggs when climate and food become less favorable.

Often coupled with protection of offspring by the mother is a reduction in progeny. We may mention the extreme case of the Glossinidae which in a life span of up to 8 months produce no more than 10–14 larvae. But such reduction of progeny is not always associated with viviparity. For example, the viviparous Strepsiptera produce up to several thousands of larvae; the released soft larvae are very vulnerable during their search for a new host and many of them die through desiccation and predation.

As this brief outline shows, viviparity among insects is an interesting and unique phenomenon coupled with a variety of adaptive features. Unquestionably, viviparous species have been reproductively successful even though many among them have a lower reproductive rate than related oviparous species.

5.6 OVIPOSITION

Deposition of a fully grown egg is affected and controlled by a multiplicity of factors in many insects. We can approach the topic from both an environmental view or from purely physiological considerations (cf. Engelmann 1970). I will attempt to do both and extract from the vast literature some of the essentials.

Females of many species begin oviposition shortly after mating. This could be caused by either mechanical or chemical stimulation of the oviposition mechanisms. For example, in several dipterans the male transfers to the female substances that may either directly or indirectly (via the neuroendocrine systems) cause oviposition (cf. Leopold 1976). These substances have been shown to be produced in the paragonia of the male *Drosphila,* the accessory glands of male mosquitoes, or the ejaculatory ducts in *Musca.* For *Drosophila* it was shown that two biologically active polypeptides are transferred to the female and that one of these—PS 2—induced egg laying in virgins (Baumann 1974). The mode of action is not entirely clear. It has also been shown in *Bombyx* that compounds of the male reproductive tract are transferred to the female during mating and induce oviposition (Yamaoka 1977). In *Bombyx* the transferred substances appear to act directly on the last abdominal ganglion and the nerves leading to the genitalia; an increased neuronal activity was recorded.

The chemical nature of the transferred and functionally active compounds and their mode of action are essentially not known. However, one interesting case was recently reported and may open up a new approach to our attempt to decipher the physiology of induced oviposition. The testes, seminal fluids, and spermatophore of *Acheta* contain prostaglandin synthetase which apparently is transferred to the female during mating (Destephano & Brady 1977). Also, injection of prostaglandin induced oviposition in virgins. In another cricket, *Teleogryllus,* similar results were obtained (Loher et al. 1981). In this species arachidonic acid, the precursor to prostaglandin, is present in the spermatheca of the female and after transfer of the prostaglandin-synthesizing complex by the male, prostaglandin is made. Mated females contain approximately 500 pg PGE_2, whereas virgins have only about 20 pg. PGs are known to enhance muscular contractions in other animals, and one could postulate that this is the mode of action during oviposition in these crickets. The action may, however, be indirect via the neuroendocrine system which has been shown, in some species, to contain myotropic hormones (cf. Engelmann 1970). Little information is available on the fine details for any species and the overall findings allow only limited speculation.

Neurohormones or the compounds transferred to the female may be sufficient to induce laying of the fully grown eggs in many species. In others a proper substrate is also required for oviposition. Many species probe the consistency of the substrate with their abdomen (e.g., sand in locusts, or leaf texture by many phytophagous species) (cf. Engelmann 1970). Chemosensors on the ovipositor are also important for parasitoids that probe potential hosts for their suitability. Casual information for many species is scattered throughout the entomological literature, but there has been no thorough analysis on all of the ramifications, particularly the CNS integration of the sensory information received.

From the ecological view we are fascinated by observations on how a species finds its host (animal or plant) and is induced to deposit its eggs. A female is attracted by host-specific factors and lays her eggs on the most suitable species. Parasitic Hymenoptera may be attracted to the host's odor (kairomone) over some distance, and on arrival contact chemoreception may play a role (Lewis et al. 1976). Whether oviposition actually follows may depend on the "quality" of the host's hemolymph (Hedgekar & Arthur 1973) or other factors, such as the presence of previously implanted eggs or developing larvae. A parasitoid may be attracted to the medium the host lives on, as in *Cothonaspis* which parasitizes *Drosophila* (Carton 1977). Both *Drosophila* and its parasitoid are attracted by ethanol emanating from fermenting fruits; they tolerate concentration up to 10%.

Many Lepidoptera whose larvae feed on plants exhibit a remarkable coevolution with that of the host species. Herbivorous insects may totally depend on one or a few plant species and many have evolved a tolerance for toxic secondary plant chemicals (Kogan 1977, Swain 1977). In order to fully understand the insect–plant relationship, for every case we need to take a holistic approach. The adults are attracted to and prefer to oviposit on a particular plant over that of another. The plant should contain the phagostimulants that induce the larvae to feed adequately; should the plant contain toxic chemicals, the species must tolerate these compounds. Adequate and suitable nutrients should be contained in the plant.

An analysis of certain particular insect–plant relationships provides us with some rather interesting insights; that is, although many times we observe what we would expect to find, in some cases just the opposite is found. For example, *Papilio machaon* of Scandinavian regions studied occurs exclusively on *Angelica archangelica*, even though other Umbelliferae exist in the vicinity. In laboratory tests it was shown that a number of other plants were acceptable for oviposition (Fig. 5.12), some even more acceptable than *A. archangelica* (Wiklund 1975). A great many additional plant species that were not attractive for oviposition could, however, support growth of the larvae. As is seen, others, functioning as oviposition substrates, were unsuitable for larval growth, since the larvae did not accept them and starved to death.

Another interesting case is the Colorado potato beetle, which is normally found on *Solanum tuberosum* but is attracted to oviposit on a variety of solanaceous plants. If offered a choice between potato and other solanaceous plants, they sometimes preferred plants that were unsuitable for larval growth or even were toxic (Hsiao & Fraenkel 1968). Obviously, ovi-

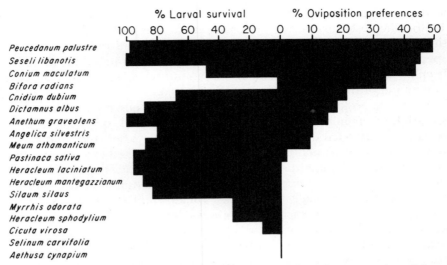

Fig. 5.12 Oviposition preferences and larval survival of *Papilio machaon* on a variety of plants. As is seen, successful larval growth is not necessarily related to preferred egg-laying sites. (Modified after Wiklund 1975.)

position preference in this case is unrelated to acceptability by the larvae. A further interesting and different case may be mentioned. *Pieris* females (the cabbage white) are attracted to the green surfaces of the host plant. However, they only oviposit when perceiving the toxic mustard oil glucoside sinigrin (Ma & Schoonhoven 1973). Sinigrin is perceived by receptors on the tarsae, as shown by electrophysiological techniques. Similarly, the dipteran *Dasyneura* is attracted to its sinigrin-containing host plant (Petterson 1976), but in this species the receptors are located on the antennae, since after their amputation the insects laid indiscriminately.

The few examples given here could be extended and the basic findings amplified significantly. It is seen that the physiology of finding suitable hosts and responding to them appropriately can be as varied and intricate as the insects themselves. Certainly, the ability to utilize a variety of plants and animals as hosts and at the same time evolve detoxifying biochemical pathways for certain toxins present has contributed to the success of various insect species. Both generalists and specialists in oviposition and feeding thrive in their respective ecological niches. Specialization obviously has not been a detriment in many cases.

REFERENCES

Albert, P. J., W. D. Seabrook, and U. Paim. 1974. *J. Comp. Physiol.* **91**: 78–89.

Alexander, R. D. 1961. *Behaviour.* **17**: 130–223.

Barfield, C. S., D. G. Bottrell, and J. W. Smith, Jr. 1977. *Environ. Entomol.* **6:** 133–137.

Barker, J. F. and W. S. Herman. 1973. *J. Exp. Zool.* **183:** 1–10.

Barth, R. H. Jr. 1968. *Adv. Reprod. Physiol.* **3:** 167–207.

Baumann, H. 1974. *J. Insect Physiol.* **20:** 2347–2362.

Beck, S. D. 1980. *Insect Photoperiodism* (2nd ed.). Academic, New York.

Beck, S. D. and J. C. Reese. 1976. Pp. 41–92 in J. W. Wallace and R. L. Mansell (Eds.), *Biochemical Interactions between Plants and Insects.* Plenum, New York.

Bedford, G. O. 1978. *Annu. Rev. Entomol.* **23:** 125–249.

Birch, M. C. 1974. *Pheromones.* North-Holland Publishing, Amsterdam.

Blum, M. S. and N. A. Blum. 1979. *Sexual Selection and Reproductive Competition in Insects.* Academic, New York.

Bodenheimer, F. S. 1943. *Bull. Soc. Fouad er Entomol.* **27:** 1–28.

Boeckh, J. and V. Boeckh. 1979. *J. Comp. Physiol.* **132:** 235–242.

Bownes, M. 1979. *FEBS Lett.* **100:** 95–98.

Bracken, G. K. 1965. *Can. Entomol.* **97:** 1037–1041.

Brunelli, M., F. Magni, and M. Pellegrino. 1977. *J. Comp. Physiol.* **119:** 15–35.

Butenandt, A., R. Beckmann, D. Stamm, and E. Hecker. 1959. *Z. Natuforsch.* **14b:** 283, 284.

Carton, Y. 1977. *Coll. Int. C.N.R.S.* **265:** 285–303.

Corbet, P. S. 1964. *Nature* **203:** 669.

Cuellar, O. 1977. *Science* **197:** 837–843.

Davis, N. T. 1965. *J. Insect Physiol.* **11:** 355–366.

Destephano, D. B. and U. E. Brady. 1977. *J. Insect Physiol.* **23:** 905–911.

De Wilde, J. and J. A. De Boer. 1961. *J. Insect Physiol.* **6:** 152–161.

De Wilde, J., C. S. Duintjer, and L. Mook. 1959. *J. Insect Physiol.* **3:** 75–85.

Dingle, H. 1978. *Evolution of Insect Migration and Diapause.* Springer-Verlag, New York.

Ejezie, G. C. and K. G. Davey. 1976. *J. Insect Physiol.* **22:** 1743–1749.

Elsner, N. and A. V. Popov. 1978. *Adv. Insect Physiol.* **13:** 229–355.

Engelmann, F. 1957. *J. Insect Physiol.* **1:** 257–278.

Engelmann, F. 1970. *Physiology of Insect Reproduction.* Pergamon, Oxford.

Engelmann, F. 1979. *Adv. Insect Physiol.* **14:** 49–108.

Erickson, J. M. and P. Feeny. 1974. *Ecology* **55:** 103–111.

Finch, S. 1971. *Entomol. Exp. & Appl.* **14:** 115–124.

Friedel, T. 1974. *J. Insect Physiol.* **20:** 717–733.

Giard, A. 1905. *Bull. Sci. Fr. Belg.* **39:** 153–187.

Greenberg, J. 1951. *J. Nutr.* **43:** 27–35.

Grison, P. 1948. *C.R. Acad. Sci., Paris* **227:** 1172–1174.

Hagan, H. R. 1951. *Embryology of the Viviparous Insects.* Ronald, New York.

Hagedorn, H. H. 1974. *Am. Zool.* **14:** 1207–1217.

Hedgekar, B. M. and A. P. Arthur. 1973. *Can. Entomol.* **105:** 787–793.

Heed, W. B. and H. W. Kircher. 1965. *Science* **149:** 758–761.

Hodkova, M. 1977. *J. Insect Physiol.* **23:** 23–28.

Hoffmann, K. H. 1974. *Oecologia* **17:** 39–54.

Hsiao, T. H. and G. Fraenkel. 1968. *Annu. Entomol. Soc. Am.* **61:** 493–503.

Huggans, J. L. and C. C. Blickenstaff. 1964. *J. Econ. Entomol.* **57:** 167, 168.

Kaae, R. S., H. H. Shorey, and L. K. Gaston. 1973. *Science* **179:** 487, 488.

Kaae, R. S., H. H. Shorey, S. U. McFarland, and L. K. Gaston. 1973. *Ann. Entomol. Soc. Am.* **66:** 444–448.

Kircher, H. W., W. B. Heed, J. S. Russell, and J. Grove. 1967. *J. Insect Physiol.* **13:** 1869–1874.

Kochansky, J., R. T. Cardé, L. Liebherr, and W. L. Roelofs. 1975. *J. Chem. Ecol.* **1:** 213–225.

Kogan, M. 1977. *Proc. 15th Int. Congr. Entomol.* **1976:** 211–227.

Labeyrie, V. 1960. *Entomophaga Mem.* **1:** 1–193.

Labeyrie, V. 1978. *Annu. Rev. Entomol.* **23:** 69–89.

Lea, A. O. 1964. *Ann. Entomol. Soc. Am.* **57:** 656, 657.

Leopold, R. A. 1976. *Annu. Rev. Entomol.* **21:** 188–221.

Lewis, W. J., R. L. Jones, H. R. Gross, Jr., and D. A. Nordlund. 1976. *Behav. Biol.* **16:** 267–289.

Light, D. M. and M. C. Birch. 1979. *J. Insect Physiol.* **25:** 161–167.

Lloyd, J. E. 1971. *Annu. Rev. Entomol.* **16:** 97–122.

Lloyd, J. E. 1973. *Environ. Entomol.* **2:** 991–1008.

Loher, W. 1962. *Naturwiss.* **49:** 406.

Loher, W., I. Ganjian, I. Kubo, D. Stanley-Samuelson, and S. S. Tobe. 1981. *Proc. Natl. Acad. Sci. USA* **78:** 7835–7838.

Loher, W. and B. Rence. 1978. *Z. Tierpsychol.* **46:** 225–259.

Ma, M.-C. and L. M. Schoonhoven. 1973. *Entomol. Exp. & Appl.* **16:** 343–357.

Macaulay, E. D. M. 1973. *Entomol. Exp. & Appl.* **16:** 48–52.

McFarlane, J. E. 1976. *Can. Entomol.* **108:** 391–394.

Marsh, D., J. S. Kennedy, and A. R. Ludlow. 1978. *Physiol. Entomol.* **3:** 221–240.

Mayer, M. S. 1973. *J. Insect Physiol.* **19:** 1191–1198.

Meikle, J. E. S. and J. E. McFarlane. 1965. *Can. J. Zool.* **43:** 87–98.

Paarmann, W. 1976. *Entomol. Exp. & Appl.* **19:** 23–36.

Pan, M. L. 1977. *Biol. Bull.* **153:** 336–345.

Papi, F. 1969. *Monitore Zool. Ital* (N.S.) **3:** 135–184.

Pappas, C. and G. Fraenkel. 1977. *Physiol. Zool.* **50:** 237–246.

Pener, M. P. 1965. *J. Zool.* **147:** 119–136.

Pettersson, J. 1976. *Symp. Biol. Hung.* **16:** 203–208.

Phillips, W. M. 1976. *Physiol. Entomol.* **1:** 223–226.

Popov, A. V. and V. F. Shuvalov. 1977. *J. Comp. Physiol.* **119:** 111–126.

Portchinsky, J. A. 1885. *Horae Soc. Entomol. Ross.* **19:** 210–244.

Regen, J. 1913. *Pflügers Arch. Ges. Physiol.* **155:** 193–200.

Roelofs, W. L. 1978. *J. Chem. Ecol.* **4:** 685–699.

Rothschild, M. and B. Ford. 1966. *Nature* **211:** 261–266.

Rothschild, M. and B. Ford. 1973. *J. Zool. Lond.* **170:** 87–137.

Roubaud, E. 1929. *C.R. Acad. Sci., Paris* **188:** 735–738.

Saunders, D. S. 1975. *Insect Clocks*, Pergamon, Oxford.

Schneider, D. 1966. *Symp. Soc. Exp. Biol.* **20:** 273–297.

Schneider, D. and E. Hecker. 1956. *Z. Naturforsch.* **11b:** 121–124.

Schneider, D., W. A. Kafka, M. Beroza, and B. A. Bierl. 1977. *J. Comp. Physiol.* **113:** 1–15.

Schooneveld, H., A. O. Sanchez, and J. De Wilde. 1977. *J. Insect Physiol.* **23:** 689–696.

Schubert, W. 1927. *Z. angew. Entomol.* **13:** 129–155.

Schwalb, H. H. 1961. *Zool. Jahrb. Syst.* **88:** 399–550.

Schwinck, I. 1954. *Z. Vergleich. Physiol.* **37:** 19–56.

Seabrook, W. D. 1978. *Annu. Rev. Entomol.* **23:** 471–485.

Shorey, H. H. and J. J. McKelvey. 1977. *Chemical Control of Insect Behavior.* Wiley, New York.

Spieth, H. T. 1974. *Annu. Rev. Entomol.* **19:** 385–405.

Steinbrecht, R. A. 1973. *Z. Zellforsch.* **139:** 533–565.

Suomalainen, E., A. Saura, and L. Lokki 1976. *Evol. Biol.* **9:** 209–257.

Swain, T. 1977. *Proc. 15th Int. Congr. Entomol.* **1976:** 249–256.

Tauber, M. J. and C. A. Tauber. 1973. *J. Insect Physiol.* **19:** 1455–1463.

Tauber, M. J. and C. A. Tauber. 1976. *Annu. Rev. Entomol.* **21:** 81–107.

Thibout, E. 1974. *Ann. Zool. Ecol. Animal* **6:** 81–96.

Thiele, H. U. 1975. *Oecologia* **19:** 39–47.

Thiele, H. U. 1977. *Carabid Beetles in their Environment.* Springer-Verlag, Berlin.

Tinbergen, N., B. J. D. Meeuse, L. K. Boerema, and W. W. Varossieau. 1942. *Z. Tierpsychol.* **5:** 182–226.

Tobe, S. S. and P. A. Langley. 1978. *Annu. Rev. Entomol.* **23:** 283–307.

Tychsen, P. H. and B. S. Fletcher. 1971. *J. Insect. Physiol.* **17:** 2139–2156.

Wigglesworth, W. B. 1936. *Quart. J. Micr. Sci.* **79:** 91–121.

Wiklund, C. 1975. *Oecologia* **18:** 185–197.

Wyatt, G. R. and M. L. Pan. 1978. *Annu. Rev. Biochem.* **47:** 779–817.

Yamaoka, K. 1977. *Adv. Invert. Reprod.* **1:** 414–431.

Zaretsky, M. D. 1972. *J. Comp. Physiol.* **79:** 153–172.

Chapter 6

Adaptations to Hazardous Seasonal Conditions: Dormancy, Migration, and Polyphenism

M. J. TAUBER, C. A.
TAUBER, and S. MASAKI

6.1 INTRODUCTION

Annual cycles in resources and unfavorable conditions characterize virtually all biological environments. Thus, a basic aspect of ecological adaptations of insects is their phenological "strategy"—that is, the set of adaptations that leads to appropriate seasonal timing of recurring events in their life cycles. Seasonality is observable at different ecological levels, and it affects ecological studies in a variety of ways. First, phenological strategies are basic components underlying inter- and intraspecific interactions at the population and community levels. Knowledge of these phenological relationships is required for efficient pest management (Chapters 23–25), specifically in predicting timing of dormancy, migration, development, and reproduction of pests and beneficial species, in selecting well-adapted species or biotypes of natural enemies for use in biological control, and in manipulating cultural practices to reduce pest damage. Second, the ecological adaptations of individuals are based on dynamic sets of physiological and behavioral responses that change dramatically throughout the seasons. An understanding of how these responses change seasonally must precede experimental studies designed to define relationships between insects and their

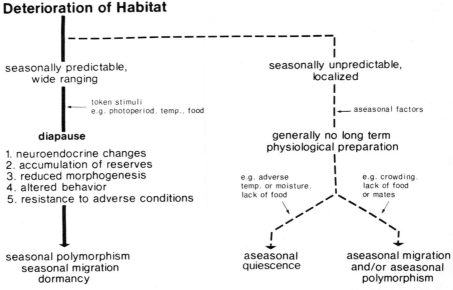

Deterioration of Habitat

seasonally predictable,
wide ranging

— token stimuli
e.g. photoperiod. temp., food

diapause

1. neuroendocrine changes
2. accumulation of reserves
3. reduced morphogenesis
4. altered behavior
5. resistance to adverse conditions

seasonal polymorphism
seasonal migration
dormancy

seasonally unpredictable,
localized

— aseasonal factors

generally no long term
physiological preparation

e.g. adverse
temp. or moisture.
lack of food

e.g. crowding.
lack of food
or mates

aseasonal
quiescence

aseasonal migration
and/or aseasonal
polymorphism

Fig. 6.1 Insect responses to seasonally predictable and seasonally unpredictable deterioration of conditions in the habitat. In general, when changes are regular and widespread, diapause-mediated responses control dormancy, migration, and polyphenism. Seasonally unpredictable or localized exigencies elicit the immediate responses of quiescence, aseasonal migration, and aseasonal polyphenism.

environment. Third, genetically controlled variation in seasonal cycles serves as building material for evolution of adaptations to the environment and thus is important in understanding species distributions, abundance, and divergence (cf. Chapter 8).

Among the characteristics that enable insects to have widespread geographic ranges and to occupy a multitude of habitats is their ability to (1) anticipate adverse seasonal conditions in order to synchronize their life cycles to the seasonal character of the environment and (2) undergo physiological and behavioral changes in order to withstand environmental extremes. The ecological manifestations of these two functions include cold, drought, and heat hardiness, seasonal polyphenism (changes in color or morphology), dormancy, and various forms of migration.

The goal of this chapter is (1) to summarize basic concepts concerning the types, functions, environmental control, evolution, and ecological significance of these seasonal adaptations and (2) in doing so, to develop the thesis that diapause is the major regulating and synchronizing mechanism underlying most phenological adaptations in insects and mites. Because migration is the subject of Chapter 7, we focus our own discussion of migration on those aspects that relate to diapause. We do not review all recent literature dealing with insect seasonal adaptations; rather we refer to selected studies that illustrate general concepts. As such, we benefit from previous reviews (Andrewartha 1952, Lees 1955, 1968, Masaki 1961, 1980, de Wilde 1962, 1970, Danilevskii 1965, Beck 1980, Danilevskii et al. 1970, Wigglesworth 1972, Tauber & Tauber 1973a, 1976a, Saunders 1976).

6.2 SEASONAL ADAPTATIONS

In general, insects are faced with two major types of environmental exigencies (Fig. 6.1). First are the *irregularly* occurring, temporary, usually localized stresses exemplified by brief periods of aseasonally high or low temperatures, drought, or depletion of food. Because of the unpredictability of these stresses, survival is dependent on the appropriateness of insects' immediate response. Insects typically respond to these unpredictable changes by becoming quiescent (torpid), or by moving away from the unfavorable area. Their second type of environmental stress is associated with *regularly* occurring, geographically widespread seasonal fluctuations in temperature, humidity or moisture, food, and other biotic factors, such as natural enemies and competitors (Fig. 6.1). These changes are usually patterned (i.e., predictable), and insects take advantage of this predictability by undergoing physiological and behavioral alterations that prepare them for the approaching season. These complex preparations constitute the *diapause syndrome*, and they form the basis for most cases of dormancy, seasonal migration, and seasonal polyphenism (cf. Chapter 7).

6.2.1 Response to Irregular or Localized Exigencies: Quiescence and Migration

Reproducing or developing insects that are suddenly confronted by an environmental exigency may either stay in place and withstand the condition until it ends or move to a new place in which to continue reproduction or development (Solbreck 1978). Both options result in a delay in reproduction and/or development; however, they are not to be confused with dormancy (Section 6.2.2.1) or seasonal migration because they are distinct both physiologically and ecologically.

When contrasting aseasonal periods of quiescence or torpor with the regularly occurring periods of dormancy, it is important to recognize both the different functions and different regulatory mechanisms involved in the two phenomena. During quiescence, growth and reproduction are delayed only until conditions become favorable; whereas during dormancy, diapause usually prevents growth and reproduction even though conditions may temporarily be favorable for these functions. Quiescence, in contrast to diapause, does not involve preparatory hormonal or physiological changes, and it usually cannot be greatly prolonged without pathological changes occurring. However, rare cases of exceptionally long periods of cold torpor are known.

Larvae of the chironomid *Polypedilum vanderplanki* undergo a state of prolonged quiescence. During the dry season, when the pools they inhabit dry up, the larvae become dehydrated and development ceases. These larvae can withstand repeated dehydration down to 8% or less of water content, and some survive storage in dry conditions for 7–10 years. Development resumes within hours of return to moist conditions (Hinton 1960).

Insects may also respond to unfavorable conditions by migrating. In this respect, two general types of migration are recognized (Solbreck 1978). The first occurs as an immediate response to unpredictable deterioration of conditions in localized areas (= Class I and Class II migration of Johnson 1969). This type occurs when the insects are in a nondiapause state and is followed by immediate resumption of feeding and reproduction when the insects encounter favorable conditions. In the seed bug *Neacoryphus bicrucis*, only adults that have fed and mated reproduce without migrating; lack of food or mates induces migration to another area where reproduction occurs (Solbreck 1978). Phenologically, this type of migration, which we term *aseasonal migration*, is equivalent to quiescence or torpor, both because of the immediacy of its control by environmental conditions and because of its ecological consequences, that is, short-term postponement of reproduction (cf. Chapter 5). Aseasonal migration thus differs from the seasonal migration that occurs as part of the diapause syndrome and usually results in long delays in reproduction.

6.2.2 Responses to Widespread, Seasonal Changes: Diapause-controlled Dormancy, Migration, and Polyphenism

In contrast to the irregularly occurring and/or localized environmental changes discussed above, most seasonal changes occur on a regular and geographically widespread basis. Insects have evolved the ability to perceive environmental cues that signal approaching seasonal changes, and they respond to these cues by undergoing specific physiological, morphological, and behavioral modifications that prepare them for approaching adverse conditions. These modifications comprise the diapause syndrome, and they form the basis for a number of phenological adaptations—including nearly all cases of dormancy, seasonal migration, and seasonal polyphenism—that require physiological, morphological, or behavioral preparations before the arrival of adverse conditions.

Considering the above, diapause is defined as a hormonally mediated state of low metabolic activity associated with reduced morphogenesis, increased resistance to environmental extremes, and altered or reduced behavioral activity. Diapause occurs during a genetically determined stage of metamorphosis and generally in response to token environmental cues that precede unfavorable conditions.

Some authors have used the term "diapause" to denote dormancy and thus both physiological and ecological connotations were included. However, we restrict the term to its physiological meaning because the physiological changes that comprise the diapause syndrome subserve other seasonal adaptations (e.g., seasonal migration and polyphenism), in addition to dormancy. We use the terms diapause-mediated dormancy, seasonal migration, and seasonal polyphenism to denote the major ecological expressions of this diapause syndrome.

Environmental cues that regulate diapause are called *token stimuli* because they, in themselves, are not favorable or unfavorable for growth, development, or reproduction, but they herald a change in environmental conditions. The most common token stimulus is photoperiod, although temperature, moisture, and biotic factors can also provide reliable cues of seasonal change. These stimuli function during periods favorable for growth, development, and reproduction and are often, but not exclusively, perceived by earlier developmental stages than those that undergo diapause (Section 6.3.2.). Once diapause has begun, growth and development are suppressed even if favorable conditions prevail.

In the past the various phenological expressions of the diapause syndrome generally have been considered in isolation. In fact, dormancy (= diapause of various authors) and migration have often been contrasted with each other as evolutionary alternatives, that is, as "escape in time" versus "escape in space." However, the intimate association between seasonal migration and dormancy (cf., e.g., Kennedy 1961, Southwood 1962,

Dingle 1978b, 1979, and Chapter 7) speaks against such a polarity. In many cases the physiological bases and ecological functions of both dormancy and seasonal migration appear to be closely related, and for both, diapause often is the major underlying adaptation.

6.2.2.1 Dormancy.

Dormancy is a general term that refers to a seasonally recurring period (*phenophase*) in the life cycle of a plant or animal during which growth, development, and reproduction are suppressed. It includes suppressed development and reproduction involving either diapause or quiescence (or both) provided they occur on a regular *seasonal* basis. Dormancy can occur during summer, fall, winter, or spring, and these periods are termed *aestivation, autumnal dormancy, hibernation,* and *vernal dormancy,* respectively. We stress that these terms do *not* indicate the organism's physiological state during dormancy; rather, they define only the seasonal timing of dormancy. The classification of dormancy was recently discussed by Tauber and Tauber (1981).

Not all species undergo dormancy. Tropical cotton stainers in the genus *Dysdercus* and tropical species of *Oncopeltus* milkweed bugs continue to grow and reproduce all year round (Southwood 1962, Dingle 1978a). If food or other requisites become unavailable, these insects either use alternate foods or migrate to new areas where growth and reproduction resumes. Temperate zone insects (e.g., *Gryllodes sigillatus* and *Myrmecophilus sapporoensis*) also may lack dormancy, but these occur in well-protected habitats (Masaki 1978). We stress that insects that totally lack dormancy are rare; and the generalization that tropical insects do not undergo dormancy or diapause is clearly questioned (Tauber & Tauber 1981).

The expression of diapause-mediated dormancy varies consideraly between species. Usually insects stop growing and feeding completely during diapause; however, diapausing embryos (e.g., the grasshopper *Aulocara elliotti*) may undergo some morphometric development (Visscher 1976). Moreover, diapausing caterpillars sometimes continue to develop at a rate that is regulated, at least in part, by token stimuli (Geyspits & Zarankina 1963, Saulich 1975).

6.2.2.2 Seasonal Migration.

Free-living insects often move from their site of reproduction to another site to undergo dormancy. These movements are well defined, have a directional component, and in some cases depend on environmentally influenced changes in wing musculature and structure (e.g., Chapter 7, and Dingle 1979). Such movements range from a few centimeters or meters to thousands of kilometers, depending on the species. In general, movements that take insects to and from their dormancy sites occur in response to token stimuli, and they usually involve the physiological and behavioral changes typical of diapausing insects (Kennedy 1961, Johnson 1969, Dingle 1978a, Rankin 1978). In view of this, we consider these movements as part of the diapause syndrome and refer to them as diapause-mediated *seasonal migration*.

A striking example of seasonal migration is seen in the monarch butterfly *Danaus plexippus*. During summer it occurs throughout most of the United States and southern Canada, and each autum large numbers fly southward to hibernate. Populations of monarchs from western United States and western Canada congregate on the Pacific coast, south of San Francisco; populations from eastern and midwestern North America move to areas of moderate climate in the Sierra Madre Occidentale of Mexico (Urquhart & Urquhart 1979). Most monarchs undergoing southward migration are in a state of reproductive diapause characterized by lack of ovarian development, increased lipid content, reduced responsiveness to reproductive and vegetative stimuli, and increased sensitivity to juvenile hormone isomers (Herman 1973). They remain in the photoperiodically and thermally regulated diapause until February when mating occurs, and mated adults disperse northward from the dormancy site (Tuskes & Brower 1978). Another long-distance migrant in which the relationship between diapause and migration has been analyzed is the milkweed bug *Oncopeltus fasciatus* (Dingle 1978a, Rankin 1978). Short day lengths maintain a high level of flight behavior in the diapausing insects, whereas long days terminate diapause, reduce movement, and promote reproduction (cf. Chapter 7).

Many insects entering diapause do not undergo long-distance migration but move from the reproductive site and seek hiding places in nearby fields and forest edges. The Colorado potato beetle, *Leptinotarsa decemlineata*, abandons the food plant, moves to edges of fields, and burrows up to 25 cm in the soil to hibernate; during this period the beetle's movements are guided by positive geotaxis and negative phototaxis (de Wilde 1954). Similarly, in late summer and early autumn the lady beetle, *Coleomegilla maculata*, moves out of corn fields and forms aggregations in litter near the field edges. Analogous patterns are found in other predaceous species (Hagen 1962, Hodek 1973), alfalfa and sweet clover weevils, and cereal leaf beetles. Diapausing adult mosquitoes sometimes seek shelter in protective places (Spielman & Wong 1973). Also, mature lepidopteran larvae (e.g. *Manduca sexta* and *Pyrrharctia isabella*) often wander several days before digging into the soil or reaching a suitable site for hibernation.

6.2.2.3 Seasonal Polyphenism.
Insects entering diapause often undergo color and/or morphological changes that provide crypsis or other forms of protection against seasonal exigencies (e.g., Shapiro 1976). We refer to these physical changes as *seasonal polyphenism*. In insect migration we stress that not all types of polyphenism are related to diapause, and here we divide polyphenism into two types (seasonal and aseasonal) that are analogous to dormancy and seasonal migration on the one hand and aseasonal quiescence and migration on the other. Since organisms generally cannot express polyphenism as quickly as they show movement or enter a state of

torpor, the distinction between these two forms of polyphenism is more difficult to perceive. However, we make the distinction on ecological grounds, recognizing that the two categories overlap considerably.

First, environmentally induced changes in form or color can occur in response to seasonally unpredictable or localized alterations in the environment. This type of polyphenism does not involve a long delay in either growth or reproduction and does not involve diapause. We refer to it as *aseasonal polyphenism* because it is triggered largely by the immediate conditions of the environment, as opposed to long-range seasonal changes. A notable example is the butterfly *Papilio polyxenes asterius* in which nondiapausing summer pupae are either green or brown depending on the color of the substrate (West et al. 1972). Other examples of aseasonal polyphenism are found in the sycamore aphid *Drepanosiphum platanoides* and the lime aphid *Eucallipterus tilliae* (Dixon 1972, Kidd 1979). In these species melanic pigmentation appears in asexual forms as a result of temperature, crowding, and/or food quality during summer, whereas photoperiod regulates the subsequent production of sexual forms that mate and produce diapausing eggs. The within-season variation in melanin deposition is a form of aseasonal polyphenism (whereas the photoperiodic regulation of diapause-producing sexual forms is an example of the second type of polyphenism—seasonal polyphenism—below). Similarly, the asexual, winged aphid forms produced in response to crowding and food quality during summer (Bonnemaison 1951, Lees 1966) are examples of aseasonal polyphenism.

Seasonal polyphenism occurs in response to seasonally predictable, regularly occurring changes in the environment. Examples include tightly spun, as opposed to coarsely spun, cocoons, increased wax secretion on pupal cuticle, sexual versus asexual forms, long versus short wings, seasonal color changes (associated with crypsis or thermoregulation), and hard versus soft egg shells. This type of polyphenism is usually induced by environmental cues—diapause-inducing token stimuli—that indicate long-term and widespread alterations in the environment. Some of these form changes have been shown to be hormonally linked to diapause (Pener 1974, Chippendale 1977, Endo, 1972).

Seasonal polyphenism in cocoon construction has reached an extreme state in some species. Prediapause larvae of the peach moth, *Carposina niponensis*, spin tough, compact, ball-shaped cocoons that are distinct from the elliptical, coarsely spun cocoons of the nondiapause larvae (Toshima et al. 1961). When diapause ends, the larvae bite through the winter cocoon and spin the nondiapause type for pupation. Thus, construction of ball-shaped winter cocoons is a highly specialized behavior associated with diapause. Similar behavior is found in certain tortricid moths that overwinter as midsized larvae. The immature larvae spin cocoons on the bark of trunks or branches only if they are entering hibernation (Oku 1966). Likewise, after hatching, prediapause spruce budworm larvae spin small cocoons in which they hibernate; such cocoons are not spun by nondiapausing larvae (Harvey 1957).

6.2.2.4 Cold Hardiness.

Hibernating animals must often tolerate extremely low temperatures that could cause injury, and insects have evolved a number of physiological and behavioral adaptations that are associated with diapause and that protect them from subzero temperatures. Two main physiological mechanisms contribute to such cold hardiness—*supercooling* (resistance to freezing by lowering the temperature at which freezing of body fluids begins) and *freezing tolerance* (survival despite freezing of body fluid) (e.g., Salt 1969, Downes 1965, Asahina 1969). Both of these mechanisms help determine an insect's overall tolerance to cold. Some insects, like the poplar sawfly *Trichiocampus populi*, have high supercooling points and can survive freezing at temperatures far below the supercooling point, while others with low supercooling points may be highly susceptible to freezing (Asahina 1969).

Gradual acclimation to low temperature often influences an insect's cold hardiness; in the adult tenebrionid, *Upis ceramboides*, acclimation increases freezing tolerance, but it has no significant effect on the supercooling point (Miller & Smith 1975). However, diapausing pupae of *H. cecropia* that have been reared at 25°C are cold hardy at temperatures down to -70°C (Asahina 1969).

6.2.2.5 Drought and Heat Hardiness.

Periods of water stress may confront aestivating insects when temperatures are high. However, drought is not restricted to summer; hibernating insects also must often cope with it, especially when temperatures go below freezing. In addition to behavioral modifications, survival during drought is accomplished by two physiological mechanisms: resistance to desiccation and/or tolerance of water loss. Some species use both mechanisms; aestivating eggs of the grasshopper *Austroicetes cruciata* have both an extremely low rate of water loss and high tolerance to reduced water content.

In some insects, especially desert inhabitants, all stages are drought hardy. But species inhabiting areas where drought is seasonal usually have one stage in the life cycle that is especially well adapted to dry conditions. In these cases seasonal drought hardiness is generally, but not always, associated with diapause. Depressed metabolism, lowered water content, and high fat content during diapause, together with increased secretion of waxy coverings, confer a high degree of resistance to desiccation. A remarkable example is the "ground pearl," *Margarodes vitium*, which forms a hard, wax-coated encysted nymph that may survive more than 10 years under dry conditions (Ferris 1919).

6.3 THE DIAPAUSE SYNDROME

We have now defined diapause and considered various adaptations that have diapause as their physiological basis. We describe here the ecophysiological characteristics of the diapause syndrome. The concept that diapause is a dynamic state that generally follows a predictable course is crucial (Fig. 6.2).

6.3.1 Stage Specificity

In contrast to adaptations such as quiescence and aseasonal migration, diapause occurs at a specific stage in the life cycle. Although diapause has been recorded in all life stages of insects: various stages of embryogenesis, all immature forms, and both sexes of adults (see list by Saunders 1976), for each species the diapausing stage is genetically fixed. In some species more than one stage may enter diapause, but in these cases, both stages are species-specific. The alfalfa snout beetle, *Otiorhynchus ligustici*, which has a 2-yr life cycle, spends the first winter as a dormant, late-stage larva and the second as a dormant adult (Lincoln & Palm 1941). Other examples include species that enter diapause at one stage in summer and another stage in winter (Masaki 1980).

   Species entering diapause as midinstar larvae tend to vary in the instar that diapauses; more than five different instars of the cricket *Pteronemobius nitidus* can overwinter (Tanaka 1978). Similar variation occurs in Lepidoptera (Sugiki & Masaki 1972) and several Odonata (Corbet 1963, Lutz 1968, Norling 1971). However, in all these cases variability in the stage undergoing diapause is species-specific and limited to certain instars.

6.3.2 Sensitive Periods

Just as diapause is expressed in a specific stage(s), the diapause-inducing stimuli are also perceived only during species-specific, genetically determined stages. The sensitive stages and the diapausing stage may be widely separated within the same generation or even between generations, or they may overlap. The sensitive period in egg diapause may range from the grandparent as in the aphid *Megoura* (Lees 1959), to various periods within the parental generation, as in the silkworm *Bombyx* (Kogure 1933), the lepidopterans *Orgyia* spp. (Kind 1969, Kimura & Masaki 1977), and the cricket *Pteronemobius* (Kidokoro & Masaki 1978), to the egg itself, as in another cricket, *Teleogryllus* (= *Acheta*) (Hogan 1960) (Fig. 6.3). For *Megoura*, photoperiod determines, not diapause itself, but the appearance of the morphs (oviparae) that lay diapausing eggs (cf. Chapter 7). Sensitivity to diapause-inducing stimuli is also species-specific in its duration—ranging from a few days in species such as *Aedes triseriatus* (Shroyer & Craig 1980) to over a month in *Chrysopa carnea* (Tauber & Tauber 1973c).

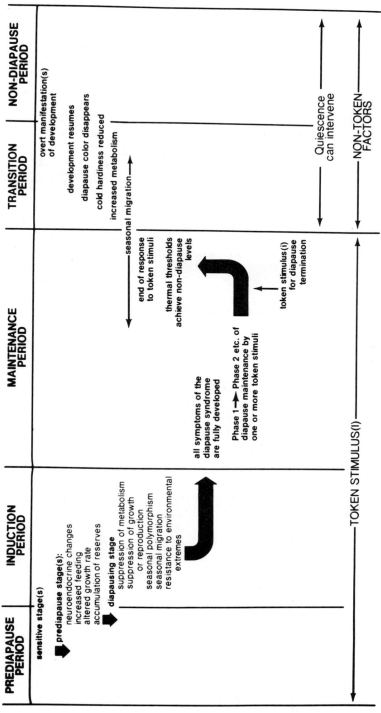

Fig. 6.2 Insect seasonal cycles comprise five distinct periods, each characterized by species-specific sets of responses. Token stimuli largely govern the first three periods, which thus constitute the diapause-regulated periods of the seasonal cycle. During the last two periods the immediate conditions of the environment (not token stimuli) regulate growth and development. Dormancy commences during the diapause induction period and ends at a species-specific time during the postdiapause transitional period. Migration to the site of dormancy usually occurs during the early part of diapause; return migration occurs near the end of diapause maintenance or after diapause had ended.

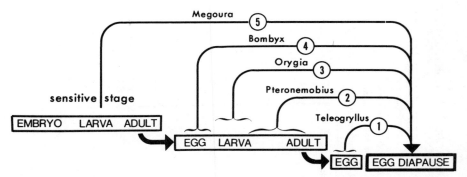

Fig. 6.3 Examples illustrating the broad range of sensitive stages associated with egg diapause. The *Teleogryllus* egg itself is sensitive to diapause-inducing conditions, and diapause occurs in a late stage of embryogenesis. In *Pteronemobius, Orygia,* and *Bombyx* various stages of the parental generation are sensitive to diapause-inducing stimuli. Finally, in *Megoura* the photoperiodic conditions of the grandparental generation determine whether or not the parental generation will oviposit diapausing eggs. Other examples are given in the text.

6.3.3 Diapause Induction and Intensification

The perception of diapause-inducing stimuli by insects results in multiple changes that are recognizable by symptoms at four general levels: neuroendocrine, physiological, biochemical, and behavioral; together these symptoms constitute the diapause syndrome. At each of these levels the various symptoms of the syndrome are expressed in a species-specific pattern that generally follows a U-shaped curve (Fig. 6.2).

Because the physiological processes occurring during diapause are poorly known, the use of "physiologically based" terms for the phases during diapause seems premature. We therefore use the terms diapause induction, intensification, maintenance, and postdiapause transition. These terms do not imply specific physiological processes during diapause. Similarly, "diapause development" (= diapause maintenance and termination) is commonly used to refer to the sequence of changes that occur once diapause is fully under way and that lead to return of the nondiapause state. We retain this term because it does not imply definition of physiological (or other) processes occurring during diapause; it merely denotes occurrence of physiological and other changes.

The neuroendocrine control of diapause has been studied in relatively few insects (Chapter 5). Nevertheless, it is apparent that diapause is a distinct physiological state with distinguishing features, not merely the cessation of growth, development, or reproduction. It is brought about not merely by absence of hormonal function; rather, full development of the syndrome is mediated through three general mechanisms: (a) a diapause factor, as in the egg diapause of *Bombyx mori* (summarized by Highnam & Hill 1977); (b) relative levels of hormones, as in *Chilo suppressalis* (cf. Yagi

& Fukaya 1974) and *Diatraea grandiosella* (Yin & Chippendale 1976); and
(c) near absence or absence of a neurohormone, as in *Hyalophora cecropia*
(Williams 1952), *Antherea pernyi* (Williams & Adkisson 1964), *Heliothis zea*
(Meola & Adkisson 1977), and *Leptinotarsa decemlineata* (de Wilde & de
Boer 1969, Schooneveld et al. 1977).

One external indicator of diapause induction is an acceleration of prediapause feeding leading to accumulation of metabolic reserves. After diapause is induced, there is a profound reduction in the insect's responses
to food (Stoffolano 1974). In some species feeding ceases, in others it
continues at a low rate, and in still others, such as the rice stem borer, *Chilo
suppressalis* (Koidsumi & Makino 1958), it occurs sporadically throughout
diapause.

The buildup of metabolic reserves during the prediapause phase may
be enhanced by breakdown of certain body tissues; in some species [e.g.,
the Colorado potato beetle (de Wilde 1970)] wing muscles degenerate and
eggs resorb during diapause induction. These changes are accompanied
by accumulation of reserves in fat body and other storage organs, including
the haemolymph (Dortland & de Kort 1978). During diapause and postdiapause periods of dormancy, these reserves serve both as a source of
energy and as a component of the protective mechanisms involved in cold
hardiness, heat tolerance, and resistance to desiccation.

Diapause induction is also characterized by changes in responses to reproductive stimuli. Although diapause can occur in either mated or unmated adults, in most species, mating does not occur during diapause (cf.
Orshan & Pener 1979). In *C. carnea* neither males nor females mate during
diapause even when paired with receptive partners (Tauber & Tauber
unpub.); whereas in the cereal leaf beetle diapausing females do not mate
but diapausing males apparently will if presented a receptive female (Connin & Hoopingarner 1971).

After diapause begins, it becomes increasingly intensified, as illustrated
in a number of ways. A commonly measured expression of intensification
is depression in oxygen consumption (e.g., Schneiderman & Williams 1953,
Siew 1966). Other expressions include decreased sensitivity to metabolic
inhibitors (de Wilde 1970), reduced RNA synthesis (cf. Wigglesworth 1972),
alteration in responses to diapause-terminating stimuli (Tauber & Tauber
1973c), altered sensitivity to diapause-terminating hormones or hormone
mimics (Bodnaryk 1977), increased resistance to temperature extremes
(Ilyinskaja 1968), alteration in thermal relations (cf. Tauber & Tauber 1976a),
and intensification of diapause color (Tauber et al. 1970).

6.3.4   Diapause Maintenance and Termination

Diapause induction and subsequent intensification are followed by diapause
maintenance during which all the species-specific symptoms of diapause
are present (Fig. 6.2). The duration of this period is highly variable among

species (and biotypes), and depending on environmental conditions before and during diapause, it may persist from several weeks to several years (Tauber & Tauber 1976a, Waldbauer 1978). Diapause may be maintained in one or more ways. Sensitivity to day length and altered thermal thresholds for diapause development are the most common (Tauber & Tauber 1976a). If more than one diapause-maintaining factor is involved, they may act simultaneously or sequentially. In the latter case it is often possible to distinguish two or more phases of diapause. In *Daphnia pulex* the autumnal phase is largely controlled by photoperiod, but in late spring, although light is necessary for diapause termination, a stimulus associated with the population's density is almost completely effective in terminating the second phase (Stross 1969b). In the carabid *Pterostichus nigrita* recognition of a two-phase diapause is made on the basis of morphological differences in the ovaries during the two phases (Thiele 1966). Photoperiod regulates both phases of this diapause. The distinction between transition from one phase to another and the termination of diapause itself may be complicated because the action of diapause-maintaining and/or diapause-terminating stimuli may be very subtle.

Apparently, aestival diapause generally requires a specific stimulus for termination, whereas hibernal diapause, which often ends "spontaneously" sometime during winter, may not require a terminating stimulus (Tauber & Tauber 1976a). Since no seasonal cue synchronizes diapause termination in the population, diapause in such species usually ends over a broad time span. Subsequent development is synchronized by factors regulating the postdiapause transitional period. We consider that, in general, diapause has ended and the postdiapause transitional phase has begun when token stimuli no longer prevent growth and development and when thermal and other responses have returned to the nondiapause state (Fig. 6.2). Thus, thermally maintained diapause ends when developmental thresholds and response curves return to normal nondiapause levels.

6.3.5 The Postdiapause Transitional Period

After diapause ends, insects enter a postdiapause transitional period of dormancy. During this period there is a species-specific, usually temperature-dependent progression in loss of diapause symptoms and in resumption of development. Rate of loss of these symptoms and rate of resumed development are generally controlled by environmental conditions (i.e., temperature, moisture, and food) that immediately affect growth and development, and not by token stimuli.

Postdiapause development may be prevented or interrupted by unfavorable conditions; in this case the insects enter the *postdiapause quiescent* period of dormancy. We consider this quiescence a part of dormancy because it is seasonally recurrent. Postdiapause quiescence is especially notable in insects that end diapause during late autumn or early winter but do not resume development until spring. In such cases the postdiapause quiescent period of dormancy precedes postdiapause development, and the insect retains most of its diapause symptoms, such as enlarged fat body, cold hardiness, and diapause color, until vernal conditions become favorable for development. During this period, continued exposure to low temperatures may aid in retaining the cold-hardiness characteristic of diapause by retarding both postdiapause development and the conversion of protective cryosubstances to other substances (Asahina 1969).

Because token stimuli do not regulate quiescence, return to favorable conditions ends postdiapause quiescence and initiates postdiapause development without delay. Thus, in *Chrysopa harrisii* photoperiodically controlled reproductive diapause ends in December; postdiapause development is inhibited until spring, but it begins immediately when temperatures rise above approximately 11°C (Tauber & Tauber unpubl.).

The term postdiapause development has sometimes been broadly interpreted (e.g., Campbell 1978) to include nondiapause development long after dormancy has ended. However, recognizing the paucity of physiological investigations of this phenomenon, we delineate the end of the postdiapause transitional period in morphological terms; that is, postdiapause development ends at the next irreversible stage in the life cycle. For eggs, it ends when the embryo reaches the next irreversible stage of embryogenesis or when hatching occurs; it does not include development of the first instar. Our criteria would exclude "stationary" or supernumerary molts that occur in some *diapausing* larvae (e.g., the southwestern corn borer). Thus, for each species, the criteria designating the end of the postdiapause developmental period must be clearly defined.

6.4   THE ENVIRONMENTAL CONTROL OF DORMANCY

We have stressed above that each species exhibits a unique set of characteristics that comprise its diapause syndrome. These species-specific characteristics (e.g., the sensitive period, diapausing stage(s), physiological expressions of diapause, diapause intensity) are primarily under genetic control. However, the degree to which each generation of a population expresses diapause usually is profoundly influenced by environmental factors, and variation in incidence of diapause between generations has led researchers to categorize diapause as either "facultative" or "obligatory." These terms are of limited usefulness (Tauber & Tauber 1981). The term *facultative diapause* denotes diapause that would be averted under appropriate environmental conditions; thus, it describes all cases of diapause in which the controlling environmental factors are known and alterable. It is used in contrast to *obligatory diapause*, which implies diapause that is expressed in every individual, in each generation, *regardless* of environmental conditions. The category of "obligatory diapause" has served largely as a catch-all for univoltine species for which the sensitive stage(s) and the dia-pause-controlling environmental stimuli have not been determined. How-ever, some authors use the term in reference to seasonal cycles in which, under normal circumstances, environmental factors interact with intrinsic characteristics of the species to produce diapause in nature, in each generation.

Generally, it is not useful to draw a sharp distinction between "faculta-tive" and "obligatory" diapause, just as it is not useful to take a one-sided view of the "nature versus nurture" problem, or to view behavioral patterns as either strictly learned (plastic) or strictly instinctive (stereotyped). The vast majority of response patterns have both genetic and environmental components. We suggest that the productive approach is to elucidate the mechanisms underlying both components and to determine how they in-teract to adapt the organism to its environment.

6.4.1 Photoperiod

Almost all environmental factors (light, temperature, humidity, rainfall, food, enemies, competitors) show characteristic seasonal patterns of change. Of these, photoperiod offers the most reliable cue to approaching seasonal changes. Except at the equator, each latitude shows a regular seasonal pattern in day length changes, and as one moves farther from the equator, these changes become more pronounced (cf. Beck 1980). Insects have com-monly evolved adaptations that take advantage of this reliability, and thus photoperiod is the primary regulator of diapause induction. Generally, temperature and other environmental factors modify the diapause-reg-ulating role of photoperiod, although sometimes they act as primary dia-pause-regulating factors (Sections 6.4.2 and 6.4.3.).

During diapause, insects continue to rely, for varying periods of time, on photoperiod as an indicator of seasonal conditions (Tauber & Tauber 1976a). In some, like *Chrysopa oculata* (Tauber & Tauber unpubl.), photo-periodic sensitivity disappears soon after diapause induction. Others, such as the European corn borer, *Ostrinia nubilalis* (McLeod & Beck 1963), the odonate *Tetragoneura cynosura* (Lutz & Jenner 1964), and *Chrysopa carnea* (Tauber & Tauber 1976a), lose this sensitivity gradually as diapause progresses.

6.4.1.1 Response Curves and the Critical Photoperiod. Insects have evolved many ways to use photoperiod as a diapause-regulating cue (cf. reviews by e.g., Danilevskii 1965, Danilevskii et al. 1970, Lees 1968, Beck 1980, Saun-ders 1976). Spectral sensitivity of insects is also covered by Norris et al. (1969), Lees (1971), and Bradshaw and Phillips (1980). Usually photoper-iodic responses of insects are categorized according to the percentage of animals entering or terminating diapause under a series of stationary pho-toperiods. For example, in the lacewing *Meleoma signoretti*, when percentage diapause is plotted against photoperiod, a characteristic response curve results. The photoperiod that elicits 50% response is called the *critical photoperiod*. For diapause induction in *Meleoma signoretti* this is between 15 and 16 hr of light/day, and for diapause termination, it shifts downward to between 12 and 14 hr/day (M. J. Tauber & C. A. Tauber 1977).

Two major types of response curves have been shown: the long-day type and the short-day type. The long-day type (Type I of Beck 1980) is characteristic of insects that reproduce, grow, and develop in response to long-day conditions and go into diapause in response to short days. The short-day response (Type II of Beck 1980) is less common and typifies insects that undergo aestival diapause in response to long day lengths (Masaki 1980). Exceptions to these generalizations are found in several species, such as the commerical silkworm in which the sensitive stages occur far in advance of the diapausing stage (see Fig. 6.3). Moreover, some insects show a combination of these two types of response curves and thus continue to develop only if day lengths fall within a narrow range between long and short days (Type IV of Beck 1980). In the bivoltine ladybird beetle *Coccinella novemnotata*, day lengths of 10, 12, and 18 hr result in a high incidence of diapause, whereas intermediate daylengths of 14 and 16 hr promote reproduction (McMullen 1967a,b).

Many texts state or imply that a population's critical photoperiod for diapause maintenance and/or termination is the same as that for diapause induction. This generalization has invariably been based on early findings with *Antheraea pernyi* in which the critical photoperiods for diapause induction and diapause termination are very similar (Williams & Adkisson 1964). In other insects for which both photoperiodic induction and termination of diapause were studied using natural populations from out of doors (e.g., *Meleoma signoretti* and the mosquito, *Wyeomyia smithii*), the two values differ (Tauber & Tauber 1976a).

6.4.1.2 Responses to Changing Day Length.

An important limitation on the usefulness of the photoperiodic response curve is that it is based on stationary light–dark cycles that do not occur in nature; it therefore does not take into account the considerable influence of changing daylength. Insect responses to photoperiodic changes fall into four categories.

1. *Species for which it is not important whether day lengths are increasing, decreasing, or stationary.* The only significant factor to these species is the duration of day length in relation to the critical photoperiod. For example, the mite *Panonychus* (= *Metatetranychus*) *ulmi* enters diapause under constant short-day conditions. Neither decreasing day lengths above the critical photoperiod nor increasing day lengths below it affect diapause induction (Lees 1953).

2. *Species that respond to a change in day length across a critical photoperiod by inducing or maintaining diapause.* Neither constant long days nor constant short days in themselves are sufficient to induce a high incidence of diapause, and if reared strictly under stationary photoperiods these insects

show a very low incidence of diapause. However, diapause is induced if the sensitive stage experiences a decrease in day length from a long day to a short day that is below the critical photoperiod. This response has been demonstrated for the bollworm *Heliothis zea* and is termed the long-day–short-day response (Wellso & Adkisson 1966, Roach & Adkisson 1970). A similar response is found in the red locust, *Nomadacris septemfasciata* (Norris 1965).

3. *Species that respond to a change in day length across a critical photoperiod by averting or terminating diapause.* If these species are reared under stationary periods, the response curve consists of an approximately straight line at 100% diapause. However, if the sensitive stage experiences an increase in day length from a short day (below the critical photoperiod) to a longer one (above it), diapause is avoided or terminated. This response is termed the short-day–long-day response.

The first study of this type of response dealt with the imaginal diapause of the univoltine delphacid *Stenocranus minutus* (e.g., Müller 1960). Ovarian development ceases at an early stage if the insect has experienced a constant, long day (16 hr or longer) throughout its previous development. Short days experienced after adult emergence allow some egg development, but full yolk formation and oviposition occurs only if the insects experience a short-day–long-day sequence. The short-day–long-day response occurs in several other univoltine species, including carabids (Thiele 1966), a leafhopper (Müller 1976), a lacewing (Tauber & Tauber 1976b), a water strider (Vepsäläinen 1974b), and a cricket (Tanaka 1978).

4. *Species with response to changes in day length without a critical photoperiod.* The above examples of response to changes in day length involve at least one critical photoperiod. The question remains whether the actual *change* in day length *alone* can induce or terminate diapause independently of a critical photoperiod. *C. carnea*, which has a critical photoperiod for diapause induction, can enter diapause in response to an abrupt decrease in day length that does not encroach on the critical photoperiod (Tauber & Tauber 1970). However, it is unlikely that it actually uses this ability to respond to changes in day lengths *per se* during natural diapause induction or termination.

6.4.1.3 Photoperiod and Postdiapause Development.

Although day length can influence prediapause and nondiapause development (e.g., Masaki 1978) and fecundity (Deseö & Saringer 1975), it has been shown not to affect postdiapause morphogenesis and reproduction in many species (cf. Tauber & Tauber 1976a). In other species photoperiod can influence postdiapause oviposition and can reinduce diapause after a period of oviposition (Hodek 1979, Pener & Broza 1971). Short day lengths can also reinduce diapause after a long period of oviposition in the Colorado potato beetle (de Wilde et al. 1959); presumably, these long-lived adults can pass a second winter in diapause. The incidence of a species-specific second diapause in *Choristoneura* spp. is influenced by photoperiodic conditions after completion of the first diapause (Harvey 1967).

6.4.2 Temperature

Temperature influences dormancy in a number of ways: (1) For some insects it is the major environmental diapause-inducing factor. (2) More commonly, it modifies, to varying degrees, the insect's response to diapause-inducing photoperiods. (3) In many species it is very important in diapause maintenance. (4) It can be an active diapause-terminating stimulus. (5) Finally, it has a major role in regulating the rate of postdiapause development.

6.4.2.1 Temperature as a Diapause-inducing Stimulus.
The function of temperature as the primary diapause-inducing factor has been demonstrated in relatively few cases. In the dipteran *Chortophila brassicae* photoperiod appears to affect diapause induction only slightly, whereas if larvae experience constant temperatures below 15°C, the incidence of pupal diapause is high. The delay in pupation is not just an effect of low temperature on development because it persists in pupae after they are transferred to 20°C (Missonnier 1963). A similar response to low temperature occurs in flies from regions near the equator where changes in day length are insignificant (Denlinger 1974). By contrast, high temperature, especially in the tropics and subtropics may function as an inducing stimulus for aestival diapause (Masaki 1980).

6.4.2.2 Thermal Modification of Photoperiodic Responses.
Temperature often interacts with photoperiod to induce diapause (Danilevskii 1965, Beck 1980, Lees 1968, Saunders 1976). In long-day insects (those that develop without diapause under long day lengths), low temperatures tend to promote diapause, whereas high temperatures tend to prevent it. In the muscoid fly *Sarcophaga argyrostomata* the critical day length is approximately 13.5–14 hr of light/day at both 15 and 20°C, whereas at 25°C all photoperiods produce almost no diapause (Saunders 1976). In short-day insects or insects that aestivate, high temperature often has the reverse role (Masaki 1980). Summer diapause occurs in pupae of the noctuid *Mamestra brassicae* when the larvae are reared under long days; high temperatures increase both the incidence and duration of this diapause (Masaki & Sakai 1965). Temperature may also drastically influence the critical photoperiod. In the noctuid moth *Acronycta rumicis*, which enters diapause under short daylengths, the critical photoperiod lengthens by $1\frac{1}{2}$ hr with each 5°C drop in the rearing temperature; that is, longer days are necessary to avert diapause at lower temperatures (Danilevskii 1965).

6.4.2.3 Thermoperiod.
Temperature, like light, has daily patterns of change, and thermoperiods (24-hr cycles in temperature) may enhance or diminish the effect of diapause-inducing photoperiods. Cold nights tend

to promote diapause induction in long-day insects. Under LD 15 : 9, the European corn borer has a high incidence of diapause under either constant 21°C or a thermoperiod of 31°C during the day and 21°C during the night. However, a day temperature of 21°C combined with a night temperature of 31°C results in a much lower diapause incidence (Beck 1980). Similar effects have been shown for insects in other orders (Saunders 1976, Beck 1980). Similarly, in the absence of light, thermoperiod simulates a photoperiodic effect in controlling diapause (cf. Saunders 1976, Beck 1980).

6.4.2.4 Temperature and Diapause Maintenance.

Thermal reaction curves for diapause development are often quite different from those for non-diapause development and growth (Andrewartha 1952, Lees 1955, Tauber & Tauber 1976a). Diapause in *N. vitripennis* terminates fastest (approximately 10 weeks) at 2°C; temperatures above 15°C and below −6°C are ineffective in terminating this diapause (Schneiderman & Horwitz 1958). Although the temperature range for rapid diapause development is usually much below that for nondiapause development, there is wide interspecific variability in this characteristic. In the grasshopper *Austroicetes cruciata* the optimum range for diapause development is 6–13°C; the lower thermal threshold for nondiapause development is above 13°C (Andrewartha 1952). In another grasshopper, *Melanoplus bivittatus*, the difference between the temperature reactions during diapause versus nondiapause development is not as great (Church & Salt 1952). In still other species, such as *Teleogryllus* spp. (Masaki 1965, Masaki et al. 1979) and *Leptinotarsa* (de Wilde 1969), responses to temperature during diapause development are similar to those during nondiapause development. Finally, in *Agromyza frontella*, diapause development proceeds faster under warm conditions than under cool, and low temperatures serve as the primary diapause-maintaining factor (Tauber et al. 1982).

Despite the above variation in thermal optima for diapause development, many texts (e.g., Gillott 1980, Borror et al. 1981) still popularize the idea that exposure to low temperature (= "chilling") is generally required for termination of hibernal diapause. This lessens our understanding of the physiological and ecological adaptations underlying diapause. The adaptive value of low thermal thresholds during diapause lies in the fact that diapause is maintained and development is prevented by warm conditions that may occur before or during winter (Tauber & Tauber 1976a). High temperatures at this time either retard or reverse diapause development.

6.4.2.5 Changing Responses to Temperature during Diapause.

The change in threshold from the low diapause level to the nondiapause level can occur relatively abruptly; in *A. cruciata* the ability to hatch at high temperatures changes suddenly between mid-May and mid-June (Birch 1942). However, in the gypsy moth, *Porthetria dispar*, the rate of diapause development under constant low temperatures decreases with time, and there is a gradual change in thermal reactions and a gradual transition from diapause to the postdiapause period (Masaki 1956). In *Tetrastichus julis, Agromyza frontella*, and other species, thermal responses return to the nondiapause level before or during early winter. At this time, low temperatures prevent development and the loss of the characteristics of dormancy (Nechols et al. 1980, Tauber et al. 1982).

6.4.2.6 Temperature as a Diapause-terminating Signal.

Because of the downward shift in the thermal reactions of insects during autumnal–hibernal diapause, it is sometimes assumed that "chilling" acts as a diapause-terminating *stimulus*. Although low temperatures are used to terminate diapause in the laboratory, as in *Hyalophora cecropia* (Williams 1956), it does not necessarily follow that "chilling" acts as a signal triggering diapause termination. It is evident in most cases that temperature acts to regulate the *rate* of diapause development (Sections 6.4.2.4. and 6.4.2.5. above), not as a specific signal to terminate diapause. The only well-studied example of temperature acting as a diapause-terminating cue is found in the cricket *Teleogryllus commodus*. In this species egg diapause persists after long periods (60–80 days) of exposure to cold conditions. Termination of a second phase of diapause requires brief (about 3 days) exposure to relatively *high* temperatures (above 20°C) (Masaki et al. 1979). In this instance high temperature acts as a diapause-terminating cue, rather than as a requirement for postdiapause development, because the lower threshold for postdiapause development is significantly lower than 20°C.

6.4.2.7 Temperature and Postdiapause Development.

After diapause ends, the temperature responses of insects are at a level characteristic for normal growth and development (Chapter 3), and given adequate nutrition and so on, temperature is usually the primary environmental factor governing the rate at which diapause characteristics (e.g., fat body, etc.) are lost and the rate at which postdiapause growth and development occur (Tauber & Tauber 1976a). As a result, thermal thresholds and temperature-dependent growth rates are often useful in predicting the timing of postdiapause events in the field—especially in computer simulations of population trends in insect pest management programs (cf. Chapters 11, 24).

Most studies using thermal accumulations to predict postdiapause events in the field are based on inaccurate or incomplete information on when diapause ends and when the postdiapause transitional phase begins. As a result they use an arbitrary date for the beginning of heat accumulation. This is inappropriate for species in which the diapause period overlaps with the time favorable for active growth, that is, those species with post-diapause developmental thermal thresholds that are lower than the temperatures that prevail after diapause ends. For example, *C. carnea* from the Ithaca, New York, area has a low (4°C) thermal threshold for postdiapause development, and this can allow postdiapause heat accumulation almost immediately after diapause ends in mid to late winter (Tauber & Tauber 1973c). A complication in calculating heat accumulations for post-diapause development is found in the wheat stem sawfly, *Cephus cinctus*, in which high temperatures immediately after diapause termination can rein-duce diapause. Moreover, temperatures during diapause, as well as the duration of diapause, may influence postdiapause thermal reactions and postdiapause developmental cycles (cf. Tauber & Tauber 1976a).

The importance of temperature in the postdiapause synchronization of vernal emergence is seen in the dragonfly *Lestes eurinus*, which overwinters in diapause, in three different larval instars (Lutz 1968). Each successive larval stage has a higher thermal threshold for development; therefore, larval development is synchronized as water temperatures rise in spring, and adult emergence occurs within a rather narrow time period.

6.4.3 Food

Food can be a reliable purveyor of seasonal cues, and insects have evolved different ways of utilizing alterations in food quality and quantity as factors regulating dormancy (cf. Masaki 1980); this may be especially true in re-gions near the equator, where rainfall varies seasonally, but where tem-perature and photoperiod show little annual change.

Food has also been shown to act as a primary diapause-regulating factor, but more often it modifies the diapause-inducing effects of another primary stimulus such as photoperiod and/or temperature. It can also influence postdiapause development.

6.4.3.1 Food as a Primary Diapause-regulating Factor.

Food has been shown to be a major diapause-regulating factor for relatively few insects, most of which undergo aestival diapause. In the *mohave* strain of *C. carnea* (native to California) both long day lengths and prey presence promote continuous reproduction. Under constant long-day conditions, lack of prey results in approximately 60% diapause; introduction of prey terminates the food-induced diapause. Thus, during summers when prey are scarce, a high proportion of the population enters a food-mediated, aestival diapause (Tauber & Tauber 1973b). An analogous response to prey apparently con-trols aestivation in *Hippodamia convergens* (Hagen 1962). Large populations of aphids allow this predator to avoid aestivation and continue reproducing throughout the summer. Aestival diapause in two Australian species, a collembolan, *Sminthurus viridis*, and a mite, *Halotydeus destructor*, is also in-fluenced by food. In these cases increased maturity of the host plant ap-parently induces diapause (Wallace 1970), but the chemical basis for this is unknown. Such effects of senescent vegetation may be related to plant growth hormones (e.g., Ellis et al. 1965, Visscher 1980).

In addition to aestivation, hibernation may also be influenced by food.

In *Chaoborus americanus* hibernal diapause terminates in response to the simultaneous occurrence of prey and long day length (Bradshaw 1970). Similarly, hibernal diapause in the Indian meal moth, *Plodia interpunctella*, terminates in response to rice bran extract (Tsuji 1966).

The role of food in morph determination in aphids and in phase determination and migration in grasshoppers is complex and has been reviewed by Blackman (1974) and Dixon (1977) (cf. Chapter 7).

6.4.3.2   Food as a Modifier of Photoperiodic Responses.

Food can modify the response to photoperiod by altering the incidence and duration of diapause (reviewed by Andrewartha 1952, Lees 1955, Tauber & Tauber 1973a, Saunders 1976). When females of the mite *Panonychus ulmi* were reared on senescent or "bronzed" foliage, they laid a high proportion of diapausing eggs even under diapause-averting temperatures and photoperiods (Lees 1953). Similarly, in *L. decemlineata* diapause is primarily induced by short day lengths, but it may also be induced by senescing potato leaves (de Wilde et al. 1969). Analogous responses are found in *P. gossypiella* (Adkisson et al. 1963), and *Hyphantria cunea* from New Brunswick and Nova Scotia (Morris 1967).

6.4.3.3   Food and Postdiapause Development.

Dependency on an external source of food for completion of postdiapause development is a species-specific characteristic. Some species that overwinter as larvae must feed prior to molting and continued development; others, for example, the "woolly bear," *Pyrrharctia isabella*, do not have this food dependency. Many mosquitoes that overwinter as adults require a blood meal after diapause, before they oviposit (Clements 1963). In other insects (e.g., *C. carnea*) food may not be required to initiate postdiapause oviposition but is required for sustained postdiapause oviposition (Tauber & Tauber 1973c).

6.4.4   Influence of Host on Parasitoid Diapause

Seasonal synchrony in the life cycles of parasitoids and their hosts is a major component of host–parasitoid relations, and this synchrony is often achieved by the timing of diapause in both host and parasitoid (Doutt 1959, Doutt et al. 1976, Tauber et al. 1983). Three types of interactions characterize the relationship between hosts and their parasitoids during diapause: (1) independent responses of the host and parasitoid to environmental cues, (2) simultaneous dependence of the parasitoid on its host's physiological state and external environmental cues, and (3) dependence of the parasitoid on the host's physiological state. These three conditions are not mutually exclusive and there is a continuum of parasitoid responses from complete independence to complete dependence on the host.

6.4.4.1 Independent Responses of Host and Parasitoid to Environmental Diapause-regulating Factors.

A number of parasitoids rely on environmental diapause-inducing signals that are independent of the physiological state of the host. In various chalcids attacking the pine sawfly (Eichhorn & Pschorn-Walcher 1976) and in the aquatic mymarid *Caraphractus cinctus* (Jackson 1963), diapause induction is relatively independent of host condition. Diapause induction in such parasitoids, as in nonparasitoids, involves critical photoperiods, response to changes in day length, and photoperiodic interaction with temperature (cf. Tauber et al. 1983). In the pteromalid *Nasonia vitripennis*, larval diapause is primarily induced by the photoperiod and temperature experienced by the parasitoid's mother (Saunders 1976). The host may exert a secondary effect through the maternal diet (host feeding). The species of blowfly the mother feeds upon influences both the incidence of diapausing *Nasonia* larvae produced and the speed with which the female switches to producing diapausing offspring (Saunders et al. 1970). Similar effects are found in the braconids *Aphaereta minuta* and *Alysia manducator* that attack various blowflies (Vinogradova & Zinovjeva 1972).

Once initiated, diapause in parasitoids may be regulated by external environmental factors independently of the host. Both the braconid *Perilitus coccinellae* and its ladybird host, *Coleomegilla maculata*, respond to photoperiod during diapause, but the photoperiodic regulation of diapause ends 3 months earlier in the parasitoid than in the host (Obrycki & Tauber 1979). In the cereal leaf beetle parasitoid, *Tetrasticus julis*, both photoperiod and temperature regulate the rate of diapause development (Nechols et al. 1980).

6.4.4.2 Dependence of Parasitoid on the Host's Physiological State and External Cues.

This type of interaction, which appears to be common, is based on a combination of the independent reactions of hosts and parasitoids to external physical conditions and some degree of parasitoid dependence on the host's diapause state. Although photoperiod, temperature, and maternal age play important roles in diapause induction in the hyperparasitoid *Catolaccus aeneoviridis*, a higher proportion enter diapause when the host, *Apanteles congregatus*, is in diapause versus nondiapause condition (McNeil & Rabb 1973).

6.4.4.3 Dependence of Parasitoid on Its Host's Physiological State.

Some parasitoids depend almost entirely on the host's condition to regulate diapause induction and/or termination. This may depend on the nutritional value of the host, as in the chalcid *Melittobia chalybii*, or on a hormonal linkage between host and parasitoid (cf. Tauber et al. 1983). Synchronous diapause termination in the tachinid *Eucaracelia rutilla* and its host, the pine looper, *Bupalus piniarius*, is maintained when there is a physical connection between the host and the parasitoid (Schoonhoven 1962).

6.4.5 Diapause Regulation in Social Insects

There are few experimental studies showing the cause–effect relationship between environmental cues and seasonal activity in social insects. Present evidence indicates that photoperiod and temperature provide the ultimate cues that synchronize seasonal cycles with external conditions. Endogenous annual rhythms may also be important in regulating diapause in social insects (Brian 1977). What appears unique to social insects is their ability to transmit photoperiodic information *between* individuals. For example, the physiological status of *Myrmica rubra* workers determines the diapause status in larvae and, in conjunction with photoperiod, regulates the diapause of queens (Kipyatkov 1976).

6.4.6 Density

In some gregarious insects and also in other arthropods from confined habitats, crowding may be significant in diapause induction. However, in most cases, the influence of crowding, or density, has not been separated from associated effects of reduced food quality and/or quantity. Despite this problem, there are some well-documented cases of density-dependent diapause. The Indian meal moth has a dual protective mechanism: a "density-independent" diapause for avoiding mortality due to seasonal environmental changes and a "density-dependent" dispause for avoiding mortality associated with increased population growth under crowded conditions (Tsuji 1966). Density effects on diapause have also been shown in *Daphnia* spp. (Stross 1969a, b) and *Naranga aenescens* (Iwao 1968).

6.4.7 Moisture

The mechanism through which moisture may influence diapause is obscure and has been studied primarily in relation to egg diapause. Designation of water absorption (or lack of it) as a diapause-regulating factor (token stimulus) has been difficult because water absorption or intake is a normal requirement for embryonic development in many species (Ando 1972). Postdiapause eggs of the corn rootworm *Diabrotica virgifera* require moisture to complete embryogenesis. When deprived of contact moisture postdiapause eggs, after some development, enter a state of quiescence that is separate from diapause (Krysan 1978). The same is true for eggs of the acridid *Locustana pardalina* (Matthée 1951). Further studies are needed with mosquitoes, such as *Aedes vexans*, that require water for initiating postdiapause embryogenesis (Horsfall et al. 1973) to determine if a second phase of diapause is ended by contact with moisture, or if moisture is a requirement for postdiapause development. An important consideration in this regard is the association of hatching by mosquito eggs with reduction of dissolved oxygen and other related factors (Gillett 1972).

Moisture may also influence diapause and postdiapause development in insect larvae, for example, the European corn borer (Beck 1980), the wheat blossom midges *Contarinia tritici* and *Sitodiplosis mosellana* (Basedow 1977), and the southwestern cornborer (Reddy & Chippendale 1973). Despite the problems in distinguishing the direct effect of moisture on diapause from that on postdiapause development, the ecological consequences are similar in both cases; development is delayed until moisture conditions become favorable.

6.5   THE GENETIC CONTROL OF DORMANCY

As expected for an adaptation with such wide-ranging effects, the genetic basis for diapause is not simple (Masaki 1961, Dingle 1974, Hoy 1978, Tauber & Tauber 1979, 1981). Several major types of genetic mechanisms help determine diapause characteristics: (1) polygenic inheritance, (2) supergenic inheritance, and (3) Mendelian inheritance; each appears to influence particular types of diapause characteristics.

6.5.1   Polygenic Inheritance

Evidence for polygenic inheritance of diapause characteristics is substantial and comes from both hybridization and selection tests. In general, the characteristics subject to polygenic control are quantitative aspects of diapause: primarily *length* of the critical photoperiod, *percentage* of diapausing individuals, and *duration* of diapause (Tauber & Tauber 1979). The critical photoperiods for diapause induction in Leningrad and Sukhumi populations of the noctuid *Acronycta rumicis* differ by about 4 hr (Danilevskii 1965). Both F_1 and F_2 hybrids, as well as backcross offspring, show critical photoperiods intermediate to those of their parents. These results are generally consistent with those for polygenic inheritance. Similar differences in photoperiodic responses exhibited by geographic strains of other species in the Lepidoptera, crickets, and lacewings are also attributable to differences in polygenes (cf. Tauber & Tauber 1981).

In some cases of polygenic inheritance, the diapause characteristics differentiating two strains appear to be, at least in part, sex-linked: for example, in *A. rumicis* and *Leucoma salicis* (Danilevskii 1965); *M. sexta* (Rabb 1969), and *P. dispar* (Lynch & Hoy 1978). Additional evidence for the polygenic mode of inheritance of diapause characteristics comes from selection experiments that altered certain quantitative aspects of diapause, for example, incidence and depth of diapause and critical photoperiods (cf. Dingle 1974, Hoy 1978, Tauber & Tauber 1979, 1981).

6.5.2 Supergenic Inheritance

In some cases the polygenes controlling diapause apparently have become closely linked and are inherited as a "supergene" (Lumme 1978). *Drosophila littoralis* has a continuous latitudinal cline in photoperiodic reactions; however, the factors controlling the critical day length segregate as single, autosomal Mendelian units. Allelic variation within the unit apparently is sufficient to form a continuous cline in photoperiodic reaction (Lumme & Oikarinen 1977).

6.5.3 Mendelian Inheritance

Mendelian inheritance of diapause characteristics has been demonstrated for two sibling species of *Chrysopa* (Tauber et al. 1977). In this case allelic differences at two autosomal loci result in gross differences in the seasonal characteristics of two species, that is, multivoltine versus univoltine life cycles. Mendelian inheritance of a switch mechanism controlling diapause and associated wing length has also been shown for a water strider, *Gerris*. Although the possibility of supergenic inheritance has not been ruled out in this case, monogenic inheritance is more likely (Vepsäläinen 1974a). The silkworm presents an interesting case in which voltinism is apparently under the control of three sex-linked alleles at one locus and paired alleles at each of three unlinked autosomal loci. The sex-linked alleles exhibit epistasis, whereas the effect of the autosomal alleles is additive (cf. Lees 1955). In hybrids various combinations of the genes are expressed as a continuous gradient—from univoltine to multivoltine life cycles.

6.6 VARIATION IN DORMANCY AND ITS EVOLUTIONARY CONSEQUENCES

6.6.1 Geographic Variation

Widespread species encounter a great variety of climatic conditions and therefore exhibit considerable variation in their seasonal cycles, a notable example being the number of generations per year (univoltine, bivoltine, multivoltine, nondiapause strains). Workers in the USSR first analyzed the nature of this geographic variation and its underlying physiological causes. They showed that there are geographic clines in the critical photoperiod for diapause induction and that these clines follow the latitudinal gradient in day length. Such clines can result in geographic variation in voltinism (cf. Danilevskii 1965) or in stabilizing a univoltine life cycle over a wide latitudinal gradient (Goldschmidt 1934, Masaki 1978).

Intraspecific clines of geographic variability have also been shown for other diapause characteristics—for example, diapause depth, temperature responses during diapause, size trends in relation to voltinism, altitudinal variation in critical photoperiod, and latitudinal variation in postdiapause developmental rates (Tauber & Tauber 1978, 1981). Most of these characteristics are *quantitative* aspects of dormancy (e.g., the *duration* of diapause and the *length* of the critical photoperiod) and most are probably based on gradients in polygene frequencies.

In contrast to clinal variation in quantitative characteristics, geographic variation can also be disjunct and involve qualitative differences in diapause characteristics. An example is found in *C. carnea*, in which the *carnea* strain of eastern and central North America does not have an aestival diapause, whereas the *mohave* strain of certain parts of western North America, has a prey-mediated aestival diapause (see Section 6.4.3).

6.6.2 Intrapopulation Variation

Genetically based intrapopulation variation in dormancy has both ecological and evolutionary importance. Ecologically, it underlies certain types of phenotypic variation that adapts populations to unpredictable or heterogeneous environments. During evolution, it serves as the building material for the diversification of seasonal cycles. As such, it also functions in the evolutionary adaptation of species and the diversification of taxa (Tauber & Tauber 1981). Therefore, an understanding of the variability in phenological responses and the underlying genetic variability is essential to a full understanding of the evolutionary ecology of any species (cf. Chapters 8, 9).

Intrapopulation variation in dormancy is expressed in a variety of ways (Tauber & Tauber 1981). We group these into two types that have both different underlying mechanisms and different ecological consequences.

6.6.2.1 Continuous Intrapopulation Variation. Continuous intrapopulation variation is expressed in quantitative aspects of dormancy and generally results in variability in the timing of phenological events—that is, variability around a mean date for diapause induction or termination. This is exemplified by the critical photoperiod which, for most species, is very responsive to selection pressure, and which shows high additive genetic variance (cf. Dingle 1974). In the milkweed bug, a long-distance migrant, genetic variability probably provides variation in the time of diapause initiation and prereproductive migration. Such genetic variance is probably maintained by seasonally fluctuating selection pressures and by interbreeding between geographic populations (Dingle et al. 1977) (cf. Chapters 7, 8).

Because continuous intraspecific variability is usually based on polygenic differences between individuals, it sometimes results in occurrence of a few nondiapausing individuals or ones with very short diapause periods. Such individuals are found in nature and laboratory stocks and have been used in selection experiments for establishing nondiapausing laboratory populations of the spruce budworm and gypsy moth.

6.6.2.2 Disjunct Intrapopulation Variation.

In contrast to continuous variation, disjunct intrapopulation variation in dormancy is either expressed in a polymorphic system or it may lead to speciation (Tauber & Tauber 1981). Disjunct variation is most easily studied in populations that have polymorphic expression of some aspect of their seasonal cycle, such as diapause induction or termination. Most of these insects live in environments characterized by unpredictable seasonal changes, and polymorphism in dormancy allows the population to spread its risks along more than one line (Masaki 1980). Perhaps the best example is the prolonged diapause exhibited by a small percentage of individuals in local populations of sawflies, Lepidoptera, and certain other insects (e.g., Powell 1974, Waldbauer 1978). These individuals probably serve to replenish the population in event of local extinction of the nondiapausing, active individuals by some catastrophe. Another example is the cecidomiid midge *Hasegawia sasacola*, which forms galls in bamboo buds. The availability of these buds varies greatly from year to year. After flowering (which occurs at unpredictable intervals of approx. 50 yr), the bamboo plants die synchronously, leaving the midge without a food source for a long time. Survival of the population depends on those individuals that survive this host-free period in a state of prolonged diapause (Sunose 1978).

Another type of polymorphism in dormancy involves within-season diapause duration. The moth *Hyalophora cecropia* exhibits bimodal and trimodal emergence during the spring and summer; probably this polymodality in emergence prevents extermination of local populations due to unpatterned catastrophes (Waldbauer 1978). Disjunct variation in termination of diapause also occurs in the predaceous mosquito, *Chaoborus americanus*, which overwinters as larvae (Bradshaw 1973). In this case polymorphism in dormancy is associated with polymorphism in larval morphology. During spring the aquatic habitat of the overwintered larvae is subject to unpredictable periods of thawing and freezing. Early developing morphs, which are large and yellow, do well when spring is mild; however, they suffer heavy mortality if spring is interrupted by freezing periods. Under cold conditions, the late-developing morphs, which are small and pale, do well.

Polymorphism in dormancy characteristics is not restricted to diapause termination; it is also seen in diapause induction. Here again, polymorphism allows the insect to overcome environmental exigencies that occur irregularly and for which there are no reliable predictive cues. When investigated, polymorphism in diapause induction will probably be shown to be common among insects in unstable or variable environments. Perhaps the most common form is polymorphism in the occurrence of aestival diapause (i.e., a part of a population enters aestival diapause). In temperate zones all, or a very high percentage of, individuals usually enter hibernation; whereas when aestival diapause occurs, it is commonly found in only part of the population. In the *mohave* strain of *C. carnea*, 40–60% of the individuals enter a food-mediated aestival diapause in response to lack of prey during hot, dry summer conditions; the others continue to reproduce even under severe drought (Tauber & Tauber 1973c, unpubl.). The density-dependent diapause of the Indian meal moth is another example. Part of a population avoids mortality due to overcrowding and lack of food by entering a density-mediated diapause, whereas the rest remain in an active state under the same conditions (Tsuji 1966).

In the cases above, the polymorphic systems appear to be stable. However, the evolutionary consequences of disjunct intraspecific variation in seasonal cycles are not restricted to stable polymorphisms. In other cases the variability may result in reproductive isolation between segments of a population, and thus may have a major role in speciation. For example, in both monophagous fruit flies (Bush 1975) and habitat-specific green lacewings (M. J. Tauber & C. A. Tauber 1977), differences in seasonal cycles are responsible, in large part, for the reproductive isolation between potentially interbreeding populations. Intrapopulation divergence in seasonal strategies may also be important to diversification and speciation in other insects (Tauber & Tauber 1981).

Clearly, insects have evolved multiple mechanisms by which they attune their life cycles to the seasonality of their biotic and abiotic environment. The richness of this diversity underscores the importance of insect phenology to many facets of ecology, including evolutionary, physiological, behavioral, and applied ecology.

ACKNOWLEDGMENTS

We thank Professor A. D. Lees, Imperial College, London, for his comments on the manuscript. We also thank James R. Nechols and E. Beth French for their help. This chapter benefited greatly from background information and knowledge gained during research sponsored by the National Science Foundation (MJT & CAT).

REFERENCES

Adkisson, P. L., R. A. Bell, and S. G. Wellso. 1963. *J. Insect Physiol.* **9:** 299–310.

Ando, Y. 1972. *Appl. Entomol. Zool.* **7:** 142–154.

Andrewartha, H. G. 1952. *Biol. Rev.* **27:** 50–107.

Asahina, E. 1969. Pp. 1–49 in J. W. L. Beament, J. E. Treherne, and V. B. Wigglesworth (Eds.), *Advances in Insect Physiology*, Vol. 6. Academic, London.

Basedow, T. 1977. *Zool. Jb. Syst. Bd.* **104:** 302–326.

Beck, S. D. 1980. *Insect Photoperiodism* (2nd ed.). Academic, New York.

Birch, L. C. 1942. *Aust. J. Exp. Biol. M. Sci.* **20:** 17–25.

Blackman, R. 1974. *Aphids.* Ginn & Co., London.

Bodnaryk, R. P. 1977. *J. Insect Physiol.* **23:** 537–542.

Bonnemaison, L. 1951. *Ann. Epiphyt. C* **2:** 1–380.

Borror, D. J., D. M. DeLong, and C. A. Triplehorn. 1981. *An Introduction to the Study of Insects* (5th ed.). Holt, Rhinehart & Winston, New York.

Bradshaw, W. E. 1970. *Biol. Bull.* **139:** 476–484.

Bradshaw, W. E. 1973. *Ecology* **54:** 1247–1259.

Bradshaw, W. E. and D. L. Phillips. 1980. *Oecologia* **44:** 311–316.

Brian, M. V. 1977. Pp. 202–206 in *International Union Study of Social Insects.* Proc. 8th Int. Cong. Wageningen.

Brown, V. K. and I. Hodek (Eds.). 1983. *Diapause and Life Cycle Strategies in Insects.* W. Junk, The Hague.

Browne, L. Barton (Ed.). 1974. *Experimental Analysis of Insect Behavior.* Springer-Verlag, New York.

Bush, G. L. 1975. Pp. 187–206 in P. W. Price (Ed.), *Evolutionary Strategies of Parasitic Insects.* Plenum, London.

Campbell, R. W. 1978. *Ann. Entomol. Soc. Am.* **71:** 442–448.

Chippendale, G. M. 1977. *Annu. Rev. Entomol.* **22:** 121–138.

Church, N. S. and R. W. Salt. 1952. *Can. J. Zool.* **30:** 173–184.

Clements, A. N. 1963. *The Physiology of Mosquitoes.* Pergamon, Oxford.

Connin, R. V. and R. A. Hoopingarner. 1971. *Ann. Entomol. Soc. Am.* **64:** 655–660.

Corbet, P. S. 1963. *A Biology of Dragonflies.* Quadrangle Books, Chicago.

Danilevskii, A. S. 1965. *Photoperiodism and Seasonal Development of Insects* (English trans.). Oliver & Boyd, London.

Danilevskii, A. S., N. I. Goryshin, and V. P. Tyshchenko. 1970. *Annu. Rev. Entomol.* **15:** 201–244.

Denlinger, D. L. 1974. *Nature* **252:** 223, 224.

Deseö, K. V. and G. Saringer. 1975. *Acta Phytopathol. Acad. Sci. Hung.* **10:** 131–139.

de Wilde, J. 1954. *Arch. Neerl. Zool.* **10:** 375–378.

de Wilde, J. 1962. *Annu. Rev. Entomol.* **7:** 1–26.

de Wilde, J. 1969. Pp. 263–284 in Woolhouse (Ed.), 1969, referenced here.

de Wilde, J. 1970. *Mem. Soc. Endocrinol.* **18:** 487–514.

de Wilde, J., W. Bongers, and H. Schooneveld. 1969. *Entomol. Exp. & Appl.* **12:** 714–720.

de Wilde, J. and J. A. de Boer. 1969. *J. Insect Physiol.* **15:** 661–675.

de Wilde, J., C. S. Duintjer, and L. Mook. 1959. *J. Insect Physiol.* **3:** 75–85.

Dingle, H. 1974. Pp. 329–342 *in* Browne (Ed.), 1974, referenced here.

Dingle, H. 1978a. Pp. 254–276 *in* Dingle (Ed.), 1978b, referenced here.

Dingle, H. (Ed.) 1978b. *Evolution of Insect Migration and Diapause.* Springer-Verlag, New York.

Dingle, H. 1979. Pp. 64–87 in R. L. Rabb and G. G. Kennedy (Eds.), *Movement of Highly Mobile Insects: Concepts and Methodology in Research.* North Carolina University Press, Raleigh.

Dingle H., C. K. Brown, and J. P. Hegmann. 1977. *Am. Nat.* **111:** 1047–1059.

Dixon, A. F. G. 1972. *J. Animal Ecol.* **41:** 689–697.

Dixon, A. F. G. 1977. *Annu. Rev. Ecol. Syst.* **8:** 329–353.

Dortland, J. F. and C. A. D. de Kort. 1978. *Insect Biochem.* **8:** 93–98.

Doutt, R. L. 1959. *Annu. Rev. Entomol.* **4:** 161–182.

Doutt, R. L., D. P. Annecke, and E. Tremblay. 1976. Pp. 143–168 in C. B. Huffaker and P. S. Messenger (Eds.), *Theory and Practice of Biological Control.* Academic, New York.

Downes, J. A. 1965. *Annu. Rev. Entomol.* **10:** 257–274.

Eichhorn, O. and H. Pschorn-Walcher. 1976. *Zeit. Angew. Entomol.* **80:** 355–381.

Ellis, P. E., D. B. Carlisle, and D. J. Osborne. 1965. *Science* **149:** 546, 547.

Endo, K. 1972. *Development, Growth and Differentiation* **14:** 263–274.

Ferris, G. F. 1919. *Entomol. News* **30:** 27. Cited by Lees, A. D. 1955. *The Physiology of Diapause in Arthropods.* Cambridge University Press, London.

Geyspits, K. F. and A. I. Zarankina. 1963. *Entomol. Rev.* **42:** 14–19.

Gillett, J. D. 1972. *The Mosquito.* Doubleday, Garden City, N.Y.

Gillott, C. 1980. *Entomology.* Plenum, New York.

Goldschmidt, R. 1934. *Bibliogr. Genet.* **11:** 1–186.

Hagen, K. S. 1962. *Annu. Rev. Entomol.* **7:** 289–326.

Harvey, G. T. 1957. *Can. J. Zool.* **35:** 549–572.

Harvey, G. T. 1967. *Can. Entomol.* **99:** 486–503.

Herman, W. S. 1973. *J. Insect Physiol.* **19:** 1883–1887.

Highnam, K. C. and L. Hill. 1977. *The Comparative Endocrinology of the Invertebrates* (2nd ed.). Edward Arnold, London.

Hinton, H. E. 1960. *J. Insect Physiol.* **5:** 286–300.

Hodek, I. 1973. *Biology of Coccinellidae.* Junk, The Hague, Prague.

Hodek, I. 1979. *J. Insect Physiol.* **25:** 867–871.

Hogan, T. W. 1960. *Aust. J. Biol. Sci.* **13:** 14–29.

Horsfall, W. R., H. W. Fowler, Jr., L. J. Moretti, and J. R. Larsen. 1973. *Bionomica and Embryology of the Inland Floodwater Mosquito, Aedes vexans.* University of Illinois Press, Urbana.

Hoy, M. A. 1978. Pp. 101–126b, in Dingle (Ed.), 1978 referenced here.

Ilyinskaja, N. B. 1968. Pp. 387–388 in *Proc. 13th Int. Cong. Entomol.* Moscow.

Iwao, S. 1968. Pp. 185–212 in *l'Effe de Groupe chez les Animaux.* 173. Colloques Internationaux du Centre National de la Recherche Scientifique, Paris.

Jackson, D. J. 1963. *Parasitology.* **53:** 225–251.

Johnson, C. G. 1969. *Migration and Dispersal of Insects by Flight.* Methuen, London.

Kennedy, J. S. 1961. *Nature* **189:** 785–791.

Kidd, N. A. C. 1979. *Entomol. Exp. & Appl.* **25:** 31–38.

Kidokoro, T. and S. Masaki. 1978. *Jap. J. Ecol.* **28:** 291–298.

Kimura, T. and S. Masaki. 1977. *Kontyu* **45:** 97–106.

Kind, T. V. 1969. *Dokl. Akad. Nauk SSSR* **187:** 517–520 (transl.).

Kipyatkov, V. Y. 1976. *Entomol. Rev.* **55:** 27–34.

Kogure, M. 1933. *J. Dept. Agr. Kyushu Univ.* **4:** 1–93.

Koidsumi, K. and K. Makino. 1958. *Jap. J. Appl. Entomol. Zool.* **2:** 135–138.

Krysan, J. L. 1978. *J. Insect Physiol.* **24:** 535–540.

Lees, A. D. 1953. *Annu. Appl. Biol.* **40:** 449–486.

Lees, A. D. 1955. *The Physiology of Diapause in Arthropods.* Cambridge Univ., London.

Lees, A. D. 1959. *J. Insect Physiol.* **3:** 92–117.

Lees, A. D. 1966. Pp. 207–277 in J. W. L. Beament, J. E. Treherne, and V. B. Wigglesworth (Eds.), *Advances in Insect Physiology,* Vol. 3. Academic, London.

Lees, A. D. 1968. Pp. 47–137 in A. C. Giese (Ed.), *Photophysiology.* Academic, New York.

Lees, A. D. 1971. Pp. 372–380 in M. Menaker (Ed.). *Biochronometry*. National Academy of Science, Washington, D.C.

Lincoln, C. and C. E. Palm. 1941. *Cornell Univ. Agric. Exp. Sta. Mem.* **236:** 1–44.

Lumme, J. 1978. Pp. 145–170 in Dingle (Ed.), 1978b, referenced here.

Lumme, J. and A. Oikarinen. 1977. *Hereditas* **86:** 129–142.

Lutz, P. E. 1968. *Ecology* **49:** 637–644.

Lutz, P. E. and C. E. Jenner. 1964. *Biol. Bull.* **27:** 304–316.

Lynch, C. B. and M. A. Hoy. 1978. *Genet. Res. Camb.* **32:** 129–133.

Masaki, S. 1956. *Jap. J. Appl. Entomol. Zool.* **21:** 148–157.

Masaki, S. 1961. *Bull. Fac. Agric. Hirosaki Univ.* **7:** 66–98.

Masaki, S. 1965. *Bull. Fac. Agric. Hirosaki Univ.* **11:** 59–90.

Masaki, S. 1978. Pp. 72–100 in Dingle (Ed.), 1978, referenced here.

Masaki, S. 1980. *Annu. Rev. Entomol.* **25:** 1–25.

Masaki, S., Y. Ando., and A. Watanabe. 1979. *Kontyu* **47:** 493–504.

Masaki, S. and T. Sakai. 1965. *Jap. J. Appl. Entomol. Zool.* **9:** 191–205.

Matthée, J. J. 1951. *Union S. Afr. Dept. Agric. Sci. Bull.* **316:** 1–83.

McLeod, D. G. R. and S. D. Beck. 1963. *Biol. Bull.* **124:** 84–96.

McMullen, R. D. 1967a. *Can. Entomol.* **99:** 42–49.

McMullen, R. D. 1967b. *Can. Entomol.* **99:** 578–586.

McNeil, J. N. and R. L. Rabb. 1973. *J. Insect Physiol.* **19:** 2107–2118.

Meola, R. W. and P. L. Adkisson. 1977. *J. Insect Physiol.* **23:** 683–688.

Miller, L. K. and J. S. Smith. 1975. *Nature* **258:** 519, 520.

Missonnier, J. 1963. *Ann. Epiphyties* **14:** 293–310.

Morris, R. R. 1967. *Can. Entomol.* **99:** 522–528.

Müller, H. J. 1960. *Verh. 11th Int. Kongr. Entomol. Wien.* **1:** 678–689.

Müller, H. J. 1976. *Zool Jb. Physiol. Bd.* **80:** 231–258.

Nechols, J. R., M. J. Tauber, and R. G. Helgesen. 1980. *Can. Entomol.* **112:** 1277–1284.

Norling, U. 1971. *Entomol. Scand.* **2:** 170–190.

Norris, K. H., F. Howell, D. K. Hayes, V. E. Adler, W. N. Sullivan, and M. S. Schechter. 1969. *Proc. Natl. Acad. Sci. USA* **63:** 1120–1127.

Norris, M. J. 1965. *J. Insect Physiol.* **11:** 1105–1119.

Obrycki, J. J. and M. J. Tauber. 1979. *Environ. Entomol.* **8:** 400–405.

Oku, T. 1966. *Kontyu* **34:** 144–153 (in Japanese).

Orshan, L. and M. P. Pener. 1979. *Physiol. Entomol.* **4:** 55–61.

Pener, M. P. 1974. Pp. 264–277 in Browne (Ed.), 1974, referenced here.

Pener, M. P. and M. Broza. 1971. *Entomol. Exp. & Appl.* **14:** 190–202.

Powell, J. A. 1974. *Pan-Pac. Entomol.* **50:** 220–225.

Rabb, R. L. 1969. *Ann. Entomol. Soc. Am.* **62:** 1252.

Rankin, M. A. 1978. Pp. 5–32 in Dingle (Ed.), 1978b, referenced here.

Reddy, A. S. and G. M. Chippendale. 1973. *Entomol. Exp. & Appl.* **16:** 445–454.

Roach, S. H. and P. L. Adkisson. 1970. *J. Insect Physiol.* **16:** 1591–1597.

Salt, R. W. 1969. Pp. 331–350 in Woolhouse (Ed.), 1969, referenced here.

Saulich, A. K. 1975. *Entomol. Rev.* **34:** 52–59.

Saunders, D. S. 1976. *Insect Clocks*. Pergamon, Oxford.

Saunders, D. S., D. Sutton, and R. A. Jarvis. 1970. *J. Insect Physiol.* **16:** 405–416.

Schneiderman, H. A. and J. Horwitz. 1958. *J. Exp. Biol.* **35:** 520–551.

Schneiderman, H. A. and C. M. Williams. 1953. *Biol. Bull.* **105:** 320–334.

Schooneveld, H., A. Otazo Sanchez, and J. de Wilde. 1977. *J. Insect Physiol.* **23:** 689–696.

Schoonhoven, L. M. 1962. *Arch. Neerl. Zool.* **15:** 111–174.

Shapiro, A. M. 1976. *Evol. Biol.* **9:** 259–333.

Shroyer, D. A. and G. B. Craig, Jr. 1980. *Ann. Entomol. Soc. Am.* **73:** 39–43.

Siew, Y. C. 1966. *Trans. Roy. Entomol. Soc. Lond.* **118:** 359–374.

Solbreck, C. 1978. Pp. 195–217 in Dingle (Ed.), 1978b, referenced here.

Southwood, T. R. E. 1962. *Biol. Rev.* **37:** 171–214.

Spielman, A. and J. Wong. 1973. *Ann. Entomol. Soc. Am.* **66:** 905–907.

Stoffolano, J. G. Jr. 1974. Pp. 32–47 in Browne (Ed.), 1974, referenced here.

Stross, R. G. 1969a. *Biol. Bull* **136:** 264–273.

Stross, R. G. 1969b. *Biol. Bull* **137:** 359–374.

Sugiki, T. and S. Masaki. 1972. *Kontyu* **40:** 269–278.

Sunose, T. 1978. *Kontyu.* **46:** 400–415.

Tanaka, S. 1978. *Kontyu* **46:** 207–217.

Tauber, C. A. and M. J. Tauber. 1977. *Nature* **268:** 702–705.

Tauber, C. A. and M. J. Tauber. 1981. *Annu. Rev. Ecol. Syst.* **12:** 281–308.

Tauber, C. A., M. J. Tauber, and J. R. Nechols. 1977. *Science* **197:** 592, 593.

Tauber, M. J. and C. A. Tauber. 1970. *Science.* **167:** 170.

Tauber, M. J. and C. A. Tauber. 1973a. *Search. (Agric.) Cornell Univ. Agric. Exp. Sta., Ithaca, N.Y.* **3:** 1–16.

Tauber, M. J. and C. A. Tauber. 1973b. *J. Insect Physiol.* **19:** 729–736.

Tauber, M. J. and C. A. Tauber. 1973c. *J. Insect Physiol.* **19:** 1455–1463.

Tauber, M. J. and C. A. Tauber. 1976a. *Annu. Rev. Entomol.* **21:** 81–107.

Tauber, M. J. and C. A. Tauber. 1976b. *J. Insect Physiol.* **22:** 331–335.

Tauber, M. J. and C. A. Tauber. 1977. Pp. 639–642 in *Proc. 12th Int. Conf. Int. Soc. Chronobiology*, The Publishing House, Il Ponte, Milan.

Tauber, M. J. and C. A. Tauber. 1978. Pp. 53–71 in Dingle (Ed.), 1978b, referenced here.

Tauber, M. J. and C. A. Tauber. 1979. *Bull. Entomol. Soc. Am.* **25:** 125–128.

Tauber, M. J., C. A. Tauber, and C. J. Denys. 1970. *J. Insect Physiol.* **16:** 949–955.

Tauber, M. J., C. A. Tauber, J. R. Nechols, and R. G. Helgesen. 1982. *Science* **218:** 690, 691.

Tauber, M. J., C. A. Tauber, J. R. Nechols, and J. J. Obrycki. 1983. Pp. 87–108 in Brown and Hodek (Eds.), 1983, referenced here.

Thiele, H. U. 1966. *Z. Angew. Entomol.* **58:** 143–149.

Toshima, A., K. Homma, and S. Masaki. 1961. *Jap. J. Appl. Entomol. Zool.* **5:** 260–269.

Tsuji, H. 1966. *Appl. Entomol. Zool.* **1:** 51.

Tuskes, P. M. and L. P. Brower. 1978. *Ecol. Entomol.* **3:** 141–153.

Urquhart, F. A. and N. R. Urquhart. 1979. *Can. Entomol.* **111:** 15–18.

Vepsäläinen, K. 1974a. *Hereditas* **77:** 163–176.

Vepsäläinen, K. 1974b. *Nature* **247:** 385, 386.

Vinogradova, E. B. and K. B. Zinovjeva. 1972. *J. Insect Physiol.* **18:** 1629–1638.

Visscher, S. N. 1976. *Cell Tiss. Res.* **174:** 433–452.

Visscher, S. N. 1980. *Experientia* **36:** 130, 131.

Waldbauer, G. P. 1978. Pp. 127–144 in Dingle (Ed.), 1978b, referenced here.

Wallace, M. N. H. 1970. *Aust. J. Zool.* **18:** 295–313.

Wellso, S. G. and P. L. Adkisson. 1966. *J. Insect Physiol.* **12:** 1455–1465.

West, D. A., W. M. Snellings, and T. A. Herbek. 1972. *J. N.Y. Entomol. Soc.* **80:** 205–211.

Wigglesworth, V. G. 1972. *The Principles of Insect Physiology* (7th ed.). Methuen, London.

Williams, C. M. 1952. *Biol. Bull.* **103:** 120–138.

Williams, C. M. 1956. *Biol. Bull.* **110:** 210–218.

Williams, C. M. and P. L. Adkisson. 1964. *Biol. Bull.* **127:** 511–525.

Woolhouse, H. W. (Ed.). 1969. *Dormancy and Survival, 23rd Symp. Soc. Exp. Biol.* Academic, New York.

Yagi, S. and M. Fukaya. 1974. *Appl. Entomol. Zool.* **9:** 247–255.

Yin, C.-M. and G. M. Chippendale. 1976. *J. Exp. Biol.* **64:** 303–310.

Chapter 7

Insect Movement:

Mechanisms

and Effects

MARY ANN RANKIN and
MICHAEL C. SINGER

7.1 INTRODUCTION

Study of insect movement impinges on many disciplines. Applied entomologists are interested, for example, in techniques for measuring or predicting movement; the data obtained may be used to locate sources of
disease vectors or to identify areas and crops that require pesticide treatment. Students of behavior may concentrate on foraging for pollen, nectar
or oviposition sites, or they may study or model behavioral decision-making
processes such as when to move, what direction to take, and when to stop.
The neural and hormonal mechanisms that underlie these decisions are the
domain of the physiologist. Situations especially amenable to study are those
in which animals can be clearly categorized as migratory or nonmigratory
and where hormonal mechanisms integrate the behaviors of the insect and
bring its various organ systems into a condition appropriate for migration.
Many environmental variables have been implicated in the induction of
long-range movement of migrant forms. Photoperiod, population density,
food quality, and weather patterns may all be involved (cf. Chapters 6, 13).
This is true whether each insect moves repeatedly or migrates only once
and whether differences between individuals which move and those which
do not are heritable or environmentally induced.

Ecologists and evolutionary biologists are interested in movement as an
adaptation resulting from (or at least correlated with) temporal and spatial
patterns of habitat suitability (cf. Chapter 14). This approach may involve
attempts to quantify the "cost" of migration in terms of lost reproduction
and hence to specify the circumstances under which migration would occur,
if the organisms were maximizing their fitness. Reciprocal interactions between movement and local population density have also received attention.

Population geneticists have been concerned with several aspects of movement. Genetic differences in tendency to move may constitute polymorphisms. Movement often results in gene flow, and gene flow may counteract
tendencies towards the development of local genetic differences. Patterns
of gene flow also affect modes of action and consequences of natural selection, and the extent to which group selection or kin selection operate
will influence many characteristics of the insect, especially behavioral ones
(cf. Chapter 9).

Many authors have argued about definitions of insect migration. We
avoid this controversy by addressing our discussion to insect movement.
This term includes a wide variety of phenomena from short-distance foraging to long-distance migration, which can comprise a separate phase in
the life cycle. Yet much of the work in this area has investigated insect
migration in species where it is a clearly defined, recognizable syndrome
(Chapter 6). We attempt here to specify when this type of migration, as
opposed to general movement, is the topic of discussion.

7.2 MEASUREMENT OF MOVEMENT OR TENDENCY TO MIGRATE

7.2.1 Analyses of Laboratory Behavior

Migrants have been identified by laboratory flight testing in several species of insects (Kennedy & Booth 1964, Dingle 1965, M. A. Rankin & S. A. Rankin 1980). Monitoring duration of tethered flight is a simple procedure which works fairly well among insects such as Hemiptera and Coleoptera, which generally move about by walking and fly only briefly or occasionally unless they are migrating. Among other insects, such as Diptera and Leptidoptera, that typically use flight as their primary form of locomotion, duration of tethered flight may not be a reliable indication of migratory behavior; that is, analyses are often difficult unless a clear dimorphism in flight activity exists within the population (M. A. Rankin & S. A. Rankin 1980) or unless other criteria besides duration of flight can be measured. For each species whose migratory behavior is being investigated using a tethered flight test, it is important to establish that such a test will identify migrants in the field. It is desirable to use, when possible, criteria such as propensity to take off, response to host plants or mates, and behavior in a free-flight system, in addition to duration of tethered flight.

Behavioral analysis of migratory behavior as compared to nonmigratory or "trivial" flight has been carefully done by J. S. Kennedy and co-workers (e.g., Kennedy 1956, 1961, Kennedy & Booth 1964, Kennedy & Ludlow 1974). Kennedy (1961) observed that migratory flight could be distinguished from appetitive flight on the basis of response to "vegetative" stimuli such as feeding, mating, or oviposition cues. On the basis of his studies on aphid flight behavior, Kennedy characterized migration as "persistent, straightened-out movement that is accompanied by and dependent upon the maintenance of an internal inhibition of those vegetative reflexes that will, eventually, arrest movement." In the black bean aphid, *Aphis fabae*, stimulating one of two antagonistic reflex systems (flying and settling) inhibited the other, and this inhibition left an aftereffect (termed *postinhibitory rebound*) on the system that had been inhibited. When the aftereffect of temporary inhibition was a net strengthening, the process was called *antagonistic induction*. When it was a net weakening, it was called *antagonistic depression*. These processes appeared to be central and with repetition could be cumulative until one activity came to replace the other entirely.

Because the completely antagonistic relationship between flight and settling in the aphid is not typical of the relationships between many of the behavioral activities of aphids, Kennedy and Ludlow (1974) analyzed two more closely related activities. Migratory flight was measured by flight upward toward a light source while the horizontally directed approach response of the flying aphid to a leaf-shaped yellow card measured targeted flight. The results of these experiments were similar in many respects to those of earlier work, again suggesting an antagonistic interaction between migratory and nonmigratory reflexes. The longer an aphid had flown before the target appeared, the more persistently it homed in on the target. This effect could be counteracted by repeated presentation of the target without allowing the aphid to settle on it. However, the visual stimulus from the target could be either excitatory or inhibitory depending usually on antecedent events in the behavioral chain. The excitatory effect generally followed an inhibitory one, but it could occur first, or even alone, and was thus a separate effect, not comparable with postinhibitory rebound. Kennedy and Ludlow (1974) suggest that such change in responsiveness is governed by antecedent reflex activity beyond the antagonistic induction and depression, implying a more complex central processing than they had previously envisioned.

7.2.2 Mark–Recapture Studies of Insect Movement

Insect movement has been studied at the population level in the field using a variety of techniques. One method of measuring distance traveled, and estimating numbers of migrants, involves recapture of marked insects at known distances from the site of initial marking capture. By numbering individuals, the migration paths of the monarch butterfly, *Danaus plexippus*, have been elucidated (Urquhart 1960, Urquhart & Urquhart 1977). A range of marking techniques is available (Southwood 1978). Where individuals are uniquely marked, movement within populations can be studied. Mass marking can be used to treat large numbers of small organisms to give information on rates of dispersal from one or more release points and rates of exchange of individuals between adjacent areas. Dobzhansky and Wright (1943) used genetically marked *Drosophila pseudoobscura* dispersing from a central release point and recaptured in traps set at different distances. Such results can be compared with those which would result from random dispersal.

The capture or marking technique may itself affect the patterns of movement being studied. Southwood (1978:328) refers to examples in a grasshopper and a carabid beetle. Singer and Wedlake (1981) compared recapture rates of the Asian butterfly *Graphium sarpedon* following two different marking procedures, only one of which involved capture of the insects. Not surprisingly, those which were marked without being captured (by creeping up on them) were later recaptured at significantly higher rates than those which had been captured, marked, and released in traditional fashion. In this context there is a distinct advantage in techniques such as used by Richardson and Johnston (1975) for *Drosophila* which were attracted to baits and marked as a result of feeding on the baits, without any handling.

Some authors have used marked insects to study rates of exchange between areas of concentration (Richards & Waloff 1954, Ehrlich 1961). Such studies show a diversity of population structures; some insects have very discrete populations with little exchange over short distances, while others with apparently similar flight capacity may be much less reticent to cross areas of unsuitable habitat and, in consequence, show more open population structures (Brussard & Ehrlich 1970). This information can also be obtained by following individuals and plotting their tracks. For organisms with small, discrete populations, the restricted nature of their movement is thereby revealed, while following individuals of more mobile species gives tracks which may be integrated to elucidate the overall pattern of movement (Baker 1978, Jones 1977).

The results of most mark–release studies give estimates of population gains and losses in the areas studied. In order to know how movement contributes to the dynamics (Chapters 11,12), we must segregate immigrations from births among the "gains" and emigrations from deaths among the "losses." This is usually impossible. Perkins (1980), for example, discusses disagreements among entomologists over the results of a pilot attempt to eradicate the boll weevil, *Anthonomus grandis,* in Mississippi. Some weevils appeared in the test area following "eradication." Without knowledge of the source of these insects, it was impossible to judge whether or not eradication of the weevils present during the treatment had been achieved. This must be a frequent problem with evaluation of pest control measures. However, there are sometimes means whereby helpful information on immigration and emigration might be measured with directional traps oriented perpendicularly to the boundaries of the study area.

Where migrants are distinguished from nonmigrants by possession of fully developed wings, this distinction may be already evident from the wing buds borne by immature stages. The categorization of immatures as presumptive alatae is helpful in assessing causes of population size change. Alatae appearing where no presumptive alatae have been censused, for example, are probably immigrants. Population crashes are more likely to represent mass emigration and less likely to have resulted from predation if a high proportion of nymphs bore wing buds just before the "crash." However, alate forms do not always emigrate (Shaw 1970). The quantification of emigration in species in which migrants are *not* physically distinguishable before they leave is more difficult than that of immigration, since once the migrant has left the study area, it usually cannot be examined. When migrants cannot be physically distinguished from nonmigrants, indirect evidence may still be obtained. Baltensweiler (1970) identified immigrant European spruce budworm moths both from the timing of their appearance relative to that of the local larvae and from their lack of aggregation behavior when caught in pheromone traps.

For species in which diffusion models of dispersal are appropriate, some measure of emigration can be made by examining the rate of decrease of recapture frequency with increasing distance from the release point and extrapolating to calculate the number which traveled further than the greatest distance sampled (Southwood 1978). However, this will not work with species that have discrete population structure and patchily distributed habitats or a marked dimorphism in flight behavior in the population. Most individuals in such populations may travel only a few meters in their lives, yet the few which emigrate may travel great distances. The distribution of distances traveled is likely to be bimodal and difficult to estimate by extrapolation of data obtained by releasing insects at a point source.

McLean et al. (1979) showed that the proportion of metal ions in plant populations and the insects which feed on them may be diagnostic of locality on a small enough scale that immigrant insects can be recognized from the ion balance, measured by neutron activation analysis, resulting from their development as larvae in some other locality. It may even be possible to identify the source of immigrants by this technique. Where new populations are founded in rapid sequence, the source may be identified from the loss of rare alleles in repeated founding events (Pashley & Bush 1979).

Large groups of flying insects have also been tracked by radar or observed from aircraft (e.g., Rainey 1974). Detection and tracking with radar has occasionally yielded useful information. Rainey (1978) and Schaefer (1979) successfully used radar, along with aircraft trapping and meteorological studies, to monitor insect movements with respect to movements of weather fronts. However, although radar is potentially very useful for such work, it has some limitations and difficulties. The data obtained are often somewhat ambiguous since identification to species of radar-detected insects is usually not possible. At low densities object size, shape, and wing-beat frequencies can be determined by radar, and these parameters can be used to determine only very generally the type of insect being observed. Successive echoes from the target may also be used to determine ground speed and direction of movement. If wind speed and direction are known, true target heading and air speed can be determined. At high densities, however, these individual measurements are more difficult to make. In combination with aircraft sampling, probable species identification can be made, but this is expensive and may not be practical. At the high sensitivities necessary for detection of small insects by radar, interference by ground clutter is a serious problem and can greatly limit resolution. Most commercially available radars have not been designed for use on targets as small as insects and thus provide only limited resolution, target identification, and detection range. Future research would greatly benefit from development of radar equipment specifically designed for insect tracking.

7.2.3 Broad-Range Surveys Using Traps

Long-term data on changes in density and distribution of over 500 moth species and over 200 aphid species have been gathered by the Rothamstead Insect Survey (RIS) (Taylor 1979). This unique set of data has been used to produce consecutive density and distribution maps for the whole of Great Britain. The data are from light traps and suction traps at standard heights above the ground. In one case, that of the hop aphid *Phorodon humili*, the maps are consistent with dispersion from localized areas which can be regarded as point sources (Taylor et al. 1979). From these data it was possible to estimate the mean distance traveled as 15–20 km and the probable limit as 100–150 km. However, maps for other species do not contain point sources and are not amenable to this analysis. Rather, they show patches of high density, low density, and apparent absence, which appear and then disappear again from year to year. Rates of change of distribution and the steepness of density contour ("roughness") are characteristic of such species. Taylor and Taylor (1977) suggest there is some predictability in these patterns, in that for each species the spatial variance of density has a characteristic value for each mean density level—a value which returns each time the same mean density returns. They argue that for this to be the case, there must be interaction between the densities in different places, and that this indicates large-scale movement of the insects. They note that classical population dynamics has concentrated on changes in natality and mortality as the causes of changes in density within a study area, neglecting effects of movement. They argue that study areas infrequently constitute isolated populations and that their data implicate patterns of movement, rather than internal changes in mortality and natality, as factors largely responsible for the documented patterns of density and distribution. The authors regard movements as density-dependent, and as direct or indirect responses to conspecific individuals, with aggregating tendency predominating at low density and dispersal at high density. They propose that these density responses provide a means of population control by density-dependent movement which results in "avoidance of overt competition."

This work is important since the evidence is unique, and the deductions made from it suggest a greater importance of movement as a cause of observed patterns of density and distribution in space and time than has commonly been recognized. But one must be cautious in extending these conclusions to insects generally, for several reasons. First, the data are gathered from a nonrandom group of insects. Light traps are often placed in suburban areas, and the moths captured are either characteristic of disturbed habitats or are outside their normal habitat. In the latter case it is not surprising that wide fluctuations in density occur, since absence in the samples may simply reflect lack of migration rather than absence in the geographical area. The tendency to sample organisms characteristic of disturbed habitats is accentuated by the location of this work in Great Britain, where there are no extensive areas of natural habitat.

The second group of species, alate aphids in flight, sometimes comprises whole populations engaged in compulsory host alternation, but at other times consists of winged forms produced in response to various environmental factors. In the latter case the density of alates bears an unknown and perhaps complex relationship to the total density, since wingless (apterous) nonmigratory forms are not susceptible to suction trapping. The known proximate factors which trigger aphid emigration sometimes include density-related factors but are also frequently unrelated to density. Photoperiod triggers production of alate gynoparae of many aphids, and food quality or quantity changes are also important (cf. Chapter 4). Sometimes food shortage results from consumption of food by conspecifics, but at other times may be entirely unrelated to intraspecific competition (e.g., Singer 1972). Clearly, for many insects, tendency to move increases with increasing density, but this relationship is not universal and may even be reversed, as in the San Francisco Bay Area ecotype of the butterfly *Euphydryas editha*, (Gilbert & Singer 1973). It is also clear from measured changes in population density and movement that, for these populations of *E. editha* which are separated by only a few meters, interpopulation movement makes a negligible contribution to population dynamics. The sole exception is the recolonizing of areas where the insect has become temporarily extinct. These studies (Ehrlich et al. 1980) show that population extinction events, which have been observed in at least five *E. editha* populations, result from death of the insects rather than from their emigration. For species such as *E. editha* (and how common they are we do not know), the classical view of changes in density being the result of changing local patterns of natality and mortality seems closer to being correct.

We disagree with Taylor and Taylor's (1977) interpretations in three further respects. First, they overextend their conclusions: their graph of monarch butterfly movement against density shows that these insects move 3000 km away from a center of abundance in response to a density of 130 units; 1600 km in response to a density of 100 units; 400 km in response to 80 units; and that at low densities they reverse the trend and move toward centers of abundance. We wonder how they manage to arrive at the same overwintering sites in years when mean abundance is different. Second, the conclusion that spatial behavior is density-dependent does not follow from the relationship between spatial variance and density. If mortality were more variable at high than at low densities (which might occur), the same phenomenon would be generated. Third, they regard movement as a population-regulating force. The validity of this argument depends greatly on scale. For some of the highly mobile species studied, the "population" could be considered to occupy the whole of Great Britain, and the patterns being studied would comprise intrapopulation movements.

We have discussed some techniques that may identify immigrants and help apportion population gains between births and immigration. The role of migrants in subsequent population increase often depends on their time of arrival, as in the bushfly (Hughes & Sands 1979, and Chapter 11).

7.3 MECHANISMS BY WHICH DENSITY OR QUALITY OF FOOD MAY AFFECT MOVEMENT

Density and food quality are considered together because they have often been studied together, and it is frequently difficult to distinguish their effects. Clearly, crowding is a major stimulus to movement in many species. However, whether the effect of high density is direct or via deterioration or depletion of food has been of interest to many investigators of mechanisms of induction of migrants. Effects of high density and/or quantity and quality of food on movement or formation of migrant morphs have been studied in many insect groups. We focus on three of these: the Orthoptera, Aphididae and Heteroptera.

7.3.1 Orthoptera

Certain locust species periodically undergo great increases in population density, accompanied by swarm formation and long-distance migration. In some species, including *Schistocerca gregaria* and *Locusta migratoria,* changes in density are also associated with phase transformation. There are two extreme forms or phases, *solitaria* and *gregaria,* which differ in physical and behavioral characteristics. *Gregaria* nymphs are usually more conspicuously colored, larger, and more active than *solitaria* nymphs. *Gregaria* females average fewer ovarioles and fewer egg pods than *solitaria* females; *gregaria* locusts are diurnal migrants; *solitaria* migrate only at night, if at all (Uvarov 1966). Ellis (1953) found that nymphs reared under crowded conditions aggregate with one another or with decoys preferentially, while nymphs reared in isolation distribute themselves away from other locusts or locustlike decoys. Crowded *Schistocerca* nymphs develop faster than isolated ones, and the presence of mature males can accelerate maturation of younger locusts (Loher 1961, Highnam & Lusis 1962).

Acceleration of maturation appears to involve a volatile lipophylic pheromone transmitted to recipients through antennal and bodily contact (Norris, 1964). More recently Nolte et al. (1973) reported that crowding increases frequency of chiasmata (Meiotic crossing-over) in certain male locusts. This is apparently due to a pheromone in the feces (locustol: 2-methoxy-5-ethylphenol) which is a degredation product of ingested lignin. It is not clear whether or not locustol is the maturation accelerating hormone of Norris (1964) and Loher (1961). Synthetic locustol does affect pigmentation and behavior, but the possibility of there being two *gregaria*-inducing pheromones has not been ruled out.

The role of high density in locust behavior is clearly very complex but has the ultimate effect of synchronizing development within the population. The question of what environmental parameters cause the initial increase in density remains controversial. It is generally accepted that weather, particularly drought, and food quality are important in such changes in abundance, but the mechanisms of these effects are still unclear (Gunn 1960, Dempster 1963, Johnson 1974). For several species of plague locusts in Africa and North America, such as *Nomadacris septemfasciata*, *Locustana Pardalina*, *Locusta migratoria*, *Melanoplus sanguinipes*, and *Camnula pellucida*, outbreaks seem to follow periods of drought (or occasionally periods of flooding) (White 1976, Pickford 1966, Symmons 1960). These are generally times in which the locusts might be supposed to be under food stress. Yet a wide variety of plant species, particularly grasses, show increases in percentage of nitrogen, especially soluble amino acids, in their aerial parts when under water stress (e.g., Kimble & MacPherson 1954, Wiggins & Williams 1955). White (1976) suggests that locust populations build up when nutrients, particularly nitrogen, in the food occasionally reach levels that support many more early-instar hoppers than is usual in conditions less stressful to food plants. Lewis (1981, 1979) showed that all developmental stages of the grasshopper *Melanoplus differentialis* prefer wilted to turgid sunflowers as food, and that those given access to wilted leaves grow faster and have higher survival and fecundity. If this effect occurs in locusts, it may well underlie the correlation between drought and population increase. Lewis' insects did not respond to nitrogen levels in their hosts.

One of the major plague locusts which does not seem to follow the pattern of migration following drought is *S. gregaria*. Outbreaks are correlated with recent high rainfall, but the habitat is so very arid that only at times of peak rainfall is there sufficient water to support the plants necessary to feed a large population of young hoppers (White 1976). White suggests that the effects of rainfall on plant quality are so ephemeral that even after heavy rain most individuals feed on drought-stressed vegetation at some time in their preadult lives. Symmons and McCulloch (1980) report that migrations of Australian plague locusts, *Chortoicetes terminifera*, also occur following rain. If young adults feed on green food, they subsequently emigrate; if not, they are likely to remain and reproduce. No inference is made about the effects of green vegetation on the density of the insects nor on the effects of density on migration in this species.

In many insects food deprivation or severe shortage increases flight activity for at least a brief period. Hoyle (1954) and Ellis and Hoyle (1954) were able to relate this response in *S. gregaria* and *L. migratoria* to a decrease in hemolymph potassium with starvation or feeding on senescent vegetation. Hoyle suggested, on the basis of experimental manipulations of dietary K^+ levels, that starvation or change in the K^+ content of the diet might influence behavior by a direct effect of K^+ on muscle fiber activity. This argument assumes the postsynaptic membrane of the muscle is unprotected from fluctuations in hemolymph potassium. Work by Njio and Piek (1977) suggests that within the lumen of the transverse tubular system of insect muscle (the effective postsynaptic membrane area) relatively high Na^+ and low K^+ levels are maintained, in contrast to high K^+ levels in the intercellular spaces and hemolymph. Thus, Hoyle's idea that, while nerve trunks are protected from high extracellular K^+ concentrations by their nerve sheaths, muscles are exposed may not be entirely accurate. Also, Chapman (1958) examined K^+ levels in fed and hungry *Nomadencis* adults and found that they varied only slightly and were not correlated with locomotor activity. Chapman's experiments suggest that the high concentrations of K^+ used by Hoyle may have been toxic to the locusts. If so, a drop in locomotor activity after ingestion of such a diet would not necessarily indicate an effect of this ion on locomotor behavior. There are many ways that insects obtain information concerning the food they ingest other than hemolymph potassium levels, and it is likely that starvation exerts its effect on locomotor activity in a less direct way in most insects.

7.3.2 Aphididae

Both food quality and population density affect production of nonsexual winged forms (alates) among aphids. Often the separate effects of these two factors are difficult to distinguish, since high density rapidly affects food quality. However, Lees (1961) was able to investigate effects of density independently of food quality by growing only one aphid on each plant and removing them from their hosts temporarily for the density treatments. "Crowded" *Megoura viciae* were placed in tubes together (2/tube) for several hours while control insects were similarly used singly. After replacement on their hosts (again only 1/plant) those which had been temporarily crowded had a higher tendency to produce alates than did the controls. The stimulus appears to be mainly tactile because a gauze barrier between two aphids inhibits the response and jostling by cockroaches will elicit it (Lees 1967). Similar effects of crowding have been shown in *Aphis craccivora* (Johnson 1965) and *Dysaphis devecta* (Forrest 1974). Forrest maintained food quality at different aphid densities by using an artificial diet.

Observations on *Elatobium abietunim* in Scotland showed that the peak number of alates produced, but not the time of their first appearance, was density related (Parry 1978). Carter and Cole (1977) suggest that alate production coincides with amino acid changes in Sitka Spruce needles, the primary host plant. Parry (1978) also showed that amino acid levels were higher during mid-May when alate production was greatest than in mid-June when no alates were being formed. However, these workers did not induce alate formation by manipulating amino acid levels so a cause-and-effect relationship, though a possibility, remains unproven. In *Macrosiphon pisum* (Schaefer 1938) and *Rhopalosiphum prunifoliae* (Noda 1956) starvation increased the proportion of alates. The same treatment increased alate production in *Aphis craccivora* (Johnson 1966) but had no effect in *Therio-aphis maculata* (Paschke 1959) or isolated *Megoura viciae*. However, when apterous *M. viciae* were kept in pairs, they produced more alatae on senescent than on young leaves (Lees 1966). A direct effect of food has been indicated in *Eucalliptus tiliae* in which the same proportion of fliers developed from both crowded and isolated aphids kept on the same leaf (Kidd 1977).

Several other factors can influence alary polymorphism in aphids. Increase in temperature suppresses alate formation in both isolated and crowded aphids (Lees 1967, and Chapter 6). Colonies of *Aphis fabae* maintained on artificial diet contained a higher proportion of apterae when tended by ants, possibly due to effects of ant mandibular gland secretions on the aphids (Kleinjan & Mittler 1975). Maternal effects are also observed. Alate parthenogenetic aphids are usually less likely to have alate offspring than are apterous forms (Lees 1966, Sutherland 1970). However, alates which have not flown may not differ from apterae in their tendency to produce alates (Burns 1972). This tendency may arise only gradually over several generations among the offspring of a particular alate aphid (MacKay 1977, MacKay & Wellington 1977).

7.3.3 Heteroptera

Food quality and quantity have been shown to be important in migration of several heteropterans, as in Dingle and Arora's (1973) study of migratory behavior in three species of African *Dysdercus*. The flight system of males of all three species matures 4–6 days after adult emergence, and flight can continue throughout life. Fed females of *D. fasciatus* do not fly but, rather, undergo flight muscle histolysis and ovarian development. Fed females of *D. nigrofasciatus* and *D. Superstitiosus* exhibit some flight at 4 days after adult eclosion, but at 6 days flight is essentially absent and flight muscle histolysis and oogenesis are in progress. Unfed females of all three species display considerable flight and do not undergo flight muscle histolysis or ovarian development. In unfed pairs copulation is reduced, permitting increased migration. The three species are thus facultative migrants using lack of food as a releasing cue. A similar situation exists in the lygaeid *Neacorphis bicrusis* in which prereproductive starvation enhances flight. Cessation of flight in the presence of ample food is associated with flight muscle histolysis such that postreproductive starvation cannot elicit further flight (Solbrek & Pehrson 1979).

Feeding on seeds is necessary for the migratory milkweed bug, *Oncopeltus fasciatus*, to complete oogenesis. Starvation or poor quality food (green plant material) results in delayed reproduction and increased migratory activity. With continued starvation the increase is transient and flight eventually declines to very low levels but can be restimulated by treatment with juvenile hormone (Rankin 1974). Even if it occurs after the initiation of reproduction, starvation stimulates a second brief period of flight (Dingle 1968, Rankin & Riddiford 1977).

7.4 EFFECTS OF PHOTOPERIOD ON MOVEMENT

In some insects migratory flight or development of migrant morphs is triggered by changes in photoperiod, often in conjunction with other environmental cues. In laboratory flight tests the duration of tethered flights in *Oncopeltus fasciatus* populations reared under long-day photoperiods is greatest 8–12 days after adult eclosion (Dingle 1966). Since the teneral period includes the first 5–6 days of the adult stage and oviposition begins about day 15 under long-day conditions, migratory flight occurs during the postteneral, prereproductive period. In this respect *Oncopeltus* is typical of migrants in which long-range movement is not interspersed with foraging (Dingle 1972, Johnson 1969). An oogenesis-flight antagonism is apparent in this species, that is, oogenesis is inhibitory to migratory flight. Photoperiod, temperature, and food quality and availability influence the amount of migratory behavior displayed by an *Oncopeltus* population. When reared in short photoperiods (12L : 12D), it undergoes an adult diapause (Dingle 1974). The delay in reproduction that results provides more time prior to egg maturation during which females may be stimulated to make long flights. It also results in a greater percentage of both sexes making a long flight (Caldwell 1974, Rankin 1974). The effect of photoperiod on reproduction and flight is mediated via the endocrine system in this species (Rankin & Riddiford 1978).

In a migrant coccinellid, *Hippodamia convergens*, the effect of photoperiod on flight behavior is more complex. This species migrates long distances to mountain aggregation sites in early or midsummer and remains there until early spring, at which time a return flight to aphid-infested fields occurs. Migratory flight is prereproductive and photoperiod seems only to influence the duration of the long flight phase when food quality is optimal. When this is so, short photoperiod greatly lengthens the period of time in which beetles will make long flights. When the animals are starved or food quality is poor under either photoperiod regime, flight activity is enhanced to quite high levels (M. A. Rankin & S. A. Rankin 1980).

Another species which shows effects of photoperiod on migration and adult diapause is the monarch butterfly, *Danaus plexippus*. Populations in northern North America undergo dramatic migrations. They fly up to 2000 mi. from Canada and the northern United States overwintering sites in southern California, Florida, and Mexico, then migrate north the following spring. The southward flight occurs in large aggregations at a leisurely pace, moving with the weather fronts. They fly primarily in the middle of the day and accumulate in roosts overnight and during bad weather. Migrators replenish their fuel reserves by feeding on nectar. Most fall migrants are thought to be newly emerged individuals in an adult reproductive diapause induced by short fall photoperiods and cool temperatures (Herman 1973, Urquhart 1960), although some oviposition does occur along the migration route where weather and nectar conditions are favorable.

Less is known about the monarch's northern flight. It is said to be more rapid; they apparently fly alone at night as well as in the day, resting singly rather than in aggregations, stopping to feed less often and for shorter periods, using the fat stored prior to their winter diapause (Urquhart 1960, Johnson 1969). Some oviposition has been noted along the route of the spring migration. It is not known whether gravid females terminate migration when they begin oviposition or continue on with the nongravid individuals. Populations in southern Florida appear to reproduce continuously (Brower 1962). The nature of the temporal relationships between reproduction and migration has long been debated and is important to understanding migration physiology in the species and to explaining apparent differences between spring and fall migrants.

Photoperiod effects on alary polymorphism have also been observed. Vepsalainen (1971) concluded that alary dimorphism and diapause in Finnish *Gerris odontogaster* are determined by a photoperiodic switch mechanism which directs development toward either a short- or a long-winged morph. Honek (1974) showed that apterous or brachypterous adults of *Pyrrhocoris apterus* are induced by short, while macropters are induced by long, photoperiods. In aphids production of the winged sexual morph (gynopara) is also induced by changes in photoperiod.

7.5 HORMONAL CONTROL OF MIGRATORY BEHAVIOR AND ALARY POLYMORPHISM

7.5.1 Migratory Behavior

At least some of the environmental parameters that affect flight behavior do so by way of the neuroendocrine system. However, a diversity of physiological mechanisms exists among the species investigated. In several insects juvenile hormone (JH) has been shown to induce flight muscle degeneration (e.g., Dingle & Arora 1973, Edwards 1970, Davis 1975, Borden & Slater 1968, Chudakova & Bocharova-Messner 1968). In *Leptinotarsa decemlineata*, however, JH is necessary for flight muscle regeneration after diapause (DeWilde & DeBoer 1969). Juvenile hormone titer analyses (DeWilde et al. 1968, DeKort et al. 1982) indicate that migration to and from the diapause site may be stimulated by intermediate titers of JH below the threshold for ovarian development but sufficient to stimulate flight muscle regeneration and migratory flight. Similarly, in *Melolontha melolontha* activity of the corpus allatum (CA) affects orientation of flight (Stengel 1974).

Several hormones are implicated in control of migratory behavior in locusts. Compared to *gregaria, solitaria* hoppers have larger prothoracic glands which persist a longer time in newly emerged adults (Carlisle & Ellis 1959). When prothoracic gland homogenates were injected into gregarious hoppers, marching was decreased (Carlisle & Ellis 1963). Thus at least one physiological difference between solitary and gregarious phase locusts might be the level of activity of this gland. Ecdysone caused a marked increase in ventral ganglion interneuron firing and a marked decrease in motor output to the extensor tibialis muscle of the metathoracic leg (Haskel & Moorehouse 1963). Moreover, implants of prothoracic glands from solitary adults into gregarious adults decreased duration of tethered flight of the host (Michel 1972). However, differences in flight behavior were not striking, no sham implants of other tissues were done, and since a decrease in activity could be caused by many factors, including injury, the results are inconclusive.

The corpora cardiaca (CC) release an adipokinetic hormone (AKH) which is necessary to mobilize fat body lipid reserves for flights longer than about 15 min in *L. migratoria* and *S. gregaria* (Goldsworthy et al. 1972, Mayer & Candy 1969). The hormone is produced by the glandular lobes, and its release, as flight begins, is governed by a double innervation from the brain via the NCCI and NCCII (Goldsworthy et al. 1973). During the initial 5–10 min locust flight speed is high but declines markedly for about 25 min until a "cruising" speed is attained. Weis-Fogh (1952) suggested that during the initial high-speed flight, carbohydrate is the predominant fuel, and the subsequent decrease in speed corresponds to a gradual switch over to utilization of lipid, a process presumably dependent upon AKH from the CC. Recent work has revealed existence of a second hyperlicemic hormone and a hypolicemic hormone in locust CC which presumably act in conjunction with AKH (Pines et al. 1981).

In addition to ecdysone and AKH, JH has been implicated in control of locust flight. Implanting CA into mature male *L. migratoria* results in increased walking speed associated with a stronger phototactic response (Cassier 1964). Allatectomized male *S. gregaria* show decreased locomotor activity, which can be reversed by CA replacement therapy (Odhiambo 1966). Flight activity does not decrease in locusts allatectomized when mature (Goldsworthy et al. 1972). Allatectomy of immature male locusts retards development of normal flight capability, possibly because JH is necessary for normal flight muscle development. Thus for a short period after operation, allatectomized locusts fly poorly compared with operated controls. Subsequent flight performance, however, is superior (i.e., flight speed on a roundabout is greater) to that of sham-operated controls. In females allatectomy increases flight speed at all ages (Lee & Goldsworthy 1976). It would be useful to determine whether JH treatment of controls would stimulate or inhibit flight. Further, given Weis-Fogh's observation that migratory flight is characterized by a drop in speed, it is not clear that increase in flight speed on a roundabout indicates an increase in migratory behavior.

More work is necessary before a firm statement can be made as to the roles of JH, adipokinetic hormones, and ecdysone in the control of locust migratory behavior.

Another migrant in which hormonal control of migratory behavior has been investigated is the milkweed bug, *O. fasciatus*. In this species both migratory behavior and oogenesis are JH-dependent. Low and intermediate titers of JH are correlated with the migratory period and higher titers with oogenesis. CA activity seems to be under photoperiodic control such that a prolonged period of intermediate JH titers, reproductive diapause, and migratory behavior occur in response to short photoperiods. Long photoperiod results in a rapid rise in JH titers to high levels, rapid onset of oogenesis, and little migratory behavior. Ovarian development appears to be inhibitory to migratory behavior (Rankin 1974, Rankin & Riddiford 1978). Application of JH mimic or implants of CA significantly increased flight behavior of males and ovariectomized females but decreased flight activity of intact females, due to induction of precocious ovarian development. Treatment with precocenes (Chromene derivatives which produce chemical allatectomy in this species) caused a significant decrease in migratory behavior of both sexes. JH replacement therapy restored flight activity and stimulated ovarian development. As oviposition began, flight activity again declined among JH-treated females. Neither CC implants nor injection of CC extract had any effect on flight. Flight behavior of precocene-treated animals improved within 1 hr after JH injection, indicating a rapid effect of the hormone on some aspect of the flight system. Recent evidence indicates that *O. fasciatus* JH may act to mobilize lipid fuel for long-duration flights (Rankin 1980, 1981).

Topical application of JH mimic to *Hippodamia convergens* also stimulates increased long-duration flight activity in both sexes. Precocene II inhibited flight as well as ovarian development, and JH replacement therapy restored both to control levels or above, indicating a JH effect on flight similar to that observed in *O. fasciatus* (S. A. Rankin & M. A. Rankin 1980).

7.5.2 Alary Polymorphism

In many species migrant individuals are not only behaviorally but also morphologically distinct from their nonmigrant conspecifics. Alary polymorphism is present to some extent in all pterygote insect orders but has been investigated in detail in relatively few. The physiological and morphological bases and their relevance to flight behavior is best known in the Hemiptera, and we therefore emphasize this group. Southwood (1961) suggested that brachypterous gerrids are either juvenilized adults (metathetely) or neotenous, a larval form with adult characters (prothetely). Wing muscles are absent in brachypters; some, though not all, macropters are migrants. The flight muscles of macropters are resorbed as the ovaries mature. There is no obvious difference in the number of eggs in short-

and long-winged females. Wing buds of both morphs appear identical in fifth instars, but there seems to be no growth of wing tissue between the penultimate and imaginal molts in brachypters (Brinkhurst 1963). Southwood suggested that brachyptery is due to either an excess of JH if metathetely, or a reduced effect of this hormone if prothetely, allowing premature metamorphosis. Since brachyptery is often associated with colder climates and higher altitudes, he concluded that low temperatures prolong the larval state and thus exposure to JH. He therefore proposed that brachypters are metathetelous adults. Indeed, short-winged adults can be produced by JH application to fifth-instar nymphs, but treatment effects are not confined to wing morphology; suppression of other adult characters in the cuticle, genitalia, and sense organs also occurs.

The key to understanding the mechanism of short-wing formation may be the apparent lack of growth of wing tissue during the final nymphal stage, which could be due to prolonged exposure to JH but might also be due to premature stimulation of the final molt by ecdysone. In many insects there are two peaks of ecdysone secretion during the final immature stage (e.g., Hoffman et al. 1974, Bollenbacher et al. 1975, Gande et al. 1979), having quite different effects: in *Manduca sexta* the first ecdysone peak triggers behavioral changes, growth of pupal structures, and insensitivity of the epidermis to JH (JH applied later has no juvenilizing effect) (Mitsui & Riddiford 1976), the second initiates the pupal molt. Ecdysone acts as both a growth hormone and a molting hormone, its effect depending on time of secretion and titer of the hormone. Thus, if the second ecdysone peak (which triggers the molt) is premature or if ecdysone titers rise more rapidly than normal in the first or a single peak, some tissues may undergo ecdysis before they have completed adult development. Wigglesworth (1940), for example, observed small imaginal wing rudiments in early fifth-instar *Rhodnius* induced to complete their metamorphosis precociously by putting them in parabiosis with late fifth-instar nymphs just entering the imaginal molt. Exogenous ecdysone treatment results in retention of juvenile characteristics in precocious adult *O. fasciatus*. Also, Truman et al. (1974) showed that epidermal tissue of *M. sexta* has a reproducible pattern in time of ecdysone sensitivity. Thus the timing of ecdysone release and the pattern of tissue sensitivity to the hormone determines which tissues would be affected by premature metamorphosis.

Thus, in polymorphic Heteroptera, macropterous versus brachypterous morphs may differ in the times of ecdysone release, the amount of ecdysone, or tissue sensitivity to the hormone. JH titer could play a role in timing of ecdysone titer changes, as it has been shown to do in *M. sexta* (Riddiford, 1980) or might indeed exert a direct juvenilizing effect. Such hormone level differences between morphs, if they exist, may depend on environmental cues, genetic programming or both.

Hormonal control of polymorphism in aphids has been extensively studied, but results are conflicting and few firm generalizations or predictions can be made as yet. Lees (1967) suggested that activity of the maternal CA influences production of alate and apterous aphids. Since apterous aphids appear to retain many larval characters (e.g., underdeveloped wings), high maternal hemolymph titers of JH were thought to stimulate aptery in the offspring. In reviewing the research reported between 1965 and 1981, one finds an array of conflicting results bearing on this hypothesis. Johnson and Birks (1960) and Lees (1975) showed that decapitation (removing among other things, the CA) of alate-producing aphids causes such headless females to produce apterous offspring. (Note this is opposite to what the JH hypothesis for hemiptera would predict.) Further, Mittler et al. (1976) showed that JH mimic treatment of alate gynoparae of *Megoura viciae* and *Aphis fabae* (raised under short photoperiod and low temperatures so they would normally have produced oviparous apterous offspring) caused them to produce alate virginoparae instead (viviparous, parthenogenetic females). Hardie (1981a) confirmed these results for *A. fabae* using the three naturally occurring JH's. Thus high maternal JH titers may induce parthenogenesis, possibly by stimulating the oogonia of presumptive oviparae to undergo division and develop into embryos instead of haploid gametes (but see Steele 1976 for another possible mechanism of control of parthenogenesis). It seems then that maternal JH levels may determine the alary morphology of offspring of gynoparae but in reverse direction to that originally suggested by Lees (1967). Lees (1980) showed that high concentrations of JH or JH mimics applied to *M. viciae* were ineffective in causing alate producers to switch to exclusive production of apterae. He concluded that JH is *not* involved in the switch mechanism. Hardie (1980, 1981b) similarly concludes that JH is not involved in the alate–apterous polymorphism in virginoparous generations of *A. fabae*. Mackauer et al. (1979) reported that precocene II applications to virginoparous *Acyrthosiphon pisum* caused a switch in production from apterae to alates, and this could be partially reversed by high doses of JHI. Precocene did not, however, have other antiallatotropic effects. It may have had additional or different effects in *A. pisum*, such as inhibition of neurosecretory cells or interference with gut secretions. Mackauer et al.'s (1979) conclusion that JH is involved in the wing development of virginoparae directly conflicts with that of Lees (1980) and Hardie, though all three studies were done with different species.

Control of wing development may also be postnatal. Several studies have involved hormone treatment of early-instar nymphs, conducted to gain an understanding of the mechanisms of apterous development, as well as the observed postnatal induction of alates. Lees (1980) found that treatment of any instar of *M. viciae* before the early fourth with JH or JH mimic induces wing deformities and/or supernumery molts. Production of malformed wings is not, however, equivalent to induction of neoteny (Lees 1975, 1980). As in the Heteroptera, Lees found many other aphid characters to be affected by exogenous JH treatment of nymphs, particularly the genitalia. He again concluded that JH is probably not the key to post-natal induction of aptery. Postnatal JH treatment of *A. fabae* presumptive gynoparae, however, did induce aptery without juvenilization if adminis-tered before the second day of the second instar (Hardie 1981a).

Clearly, exogenously applied JH may affect adult morphology and re-productive form in aphids, but whether or not JH is a major determining factor of wing development under natural circumstances is not clear nor is the time of its putative action, whether pre or postnatal. Indeed there may be substantial species differences in the role of JH and/or the timing of its action in the control of aphid alary polymorphism.

7.6 EFFECTS OF WEATHER ON MOVEMENT

The layer of air that extends from ground level upward through increasing speeds of the wind profile to the level at which the wind speed comes to equal the insect's independent flight speed has been called *boundary layer*. It is only within this layer that the insect can control its course. Above that level the flier's track can only be more or less downwind whatever its orientation. Even within the boundary layer "movement will still be down-wind unless some other strong directional element enters the field of ori-entation" (Taylor 1965). Such "directional elements" might be orientation to food or oviposition sites, optomotor responses to movement, or airspeed-regulating reactions to wind. There will also be an upper height limit (varying with different air and wind speed conditions) above which the given ground pattern can no longer be resolved by the insect's eye. This limit has been termed the *maximum compensatory height* (Kennedy 1951, Johnson 1969).

Many insects in migratory phase fly up and through their boundary layer and may be carried long distances by downwind displacement. Indeed one difference between a migrant and a nonmigrant may be in its orien-tation after takeoff, that is, whether it stays within its boundary layer, orienting to stimuli on or near the ground or flies upward, past its boundary layer.

In several insects in which wind effects have been studied, strong wind seems to inhibit takeoff but light winds often stimulate it. *S. gregaria* on the ground head into wind before they make the jump at takeoff that releases tarsal inhibition and allows the wings to flap. This upwind attitude gives maximum lift at takeoff and is adopted in response to flow of air over sense organs on the head, anal cerci, thorax, and abdomen. In fast winds locusts cling tightly to the substrate and do not fly spontaneously at wind speeds above 10 ft/sec (Rainey et al. 1957). Strong winds do not inhibit takeoff of migrant aphids and many Diptera. This ability to take off in strong wind may facilitate wind-borne transport.

Winds may also cool an insect and thus inhibit flight. Most insects have a thoracic temperature threshold for flight, which may be considerably above the ambient temperature. Warming of small insects by insolation is often quite rapid, but larger species reach their thresholds more slowly and often require more power per unit wt. for takeoff than do small species. Consequently, warming behaviors are common. For locusts to attain characteristic flapping flight, wing muscle temperature must be between 25–40°C. Body temperature depends upon air temperature, radiation, convection, evaporation, and metabolic heat. Warming in some species is facilitated by sun basking, vibrating the wings or wing muscles, or respiratory movements of the abdomen (Johnson 1969). These warmup behaviors are often seen in cryptic insects when disturbed at rest. Once the insect is warm, it flies in search of a new resting site. Low air temperature may limit the length of such flights, and insects may alight repeatedly to regain their operating temperature. Stratification of air temperature has been shown to be correlated with the vertical distribution of insects (Rainey 1974).

As air is warmed by conduction from the earth's surface, it becomes less dense than the air above it and rises in a thermal convection current while cooler air descends to take its place. Such warm air continues to rise until it reaches a point at which its density is the same as that of the surrounding air. Thus, ground air can provide thermal up-currents capable of rising to a sharply defined ceiling, the isothermal layer. For an insect riding these up-currents, the maximum attainable height should be the level at which the vertical velocity of the up-current has decreased to a value just equal and opposite to the "sinking speed" of the insect relative to the air. In one locust swarm the base of the isothermal layer was only 130 m above the topmost recorded locusts, suggesting that they had indeed ridden thermal up-currents (Rainey 1974). As night falls these up-currents collapse, and locusts are often observed descending to within a few meters of the ground (Rainey, 1958). Other insects also seem to use thermals in this way (Rainey 1974). Some night-flying migrants, however, such as *Aiolopus simulatrix*, apparently fly almost or entirely on their own power or even against air currents (Rainey & Joyce 1972) (cf. Chapter 13).

Based on extended studies R. C. Rainey and co-workers suggest that movements and distributions of desert locusts (and many other insects) on scales of hundreds or thousands of kilometers are not merely correlated with but are to a very large extent determined by low level wind fields (Rainey 1974, Rainey & Aspliden 1963, Taylor 1979). Radar observations have confirmed this hypothesis (e.g., Schaefer 1970). Indeed, wind trajectories have been used successfully to find the source of a migrant swarm of insects and to predict the next settlings, for example, for *S. gregaria*, *Spodoptera exempta*, *Agrotis ipsilon*, *Alabama argillacea*, and many hemiptera (Rainey 1974). Downwind movement of locusts takes them toward areas of low level wind convergence, often associated with rain (Waloff 1966).

Swarms tend not only to arrive with seasonal rains in their major breeding areas but also to arrive in association with exceptional rains in areas that are normally arid. Being carried by winds into zones of convergence has the effect of concentrating the migrants, bringing them together for reproduction, and rendering them vulnerable to human control measures (Rainey 1974).

Riordan (1979) assessed the atmospheric transport possibilities for insects in the southeastern United States. The air arriving over a North Carolina sampling area in April, 1978, had passed over the Gulf Coast only 15–18 hr previously and had experienced a net displacement of 300 km during that time. Transport of an airborne but passive insect population across the entire southeastern United States could thus have occurred within 2–3 days or less, the only restriction being limitations in the length of time the insects could remain aloft. The end point of a wind-borne insect's trajectory depends strongly on its altitude. Upper level wind is stronger than that at 10 m and is more predictable in its direction. Since both wind velocity and direction change with altitude, there is uncertainty in predicting the paths of passive migrants. However, this phenomenon may afford a wind-borne organism some degree of choice in final destination. If, like a balloonist, an insect could adjust its flight level, then it could control its direction and speed of transport. Furthermore, surface weather conditions are correlated with wind direction. In most of the north temperature zone, temperatures are significantly warmer during southerly than during northerly winds. Riordan (1979) suggests that if a passive organism were to become airborne only during warm periods, it would be carried northward. When surface temperatures are 23°C or higher, the probability that winds aloft in April over North Carolina are southerly is 0.81 (Riordan 1979). Thus, although insects may use the wind for transport, they are not at its mercy. If choices can be made concerning which day to take off and at what altitude to fly, considerable control might be exerted by "passive" insect migrants over the direction and speed of their movement. An example is given by Hagen (1962) for *Hippodamia convergens* (cf. Chapter 13).

Effects of atmospheric pressure on flight, although probably important to insects in choosing which days to fly, have been little studied. However, Wellington (1946) studied the effects of pressure on kinetic and tactic responses of some Diptera. A steady rise in pressure seemed to reduce takeoff activity while a rapid decrease from 1000 to 985 millibars increased activity. Similarly, Broadbent (1949) found that frequency of aphid takeoff increased as atmospheric pressure dropped. Edwards (1961) studied takeoff of *Calliphora vicina* subjected to natural and artificial changes in pressure. Takeoff activity increased by 60–90% when pressure fell steadily over a period of several hours, but a comparable rise in pressure had no detectable effect.

7.7 MOVEMENT AND NATURAL SELECTION

Movement plays various roles in insect life histories. Where movement is for foraging (searching) for food, mates, or oviposition sites, the consequences of different movement patterns (in terms of rates of resource finding) can be modeled and compared with observed success rates (Parker 1978, Hubbard & Cook 1978, Jones 1977, Heinrich & Raven 1972). Where alternative search behaviors exist, advantages of performing one rather than the other may depend on the proportion of insects currently performing each behavior. Parker's work on dung fly males searching for females shows this is true and that males come close to optimizing their strategies since the rates of encounter with females are similar for those that wait on dung and those that search in the surrounding grass.

Studies of foraging are concerned with rules that govern search behavior, such as where to search, how long to spend in each "patch," when to leave a patch, which direction to travel, and so on. Parker (1978) discusses how patch leaving (emigration) should relate to rates of resource location on the patch and predicted rates elsewhere, while Baker (1978) gives extensive data on directions of travel. As Baker (1978) notes, animals that move may retain the ability to return to their "pre-migration familiar area" (exploratory movement) or may lack this ability and be forced to search until they die or locate a new habitat. Migration that removes the insect from its natal population can be considered from the point of view of its effects on the population left behind or from that of its effects on the fitness of the individuals that leave. If the function of emigration is viewed as "self-regulation" of population size in relation to resources, and if those insects that leave reduce their own fitness by doing so, as Wynne-Edwards (1962) proposed, then such behavior could only evolve by group selection. In this case, group selection would constitute differential survival and reproduction of two or more populations which differ in their tendency to produce emigrants. Although the power of group selection to oppose selection at the individual level is still strongly debated (cf. Chapters 9, 19), and many models of group selection are pertinent to only a narrow range of behaviors, some workers seem to assume a group selection basis for the evolution of migration. They expect migration to evolve so as to maximize population stability (e.g., Myers 1976). To the extent that group selection does operate, it might favor evolution of environmentally triggered emigration over genetic polymorphism for the ability to migrate since the former allows a more rapid and complete response at the population level without the loss of the migrant genotype(s) from the population. The power of group selection is itself much influenced by migration and by the influence of migration on deme structure and gene flow (e.g., Wilson 1977).

We think it unlikely that group selection is important in the evolution of insect migration since insect populations do not characteristically defend their resources. Where "population dynamic" altruism evolves to preserve a resource base, a population of altruists is vulnerable to the appearance of individuals with genes causing them to exploit the resources more rapidly. Where resources are undefended, this vulnerability must be extreme. Kin selection is even less likely to be important in this context since, if this form of selection is to lead to evolution of altruistic emigration from a population where resources are in short supply, two conditions must be met. First, the resources given up by the emigrants must be disproportionately allocated to their relatives among those that remain behind. Second, the loss in fitness of emigrants resulting from the surrender of these resources must be less than the gain in fitness of nonmigrants resulting from use of the same resources. We think these conditions are not likely to be met in most nonsocial insect populations.

Authors who take the individual selection approach, such as Parker (1978) and Baker (1978) discuss effects of migration on fitness of the individual and hence try to describe circumstances under which some cost benefit analysis on this basis would favor emigration. This necessitates some estimate of cost of migration in terms of fitness. Such might stem from reduced fecundity due to energy costs of flight (quantified by Roff 1975), hazards encountered in migration, including risk of failure to find suitable habitat (Baker 1978), loss of time before reproduction, or other problems such as loss of fighting ability by male fig wasps encumbered with wings (Hamilton 1979). For individual selection to favor migration, these losses must, on the average, be outweighed by gains associated with the change in location achieved. Baker (1978) discusses this question in detail.

7.7.1 Migration and Habitat Stability

Southwood (1977) introduced the idea of the importance of the relationship between generation time (τ) and the length of time for which the habitat remains suitable (H) (cf. Chapter 12). Clearly, where $\tau > H$, the insects die before they become adult, if the adult is the only mobile stage. Where $\tau = H$, migration of the adult is obligate, and where $\tau < H$, migration can be facultative. Other workers have developed this line further. Dingle (1979) has measured greater migration tendency in *Oncopeltus* from temperate latitudes than in those from supposedly more stable tropical habitats.

Many authors have noted a correlation between habitat permanence and flightlessness (Putshova 1971, Andersen 1971, Brinkhurst 1963). Denno (1978) and Denno and Grissell (1979) observed that populations of the multivoltine planthopper *Prokelisia marginata* contain two wing forms, flightless short-winged brachypters and long-winged macropters with fully developed wings capable of flight. *P. marginata* feeds selectively on the grass *Spartina alterniflora*, which dominates the vegetation of Atlantic and Gulf Coast tidal salt marshes. Along most of the Atlantic seaboard *S. alterniflora* occurs in stable patches, which provide life history requisites year round, and unstable patches, which disappear during winter. The proportion of macropters in the planthopper populations is significantly correlated with the proportion of unstable patches in the local environment. A dimorphism is maintained along most of the Atlantic Coast, apparently because brachypters remain and effectively exploit stable patches and macropters colonize and subsequently escape unstable habitats (Denno & Grissell 1979). The percentage of macropters is positively correlated with levels of crowding incurred during the nymphal stage, suggesting that wing form is determined by a developmental switching mechanism triggered by various environmental cues that predict habitat deterioration (Denno & Grissell 1979).

Vepsalainen (1978) has studied alary polymorphism with respect to habitat and life history strategies of several Finnish and Hungarian water striders (*Gerris* spp.). Species occurring in the most permanent habitats were almost completely apterous. When habitats are extremely unstable, migration and habitat selection between each generation is the only means to cope with the changing environment and the species are entirely long winged (e.g., *G. thorcicus*).

Solbreck (1978) considered that where habitats fluctuate in phase with one another, emigrants are unlikely to meet success, and diapause would seem a more appropriate response to adversity. The lower the correlation between such fluctuations in different habitats, the more advantageous migration becomes. Habitat quality may also be important in influencing the relative timing of migration and reproduction.

Although logical, these ideas are difficult to test since quantitative data on habitat stability are not easily obtained. Furthermore, some of these predictions may need to be modified since Hamilton and May (1977) have modeled theoretical situations in which even in a completely stable habitat, production of fractions of both dispersing offspring and sedentary offspring gives higher fitness than either of the extreme alternatives (all sedentary or all dispersing).

Isolated habitats such as island or high-altitude environments have also been suggested as places where emigration should be very low. If the probability of finding a suitable new habitat under these circumstances is low, selection should favor evolution of flightless forms. Some authors have cited exposure to very high winds as a reason for the high proportion of flightless species on oceanic islands and alpine meadows, but these ideas have been disputed and the question remains unresolved (Darlington 1943, Carlquist 1974).

7.7.2 Migration and Reproduction

We have noted that migration involves a cost in reproduction which must be balanced by the reproductive advantages of the new environment. The numerous observations of flight muscle histolysis accompanying reproduction support the idea that energy investment in a migrant, as well as the space devoted to flight muscle, is switched to reproduction in a nonmigrant (e.g., Dingle & Arora 1973, Andersen 1973, Larson 1970). Dixon and Wratten (1971) compared laboratory fecundities of alate and apterous aphid morphs of the same weight. Over the entire life span the net reproductive rate of alates was only 70% that of apterae (note that the alates did not actually make a migratory flight). However, since the alates produced a larger proportion of early offspring, the intrinsic rate of increase of the two morphs was actually nearly equal. Long-distance flight requires fuel that might go to reproduction in a nonmigrant. Thus, decreased total fecundity might be expected in a migrant, other factors being equal. Roff (1975, 1977) found that flight significantly reduced the eggs laid per day by *Drosophila melanogaster*.

When migration is prereproductive, onset of reproduction may be delayed in migrants compared to nonmigrants (e.g., Mochida 1973, Waloff 1973). In *Oncopeltus*, however, age of first reproduction is similar in migrants and nonmigrants. In the cricket *Pteronemobius taprobanensis* peak oviposition occurs 15 days later in long-winged than in short-winged individuals (presumed migrants and nonmigrants, respectively) (Tanaka 1976). The long-winged forms shed their wings as oogenesis begins. If the wings are removed from these insects, flight muscles degenerate and there is no delay in oviposition.

While some workers stress the reproductive immaturity of many migrants (Johnson 1969), others point out that reproduction during migration distributes eggs over a wide area (Baker 1978). Fisher and Greenbank (1979) noted that spruce budworm moths reproduce both before and after migration. These different patterns raise the question: When should the insect carry with it the resources for reproduction and when should it gather these in the new habitat? What is optimal should be largely influenced by relative habitat quality. Dixon (1976) showed, for example, that fall-produced winged females of the aphid *Rhopalosiphum padi* migrate from grasses to *Prunus* trees and there produce 40% of their offspring in the 24 h following arrival. Summer-produced alate *R. padi* migrate from crowded conditions on grasses to colonize new growth on other grasses and do not reproduce in the first 24 h following arrival. Rather, reproduction builds up more gradually. Fall *R. padi* are leaving a temporarily suitable habitat and entering one less favorable in the short term but necessary in the long term because frost-resistant eggs will be produced (Singer, unpubl.). Summer migrants leave adverse conditions and search for more suitable ones. Thus, the differences in their strategies are logical since each form gathers resources for reproduction in the most favorable habitat. Where the food is gathered before migration, the migrant is reproductively mature and ready to reproduce immediately on arrival. Of course, these arguments apply only to insects that feed as both larvae and adults.

7.7.3 Genetic Versus Environmental Switching Mechanisms

When times are hard, it is frequently observed that some members of a population emigrate and others remain. It is not necessary to postulate genetic differences between the individuals that leave and those that stay since there will be stochastic variation in the way they experience their environment or perceive population density (Lomnicki 1978). If the emigration is in response to high density, those that leave may reduce the density below the "emigration threshold" for even genetically similar individuals which happened to remain longer. However, in some cases it is clear that differences in movement tendency are genetic. Heritable differences in flight behavior have been measured between physically similar individuals (Caldwell & Hegmann 1969, Dingle et al. 1977, Rose 1972), and genetic polymorphism for presence or absence of wings can occur. In the weevil *Sitona hispidula*, individuals from a macropterous population were crossed with ones from a 95% brachypterous population (Jackson 1928). The proportions of long- and short-winged forms in the F_2 progeny suggested that wing length was a simple Mendelian character, with the brachypterous morph dominant. Similar results were obtained by Stein (1973) in another weevil, *Apion virens*, and by Lindroth (1946) in a carabid *Pterostichus anthrocinus*. In contrast the genetics of wing polymorphism in *Melanogryllus* sp. does not suggest a simple Mendelian relationship (Sellier 1954).

Several studies have attempted to determine the mode of inheritance of various morph types in gerrids. Most investigators (Poisson, 1924; Guthrie, 1959; Brinkhurst, 1959; Vepsalainen, 1978) have concluded that morph determination is controlled by a single locus, with the allele for the short-winged or wingless morph dominant to the allele for the long-winged morph. However, as pointed out by several workers (Harrison, 1980; Dingle, 1982; Zera, et al. 1983), there are serious problems with each of these studies. In addition to technical errors such as inadequate control of rearing photoperiod and temperatures or previous mating experience, morph ratios of many crosses were inconsistent with the single locus model proposed by the investigators. Vepsalainen (1978) explains some of the inconsistencies in his data by proposing that in some gerrids both genetic and environmental factors can determine wing length, but genotype determines the levels of environmental cues which are effective. Recently, however, in a more carefully controlled study, Zera et al. (1983) investigated the genetic basis of morph determination in the waterstrider, *Limnoporus canaliculatus*. Adults were pair crossed and resultant broods were split between two photoperiods. Under both photoperiods there was a strong genetic component to morph determination, but the mode of inheritance was clearly more complex than a single locus with two alleles. Both winged and wingless adults were produced in nearly all winged × winged and wingless × wingless crosses.

There have been numerous attempts to document fitness differences between the long-winged and short-winged or wingless morphs in various gerrid species. Most studies have focused on trade-offs between reproductive and dispersal ability, and the results of these studies are summarized in Vepsalainen (1978). Other than differences in overwintering ability (Ekblom, 1949), no fitness differences have been documented. However, in all cases, the sample sizes employed in these experiments were so small that only enormous differences would have been detected. In a recent study in *L. canaliculatus,* using larger sample sizes, Zera (1983) demonstrated highly significant differences between the long-winged and wingless morphs in three characteristics. Winged morphs exhibited substantially greater survivorship than wingless morphs under overwintering conditions. On the other hand the wingless morph exhibited a significantly faster rate of development from egg hatch to adult eclosion. Wingless females also laid a significantly greater number of eggs than winged females, and the difference was particularly marked during the early stages of egg laying. Similarly, Stein (1977) showed that the proportion of an alfalfa feeding curculionid beetle with short wings increased year by year after the founding of new populations. Since the differences in wing length in this species are genetic, the change in frequency is likely to stem from a reproductive advantage of the short-winged over the long-winged form.

The extent to which movement results in gene flow is variable and dependent on ecological (especially phenological) factors (Gilbert & Singer 1973, Richardson & Johnston 1975). Gene flow interacts with selection and drift to affect genetic differentiation of populations. Although these interactions are usually too complex to allow conclusions about the nature of the migration to be drawn from genetic data, attempts have been made to do so. For example, Jaenike and Selander (1980) conclude that the genetic similarity of the fall webworm, *Hyphantria cunea,* on two different hosts indicates that these hosts do not harbor separate populations with restricted gene flow between them. As a counter example, Ehrlich and White (1980) found strong genetic similarity between populations of *Euphydryas* in Colorado and California. These populations are not likely to have exchanged genes for at least 7000 generations. Attempts to interpret genetic differences as results of lack of gene flow tend to ignore the power of natural selection to produce genetic differentiation in the face of moderate gene flow or to maintain similarity in the absence of it. Nonetheless, Eanes and Koehn (1978) and Eanes (1979) showed that genetic differences build up in summer populations of the monarch butterfly and are obliterated by the fall migration.

ACKNOWLEDGMENT

The authors wish to gratefully acknowledge the kind assistance of Dr. W. D. Hamilton in the preparation of the manuscript.

REFERENCES

Andersen, N. M. 1971. *Proc. 13th Int. Cong. Ent. (Moscow)* **1:** 469, 470.

Andersen, N. M. 1973. *Entomol. Scand.* **4:** 1–20.

Baker, R. R. 1978. *The Evolutionary Ecology of Animal Migration.* Holmes & Meier, New York.

Baltensweiler, W. 1979. *Proc. 1st European Cong. Entomol.*

Barton Browne, L. (Ed.). 1974. *The Experimental Analysis of Insect Behaviour.* Springer-Verlag, New York.

Bollenbacher, W. E., W. V. Vedeckis, L. I. Gilbert, and J. D. O'Connor. 1975. *Devel. Biol.* **44:** 46–53.

Borden, J. H. and C. E. Slater. 1968. *Z. Vergl. Physiol.* **61:** 366–368.

Brinkhurst, R. O. 1959. *J. Anim. Ecol.* **28:** 211–230.

Brinkhurst, R. O. 1963. *Proc. Roy. Entomol. Soc. Lond. (A)* **35:** 91, 92.

Broadbent, L. 1949. *Ann. Appl. Biol.* **36:** 40–62.

Brower, L. 1962. *Ecology* **43:** 549–552.

Brussard, P. F. and P. R. Ehrlich. 1970. *Nature* **227:** 91, 92.

Burns, M. 1972. *Ent. Exp. & Appl.* **15:** 319–323.

Caldwell, R. L. 1974. Pp. 304–316 in Barton Browne (Ed.), 1974, referenced here.

Caldwell, R. L. and J. P. Hegmann. 1969. *Nature* **223:** 91, 92.

Carlisle, D. B. and P. E. Ellis. 1959. *Comp. Rend. Acad. Sci., Paris* **249:** 2059, 2060.

Carter, C. I. and J. Cole. 1977. *Ann. Appl. Biol.* **86:** 137–151.

Cassier, P. 1964. *Comp. Rend. Acad. Sci. Paris* **258:** 723–725.

Carlquist, S. 1974. *Island Biology,* Columbia University Press, New York.

Chapman, R. F. 1958. *Anim. Behav.* **6:** 60–67.

Chudakova, I. V. and A. Bocharova-Messner. 1968. *Akad. Nauk. SSSR Dokl. Biol. Sci.* **179:** 157–159.

Darlington, P. J. Jr. 1943. *Ecol. Monogr.* **13:** 39–61.

Davis, N. T. 1975. *Ann. Entomol. Soc. Am.* **68:** 710–714.

Dempster, J. P. 1963. *Biol. Rev.* **38:** 490–529.

Denno, R. F. 1978. *Can. Entomol.* **110:** 135–142.

Denno, R. F. and E. E. Grissell. 1979. *Ecology* **60:** 221–236.

DeKort, C. A. D., M. A. Khan, B. J. Bergot and D. A. Schooley. 1982. *J. Insect Physiol.* **23:** 471–474.

DeWilde, J. and J. A. DeBoer. 1969. *J. Insect Physiol.* **15:** 661–667.

DeWilde, J., G. Staal, C. A. D. DeKort, A. DeLoof and G. Baard. 1968. *Proc. K. Ned. Akad. Wetensch. Ser. C. (Amsterdam)* **71:** 321–326.

Dingle, H. 1965. *J. Exp. Biol.* **42:** 269–283.

Dingle, H. 1968. *J. Exp. Biol.* **48:** 175–184.

Dingle, H. 1972. *Science* **175:** 1327–1335.

Dingle, H. 1974. *Oecologia* **17:** 1–10.

Dingle, H. 1978 (Ed.). *Evolution of Insect Migration and Diapause.* Springer-Verlag, New York.

Dingle, H. 1979. Pp. 64–87 in Rabb and Kennedy (Eds.), 1979, referenced here.

Dingle, H. 1982. *Ent. Exp. Appl.* **31:** 36–47.

Dingle, H. and G. Arora. 1973. *Oecologia* **12:** 119–140.

Dingle, H., C. K. Brown and J. P. Hegmann. 1977. *Am. Natur.* **111:** 1047–1059.

Dixon, A. F. G. 1976. *J. Anim. Ecol.* **45:** 817–830.

Dixon, A. F. G. and S. D. Wratten. 1971. *Bull. Entomol. Res.* **61:** 97–111.

Dobzhansky, T. and S. Wright. 1943. *Genetics* **28:** 304–340.

Eanes, W. F. 1979. Pp. 88–103 in Rabb and Kennedy (Eds.), 1979, referenced here.

Eanes, W. R. and R. K. Koehn. 1978. *Evolution* **32:** 787–797.

Edwards, D. K. 1961. *Can. J. Zool.* **39:** 632–635.

Edwards, F. J. 1970. *J. Insect Physiol.* **16:** 2027–2031.

Ehrlich, P. R. 1961. *Science* **134:** 108, 109.

Ehrlich, P. R., D. D. Murphy, M. C. Singer, C. B. Sherwood, R. R. White and I. L. Brown. 1980. *Oecologia* **46:** 101–105.

Ehrlich, P. R. and R. R. White. 1980. *Am. Natur.* **115:** 328–341.

Ekblom, T. 1949. *Not. Entomol.* **29:** 1–15.

Ellis, P. E. 1953. *Behaviour* **5:** 225–260.

Ellis, P. E. and G. Hoyle, 1954. *J. Exp. Biol.* **31:** 271–279.

Fisher, R. A. and D. O. Greenbank. 1979. Pp. 220–231 in Rabb and Kennedy (Eds.), 1979, referenced here.

Forrest, J. M. S. 1974. *J. Entomol.* **48:** 171–176.

Gande, A. R., E. D. Morgan and I. D. Wilson. 1979. *J. Insect Physiol.* **25:** 669–675.

Gilbert, L. E. and M. C. Singer. 1973. *Am. Natur.* **107:** 58–72.

Goldsworthy, G. J., A. J. Coupland and W. Mordue. 1973. *J. Comp. Physiol.* **82:** 339–346.

Goldsworthy, G. J., R. A. Johnson and W. Mordue. 1972. *J. Comp. Physiol.* **79:** 85–96.

Gunn, D. L. 1960. *Annu. Rev. Ent.* **5:** 279–291.

Guthrie, D. M. 1959. *J. Anim. Ecol.* **28:** 141–152.

Hagen, K. S. 1962. *Annu. Rev. Ent.* **7:** 289–326.

Hamilton, W. D. 1979. Pp. 167–200 in M. S. Blum and N. A. Blum (Eds.), *Sexual Selection and Reproductive Competition*. Academic, New York.

Hamilton, W. D. and R. M. May. 1977. *Nature* **269:** 578, 579.

Hardie, J. 1980. *Nature* **286:** 602–604.

Hardie, J. 1981a. *J. Insect Physiol.* **27:** 257–266.

Hardie, J. 1981b. *J. Insect Physiol.* **27:** 347–356.

Harrison, R. C. 1980. *Annu. Rev. Ecol. Syst.* **11:** 95–118.

Haskel, P. T. and J. E. Moorhouse. 1963. *Nature* **197:** 56–58.

Heinrich, B. and P. H. Raven. 1972. *Science* **176:** 497–502.

Herman, W. S. 1973. *J. Insect Physiol.* **19:** 1883–1887.

Highnam, K. C. and O. Lusis. 1962. *Quart. J. Microsc. Sci.* **103:** 73–84.

Hoffman, J. A., J. Koolman, P. Karlson, and P. Joly. 1974. *Gen. Comp. Endocrin.* **22:** 90–97.

Honek, A. 1974. *Vestnik Seskoslovenske Spolecnosti Zoologicke* **38:** 241, 242.

Hoyle, G. 1954. *J. Exp. Biol.* **31:** 267–270.

Hubbard, S. R. and R. M. Cook. 1978. *J. Anim. Ecol.* **47:** 593–606.

Hughes, R. D. and P. Sands. 1979. *J. Appl. Ecol.* **16:** 117–139.

Jackson, D. J. 1928. *Trans. Roy. Soc. Edinburgh.* **55:** 665–735.

Jaenike, J. and R. K. Selander. 1980. *Ent. Exp. & Appl.* **27:** 31–37.

Johnson, B. 1965. *Ent. Exp. & Appl.* **8:** 49–64.

Johnson, B. 1966. *Ent. Exp. & Appl.* **9:** 213–222.

Johnson, B. and P. R. Birks. 1960. *Ent. Exp. & Appl.* **3:** 327–339.

Johnson, C. B. 1974. Pp. 279–443 in E. Rockstein (Ed.), *Physiology of Insecta*. Academic, New York.

Johnson, C. G. 1969. *Migration and Dispersal of Insects by Flight.* Methuen, London.

Jones, R. E. 1977. *J. Anim. Ecol.* **46:** 194–212.

Kimble, A. R. and H. T. MacPherson. 1954. *Biochem. J.* **58:** 46–50.

Kennedy, J. S. 1951. *Phil. Trans. B.* **235:** 163–290.

Kennedy, J. S. 1956. *Biol. Rev.* **31:** 349–370.

Kennedy, J. S. 1961. *Nature* **189:** 785–791.

Kennedy, J. S. and C. O. Booth. 1964. *Ann. Appl. Biol.* **41:** 88–106.

Kennedy, J. S. and A. R. Ludlow. 1974. *J. Exp. Biol.* **61:** 173–196.

Kidd, N. A. C. 1977. *Ent. Exp. & Appl.* **22:** 251–261.

Kleinjan, J. E. and T. E. Mittler. 1975. *Ent. Exp. & Appl.* **18:** 384–388.

Larson, O. 1970. *Entomol. Scand.* **1:** 227–235.

Lee, S. S. and G. J. Goldsworthy. 1976. *Acrida* **5:** 598–611.

Lees, A. D. 1961. *Symp. Roy. Entomol. Soc. Lond.* **1:** 68–79.

Lees, A. D. 1966. *Adv. Insect Physiol.* **3:** 207, 278.

Lees, A. D. 1967. *J. Insect Physiol.* **13:** 289–318.

Lees, A. D. 1975. *Proc. Roy. Soc. Lond. (A)* **39:** 59–64.

Lees, A. D. 1980. *J. Insect Physiol.* **26:** 202–207.

Lewis, A. C. 1979. *Ent. Exp. & Appl.* **26:** 202–207.

Lewis, A. C. 1981. *Ph.D. dissertation,* University of Texas, Austin, Tex.

Lindroth, C. H. 1946. *Heriditas* **32:** 37–40.

Loher, W. 1961. *Proc. Roy. Soc. (B)* **153:** 380–397.

Lomnicki, A. 1978. *J. Anim. Ecol.* **47:** 461–475.

Mackauer, M., K. K. Nair and G. C. Unnithan. 1979. *Can. J. Zool.* **57:** 856–859.

MacKay, P. A. 1977. *J. Insect Physiol.* **23:** 889–893.

MacKay, P. A. and W. G. Wellington. 1977. *Res. Popul. Ecol.* **18:** 195–209.

Mayer, R. J. and D. J. Candy. 1969. *J. Insect Physiol.* **15:** 611–620.

McLean, J. A., R. F. Shepher and R. B. Bennett. 1979. Pp. 369–381 in Rabb and Kennedy (Eds.), 1979, referenced here.

Michel, R. 1972. *Gen. Comp. Endocrin.* **19:** 96–101.

Mitsui, T. and L. M. Riddiford. 1976. *Dev. Biol.* **54:** 178–186.

Mittler, T. E., S. G. Nasser and G. B. Staal. 1976. *J. Insect Physiol.* **22:** 1717–1725.

Mochida, O. 1973. *Trans. Roy. Entomol. Soc. London* **125:** 177–225.

Myers, J. H. 1976. *Oecologia* **23:** 255–269.

Njio, K. D. and T. Piek. 1977. *J. Insect Physiol.* **23:** 919–929.

Noda, O. 1956. *Mem. Ehime Univ., Sect. 2, Ser B* **2:** 61–70.

Nolte, D. J., S. H. Eggers and I. R. May. 1973. *J. Insect Physiol.* **19:** 1547–1554.

Norris, M. J. 1964. *Nature* **203:** 784–785.

Odhiambo, T. R. 1966. *J. Exp. Biol.* **45:** 51–63.

Parker, G. A. 1978. Pp. 214–244 in J. R. Krebs and N. B. Davies (Eds.), *Behavioural Ecology, An Evolutionary Approach.* Blackwell, London.

Parry, W. H. 1978. *Ann. Appl. Biol.* **89:** 9–14.

Paschke, J. D. 1959. *Univ. Calif. Publs. Entomol.* **16:** 125–180.

Pashley, J. D. and G. L. Bush. 1979. Pp. 333–341 in Rabb and Kennedy (Eds.), 1979, referenced here.

Perkins, J. H. 1980. *Science* **207:** 1044–1049.

Pickford, R. 1966. *Can. Entomol.* **98:** 158–169.

Pines, M., A. Tietz, H. Weintraub, S. W. Applebaum and L. Josefsson. 1981. *Gen. & Comp. Endocrin.* **43:** 427–431.

Poisson, R. 1924. *Bull. Biol. Fr. Beig.* **58:** 49–305.

Putshova, I. V. 1971. *Ent. Obozr.* **50:** 537–549 (English trans., Pp. 303–309).

Rabb, R. L. and G. G. Kennedy (Eds.). 1979. *Movement of Highly Mobile Insects: Concepts and Methodology in Research.* North Carolina State University Graphics, Raleigh, N.C.

Rainey, R. C. 1958. *J. Sci. Food Agric.,* pp. 9677–9692.

Rainey, R. C. 1974. *Annu. Rev. Entomol.* **19:** 407–439.

Rainey, R. C. 1978. Pp. 35–48 in Dingle (Ed.), 1978, referenced here.

Rainey, R. C. and C. I. H. Aspliden. 1963. *Anti-Locust Mem.* **7:** 54–103.

Rainey, R. C. and R. J. V. Joyce. 1972. *Proc. Int. Aerosp. Instrum. Symp. 7th Cranfield 1972,* Pp. 81–84.

Rainey, R. C., Z. Waloff and G. Burnett. 1957. *Anti-Locust Bull.* **26.**

Rankin, M. A. 1974. Pp. 317–328 in Barton Browne (Ed.), 1974, referenced here.

Rankin, M. A. 1980. *J. Insect Physiol.* **26:** 67–74.

Rankin, M. A. 1981. *Am. Zool.* **21:** 918.

Rankin, M. A. and S. A. Rankin. 1980. *Biol. Bull.* **158:** 356–369.

Rankin, M. A. and L. M. Riddiford. 1977. *Gen. & Comp. Endocrin.* **33:** 309–321.

Rankin, M. A. and L. M. Riddiford. 1978. *J. Insect Physiol.* **24:** 31–38.

Rankin, S. A. and M. A. Rankin. 1980. *Physiol. Entomol.* **5:** 175–182.

Richards, O. W. and N. Waloff. 1954. *Anti-Locust Bull.* **17.**

Richardson, R. H. and J. S. Johnston. 1975. *Oecologia* **20:** 287–299.

Riddiford, L. M. 1980. *Annu. Rev. Physiol.* **42:** 511–528.

Riordan, A. J. 1979. Pp. 120–132 in Rabb and Kennedy (Eds.), 1979, referenced here.

Roff, D. 1975. *Oecologia* **19:** 217–237.

Roff, D. 1977. *J. Anim. Ecol.* **46:** 443–456.

Rose, J. W. 1972. *J. Anim. Ecol.* **41:** 589–609.

Schaefer, G. W. 1938. *J. Agric. Res.* **57:** 825–841.

Schaefer, G. W. 1970. *Proc. Int. Study Cong. Curr. & Fut. Probl. Acridol., Lond. 1970,* Pp. 370–438.

Schaefer, G. W. 1979. *Phil. Trans. Roy. Soc. Ser. B.* **287:** 215–221.

Sellier, R. K. 1954. *Ann. Sci. Nat. 11th Ser.* **16:** 595–740.

Shaw, M. J. P. 1970. *Ann. Appl. Biol.* **65:** 191–196.

Singer, M. C. 1972. *Science* **173:** 75–77.

Singer, M. C. and P. Wedlake, 1981. *Ecol. Entomol.* **6:** 215–216.

Solbreck, C. 1978. Pp. 195–217 in Dingle (Ed.), 1978, referenced here.

Solbreck, C. and I. Pehrson. 1979. *Oecologia* **43:** 51–62.

Southwood, T. R. E. 1961. *Proc. Roy. Entomol. Soc. Lond.* (A) **36:** 63–66.

Southwood, T. R. E. 1977. Pp. 471–493 in V. Labeyrie (Ed.), *Comportmement des Insectes et le Milieu Trophique,* CNRS, 265, Paris.

Southwood, T. R. E. 1978. *Ecological Methods with Particular Reference to the Study of Insect Populations.* Chapman and Hall. London.

Steele, C. G. H. 1976. Pp. 117–120 in M. Luscher (Ed.), *Phase and Caste Determination in Insects.* Pergamon, Oxford.

Stein, W. 1973. *Z. Ang. Entomol.* **74:** 62, 63.

Stein, W. 1977. *Z. Ang. Entomol.* **83:** 37–39.

Stengel, M. 1974. Pp. 297–303 in Barton Browne (Ed.), 1974, referenced here.

Sutherland, O. R. W. 1970. *J. Insect Physiol.* **16:** 1349–1354.

Symmons, P. 1960. *Bull. Entomol. Res.* **50:** 106–109.

Symmons, P. M. and L. M. McCulloch. 1980. *Bull. Ent. Res.* **70:** 197–201.

Tanaka, S. 1976. *Kontyu* **44:** 327–333.

Taylor, L. R. 1965. *Proc. N.C. Branch Entomol. Soc. Am.* **20:** 9–19.

Taylor, L. R. 1979. Pp. 148–187 in Rabb and Kennedy (Eds.), 1979, referenced here.

Taylor, L. R. and R. A. J. Taylor. 1977. *Nature* **265:** 415–421.

Taylor, L. R., I. P. Woiwod and R. A. J. Taylor. 1979. *J. Anim. Ecol.* **48:** 955–986.

Taylor, R. A. J. 1979. *J. Anim. Ecol.* **48:** 577–602.

Truman, J. W., L. M. Riddiford and L. Safranek. 1974. *Dev. Biol.* **39:** 247–267.

Urquhart, F. A. 1960. *The Monarch Butterfly*. University of Toronto Press, Toronto.

Urquhart, F. A. and N. R. Urquhart. 1977. *Can. Entomol.* **109:** 1583–1589.

Uvarov, B. P. 1966. *Grasshoppers and Locusts*, Vol. 1. Center for Overseas Pest Research, Cambridge University Press, London.

Vepsalainen, K. 1971. *Ann. Acad. Sci. Fenn. A, IV Biologica* **183:** 1–25.

Vepsalainen, K. 1978. Pp. 218–253 in Dingle (Ed.), 1978, referenced here.

Waloff, N. 1973. *J. Appl. Ecol.* **10:** 705–730.

Waloff, Z. 1966. *Anti-Locust Mem.* **8**.

Weis-Fogh, T. 1952. *Phil. Trans. Roy. Soc. Lond. Ser. B.* **237:** 1–36.

Wellington, W. G. 1946a. *Can. J. Res.* **24:** 51–70.

Wellington, W. G. 1946b. *Can. J. Res.* **24:** 105–117.

White, T. C. R. 1976. *Oecologia* **3:** 119–134.

Wiggins, L. F. and J. H. Williams. 1955. *J. Agr. Food Chem.* **3:** 341–345.

Wigglesworth, V. B. 1940. *J. Exp. Biol.* **17:** 201–222.

Wilson, D. S. 1977. *Am. Natur.* **111:** 157–185.

Wynne-Edwards, V. C. 1962. *Animal Dispersion in Relation to Social Behavior*. Oliver & Boyd, Edinburgh.

Zera, A. J. In press. *Evolution*.

Zera, A. J., D. J. Innes and M. E. Saks. 1983. *Evolution.* **37:** 513–522.

Chapter 8

Adaptations
of Insects
to Modes of Life

LEOPOLDO E. CALTAGIRONE

*O*rganisms are adapted to the conditions in which they live.

VERMEIJ 1978

8.1 INTRODUCTION

The above quotation is a tautology. As such it brings into focus the enormity of the task when attempting to summarize in one chapter the adaptations of the largest and most diversified group of organisms on earth: the Insecta. The product of some 350 million years of evolution, the insects occupy as diverse environments as the ice-capped poles, tropical rain forests, fast flowing, cold rivers, swamps, and deserts where it never or very seldom rains. Their size varies from that of the tiniest trichogrammatid wasps, only a fraction of a millimeter long, to the giant walking sticks which can reach some 30 cm in length. Equally varied are their nutritional adaptations. There are those that are specific or narrowly oligophagous, like the yucca moths whose larvae are restricted to the developing fruits of species of *Yucca*; those that are broadly oligophagous or polyphagous, like the migratory locust; and those that belong to categories in between. But the habit of eating plants and substances of plant origin is far from being the one that satisfies the nutritional needs of all of the insects; on the contrary, a great number of species are carnivorous, feeding on other animals, including other insects. A number of insects are able to utilize food to which they were never exposed, as demonstrated in those instances in which native insects successfully use exotic plants or animals as food or, conversely, exotic insects that feed on native plants or animals.

In their struggle to eat, to avoid being eaten, and to leave descendants, insects, like other living organisms, have adapted themselves in a myriad of ways, as indicated by the intricacies of their morphology, physiology, and behavior. The hard, strong exoskeletons of some beetles make the fragile body of a mayfly look even more delicate by comparison. The complexity of antennae such as those of green lacewings with their thousands of chemoreceptors contrasts with the apparent simplicity of antennal structure in immature stages of some endoparasitic forms. The short life cycle of some drosophilid fruit flies, which can be less than a week, seems even shorter when compared with that of the 17-yr cicada. The majority of insects are zygogenetic, that is, eggs produced by females need to be fertilized by sperm produced by males in order to develop into new individuals; however, many insects are parthenogenetic, that is, unfertilized eggs develop into females; there are no males in these species. In a number of species, such as aphids and cynipid wasps, zygogenetic generations alternate with parthenogenetic ones. In the zygogenetic species a most important requirement for successful reproduction is fertilization of the eggs and, because in insects fertilization is internal, a number of adaptations have developed to ensure that the sperm will reach the egg in the female. This has resulted in many mating tactics of varying degress of intricaty.

Adaptations to avoid being eaten are legion and vary from escape movements, such as dropping to the ground at the slightest disturbance and becoming totally inconspicuous, to aposematic coloration reinforced with chemical defenses, to mimicry and camouflage. Because predation and parasitism (carnivory) are very prevalent in the insects and constitute sustained selective forces impinging upon the populations of both the eater and the eaten, adaptations leading to improved chances of survival of both groups have developed rather quickly, by comparison to the evolution of some traits that do not involve prey–predator or host–parasite interactions.

In the following sections I discuss some adaptations that have resulted in the insects' modes of life as we know them. I am well aware that the subject is too varied and extensive to be covered adequately in a single chapter by a single author. But if readers become interested enough to seek additional information in the references cited in this and other chapters, thus becoming better acquainted with the rich variety of insect adaptations, I would consider my attempt a success.

The scope of coverage in this chapter is somewhat biased toward coverage of areas involving parasitoids and entomophagous predators, by virtue of my own area of expertise. This is justified in part by the fact that adaptations of insects to the physical conditions of their environments are covered in Chapters 6, 7, and 13, and adaptations for interspecific competition in Chapter 15 and to a degree Chapter 16, involving such adaptations as resource finding ability, efficiency of resource use, rate of development, aggregative tendency, and the like. Moreover, there is some overlap, from somewhat different points of view and intent of the chapters, with the chapters on food (Chapter 4), reproduction (Chapters 5 and 17), and evolutionary processes (Chapter 9). The many specialized adaptations for aquatic life are largely neglected. The texts of Merritt and Cummings (1978) and Usinger (1963) on this subject should be consulted.

8.2 ADAPTATIONS FOR OBTAINING FOOD

Insecta have developed numerous adaptations that facilitate obtaining the necessary nutrients from a large variety of sources in a large variety of habitats. All organic matter, regardless of the form in which it is available—plants and animals and all their derivatives, whether the result of decomposition or manufacture—is susceptible to being used as food by insects (Chapter 4). The ability to extract nutrients from the various textures in which organic matter occurs is primarily due to the variety of mouthparts found in insects. Companion to the various types of mouthparts are the morphological and physiological adaptations of the digestive tract. A discussion of a representative sample of these adaptations is not possible within the limits allowed, so I will summarize a few selected cases. Additional material can be found in the references of this and other chapters in this volume.

8.2.1   Types of Mouthparts and Related Digestive Features of Insects

The most generalized, and probably the most primitive type of mouthparts is the mandibulate. It is found in cockroaches, grasshoppers, beetles and wasps, among other insects. The mandibles are effectively used to cut food into pieces small enough to be ingested. The larvae of mosquitoes have mandibulate mouthparts, but they do not use their mandibles and maxillae to grasp and tear food. Instead, they feed on small particles (bacteria, algae) dispersed in the water. To gather this food the larvae have tufts of hair, the feeding brushes, in the front of the head. The brushes can be spread out or closed in, and when these alternating movements are continuous, a current toward the mouth is created. The edges of the mandibles and the maxillae are lined with fine, closely set setae which capture the food particles. Thus, mosquito larvae gather food in essentially the same manner as baleen whales, basking sharks, phytoplankton-eating tilapias (fish), flamingoes, and other filter feeders.

The other types of mouthparts are derived from the mandibulate type and are essentially adapted for ingesting liquid nutrients. Insects that feed on plant fluids from the xylem (e.g., cicadas) are faced with the problem of excessive intake of water. The xylem is poor in amino acids and comparatively rich in inorganic salts; consequently, as noted in Chapter 4, the insects have to ingest large amounts of sap to get enough amino acids. The digestive tract has developed in such a way that excess water and salts are transferred from the anterior midgut to the posterior midgut and hindgut. The posterior midgut loops cephalad against the anterior midgut which wraps around the posterior midgut and the proximal end of the Malpighian tubules. This is the *filter chamber*. Salts are eliminated through the Malpighian tubules and water passes directly to the posterior midgut. Nutrients are concentrated and absorbed in the middle midgut. In some sap feeders, for example, aphids, soft scales, and whiteflies, the fluids flow constantly, and are mostly eliminated as honeydew.

8.2.2   Securement and Conservation of Water

In the process of digesting food insects, like other animals, are confronted with the problem of maintaining an adequate amount of water for an efficient metabolism, including excreting nitrogenous wastes. The majority of insects dispose of waste nitrogen in the form of uric acid, which is crystalized in the hindgut where water is absorbed back into the hemolymph. The nontoxic crystals of uric acid are excreted as part of the feces.

This adaptation, together with the ability to use water produced in the metabolic processes, makes it possible for species such as the mealworm, *Tenebrio molitor*, to subsist on very dry food. Mealworm larvae when living on dry flour conserve most of the water used during digestion and that produced in the metabolic process, excreting very dry, powdery feces (Wigglesworth 1964, Chapman 1971) (cf. Chapter 4).

Most insects obtain the water they need either from the food they ingest or drink it as free water. Some tenebrionid beetles of desert areas manage to obtain water from the dry atmosphere that surrounds them. L. J. and M. Milne (1980) observed the process in the Namibian desert. The beetles are dull black, with hind legs disproportionally long. This allows them to stand or walk with their body at an angle (45° or more), their heads close to the ground. The individual they observed was standing head down in the faint predawn light, and presumably heat was radiating off its body. "Moisture from the saturated air condensed on the insect's back." As droplets grew larger, they coalesced and trickled down to the insect's mouth. The Milnes are of the opinion that this method of obtaining water in dry areas is the result of a combination of morphological characters (long hind legs) and behavioral traits (standing or walking head down), and that this is one of the functions of similar structures and behavior seen in similar tenebrionids in south western United States and Libya.

8.2.3 Scramble and Contest Competition

Food is a finite resource which, depending upon the number of individuals requiring it, may become scarce and the reason for competition. Whether intra- or interspecific, competition can result in none of the individuals obtaining even the minimum of the resource required (pure *scramble* competition) or only one or a few individuals obtaining the amount of resource needed (*contest* competition). In most phytophagous insects acute competition for food seems to be rare. In most cases the population density is reduced by mechanisms such as dispersal, or by the action of natural enemies—parasitoids, predators, and pathogens—before the density becomes high enough to trigger much competition (cf. Chapter 15).

Because there are no victors in pure scramble competition, adaptations could hardly arise that are applicable to this hypothetical type of interaction. Nevertheless, adaptations for quickly finding patches of an ephemeral resource and for its rapid utilization by a highly gregarious progeny are commonly associated with a scramble type of resource utilization, and some of these patches of resource may be so overpopulated that few or none of the larvae can mature. Nicholson (1955) studied the effect of food shortage (scramble competition) on laboratory populations of the sheep blowfly, *Lucilia cuprina*. When 30 larvae were given 1 g of steer's brain, 16.5 adults were obtained. As density was increased, the issuing adults were fewer and smaller, until none was able to develop when density reached 200 larvae.

At 180 larvae only a very few adults were produced and they had a very low fecundity.

Price (1975) considers scramble as "the most common form of competition in insects." In some groups, under certain conditions, scramble competition does occur frequently, though locally. For example, when the California red scale, *Aonidiella aurantii*, develops to high densities on individual leaves of orange and other citrus, it causes premature dropping of the leaves and the death of all the still-feeding scales on them (my observation). The fatal outcome for the individuals involved may be due not only to the depletion of food but also to reaction of the plant to saliva injected by the insects or to other disoperative effects (Allee et al. 1949).

In solitary endoparasitoidism, that is, the condition in which a single parasitoid can develop within a host, there are situations when *contest* competition, either intra- or interspecific, occur. Ovipositing females are able to discriminate, to a greater or lesser degree, between a host that is not parasitoidized (suitable) and one that is already parasitoidized by either a conspecific or a different species (unsuitable). When there are too many parasitoids to be distributed among the available hosts, restraint against ovipositing in an already parasitized host breaks down and oviposition takes place, occasioning superparasitoidism (all parasites of the same species) or multiple parasitoidism (parasites belong to two or more species). The elimination of supernumerary parasitoids (all but one) commonly occurs by direct combat among the parasitoids within the host. First-stage larvae of a great number of hymenopterous parasitoids have sharp mandibles, a heavily sclerotized head, and a highly mobile body which renders these larvae well adapted for combat. Supernumeraries can be also eliminated by physiological suppression (e.g., depletion of oxygen or secretion of toxin by the victor) or otherwise rendering the host unsuitable for the losing competitor. Competition in endoparasitoids has been thoroughly discussed by Fisher (1961) and Salt (1961).

It is intriguing that in species of *Copidosoma*, *Litomastix*, and *Pentalitomastix*, and probably in other Copidosomatini (Hymenoptera:Encyrtidae) there are two types of larvae, one of which is apparently a defender of the other against competitors (Cruz 1981). All species of Copidosomatini that have been studied are polyembryonic, that is, a single egg develops into two or more adults. The parasitoid lays the eggs in the host's eggs. Embryonic development is very slow, remaining as a growing mass of undifferentiated cells inside the developing host larva. Eventually larvae are formed that consume all of the host's tissue by the time the host larva has reached the last stage. The number of adult parasitoids developing in a single host varies according to the species: from about 35 in *Copidosoma desantisi* developing in *Phthorimaea operculella* (Lepidoptera:Gelechiidae) to about 3000 in *Litomastix maculata* developing in *Trichoplusia ni* (Lepidoptera:Noctuidae). Early in the embryonic development of the parasitoid, while the great majority of the embryonic mass is undifferentiated, a few groups of cells differentiate rapidly into fully formed larvae that promptly set themselves free in the host. These larvae have very sharp mandibles, a heavily sclerotized head, and are very active. Invariably they eventually die without developing. They are continually formed until the normal larvae, which will develop into adults, dissociate from the polygerm. Because they appear very early in the development of the polygerm, these larvae are called precocious (Doutt 1952, Cruz 1981). Experimenting with an undescribed species of *Pentalitomastix*, which is parasitoidic on *Anagasta kuehniella* (Lepidoptera: Pyralidae), Cruz (1981) demonstrated that when an *Anagasta* larva was multiple parasitoidized, interspecific competition occurred, the precocious larva invariably being the winner. It is conceivable that the same function is performed by these larvae in the other species where they are found. The larvae represent a successful "altruistic" adaptation to protect their siblings during their protracted embryonic development when they are not equipped to defend themselves.

Normally, under natural conditions, populations occur in low enough densities as to avoid intense resource competition. But experimentally, very dense populations can be induced and competition studied. According to Marshall (1981), Mock (1974) observed that when cattle were prevented from grooming, populations of the louse *Damalinia bovis* rose to levels of 80 adults and hundreds of eggs and nymphs per cm^2 in certain favored places, and an average of 20 adults/cm^2 over the entire body. At these high densities strong competition for food and space occurs. Under artifically induced high densities of Anoplura, Cimicidae (Hemiptera) and Nycteribiidae (Diptera) abnormally frequent copulations occur which may result in death of females (Buxton 1947, Ryberg 1947, Usinger 1966).

In any event a mechanism to avoid intense competition and to locate new food resources is dispersal. Species that exhibit this mechanism are adapted to increase their dispersive proclivity when the density increases. In delphacid Homoptera wing polymorphism is determined by environmental factors acting on the nymphal stages (Denno 1976, Kisimoto 1956, 1965, Mochida 1973, Raatikainen 1967). Mori and Kiritani (1971) determined that crowded conditions during the nymphal stage of the brown plant hopper *Nilaparvata lugens* induced a progressive increase in the proportion of macropterous females in subsequent generations. A number of other homopterans are similarly adapted (cf. Chapter 7).

Herbivory itself may induce changes in the quality of the foliage in certain plants, rendering them less suitable for development of various herbivores. Heavy defoliation of red oak, *Quercus rubrum*, by the gypsy moth, *Lymantria dispar*, induces high concentration of tannin and phenolic compounds on damaged and replacement leaves of defoliated trees to levels above those known to retard development of the gypsy moth larvae (Schultz & Baldwin 1982). It is conceivable that a similar change in foliage quality was the reason that gypsy moth larvae developed more slowly and produced lighter weight pupae, as compared with control individuals, when fed foliage from previously defoliated black oak (*Q. nigra*); the larvae also suffered an 80% nondisease mortality after the third instar (Wallner & Walton 1979, Richerson et al. 1978). This change in the chemistry of the host plant increases the severity of competition in the gypsy moth.

8.2.4   Ecological Segregation and Coexistence

Competition can affect the area occupied by a population. When the requisites of two species are such that the areas where they can live overlap more or less, and there is competition, the possible area of overlap is reduced to a minimum. This is an expression of the general phenomenon of *ecological segregation*, according to the specific adaptations of the species concerned. If in certain segments of the perimeter of the area occupied by a population there are no competing species, the area occupied is extended in that part of the perimeter (Margalef 1977).

In England various species of ants colonize tree stumps. Generally there is no interspecific competition and *Myrmica rubra* can be found all over the wood, although it is more densely grouped on the south-facing side of the stumps. But when the same stump is also colonized by *Formica fusca*, which is a dominant species that inhabits the south-facing side, *M. rubra* is restricted to the north-facing side; thus both species coexist in the same stump. Occasionally *M. scabrinodes* competes with *M. rubra* and displaces it. *F. fusca* may also displace both species of *Myrmica*. The extent of competition in these and other ants (e.g., *Acanthomyops*) can be noticed in the changes of the territorial boundaries of adjacent ant hills (cf. Chapter 18).

Ecological displacement (segregation) in a wide area has occurred in southern California where three species of aphelinids (Hymenoptera), imported for control of the California red scale, segregated themselves into different areas according to their adaptability to each area. The first species, *Aphytis chrysomphali*, was imported from the Mediterranean area before 1900. It was successfully established and became the most common parasitoid in the area occupied by the scale. In 1948 *A. lingnanensis* was imported from southern China and colonized profusely. By 1958 *A. chrysomphali* had been almost completely displaced from most of the area. In 1956–57 *A. melinus* was imported from India and Pakistan. By 1961 *A. melinus* had displaced *A. lingnanensis* from the interior areas where it accounted for 94–99% of the total *Aphytis* populations (DeBach & Sundby 1963). The adaptations accounting for these consequences are debated and still under study.

On the other hand in the galls incited by the sawfly *Pontania pacifica* on the leaves of the arroyo willow (*Salix lasiolepis*) in California, there are several species that utilize the gall tissues as food. The largest portion of the gall is consumed by the sawfly larva. The cosmopterygid moth *Batrachedra salicipomonella* and the weevil *Anthonomus sycophantha* use the tissues of this and similar galls as food, so they are dependent upon the gall makers for their dietary requirements. Both the moth and the weevil commonly develop in galls vacated by the full grown sawfly larva. But it is not rare to find galls simultaneously inhabited by any two of these three species. In these cases contest competition ensues. Whenever the sawfly is involved in the competition, it is the loser. When the moth and the weevil compete, the weevil is the winner (Caltagirone 1964). In this case competition does not result in displacement of populations (ecological segregation). Populations of the three species *coexist* in the same environment, and it is common to find individuals of the three species in galls on the same leaf. The frequency and intensity of the competition is probably dependent on the relative densities of the populations involved.

8.2.5 Specialized Adaptations of Predatory Insects

Carnivory, in its varied expressions from predation to parasitoidism, has provided the basis for a large number of adaptations. A predatory insect is one that consumes more than one prey individual to complete its development. This habit occurs in the majority of insect orders. A species can be predaceous as larvae and/or adults, and their prey consist of a variety of groups: other insects, mites and spiders, molluscs, fish, and tadpoles, among others. Predators occur in a variety of environments and their adaptations to the predaceous mode of life constitute a broad spectrum of morphological, physiological, and behavioral traits that bear witness to the adaptive plasticity of insects as organisms.

Preying mantises are well-known predators. Their prey include aphids and other similar soft-bodied insects for the first-instar nymphs and insects of any size and consistency for the adults. Mantids exhibit obvious adaptations to the predaceous mode of life. Perhaps the most conspicuous is the first pair of legs, which are raptorial, with strong coxae and femora and spiny tibiae that can be folded tightly against the corresponding femora. These legs are adapted to catch and hold prey. The highly mobile head— mantids can turn their heads over their shoulders—is provided with large eyes and powerful mandibles. If a prey is spotted, but beyond reach, the mantid approaches it very deliberately, carefully taking each step, causing no major disturbance; the target is always kept within the field of vision, which is accomplished by turning the head in the required direction and adjusting the course of displacement as needed. During this process the front legs are cocked. As soon as the intended victim is within reach, it is caught by extending the legs very rapidly, grabbing the prey, and retracting the legs with the prey in them. The whole sequence is almost instantaneous. Consumption of the prey begins immediately. If a mantis is very hungry and is consuming a prey, and another potential prey lands within reach, the predator will try to catch it with one front leg only, holding the partially consumed one with the other. Frequently the attempt is successful.

Many species of mantids have the general color of the environment in which they live. This and their ability to hold themselves very still make it possible for prey to land within reach. Some tropical mantids take the shape and color of flowers which attract other insects that are potential prey, an obvious adaptation to facilitate predation. It also serves the purpose of avoiding being preyed upon (see below).

A similar method of catching prey is used by dragonfly naiads. They are aquatic and feed on other aquatic insects, tadpoles, crustaceans, and small fish. Adaptations that make it possible for the naiads to catch prey include a highly modified labium. This structure is very long, hinged across the middle, and the labial palpi are modified into two sharp, opposable claws. At rest the labium is folded under the head. The naiads also have well-developed eyes. When hungry, a naiad moves slowly toward a spotted target or waits until a potential prey comes within reach. It strikes by quickly extending the labium, grabbing the prey with the labial claws, and bringing it (the prey) against the mandibles and maxillae by folding back the labium.

Adult dragonflies are also predators, feeding on a variety of flying insects which are caught in flight. The adult dragonfly has large eyes; a thorax that has been modified in such a way that the legs are clustered near the head, close to the mouthparts. These legs are not suited for walking; they are used for perching and to catch prey.

The larvae of green and brown lacewings (Neuroptera: Chrysopidae and Hemerobiidae) prey on soft-bodied insects such as aphids, mealybugs, and small lepidopterous larvae. These predators have their mouthparts adapted for catching and consuming prey. The mandibles are sickle shaped and grooved on the inner side. The lacinia is tightly appressed to this groove, forming a canal through which the fluids of the prey are sucked. A prognathous head and an elongate, agile body with strong thoracic legs form a set of adaptations that facilitate catching prey.

The larvae of tiger beetles (Cicindelidae) spend their lives in burrows dug in the soil or plants. At the entrance of these burrows they wait for their prey, which include terrestrial crustaceans, centipedes, spiders, and insects. Tiger beetle larvae exhibit remarkable morphological adaptations: the mandibles are sharp and curved; and head and pronotum are heavily sclerotized and form dorsally a somewhat circular plate positioned at an angle with respect to the rest of the body, which is flexible and S-shaped; the tarsal claws are sharp; the fifth segment bears dorsally a hump on which there are heavily sclerotized, forward-directed hooks. The larvae of tree-inhabiting species lie in their burrows with the head in the entrance, mandibles wide open, waiting for prey. When a prospective prey crawls within range, "the predator suddenly throws his head forward, simultaneously emerging a short distance out of the entrance and snapping the mandibles shut upon the prey, then backing into the burrow again" (Balduf 1935).

Enock (1903), as cited by Balduf (1935), observed the larva of *Cicindela campestris*, which is a soil-inhabiting species. He discovered that the tarsal claws were driven at right angles into the sides of the gallery, the dorsal hump wedged up close to the thorax with the spines driven at right angle into the opposite side of the gallery, with the anal spines driven into the wall also. "In this manner the larva was held securely in place in its peculiar zigzag form, safeguarding it especially against possible forced removal should large prey be seized. Yet, by releasing the anal spines and straightening the abdomen it could drop down to the bottom of the burrow instantly" (Balduf 1935).

The catching of prey has reached a high level of sophistication in the predaceous lampyrid beetles in the genus *Photuris*. In the family Lampyridae males and females find each other by emitting light signals. Flying males emit light bursts according to highly specific patterns. The females, usually stationary, respond to the flash patterns of males of their own species with specific patterns of their own. In turn the flying male responds to the appropriate female signal by approaching, landing, courting, and mating. Most adult fireflies are short lived and do not feed at all, but females of the genus *Photuris* have been found to feed on other beetles, including males of the lampyrid genus *Photinus*. Upon perceiving a flashing male *Photinus*, a female *Photuris* responds with a flash that mimics that of the female *Photinus*. As the male *Photinus* approaches, the female *Photuris* even reduces the intensity of her flashes, thus resembling closely the weaker signals of the smaller female *Photinus*. After landing the hapless male is seized and eaten by the *Photuris* female. In response to males of their own species, of course, the female *Photuris* gives a flash response quite different from that of *Photinus* (Lloyd 1965, 1975).

Certain predators increase their success in catching prey by camouflaging themselves. Eisner (1982) reports an interesting case. The wooly alder aphid, *Pociphilus tesselatus*, is mutualistic with the ant *Camponotus noveboracencis*: the ants defend the aphids from intruders and consume the honeydew secreted by the aphids. The ants prevent wasps from getting to the honeydew and attack any intruder in the aphid colony; they attacked the tools the observers used to scare the ants. In the colony, larvae of *Chrysopa slossonae*, which looked just like the aphids, were preying on the aphids without being noticed by the ants nearby. The predator camouflages itself with waxy strands which it plucks from the aphids with its mandibles and places them on its dorsum until totally covered, assuming the appearance of the prey. When naked predator larvae were released in an aphid colony, the first priority for the predator was to rebuild the waxy coating. If naked predator larvae were released in an ant-tended colony, the predators were promptly discovered by the ants and removed.

Comprehensive discussions on predatory insects, their biologies, adaptations, and behavior are given by Balduf (1935, 1939), Clausen (1940), and Sweetman (1958).

8.2.6 Specialized Adaptations of Parasitoids

Many carnivorous insects develop in or on other arthropods, maintaining
a rather lengthy trophic association with the food, the host, at least during
the larval stage. These insects are called parasitoids, protelean parasites,
or simply parasites. Differing from predators, which normally require more
than one individual of the prey to complete larval development, parasitoids
utilize only one individual of the host. Parasitoidism is found only in the
Hymenoptera, Diptera, Strepsiptera, Coleoptera, and Lepidoptera, the great
majority of species belonging in the first three orders. Most of the hosts
belong in the Insecta (most orders) and the Arachnida (spiders and ticks).

Detailed information on insect parasitoids is found in Askew (1971),
Clausen (1940), and Price (1974).

The diversity of hosts and environments where parasitoids are found
suggest that a number of adaptations have made these trophic associations
possible. In most cases the females are the ones that locate the hosts that
will receive the eggs or larvae from which the next generation will develop.
Depending upon the species, the host may be located by sight or sound or
recognized by means of the antennae or ovipositor. In a number of parasitoids the eggs are deposited away from the host, which is eventually
contacted through various mechanisms.

Females of *Perilitus coccinellae*, a parasitoid of various lady beetle adults,
recognize and pursue moving red-colored lady beetles. Some tachinid flies
(e.g., *Compsilura concinnata*) react to oviposit only when the host moves. Flies
in the families Pipunculidae, parasitic on leafhoppers and spittlebugs, and
Conopidae, which parasitize adult bees or wasps, use their vision to locate
their host. The adults have well-developed eyes.

The tachinid fly *Euphasiopteryx ochracea* is parasitic on adult crickets.
When a cricket sings, the female fly is attracted to it and parasitizes it. Its
ability to locate the host through sound was demonstrated when a recording
of *E. ochracea* were attracted to the speaker from which the sound came
and they larviposited on a dead cricket that was tied to the speaker (Cade
1975).

By far the most common way for a gravid female parasitoid to locate
the host is by using the antennae. When a hymenopterous parasitic female
is searching for a host, she usually scans the substratum with the antennae;
if a potential host is present, she will locate it by following chemical trails
left by the host. In the coevolution between some parasitoids and their
hosts, the parasite has developed the ability to recognize chemical substances normally produced by the host. A number of parasitoids utilize
these substances, called kairomones, as cues to locate hosts (Weseloh 1981)
(cf. Chapters 4 and 5).

Some tachinid flies produce very small eggs which have a hard, smooth
exochorion. These eggs are deposited on vegetation away from the host.
The fly larvae will hatch only if the eggs are ingested by hosts. Hatching

takes place as soon as the parasitoid egg reaches the midgut. The maggot bores through the midgut into the host's body cavity where it will complete development. *Cyzenis albicans*, a parasitoid of the winter moth, *Operophtera brumata*, belongs in this group. Other tachinids, among which are species in the genera *Peleteria* and *Archytas*, lay fully incubated eggs away from, but in the vicinity of, hosts. The parasitoid larva hatches immediately but remains attached to the substratum by the partially shed chorion. The larvae take an erect stance and sway about. If a caterpillar passes close enough, the maggot will attach to it and eventually will bore its way into the host.

The Acroceridae, a group of flies that parasitize spiders, lay eggs in sites apparently unrelated to the presence of the host. The larvae hatch and move about, frequently stopping and standing erect swaying. When a spider passes within reach, the larva gets on and eventually penetrates the host through a membranous area. A way of reaching their food very similar to that of the Acroceridae is found in wasps of the genus *Perilampus*. Species in this genus are primary parasitoids (i.e., parasitoids of a primary consumer or predatory arthropod) or secondary parasitoids (i.e., a parasitoid of a parasitoid). *P. chrysopae*, a primary parasitoid of larvae of the green lacewing, *Chrysopa carnea*, lays its eggs on vegetation, regardless of whether or not the host is present. The newly hatched larvae are heavily sclerotized and very active. From time to time they stop, stand erect, and sway. If a green lacewing larva passes by, the *Perilampus* larva will climb on the host and take a position along a seta with its head on the host's integument. It will remain in this position until the lacewing larva has spun its cocoon. Only then will it penetrate the host's integument and develop as an endoparasitoid. If the parasitoid larva reaches a host that is younger than full grown, when the host molts the parasitoid abandons the exuvium at an appropriate moment and again takes the usual position on the new integument.

Among various groups of parasitoids of insect eggs that are laid in masses, there are species in which the female is transported by the adults of the host species. The advantage of this adaptation is that the parasitoid, which has a low reproductive capacity, can distribute its progeny more efficiently. No time is spent looking for host eggs because the females are there when the eggs are being laid. Phoresy of adult females of egg parasitoids has been reported in the hymenopterous families Scelionidae, Trichogrammatidae, Encrytidae, Eupelmidae, Eulophidae, and Torymidae (Clausen 1976). In some cases, such as in the scelionid *Mantibara manticida*, the parasitoid is totally dependent on the phoretic relationship with the adult host: the parasitoid sheds its wings shortly after attachment to the host. *M. manticida* is parasitic on egg masses of *Mantis religiosa*. The females attach themselves to adult mantises, either males or females, but those attached to males will transfer to females when mating occurs. In addition to the advantage of not having to search for hosts, the parasitoid can oviposit in newly laid, not yet hardened egg masses.

The physiological and behavioral adaptations for obtaining the required kind and/or amount of food can reach, in some insects, levels of complexity that defy a complete evolutionary interpretation of the process. The sphecid wasp *Solierella pekhami* female hunts for older nymphs of the chinch bug *Nysius raphanus*, paralyzes them, and carries them to a nest (any crevice that can give some protection). Some 10–15 nymphs are stored, and an egg is laid on one of them. The larva feeds on the paralyzed prey. Females of the chrysidid wasp *Pseudolopyga taylori* search for first- or second-stage nymphs of *N. raphanus* in which a single egg is laid. The chrysidid larva hatches and remains in the bug nymph apparently without causing any damage and without developing beyond the first-stage larva. If a *Nysius* nymph containing a *Pseudolopyga* larva is paralyzed by *Solierella* and taken to the nest, the chrysidid develops fast, emerges from its host's skin, searches for the *Solierella* egg, which is promptly destroyed, and then completes its development feeding on the paralyzed *Nysius* nymphs. If a bug nymph containing a *Pseudolopyga* larva is not paralyzed, the nymph continues developing normally to the adult stage, and the chrysidid dies (Carrillo & Caltagirone 1970).

Parasitoids, especially endoparasitoids (i.e., those that complete their larval stages inside the host) exhibit a number of adaptations that have resulted in their specialized mode of obtaining food.

The majority of insects in fact, including many parasitoids, lay eggs that contain enough nutrients for the embryo to develop into a fully functional first-stage larva. But in a number of endoparasitoids the egg contains very little or no yolk. In these the chorion is shed very early, and the embryo is surrounded by a membrane, the cells of which absorb nutrients from the host and transfer them to the embryo. The origin of this membrane can be the serosa (species with meroblastic cleavage) or the polar bodies (species with holoblastic cleavage) (Tremblay & Caltagirone 1973). The fate of the trophic membrane is known in very few species. In *Perilitus coccinellae*, after the embryo completes development, the cells disperse in the host and grow very large, and the parasitoid larva feeds on these cells. But in the ichneumonid *Larthrostizus euurae*, an endoparasitoid on certain gall-inducing sawflies, after the embryo frees itself from the membrane, the host encapsulates the membrane, which remains in the host as a melanized body (Caltagirone 1964).

Respiration is an oxygen obtaining function for which parasitic insects, in fact all insects, have developed admirable adaptations. The open tracheal system found in the vast majority of insects is somewhat modified in those that live in special situations, as do parasitoids and aquatic insects. In aquatic insects tracheal gills, air bubbles, and plastrons are devices that permit gaseous exchange. In endoparasitoids various types are found. In many of the Hymenoptera respiration is through the cuticle (cutaneous). In some encyrtids (Hymenoptera) that are parasitoidic on soft scales and mealybugs, the larva respires through a special device in the egg. Along one side of the egg pedicel and extending onto the main body there is a spongy area, the aeroscopic plate. When the egg is laid, a portion of the pedicel with the aeroscopic plate is left to protrude outside, while the rest of the egg is inside the host. When the larva hatches, it remains attached to the chorion with the caudal pair of spiracles appressed to the aeroscopic plate. After each molt, the old cuticle is pushed back and the new spiracles become attached to the plate.

Tachinid fly larvae, which are all endoparasitoidic, have only one pair of functional spiracles, the anal pair. To respire, the parasitoid attaches itself either to the integument or to a tracheal trunk, and with its spiracular teeth wounds the area of attachment. This triggers a reaction by the host, and hemocytes accumulate around the wound; melanization also takes place. These structures induced by the parasitoid and formed by the host are known as respiratory funnels. A single host may exhibit several respiratory funnels induced by a single larva as a result of the parasitoid changing feeding sites.

In the nemestrinid flies *Trichopsidea* and *Neorhynchocephalus*, which are endoparasitoidic on grasshoppers, the respiratory adaptation consists of a single, very long respiratory tube attached at one end to the host's integument and at the other to the parasitoid's caudal spiracles, thus the larva has access to atmospheric air (Greathead 1958, Spencer 1958).

8.3 ADAPTATIONS TO AVOID BEING EATEN

"Natural history, to a great extent, is a tale of different adaptations to avoid predation" (Gould 1977). As consumers, phytophages and carnivores exert a great selective pressure on the populations subjected to their actions. The result is a diversity of adaptations that tend to reduce the effect of the eater on the eaten.

Insects have developed various strategies to minimize the effect of carnivores on them. These strategies have been grouped in a number of categories (Blum 1981, Eisner 1970, Mathews & Mathews 1978, Milne & Milne 1980, Robinson 1969), examples of which are given here.

8.3.1 Crypsis

This category includes all those adaptations that render the prey undiscernible from the background. These are (1) homochromy, (2) countershading, and (3) disruptive coloration. In the majority of the cases the camouflaged insect feeds at night while during the day it remains motionless, thus avoiding detection by visually hunting predators, movement being one of the conditions that renders a camouflaged target visible.

Many species of moths that are active at night are quite undiscernible during the day. One of the better known cases of homochromy in moths and of the conditions that make this kind of camouflage effective is that of the European pepper moth, *Biston betularia*. The moths are white or creamy-white in color, with dark markings scattered over the wings and rest of the body. When they rest on lichen-covered trunks, they blend perfectly with the background. In a normal population occasionally a dark, almost black individual appears, but these phenotypes are not abundant. Environmental changes, such as industrialization in England, resulted in a change in color of the adults in the vicinity of factories: the light-colored form became scarce, while the dark-colored form became very abundant; by late 1895 these populations consisted mostly of black forms. The black form is dominant, but because it is conspicuous to predators when resting on light colored backgrounds (formerly the normal), they were easily decimated. With the onset of industrialization, pollutants killed lichens on the trees and rendered the bark sooty dark. The dark forms of *B. betularia* were then camouflaged while the light forms were exposed to predators. This case is illustrative of the way natural selection operates, in this case through predatory birds, and also of the value of camouflage as protection against predators.

A number of species in various taxonomic groups match the color of the background against which they remain motionless during the day, becoming inconspicuous to predators. Noctuids, pyralids, and other moths exhibit the color and patterns of the background on which they rest. Mantids, walking sticks, long-horned grasshoppers (katydids), and membracids resemble leaves, twigs, or spines. Individuals of the same population may adopt a different color according to color changes in the background. The African grasshopper *Ruspolia differens* is a large insect (about 6 cm long) that remains hidden in grass during the day. Two distinct color forms are the most abundant: solid green (63.34%) and brown (33.38%); the other colors or patterns are scarce: purple-striped green (2.89%), purple-headed brown (0.05%), purple-headed green (0.32%), and purple striped (0.02%). The colors in the grasses where the grasshoppers live can be similarly ranked, green being by far the most abundant, followed by brown and finally various patches of purple on either leaves or stems. In the area where this grasshopper occurs a number of predators, including at least 16 species of birds (among which are kites, herons, storks, cranes, kingfishers, shrikes, crows, starlings, sparrows, and chicken), rats, jackals, civets, genets, mongoose, dogs, and cats are prevalent. This complex of predators suggests that polymorphism to match the background in *Ruspolia differens* evolved as a mechanism to avoid predation through camouflage.

Countershading, that is, different colors of certain parts of the body, according to the intensity and direction of the light, rendering the individual inconspicuous, is most commonly found in mammals, birds, fish, and insects. Organisms that are countershaded are darker on the sides that are directly exposed to light. Fish, except for deep bottom dwellers, are normally dark grey on the dorsum, while ventrally they are much paler, making it difficult for predators to distinguish them from their background.

Similarly, insects that exhibit this adaptation orient themselves with the darker areas toward the light, as do larvae of the sphingid moth *Smerinthus ocellata* whose ventral side is darker. These larvae rest on the underside of twigs, their ventral side toward the light, producing a flattened, leaflike image that blends with its surroundings.

Startling coloration is another mechanism that results in avoiding predation. Some cryptic colored insects have conspicuous patches of vivid colors which are exhibited only when disturbed. Some moths have brightly colored hind wings which, when the insect is at rest, are covered by the cryptic colored fore wings. If disturbed, as when discovered by a predator, the moth suddenly exposes the hind wings, dramatically changing its appearance. This causes the predator to hesitate, giving the intended prey the opportunity to escape, or the predator may then look for prey elsewhere. In grasshoppers it is common to find species that are perfectly camouflaged, having brightly colored hind wings. When the grasshopper is disturbed (e.g., by a searching predator) the hind wings are exposed at the same time the grasshopper takes to the wing. Presumably the predator concentrates on the bright colored areas, which vanish when the grasshopper conceals its hind wings just before landing.

8.3.2 Mimicry

Insects that are cryptic convey the message that "they are not there." The message from those that are mimics is that "they are different than what they really are." Mimicry has been extensively studied; among the publications are those of Alcock (1975), Mathews and Mathews (1978), Owen (1980), Pasteur (1982), Rettenmeyer (1970), and Wickler (1968).

Insects that exhibit mimetic adaptations can be placed in either of two major groups: those that mimic inedible objects and those that resemble species that are unpalatable.

Twigs, leaves, thorns, and bird droppings are among the inanimate, inedible (for their predators) objects mimicked by insects. Some geometrid larvae resemble leafless twigs to the smallest detail; some noctuid adults sit motionless during the day seeming to be a part of the twig where they rest. Walking sticks (phasmids) resemble twigs, and katydids and some praying mantids and moths resemble leaves so perfectly that some of them have spots that simulate disease symptoms; others have the margins indented as if they had been fed upon by a phytophage.

In all cases involving mimicry of inedible objects, as are also those involving crypsis, the success of the mimic depends on its ability to stand still in daylight. The most perfect mimic can be promptly discovered if it moves within the visual range of a hunting predator. The prevalence of these mimics in a variety of environments is a strong indication of effectiveness of these adaptations in protecting the mimics from predators that locate their prey by sight.

There are insects that are brightly colored—orange and black butterflies, red and black lady beetles, yellow and black wasps and caterpillars—thereby making themselves conspicuous in their surroundings, yet predators in general avoid them. These insects are poisonous, unpalatable or otherwise noxious to predators. Accumulated evidence indicates that naive predators accept these insects initially, but reject them before swallowing or become ill after ingestion. Predators quickly learn to associate appearance, especially color, with unpalatability and soon these aposematically colored forms are not preyed upon by experienced predators. Such insects may become models in mimetic adaptation, as with the monarch butterfly and most acraeid butterflies.

When handled, acraeid butterflies exude large quantities of a foamy substance that smells like hydrogen cyanide. This substance has been found in the secretion of *Acraea entedon*. Most acraeids are orange and black or black and white in color. They are slow fliers and seldom take evasive action when threatened.

Monarch butterflies (Danaidae) feed as larvae on milkweeds (Asclepiadaceae), many of which contain cardiac-active steroids in the form of glycosides or cardenolides, which in addition to being heart stimulants induce vomiting, diuresis, central nervous system disorders, and muscle spasms. Clayton (1970) wrote: "It seems clear that these highly unpleasant effects are protective against grazing animals." These compounds, stored in the larvae, are transferred to the monarch pupae and adults. When a naive predator ingests a monarch, it becomes ill and vomits. Monarchs are conspicuously colored orange and black or white and black.

Because predators learn to associate color and appearance with unpalatability, a number of species that are palatable but resemble (mimic) unpalatable species, such as the Acraeidae and Danaidae mentioned above, are protected from experienced predators. This is known as Batesian mimicry, found in many orders of insects. These mimics commonly resemble bees, wasps, inedible butterflies, or beetles. Ladybird beetles (coccinellids) and leaf beetles (chrysomelids) are inedible and exhibit conspicuous colors, usually with contrasting spots. A group of Philippine cockroaches in the genus *Prospoplecta* mimics these beetles, having undergone profound modifications in accomplishing the similarity. To simulate the short, rounded form of the ladybirds, the large hind wings of the roaches are rolled and folded in a manner unparalleled in other insects. The Lepidoptera abound with Batesian mimics. A well-known case is that of the palatable swallowtail butterfly, *Papilio dardanus*, a widespread African species. In many of its populations the females are polymorphic, each morph being a mimic of a locally unpalatable butterfly of another genus, either *Danaus* or *Amauris* (Danaidae). However, we note that in California 47% of the population of *Danaus plexippus* and in Ghana 84% of the *D. chryssippus* feed on nontoxic plants and are quite palatable (Turner 1977).

The advantage of Batesian mimicry to the mimic has been experimentally demonstrated. When toads (*Bufo terrestris*) were fed live bees, they were exposed to the noxious stings. After the experience these toads avoided the palatable drone fly *Eristalis vinetorum*, which mimics bees. If the toads were fed bees from which the sting had been removed, the toads subsequently ate the drone flies quite readily (Brower & Brower 1962). Similar results were obtained when blue jays were tested with unpalatable models (monarchs) and palatable mimics (viceroy butterflies) (Brower 1958; cf. Ricklefs 1979).

When a species obtains nourishment from both toxic and nontoxic plants, a certain proportion of a population (those that developed on nontoxic plants) are mimics of those that have become toxic. As noted above, monarch caterpillars feed on milkweeds (*Asclepias*). Of the species in this genus *Asclepias humistrata* and *A. curassavica* produce cardiac glycosides that make adult monarchs noxious to birds. Larvae fed on other species of *Asclepias* produce perfectly palatable adult monarchs. Where Asclepias that do and do not produce cardiac glycosides occur together, as in Florida, monarch populations may consist of both palatable and unpalatable individuals. The palatable ones are mimics of their unpalatable conspecifics (automimicry) (Brower 1969, 1970, Brower et al. 1968, cited by Ricklefs 1979).

When groups of unrelated species, all unpalatable, possess the same conspicuous warning coloration, they are referred to as Müllerian mimics. In Batesian mimicry the mimics derive the greatest protection when the model is considerably more abundant than the mimic. In these latter circumstances the probability that naive predators will catch unpalatable specimens prior to a palatable one is large; thus it learns to avoid the model (and the mimic) before taking a toll from the mimic. On the other hand, in Müllerian mimicry all species share in the learning process proportionally to their frequency, so the cost of predator learning to each mimic is much less than if the species involved were different in their coloration.

8.3.3 Chemical Adaptations

A number of insects have been successful at exploiting their ability to produce and use natural products to avoid being eaten by predators. Comprehensive treatments of this subject are those of Blum (1981) and Eisner (1970).

Natural products contained in saliva have been reported as having deterrent effects on predators. The assassin bug *Platymeris rhabdamanthus* produces saliva that contains a trypsinlike protease, hyaluronidase, and phospholipase. When disturbed, the bug can eject saliva at a distance of about 30 cm (12 in.). Depending upon the degree of excitability, the bug can spit once or twice, up to 15 times, at a rate of 3–5 spits/sec. The aim is very accurate over a firing arc of about 65°, which results from the ability to deflect the penultimate segment of the rostrum. If the secretion gets to the eyes or nose of a predator, intense local pain, vasodilation, and edema result (Edwards 1961, 1962, cited by Blum 1981).

An example of the refinement of the biogenesis of some defensive chemicals is the production of 1,4-quinones in the ground beetles *Brachinus crepitans* and *B. explodens* (Carabidae). The paired pygidial glands consist of the gland proper, a reservoir, and a highly sclerotized firing chamber. The gland produces two initially nonreactive compounds which are stored in the reservoir in solution: 10% hydroquinone and 25% hydrogen peroxide. When the beetle is threatened, the mixture is forced into the firing chamber where glands in its walls secrete catalases and peroxidases, which decompose the peroxide into oxygen and water and oxidize the hydroquinone to 1,4-quinone which is forcibly expelled by the generated oxygen. Because the reaction is exothermic, the expelled solution can reach a temperature of 100°C (212°F) (Thiele 1977). Production of 1,4-quinone is found in other carabids and in various termites, grasshoppers, and other beetles (tenebrionids, staphylinids) (Blum 1981).

Saliva, which has the property of lowering the surface tension of water, makes it possible for the water strider *Velia capraii* (Hemiptera: Veliidae) to escape predators. When disturbed, the water strider ejects saliva through the rostrum onto the water surface behind. As a result of lowering of the surface tension, the rapidly receding water propels the bug for a distance of 10–25 cm. (4–10 in.) (Blum 1981). If the cause of disturbance is a predator, the fast displacement may make it possible for the water strider to escape safely (cf. Chapter 15).

Blood as a deterrent against predators is of common occurrence. In some cases, for example, in larvae of the chrysomelid beetles *Diabrotica undecimpunctata* and *D. balteata*, it acts as a purely physical deterrent by its copiousness and rapid clotting properties. In other cases, such as blister beetles, ground beetles, and lady beetles, the blood is enriched with natural products produced by the insects themselves (cantharidin, alkaloids) or contains sequestered natural plant products such as cardenolides in the milkweed bug, *Oncopeltus fasciatus*. These products render the secreted blood highly repellent to predators.

Other secretions used by insects to escape predators include entangling saliva (syrphid fly larvae against ants), viscous abdominal secretions (cockroaches against ground beetles, ants, and centipedes), secretions that become viscous shortly after being ejected (termite soldiers against predaceous ants), cornicle secretion (aphids not tended by ants against predators), and froths containing cardenolides (aposematic colored grasshoppers).

8.3.4 Parental Care

Parental care of the progeny (eggs, larvae) is present in social insects (Chapter 18) and various groups of nonsocial insects. This seems to be an adaptive behavior to protect the progeny from predators. Female water bugs (*Bel-*

ostoma) glue their eggs on the back of males. Until the nymphs hatch, the males fight off any predator that may threaten the progeny. Maternal care of eggs and larvae has been amply documented in the bethylid wasps *Goniozus* and *Prosierola* (Clausen 1940, Doutt 1973, Malyshev 1968). The behavior of the females of these gregarious ectoparasites of moth larvae is similar in the various species that have been studied: the female paralyzes a host, lays eggs on it, and remains with her progeny until the larvae are full grown and start spinning their cocoons. Doutt (1973) observed that the female of *Prosierola bicarinata* would aggressively attack any moving object nearby, such as the tip of a brush or needle. The attack involved attempts to bite and sting the object. This behavior possibly protects the progeny from predators, parasites, or competitors (same or different species).

It is logical to conclude that a female that invests considerable time caring for her progeny is spending less time in distributing her progeny. A study by Bristow (1983) strongly suggests that females of the treehopper *Publilia reticulata* have solved this dilemma by delegating certain "maternal duties" to tending ants. *P. reticulata* feeds on ironweed (*Vernonia sp.*). Eggs are laid in leaves and the females remain with their eggs and young nymphs. This parental care, found also in other species of treehoppers, results in protection of the progeny from parasites and predators and in the maintenance of the aggregation. Both adult females and nymphs produce honeydew, avidly sought by ants. In colonies tended by ants the female treehopper left the colony about 6 days after nymphs hatched, becoming free to produce additional colonies, while in colonies not tended by ants the female remained with the nymphs for about 32 days after the nymphs hatched.

8.4   ADAPTATIONS FOR REPRODUCTION

The great majority of insects reproduce sexually, each individual developing from a zygote (fertilized egg), the result of the fusion of two gametes—an egg and a sperm—each produced by a different individual, a female and a male, respectively. However, in some Iceryini (Homoptera Margarodidae), especially in the cottony cushion scale, *Icerya purchasi*, both eggs and sperm are produced in the same individual (hermaphrodite) in which fertilization takes place (Pierantoni 1913, Hughes-Schrader 1925, 1927). The hermaphrodites in this species are similar to the females in other Iceryini. Occasionally males appear in *I. purchasi*; apparently they are the result of eggs failing to get fertilized. These males are capable of copulating with hermaphrodites, but hermaphrodites are incapable of fertilizing one another (White 1973). *I. bimaculata* is also hermaphroditic, but some individuals seem to be pure females (Hughes-Schrader 1963).

Because fertilization of the egg is internal in the female, sperm must reach the *spermatheca*—a sperm storage pouch in the usual female reproductive system. In the large majority of species transfer of sperm takes place directly into the female genitalia during copulation. However, in Odonata transfer of sperm is indirect. An accessory genitalia, which temporarily stores and transfers sperm to the females, is present on the second and third abdominal sternites of males. Prior to copulation the male transfers sperm through the gonopore in the eighth abdominal sternite to the accessory gentialia; then he grasps the female by the occiput or the thorax (depending on the group) with claspers at the tip of his abdomen; then the female bends her abdomen downward, forward, and upward, bringing her genitalia in contact with the accessory genitalia of the male. Copulation and transfer of sperm then takes place. In the rest of the insects the sperm is transferred to the female directly through the male genitalia. In four families of the Cimicoidea (Hemiptera)—Cimicidae, Anthocoridae, Polyctenidae, and some Nabidae—insemination does not take place during normal copulation; instead, it is traumatic: the integument of the female is pierced by the male genitalia and the sperm is injected into the hemocoel, eventually reaching the eggs. The mechanisms of traumatic insemination in the Cimicoidea vary from species in which the integument may be pierced at almost any place in the abdomen (e.g., *Primicimex cavernis*) to those in which the piercing of the integument occurs at a specialized organ, the *spermalege*. In *Stricticimex brevispinosus* the spermalege is connected to the oviducts. The sperm migrates through this connection to the female reproductive system. In the bed bug, *Cimex lectularius*, the sperm is injected through a spermalege that has no connection with the female reproductive system (Carayon 1966; cf. Engelmann 1970). (Cf. Chapter 5)

In the Strepsiptera, which are all endoparasitic on insects, mature females remain in the host with only the cephalic end protruding outside. Ventrally there is a longitudinal brood chamber that has a single opening to the outside. Internally there are a series of genital canals that open into the brood canal. Eggs are produced in the ovaries and float in the hemocoel. Male strepsipterans are winged and short lived. During their short life they search for hosts containing females. Copulation takes place through the opening of the brood chamber into which sperm is deposited. Spermatozoa penetrate into the hemocoel through the genital canals and fertilize the eggs. Upon completion of embryonic development the larvae migrate to the brood chamber through the genital canals and out through the brood chamber opening.

In the rest of the sexually reproducing insects transfer of sperm to the female during copulation (insemination) results in accumulation of spermatozoa in the spermathecae (above), which is connected with the median oviduct. Fertilization, that is, penetration of a spermatozoan into the egg and eventual zygote formation, takes place as the egg passes from the ovary into the median oviduct through the area into which the spermathecal duct opens. In these species mating is indispensable for fertilization to occur.

Parthenogenesis, or complete thelytoky—the form of reproduction in which females produce female progeny without fertilization—has arisen several times in groups of insects that normally reproduce sexually (cf. Chapter 5). Thelytokous species and/or "races" are found in several of the orthopteroid orders and in Lepidoptera, Homoptera, Coleoptera, and Hymenoptera. In some aphids and gall wasps (Cynipidae) thelytoky alternates with sexual reproduction in a more or less regular fashion. This is called cyclical thelytoky or heterogony (White 1973).

Some sexual species are capable of producing thelytokous populations under as-yet-undetermined circumstances. *Venturia* (= *Nemeritis*) *canescens*, an endoparasitic ichneumonid, is thelytokous in the Nearctic region. In Spain there is a sexual population of *V. cansecens* in which the sex ratio is about 1:1. Specimens of this population were imported to California and mass produced in the insectary. After the third-generation males were no longer produced; the population continued reproducing parthenogenetically (this author, unpubl.). The same change is occurring in a species of *Trathala* I collected in southern Spain. The field-collected specimens exhibited a sex ratio of about 1:1. The sex ratio of the first few generations was strongly biased in favor of males; however, later generations showed a predominance of females. At present the laboratory colony consists of mostly females (L. Caltagirone & L. Etzel, unpubl.). These populations have the capability of producing parthenogenetic individuals, and under certain conditions these individuals produce descendants that eventually eliminate the sexually reproducing ones. This is in agreement with Maynard Smith's (1971, 1978) theoretical analysis of these two reproductive strategies. However, the conditions prevailing in Spain do not favor expression of parthenogenesis, and the species remain sexual.

In a few groups of insects males develop parthenogenetically from unfertilized eggs, while females develop from fertilized eggs. This mode of reproduction, known as arrhenotoky, haplo-diploidy (males are haploid and females diploid), or haploid parthenogenesis, is found in the Thysanoptera, Homoptera, Coleoptera, and Hymenoptera. In the Homoptera haplo-diploidy is found in the Iceryini and some whiteflies. The males of armored scales are haploid, but they originate from fertilized eggs in which the paternal chromosomes are discarded early in development (Brown 1963, Brown & Bennett 1957, Brown & McKenzie 1962). In the Thysanoptera and Hymenoptera haplo-diploidy seems to be the rule, except for the thelytokous species (White 1973).

In arrhenotokous species females are produced only be mated females, consequently mating is essential for reproductive continuity. Unmated females produce males only, while mated females produce either only females or both sexes. In species in which mated females lay only fertilized eggs, unmated females are indispensable.

The aphelinid *Prospaltella porteri* is an endoparasite whose progeny develop in two taxonomically different hosts: female larvae develop in nymphs of the greenhouse whitefly, *Trialeurodes vaporariorum* on cucurbits, while male larvae develop in eggs of the noctuid (Lepidoptera) *Rachiplusia nu* on alfalfa. Unmated females visit alfalfa where they find suitable hosts for development of unfertilized eggs and where there are adult males with which to mate. After mating these females are no longer attracted to alfalfa and noctuid eggs but to cucurbits and whitefly nymphs (Rojas 1968). A similar case involving *Prospaltella* sp. has been reported by Gerling (1967).

In the aphelinid *Coccophagoides utilis*, an endoparasite of the olive scale, *Parlatoria oleae* (Homoptera: Diaspididae), mated females search for scales into which they lay fertilized eggs that develop into females. Unmated females also search for scales but select as oviposition sites only those that contain a female larva of its own species (*C. utilis*) upon which an unfertilized egg that will develop into a male will be laid. Thus, the males are parasitic on females of their own species (Broodryk & Doutt 1966). Zinna (1962) reported a similar biology in *Coccophagoides similis*.

Although in the cases just described it is theoretically possible for a population to reach the point of extinction if at some time only unmated females oviposit, in reality the host–parasitoid complexes adjust themselves in such a way that there always seem to be enough unmated females to produce males and enough males to inseminate females.

In most insects, as in most sexual species, males and females are produced in approximately equal numbers. However, a number of them, representing diverse taxonomic groups, normally exhibit a sex ratio that is biased in favor of females. This deviation from the 1:1 sex ratio, apparently in conflict with the widely accepted explanation as to why in a sexual population the number of females should be approximately equal to that of males (Fisher 1958), has been explained by Hamilton (1967). These species show the following characteristics:

1. reproduction is haplo-diploidic;
2. females are decidedly more numerous;
3. the immature stages are gregarious or closely clumped;
4. there is at least one male in each clump;
5. male adults emerge first;
6. males mate many times in rapid succession;
7. mating takes place immediately after females emerge or even before;
8. males stay with the clump, sometimes being territorial; and
9. the amount of sperm received by each female is sufficient to fertilize all or most of the eggs.

In populations in which the sex ratio is about 1:1 it seems logical to assume that every individual has an equal chance of mating and that competition for mates is not common, but this is not so. There are species in which competition for females has led to the development of very elaborate mating behavior (cf. Chapter 17).

Where they arrive before females, male dragonflies establish territories and chase away competitors. Competition for territory results in reduction of potential density at breeding sites, and dispersal of males from these to other sites (Moore 1964). The size of the territories depends on the microtopographic areas, delimited by features such as overhanging shrubs, and on population density. Kormondy (1961) observed territoriality in *Tetragoneura cynosura* and noticed that at the time of greatest density males flew at intervals of 10–30 ft, 3 ft above water and 1–3 ft away from shore. In these territories males will mate. In some cases an area may contain individuals of different species, males of each of them establishing their own territories over the same pond (Corbet 1962). When a territory occupied by *T. cynosura* is entered by an individual of the same species, the owner of the territory responds quickly, exhibiting an exploratory behavior. If the intruder is a male, aggressive behavior is displayed; if it is a female, copulatory activity takes place. Aggressive behavior in the dragonfly *Plathemis lydia* consists of the intruder being chased away, the pursuer flying with his abdomen raised while the intruder flies away with his abdomen lowered. This continues for 8–16 m, then the roles are reversed, the pursuer becoming the pursued. This routine is repeated back and forth several times until the newcomer settles in a nearby site (Corbet 1962).

In the Odonata competition occurs not only for territories in which to mate and where the female will oviposit but also involves mechanisms by which the sperm of a previous male is removed before copulation, thus ensuring that the eggs that will be laid will be fertilized by the sperm of the last male to mate with a female. This is the case in the damselfly *Calopteryx maculata*. In this species the male uses the penis first to remove any sperm that may be in the female genitalia and then to transfer new sperm (Waage 1979).

The pteromalid *Cheiropachus colon* is a parasitoid of the larvae of the shothole borer, *Scolytus rugulosus*, and other scolytid beetles. The larvae pupate in the galleries carved by the host. When ready to emerge, the adult cuts a hole through which it exits. If the emerging adult is a female, as soon as the bark is broken the males in the vicinity congregate around the hole, touching it with the antennae and pushing the others away. This strongly suggests that emerging females produce a sexual pheromone. I have observed as many as 20 males converging at an exit hole, each struggling to get onto it and trying to stay there, waiting for the female to emerge. Russo (1938) reports that as the female is emerging a fight ensues among the males, and when the most able among them succeeds in grabbing her and positions himself on her dorsum, the other males abandon the struggle. My own observations are slightly different: as the emerging female enlarges the hole, the intensity of the struggle for the female increases. As soon as the female puts her head through the exit hole, the closest male grabs her with his mandibles and legs. At that moment all struggle by the other males ceases.

Internecine activity among males has been reported in two species of *Melittobia* (Hymenoptera: Eulophidae) by Mathews (1975). Both species are gregarious parasitoids of solitary wasps. Up to 150 parasitoids, all progeny of a single female, may develop in a single host. In *M. chalybii* from 3 to 4% of the progeny are males. When males encounter each other they fight; although death does not occur, disabling damage usually results. *Melittobia* sp. adopts a different strategy. The first male to emerge seeks out male pupae and systematically decapitates them. No male was observed attacking females. This aggressive behavior, which allows the victor to mate with all the females in the brood, takes place in spite of the fact that the males are all brothers.

In the dung fly *Scatophaga stercoraria*, if a female mates repeatedly, sperm from the different males is stored in the spermatheca, but most of the eggs she will lay in a batch will be fertilized by sperm from the last male with which she mated immediately before ovipositing. Males in this species gather on fresh cow pads and compete for females that come to oviposit. After mating the male withdraws his gentialia but remains in possession of the female while she lays her batch of eggs, fighting off other males that attempt to mate with her. If a male is successful in displacing a guarding male, he immediately mates with the female and keeps hold of her until she finishes ovipositing.

A number of other adaptations that increase the probability that the reproductive potential of individuals is realized have evolved in the insects. A more extensive discussion of the subject is found in Engelmann (1970), Halliday (1980), and Mathews and Mathews (1978) (cf. Chapter 5).

8.5 CLOSING REMARKS

As emphasized in Chapter 23, humans frequently find themselves in antagonistic relationships with insects. The same food or the same sources of food that humans require are, in many instances, also required by insects. It is widely recognized that insects, together with plant pathogens and weeds, take an important proportion of the food and other commodities that humans need or demand. Insects also interfere with human welfare through the various species of medical importance which feed on blood and other fluids, causing discomfort, or in addition, and more important, vector organisms that cause disease. So, although there are many species that from the human point of view are beneficial (e.g. . . . pollinators, enemies of pests), it is understandable that during the extended period in which humans have been adversely affected by insects, procedures to fight them have been developed. By and large, and in spite of dramatic failure in some instances, the effort to stay ahead of pestiferous insects can be considered successful. However, the ability of insects to adapt themselves to various and varying modes of life and to effect adaptive changes rather rapidly should make us reject the notion that we can eliminate insects as competitors.

The case of the pepper moth, described above, is a sobering reminder of the ability of insects to adapt. Another one, and of a more ominous nature, is their developing of resistance to many of the chemicals that have been in use since the beginning of what may be called the modern period in the fight against pestiferous insects. The synthetic organic pesticides of the mid 1940s and 1950s were considered by many crop protectionists as the tools that were going to make it possible to keep insect pests under control permanently. The insects, through their adaptability, proved the optimism wrong. Up to 1980 resistance in insects to one or more pesticides had been reported in 375 species. Based on the total number of arthropods that have been reported as resistant (428 species), there has been a 2.65-fold increase in the decade from 1971 to 1980 (Georghiou 1982) (cf. Chapter 9).

Whenever new tactics for crop protection are developed or any procedure involving change in environmental conditions (creating new agricultural varieties, changing cultural practices, or damming rivers) is considered, the great adaptability to change that insects possess must be kept in mind.

REFERENCES

Alcock, J. 1975. *Animal Behavior—An Evolutionary Approach*. Sinauer, Sunderland, Mass.

Allee, W. C., O. Park, A. Emerson, T. Park, and K. P. Schmidt. 1949. *Principles of Animal Ecology*. Saunders, Philadelphia.

Askew, R. R. 1971. *Parasitic Insects*. Elsevier, New York.

Balduf, W. V. 1935. *The Bionomics of Entomophagous Coleoptera*. Swift, New York.

Balduf, W. V. 1939. *The Bionomics of Entomophagous Insects, Part II*. Swift, New York.

Blum, M. S. 1981. *Chemical Defenses of Arthropods*. Academic, New York.

Bristow, C. M. 1983. *Science* **220:** 532–533.

Broodryk, S. W. and R. L. Doutt. 1966. *Hilgardia* **37:** 233–254.

Brower, J. V. Z. 1958. *Evolution* **12:** 32–47, 123–136, 273–285.

Brower, J. V. Z. and L. P. Brower. 1962. *Am. Natur.* **96:** 297–308.

Brower, L. P. 1969. *Sci. Am.* **220:** 22–29.

Brower, L. P. 1970. Pp. 69–82 in K. L. Chambers (Ed.), *Biochemical Evolution (Proc. Annu. Biol. Coll. 1968)*. Oregon State University, Corvallis.

Brower, L. P., W. N. Ryerson, L. L. Coppinger, and S. C. Glasier. 1968. *Science* **161:** 1349–1351.

Brown, S. W. 1963. *Chromosoma* **14:** 360–406.

Brown, S. W. and F. D. Bennett. 1957. *Genetics* **46:** 510–523.

Brown, S. W. and H. L. McKenzie. 1962. *Hilgardia* **33:** 141–170.

Buxton, P. A. 1947. *The Louse*. Arnold, London.

Caltagirone, L. E. 1964. *Ann. Entomol. Soc. Am.* **57:** 279–291.

Cade, W. 1975. *Science* **190:** 1312, 1313.

Carayon, J. 1966. *The Thomas Say Foundation* **7:** 81–166.

Carrillo, J. L. and L. E. Caltagirone. 1970. *Ann. Entomol. Soc. Am.* **63:** 672–681.

Chapman, R. F. 1971. *The Insects—Structure and Function.* Elsevier, New York.

Clausen, C. P. 1940. *Entomophagous Insects.* McGraw-Hill, New York.

Clausen, C. P. 1976. *Annu. Rev. Entomol.* **21:** 343–368.

Clayton, R. B. 1970. Pp. 235–279 in E. S. Sondheimer and J. B. Simeone (Eds.), *Chemical Ecology.* Academic, New York.

Corbet, P. S. 1962. *A Biology of Dragonflies.* Witherby, London.

Cruz, Y. P. 1981. *Nature* **294:** 446, 447.

DeBach, P. and R. A. Sundby. 1963. *Hilgardia* **34:** 105–166.

Denno, R. F. 1976. *Ecol. Entomol.* **1:** 257–266.

Doutt, R. L. 1952. *Can. Entomol.* **84:** 247–250

Doutt, R. L. 1973. *Ann. Entomol. Soc. Am.* **66:** 486, 487.

Edwards, J. S. 1961. *J. Exp. Biol.* **38:** 61.

Edwards, J. S. 1962. *Proc. 11th Int. Congr. Entomol.* **3:** 259.

Eisner, T. 1970. Pp. 157–217 in E. S. Sondheimer and J. B. Simeone (Eds.), *Chemical Ecology.* Academic, New York.

Eisner, T. 1982. *BioScience* **35:** 321–326.

Engelman, F. 1970. *The Physiology of Insect Reproduction.* Pergamon, Oxford.

Enock, F. 1903. *Proc. Entomol. Soc. London.* xv–xix

Fisher, R. A. 1958. *The Genetical Theory of Natural Selection.* Dover, New York.

Fisher, R. C. 1961. *J. Exp. Biol.* **38:** 605–628.

Georghiou, G. P. 1982. *Proc. Int. Workshop Resistance to Insecticides Used in Public Health and Agriculture.* Nat. Sci. Council Sri Lanka, Colombo, pp. 46–62.

Gerling, D. 1967. *Ann. Entomol. Soc. Am.* **60:** 1306–1321.

Gould, S. J. 1977. *Ever Since Darwin. Reflexions in Natural History.* Norton, New York.

Greathead, D. J. 1958. *Proc. Entomol. Soc. Lond., A* **33:** 107–119.

Halliday, T. 1980. *Sexual Strategy.* University of Chicago Press, Chicago.

Hamilton, W. D. 1967. *Science* **156:** 477–488.

Hughes-Schrader, S. 1925. *Z. Zellforsch.* **2:** 264–292.

Hughes-Schrader, S. 1927. *Z. Zellforsch.* **6:** 509–540.

Hughes-Schrader, S. 1963. *J. Morph.* **113:** 173–184.

Kisimoto, R. 1956. *Nature* **178:** 641, 642.

Kisimoto, R. 1965. *Bull. Shikoku Agric. Exp. Stn.* **13:** 1–106.

Kormondy, E. J. 1961. *J. N.Y. Entomol. Soc.* **69:** 42–52.

Lloyd, J. E. 1965. *Science* **149:** 653–654.

Lloyd, J. E. 1975. *Science* **197:** 452–453.

Malyshev, S. E. 1968. *Genesis of the Hymenoptera and the Phases of Their Evolution.* Methuen, London.

Margalef, R. 1977. *Ecologia.* Omega, Barcelona.

Marshall, A. G. 1981. *The Ecology of Ectoparasitic Insects.* Academic, New York.

Mathews, R. W. 1975. Pp. 66–86 in P. W. Price (Ed.), *Evolutionary Strategies of Parasitic Insects and Mites.* Plenum, New York.

Mathews, R. W. and J. R. Mathews. 1978. *Insect Behavior.* Wiley, New York.

Maynard Smith, J. 1971. Pp. 163–175 in E. G. Williams (Ed.), *Group Selection.* Aldine-Atherton, Chicago.

Maynard Smith, J. 1978. *The Evolution of Sex*. Cambridge University Press, Cambridge.

Merritt, R. W. and K. W. Cummings (Eds.). 1978. *An Introduction to the Aquatic Insects of North America*. Kendell/Hunt, Dubuque, Iowa.

Milne, L. J. and M. Milne. 1980. *Insect Worlds*. Scribner's, New York.

Mochida, O. 1973. *Trans. Roy. Entomol. Soc. Lond.* **125:** 177–225.

Mock, D. E. 1974. *The Cattle-biting Louse, Bovicola bovis (Linn.). I. In Vitro Culturing, Seasonal Population Fluctuations, and Role of the Male. II. Immune Response of Cattle*. Ph.D. dissertation, Cornell University, Ithaca, N.Y.

Moore, N. W. 1964. *J. Anim. Ecol.* **33:** 49–71.

Mori, A. and H. Kiritani. 1971. *Jap. J. Ecol.* **21:** 146–152.

Nicholson, A. J. 1955. *Aust. J. Zool.* **2:** 9–65.

Owen, D. 1980. *Camouflage and Mimicry*. University of Chicago Press, Chicago.

Pasteur, G. 1982. *Annu. Rev. Ecol. Syst.* **13:** 169–199.

Pierantoni, U. 1913. *Arch. Zool. Ital.* **7:** 27–49.

Price, P. W. (Ed.). 1974. *Evolutionary Strategies of Parasitic Insects and Mites*. Plenum, New York.

Price, P. W. 1975. *Insect Ecology*. Wiley, New York.

Raatikainen, M. 1967. *Ann. Agr. Fenn.* **6:** 1–149.

Rettenmeyer, C. W. 1970. *Annu. Rev. Entomol.* **15:** 43–74.

Richerson, J. V., E. A. Cameron, D. E. White, and M. Walsh. 1978. *Ann. Entomol. Soc. Am.* **71:** 60–64.

Ricklefs, R. E. 1979. *Ecology*. Chiron, New York.

Robinson, M. H. 1969. *Evol. Biol.* **3:** 225–259.

Rojas, S. 1968. *Revista Chilena Entomol.* **6:** 123–125.

Russo, G. 1938. *Boll. R. Lab. Entomol. Agr.* (Portici) **2:** 1–418.

Ryberg, O. 1947. *Studies of Bats and Bat Parasites*. Bökforlaget Svensk Natur, Stockholm.

Salt, G. 1961. *Symp. Soc. Exp. Biol.* **15:** 96–119.

Schultz, J. C. and I. T. Baldwin. 1982. *Science* **217:** 149–151.

Spencer, G. J. 1958. *Proc. 10th Int. Congr. Entomol.* **4:** 503–509.

Sweetman, H. L. 1958. *The Principles of Biological Control*. Wm. C. Brown, Dubuque, Iowa.

Thiele, H.-U. 1977. *Carabid Beetles in their Environments*. Springer-Verlag, Berlin.

Tremblay, E. and L. E. Caltagirone. 1973. *Annu. Rev. Entomol.* **18:** 421–444.

Turner, J. R. G. 1977. *Evol. Biol.* **10:** 163–206.

Usinger, R. L. (Ed.). 1963. *Aquatic Insects of California*. University of California Press, Berkeley.

Usinger, R. L. 1966. *Monograph of Cimicidae*, Vol. 7. Entomological Society of America, Thomas Say Foundation, College Park, Md.

Vermeij, G. J. 1978. *Biogeography and Adaptation*. Harvard University Press, Cambridge.

Waage, J. K. 1979. *Science* **203:** 916–918.

Wallner, W. E. and G. S. Walton. 1979. *Ann. Entomol. Soc. Am.* **72:** 62–67.

Weseloh, R. M. 1981. Pp. 79–95 in D. A. Nordlund, R. L. Jones, and W. J. Lewis (Eds.), *Semiochemicals—Their Role in Pest Control*. Wiley, New York.

White, M. J. D. 1973. *Animal Cytology and Evolution*. Cambridge University Press, Cambridge.

Wickler, W. 1968. *Mimicry in Plants and Animals*. McGraw-Hill, New York.

Wigglesworth, V. B. 1964. *The Life of Insects*. New American Library, New York.

Zinna, G. 1962. *Boll. Lab. Entomol. Agr. F. Silvestri (Portici)* **20:** 73–184.

Chapter 9

Evolutionary Processes in Insects

GUY L. BUSH and
MARJORIE A. HOY

9.1 INTRODUCTION

Insect ecology is becoming inextricably bound with the study of the evolution and population genetics of insects. The diversity and variety of insect species, in both numbers of individuals and numbers of habitats exploited, emerges from a study of insect ecology and inevitably leads to the question of how? and why?. Insects comprise about 80% of the described animals in the world; over 900,000 described species as contrasted to 60,000 Chordata. The incredible diversity in form and function exhibited by insects invites enthusiasm, amazement, and sometimes, awe.

What are the attributes of insects that have contributed to their evolutionary success as measured by numbers of species, numbers of individuals, and numbers of habitats exploited? The Insecta represent one of the most complex arrays of evolutionary experimentation among eukaryotic organisms. Insects exhibit metameric segmentation with appendages capable of modification, and this, coupled with the mechanical advantages of an exoskeleton, has been exploited in diverse ways resulting in evolutionary novelty. Wings promoted widespread dispersal and escape and increased the range of feeding and breeding opportunities. A most intriguing adaptation is complete metamorphosis in higher insects, which has opened up a variety of habitat and food possibilities in which immature forms and adults often exploit radically different habitats and thus avoid competition. Insects tend to be small, have a high reproductive rate, a short generation time, diverse modes of dispersal, and diverse genetic systems. All of these attributes, coupled with their fantastic diversity, makes insects highly desirable experimental organisms for the study of evolutionary processes. It is not surprising, therefore, that insects such as *Drosophila, Tribolium, Habrobracon,* and others have contributed fundamentally to our understanding of ecology, genetics, behavior, and evolutionary theory.

Evolution by *natural selection* is an elegantly simple principle first proposed by Darwin (1859). Only two conditions are necessary for evolution to occur: *reproduction* and *hereditary variation* such that it influences the success of reproduction. So, evolution is the result of differential reproduction. As a result of natural selection, the most fit and best adapted individuals survive to reproduce and thus transmit their genes to the next generation. Although the principle is simple, the processes involved in adaptive evolution are complex and are treated in the following sections.

9.2 MATERIAL BASIS FOR EVOLUTION—GENETIC VARIATION AND REPRODUCTION

Fundamental to understanding the evolutionary process in insects is an appreciation of the underlying genetic basis for adaptation. Ultimately, the raw material on which natural selection acts is genetic variation, generated and maintained in a variety of ways. We first consider the origin and types of variation and then examine how it is molded by selection in the course of adaptation.

9.2.1 The Origin of Genetic Variability

Adaptation and evolutionary change ultimately arise by action of natural selection and stochastic events on genetic variability. What is the nature of this variability and how is it expressed? To understand the role genetic variation plays in the adaptive process, we must understand the genetic material itself.

9.2.1.1 The Nature of the Genetic Material. In recent years our concept of what constitutes a gene has changed rapidly and is still not resolved. Recent advances in molecular biology have revealed that the genetic material in eukaryotes is far more complex and diverse in its structure, organization, and function than previously imagined. Part of the genome consists of *unique* nucleotide sequences that probably encode for most of the genes eventually translated into enzymatic and nonenzymatic proteins. Other nucleotide sequences, which code for a protein or ribosomal RNA that is produced in large amounts, may be *repeated* many times. Nucleotide sequences that give rise to a diffusible product are called *structural genes*. In the case of a protein-producing locus, one structural gene encompasses one polypeptide unit. This polypeptide may represent a functional protein or serve as a subunit incorporated into a more complicated protein molecule (Lewin 1980). *Regulatory genes* control *timing* and *rate* of structural gene expression. Regulatory genes may or may not produce a gene product. Even when a gene product is produced, it may not be translated into a protein but serves as a nuclear RNA molecule of special function in gene expression that never finds its way into the cytoplasm. Unraveling the structure and function of the regulatory process is now a most active area of research.

Unique nucleotide sequences coding for proteins may constitute a relatively small part of the total DNA in eukaryotes. The rest, sometimes collectively called *heterochromatin*, may represent up to 80–90% of the genome. Heterochromatin is made up of highly reiterated, short to moderately long repetitive sequences. Some sequences may be repeated millions of times and dispersed throughout the chromosomes in discrete blocks at specific sites. The genetic role of heterochromatin is complex and poorly understood. At one time heterochromatin was regarded as inert and of little importance in genetic expression. Indeed, some molecular geneticists still regard most of it as "parasitic," or "selfish," DNA with no function other than its own replication and survival. Today we know that it is directly involved in such diverse processes as chromosome pairing, recombination, chromosome rearrangement, gene regulation and developmental programming, speciation, and macroevolution (Bonner 1982), with the list of function and evolutionary role increasing as new discoveries are made.

Extrachromosomal genes represent a fourth class that reside outside the nucleus as circular DNA in eukaryotic organelles such as mitochondria and chloroplasts. In many respects these genes resemble those of prokaryotes, and there is convincing evidence that they represent highly modified symbiotic microorganisms acquired during early stages of eukaryotic evolution (Margulis 1981). Characterization of these genes is now advanced. Because of their small size and the ease with which they can be isolated and their DNA sequence, they are becoming increasingly important as tools in a wide range of molecular evolutionary studies ranging from tracing lineages and polygenetic relationships among taxa by restriction enzyme mapping to establishing rates of speciation (Brown 1981).

9.2.1.2 The Organization and Function of Structural Genes.

Structural genes in eukaryotes are similar to genes in prokaryotes only in the general pattern of their expression. That is, the DNA code for protein is transcribed to an RNA copy of *messenger* RNA (m-RNA). This m-RNA is carried out of the nucleus and translated on ribosomes while interacting with different transfer RNAs (t-RNA) that order amino acids into a specific sequence along the m-RNA molecule. Eukaryote genes, however, are now known to be quite different from those found in prokaryotes in organization, regulation, process, and order on the chromosome (Lewin 1980).

Eukaryotic structural genes frequently consist of far more nucleotides than are ultimately translated into protein. The transcribed nucleotide sequences that code for protein (*exons*) may be interrupted by a sequence of transcribed, but nontranslated, sequences of nucleotides called *introns* which are enzymatically excised before translation. The functional role of introns is unclear, but they have been implicated in recombination, chromosome rearrangements, genetic control of cell differentiation during embryogenesis, and other cellular and immunological processes. They may also facilitate reshuffling of exons to create new genes (Gilbert 1978). A gene coding for a polypeptide unit also includes long stretches of nucleotide sequences at either end that do not produce a gene product.

A recent startling discovery is the presence of a previously unrecognized class of DNA sequence called a *transposon*. This special DNA sequence, sometimes referred to as a "jumping gene," was first described in corn by McClintock (1951). Transposons, under certain conditions, can jump from one chromosome to another and may be a major mechanism of chromosomal mutations and rearrangements. They may not only cause mutations but also alter the pattern of gene expression and evolution (Bonner 1982).

Regulation of eukaryote genes may be influenced by the presence of histone and nonhistone proteins coupled with the chromosome. Also, gene control may be altered by diminution, amplification, rearrangement, and modification or modulation through transcriptional, posttranscriptional, and translational control (Brown 1981). Indirect evidence suggests that such proteins (not present in prokaryotes) are somehow involved in regulating gene expression, but their mode of action is not clear. Although we lack a clear understanding of how gene regulation works in eukaryotes, it is clear that eukaryotes have evolved an extremely flexible regulatory apparatus. Without such flexibility the complex organ systems and way of life of higher organisms, particularly animals, could not have evolved. This flexibility also provides many avenues of adaptation to cope with new environmental situations.

9.2.1.3 How is the Genetic Material Changed? If genetic variation is the raw material for evolution, how does it arise? By the genetic material we include both structural and regulatory genes as well as the various kinds of heterochromatic DNA. Changes in chromosomal and extrachromosomal DNA or *mutations* fall into two categories: those that result in the alteration of a single nucleotide or *point mutation* and *chromosome mutations* that involve a change in a gene segment or large blocks of DNA (Table 9.1). The effects on fitness of point and chromosome mutations vary greatly. Some may be essentially neutral while others are lethal. Point mutations of structural and control genes, as well as heterochromatin, may result in a limited change in expression of one gene and possibly genes with which it interacts. Chromosomal mutations, however, can result in more extensive reorganization of the genome that alters the pattern of gene expression over many loci simultaneously (Bush 1981). Chromosome rearrangements thus appear to be of prime importance in implementing major adaptive shifts and evolutionary novelty (White 1978). Almost all species, even those most recently

Table 9.1   Types of mutation in eukaryote nuclear DNA

A. Point or allelic mutations causing alterations in:
  1. Structural genes (exons or transcription units) specifying protein
  2. "Operator" genes adjoining a structural gene
  3. "Control" genes
  4. Intron units
  5. "Heterochromatin"
B. Chromosome mutations (involving a gene segment or large blocks of DNA):
  1. Changes in amount of DNA
    a. duplication and deletion of structural, operator, and control genes
    b. addition and loss of heterochromatin
    c. additions or deletions of whole chromosomes (aneuploidy) or complete chromosome sets (polyploidy)
  2. Alteration of gene arrangement within a chromosome (inversions)
  3. Shift or exchange of one or more genes or gene segments between non-homologous chromosomes (translocations)

evolved, differ in the karyotype and overall genetic architecture in some unique way.

9.2.2 A Basis for Evolution—Sex and Reproduction

Although mutations are the raw material on which natural selection acts, genetic variation is affected by several other processes such as *recombination, genetic drift,* and *inbreeding.* Another factor of profound influence on the pattern of genetic variation is the mode of reproduction. Insects and other arthropods have evolved a most diverse array of reproductive mechanisms controlling or affecting the level of genetic variation.

9.2.2.1 Sexual Versus Asexual Reproduction.   Most insects reproduce sexually and are diploid. Many diploid species, however, contain polyploid tissues (i.e., with more than two sets of chromosomes) or tissues with polytene chromosomes that consist of multiple copies of a chromosome united together in a single giant chromosome. Meiosis in diploid species results in production of haploid eggs and sperm. Mating commonly results in transfer of sperm to the female where it is stored and released to fertilize eggs as they pass out the genital tract. In such species recombination occurs each generation by a variety of mechanisms including crossing over, random assortment of chromosomes during meiosis, and by recombining male and female chromosome sets at fertilization.

Parthenogenesis, development of eggs without fertilization, occurs in several forms in insects and has different consequences with respect to genetic variation. *Arrhenotoky* (haplo-diploidy) is the parthenogenetic production of haploid males, while fertilized diploid eggs develop into females. In this case genes are effectively dominant in the male and thus exposed to direct selection every generation. Concealed variability, except for attributes limited to the females, should be reduced vis-á-vis diplo-diploid species. *Thelytoky* is asexual production by way of females only; sterile or fertile males are produced very rarely or not at all. Their genetic variability is predominantly limited to production of mutations, as recombination has been forfeited.

Several cytological mechanisms have been described for parthenogenetic arthropods (White 1973). These include *apomictic* parthenogenesis, in which the egg fails wholly, or in part, to undergo meiosis, so that no reduction in chromosome number results. Since apomictic reproduction is asexual, new gene combinations are unlikely except for new mutations. In *automictic* parthenogenesis the egg undergoes meiosis, and the diploid condition is restored in a variety of ways (i.e., through fusion of two cleavage nuclei, two polar bodies, or a polar body with the egg pronucleus, Tremblay & Caltagirone 1973). Some genetic recombination may be possible.

Parthenogenesis may be faculative or obligatory within a species. It is obligatory in most thelytokous species–males are very rare or lacking. Cyclical parthenogenesis involves alternation of parthenogenetic and sexual generations. Aphids, cynipid wasps, and certain Diptera exhibit this and have very complex life cycles.

9.2.2.2 Mechanisms of Sex Determination. Sex determination in diplo-diploid species is usually the result of genic balance, with the two sexes having different combinations of sex chromosomes. (Sex chromosomes are chromosomes containing the genetic information responsible for sexual differentiation in diploid organisms.) In many cases one of the sex chromosome homologues is visibly different. The sex that carries the heteromorphic chromosome is the heterogametic sex (i.e., XY or ZW), while the other sex is homogametic (XX or ZZ). In *Drosophila* a double dose of X chromosome determines femaleness, while the autosomes contain male-determining genes. Thus, XY or XO *Drosophila* individuals are males because they have one X only; XXY or XX individuals are females, assuming the normal (diploid) complement of autosomal chromosomes is present.

The genic balance theory of sex determination is untenable with haplo-diploid species because the sex-determining chromosomes will have the same relative frequency in both sexes. Several hypotheses currently attempt to explain sex determination in haplo-diploid species (Crozier 1977), and other mechanisms of sex determination have been reviewed (White 1973). Environmental and physiological factors also affect sex determination and differentiation in insects (Bergerard 1972). The evolutionary implications of sexual reproduction have been discussed by Mittwoch (1967), Williams (1975), and Maynard Smith (1978, cf. Chapter 17).

9.2.2.3 Inbreeding, Genetic Drift, and Gene Flow. In sexually reproducing species population structure, patterns of dispersal, and stochastic factors can greatly influence gene frequency. The number and sex of breeding individuals, for instance, may result in radical departures from random mating. In parasitic Hymenoptera brother–sister matings are quite common. In other groups the breeding structure is highly polygynous and a few males may fertilize most of the females. Counts of individuals may be misleading in such cases, as the effective population number (N_e) may be smaller than the apparent numbers would indicate. Thus, in small populations if one male inseminates 100 females, N_e is slightly over 4 because it is proportional to the harmonic mean of N_m and N_f and is strongly influenced by the smaller values (see Crow & Kimura 1970). Under such conditions new mutations can be fixed rapidly through selection, inbreeding, and drift.

Effects of inbreeding and drift are often confused. The immediate outcome of *inbreeding* is to alter genotype frequencies but not gene frequencies. Thus, inbreeding tends to increase the frequency of homozygotes and reduce heterozygotes, each generation making it easier for selection to act on concealed homozygotes. *Genetic drift*, however, can change both genotype and gene frequencies. Genetic drift occurs primarily in small populations and results from genetic "sampling error." When the number of breeding individuals is small, some alleles may be lost or fixed in the population by chance alone, thus reducing genetic variability.

Population structure and the pattern of geographic distribution may also affect gene frequencies and the level of genetic variation within a species. Gene flow by immigration or emigration between populations may increase or decrease the number of alleles, with the magnitude and rate of change determined by the number of migrants involved, size of the population affected, and intensity of selection on new alleles (cf. Crow & Kimura 1970).

9.3 MEASURING GENETIC VARIATION—METHODS AND IMPORTANCE

We have seen that evolution by natural selection arises from genetic variation and sexual reproduction. Measuring and characterizing genetic variation and changes in natural populations has therefore taken on an ever more important role in understanding insect adaptation and evolution. Ideally, we need a cheap, rapid, and simple method for surveying allele frequencies of specific genes of adaptive importance. Thus, if we are studying the genetics of host plant selection by insects, we might want to examine genes controlling the chemoreceptors of different races of insects with differing responses to host plants. Although spectacular advances are being made in molecular genetics, it is not yet possible to provide nucleotide sequences for such genes from large numbers of individuals. For population studies we are more limited to what kinds of genes and gene products we can study.

9.3.1 Gel Electrophoresis

Major advances in our knowledge of the genetic structure of insect populations, however, have been made in recent years through use of a cheap and sensitive method, gel electrophoresis. Allelic forms of protein, differing by only a single amino acid, can often be identified if the substitution alters the overall electrical charge of the molecule. Over 80 proteins can now be routinely studied by gel electrophoresis in some animals. A major limitation of the technique is that only genes coding for soluble proteins can be studied. Membrane, chromosomal, and other proteins cannot be sampled. Furthermore, genetic variation in *regulatory* genes is not amenable to gel electrophoresis, and yet such genes are probably as important from an adaptive and evolutionary standpoint as structural genes. A third limitation stems from the fact that gel electrophoresis underestimates soluble protein variation, as only variants that alter the overall electric charge or configuration of the protein can be detected. Despite this, gel electrophoresis is superior to older methods used to measure genetic variation, which were time consuming and usually restricted to visible morphological characteristics and cytogenetic markers (e.g., chromosome rearrangements). More sophisticated biochemical techniques are also available which provide considerably more detailed information than one-dimensional gel electrophoresis. Two-dimensional electrophoresis, microcomplement fixation, restriction enzyme analysis, and DNA and protein sequencing, however, are more costly and time consuming. A detailed discussion of these methods and those used in insect systematics and biology is presented by Bush and Kitto (1978).

9.3.2 Chromosomal Organization

Cytogenetic and biochemical approaches to the study of chromosome structure have application in the study of genetic variation. Various cytogenetic methods have been developed to establish specific attributes such as chromosome number and form of the chromosome complement or *karyotype* of an organism. Cytogenetics has been the conventional way of looking at the gross anatomy of chromosomes since the turn of the century. Until about 1955, methods remained relatively unchanged and centered primarily on conventional DNA and chromatin stains and the light microscope. In recent years resolution has been greatly increased, and our knowledge of cytogenetic variation enhanced through use of the electron microscope and sophisticated staining techniques. New staining methods permit the cytogeneticist to resolve fine details of chromosome structure and to examine the distribution of heterochromatin and euchromatin. These methods make it possible to establish the presence of very small translocations and inversions, and thus to trace the pattern of karyotypic evolution with greater accuracy (Blackman et al. 1980, Schultz-Schaeffer 1980).

At the biochemical level gene sequencing methods are being used to map chromosome structure and establish the function of various parts of the eukaryotic chromosome, such as the centromere, satellite DNA, and unique sequence DNA. Molecular genetics, therefore, is an increasingly important tool. The methods employed [e.g., DNA sequencing, microcomplement fixation (MCF), DNA hybridization, or gel electrophoresis] will depend on what level of genetic delineation is required. Gel electrophoresis is currently the most inexpensive and effective method for routinely studying the genetic structure of populations, clines, races, species, and closely related genera. For evolutionary studies on higher categories, MCF and DNA hybridization are more appropriate. As techniques improve, DNA and protein sequencing may be used to resolve problems at all systematic levels. Only DNA sequencing offers the capability of establishing the fine details of genetic evolution at the molecular level of the gene, as nucleotide sequences of specific genes and chromosome regions can be compared between species.

9.4   INTRASPECIFIC VARIATION

9.4.1   Genetic Variation Within and Between Insect Populations

The advent of gel electrophoresis and molecular genetics brought into question some long-held theories on the genetic structure of animal and plant populations. Most importantly, the level of genetic variation maintained in natural populations, at least in the structural genes amenable to study, is considerably higher than originally predicted (Lewontin 1974). This has divided population geneticists into two schools. On the one hand the *neutralists* feel that most variation expressed electrophoretically has a minimal effect on fitness and exists in a population because of a combination of factors such as mutation, finite population size, migration, and drift (Nei 1975), with selection playing only a minor role in maintaining different electrophoretic alleles. The opposite view is held by the *selectionists* who believe that most, if not all, electrophoretic variation is maintained by some form of "balancing selection" such as *overdominance* (superiority of the heterozygote), *density-dependent selection, differential selection* between two sexes or life stages, or variable selection over time and space in a heterogeneous environment (Futuyma 1979). Both groups assume that many deleterious substitutions are eliminated by natural selection and that advantageous mutations have a role in protein evolution (Wilson et al. 1977). The controversy is over how the polymorphisms are maintained, and this problem is yet to be resolved.

9.4.2   Factors Affecting Genetic Variation

Environmental heterogeneity is one attribute that has been given particular attention recently as an important factor in maintaining genetic polymorphism in natural populations. Because agricultural practices, such as planting monocultures with little genetic variation, often have profound effects on environmental heterogeneity and the distribution and importance of beneficial and pest insects, we examine a few examples (cf. review by Hedrick et al. 1976). The best examples come from associations between particular genotypes and environmental parameters. For instance, there are now several well-studied cases in both plants and animals that have demonstrated clinal, temporal, and density-dependent variation, as well as "area effects" due to environmental heterogeneity, in allele frequencies (Hedrick et al. 1976).

At another level genetic perturbation studies of natural populations of *Drosophila* have shown that a suite of factors may affect allele frequencies at single loci or over all levels of genetic heterozygosity (Barker & East 1980). Changes in frequency of alcohol dehydrogenase alleles can also be initiated by altering concentrations of various alcohols in the larval food media (van Delden et al. 1975), and the proportion of heterozygous loci per individual can be raised or lowered by increasing or decreasing the number of environmental variables (Hedrick et al. 1976). Levels of heterozygosity can likewise affect survival. Populations heterozygous for the ADH locus appear to be better adapted to different environments than monomorphic ones (Bijlsma-Meeles & van Delden 1974). These experiments have important implications in mass-rearing insects for sterile insect releases, biological control, or pesticide screening. If levels of genetic variation are not maintained at near normal levels, characteristics may be radically altered. This evidence for adaptive changes supports the selectionist theory for maintenance of genetic variation in nature. The gene–environment relationship, however, is not a simple one and caution should be the rule in any attempt to establish cause-and-effect associations.

9.4.3 Chromosomal Variation

Chromosomal variation in natural animal populations arising from chromosome rearrangements (Table 9.1) has been extensively investigated and thoroughly reviewed by White (1973, 1978). *Drosophila* has probably contributed more to this field than any other organism. *Drosophila* (many Diptera) have giant banded polytene chromosomes in various tissues which permit identification of small chromosomal rearrangements, deletions, and duplications. Recent developments have also enhanced the resolution of chromosome studies of non-Dipteran species. With special chromosome banding techniques, many types of chromosome rearrangements can now be recognized (Blackman et al. 1980).

Certain chromosomal polymorphisms in populations are common in some organisms, but rare in others. Paracentric inversions (ones that do not include the centromere) in some *Drosophila* and trimerotropin grasshopper species are widespread, while pericentric inversions, translocations, and changes in chromosome number are encountered in these insects only as rare mutants. In general, chromosomal polymorphisms are less common in other insect groups because they generally reduce the fecundity of heterozygotes. The conventional interpretation of why such chromosomal polymorphisms exist in certain populations is that they tie up blocks of coadapted gene complexes with superior fitness conferred on the heterozygotes (Mayr 1963). Although some laboratory experiments appear to support this view, alternative explanations have been offered (Bush 1982). It can be argued that the inversion polymorphisms found in *Drosophila* maintain a regulatory polymorphism rather than genetic variability or co-adapted complexes.

9.5 EVOLUTIONARY PROCESSES—GENETIC VARIATION BETWEEN SPECIES

9.5.1 Species and Speciation

Before we can consider genetic variation between species and its ecological and evolutionary implications, we must first understand what we mean by the term *species*. What constitutes a species and how species originate are actually one and the same question. How a species is defined will in turn profoundly affect the way an investigator interprets the way new species arise and vice versa. It is for this reason that so much controversy on this subject has been generated over the years. The conventional and most widely accepted definition of a species is based on the "*biological species concept.*" According to Mayr (1963), "species are groups of actually or potentially interbreeding natural populations which are reproductively isolated from other such groups." Although this is a handy operational definition for sexually reproducing eukaryotes, it cannot apply to asexual organisms and is difficult to use objectively in cases involving geographically isolated populations of closely related taxa. Speciation is the development of separate evolutionary lineages over time. If *sympatric* (e.g., their home ranges overlap), taxa must be sufficiently reproductively isolated from one another to indefinitely maintain separate identities. If they coexist in the same area yet retain separate gene pools, irrespective of how phenetically similar they might be, they may be recognized as species. When dealing with closely related, but *allopatric* (geographically isolated) taxa identification of species becomes more subjective. If reproductive isolation can be demonstrated by carefully controlled hybridization tests, species status can be conferred. However, when hybrids and backcrosses are normal, such cannot be inferred. Correct mate selection in nature may depend on certain premating isolating mechanisms such as habitat selection or some other ecological or behavioral cue not furnished in laboratory tests. Species that would never mate in nature may sometimes be artificially hybridized.

Strong morphological differences sometimes encountered between geographic isolates may also give a false impression. Evidence suggests that there is no direct cause-and-effect relationship between the level of genetic and morphological divergence in the degree of reproductive isolation (Wilson 1976). Therefore, taxonomic treatment of many allopatric demes is, at best, an educated guess. Whether two or more populations represent identifiable, independent evolutionary lineages will thus depend on the quality and extent of evidence mustered. Changes that result in divergence of separate lineages are attributed to complex interactions between spatial, temporal, ecological, stochastic, and genetic factors that affect epigenetic, physiological, behavioral, and morphological traits. Because the processes involved are diverse, and rather poorly understood, an unequivocal solution to the species problem frequently voiced by evolutionary biologists is unlikely in the near future.

9.5.2 Methods for Measuring Genetic Differences Between Species

The most direct way to establish if related populations represent two or more species is to study their genetic structure. Because species are reproductively isolated from one another, it seems logical that a measure of genetic differentiation would resolve their status. Unfortunately, this is not necessarily the case. Ideally, we would like to compare nucleotide sequences of major sets of genes on a population basis, but this day has not yet arrived. Indirect methods such as gel electrophoresis, microcomplement fixation, restriction enzyme mapping, and polynucleotide hybridization techniques must therefore be employed. These can only give a rough approximation of the true variability, but they are sufficiently accurate for resolving many population–species problems. Gel electrophoresis is currently the method of choice for most population studies because it is simple and inexpensive. Genetic differences expressed electrophoretically can be used to establish the degree of genetic divergence between populations, species, and sometimes genera.

Several numerical approaches for quantifying electrophoretic data have been developed. Two commonly used are modifications of techniques developed by numerical systematists for analysis of morphological traits. One, called *phenetics*, involves comparisons of taxa on the basis of similarities (Sneath & Sokal 1973). *Cladistics*, on the other hand, establishes phylogenetic relationships on the basis of shared, derived traits (Wiley 1981). Both approaches have their advantages and adherents (Mayr 1981), but cladistic methods may be more appropriate for generating phylogenetic trees over those that employ phenetic methods (Wiley 1981).

The lesson from these studies on genetic variation in and between populations and species is that genetic differences expressed electrophoretically are not necessarily directly correlated with the degree of morphological differentiation. In certain groups large phenotypic differences may be accompanied by little or no measureable genetic divergence and vice versa. Allozymes, and the structural genes they represent, therefore appear to have little to do (at least directly) with evolution of the phenotypic traits used in systematics to establish taxonomic groupings.

Although cladistic methods are increasingly being applied to molecular data, most electrophoretic results in the past were analyzed by phenetic methods to establish differences between species. A similarity coefficient (I) is generated using the allelic variations expressed electrophoretically (Nei 1975). The coefficient of similarity may range from 1 (complete concordance) to 0 (no alleles shared). Thus, if two populations have an I value of 0.99, they share about 99% of their alleles. As the phenetic approach provides useful information on the overall genetic similarity between related taxa it is used here to illustrate some problems of relying too heavily on genetic data to establish species boundaries.

9.5.3 Examples of Genetic Differences Between Species

Much has been written on the origin and evolution of genetic differences between species, hard data on the degree and kind of genetic differences is meager. It is based almost entirely on electrophoretic studies and is reviewed by Nevo (1978), White (1978), and Ayala (1982). In insects comparative estimates are available only for *Drosophila*. This has led to an unbalanced view of the significance of genetic differences between species as this sample hardly reflects the genetic diversity known to occur in the Insecta. There are at least three problems in using electrophoretic data to estimate genetic distances between species. First, the DNA coding for structural genes that produce protein represents only about 10% of the total found in a eukaryote. As yet we have no way of establishing the level of variation or its function in the remaining 90%. A second source of error results from the nature of the genetic code. About 30% of the changes in structural DNA cause no amino acid sequence change in the resultant proteins. Finally, only about 40% of amino acid substitutions alter the overall charge of a protein and thus can be detected by electrophoretic methods.

Since *Drosophila* represents the most thoroughly analyzed group of insects, we will closely examine their pattern of genetic divergence. Dobzhansky et al. (1977) compared the genetic distance between taxa at different levels of evolutionary divergence in the *Drosophila willistoni* group, ranging from local populations to nonsibling species. Their findings are summarized in Table 9.2. Caution should be observed in interpreting these results, particularly with respect to subspecific and semispecies categories.

Table 9.2 Genetic identity and genetic distance between various levels of evolutionary divergence in the *Drosophila willistoni* group[a]

Level of divergence	Identity	Distance
A. Local populations	0.970 ± 0.006	0.031 ± 0.007
B. Subspecies	0.795 ± 0.013	0.230 ± 0.016
C. Semispecies	0.798 ± 0.026	0.226 ± 0.033
D. Sibling species	0.563 ± 0.023	0.581 ± 0.039
E. Full species	0.352 ± 0.023	1.056 ± 0.068

[a]After Dobzhansky et al. (1977).

Both are operational taxonomic terms that are ill defined from a genetic and evolutionary standpoint. *Subspecies* represent geographic races of a species, each recognized on the basis of some phenotypic characteristic. As there is no test of sympatry, the decision to designate geographically isolated populations subspecies or species is usually made on purely subjective grounds and is open to interpretive error. In certain cases subspecies may represent distinct species. *Semispecies* are taxa on the borderline between subspecies and species. They qualify as species under some but not under other criteria. Again, classification is based on subjective interpretations. As pointed out by Bush (1975) and White (1978), most, if not all, the *willistoni* group semispecies could be regarded as distinct species. Each is characterized by its own set of unique chromosome inversions. Furthermore, several are sympatric. The Andean-Brazilian, Amazonian, and Interior populations coexist over a broad area. They do so without losing their genetic and phenotypic identity. They pass the test of sympatry and should be recognized as species. The reasons for conferring semispecies status apparently rests on the fact that under laboratory conditions they are not fully reproductively isolated. Such tests, as already noted, circumvent many natural barriers to gene flow and fail to account for those reproductive isolating mechanisms based on habitat selection and other environmental factors.

The genetic distances given in Table 9.2, therefore, do not provide an accurate picture of the kinds of genetic differences in structural genes one might necessarily expect between local populations, geographic races, and species. There is, in fact, mounting evidence from a variety of studies on vertebrates and invertebrates that genetic distances as measured by electrophoresis bear little if any direct relationship to the genetic changes involved in speciation (Nevo & Cleve 1978). The genetic similarity between some species of rodents, for instance, may be as high as 0.98, or as low as 0.63 between some local populations of the same species (Patton & Yang 1977). In some groups morphologically distinct species placed in separate *families* may have genetic distances no greater than those found between some sibling species (Wilson et al. 1977). For the reason behind the poor concordance between genetic divergence, morphological differentiation, and reproductive isolation, we must consider other types of genetic change such as those involved in altering patterns of gene regulation (Bush 1982).

9.5.4 Evolutionary Rates and Phylogenetic Trees

9.5.4.1 The Molecular Clock.
Comparative biochemists have discovered that macromolecular sequences or proteins and nucleic acids evolve at relatively constant rates within major groups, and thus can be used as evolutionary clocks. Although the factors underlying this constancy are not understood, data indicate that variation in the rate of this clock within a given protein class is about twice that of a radioactive decay clock. This provides a better degree of accuracy than most fossil records available for

plants and animals. Wilson et al. (1977), review the evolutionary clock and its implications. The clock appears to be geared to years rather than generation time and provides an evolutionary time range for groups that lack or have a poor fossil record.

9.5.4.2 Molecular versus Morphological Phylogenies.

Generally, there is a fair concordance between molecular- and morphologically based phylogenies. Whereas macromolecular evolution within a group appears to be rather constant over time, the rate of morphological evolution is more variable, proceeding through long periods of morphological statis punctuated by rapid changes in morphology and way of life (Gould 1982). Furthermore, the rate and pattern of morphological change appear to differ greatly between major groups. Mammals have evolved much faster than some nonmammalian vertebrates such as amphibians. Phylogenies based strictly on macromolecular sequences, therefore, apparently reflect a better approximation of evolutionary relationships (Wilson et al. 1977) than those based strictly on morphological traits. Classifications based strictly on morphological criteria, on the other hand, do not necessarily provide close concordance with molecular-based, phylogenetic relationships. Because radical morphological changes can occur over short periods of time, certain taxa that appear to be distantly related on morphological (or other biological) grounds may have diverged from one another fairly recently and be more closely related genetically than one of them is to a third species with which it is morphologically very similar, or even a sibling species.

9.5.4.3 Regulatory versus Structural Gene Evolution.

Why does such a difference in the evolutionary pattern of morphological and genetic traits exist? Structural gene mutations appear to play a secondary role in adaptive radiation and speciation. Although these mutations may alter fitness of individual proteins, only a small minority appear to be involved in phenotypic change (Wilson et al. 1977). Major adaptive advances, such as acquisition of a novel metabolic activity or alteration of developmental pathways and ontogenetic development, appear to depend initially on changes in activity of rate-limiting proteins. Experimental studies on microbes and eukaryotes show that such changes are usually the result of mutations in control genes, in genes that exert control at levels other than transcription, or from chromosomal mutations that alter the arrangement of genes (Bonner 1982).

9.6 MODES OF SPECIATION IN INSECTS

There has been a quiet revolution over the past 10 years in our appreciation of how new species arise. As recently as 1963, with the publication of Mayr's elegant *Animal Species and Evolution*, most evolutionary biologists became convinced that in sexually reproducing animals speciation only occurs as a result of genetic differences accumulated during periods of complete geographic isolation (*allopatric* speciation—Table 9.3). This divergence must be sufficient to insure that hybridization between populations, if it occurs, does not result in the breakdown of reproductive isolation. Recently several authors have questioned the universality of allopatric speciation in animals (Bush 1975, Endler 1977, White 1978, Templeton 1981). Alternative models have been proposed that do not require geographic isolation.

Speciation in animals may be operationally grouped into three categories based on presence or absence of geographic barriers to gene flow and mode of reproduction. These are allopatric, sympatric, and asexual speciation. Other schemes have been proposed by Endler (1977), White (1978), and Templeton (1981); they differ primarily in the way modes of nonallopatric speciation are grouped. We prefer to place them all under the general heading of sympatric speciation, with the exception of cases involving animals with asexual reproduction. While allopatric speciation requires complete geographic isolation of populations as a prerequisite, sympatric speciation occurs in situations where the ranges of two or more populations overlap in such a manner that interbreeding could occur unless prevented or greatly reduced by some form of genetically based isolating mechanism. The various forms of sympatric speciation listed in Table 9.3 represent different degrees and patterns of gene flow between diverging populations during speciation. They all share one thing: at no stage in speciation is the potential contact broken between parent and daughter populations.

9.6.1 Allopatric Speciation

Allopatric speciation was discussed in detail by Mayr (1963), Dobzhansky et al. (1977), and Carson (1975). Two types, the classical and founder principle models, have been proposed.

The *classical model* involves subdivision of a species into two or more large

Table 9.3 Modes of speciation in animals

A. Allopatric speciation (Mayr 1963)
   1. Without population bottleneck (classical model)
   2. With population bottleneck (founder principle)
B. Sympatric speciation (Bush 1975, Endler 1977, White 1978)
   1. By clinal and area effects (parapatric speciation)
   2. By divergent selection (competitive speciation)
   3. By chromosome rearrangement (stasipatric speciation)
   4. By polyploidy
   5. Resulting from intraspecific hybridization
C. Asexual speciation (White 1978)

geographically isolated populations. Divergence in such large populations occurs as a result of genetic changes incorporated in response to environmental changes that occur over time in the different geographic regions. First, the majority of these new allels would exist as balanced polymorphisms for varying lengths of time. As such, they must be what Mayr called "good mixers," they interact with other alleles throughout the genome without reducing average overall individual fitness. Mutations that appreciably reduce fitness of the heterozygote are not likely to be incorporated into the gene pool except under special conditions. Allopatric speciation by way of large, isolated populations is thus a slow process that requires long periods of isolation, and reproductive isolation would seldom be perfected when previously isolated populations reestablish contact.

A more rapid form of allopatric speciation, *the founder principle,* was proposed (Mayr 1963) that results from severe genetic bottlenecking in small, geographically isolated populations. Such populations may be founded by a small number of individuals that carry with them only a fraction of the genetic variation in the parent population. While the population is small, both chance and selection will play important roles in rapidly altering gene frequency. Mayr (1963) and Carson (1971) suggested that the combined effect of these two evolutionary forces will lead to a genetic revolution in the small, isolated population. Genetic divergence sufficient to insure reproductive isolation may then arise rapidly.

Both modes of allopatric speciation have apparently occurred throughout the animal and plant kingdoms. However, evidence in support of both types is circumstantial. Examples offered in their support are drawn from species that have already speciated, as no one has witnessed either type of speciation event in nature. Bush (1975), Endler (1977), and White (1978) noted that many cases offered as proof of the allopatric mode can be equally well explained by one of the sympatric modes discussed below.

9.6.2 Sympatric Speciation

9.6.2.1 Clinal and Area Effects. This form of sympatric speciation has been called *parapatric speciation,* and involves fragmentation of a species in an originally continuous cline or patchy environment into two or more species as a result of strong selection in the vicinity of an ecological boundary (White 1978). Both spatial segregation and spatial differentiation initiate the process and lead to evolution of isolating mechanisms between groups of geographically distinct, but contiguous, populations (Endler 1977). Three basic questions must be resolved in order to establish if it occurs in nature: (1) Can sharp genetic boundaries arise within a spatially and genetically continuous series of populations? (2) Can the resulting step clines give rise to hybrid zones? (3) Can reproductive isolation evolve in such zones? Direct field studies on plants and animals favor an affirmative answer to the first two questions, but as in the case of allopatric speciation there is only circumstantial evidence and theoretical arguments that the process of genetic differentiation in parapatric situations will eventually lead to speciation.

It is clear that contiguous populations occurring across a spatially abrupt environmental gradient or *ecotone* can diverge genetically to a considerable degree in a short time. Populations of *Agrostis* grasses, for instance, have become adapted to growing in contaminated soils of abandoned heavy metal mines in England. These mine-tiling races show marked differences in flowering time, amount of selfing, and other traits from the parent population on normal soil only inches away. Selection has favored evolution of a narrow hybrid zone resulting in greatly reduced gene flow between the two races. Of key importance in the *Agrostis* example is that the rate of gamete and seed dispersal is low and divergent selection in the two habitats is high enough to favor development of some reproductive isolation and a narrow, primary hybrid zone. Other examples and theoretical models are presented by Endler (1977) and White (1978). In no case has it been unequivocally demonstrated, however, that parapatric divergence has resulted in complete reproductive isolation. This question cannot be resolved until we have a better understanding of the relationship between gene flow and divergent selection across an ecotone. Current theoretical models, based on one or two locus situations, do not reflect the complicated, multilocus interaction occurring in natural populations. Such interactions can have unpredictable pleiotropic effects that could affect reproductive isolation and other adaptive processes. We must therefore withhold judgment on whether or not parapatric speciation can occur, although the evidence strongly suggests that it can.

9.6.2.2 Competitive Speciation. Another form of sympatric speciation involves divergence of two populations as a direct result of competitive interaction between genotypes adapted to different zones. It appears to be a common mode of speciation in several parasitic insect groups (Bush 1975) and possibly many other insects as well. Sympatric speciation in parasitic insects was reviewed by Price (1980).

Competitive speciation can occur rapidly and apparently with minimal gene changes. The best evidence supporting this comes from studies on insect pests that have formed new host races on introduced plants. The original host of the North American fly, *Rhagoletis pomonella*, is the hawthorn fruit (*Crataegus* spp.). Around 1860 it was found infesting apple, introduced over 200 years earlier. More recently, it established a population on sour cherry in a small area of Door County, Wisconsin. All three host races now coexist in this area. Each fly population has a different emergence time and pattern of host utilization, suggesting that the three races are biologically distinct. In these flies males and females meet and mate on the host fruit. In a series of papers Bush and colleagues (see Bush 1975 for summary and references) showed that host selection is determined mainly on the basis of chemical cues. Hybridization experiments with a related fly species indicated that host selection and larval survival were controlled by only a few loci, suggesting that a host shift may occur through genetic changes that affect recognition of hosts, (perhaps involving chemoreception) and survival. Changes in host selection and survival genes could follow the gene-for-gene coevolved genetic system as seen for wheat and the Hessian fly (cf. Day 1974). Newly established host races have been observed in other insect pests (reviewed by Diehl & Bush 1983).

9.6.2.3 Stasipatric Speciation.

The major differences between stasipatric speciation and clinal and area effect speciation is in the way reproductive isolation is attained. In the latter it arises by accumulation of genic differences between parapatric populations across an ecotone. *Stasipatric speciation* is initiated by a chromosomal rearrangement that reduces fecundity when heterozygous, and when homozygous, produces a new adaptation. These mutations may arise anywhere within the range of the ancestral species (White 1978). This mode of speciation appears to be restricted to taxa that are subdivided into many relatively small, semiisolated populations (Bush et al. 1977). As with parapatric speciation, stasipatric speciation is most common in organisms with low vagility or that have breeding systems that promote inbreeding. Small effective population size, coupled with inbreeding and drift, are a prerequisite for fixing chromosome rearrangements that reduce viability in the heterozygote (Lande 1979).

Stasipatric speciation may therefore be widespread in certain arthropods and other invertebrates. White (1978), who developed the concept of speciation by chromosome rearrangement to explain the origin and evolution of moribine grasshoppers of the genus *Vandiemenella*, discusses several striking cases in other grasshopper genera, phasmids, beetles, and isopods. Chromosome evolution and speciation have also been rapid in some parasitic Hymenoptera where inbreeding, including brother–sister mating, is widespread (Askew 1968, Goodpasture & Grissell 1975). Because this mode has probably played an important role in the evolution of new host races and sibling species of parasitoids, attempts to use parasitoids as biological control agents should be accompanied by careful screening at both the genetic and chromosomal level to establish the taxonomic status of each population sampled.

9.6.2.4 Polyploidy and Hybridization.

These two factors have contributed significantly to multiplication of species in plants but are of little or no significance in animals. Polyploidy involves multiplication of complete sets of chromosomes. It can occur either as a result of *autopolyploidy* in which there is a multiplication of a chromosome set within a single individual or by *allopolyploidy* in which chromosome sets from two related species are joined as a result of the hybridization, then doubled during cell division by various mechanisms. Only the latter appears to be of significance in annual plants and possibly a few animal groups. The result is instantaneous speciation in which a new sexually reproducing species occurs in one generation. Interspecific hybridization has also been invoked to explain cases of speciation in both plants and animals in which no increase in chromosome number has been found. No example of polyploidy as a mechanism for speciation in sexually reproducing animals has been confirmed or supported.

9.6.2.5 Asexual Speciation.   Many species of animals (and plants) reproduce asexually. Males are either nonexistent or rare and nonfunctional. Asexual reproduction occurs in almost all insect orders, including parasitic Hymenoptera, aphids, cockroaches, and flies. In an asexual species there is a common phenotype. There is no recombination through sexual reproduction, although a considerable amount of genetic and chromosomal variations may be accumulated by mutations. There are a variety of ways, including interspecific hybridization and polyploidy, that may give rise to an asexual taxa (see review by White 1978).

9.7 COEVOLUTION

Another aspect of insect evolution of major significance and interest to ecologists is *coevolution,* that is, reciprocal evolution arising from pressures that occur between different species which have close ecological relationships with one another (Ehrlich & Raven 1964, Gilbert & Raven 1975). One should remember that the coevolutionary process ultimately involves selective forces interacting between species at the genetic level. Thus, genomes of unrelated organisms are coevolving as an outcome of their close, and frequently, obligate association. The best examples at the genetic level are those worked out for host–parasite interactions. In insects the genetics of host race formation in the Hessian fly provides an interesting case. The larvae mine stems of wheat. Since its introduction into North America in the early 1800s, it has repeatedly formed races on new cultivars selected for resistance to its attack. Larvae usually die in resistant varieties but cause stunting in susceptible ones. Controlled test crosses and oviposition experiments have shown that resistance is conferred by a series of dominant genes in the wheat varieties and virulence (ability to overcome resistance) is the product of a series of recessive genes in the fly. Each cultivar has a unique resistant gene and can be attacked by only one race of the fly, one that is homozygous for a specific virulent gene (Stebbins et al. 1980). This gene-for-gene coevolved genetic system is typical of a range of host–parasite interactions (Day 1974).

In some cases host–parasite interactions are complicated by bacterial and viral symbionts. The genetic interactions involved, however, are not known. A few examples will suffice to illustrate the high degree of coevolutionary interaction. Many braconid and ichneumonid parasitoids, for instance, transmit highly specialized bacculoviruses and other viroidlike particles when depositing eggs in their host (Edson et al. 1981). The viruses

appear to have an important commensal role with the parasitoid and somehow contribute to survival of parasitoid larva. Many insects also harbor
highly host-specific, symbiotic bacteria, yeasts, and protozoans that are finely
tuned to their hosts' biochemistry (Buchner 1965). Cockroaches, for instance, have intracellular "bacteroids" that are transmitted in the egg cytoplasm. Roaches cannot survive without their symbiotes to provide several
essential amino acids and other biochemicals (cf. Chapter 4).

Coevolution also may occur between plants and herbivores. Special morphological features (spines, pubescence) and chemicals (nutrient-poor sap,
secondary plant substances, etc.) may lead to adaptation in the herbivore
to circumvent the host's defenses. Similar patterns of coevolution may also
be found in certain predator–prey, plant–pollinator relationships, and
mimicry complexes (Gilbert & Raven 1975). Coevolved systems are widespread in arthropods and frequently involve unrecognized relationships.
There are probably few insects that do not harbor some viral or bacterial
endosymbionts. The function of the vast majority of these is unknown but
of great importance to understanding how certain kinds of adaptive variation are maintained in nature. Genetic aspects of host–parasite interactions, for instance, are essential to proper management of some insect pests.
Certain aspects of community ecology can only be understood in the light
of such closely interacting systems.

9.8 IMPLICATIONS FOR APPLIED ENTOMOLOGY

9.8.1 The Role of Genetic Variability

Extensive intra- and interpopulational genetic variability exists in arthropod
populations and has important implications for applied entomology. Pest
species may be able to respond rapidly to environmental changes, including
those caused by man. A most obvious (and most economically important)
example of rapid microevolutionary change is the development of resistance to pesticides.

Resistance is the development in a strain of the ability to tolerate toxicant
levels that are lethal to most individuals in a normal population of the
species. Natural *tolerance* is distinctly different; it is the preadapted tolerance
shown by some species to some insecticides. Multiple resistance has sometimes been induced by simultaneous or successive exposure of a population
to two or more insecticides. Development of pesticide resistance is a classic
example of microevolution in which a toxic chemical acts as a selective
agent to increase the frequency of the genes responsible for survival. Over
250 agricultural pest species and 100 species of medically important arthropods are known to be resistant to one or more major insecticides (Brown
& Pal 1971). Resistance is a preadaptive trait, or arises *de novo* by mutation.
There is no evidence that pesticides induce changes in the DNA responsible

for the resistance. Mechanisms of resistance vary and include such attributes as activation and detoxification, reduced penetration and transport, larger capacity for storage or faster excretion, and reduced target-site sensitivity. Pesticide resistance may also involve changes in host or habitat preference, leading to "behavioral resistance." The genetic basis of pesticide resistance has been widely studied (e.g., Crow 1957), but controversy still exists about the relative importance of monofactorial and multifactorial determinants due, in part, to problems of methodology. Inheritance of resistance was initially attributed to multiple genes, but studies soon indicated that DDT resistance was determined by a single major gene. Estimation of mortality at single "diagnostic" doses tends to result in the conclusion that the trait is a monofactorial one, whereas use of multiple doses tends to lead to a conclusion of polyfactorial inheritance. Crow emphasized that monofactorial inheritance is best proven by isolation of the factor through repeated backcrosses, coupled with selection. Resistance genes have been located on specific chromosomes in house flies, mosquitoes, and *Drosophila melanogaster*, and these studies have shown that resistance is usually determined by genes located on all the chromosomes, although major genes often account for the majority of the effect. Furthermore, several resistance mechanisms may coexist within a population, and variability between populations is common. Increased rate of detoxification and an altered site of action account for the majority of resistance mechanisms.

Devising methods to avoid or retard development of resistance is of high priority for pest management specialists. This can be achieved in several ways: reducing selection pressure through use of reduced rates and numbers of pesticide applications, and restricting applications to a small portion of the population. An alternative strategy of effecting complete mortality, so as to preclude selection, is operationally unfeasible. However, development of resistance is unpredictable. Some species have undergone extensive selection and never developed resistance, presumably due to lack of appropriate preadaptive alleles.

Other types of resistance may develop in insect populations. The sterile insect autocidal technique requires that the insect can be mass reared, that sterility can be induced without adverse effects upon competitive ability and mating efficiency, and that population density estimates are available for low points in the population cycle to allow the calculation of the appropriate release ratio (Pal & Whitten 1974). Several components of a genetic control program are vulnerable to genetic responses by the target pest. Thus, insects that became able to reproduce asexually become invulnerable to autocidal techniques. Parthenogenesis is present in diverse insect groups and has developed many times (White, 1973). Moreover, parthenogenetic individuals occur at low, but regular rates in many sexually reproducing species, and it could be the ultimate escape from autocidal control programs.

Development of assortative mating, due to differences in behavior, diel periodicity, pheromones, and so forth, may provide another escape from autocidal control. If divergent selection were sufficiently intense, distinct races might evolve rapidly. The propensity to mate varies within and between populations, and laboratory experiments may not expose this if one of the pest "species" is not included or if forced mating under laboratory conditions overcomes natural mating barriers.

There are other possibilities for evolutionary change to disrupt insect control programs. Control of populations by use of pheromone traps or "mating disruption" techniques could also fail due to genetic changes. Nonresponsive insects could be selected for, particularly if visual and other cues are also critical for mating success. Genetic variability also exists for the use of, and metabolism of, juvenile hormone in insects (Templeton & Rankin 1978). Thus, these "third-generation" pesticides are vulnerable to development of resistance. The evolutionary potential of insects is generally underestimated by economic entomologists, but perhaps we can learn from the paradigm of resistance.

9.8.2 Genetic Variation and the Origin of New Biotypes

The appearance of new *biotypes* specializing on plant or animal hosts have frequently caused major problems. New host races have developed rapidly on a number of plants. The hawthorn fly, *Rhagoletis pomonella*, shifted to cultivars of apple in about 1860 and later to domestic sour cherries (Bush 1975). A similar pattern of host shifts has been noted in the codling moth which normally infests apple. This species has now established distinct biotypes on walnuts and plums in California, with associated changes in host preference and numbers of generations (Phillips & Barnes 1975). The genetics of host race formation are discussed in Bush and Diehl (1982). Permanent shifts appear to require genetic changes in genes affecting host selection or recognition and survival. Little work has been done on the genetics of host selection, but there are several detailed studies on survival genes. For example, the Hessian fly has established genetically distinct strains on specific cultivars of wheat. The coevolved gene-for-gene system of resistance and virulence (survival) genes is now being used experimentally to exterminate cultivar races of the fly (Foster 1977). Other examples of race formation based on a shift in seasonal cycle are discussed by Tauber and Tauber (1981).

The speed with which insects can adapt to a new habitat or host is exemplified experimentally by Templeton (1979) who described a "genetic revolution" in *Drosophila mercatorium*. Given drastically different environments, a colonizing population can rapidly evolve a new set of balanced genes that are incompatible with the old gene complex. This genetic revolution can be manifested in terms of morphology, development, physiology, life history, and behavior. These alterations can be so drastic that "new species" can develop, complete with pre- or postmating isolating mechanisms in a very short space of time. In terms of pest control an exotic pest could evolve in unpredictable fashion in the new environment. Rapid shifts to new hosts can be expected as well. The genetic revolution may be due to genes that have fundamental regulatory roles as in juvenile hormone function or metabolism (Templeton & Rankin 1978). This revolution may depend upon a small number of loci, possibly as few as four.

The implications of host shifts and race formation for economic entomologists extend beyond changes in host plant choice by pest species. Biological control of weeds with phytophagous insects is predicated on the notion that host specificity will remain stable (Andres et al. 1976). Screening of host plant choices under insectary conditions is well known to overestimate host plant acceptability since the insect may never choose the host under field conditions. To date, we have no documented case in which basic host plant shifts have occurred after introduction of insects for weed control, due to the careful screening of such agents. However, it appears likely that such screening will never give total assurance of safety due to the unpredictable likelihood of a genetic change in the introduced insect.

9.8.3   Genetic Variability and Insect Colonization

9.8.3.1   Biological Control.   Intra- and interspecific variability in insects considered for use in biological control innoculative release is widely recognized as potentially highly important. Establishment and efficiency of exotic natural enemies depends on many factors, including the initial sampling method of the population(s), rearing procedures, release strategies, and a host of other ecological and geographic variables. Since most foreign exploratory trips are of limited time and duration, it is unlikely that collections will include the full array of natural variation even though there is a strong tendency to include as many natural enemies as possible. Collections may undergo bottlenecking in the quarantine facility. Unfortunately, strategies for maintaining extensive genetic variability are not commonly practiced in biological control programs. Individual collections of a species should be taken at different sites, times of day, and seasons, if at all possible. These collections should be kept separate in quarantine and subsequent rearing, with each strain released individually. Collecting and maintaining several small colonies of different strains should maintain more variability than a single, pooled, large colony.

9.8.3.2   Genetic Problems in Mass-Rearing Programs.   With the advent of sterile insect release methods for pest control, a new era in mass rearing of insects began. Suppression or eradication of a major pest species is achieved by flooding the population with infertile, but still sexually competitive, males mass reared in a factory. The technique was demonstrated first by eradicating the screwworm fly, *Cochliomyia hominivorax*, from Florida and the southeastern United States. Later, it was suppressed in Texas and

the southwest, and now attempts are being made to push it back to southern Mexico. A major difficulty has been in maintaining tight quality control over the factory-reared flies. Various factors such as larval diet, temperature fluctuations, and other factory conditions and practices can drastically reduce competitive ability. Most of these effects can be reversed, but if the cause goes unnoticed, selection alters the genetic structure of the factory population, and the effect may be irreversible, so that a new strain must be introduced. This has occurred on several occasions over the course of the screwworm eradication program, and it is instructive to look at this example in detail (Bush 1979).

Success of the program depends on the ability of released flies to compete with wild flies for mates. Although the program experienced stunning initial success, it ran into difficulties in Texas in 1972 when reported infestations jumped from 473 to over 95,000 and remained high in subsequent years, despite the fact that the number of sterile flies released rose from less than 2 billion per year at the start in 1962 to about 10 billion in 1974 (Bush 1978). Scientists at the screwworm facility suspected that factory conditions might be inadvertently selecting a strain of flies with reduced competitive ability. To ascertain if genetic drift or selection was occurring in the factory, genetic variation in factory and wild populations was studied. The objective was to see if the processes of colonization and factory rearing had any detectable genetic effect on the flies. Natural populations of the screwworm are low, ranging from 100 to 200 per mi.^{2}; thus, released flies must disperse and remain active at appropriate times if they are to find suitable living conditions and mate in competition with wild males. Therefore, the genetic study, using standard gel electrophoresis techniques, concentrated on glycolytic enzymes involved with flight, although others were examined as well.

One enzyme, α-glycerophosphate dehydrogenase (GDH) is of particular interest because something is known concerning the adaptive function of its allelic forms. This enzyme plays a key role in regulating energy flow in the flight muscle of insects during flight, as it governs transfer of reducing equivalents from cytoplasmic NADH to the mitochondrial electron-transport chain by way of the glycerol phosphate shuttle. The amount of GDH present in flight muscle greatly affects flying ability, and mutants which lack GDH cannot fly. Because of its central importance in regulating energy flow, allelic variants of this enzyme are under tight natural selection (cf. Bush et al. 1976 for details). Variation in these enzymes appears to be maintained by a combination of spatial and temporal patterns of environmental variation in temperature.

Upon electrophoretic analysis, all factory samples examined differed significantly from wild flies in the frequency of alleles at almost all the electrophoretically detectable loci examined. This was particularly true of loci controlling flight activity, such as GDH, which showed the most dramatic change (Bush et al. 1976). In the screwworm the GDH enzyme exists in two forms. Almost all factory flies were homozygous for electromorph GDH_2, while this form of the enzyme was extremely rare in wild Texas populations, which typically retained the alternate form, GDH_1. Therefore, the difference in GDH between wild and factory flies provided an important clue to the problem. Systematic sampling of a new factory strain was undertaken, from its introduction in the laboratory to its eventual adaptation to factory mass-rearing conditions. As soon as the new strain was introduced into the factory, GDH_2 began to increase in frequency and GDH_1 to decrease. Within six months, GDH_2 had become about as common as in previous factory-adapted strains. But does the GDH_2 form of the enzyme affect flight activity?

Research on the function of the various forms of GDH in other insects, such as butterflies and fruit flies, suggested that each form of the enzyme functioned satisfactorily only within a specific, but different, temperature range (cf. Johnson 1974). To test whether the two forms of the GDH enzyme were affected by temperature, both forms were isolated and purified. It was then established that they did indeed have quite different temperature ranges within which they showed optimal activity (Kitto et al. 1976). The factory-type enzyme (GDH_2) was less active in the temperature ranges experienced in nature in the morning when flies were active and mating. The high constant temperature used to speed development in the factory apparently exerted a strong selective force favoring GDH_2 over GDH_1 (Bush et al. 1976). This suggested that the competitive ability of the fly in nature would decrease as the frequency of GDH_2 increased because factory flies would have to cope with a wide temperature range in nature and individuals which lacked the GDH_1 enzyme simply could not fly as well at the appropriate time as their wild cousins.

Although the mating behavior of wild flies has never been observed in detail under natural conditions, studies on released factory flies suggest that at least part of the mating activity occurs in the air and possibly at specific sites which require normal flight and behavioral response (Guillot et al. 1978). Thus, the "lazy" factory males would be at a disadvantage competing for mates. In fact, a USDA team found that wild females were attracted to wounds from early morning to late afternoon. Factory females were not active until early afternoon. They apparently could not get their flight muscles operating for lack of sufficient energy. Most wild females therefore were probably inseminated by wild males before the factory flies become active enough for sexual activity. Also, the lazy flies might be subject to higher predation, further reducing their efficacy as biological control agents.

In 1977 the defective factory strain was eliminated and replaced by a new strain with the defective allele at low frequency or absent. Care was taken not to mix the old and new strains. Infestation levels that summer dropped to pre-1972 levels and, combined with adverse climatic conditions for the fly during the spring and summer, excellent control was regained in Texas, although difficulty is still being experienced elsewhere and further outbreaks have occurred. Recent cytogenetic and morphological studies also indicate that this fly may be subdivided into rather distinct geographic races, further complicating control efforts by the sterile insect release method (Richardson et al. 1982).

The problems encountered in 1972 could have been averted if the factory had had a sound quality control program to monitor factory populations for harmful genetic changes. Some effort had been made to measure effects of diet and other rearing conditions on fecundity, development time, longevity, and so on, but no tests were conducted to measure known adaptive, ecological, and behavioral traits pertinent to the fly's survival in nature. Therefore, when the flies failed to perform, it was difficult to act because insufficient effort had been devoted to basic ecology and behavior, so that no one knew what might be causing the problem.

Although the screwworm example serves as a model, the effect of selection on several other species during mass rearing or laboratory colonization has been established or inferred for a diverse array of biological traits important to a pest control program. These include phototactic behavior and vision (Markow 1975, Goodenough et al. 1977), locomotory behavior (Chabora 1969), oviposition, premating, and mating patterns (Rossler 1975), and diapause (Hoy 1978). Work of Homyk and Sheppard (1977) exemplifies the ease with which behavior traits can be selected. They were able to recognize and establish 48 behaviorally different strains of *Drosophila melanogaster*.

9.8.4 Genetic Improvement of Beneficial Insects

Genetic improvement of beneficial parasitoids and predators has been discussed for more than 60 years because of the successes achieved by plant and animal breeders and with silk worms and honey bees (Rothenbuhler 1979, Yokoyama 1973). A number of beneficial insect species have been targets of artificial selection programs (reviewed by Hoy 1976, Messenger et al. 1976). Genetic improvement of beneficial insects can be accomplished in several ways. One method suggested involves increasing the genetic variability of the beneficial species through laboratory hybridization of different strains, with the assumption that natural selection will effect a desired improvement after field release. Classical biological control, where limited numbers of an exotic species are introduced into new habitats to establish and effect control of a pest, is plagued with the problems of "adequate" sampling or securing of exotic populations in space and time, and with maintenance of the sometimes limited genetic variability achieved in foreign collections during quarantine processes, insectary rearing, and colonization. Thus, hybridization of strains, either before or after release, may be a way of increasing or maintaining sufficient genetic variability to obtain establishment and/or improve the efficacy of beneficial species (Hoy 1976, 1978).

Strain hybridization could be used for inundative releases of parasitoids or predators where establishment need not be permanent. Hybridization of standard insect strains could result in insects of uniform quality. Quality control in insectary rearing is a problem that has gained increased recognition as having both genetic and environmental components, and hybridization may provide one solution.

Selective breeding offers another method whereby beneficial insects may be genetically "improved" (Hoy 1976, 1979, Roush 1979). Artificial selection is the more controversial of the genetic improvement techniques suggested and has been demonstrably effective under field conditions with pesticide-resistant predators of spider mites (Hoy 1982, Hoy et al 1983, Roush & Hoy 1981). Whether this success can be achieved with other biological control agents remains to be seen. The problem seems complex since a suite of desirable attributes must be maintained during the course of a laboratory selection program in which the beneficial insects are expected to be the effective agents. The mode of inheritance of relatively few desirable attributes are known, making selection inefficient. Sufficient genetic variability must be available to permit a response to selection. Some have suggested that after the release, "improved" strains will revert or be swamped genetically by the wild strain, thus, nullifying the extensive (and expensive) efforts of the program. Most of these problems might be reduced or eliminated if the problem is carefully defined (Hoy 1976, 1979). Most critics of genetic improvement are thinking about the genetic improvement of a parasitoid or predator that is to be released and permanently established in the environment. That is a difficult goal, which might have a limited success rate. However, "genetic improvement" may be feasible if more limited goals are sought. Not all attributes of the "wild" insect need be maintained, nor may they even be desirable. For example, obligate diapause in the silk worm is undesirable if two harvests per year are desired, and normal flight, dispersal, habitat finding, and fastidious food habits are undesirable under factory conditions. Parasitoids released into glass houses or field crops for temporary suppression of pests need not disperse, diapause, or find the appropriate habitat since they can be delivered to it. High fecundity, pesticide resistance, adequate sex ratios, and high parasitization or predation rates are desirable and are attributes that are amenable to selective breeding (Hoy 1976, 1979).

9.9 CONCLUSIONS

Population and molecular genetics have made rapid advances in recent years, shedding new light on the process of adaptation and evolution. Contributions at the molecular level have been particularly rewarding with respect to improving our knowledge of how genes function and are organized on the chromosome. More sophisticated biochemical tools are also now available for examining the pattern of genetic variation at the protein and nucleotide level. We are therefore beginning to perceive what constitutes the genetic machinery that programmes the development and other cellular functions that have long been viewed as a "black box" by population geneticists. Knowing how the genome is put together and how it works removed considerable uncertainty on how certain adaptations might evolve.

Although the molecular biologist may provide us with the mechanisms of gene action, the population biologist and naturalist will have the important responsibility of integrating this new genetic knowledge into a modern synthesis of evolutionary biology. The applied biologist also has much to gain from these advances as they open up many new roads to pest management and the development of resistance in crops and domestic animals.

REFERENCES

Andres, L. A., C. J. Davis, P. Harris, and A. J. Wapshere. 1976. Pp. 481–499 in Huffaker and Messenger (Eds.), 1976, referenced here.

Askew, R. R. 1968. *Evolution* **22**: 642–645.

Atchley, W. R. and D. S. Woodruff (Eds.). 1981. *Evolution and Speciation: Essays in Honor of M. J. D. White.* Cambridge University Press, Cambridge.

Ayala, F. J. (Ed.). 1976. *Molecular Evolution.* Sinauer, Sunderland, Mass.

Ayala, F. J. 1982. Pp. 60–82 in Milkman (Ed.) 1982, referenced here.

Barker, J. S. F. and P. D. East. 1980. *Nature* **284**: 166–168.

Bergerard, J. 1972. *Annu. Rev. Entomol.* **17**: 57–74.

Bijlsma-Meeles, E. and W. van Delden. 1974. *Nature* **247**: 369–371.

Blackman, R. L., G. M. Hewitt, and M. Ashburner. 1980. *Insect Cytogenetics.* Blackwell, Boston.

Bonner, J. T. 1982. *Evolution and Development.* Springer-Verlag, New York.

Brown, A. W. A. and R. Pal. 1971. *Insecticide Resistance in Arthropods.* Monograph Series World Health Organization #38. World Health Organization, New York.

Brown, D. D. 1981. *Science* **211**: 667–674.

Brown, W. M. 1981. *Annu. N.Y. Acad. Sci.* **361**: 119–134.

Buchner, P. 1965. *Endosymbiosis of Animals with Plant Microorganisms.* Wiley, New York.

Bush, G. L. 1975. *Annu. Rev. Ecol. & Syst.* **6**: 339–364.

Bush, G. L. 1978. Pp. 37–47, in Richardson (Ed.), 1978, referenced here.

Bush, G. L. 1979. Pp. 145–152, in Hoy and McKelvey, Jr. (Eds.), 1979, referenced here.

Bush, G. L. 1981. Pp. 201–218, in Atchley and Woodruff (Eds.), 1981, referenced here.

Bush, G. L. 1982. Pp. 119–128, in Milkman (Ed.), 1982, referenced here.

Bush, G. L., S. M. Case, A. C. Wilson, and J. L. Patton. 1977. *Proc. Natl. Acad. Sci. USA* **74**: 3942–3946.

Bush, G. L. and S. R. Diehl. 1982. In Visser and Minks (Eds.), referenced here.

Bush, G. L. and G. B. Kitto. 1978. Pp. 89–118 in Rombergus et al. (Eds.), 1978, referenced here.

Bush, G. L., R. Neck, and G. B. Kitto. 1976. *Science* **193**: 491–493.

Carson, H. L. 1971. *Stadler Genet. Symp.* **3**: 51–70.

Carson, H. L. 1975. *Am. Natur.* **109**: 83–92.

Chabora, P. C. 1969. *Evolution* **23**: 65–71.

Crow, J. F. 1957. *Annu. Rev. Entomol.* **2**: 227–246.

Crow, J. F. and M. Kimura. 1970. *An Introduction to Population Genetics.* Harper & Row, New York.

Crozier, R. H. 1977. *Annu. Rev. Entomol.* **22**: 263–288.

Darwin, C. 1859. *On the Origin of Species by Means of Natural Selection* (1st ed.). Murray, London.

Day, P. R. 1974. *Genetics of Host-Parasite Interactions.* W. H. Freeman. San Francisco.

Diehl, S. R. and G. L. Bush. 1984. *Annu. Rev. Entomol.* **29**: 471–504

Dingle, H. (Ed.). 1978. *Evolution of Insect Migration and Diapause.* Springer-Verlag, New York.

Dobzhansky, T., F. J. Ayala, G. L. Stebbins, and J. W. Valentine. 1977. *Evolution.* W. H. Freeman, San Francisco.

Edson, K. M., S. B. Vinson, D. B. Stoltz, and M. D. Summers. 1981. *Science* **211**: 582, 583.

Ehrlich, P. R. and P. H. Raven. 1964. *Evolution* **18**: 586–608.

Endler, J. A. 1977. *Geographic Variation, Speciation, and Clines (Monographs in Population Biology 10).* Princeton University Press, Princeton, N.J.

Foster, J. E. 1977. *J. Econ. Entomol.* **70**: 775–778.

Futuyma, D. J. 1979. *Evolutionary Biology.* Sinauer, Sunderland, Mass.

Gilbert, L. E. and P. H. Raven. 1975. *Coevolution of Animals and Plants.* University of Texas Press, Austin, Tx.

Gilbert, W. 1978. *Nature* **271**: 1978.

Goodenough, J. L., D. D. Wilson, and H. R. Agee. 1977. *J. Med. Entomol.* **14**: 309–312.

Goodpasture, C. and E. E. Grissell. 1975. *Can. J. Genet. & Cytol.* **17**: 413–422.

Gould, S. J. 1982. Pp. 83–104, in Milkman (Ed.), 1982, referenced here.

Guillot, F. S., H. E. Brown, and A. B. Broce. 1978. *Annu. Entomol. Soc. Am.* **71**: 199–201.

Hedrick, P. W., M. E. Ginevan, and E. P. Ewing. 1976. *Annu. Rev. Ecol. Syst.* **7**: 1–32.

Homyk. T. J. and D. E. Sheppard. 1977. *Genetics* **87**: 95–104.

Hoy, M. A. 1976. *Environ. Entomol.* **5**: 833–839.

Hoy, M. A. 1978. Pp. 101–126 in Dingle (Ed.), 1978, referenced here.

Hoy, M. A. 1979. Pp. 104–115 in Hoy and McKelvey (Eds.), 1979, referenced here.

Hoy, M. A. 1982. Entomol. Expt. & Appl. **32**: 205–212.

Hoy, M. A., P. H. Westigard, and S. C. Hoyt. 1983. J. Econ. Entomol. **76**: 383–388.

Hoy, M. A. and J. J. McKelvey, Jr. (Eds.). 1979. *Genetics in Relation to Insect Management.* Rockefeller Foundation Working Papers Series. Rockefeller Foundation, New York.

Huffaker, C. B. and P. S. Messenger (Eds.). 1976. *Theory and Practice of Biological Control.* Academic, New York.

Johnson, G. B. 1974. *Science* **184**: 28–37.

Kitto, G. B., R. Neck, and G. L. Bush. 1976. *Proc. Fed. Am. Soc. Expt. Biol.* **35**: 1858.

Lande, R. 1979. *Evolution* **33**: 234–251.

Lewin, B. 1980. *Gene Expression*, Vol. 2, *Eukaryotic Chromosome* (2nd ed.). Wiley, New York.

Lewontin, R. C. 1974. *The Genetic Basis of Evolutionary Change.* Columbia University Press, New York.

Margulis, L. 1981. *Symbiosis in Cell Evolution.* Freeman, San Francisco.

Markow, T. A. 1975. *Behav. Genet.* **5**: 339–350.

Maynard-Smith, J. 1978. *The Evolution of Sex.* Cambridge University Press, Cambridge.

Mayr, E. 1963. *Animal Species and Evolution.* Harvard University Press, Cambridge, Mass.

Mayr, E. 1981. *Science* **214**: 510–516.

McClintock, B. 1951. *Cold Spring Harbor Symp. Quant. Biol.* **16**: 13–44.

Messenger, P. S., F. Wilson, and M. J. Whitten. 1976. Pp. 209–231 in Huffaker and Messenger (Eds.), 1976, referenced here.

Milkman, R. (Ed.). 1982. *Perspectives on Evolution.* Sinauer, Sunderland, Mass.

Mittwoch, U. 1967. *Sex chromosomes.* Academic, New York.

Nei, M. 1975. *Molecular Population Genetics and Evolution.* Elsevier, New York.

Nevo, E. 1978. *Theor. Pop. Biol.* **13:** 121–177.

Nevo, E. and H. Cleve. 1978. *Nature* **275:** 125–126.

Pal, R. and M. J. Whitten. 1974. *The Use of Genetics in Insect Control.* Elsevier/North Holland, Amsterdam.

Patton, J. L. and S. Y. Yang. 1977. *Evolution* **31:** 697–720.

Phillips, P. A. and M. M. Barnes. 1975. *Annu. Entomol. Soc. Am.* **68:** 1053–1060.

Price, P. W. 1980. *The Evolutionary Biology of Parasites (Monographs in Population Biology 15).* Princeton University Press, Princeton, N.J.

Richardson, R. H. (Ed.). 1978. *The Screwworm Problem: Evolution of Resistance to Biological Control.* University of Texas Press, Austin.

Richardson, R. H., J. R. Ellison, and W. W. Averhoff. 1982. *Science* **215:** 361–369.

Rombergus, J., R. Forte, L. Knutson, and P. Lent. (Eds.). 1978. *Beltsville Symposia in Agricultural Research,* Vol. 2. *Biosystematics in Agriculture.* Allanheld, Osmern Company, Montclair, N.J.

Rossler, Y. 1975. *Annu. Entomol. Soc. Am.* **68:** 187–291.

Rothenbuhler, W. C. 1979. Pp. 84–92 in Hoy and McKelvey (Eds.), 1979, referenced here.

Roush, R. T. 1979. Pp. 97–105 in Hoy and McKelvey (Eds.), 1979, referenced here.

Roush, R. T. and M. A. Hoy. 1981. *J. Econ. Entomol.* **74:** 138–141.

Schultz-Schaeffer, J. 1980. *Cytogenetics: Plants, Animals, Humans.* Springer-Verlag, New York.

Smith, R. F., T. E. Mittler, and C. N. Smith (Eds.), 1973. *History of Entomology. Annual Reviews, Inc.,* Palo Alto, Calif.

Sneath, P. H. A. and R. R. Sokal. 1973. *Numerical Taxonomy.* Freeman, San Francisco.

Stebbins, M. B., F. L. Patterson, and R. L. Gallun. 1980. *Crop Sci.* **20:** 177–180.

Tauber, C. A. and M. J. Tauber. 1981. *Annu. Rev. Ecol. & Syst.* **12:** 181–308.

Templeton, A. R. 1979. *Genetics* **92:** 1283–1293.

Templeton, A. R. 1981. *Annu. Rev. Ecol. & Syst.* **12:** 23–48.

Templeton, A. R. and M. A. Rankin. 1978. Pp. 83–112 in Richardson (Ed.), 1978, referenced here.

Tremblay, E. and L. E. Caltagirone. 1973. *Annu. Rev. Entomol.* **18:** 421–444.

Van Delden, W., A. Kampurg, and H. van Dijk. 1975. *Experientia* **31:** 481, 419.

Visser, J. H. and A. K. Minks (Eds.). In press. *Fifth International Symposium on Insect-Plant Relationships.* PUDOC, Wageningen.

White, M. J. D. 1973. *Animal Cytology and Evolution.* (3rd ed.). Cambridge University Press, Cambridge.

White, M. J. D. 1978. *Modes of Speciation.* Freeman, San Francisco.

Wiley, E. O. 1981. *Phylogenetics: The Theory and Practice of Phylogentic Systematics.* Wiley, New York.

Williams, G. C. 1975. *Sex and Evolution.* Princeton University Press, Princeton, N.J.

Wilson, A. C. 1976. Pp. 225–235 in Ayala (Ed.), 1976, referenced here.

Wilson, A., S. S. Carlson, and T. J. White. 1977. *Annu. Rev. Biochem.* **46:** 573, 639.

Yokoyama, T. 1973. Pp. 267–284 in Smith et al. (Eds.), 1973, referenced here.

Chapter 10

Biogeography and Evolutionary History: Wide-Scale and Long-Term Patterns of Insects

EUGENE MUNROE

10.1 INTRODUCTION

All ecological processes take place in space and time, but for many of the problems studied by ecologists these dimensions provide only a theater. Component organisms are the same or replaceable and interact directly or at short range. Environmental parameters are repeatable, often cyclical, and in many cases controllable by the investigator. In contrast, the present chapter deals with distances so large and times so long that organisms differ from place to place because of isolation and environmental differences, and over time because of immigration, extinction, and evolutionary change. We look first at geographical patterns of present-day species and faunas. These can be observed directly, though explaining them may lead to speculation and controversy. Then we glance more briefly at changes in insects and their environment over geological time, known from an imperfect fossil record and supplemented by indirect evidence. Finally we touch on the profound effects of humans on insect faunas and habitats.

10.2 PROBLEMS OF INSECT GEOGRAPHY

Insects, like other organisms, differ greatly in different parts of the world. The differences are determined partly by environment—insects of deserts are not much like those of rain forests—and partly by isolation and evolutionary divergence—African insects are very different from those of South America, even in comparable environments. Geographical units, large and small, have characteristic insect faunas, which may be rich or poor in species, may have a broadly representative or narrow and disharmonic spectrum of higher groups, and whose taxa may be closely related to those of one or more neighboring units. Taxa, in turn, have characteristic geographical ranges. These differ greatly in character. Some are wide, some restricted; some are continuous, some fragmented; some are unique, some belong to classes of similar ranges. Few ranges are cosmopolitan; most center on particular geographical areas. It is the task of entomogeography to describe and classify these similarities and differences and to try to explain them.

This might seem an impossible enterprise, and in complete detail perhaps it is. However, there are many regularities in insect distribution, which permit different approaches to classification and explanation. These can begin with the fauna, and by assessing differences divide the world into biogeographical regions, subdividing these in turn into provinces and smaller divisions. Though boundaries of divisions are not always sharp, and though somewhat different systems may be appropriate for different organisms, the value of this method is shown by the fact that the regions and many of the provinces and subdivisions proposed by such authors as Sclater (1858) and Wallace (1876) are still in use with little change (Table 10.1). Boundaries between divisions generally correspond either to physiographic barriers or

Table 10.1   List of entomogeographic regions of the world, with partial list of subregions; Wallace's realms listed for comparison

Contemporary	Wallace
Holarctic	
Palaearctic	Palaearctic
Nearctic	Nearctic
Neotropical ⎫	Neotropical
Chilean ⎭	
Afrotropical	Ethiopian
Oriental ⎫	
Indo-Malayan	
Wallacean	
Papuan	
Micronesian	Australasian
West Polynesian	
East Polynesian	
Australian	
New Zealandic ⎭	
Antarctic	

to climatic gradients. Differences in divisional faunas may be determined by environment, history, or both. The Holarctic region, comprising lands of the North Temperate zone, differs strongly in its biota from adjacent tropical regions, even though physiographic barriers to interchange of organisms are relatively weak. Old and New World temperate biotas are relatively similar, especially in the north. Differences increase to the south, where ocean barriers are older and wider. The three main tropical regions, Neotropical, Afrotropical and Indo-Australian, have widely different biotas. Even though their climates are much the same, the oceans and tracts of nontropical land that separate them are virtually uncrossable and have been so for long enough to permit major evolutionary divergence. South Temperate regions, now separated by equally wide oceans, show some faunal relationships that seem to stem from former direct land connections (Brundin 1975). Island faunas have special characteristics, such as poverty, disharmony, and endemic radiation, that have made them particularly interesting from the time of Wallace (1880) to the present.

In studying faunas we try to account for both resemblances and differences. If these can be explained by present environmental factors, there is no need to invoke past history, though the wider question may be raised of how the environment has originated or changed with time. In the absence of such explanation we must look for historical processes that account for observed facts. Both current and historical factors are important in most actual problems.

A second approach to biogeographic analysis begins with the geographical range of a taxon. Here we deal with individual distributions rather than broad averages. This has the advantage that we can hope to trace specific environmental limitations and sequences of historical events. It has the disadvantage that the law of small numbers operates, and highly improbable features may be found in any particular case. The answer to this difficulty lies in detailed analysis, coupled with comparison of different cases. Hultén's (1937) comparison of concentric equiformal ranges and vicariance geographers' matching of phylogenetic and geomorphological cladograms (Nelson & Platnick 1981) are two examples of such comparative analyses. As in the study of faunas, both present ecology and past history must be considered in explaining ranges of single taxa or classes of taxa. Faunistic and taxon-by-taxon approaches can provide mutual support, the fauna showing frequencies and therefore likelihoods of particular patterns and individual distributions revealing features and sequences that might be lost in statistics of the whole fauna.

Whether we study faunas or single taxa, basic data consist of (1) observations of occurrence of insects; (2) observations and deductions on their environmental needs; (3) synopses of geographical distributions of environments; (4) information on processes and pedigrees of insect evolution in the past, including changes in geographical range; and (5) information on the nature, sequence, and timing of past geographical and environmental changes. Unfortunately there are serious gaps in our knowledge in all these categories. Even in better known regions many insect species are as yet undescribed (Danks 1979, p. 242), and classification and phylogeny are only sketched out for most groups. Ecological limiting factors are experimentally determined for relatively few species, and charting the natural distribution of these factors is difficult. Our knowledge of the past is far less complete than of the present. It is not surprising that entomogeography, and indeed biogeography in general, has remained a field notorious for speculation and controversy.

Nonetheless there are encouraging developments. Over the last three decades there has been enormous progress in knowledge of species and their distributions. Classification has been revolutionized for some groups, and similar progress is being made for many more. Ideas of evolution, genetics, and species structure are developing rapidly. Computers enable us to address environmental and distributional problems that would formerly have been totally intractable. New dating methods have given us a time scale extending back to the origins of life. New concepts of geophysics have clarified major aspects of earth history, particularly movement of continents, mountain building, climatic events, changes in magnetic field, and the possible influence of catastrophes: meteors and other impacting bodies (Clube & Napier 1982) and massive volcanic eruptions (Francis

1983). Perhaps the best sign of all is an increasing rigor in argumentation, stimulated chiefly by the encounter between phenetic and phylogenetic systematists, but extending to better defined hypotheses and more explicit and quantified statements throughout the field.

10.3   RANGES

10.3.1   General Features of Ranges

The *range* of a taxon is that part of the earth's surface on which individuals of the taxon occur. Though ranges are often shown as plane surfaces with definite boundaries, this view is too simple. Individuals are almost always dispersed and can be instantaneously represented more accurately by a pattern of dots or patches. Over time individuals move and populations extend or relocate, so that a more or less continuous figure is formed. However, this figure has variations in density of individuals and may have discontinuities in time or space; toward its edge it may grade into a zone of outliers and strays where probability rather than density may be the appropriate measure. Strictly, the figure is spheroidal and distorted by local topography rather than flat, and three-dimensional rather than plane, as many organisms can fly, burrow, climb, swim, or drift and even the most surface-bound ranges have measurable thickness. The importance of these qualifications varies from problem to problem, but they must always be kept in mind.

Considering for simplicity the plane approximation, we see that the instantaneous pattern of individuals may be random, in which case it has a Poisson statistical distribution among samples of habitat, or it may depart from randomness in the direction of even spacing or of clumping. Even spacing is rare in nature, though the tendency might be expected in territorial organisms or in sessile, competitive ones. Purely random distributions are also rare, partly because of patchiness of suitable habitat and partly because of behavioral, reproductive, and developmental influences toward aggregation. These matters are discussed more fully in Chapters 7, 11, and 18 and by Pielou (1975). Pielou notes that both pattern and texture of patchiness may vary, and that individuals, groups of individuals, and habitat patches can be assumed to have various random or nonrandom statistical distributions (cf. Chapter 14). Resultant problems of modeling, sampling, and interpretation are beyond the scope of this chapter and must be considered in detailed biogeographical studies. Density and mobility of individuals vary with season and from year to year, so that temporal as well as spatial patchiness is to be expected. Particularly well defined patches form colonies, islands (geographical or ecological), or larger segments of disjunct ranges. We return to these in discussing configurations and dynamics of ranges below.

Ranges of species are perhaps never worldwide. They are restricted first by occurrence of suitable habitat. For most insects this means emergent land, though some need fresh water for part of the life cycle and a few are marine. Further restrictions are imposed by climate, by geographical barriers such as oceans, mountain ranges, and deserts, and by biological competitors or natural enemies. Even on land relatively few species are nearly cosmopolitan, and the number considered so is decreasing as species classification is refined (Munroe 1973). Habitable regions are often discontinuous, so that taxa may be restricted to particular continents, islands, or other ecological compartments, separated from like regions elsewhere by *barriers*, physical or ecological. As already mentioned, this is an extension of the concept of patchiness.

Barriers that limit taxa and populations may be sharp or diffuse. Rarely are natural populations as sharply walled off as, say, a culture of *Drosophila* in a bottle. The degree of penetrability depends on the nature of the barrier and the vagility and viability of the organism. Consider a species with a breeding range ending at a boundary beyond which individuals can survive but not reproduce. Let individuals move at random into the nonbreeding zone. The result is a diffusion process, linear if the boundary is straight and indefinitely long, two-dimensional if organisms are radiating from a circumscribed source. Pielou notes that these cases are described by the appropriate Fokker–Planck equations, giving the probability $\phi(x, t)$ or $\phi(x, y, t)$ that at a given time an individual will be at a position in the line or the plane from the starting point. These equations correspond to normal or two-dimensional joint normal distributions, respectively. They are time-dependent and relate to spread of an initial population without reinforcement or attrition. A term can be introduced to take account of drift in a given direction, such as might be caused by a steady wind or current. With drift, probability distribution curves are skewed, and in two dimension their contours are elliptical rather than circular.

Populations of living organisms are subject to attrition, for instance by death or the termination of a vagile stage. They may also be reinforced by reproduction at the source or en route. Pielou (1969, p. 133) considers organisms dispersing at random in the plane from an initial cluster and stopping at random times. The pattern of stopping points is radially symmetrical, with density diminishing outward from the center of diffusion according to the Bessel function $K_0\rho$, where $\rho = 2r\sqrt{\lambda}/a$, with λ the reciprocal of the expected travel time of an individual and a the root mean square displacement. The shape of the curve depends only on the ratio $2\sqrt{\lambda}/a$ and not on the magnitudes of λ and a, so that a slowly diffusing population with a long mean travel time might have the same distribution as a quickly diffusing one with a short mean travel time. This distribution assumes that λ is constant, whereas in actual populations in a uniform environment λ increases and a decreases with time, according to specific mortality, maturation, and exhaustion curves. The resulting mathematics, though more complex, may not be intractable with adequate empirical data. The general effect is to reduce displacement from the center and to truncate the periphery of the curve.

So far we have considered diffusion of a population cluster dispersing with time. Suppose, however, that population density is kept constant at the point of origin, for instance by continuous reproduction. Neglecting attrition and boundary effects, population density will decline exponentially with distance from the source and will have a constant expected value at any given distance. As densities become low, the stochastic element becomes large, and occurrence of an individual in any small area and time interval becomes a random event, though with defined expectation.

If the population is reinforced by reproduction en route, then its rate and probability of dispersal are raised, and it can spread indefinitely. Skellam (1951) showed that over terrain permitting establishment and reproduction, the radius of spread is proportional to time. Such is the case of an organism that has penetrated a barrier or been introduced into a previously inaccessible part of its potential range. MacArthur and Wilson (1967) showed the importance of stepping-stone islands (on which populations of species can be regenerated) in long-distance dispersal to islands.

As well as being diffuse, range limits may change with time. Weather and biotic environment vary daily, seasonally, and from year to year. Climate fluctuates over longer periods (Goudie 1977). Over very long times mountain ranges rise and are degraded, continents move, and seas grow and shrink. Organisms themselves change by adaptation to previously inoccupable territory or by evolution to different competitive levels or life strategies. For most species vagility is greater in some parts of the life cycle than in others. In winged insects adults usually disperse best, but sometimes adults are sluggish and early stages spread more easily, as with ballooning caterpillars or spiders. Ranges may be altered sporadically or seasonally by directed mass movement or *migration* (Johnson 1969). This may involve both sexes and result in temporary or even permanent establishment of breeding populations, as in the monarch butterfly, *Danaus plexippus*; alternatively, migrants may be males or move to areas where, or at seasons when, reproduction is not possible.

Shapes and sizes of ranges are almost infinitely varied. They are influenced by geographical and ecological features and by past history. Some ranges are large, some small; some are compact, others are extended, branched, fragmented, or widely disjunct. The nature of the range influences the evolution of the taxon that inhabits it. Compact ranges inhibit geographical differentiation; extensive or compartmented ranges promote it. Extreme habitats lead to specialization, sometimes of characteristic kinds. Insular and other protected habitats may permit lowering of defenses against predation and competition. Extreme isolation may favor adaptive radiation. Constraints of geography and climate lead to repetition of range patterns in different taxa, so that range types and *suture zones*, characterized by high density of range boundaries, can be recognized (Remington 1968, Scudder 1979).

10.3.2 Determinants of Ranges

Ranges of taxa are limited by the distribution of conditions under which individuals can become established, survive, and prosper. This distribution in turn results from an interaction of *intrinsic* and *extrinsic* determinants— the environmental tolerances and dispersal capacities of organisms constituting the taxon and the global pattern of environments falling within these capacities. Both intrinsic and extrinsic determinants can change, as evolving or fluctuating genotypes alter the needs and abilities of organisms and climatic, biotic, or physiographic developments modify environments and create or destroy barriers to dispersal.

Extrinsic determinants have been discussed by many authors (e.g., Watts 1971). Numerous conditions must be satisfied for a population or taxon to survive; collectively these define its *niche* (Chapter 2). The geographical projection of the niche of a taxon defines its *potential range* (*potential area* of Good 1931). This is the maximum range the taxon could occupy without intrinsic change in tolerance or extrinsic change in environment. A taxon need not occupy its entire potential range: it may either not have had time to spread that far, or it may be prevented from doing so by a geographical barrier. In the latter case the part of the potential range to which the taxon is confined may be called the *accessible range*. Potential and accessible ranges are both subject to marginal diffuseness and temporal fluctuation, as is the actual range.

Mapping of potential ranges was approached simplistically by Cook (1924), who plotted monthly temperature and precipitation tolerances for the pale western cutworm, *Agrotis orthogonia*, and projected these geographically to predict potential outbreak areas for the pest. Though Cook's approximation worked satisfactorily, many more factors are needed for full definition of range limits. These factors may be either physicochemical or biotic, and one group may influence the other. Climate, for instance, affects vegetation, and vegetation in turn influences soil formation.

Among physicochemical factors the most significant are those with global gradients (e.g., temperature and seasonality) or regional contrasts (e.g., rainfall and mineralogy). A full enumeration is not needed, but we note first factors of the *medium* or *substrate*, including air and fresh, brackish, and salt water of various types and many kinds of rocks, soils, organic structures, and detritus. A series of factors governs *productivity*, including: incident radiation, mainly solar, about 10 times as great in the tropics as near the poles; temperature; available water; oxygen; nutrients; toxic substances; and exposure. *Climate* is determined by interaction of temperature and moisture in a global framework of gradients and circulation mediated by winds and currents and controlled by topography and seasons. The climatic classifications of Köppen and Thornthwaite, based on temperature and moisture regimes and calibrated with respect to major vegetation types, are summarized by Watts (1971). Parallel effects of latitude and altitude on temperature, length of season, and phenological dates are well known (Hardwick 1971).

Biotic factors may be positive, supplying food, habitat, or some other essential, or negative, taking the form of competitors or enemies. Among positive factors food takes first place. Insects are never primary producers: they depend on plants either directly or through a phytophagous food level. Their distribution is therefore closely related to that of plants. The relationship may be a general one, to coniferous or broad-leaved forests, grasslands, bogs, and so on, or it may be highly specific, the papilionid tribe Troidini being almost limited to *Aristolochia* as hosts (Munroe 1961) and the different species of *Calligrapha* leaf beetles being specific to different trees and shrubs (Brown 1945). Dependence on animal hosts may likewise be more or less specific; some fleas are rather general feeders, whereas others occur only within the geographical ranges of particular hosts (Holland 1949). Plants or animals may also provide shelter, substrate, or suitable microclimates without which various insect taxa could not survive. Many examples come to mind. Aphids and other Homoptera and lycaenid butterfly larvae often depend on symbiosis with, or protection from, ants; the presence of a single ant species may be necessary for survival. Suitable model species may be essential for mimics. Beetles, flies, and moths may require the excrement of specific kinds of mammals. *Sarracenia* pitcher plants provide reservoirs for *Wyeomyia* mosquito larvae.

Biotic agents with negative effect include predators, parasites, pathogens, and competitors. While evidence of exclusion of particular species by these agents is rather scanty, there are many instances of release of insects to high population levels following introduction to areas where their normal parasites and predators are absent. Success of species in impoverished habitats such as bogs may often be due to relatively poor performance of competitors rather than to a favorable character of habitat *per se*.

Intrinsic determinants of ranges depend on genetic constitution of individuals and populations and on the phenotypic repertoire of responses and tolerances to environment determined by this constitution. This is discussed in Chapter 9 and touched on only briefly here. Species and populations are neither homogeneous nor fixed in their genotypes, though clearly variability at any one time is rather limited. Opinions differ as to what proportion of genetic differences affect fitness; certainly many do. Numerous examples are known of species with adaptive genetic differences in different parts of their geographical ranges: differences may concern climatic tolerance, host preferences, protective resemblance to different backgrounds or models, and the like (Chapter 8). The general tendency is to improve chances of survival locally and to expand the range as a whole by permitting adaptation to different conditions in different parts. On the other hand, loss of alleles, for instance by the founder effect in colonist populations, by random fixation in small isolates, or by selection in harsh or marginal environments, or even a shift to genotypes with less flexible responses to changing conditions, may lead to contraction of the potential range of the population or species and to exclusion from formerly habitable territory.

The determinants of ranges, extrinsic or intrinsic, biotic or physico-chemical, clearly do not work in isolation. The *integration of factors* is complex and can be represented as a multidimensional matrix or environmental space, giving viabilities or survival probabilities for each factor measured against the others (Whittaker 1972). Projected geographically, resultant viabilities determine potential range. As individual factors normally have nonlinear effects on viability, and as different factors are not independent but affect one another's action, the mathematics rapidly become complex or intractable as factors are added. Cyclical and erratic changes in environment, differential responses of life stages, and seasonal and spatial packing of species and their life cycles all contribute to complexity. Selection of key factors for study that account for most of the variability is often the practical answer and accounts for the success of Cook's study cited above.

10.3.3   Dynamics of Ranges

Much of the interest of ranges attaches to their changes with time—spread of successful taxa, shifts as environments alter, and relict and disjunct configurations developed as formerly widespread taxa declined and retreated.

The range of a taxon is always limited by the extent of its potential range. The potential range may be divided into segments that are, at least within reasonable probabilities, inaccessible to one another. In such a range the segment in which a taxon originates may be called its *accessible range*. Within the accessible range a species may originate in one of two modes: either as a population occupying the whole accessible range, for instance a vicariant differentiating from a more widely distributed ancestor, or locally in some part of the accessible range, whether sympatrically by ecological isolation or some genetic or chromosomal event, parapatrically by differentiation along a gradient, or allopatrically in response to temporary or imperfect geographical isolation. If the species initially occupies only part of the accessible range, it will spread to fill it. In uniform terrain radius of spread tends to be proportional to time, following Skellam's model for reproducing populations (Section 10.3.1 above). Rate of spread depends on vagility and on environment traversed. Factors that alter vagility (e.g., prevailing winds) or survival and reproduction (e.g., unfavorable ecology) distort the pattern of spread.

Once the accessible range has been filled, provided the species does not split into vicariant daughter species, further change can take place in actual range only as a result of change in potential range. Such change may have direct effects, as when the potential range is altered by shift in climatic zones, emergence or submergence of land, and glaciation or deglaciation. Potential and actual ranges may increase, decrease, or change position without changing size. Such shifts can be very dramatic. In the Upton Warren Interstadial in Pleistocene England, an insect assemblage of boreo-montane facies was replaced by an austral one, indicating a change from conditions colder to warmer than the present, in a period of about 1000 yr. The beetle *Aphodius holdereri* found in glacial Britain is now restricted to high plateaus of Tibet (Coope 1979). Many comparable changes are known.

Because of accidents of geography, changes in potential range may also have large indirect effects. These come about when a formerly continuous accessible range is divided, for instance if a land bridge becomes impassable by subsidence, or conversely when an accessible range is enlarged by break-down of a barrier. The changes in connections between neighboring continents during the Cenozoic (Matthews 1980) produced many examples, as witness the mixing of North and South American faunas following estab-lishment of the Panama isthmian link. Another type of geographical ac-cident traps a taxon or population in a shrinking accessible range from which there is no egress, even though the potential range as a whole only changes position. Antarctic taxa must have been in this position as glaciers occupied their continent, as for example insects associated with Antarctic *Nothofagus* populations (Schlinger 1974). Such obliteration of local sections of the potential range accounts for many cases of disjunction and localized relict populations.

Genotypic change in a population, species, or higher taxon can cause similar changes in its potential range, and parallel effects on its actual range can be expected. Diversification of genotypes may mean that more habitats can be occupied and that the potential range expands. Specialization may lead to withdrawal from certain habitats and perhaps from part of the geographical range. If diversification proceeds to the extent of speciation, the daughter species may have smaller potential ranges than the parent or the potential ranges may overlap or coincide. In such cases the range of the original species is inherited by a monophyletic group of descendant species or higher taxa.

Because of the probabilistic nature of range boundaries, it is rarely safe to assume that barriers are absolute or that accessible ranges are perfectly sealed off. There are numerous records of insects at sea (Williams 1958), in the air (Gressitt & Yoshimoto 1963), arriving on such remote islands as Norfolk and New Zealand (Holloway 1977, Fox 1978), and flying in num-bers to altitudes above snow line (Mani 1962). Presence of endemic stocks in islands as remote as Hawaii argues for occasional successful penetration of broad oceanic barriers (Zimmerman 1948, 1970). The relative impor-tance of such events is a matter of heated controversy (Nelson & Rosen 1981).

10.4 FAUNAS

10.4.1 Main Problems and General Features of Faunas

A *biota* is the set of species that live together in one place or area. Its animal part is the *fauna* and its insect part the *entomofauna*. The idea of a fauna is most meaningful when it relates to a topographically or ecologically well-defined area, but for funding, logistic, or jurisdictional reasons a political unit is frequently made the basis. We have seen that faunas differ greatly from place to place and over even geologically short intervals of time. We have seen also that faunal make-up reflects both current ecology and past evolutionary and distributional history. How should we describe and compare faunas, how should we separate ecological from historical effects, and how can we trace the evolutionary and geographical course of the latter?

Faunas are generally specified by a list of species. To be meaningful, this list should be drawn from within spatial, temporal, and taxonomic limits (Pielou 1975). The number of species in the list is an elementary measure of *diversity*. It is referred to here as *size* of fauna. Species group into higher taxa, which have diversities and distribution patterns of their own. Both species and higher taxa can link or differentiate faunas. Species of one fauna are present or absent in another. If present they may show greater or lesser genetic differences; if absent they may be represented by similar species (*vicariants*) or unrepresented. Higher taxa can have comparable resemblances or differences; in addition they can differ in numerical and proportional representation and in identity and ecological and geographical relationships of component species.

The faunal list can be refined by distinguishing permanent residents from various classes of temporary visitors. In the intensively studied British fauna, accidental or recurrent wanderers come from continental Europe or even Africa and America. There are also migrants, some coming annually, others only in favorable years. Immigrants may reproduce locally, sometimes until the onset of winter, sometimes for a period of years, and of course a certain fraction become permanently established.

More complicated indices of diversity, which take into account relative quantities of species, are discussed by Pielou (1975). Though important for comparison of samples and communities, these are of less interest in faunal studies.

It might seem easy by comparison of species lists to decide whether one pair of faunas is more or less similar than another, but complications soon appear. The number of species in common tells us something, but not much. The proportion of species in common is more informative, but ambiguous. Consider a fauna that has no species not also present in a second neighboring fauna. If the faunas are of the same size, their species lists are identical. If the second is larger, only N_1/N_2 of its species are shared with the first, where N_1 and N_2 are the respective faunal sizes. It may be argued that the smaller fauna is only a sample drawn from the larger and that zoogeographically the two are the same. However, suppose that the small Fauna 1 shares the same list of species with a Fauna 3 that has different unshared species, as might happen if all the shared species were vagile, opportunistic, and ubiquitous. Fauna 1 can clearly not be the same as both Fauna 2 and Fauna 3. A better measure is $N_c/(N_1 + N_2 - N_c)$, where N_c is number of species common to the two faunas. This measure applies to faunas of different sizes, to faunas that each have unshared species, and is readily generalized to compare more than two faunas. A more sophisticated index is Preston's coefficient of dissimilarity, z, given by the resemblance equation

$$x^{1/z} + y^{1/z} = 1,$$

where x is the proportion of the joint Fauna found in Fauna 1 and y that in Fauna 2. Holloway and Jardine (1968) call this "the only coefficient we have seen that has a sound statistical basis."

These and similar indices can be examined intuitively or contoured or subjected to cluster analysis to yield patterns of resemblance and difference over a selected area (Huheey 1965, Holloway & Jardine 1968). Usually areas of relatively uniform fauna are separated by suture zones (Remington 1968, Scudder 1979) of steep faunal change. Where there are sharp faunal barriers, as between islands, faunal composition may change even more abruptly. On the other hand in extended environmental or dispersal gradients change may be gradual—continuous if habitats are uninterrupted, stepwise if they are compartmented (Pielou 1977, Bond 1979).

Faunas naturally tend to be most similar to their nearest neighbors. Patterns of similarity depend on geography, ecology, and past history. The Virgin Islands share most of their fauna with the nearby large island Puerto Rico, with which they were united in the Pleistocene (Carey 1972), whereas the Bahamas, isolated from other land masses in the Pleistocene but standing at a distributional crossroads, have species of Cuban, Hispaniolan, and North American affinity (Clench 1977). Both faunas lack many species found in the larger and ecologically more diverse Greater Antilles.

Interfaunal relationships are evidence of past faunal connections and of movements of species. Commonly species in a fauna can be assigned to *elements* according to their extralimital distribution patterns. By inspection Mayr (1944) distinguished 15 elements in the bird fauna of Timor. Holloway and Jardine (1968) used clustering of dissimilarity coefficients to identify nine widespread and several endemic elements among Indo-Australian birds; butterflies had a similar but simpler pattern, bats a different one. In both studies most of the elements reflected distributional history but some were ecologically constrained. Holloway (1973) emphasized ecological characteristics of elements in the Indian butterfly fauna. In the mostly postglacially formed insect fauna of Canada elements are largely ecologically defined, though history of spread from Pleistocene refugia is also important (Scudder 1979).

Further dimensions can be added by analysis at different taxonomic levels. Genera do not necessarily show the same geographical groupings as species or subspecies nor families and orders the same as genera. Higher rank of average taxonomic difference between faunas indicates longer or more effective isolation or both, especially where direct vicariants are concerned (Munroe 1957). Within genera, centers of species richness and character state diversity can be identified, and can be used to locate faunal centers, zones of hybridization, and mixing and faunal refugia (Holloway 1969, 1970, Brown et al. 1974). These concepts lead naturally to historical analysis, but first we must briefly consider world patterns of distribution.

10.4.2 Global Patterns of Faunas

The distribution of land masses and that of climatic gradients determine the broad outlines of insect geography. Species diversity generally increases from high to low latitudes. Antarctica has about 140 known species of land arthropods, of which 56 are parasites. Of the free-living species 64 are mites, 19 are Collembola, and 2 are Diptera (Gressitt 1971). In the Arctic only 28 species of insects and relatives are known from inhospitable Ellef Ringnes Island; about 250 insects and spiders are recorded from the more northerly but more protected Hazen Lake, Ellesmere Island (Scudder 1979). Canada as a whole has over 33,000 known species; the number in a comparable tropical area would be much higher. Though this trend is common, not all groups of animals follow it (Fischer 1960). Though a majority of American butterfly genera are tropical, nearly 10% are northern and a number are southern (Hovanitz 1958). The reasons for latitudinal diversity gradients are not well understood. Among factors possibly responsible are productivity, predictability, stability, habitat diversity, and length of undisturbed development. These are reviewed by Pielou (1975, 1979), but without conclusive result.

Other general diversity gradients exist, for instance a decline from rain forest to desert and an increase from uniform to dissected habitats. Species diversity tends to decline with altitude, as with latitude—for example, in Scottish carabid beetles (Greenslade 1968) and West Indian butterflies (Munroe, unpubl.). The well-known gradient from continents to islands of various sizes and distances is discussed in Section 10.4.3.

Major faunal regions have been outlined (Introduction and Table 10.1). Some further details follow. The Holarctic region is latitudinally zoned from rock desert and tundra in the north through coniferous forest to longitudinally compartmented southern formations ranging from deciduous forest through parkland or sclerophyll scrub to grassland, steppe, and desert. The major disjunction between Eurasia and North America is hardly more important than those between eastern and western United

States or Europe and temperate East Asia. Disjunction has led to extensive vicariance of insects not confined to one type of habitat, but found in forest, aquatic, steppe, and montane groups. Mountain chains in the New World run mainly north and south, in Europe and Asia mostly east and west. They have therefore tended to act as corridors for cool-adapted insects in America and barriers for warm-adapted forms in Eurasia. Some groups are disproportionately numerous in species in each main division: *Parnassius* in the Palearctic mountains; *Abraxas* and allies in China and Japan; Satyridae, plebejine Lycaenidae, and *Zygaena* in Europe; and *Catocala* in North America. Infiltrators from adjacent tropical continents have had distinctive influences on the various southern sectors.

The tropical regions are ecologically not dissimilar, having rain forest, savanna, desert, and montane habitats, but faunistically they differ widely, having few species and only a modest proportion of genera in common, though often sharing families, subfamilies, and tribes. Tropical America and Africa are each mainly composed of a single land mass, though the former has the Antilles and the latter Madagascar and the Mascarenes and Seychelles as island satellites. Both have north–south trending mountains that have permitted travel of northern and southern cool-adapted insects and that have also developed striking endemic faunas. In tropical America montane faunas are particularly rich, and a high proportion of species and genera are endemic in the Cordillera and the Brazilian and Guianan mountain systems. In Africa grassland and steppe faunas are well developed, with many species of tenebrionid beetles, crambine moths, and other characteristic forms. Both the Antilles and Madagascar have interesting but relatively poor and disharmonic faunas with many endemic species and genera (Gill 1978, Paulian 1961).

Biogeography in the Oriental region is dominated by division into islands and the distribution of mountain ranges. Vicariance between islands and strong centers of endemism and species and genus richness on the larger islands and mountain masses make for a complex and diverse fauna. The richest faunas and most important centers of endemism are in the Indo-Malayan and Papuan subregions, the former with multiple centers, the latter with its main center in New Guinea. Wallacea has a mixture of Indo-Malayan and Papuan stocks, with secondary centers of endemism, on larger islands, especially in the Philippines and Sulawesi. In the Micronesian and Polynesian subregions the setting is entirely insular (Gressitt 1954, 1956, 1963). "High" islands with persisting mountains have much richer faunas than flat atolls, which are subject to flooding and wind damage during storms and desiccation in between. Faunas become poorer in taxonomic stocks but richer in endemic species per stock with increasing distance eastward. In the East Polynesian subregion some stocks are of American or indeterminate origin (Zimmerman 1948, 1970, Carlquist 1980). Regions of the South Temperate zone differ greatly in their faunas, though all have species or groups, often of archaic type, that show clear relationships to those of other southern regions. Australia–New Zealand, Australia–South America, and New Zealand–South America links are known (Illies 1965, Schlinger 1974, Howden 1981). Chile has desert, steppe, montane, and several types of forest habitats, the cool temperate forests with *Araucaria* and *Nothofagus* being particularly interesting (Peña 1966). Australia has a dry interior and relatively moist coastal zones, especially in the east and southwest (Cracraft 1982). Papuan species enter the fauna, particularly in the Queensland rain forests, but extend down the east coast and occur more sparingly elsewhere. The greater part of the Australian entomofauna consists of species groups and genera peculiar to the region but not grossly different from members of the same families in the Oriental region. Most of these are probably colonists from the north but of greater age than the Papuan element. Finally there is a substantial element of very peculiar, often primitive, endemic, relict, or southern forms; these are concentrated in the southeast, including Tasmania, and the southwest (Mackerras 1970, 1974). The New Zealand insect fauna is smaller than that of Australia, is more disharmonic, and has large concentrations of species in certain genera. Most of the species are adapted to moist forest, grassland, montane, or aquatic habitats. Gaskin (1970) says there is a small, archaic, possibly endemic element of uncertain origin, a strong but disharmonic Australian element, a substantial Oriental element, a small Holarctic element, and a large, heterogeneous southern element of varying age. There are strong indications of faunal interchange with temperate South America in some of the southern groups (Brundin 1975, Gaskin 1975). Southern Africa, too, has important elements in common with the other temperate continents, but the differences are far larger. Geology suggests that separation of Africa from Antarctica occurred earlier and that the southern elements have had varied origins. Whatever their derivation they have penetrated far northward along mountain chains, mingling in equatorial Africa with southward-traveling northern forms.

In addition to faunas already considered, there are a number of oceanic islands or island groups, with little in common except their isolation. Examples are Bermuda, the Azores, the Canaries, Madeira, and the Cape Verdes in the North Atlantic; Ascension, St. Helena, Fernando de Noronha, Martin Vas, and Gough in the South Atlantic; the Chagos, Cocos, and Christmas Islands in the Indian Ocean; Galapagos, Cocos, and Juan Fernandez in the East Pacific; and South Georgia, Heard, Kerguelen, Amsterdam, Marion, Crozet, Macquarie, Campbell, and the Australs in the subantarctic. These agree in having restricted waif faunas, with endemism at the species or species group level, and with various adaptations to exposed insular life.

10.4.3 Dynamics of Faunas

Before examining the history of insect faunas, we must give some attention to how they are formed and how they change with time. Faunas are assemblages of coexisting species. These species are either *autochthonous* (i.e., of local origin) or *allochthonous*, having immigrated from elsewhere. Either kind may be *endemic* (i.e., confined to the geographical unit under study) or *apodemic*, occurring in other units also. If apodemic it may be *native* or *introduced*. The same classification applies to higher taxa also. An endemic species that formerly had a much wider range is called a *relict*; an area with many relict species or taxa is a *refugium*.

Faunas can originate with a complement of species if they are formed by splitting from some larger faunal unit or by movement of an environmental zone, shifting many potential ranges at once. Or they may originate by individual movement of species into a previously unoccupied habitat. They add species either by autochthonous speciation or by immigration from elsewhere. They lose species by extinction, either of the whole species or of its local population. As we have seen, immigrant species may arrive from different sources, according to what routes are open and how difficult they are. Autochthonous species may evolve by several different modes (Scudder 1974); some of these are sympatric, some allopatric. Daughter species may share the habitat with parent and sister species, or one or more may supplant or exclude the rest. Only if more than one remains is there a gain in species number.

Size and composition of a fauna depend on interplay of these three processes, the positive factors immigration and speciation opposing the negative factor extinction. If immigration and speciation together outweigh extinction, the fauna grows. If extinction predominates, the fauna shrinks. If immigration plus speciation balance extinction, then the size of the fauna, though not necessarily its composition, is in equilibrium. As these processes are stochastic, there will be deviations from expected values, and perhaps overshooting and oscillation if there is a tendency toward equilibrium. In particular a sudden rush of immigrants may overload the fauna. Origin of a fauna by splitting of a fragment from a large species-rich habitat may constitute a special case of such overloading.

Composition of the fauna is affected in a different way by the balance of factors. Speciation tends to increase autochthonous and endemic species, immigration to increase allochthonous and apodemic ones. If immigrant species come from more than one source, they add to the fauna in proportion to immigration rates from the respective sources. Extinction may affect species from different sources impartially, or autochthonous species and immigrants from various sources may have different expectations of survival, in which case their relative representation will be affected. High values for immigration and speciation offset by high extinction rate cause rapid turnover and low mean age of species in the fauna. The longer these processes operate, the more any original fauna will be modified, by alteration or splitting of species through evolution and by substitution due to immigration.

The form these relationships take in the real world has been investigated and debated intensively since the proposal of a quantitative model by

MacArthur and Wilson (1963, 1967). Developments have been reviewed comprehensively though unsympathetically by Gilbert (1980) and more analytically by Pielou (1979). MacArthur and Wilson sought to explain an observed relationship between number of species in an island fauna and area of the island, which took the form $S = CA^z$, where S is number of species, A is area, and C and z are empirical constants. They hypothesized that island faunas are approximately in equilibrium between immigration of unrepresented, and extinction of, represented species; they supposed the contribution of speciation to be negligible. In a constant flux of wandering species immigration to a given island will decline as number of species increases because (1) more species will be established already so that their arrival will not count and (2) the more abundant species tend to become established first so that the mean rate of arrival of the rest tends to decline. On the other hand, rate of extinction increases with number of species because (1) space and resources are partitioned among more species so that the population of each one is smaller and its probability of extinction in a given time correspondingly larger and (2) there are more species so that even with extinction probability unaltered for each, it is increased for the set. If these functions are definable at all, therefore, they tend toward equilibrium. If immigration rate is decreased, as by considering a more remote island, or extinction rate increased, as in a smaller island, the number of species at equilibrium is smaller, agreeing with observation.

A number of consequences can be deduced from the theory if it is accepted as correct. For details the reader is referred to MacArthur and Wilson (1967) and subsequent literature. By suitable assumptions the theory ought to be adaptable to cases where speciation is important, to ecological and other conceptual "islands," to productivity and other habitat gradients, and to continental faunas in steady and changing environments. Though there is some experimental and considerable observational support for it, serious doubts remain. Among the most important are whether available data support the theory in any but selected cases, whether natural variations are not so large as to obscure underlying trends, and whether the theory may not be so flexible through supplementary assumptions as to become an unverifiable ad hoc formulation. The present view seems to be that equilibrium theory should not be written off, but that it needs substantially more evidence and refinement for acceptance.

10.5 INSECTS, FAUNAS, AND ENVIRONMENTS OF THE PAST

10.5.1 Problems and Evidence

We have emphasized the role of history in forming present distributions, faunas, and ecosystems. How has that history unfolded? The earth is about 4.6×10^9 yr old. Probable microorganism fossils are known from 3.5×10^9

B.P., and the first insects from Lower Devonian, about 3.8×10^8 B.P. Our samples of the Recent fauna come from less than one-millionth of recorded insect history and less than one ten-millionth of the age of life and of the earth. How can we get access to this long period? The surprising thing is that it is possible to develop some kind of a historical picture, but the fragmentary and often speculative nature of the evidence should prepare us for major amplifications and revisions, such as have indeed occurred during the last few years.

Evidence on ages comes primarily from radioisotope decay measurements, supplemented by comparative stratigraphy (on which relative ages were first established) and astronomical theory and data. For late periods some additional sequences can be calibrated, such as tree-ring widths, volcanic ash deposits, and geomagnetic field reversals. Past environments are attested by chemical and physical features of earth and rock deposits and by the nature of included fossils. Fossils give information on the evolution of life, but some modern workers think that more can be deduced from comparative study of existing forms (Cracraft & Eldredge 1979). Distribution of land and water was classically studied from distribution of deposits combined with similarities or differences of contemporary fossils, but paleomagnetic studies now give latitudes with respect to the magnetic and, no doubt, the rotational poles. Plate tectonics have made it possible, by examination of ridges, fractures, rifts, bands of remanent magnetism, ophiolites, and other structures and evidence, to learn much about sea-floor spreading, movements of continents, and genesis of mountain ranges and climates, particularly since late-Paleozoic time. But it must be emphasized that resulting pictures are fragmentary and that single "snapshots" in textbooks may represent millions or tens of millions of years of varied history.

Insect fossils are not easily preserved; the useful deposits are few and unrepresentative. Rock fossils go far back in time. Some fine-grained sediments preserve exquisite detail, but most fossil insects are fragmentary, and often different observers interpret structures differently. Amber, dating back as far as early Cretaceous, gives much better detail. In favorable aspect, amber specimens can be as informative as microscope slides. Insects, unaltered but usually fragmentary, are preserved in late-Cenozoic peats. Specimens can often be matched with identical or closely related existing species.

10.5.2 History of Insects and Insect Faunas

Only the broadest historical outlines can be touched on. For more detail see Hennig (1981) and Matthews (1979, 1980). A time scale is given in Table 10.2. Arthropods appeared at the beginning of the Cambrian, 5.8×10^8 B.P., but the first known insects are from the Lower Devonian Rhynie Chert of Scotland, formed in a deltaic marsh with psilophytes and very early vascular plants, at about 4.2×10^8 B.P. *Rhyniella*, accepted by Hennig as a

Table 10.2   Geological eons, eras, periods, epochs and absolute dates

Eon	Era	Period	Epoch	Time before present (yr × 10⁶)
Phanerozoic	Cenozoic	Quaternary	Holocene	0.01
			Pleistocene	2
		Tertiary	Pliocene	7
			Miocene	26
			Oligocene	36
			Eocene	55
			Paleocene	65
	Mesozoic	Cretaceous		135
		Jurassic		200
		Triassic		240
	Paleozoic	Permian		285
		Carboniferous		375
		Devonian		420
		Silurian		450
		Ordovician		520
		Cambrian		580
Proterozoic				2600
Archaeozoic				4500

collembolan (Entognatha), and *Rhyniognatha*, known from a pair of jaws thought to belong to a dicondylian ectognathan, are the two insect genera. There are also mites and spiderlike arachnids and branchipod Crustacea. All the forms are minute. The insects show no evidence of wings. The mite has what appear to be piercing mouthparts. There is stem injury, possibly arthropod caused, and a group of arachnids was found in sporangia (McKerrow 1978).

The Rhynie, in which conditions were exceptionally favorable for preservation of minute forms, is followed by a long gap in the insect record. The next authenticated fossils are in the Upper Carboniferous, beginning at about 3.7×10^8 B.P. Here winged insects are abundant and diverse, including in the lowest strata Protodonata and varied genera of uncertain placement, soon augmented by Palaeodictyoptera, Protephemeroptera, cockroachlike insects, and others. Giant insects and neopterous wing folding had developed; insects were present on the principal continents and had ecologically diversified faunas. Identifiable lineages tended to be less advanced in organization than modern forms: Protephemeroptera and Permoplectoptera had fully developed hind wings resembling the forewings; in Odonata and precursors the nodus, discoidal cell, and attachment of anal veins to cubitus developed step by step as the Paleozoic progressed (Hennig 1981). Paleozoic pterygote nymphs had articulated, movable, veined wings and gradual metamorphosis with a succession of subimaginal instars. Both adults and nymphs often had veined and sometimes articulated wing rudiments on the prothorax. Mayfly nymphs had nine pairs of abdominal gills, with plates which may have been homologous with thoracic wings (Kulakova-Peck 1978).

The Permian period was a time of diversification and change. Holometabola, including beetles, Mecoptera, and others, and Homoptera appeared. Lower Permian insects of Kansas were smaller than those of the Carboniferous of Kansas and Europe, but numerous in individuals and species. In Siberia the fossil record continues directly on that of the Carboniferous, but there are many more individual fossils and species in the Permian. Also the fauna changes, cockroachlike forms disappearing and being replaced by beetles and others. The Permian was a time of cyclical recession and advance of seas and of union of continents into a single Pangea, the northern part of which was equatorial, the southern part polar and at times glaciated. Insect faunas in different deposits are markedly divergent, but geographical and temporal patterns have not been satisfactorily resolved.

The close of the Permian, at about 2.85×10^8 B.P., marks one of the great changes in the world's biota. As in many other organisms, insects lost numerous major and previously successful groups. Palaeodictyoptera and Megasecoptera disappeared, Protodonata, Protoperlaria, Archimylacrididae, and Glosselytrodea and others were reduced to few and perhaps incorrectly identified Mesozoic survivors. The Triassic insect fauna is already surprisingly modern, but mainly by subtraction of Paleozoic groups and expansion of surviving stocks rather than by appearance of new groups. These indeed developed during the Mesozoic, but apparently gradually— Hymenoptera in the Jurassic, Lepidoptera and ants in the Cretaceous, and so on. These are of course the times of occurrence of recognized fossils, and the origins of the groups may have been earlier. The reasons for late-Permian extinctions are obscure. Some authors prefer catastrophic explanations; others invoke changes in climate.

In the Cretaceous (late Mesozoic) and Cenozoic we see the gradual development of modern conditions. We do not know much about the Cretaceous to Tertiary change—so important for many marine organisms and for terrestrial reptiles. Cretaceous amber insects are more different from their modern counterparts than are those of the Oligocene Baltic amber. Many if not most existing insect families were in existence by the end of the Cretaceous (Matthews 1979), whereas in the Baltic amber a high proportion of species belong to modern genera. The few known Eocene butterflies are referable to existing families but extinct genera (Durden & Rose 1978), whereas Oligocene butterflies from Colorado are close to present forms (Brown 1976).

Through most of the Mesozoic Pangea remained united, though largely bisected into northern Laurasia and southern Gondwana segments by the transverse Tethys Sea. In late Jurassic or early Cretaceous the world continent began to fragment, as present patterns of spreading axes and lithospheric plate movements became established. Pangea broke up, the southern continents separated from Antarctica, and the Atlantic Ocean opened. Though this was the broad picture, there are many points of uncertainty (Hallam 1981). Not surprisingly, older insect groups such as Plecoptera appear to reflect these events in their distributions (Illies 1965), whereas younger ones do not.

Biogeographical events since fragmentation of Pangea have been dominated by the sequence of splitting and paths of divergence of new regional continents. Raven and Axelrod (1975) present a general account, emphasizing Tertiary independence of South and North American biotas, initial integration of African and South American blocs in the West Gondwana continental speciation center, and later association of the now separate African fauna with Europe, Temperate Asia, and, indirectly, North America. Subsequent studies show that there was a widespread North American tropical fauna up to the latter part of the Eocene; temperature differences between seasons were much smaller; tropical foliage characteristics at high latitudes suggest that the earth's axis of rotation was inclined much less than now, so that there was no extreme seasonal variation in day length at high latitudes as at present (Wolfe 1978). A marked decline of temperature at the end of the Eocene ushered in a time of fluctuating mean temperatures and much wider differences between annual maxima and minima. This was followed by the onset of glaciation and the alternating glacial and interglacial stages of the Quaternary. The latter resulted in profound and repeated changes in ranges of mammals, plants, and insects and disruption and mixing of habitats (Kurtén 1968, 1971, Coope 1979, Matthews 1979, 1980, Van Devender & Spaulding 1979). Reestablishment of faunal connections between North and South America by island arcs about 9×10^6 B.P. and by a land bridge about 3×10^6 B.P. led to extensive mixing of biotas (Marshall et al. 1979) and perhaps to extinction of many forms by competition. Glaciation and associated refrigeration in temperate latitudes and desiccation in the tropics no doubt effected the extinction of many more species, but the high proportion of large forms among the mammals exterminated suggests that the most important agent may have been man in pursuit of food and fur.

10.6 THE EFFECTS OF HUMANS

Although humans have changed life for other species as much as any environmental revolution since the emergence of organisms from the sea, their influence has mostly differed from that of natural phenomena in degree rather than in kind. Humans have competed more effectively, transformed environments more quickly, and transported species farther and faster than prehuman agencies would have done. These processes were not new, but they had never been carried so far so quickly before.

Early humans were hunters and gatherers. At about 1.75×10^6 B.P. *Australopithecus* in Africa was using chipped pebbles as tools and making circular stone shelters. *Homo erectus* at over 5×10^5 B.P. was using fire. Advanced weapons, and especially burning to drive game, permitted hunting peoples to deplete the large-mammal fauna and destroy wooded and scrubby environments. Beginning about 60,000 B.P. approximately 40% of African large-mammal genera became extinct, probably at the hands of man. In Europe about 30,000–10,000 B.P., after the rise of *Homo sapiens* hunting cultures, about half the species of large mammals were exterminated. In America extinction was more rapid and more severe, and about 70% of large mammals disappeared in a couple of millennia following the advent of Amerindian hunters (Kurtén 1971). Similar disasters have befallen the faunas of South America, Madagascar, Australia, New Zealand, and many continental and oceanic islands. Surviving large mammals have almost all decreased in size, probably in response to a decreased share of resources.

Agriculture, of course, led to more extensive inroads on natural habitats. Forests were cleared, fields cultivated, marshes drained, at first on a small scale, then farther and faster as populations increased and farming became more organized and technology advanced. Today whole habitats have been destroyed and the fringes of wildlands are being pushed back ever more rapidly.

All these developments, while destructive to many species, are beneficial to a few. Useful and ornamental plants and those that thrive as weeds or in disturbed environments have greatly expanded populations and habitats. So too do domestic animals and pests, opportunists, and commensals that accompany households and people or their crops, flocks, or pets. Large numbers of species have been introduced accidentally or on purpose to lands far from their original homes. Often the introductions have proved nearly as destructive as man himself. These matters are dealt with in other chapters. See also the useful symposium chaired by Downes (1980).

Perhaps the final extreme of habitat destruction is the building of cities, roads, and other paved structures (Gottman 1964). On territory so preempted no natural populations survive.

REFERENCES

Bond, J. 1979. *Proc. Acad. Nat. Sci. Phil.* **131:** 89–103.

Brown, F. M. 1976. *Bull. Allyn Museum* **37:** 1–4.

Brown, K. S. Jr., P. M. Sheppard, and J. R. G. Turner. 1974. *Proc. Roy. Soc. Lond. B.* **187:** 369–378.

Brown, W. J. 1945. *Can. Entomol.* **77:** 117–133.

Brundin, L. K. 1975. *Mem. Mus. Hist. Natur. Paris* (ns. A. Zool.) **88:** 19–27.

Carey, W. M. 1972. *Carib. J. Sci.* **12:** 79–89.

Carlquist, S. 1980. *Hawaii: A Natural History* (2nd ed.). Pacific Tropical Botanical Garden, Lawai, Kaui, Hawaii.

Clench, H. K. 1977. *Ann. Carnegie Mus.* **46:** 173–194.

Clube, V. and B. Napier. 1982. *The Cosmic Serpent; a Catastrophist View of Earth History.* Faber, London.

Cook, W. C. 1924. *Ecology* **5:** 60–69.

Coope, G. R. 1979. *Annu. Rev. Ecol. Syst.* **10:** 247–267.

Cracraft, J. 1982. *Am. Zool.* **22:** 411–424.

Cracraft, J. and N. Eldredge. 1979. *Phylogenetic Analysis and Paleontology.* Columbia University Press, New York.

Danks, H. V. (Ed.). 1979. *Mem. Entomol. Soc. Can.* **108:** 1–573.

Downes, J. A. (Chair). 1980. *Can. Entomol.* **112:** 1089–1238.

Durden, C. J. and H. Rose. 1978. *Pearce–Sellards Ser., Texas Mem. Mus.* **29:** 1–25.

Fischer, A. G. 1960. *Evolution* **14:** 64–81.

Fox, K. J. 1978. *New Zealand Entomol.* **6:** 368–379.

Francis, P. 1983. *Sci. Am.* **248:** 60–69.

Gaskin, D. 1975. *Mem. Mus. Natl. Hist. Nat. (Paris)* (n.s., A, Zool.) **88:** 87–97.

Gaskin, D. E. 1970. *Geogr. Rev.* **60:** 414–434.

Gilbert, F. S. 1980. *J. Biogeogr.* **7:** 209–235.

Gill, F. B. (Ed.) 1978. *Zoogeography in the Caribbean. The 1975 Leidy Medal Symposium.* Academy of Natural Science, Philadelphia, Pa.

Good, R. d'O. 1931. *New Phytol.* **30:** 149–171.

Gottmann, J. 1964. *Megalopolis.* MIT Press, Cambridge, Mass.

Goudie, A. 1977, *Environmental Change.* Oxford University Press, Oxford.

Greenslade, P. J. M. 1968. *Trans. Roy. Entomol. Soc. Lond.* **120:** 39–54.

Gressitt, J. L. 1954. *Insects of Micronesia. 1. Introduction.* Bishop Museum, Honolulu.

Gressitt, J. L. 1956. *Syst. Zool.* **5:** 11–47.

Gressitt, J. L. (Ed.). 1963. *Pacific Basin Biogeography, A Symposium.* Bishop Museum, Honolulu.

Gressitt, J. L. 1971. *Pacific Insects Monogr.* **25:** 107–178.

Gressitt, J. L. and C. Yoshimoto. 1963. Pp. 283–292, in *Pacific Basin Biogeography.* Bishop Museum, Honolulu.

Hallam, A. 1981. Pp. 303–330 in Nelson and Rosen (Eds.), 1981.

Hardwick, D. F. 1971. *Can. Entomol.* **103:** 1207–1216.

Hennig, W. 1981. *Insect Phylogeny* (transl. and ed. by A. Pont, rev. notes by D. Schlee). Wiley, Chichester.

Holland, G. P. 1949. *Dom. Canada, Dept. Agr. Tech. Bull.* 70.

Holloway, J. D. 1969. *Biol. J. Linn. Soc.* **1:** 373–385.

Holloway, J. D. 1970. *Biol. J. Linn. Soc.* **2:** 259–286.

Holloway, J. D. 1973. Pp. 473–499 in H. S. Mani, *The Biogeography and Ecology of India.* W. Junk, The Hague.

Holloway, J. D. 1977. *The Lepidoptera of Norfolk Island: Their Biogeography and Ecology.* W. Junk, The Hague.

Holloway, J. D. and N. Jardine. 1968. *Proc. Linn. Soc. Lond.* **179:** 153–188.

Hovanitz, W. 1958. Pp. 321–368 in *Zoogeography*. AAAS, Washington, D. C.

Howden, H. F. 1981. Pp. 1009–1035 in A. Keast (Ed.), *Ecological Biogeography of Australia*. W. Junk, The Hague.

Huheey, J. E. 1965. *Am. Midl. Natur.* **73:** 490–500.

Hultén, E. 1937. *Outline of the History of Arctic and Boreal Biota during the Quaternary Period*. Bokförlags Aktiebolaget, Thule, Greenland.

Illies, J. C. 1965. *Annu. Rev. Entomol.* **10:** 117–140.

Johnson, C. G. 1969. *Migration and Dispersal of Insects by Flight*. Methuen, London.

Kulakova-Peck, J. 1978. *J. Morphol.* **156:** 53–123.

Kurtén, B. 1968. *Pleistocene Mammals of Europe*. Weidenfeld & Nicholson, London.

Kurtén, B. 1971. *The Age of Mammals*. Columbia University Press, New York.

MacArthur, R. H. and E. O. Wilson. 1963. *Evolution* **17:** 373–387.

MacArthur, R. H. and E. O. Wilson. 1967. *The Theory of Island Biogeography*. Princeton University Press, Princeton, N. J.

Mackerras, I. M. 1970. Pp. 187–203 in *CSIRO, The Insects of Australia*. Melbourne University Press.

Mackerras, I. M. 1974. Pp. 29–31 in *CSIRO, The Insects of Australia*, Supplement. Melbourne University Press.

Mani, M. S. 1962. *Introduction to High-Altitude Entomology*. Methuen, London.

Marshall, L. G., R. F. Butler, R. E. Drake, G. H. Curtis and R. H. Tedford. 1979. *Science* **204:** 272–279.

Matthews, J. V. Jr. 1979. *Mem. Entomol. Soc. Can.* **108:** 31–86.

Matthews, J. V. Jr. 1980. *Can. Entomol.* **112:** 1089–1103.

Mayr, E. 1944. *Bull. Am. Mus. Nat. Hist.* **83:** 123–194.

McKerrow, W. S. (Ed.). 1978. *The Ecology of Fossils*. MIT Press, Cambridge, Mass.

Munroe, E. 1957. *Science* **126:** 437–439.

Munroe, E. 1961. *Can. Entomol.* (Suppl.) **17:** 1–51.

Munroe, E. 1973. *Can. Entomol.* **105:** 177–216.

Nelson, G. and N. I. Platnick. 1981. *Systematics and Biogeography, Cladistics and Vicariance*. Columbia University Press, New York.

Nelson, G. and D. E. Rosen (Eds.). 1981. *Vicariance Biogeography, a Critique*. Columbia University Press, New York.

Paulian, R. 1961. *Faune de Madagascar* **13:** 1–442.

Peña, G. L. E. 1966. *Postilla* **97:** 1–17.

Pielou, E. C. 1969. *An Introduction to Mathematical Ecology*. Wiley, New York.

Pielou, E. C. 1975. *Ecological Diversity*. Wiley, New York.

Pielou, E. C. 1977. *J. Biogeogr.* **4:** 299–311.

Pielou, E. C. 1979. *Biogeography*. Wiley, New York.

Raven, P. H. and D. I. Axelrod. 1975. *Am. Sci.* **63:** 420–429.

Remington, C. L. 1968. *Evol. Biol.* **2:** 321–428.

Schlinger, E. I. 1974. *Annu. Rev. Entomol.* **19:** 323–343.

Sclater, P. L. 1858. *J. Proc. Linn. Soc. Lond., Zool.* **2:** 130–145.

Scudder, G. G. E. 1974. *Can. J. Zool.* **52:** 1121-1134.

Scudder, G. G. E. 1979. *Mem. Entomol. Soc. Can.* **108:** 87–179.

Skellam, J. G. 1951. *Biometrika* **38:** 196–218.

Van Devender, T. R. and W. G. Spaulding. 1979. *Science* **204:** 701–710.

Wallace, A. R. 1876. *The Geographical Distribution of Animals* (2 vols.). Macmillan, London.

Wallace, A. R. 1880. *Island Life*. Macmillan, London.

Watts, D. 1971. *Principles of Biogeography*. McGraw-Hill, London.

Whittaker, R. H. 1972. *Taxon* **21:** 213–251.

Williams, C. B. 1958. *Insect Migration*. Macmillan, London.

Wolfe, J. A. 1978. *Am. Sci.* **66:** 693–703.

Zimmerman, E. C. 1948. *Insects of Hawaii. Introduction*. University of Hawaii Press, Honolulu.

Zimmerman, E. C. 1970. *Biotropica* **2:** 32–38.

Section III
NATURAL
CONTROL OF
INSECT
POPULATIONS

The focus now is on insect populations, in contrast with the previous section devoted largely to the ecological attributes of individual insects and different species. Data of central focus become numbers of individuals in specific populations, and it is often critical to partition these numbers into groups according to age and other qualitative attributes (phenotypic and genotypic). We are primarily interested in discerning and explaining spatial and temporal patterns of these groups of individuals within the heterogeneous environments in which they live. What are the basic ecological processes and general principles underlying the ebb and flow of populations within and among habitat patches? But further, what are the species-specific intrinsic and extrinsic factors which explain unique population behavior and account for the great variation among species and populations in the degree of stability they exhibit in time and space?

Pragmatically and theoretically, one of the most significant and enduring questions in populational research is how to define a population and how to set the spatial and temporal dimensions of its environmental arena. What assumptions are implicit in each arbitrary definition and how do these assumptions vary as we shift focus from a very local to a wide-area, multideme population?

From a global perspective, individuals of each species occur in aggregations within widely distributed environmental patches in which species-specific mixes of requisites occur. Except in rare cases, practicality obviates field study of such total populations and differentially restricts the duration of studies. Consequently, empirical studies are restricted to fractions of species populations in arbitrarily defined space–time arenas. At the most, such studies are but brief "film clips" of population behavior which is everchanging due to open-ended ecological and evolutionary processes.

Other factors impeding development of truly holistic population ecology are the large number of species, diversity of their life-styles, and disparity among their scales of reference. Consequently, the dynamics of but few species have been investigated in depth, and there is a tendency to select species for study on the basis of convenience and personal preference rather than ecological representativeness. Extremely small, highly mobile, and cryptic species as well as those with long (compared with grant periods and human longevity) life cycles are poorly represented.

In spite of the difficulties enumerated above, interest in population research has intensified and major contributions toward a sound theoretical frame for such population biology continue. Many of these contributions have entailed studies of insects utilizing the *life system* concept, described in Chapter 11. This approach provides a framework for viewing the interrelationships between genetic and environmental factors limiting and regulating population behavior as reflected by data on births, deaths, and movements in and out of a population.

There are two interrelated aspects of population dynamics: change in numbers and change in qualities. To date, however, relatively little attention has been given to the reciprocal relationship between the quantitative and qualitative characteristics of populations, perhaps because assumptions of equivalence among individuals simplify numerical assessments of populations and because subtle but significant qualitative changes are difficult to detect, particularly in the traditionally short-term population studies. While acknowledging this disproportionate representation of numbers versus qualities in empirical studies, the authors of Chapter 12 present a useful synthesis of natural control theories. They retain the much debated concepts of density-independent and density-dependent action of life system components and processes: the former predominantly setting ultimate limits on populations and the latter principally responsible for restricting (regulating) population fluctuations within such limits. Within the large number of species with diverse life-styles, those associated with relatively unpredictable and only temporarily available resources may seldom experience marked and obvious density-induced suppression. In contrast, density-dependent feedback mechanisms are more frequently engaged in the regulation of species within highly predictable, more stable environmental arenas. These concepts are not offered in contradiction to ideas of "spreading the risk" of extinction (den Boer 1968) in heterogeneous environments through genetic and phenotypic variability.

The remaining chapters of this section focus more sharply on the role of major life system components in population dynamics of insects. A particularly illuminating analysis of the response of insects to weather patterns and processes at different scales is presented in Chapter 13. Similarities noted in patterns of frost in a footprint, valley and large depressed plain are related to the same processes, radiant cooling and cold-air drainage, which in turn are triggered by the same regional weather. To quote the authors, "if we are to understand fully the impact of weather on an insect, we must be able to project our imaginations up- and down-scale well beyond the limits our own size imposes". In developing this theme, the authors present excellent examples of how different species of insects with different scales of reference are affected by and adjust to micro-, meso-, and macroscale weather patterns.

The theme of Chapter 14 is that spatial and temporal patterns of populations reflect in large measure variations in environmental heterogeneity. From man's view, the structure of nature imposes various levels of heterogeneity from gross to fine: for example, the mix of biomes; disturbed (e.g., cultivated) and undisturbed areas (fields) within each biome; and the variation in topography, soil, and biota within each field. However, to appreciate insects' responses to the number, size, kind, and arrangement of heterogeneous units as we view them, we must, as prescribed in Chapter 13, adjust scales of reference and become specific in defining resources and hazards. The authors' discussion of patchily distributed resources in the "insect view" context provides insights on the role of ecological availability of resources in the population biology of insects as mediated through behavioral and genetic processes.

Chapter 15 takes a detailed look at the theory of interspecific competition in insects and relates what is known about cases of interspecific competition in nature to those theories. Chapter 16 summarizes the vast literature on insect predation (including parasitoids) in terms of defining principles. This includes not only the traditional predator–prey theory but also the emerging field of foraging theory. Chapter 17 is again a refreshing and challenging account and interpretation of the question "Do populations self-regulate?" Lastly, Chapter 18 deals with population regulation in social insects. It is something of a first in a major attempt to survey and analyze the evidence concerning if, and how, social insects may be regulated or controlled.

REFERENCE

den Boer, P. J. 1968. *Acta Biotheor.* **18**: 165–194.

Chapter 11

Short-Term Patterns of Population Change: The Life System Approach to their Study

R. D. HUGHES, R. E. JONES, and A. P. GUTIERREZ

11.1 INTRODUCTION

In this chapter we relate the life history properties and processes of insects (earlier chapters) to variations of species population structure and abundance in time and space. We are not concerned with changes occurring as a result of long-term evolution (Chapters 9 and 10) but rather with the normal seasonal changes within years and the less regular differences between years, including those due to succession. Methods are described for establishing the essential features of local populations to develop a fuller understanding of the population dynamics of the species. Techniques for population dynamics analysis can be used in a variety of ways, but the life system approach outlined here provides an intuitively satisfying framework for such studies.

11.2 WHAT IS A LIFE SYSTEM?

Early entomologists observed that pest numbers were liable to dramatic change and recognized that a complex of physical and biological factors was involved. Realizing that these factors and the pest species were webbed together in a complex ecosystem, Tansley (1935) noted the need for comprehensive studies but gave little guidance on the scope and extent of the necessary investigations. With development of more quantitative approaches in the 1940s, some attempts were made to identify those parts of an ecosystem relevant to a particular species (e.g., "single-species ecosystem" of Huffaker 1970). Clark et al. (1967) were the first to describe explicitly as an *analyzable* subsystem the web of relationships directly associated with numerical changes of a species population. The "life-system" they described comprised "a subject population and its effective environment, which includes the totality of external agencies influencing the population. . . ." This definition directs attention to the ecosystem of which the life system forms the part "which determines the existence, abundance and evolution of a particular population." They saw the main purpose of the life system idea as emphasizing the fundamental nature of the link between a population and its environment—that a population cannot be considered alone and that a population study cannot reach equally into all aspects of all organisms in the larger community.

11.2.1 Life System and the Concept of Population

To serve the purposes of particular biologists, the word *population* has been applied to entities ranging from several individuals seen together to the whole species. Concepts of number, contact, place, and time are implicit. There is, however, no agreed-upon definition. It is "an entity like family. . ." (Nicholson 1957); in practice the problem is to define its limits (cf. Chapters

310

12, 23). Ecologists usually mean a group of conspecific individuals occupying a particular place. The life system approach allows a population to be defined by the spatial limits of its effective environment.

To investigate the population dynamics of an organism, it is logical to study a population which is substantially independent of other populations of the species. For any locality one can envisage the spatial limits of the population directly involved in the abundance and persistence of a species occurring there. Depending on the vagility of the organism, those limits could be extensive or confined. Ideally, the spatial limits of a study should minimize the number of direct relationships involving the subject species that cannot be analyzed because they operate from outside the area. The spatial limits of some species (e.g., some locusts) would be impractically extensive. Thus, the purpose of the study and the resources available to it will determine how much of the life system description will be unexplained input, such as the number and age structure of immigrants (cf. Chapters 2, 12, 23).

11.2.2   Life System and Ecosystem

An ecosystem is made up of the interconnecting life systems of all the species in it. Components of ecosystems included in analyses of life systems vary from species to species according to the specific requisites of the target species and the nature of the habitats in which these requisites exist. The basic life history characteristics, which in some details are unique for each species, mediate those relationships that modify the target population's development, survival, behavior, and fecundity. Those relationships comprise, at the least, the trophic interactions with food (Chapter 4), allies (Chapter 12), and natural enemies (Chapters 12, 16, 22) and the physical and chemical relationships to weather (Chapters 3, 6, 7, 13) and substrates. Although up to now we have been considering single species as subjects for quantitative study, the methodology employed can also be focused equally well on other parts of an ecosystem. In agriculture we are often interested in relationships between crop and pest or between pest and natural enemy. These interfaces may be approached similarly to the study of a single species but usually with less emphasis on peripheral components.

11.2.3   Why Study Life Systems?

As humans are usually only interested in one or a few species in a given ecosystem, the best research strategy is to study as little as possible of the total complexity to achieve the necessary understanding. Nevertheless, the ecosystem should be reviewed before planning a study. As Evans (1976) noted, ". . . to start from this comprehensive viewpoint makes clear there is no *a priori* reason for excluding particular elements. . . ." The life system idea offers guidance to the scope of necessary investigations. Knowledge of a variety of insect life systems is now available to initiate life system scenarios for most sorts of insects, as guides to the study of other species.

Predicting the effects of any manipulation of an ecosystem upon a particular population requires thorough knowledge of its life system. Indeed, the predictive power of a life system description is the ultimate test of its value. The life system also offers an unbiased frame of reference within which to examine a priori principles or preconceived notions of the importance of certain factors. If successful, such a study will provide an evaluation of the importance of various factors. With some acquaintance with a life system, it is possible to concentrate on particular aspects of an insect's behavior without losing sight of the ecological context which gives the behavior its adaptive significance. The life system is also one way of looking at the role of particular species in a larger ecosystem. Studies of connected life systems may even be a good way to start investigating ecosystem dynamics (Whittaker 1965, Gilbert & Gutierrez 1973).

Although the concept of life system is applicable to any species, numerical population studies in the field may be precluded by features of an insect's biology or habitat. Particular species or stages of development may be too mobile (above) or extremely difficult to sample quantitatively, and the necessary effort would exceed available resources (e.g., some forest insects).

11.3 COMPONENTS OF A LIFE SYSTEM

11.3.1 Life History Characteristics and Phenology

The basic information required for a successful life system study is a thorough knowledge of the life history of the subject species and other important species in its life system. This includes a knowledge of the annual timing of those life histories, that is, their phenology (Chapter 6). Where a species has nonoverlapping generations and a homogeneously aged cohort may be observed as it develops (e.g., temperate region organisms producing only one generation each year), field observations may provide many of the necessary data. When generations overlap, generation times and life history patterns can no longer be observed so clearly. In either case, however, it is desirable also to examine the development and reproduction in the laboratory, both to obtain more precise measurements of characteristics and to examine factors affecting the life history, such as food quality and quantity, population density, and photoperiod. These are considered later (Chapters 3, 4, 6, 7, 12, 13, 14). But temperature must be examined here, since it is fundamental to considering such attributes as generation times and development.

11.3.1.1 Temperature and Development. Because temperature effects on physiological rates are so fundamental to understanding insect life history patterns, life history characteristics of an insect (e.g., development rates, age-specific fecundities) are commonly evaluated in different temperature regimes in laboratory studies so that effects on developmental patterns may be determined. Insects, and most interacting species, show a change in developmental rate with temperature, of the general form shown in Fig. 11.1 (cf. Howe 1967, and Chapter 3).

Below some threshold temperature t no development occurs. Just above the threshold development rate increases gradually. Then there is a range of temperatures over which development rate increases linearly with temperature. Finally, at temperatures close to the upper lethal limit the rate again declines. For years workers have fitted appropriate curvilinear relationships to describe an insect's response to a broad spectrum of temperatures (e.g., Huffaker 1944, Andrewartha & Birch 1954, Stinner et al. 1975, Bailey 1976, Taylor & Harcourt 1978). However, for many life system studies it may prove sufficient to fit a straight line to the midsegment of the curve. From this it is possible to estimate both the x intercept (i.e., the "notional" threshold temperature t) and the reciprocal of the slope, which gives the total number (K) of day-degrees ($D°$) above the threshold required to complete the developmental stage studied ($K\ D°_t$). Campbell et al. (1974) described the estimation of t and K in detail. This linear description *usually* suffices because rather mobile organisms do not often stay in areas which expose them for long periods to daily temperatures close to their upper lethal limit, and the error introduced by ignoring the small departures for development below the range of the linear relationship is generally small. But if the temperatures experienced remain either very high or close to the threshold for long periods, then a more precise curvilinear description may be necessary (cf. Sharpe et al. 1977).

Different insect stages may have different development thresholds, as in the cabbage butterfly (Jones & Ives 1979) and the coccinellid, *Coleomagilla*

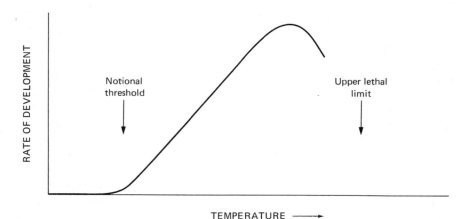

Fig. 11.1   General form of change in development rate of insects with temperature.

maculata (Obrycki & Tauber 1978) (cf. Chapter 3). Different processes may also have different thresholds; for example, that for egg laying is lower than that for ovarian development in the bushfly (Hughes et al. 1972). Geographic variation in thresholds and development times has been demonstrated, for example, for the cabbage aphid and its parasite in Australia, U.K., Canada, and the United States (Campbell et al. 1974). As expected, populations from cooler climates had lower development thresholds.

Calculation of the developmental threshold then allows the age of the insect, and the times required to complete different parts of its life cycle, to be expressed in terms of physiological time ($D°_t$), simplifying subsequent work. *Pieris rapae* eggs, for example, require 50 $D°_{10}$ to hatch. In calendar time, they require $50/(T - 10)$ days—for example, 5 days at 20°C or 10 days at 15°C. Such calculations can be applied to field conditions but fluctuation of air temperature, particularly if below the threshold, has caused difficulty. Frazer and Gilbert (1976) give an algorithm for calculating the physiological time, above a specified threshold, elapsing in the field from daily maxima and minima using the sine curve approximation for the pattern of temperature change as derived by Morris and Bennett (1967), allowing prediction of field generation times from daily temperatures. Automatic recording of physiological time is now possible (Scopes & Biggerstaff 1977). It may be necessary to allow for differences between the temperatures in the insect's microhabitat and those recorded by standard meteorological equipment. In some cases—for example, for ovarian development of *Lucilia cuprina* (Woodburn et al. 1978)—it was also necessary to allow for solar radiation on the insect's body (cf. Gutierrez, Denton et al. 1974).

Since both organisms are reacting to temperature change, it appears that there are few problems for an insect in synchronizing its phenology with that of its prey (or host plant), even though they may have different development thresholds (Gutierrez, Nix et al. 1974, Gutierrez, Havenstein et al. 1974). Accumulations of day-degrees above thresholds that are not too different for the two species are highly correlated most of the active season. An insect inhabiting a warm-blooded animal usually has a temperature that closely follows that of its host, and timing is no problem, as in nycteribiid flies on bats (Funakoshi 1977). Synchronization of the phenology of an insect to a precise physical cycle, other than temperature change, can be more complex, as with the tidal cycle of *Culicoides* emergence (Kay 1972). Annual cycles do not usually involve such precise requirements, and synchrony is usually accomplished by introducing specialized delays of development, such as diapause, which are terminated by appropriate stimuli (Chapter 6).

Especially for temperate climate insects with a fairly precise beginning and end to their growing season, and with one one or a few generations each year, development and timing studies, combined with an early-season sample to establish a "fix" on the initial age structure of the population, may be sufficient to predict the timing of population events through the season, if not their magnitude (e.g., Gage et al. 1976 for Canadian grasshoppers).

Rearing insects at several different temperatures may also reveal that the temperature affects other life history characteristics besides development. A number of behavioral processes may have thresholds lower (or higher) than those for development (e.g., oviposition in bushflies and flight in several species). Especially at very high or very low temperatures, relative longevity (in physiolgical time) may be affected.

11.3.1.2 Delays in Development. Once the basic life history pattern of the species within its natural range of experienced temperatures is established, the roles of factors which delay development and modify patterns may be evaluated. Among the factors intrinsic to the population, dormancy mechanisms, particularly diapause, is of basic importance because its termination and initiation set the beginning and end of the insect's growing season. Its termination is also a difficult event to predict (Chapter 6). Quiescence, the reversible response to temporarily adverse conditions of temperature, dryness, or lack of food (cf. Chapter 6), can give rise to variable delays in development. Other developmental delays may also occur. The necessity to search for sparse resources (food, mates, oviposition sites) may delay reproduction, as in Kitching's (1977) studies, which showed that this exerts a substantial influence on sheep blowfly reproductive rates. Differences in food quality, photoperiod, humidity, population density, and other factors may modify juvenile or adult development or adult size and fecundity. In some cases (e.g., aphids and locusts) similar influences may induce basic changes in morphology, behavior, and reproduction. Gutierrez et al. (1981) developed a general model for insect growth and development based upon food supply–demand relationships.

The basic life history pattern is seldom excessively difficult to document, but the range of environmentally induced variations on it may be enormous and may be confounded with genetic variation in the perception of the researchers. It is sensible to restrict attention to those factors most likely to significantly affect the organism in the field. The corollary is that most progress is usually made when field and laboratory studies proceed together and field experience is used to dictate the factors to be closely examined.

Genetic variability in characteristics may also occur but has received little attention except in relation to diapause; Morris (1971) showed that effects of annual selection on the timing of onset of diapause may markedly alter the phenology and intrinsic vigor of populations of *Hyphantrea cunea*. Dingle and co-workers (cf. Dingle 1978, and Chapter 7) identified an adaptive genetic component in the diapause (and migratory) behavior of milkweed bugs. It would seem that such should be true of other life history characteristics.

There are cases where nongenetic maternal quality and "birth-order" effects exert considerable influence on survival, development, vigor, or behavior of an insect's offspring. The best studied example is the tent caterpillar, where polymorphisms in activity and vigor exert profound influences on the insect's population dynamics (e.g., Wellington 1957, 1960, 1964, 1965, and Chapters 13, 17). As a result of its differential allocation of nutrients between eggs, there is a gradient of decreasing activity and vigor associated with birth order in a female's progeny. Also, more active, vigorous females, since they have eaten more as larvae, produce a higher proportion of active and vigorous offspring than do less active females. These variations in vigor affect feeding and development rates, dispersal, and susceptibility of larvae to weather, predators, parasites, disease, and starvation. Being fundamental to understanding the *Malacosoma* life system, these variations were described and modeled by Wellington et al. (1975). Effects of female nutrition on the quality of progeny were also described in spiders (Turnbull 1962), and birth-order effects have been recorded in a number of other species, notably aphids (MacKay & Wellington 1975). Such effects may be of general occurrence, though probably seldom as marked as in *Malacosoma*.

11.3.2 Trophic Relationships

11.3.2.1 Insects and Their Food. The Insecta exploit many kinds of foods— detritus, plants, other animals, or their products. The quality and quantity of available food often exerts profound influences (Chapters 4, 14); indeed White (1978) for insects (and Lack 1966 for birds) argues that the most important factor limiting abundance of an animal species is shortage of nitrogenous food, especially for young individuals and breeding females. The argument applies clearly to many plant feeders (Chapters 15, 22) and detritivors but also to predators (Chapters 12, 16). White notes that most plant tissue carries carbohydrate in abundance but is low in N; so many herbivores feed selectively on tissues with the highest N content: that is, cambium, new growth, senescing tissues, fruits, and seeds. Availability of quality food is seasonal and weather-dependent in most wild plant populations and generally coincides with a time when insect fecundity and juvenile survival are high enough to permit population increase. Eucalyptus psyllids (White 1969, 1970) are good examples, as are many other insects, including dung-breeding flies whose rates of increase are tied closely, though indirectly, to pasture quality (Greenham 1972). Recent plant–insect coevolutionary studies have shown that plants also exhibit active defences [e.g., resinosis (McClure 1977)] against insects that feed on them, which may reduce the quality of available food or, alternatively, reduce the time that high-quality food is available. Feeny (1970) showed that the tannins which oak leaves store increasingly as they age progressively reduce growth rates, survival, and fecundity of the wintermoth. Thus, a life system analysis of a herbivorous insect (or detritivore) will generally require attention to temporal changes in both quality and quantity of available food (cf. Chapters 4, 14, 23).

Caughley (1976) recalled the useful distinction between *nonreactive systems*, in which the herbivore does not, or cannot, alter the rate of growth of its food supply (consumers of dead or senescing tissue provide good examples), and *reactive systems,* in which the herbivore, depending on its density, can and does alter the rate of growth of its food (cf. Chapter 12). The relevance of this distinction to a life system study lies in the level of detail at which the food itself must be studied. A nonreactive system may require only a documentation of the seasonal spatial, and weather-related changes in food quality and abundance, perhaps using bioassay techniques to assess effects of food quality. The bushfly is a good example, and some studies of aphids (e.g., Gilbert & Gutierrez 1973, Gutierrez, Morgan et al. 1971, Gutierrez, Havenstein et al. 1974) have used this restricted approach. A reactive system, however, may require that the dynamics of the exploited species be studied in just as much detail as the exploiter species (Gutierrez et al. 1975, Wang et al. 1977). Indeed, for herbivorous pests, it may be best to make the food plant itself the subject species. This is a simpler problem for crops than for other plants, since crop population recruitment and growth are under human control (but not that of *wild* plants) so the recruitment process for *crop* plant species need not be studied.

Food quality, quantity, and even diversity may affect almost any aspect of an insect's dynamics, from survival and longevity to development rate, diapause, fecundity, and dispersiveness. A range of such effects is described in Chapters 4, 6, 7, and 14 and by Watson (1970), Peters and Barbosa (1977), Labeyrie (1978), and others. Evaluation of the processes and magnitudes of these nutritional effects (Chapters 4, 14, 23) in natural populations often forms the bulk of a life system study.

11.3.2.2 Predators and Parasitoids.

Predator–prey (including parasitoid–host) relationships have been of intense theoretical interest for many years, and since Holling's classic analysis of functional responses in predators (e.g., Holling 1959, 1961, 1963, 1965, 1966) they have frequently been studied in the laboratory. By comparison, field studies on insects which include detailed analyses are rare, and there are few, if any, full life system studies in which the subject species is a predatory insect or parasitoid. This reflects the difficulties of carrying out such studies in the field rather than their lack of intrinsic interest or importance. The object of studying a predator–prey relationship within a life system analysis is to determine the effects of the interaction on the survival and reproductive rates of the subject species in all circumstances normally encountered in the field. If the predator feeds exclusively on the subject species, it may be possible and desirable to document the life systems of both species. If the predator feeds on a wide range of prey, then such detail, however desirable, is likely to require more time than an ecologist's working lifetime. The potential benefits of documenting both species should, however, be carefully considered. If the prey species is the main concern and predators or parasitoids have little impact on it—that is, do not drive down prey numbers or act as major controlling agents and if this were fully known for both endemic and epidemic states—detailed study of either the predation process or the enemy's (enemies') life system(s) would be unnecessary. The first problem, therefore, involves evaluation of the enemy's impact on the prey population, in both the short term and long term. This may present difficulties. Southwood (1966) described a range of techniques, while Hodek et al. (1972), Kiritani and Dempster (1973), and Bottrell and Huffaker (1974) review methods found to be useful for various true predators—ranging from labeling with isotopes, precipitin tests, and use of trap plants to predator exclusion or interference experiments, and field caging trials. Often, the difficulty is not so much in evaluating the consumption capacity of a given population of predators as it is in determining how many of a given predator species and how many species are present and their "switching" behavior (Gilbert et al. 1976). Insect parasitoids present fewer problems in impact evaluation because it is generally possible to estimate by direct observation the number of deaths they caused (cf. below).

In several modeling studies of aphid–parasitoid interactions, where the parasitoids had no major impact, satisfactory prediction of parasitization rates throughout a given season was obtained by assuming random search by the parasitoids (i.e., Thompson's 1924 formula, although this is not an accurate model of the behavior). According to this model, the proportion of susceptible prey escaping attack during any time period is $\exp(-kb/a)$, where k is the number of prey each predator is able to attack during a specific time period, b is the density of parasitoids, and a is the density of susceptible hosts (Hughes & Gilbert 1968, Gilbert & Gutierrez 1973, Frazer & Gilbert 1976). The same or equally crude methods may be used for true predators, if their impact on the prey is small. But if the predators exert major effects, it usually would be necessary, as Frazer and Gilbert (1976) found for coccinellids preying on pea aphids, to develop a more realistic description of predator behavior and to include a study of predator reproductive success (Gilbert et al. 1976). Even cruder assessments may suffice if the goal is to assess only roughly the general impact of the predators rather than to produce a comprehensive life system model. In studying mite interactions, Readshaw (1975) used a modified Lotka–Volterra formulation which, as he noted, possessed manifest defects. It was nonetheless sufficient to show that the predator was indeed able to account for observed reductions in the prey's increase.

More comprehensive analyses require great effort. Solomon (1949) and others since have divided the predator–prey relationship into two components: (1) the functional response, which is defined as the change in attack rate per predator with variation in prey density and (2) the numerical response, which is the change in predator numbers with variations in prey density. Several workers have noted that this division fails to accommodate comfortably certain phenomena since shown to be important—especially the relative distributions of predator and prey and environmental factors. Hassell et al. (1976) and Beddington et al. (1976) consider a more useful distinction to be that between factors affecting prey death rate and those affecting the predator's rate of increase; we follow this distinction.

There now exists a formidable array of theoretical algebraic and simulation models for examining most predatory–prey factors (cf. Chapter 16). Here we simply list known effects which may strongly affect an outcome of predator–prey interactions (with emphasis on field studies) and consequently need to be considered (cf. Chapters 12, 16).

Factors Affecting the Prey Death Rate

1. *The functional response.* Holling (1965) identified three basic types of functional response; the first is a special case rarely relevant to insects; the second and third are shown in Fig. 11.2. Most invertebrate predators studied exhibit a type II response, although Hassell et al. (1977) suggest that this may reflect the restricted range of prey densities over which studies have been made, and that type III responses, which are considered to require a learning component, may be more common than usually supposed. The importance of this difference is that, given a type III response, each predator will act in a density-dependent—and hence stabilizing—fashion on the prey population over some part of the range of prey densities, while this is not true of a type II response (cf. Chapter 16), where for a given number of predators the risk to each prey individual continuously decreases as prey density increases. Holling described the type II response with the familiar "disc equation." For technical reasons (Rogers 1972), or in order to use parameters more easily evaluated in the field (Frazer &

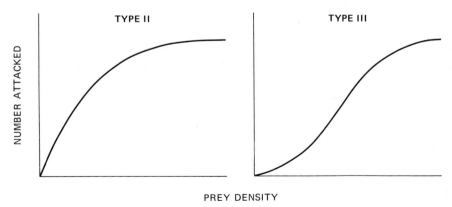

Fig. 11.2 Types of functional response relevant to insects: Holling's types II and III.

Gilbert 1976), other formulations have also been used to describe type II responses (cf. Chapter 16).

2. *Temperature.* Predators and parasitoids tend to have activity and development thresholds higher than their hosts (Campbell et al. 1974, Neuenschwander et al. 1975, Jones & Ives 1979). This may exert profound effects on predator–prey relationships; Frazer and Gilbert (1976) showed that a given density of coccinellids might drive down pea aphid numbers in warm weather but be unable to prevent increase in cooler weather. Messenger (1964) observed similar phenomena in parasitoid–aphid relationships; such are probably common, affecting both predator–prey and plant–herbivore relationships.

3. *The aggregation response.* A common response by predators to spatial variations in prey density is to aggregate in areas of high prey density. In the coccinellid–aphid study cited above, while there was a continuous flux of beetles into and out of the field, they tended to stay longer when aphid densities were high, and predator densities then increased. Conversely, numbers declined (again) when aphid densities became low (Ives 1981). This response was also found by Hughes and Mitchell (1960) for carabids preying on cabbage rootfly eggs and by Turnbull (1964) for predation by spiders. Some predators, such as syrphid larvae (see Jones 1977a), *Stethorus* and predatory mites (Fleschner 1950), and parasitic wasps (cf. Doutt 1964, Waage 1978, Wylie 1958), show an increased turning rate just after contact with prey, which tends to keep them in areas of high prey density. In species with predatory larvae, the adults may lay more eggs in areas of high prey density; for example, *Chrysopa* adults feed on and settle in response to honeydew excreted by the prey of the larvae (Hagen et al. 1970). Hassell and May (1974) showed theoretically that an appropriate aggregation response can stabilize a predator–prey relationship (cf. Chapter 16).

4. *Interference.* In laboratory trials with standardized prey densities, at high density predators may decrease their individual attack rates; this is common in parasitoids (e.g., Ullyett 1950, Hassell & Huffacker 1969, Beddington 1975, Hassell 1971, 1976, 1978, and work cited therein). For statistical reasons interference is difficult to establish in the field (Hassell & Varley 1969). Griffiths and Holling (1969) suggest that it will rarely be important at parasite densities encountered in the field. Broadhead and Cheke (1975), however, describe a case (mymarid parasites of *Mesopsocus*) where significant interference appears to exist in the field and to affect the population dynamics of the species concerned (cf. Chapter 16). Host marking by parasitoids suggests significant interference and evolution of behavior to minimize its adverse effects.

5. *Presence of alternate prey.* The existence of switching (Murdoch 1969)—that is, disproportionate attention by a predator to more abundant prey in a multiprey system—has been demonstrated for predatory insects,

but obvious logistic problems tend to discourage detailed analysis of multiple-prey (and multiple-predator) systems in the field. Whittaker (1973) presents evidence of a change in attack rate by a parasitoid with changes in relative numbers of two hosts (cf. Chapter 16).

Factors Affecting Rates of Predator Increase. Beddington et al. (1976) argue that the overall rate of increase of a predator population will depend on three main components: the duration of each instar, survival rates within instars, and adult fecundity. They show that, predictably, all three components may be affected by the numbers of prey consumed, and, we add, the numbers consumed will normally be a function of certain inherent traits, such as searching capacity, synchronizations, and so forth (cf. Chapter 16). The same considerations apply here as to any insect population study.

11.3.2.3 Pathogens and True Parasites. Insect diseases may have significant roles in their hosts' population dynamics. Aphids may be severely reduced by fungus disease when their numbers are high and the weather (wet) favors fungus reproduction and spread. True parasites rarely kill their victims but presumably reduce their vigor. Those organisms that spay or castrate their hosts can also be important by reducing reproductive potential without withdrawing the host from the life system. The nematode *Heterotylenchus autumnalis* infests the female facefly, effectively spaying her without reducing her urge to join in the oviposition behavior of other flies. Thus, "packets" of worm larvae are deposited with facefly eggs and the nematode life cycle is restarted in newly hatched fly larvae (Stoffolano 1970).

11.3.3 Competitive Relationships

In a life system study consideration of competitive interactions within and between species will usually develop from analyses of relationships between the subject species and the resource utilized. Higher densities or restricted resources often have three related effects on population increase parameters of individuals—increasing development time, reducing survival, and reducing eventual size and thus fecundity—although the density threshold for each effect may differ (Gutierrez & Baumgaertner In press).

Two types of competition are usually distinguished, and they may differ in their effects on an insect's population dynamics. *Exploitation competition* occurs when each individual consumes the resource independently of impending behavioral activities of other individuals—that is, competition acts simply via reduction in the available resource. In the case of pure "scramble" competition of this type, all individuals rather equally share the resource, and if the resource is insufficient for completion of development of all individuals, few, if any, survive. The classic example of blowfly larvae on carrion was demonstrated by Nicholson (1954). Good examples of exploitation competition by herbivores are given by Harcourt (1971) for Colorado potato beetle and Dempster (1971) for cinnabar moth (cf. Chapters 12, 15).

Interference competition is a result of some sort of "contest" between individuals wherein one behaviorally acts on another individual in such a way as to reduce the latter's access to a common required resource. This may be achieved in several ways: Many hymenopteran parasitoids mark a host they have attacked with a pheromone, and marked hosts are then avoided by other individuals. Cannibalism (e.g., by a larva of codling moth on others in an apple, Geier 1964) is an effective extreme form of interference. Territoriality in species such as dragonflies (Chapter 15) and use of a sphragis by some male butterflies to prevent remating by other males represent interference competition for mates. The interspecific form of interference competition is also common. Some ants fight and kill competing ants of other species, and this has led to spectacular faunal changes following invasions of new habitats by certain species (cf. Chapters 15, 18). With interference competition, a few individuals may appropriate the entire supply of a resource and are thus likely to survive a severe shortage, thus avoiding complete population collapse, as may occur with scramble competition.

In some cases high population densities or restricted resources lead to a developmental switch, resulting in morphologically and physiologically different types of individuals that are usually more dispersive (cf. Chapter 7). There is a trigger mechanism for the switch, a response to incipient or "anticipated" crowding, which means that individuals moving out of the population avoid the worst effects of such competition. Another view for certain cases is that evolution has produced the triggered response(s): to disperse in order to exploit *new* resources rather than to *avoid* lack of resources in the old site. Obvious examples are the induction of the *gregaria* phase in locusts, of winged forms in aphids, and of macropterous forms in plant-hoppers (cf. Chapters 6, 7).

11.3.4 Mutualistic Relationships

Mutualistic relationships—especially ones concerning pollination and nectar feeding (Chapter 20) and ant–plant (Chapter 18) and plant–herbivore (Chapter 8) interactions—are receiving increased attention because of their coevolutionary interest. The obvious relationships of ants with honeydew-producing insects such as aphids and scales have long been known (Chapter 4), but more cryptic relationships are now being recognized, such as the adaptive timing of attraction of predatory ants by wild cherry when shoot growth is threatened by caterpillars (Tilman 1978). Few such relationships have been studied in a life system context. An exception involves burying beetles and their phoretic mites as described by Springett (1968). The mites leave the beetle when it arrives at the corpse; they then seek out and consume blowfly eggs. If the eggs were to hatch, the beetle larvae would suffer severe competition from the blowfly larvae.

11.3.5 Social Relationships

Subsocial behavior is common among insects and can have important population consequences, as Carne (1969) showed for the defensive group activity of a species of sawfly. Insect sociality is dealt with in Chapter 18; we mention it here only to note that a life system study of a social species presents a whole suite of problems different from those posed by nonsocial insects, not the least of which is the extreme difficulty of obtaining more than one population estimate from a colony because sampling either destroys or may gravely disturb the insects and their microhabitat (Benois 1972). Yet, Chapter 18 presents some useful techniques. The life system approach, however, may be applied to "populations" of colonies as, for example, in the studies of colony size and survival in meat ants by Greaves and Hughes (1974) and Greenslade (1975).

For social insects, the methodology of field studies of colony structure, size, development, and survival are analogous to those of studies on these features for individuals in solitary insects. The effective environment affects each colony as a whole to modify its external relationships (e.g., Klimetzek 1970), but within the colony social activities and structure tend to ameliorate the effects of environmental variability.

11.3.6 Relationships with the Physical Environment

Almost every aspect of the life history, biology, and behavior of an insect, and many relationships to the biotic environment, are modified by variation in weather factors—wind, radiation (e.g., photoperiod), temperature, humidity, and rainfall. Thus, climate and weather have a profound influence on the populations of most insects, either directly or through their influence on other organisms in the life system (cf. Chapters 6, 7, 12, 14, 23, but especially 13).

Many insects are less active in high winds or under overcast conditions. In such circumstances, the cabbage butterfly will neither fly nor lay eggs, and the frequency of overcast days may be a major factor limiting population growth (Dempster 1967, Gossard & Jones 1977). Other wind and radiation effects are difficult to distinguish from, or act via, their effect on environmental temperature. An exception is the use of photoperiod as a token stimulus for certain types of development or behavior (e.g., Chapters 6, 7).

The effects of temperature and other factors on phenology have already been discussed. Insects may be larger and more fecund when they develop at low temperatures, as in the cabbage butterfly (Jones 1974, and refs. therein) and pea aphid; or the reverse, as in the bushfly (Hughes et al. 1972). Effects of temperature on oviposition and other behavior, besides general effects on activity, are also common. Trophic relationships may be affected by temperature, particularly, as noted above, where an insect's food supply or its natural enemy has a different temperature threshold for development. This can be important for the persistence of a life system. Thus, the biennial kale, *Brassica oleracea* var., has a lower threshold than the cabbage aphid. If a plant is badly damaged by this aphid in autumn, it can still grow away from the infestation through the winter and flower and seed successfully in the spring (Hughes 1963). It also became clear that the hold of the aphid's parasite, *Diaeretiella rapae*, on the increase of the aphid population in autumn was released over winter (Clark et al. 1967).

Humidity variations may have marked effects on the life systems of insects, although often the mechanism involved is not clear (cf. Chapter 6). In a study of the spotted alfalfa aphid, Bishop and Crockett (1961) demonstrated that high humidity reduced the insect's reproductive performance. Humid conditions are favorable for fungi pathogenic to insects. Dry conditions directly increase stress in many insects, resulting in increased mortality rates. Such conditions may reduce the moisture content of food, rendering it less "available" to an insect, as in leaf-feeding Lepidoptera and stored grain insects (Khare & Agrawal 1970). This may result in a switch to other food resources. Holdaway (1932) showed an increase in egg cannibalism in *Tribolium* when humidity was low (cf. Fig. 12.5). Sap-sucking insects are often unable to feed on wilted plants and usually leave them.

Rainfall may kill insects by drowning as with *Pieris rapae* caterpillars (Harcourt 1961) and by direct mechanical damage, as with the loss of over half a cabbage aphid population during a rainstorm (Hughes 1963); hail would probably have an even more marked effect.

Indirect effects of weather via effects on other organisms in the life system are most important when they affect the growth, quantity, and quality of plant food. Such indirect effects on food plants have been studied for the cowpea aphid (Gutierrez, Havenstein et al. 1974; Gutierrez, Nix et al. 1974), cotton herbivores (Gutierrez et al. 1975, Gutierrez et al. 1976, Gutierrez et al. 1977), and alfalfa weevils (Gutierrez et al. 1976). Effects of weather on the dung breeding bushfly act via the growth of plants, which are eaten and degraded into the dung of herbivores (Sands & Hughes 1976, Hughes 1977a).

Besides weather, temporal and spatial variation in applied or substrate features of the environment can have large effects. With aquatic insects, flooding, or drying out, or levels of pollutants may greatly affect survival and reproduction, as in simuliids (Rhum 1970). Flooding may also affect soil insects—for example, newly pupated *Lucilia cuprina* are very sensitive to waterlogging (A. R. Gilby pers. comm.). For soil insects, vertical and horizontal gradients of temperature, moisture, and gases are important, in particular, the time lags in diurnal temperature cycles at different depths. The effect of patchy occurrence of favorable soil substrates is well known (cf. Chapter 14, 23). Dung beetle reproduction is dependent upon patches of suitable soils (Nealis 1977). For insects living in situations affected by man's manipulations (e.g., in and near agro-ecosystems), the direct and indirect effects of those manipulations, including pesticides, are an integral part of the life systems (cf. Chapter 23).

11.3.7 Movements and the Spatial Structure of the Environment

Appropriate habitats for an insect species usually occur in patches. Especially when the patches are short lived, their distribution in space and time is crucial since the insects must find their way from patch to patch (Chapters 7, 12, 14, 23). Dealing with dispersal and migration often presents difficult problems, especially when movements occur on a large scale. The difficulties are twofold. The first is that of observing and tracking movements in small organisms like insects. It is often not too difficult to evaluate the intrinsic and extrinsic factors which initiate dispersal or to specify the stimuli which may act to terminate it, but to track the movement itself, let alone establish the factors which govern its progress, may require enormous effort and ingenuity—indeed may be impossible with available techniques. Relatively small-scale movements in amenable species are less problematic, since they may be studied by direct observation (e.g., Jones 1977a) or by mark-recapture techniques (e.g., Gilbert & Singer 1973), but large-scale movements usually require indirect and inferential methods (e.g. the bushfly, described in detail later).

The second problem is that of integrating the process of large-scale movement into the life system description, especially if the description takes the form of a computer simulation (cf. Gutierrez, Nix et al. 1974). The necessity to keep track of events in a large number of subpopulations, and to exchange individuals between them, is liable to result all too quickly in excessive use of computer time and storage space. An attempt at such integration has nonetheless been partially successful for spruce budworms in New Brunswick, Canada (Clark et al. 1978a, and Chapter 25).

11.4 STUDYING AN INSECT LIFE SYSTEM

There are three facets in studying an insect life system: (1) the analysis of the life history, phenology, and other biological features of the species; (2) the significant relationships have to be identified, described, and if possible quantified; and (3) the functioning of the system as a whole must be interpreted for comprehensive understanding. All three are involved throughout the study, although initially emphasis is on the first two, and finally on the last. Field studies should always be the guide to important relationships to be investigated in detail (Price 1971). By observing successive stages of the life history in relation to environmental factors, likely relationships with weather, food, and other organisms become apparent. The phenology and obvious population changes in relation to changes in the environment will also guide the investigation. Existence of each suggested relationship should if possible be confirmed by experiment or planned observation, for example, the effect of cannibalism as shown by Nakamura (1976a,b). At that stage possible interactions between relationships should be considered.

11.4.1 Methodology and Data Collection

There are several aspects to describing a population and its environment. These are discussed below and in many chapters of this book.

11.4.1.1 Review of Available Information   It is appropriate initially to assemble all available information on the insect. This should be more than a search of the *ecological* literature. So many subjects are peripheral to ecology that all papers on the insect should be scrutinized for relevant biological observations; for example, physiological relationships, genetic variability, geographic distribution. Communication with previous workers on the insect and other entomologists and naturalists should be made, as these people may have a broader view of the species than is recorded in specialist papers. The major use of such information is to get a general idea of the biology and ecology of the insect, including indications of likely relationships and some idea of their form. Although relationships may have been quantified, existing data should not be used directly without careful consideration of their experimental basis.

11.4.1.2 Identity of Material.

The Study Species.   All life stages and all types of individual polymorphism should be collected and described so they can be subsequently identified in the field. It is always desirable to have a large sample of adults checked by a taxonomist at the start of the study. This will avoid making undifferentiated observations on a mixture of species, a situation usually impossible to do anything about afterwards unless large sample series of adults have been collected at regular intervals throughout the study. Live samples of each immature stage should be reared through to confirm their identity.

Food Species, Allies, Competitors and Natural Enemies.   Organisms in these categories should be collected and described, and for the same reasons. Collections should include all relevant stages. If the insect's food is a plant species, all relevant plant stages should be collected and described.

11.4.1.3 Field Studies.   The purpose of field studies is largely to provide a data base to show the nature and timing of changes in the population and its environment. Good serial estimates of population numbers and composition, associated with estimates of environmental changes, are doubly valuable, first, in identifying relationships and, second, in testing the system description. Furthermore, data not at all used in developing a model of the life system allow a proper test of its validity and predictive ability.

Definition of Study Area. Clark et al. (1967) suggest that the spatial delimitation of a study population is "a matter of purpose and convenience." Abundance, a general term, is usually estimated in well-defined study areas, and expressed as numbers of individuals per unit of habitat (i.e., as a density). Abundance, however, can refer to the numbers of a local population or a whole species without regard to specific density. If the study habitat is uniform, determination of population density is straightforward. Where the area has a patchy structure in which the subject occurs at varying density, stratified sampling is needed to get an overall estimate of density. The relevance of the sampled area to the whole spatial environment of the system must be considered (cf. Price 1971). Investigations in an area encompassing only part of an insect's life system are liable to give undue weight to some system components. Unless considerable information is available on the biology and distribution of the species it may be better to delay definitive population studies until the relevant information has been obtained. In any case, it may prove necessary to have study areas in more than one locality as was shown by Whittaker (1971) in his study of *Neophilaenus*. Interlocality relationships may become a major aspect of such a study.

Development of Population Sampling. It is desirable to follow (and sample) individuals through all life history stages. Abundance data are collected by sampling, using appropriate methods for each stage (cf. Southwood 1966, 1978). All similar individuals should be equally likely to appear in a sample, so that ideally a population sample would be equivalent to enclosing a portion of habitat and counting all insects in it. Methods to do this have been developed for a variety of insects, but less adequate sampling may have to be employed for some species or stages. Methods dependent on variable individual behavior, that is, most forms of trapping, produce suspect data unless obvious biases (e.g., temperature effects on activity) are corrected for. Thus, Woodburn et al. (1978) corrected trap catches of sheep blowflies for both temperature and radiation effects on flight activity.

Reconciliation of density estimates of a species obtained by different sampling methods may pose problems, as many comparisons of those densities with densities of other organisms in the life system, at least some stages of which must be sampled in different ways.

All insects, as species, are patchily distributed. But within a prescribed study area, insects which are more patchily distributed pose greater sampling problems than those more uniformly spaced. Also, species vary in the distributional patterns of their patchiness, and this may bias estimates of population density. Furthermore, successive stages of development may exhibit different degrees of patchiness. *Nezara viridula* on rice (Kiritani 1971) is a good example of both eggs and first-instar nymphs showing clumped distributions, but later instars become more dispersed. Thus, variations in patchiness may increase problems of comparing successive density estimates.

The aim of sampling a population should be to describe its state at successive instants of time and so to derive the patterns of change. Its state can be described at different levels of detail. An estimate of total number of individuals is the basic requirement, but the more that is known about the qualities of the individuals the better. Composition of the population in terms of frequencies of different polymorphic forms (sexes, etc.) and stages of development (instars) should be recorded. Further information on reproductive status, diapause status, calendar age, feeding history, and so forth can be gained by more detailed examination (e.g., dissection) of a valid subsample. The individual members should be described as fully as the program permits. Depending on their relative importance to the life system, individuals of associated species will usually be described in less detail. Numbers of individuals observed to have been killed by natural enemies are often recorded from evidence seen in the samples. Examples include the corpses (mummies) left from attacks by pathogens or parasitoids or damage to egg chorions and other remains recognizably caused by predators.

Several guidelines help achieve a useful description of population change. Sampling should be frequent enough to avoid missing a change significant to the study. Generally it is better to sample at regular time intervals, but seasonal differences in likely rates of change may allow safe economies in effort. There may be little point in regularly sampling a population entirely in diapause. Insect activities in general slow down in winter and sampling frequency can be reduced. But even in winter or during aestivation the time of occurrence and the cause of a sudden catastrophic mortality may be obscured by prolonging the sampling interval. This could be important where the mortality is either the result of an irregular physical event [e.g., intense frost (Ricou 1967), heavy rain, etc.] or action of a warm-blooded animal.

Records of Environmental Conditions.   It is good to know the history of the study area. Previous history is often reflected in the current distribution of the plants and insects. A diary should be kept of any treatments of the study area (e.g., burning, cutting, application of fertilizer, pesticide, etc.).

A record of long-term average weather in the study area or nearby is needed to plan experiments under relevant conditions. Unless a well-equipped meteorological station is operated very close to the study area, basic meteorological recording equipment should be operated on the site throughout the study period. Comparison of a full year's basic data from the study area and from a regular meteorological station in the vicinity may show that conditions in the study area can be predicted from the standard records. Because rate of insect development varies markedly with temperature, daily estimates at least of average temperatures are essential. Other weather factors should be measured as frequently and accurately as their relevance suggests. Microhabitat conditions (e.g., water and air pollution, water levels, moisture content, etc.) should be monitored if circumstances so indicate.

Behavior Patterns. Behavior which affects population dynamics may need to be monitored in the field. Daily rhythms of insect behavior may affect not only population changes but one's capacity to observe such. It is thus desirable to attempt sampling or other observations over a range of times during the day. The most likely behavior to affect population change is movement of individuals in and out of the study area. Some idea of the scale and scope of such movements—for example, proportion of individuals involved, distances travelled, and net displacement in relation to the study area—is probably essential to all field population studies. Other forms of behavior which may need field study include, for example, those affecting either contest for food, predator–prey relationships, or the spatial and temporal distributions of various stages of the insects. An example is the drift downstream of stonefly larvae and the subsequent return flight of adults, as shown by Madsen and Butz (1976a,b).

11.4.1.4 Laboratory (Insectary) Studies.

At least some aspects of the parameters involved in population changes—growth, development, fecundity, birth rates, survival and death rates, and localized movement—can be studied in the laboratory to elucidate their inherent interactions with climatic variables and resource supplies. Obviously, important processes should be observed under a range of experimentally controlled conditions that provide opportunity to determine numerical relationships rather precisely. Ability to vary experimental conditions over the full range likely to occur in the field means that the study is not wholly limited to the conditions seen in the environment over the study period. These studies are only *part* of an ecological study, and because of the scale and scope of such work one cannot be as rigorous as in a physiological study. Experience with other life systems may be used to justify emphasis on particular factors in the early stages of a study, but before the importance of particular factors is claimed, it is best to review the rigor of any laboratory studies. Factors affecting phenology and/or the additive processes of insect populations generally include ambient temperature, food quality, and population density as it affects food supply, behavior, and so on. Different factors may be involved in different species. Subtractive processes are affected by the same environmental factors, particularly by extremes of climatic variables. Laboratory studies of population processes usually include initial experiments with each factor as the sole variable, the other conditions being held constant. Further (factorial) experiments with more than one variable can be particularly useful and do, of course, indicate interactions between factors.

Problems and Pitfalls in Laboratory Studies.   Many problems in the experimental approach to population ecology have been caused by using poorly planned experimental conditions and/or careless interpretation. A major problem is enclosure of insects. In any cage, movement is limited, the resources have been enclosed too, there is an impediment between the insects and the imposed treatment, and the latter often has an unnatural directional component. Abnormal behavior, resulting in either reduction or increase of, for example, mating, feeding, and so forth, may result from restricted movement. Conditions may be either more or less favorable than in the uncaged condition and more favorable than the applied level of a treatment might suggest because the resources are placed unnaturally close at hand and the cage may provide spatial variation in receipt of the imposed treatments. The manner of applying a treatment may greatly affect behavior; for example, the timing of protein feeding by female bushflies could be varied by changing the placement of food in relation to other attractive stimuli (Jones and Walker 1974). In general, enclosure of insects for experimentation should aim at minimal differences from the field. This means using cages as large as possible, with the physical barrier minimal for containing the insects. Since the caged conditions may alter the results, conduct of the experiment using different types of enclosure could, if the results are consistent, increase confidence in the interpretations. Sometimes the treatment can be applied in the absence of caging and the experiment reenclosed for subsequent observation.

The treatments imposed should encompass the normal range of field variation of the factor. In general, nonvariable conditions of a factor should be held near the long-term field means. The possibility of a switch in behavior triggered by a change in a factor must be considered. For example, a photoperiod relevant to the other conditions of the experiment should be selected.

1.   *Growth, development, and survival.*   Defining the start and end of a stage of development is often a major problem. It may be difficult to synchronize the start of development for a cohort of individuals: a collection of eggs will have a spread of ages reflecting the period of time since each egg was laid. Ecdysis or other clear morphological change is usually used to mark the completion of stages. Once the adult stage is reached, further development can rarely be observed directly. Time of first flight, egg laying, or larviposition may be recorded. Otherwise, sample dissections must be used to show stages of reproductive—or other internal—development. Unless individually marked insects are followed continuously, development periods should be estimated from the start of observations (less half the age range of the individuals used) to the average recorded time of completion of the stage (less half the time interval between the last two observations). Completion of development by half the individuals can be found more accurately using probit analysis (Finney 1971).

2. *Fecundity and birthrates.* Problems may arise here depending on whether the insects produce their progeny all at once, at a steady rate, or in discrete batches. As birthrate is normally construed as number of live births/female, a semantic problem arises where eggs are laid. Although intrinsic egg fertility may be less than 100% it is probably best to start with the eggs and consider fertility as part of survival. Careful attention to the conditions used is important when estimating fecundity. Oviposition may be dependent on the triggering effect of mating, substrate, or other environmental conditions.

3. *Survival rates.* Immature insects are most vulnerable during and right after transitions between stages. As a result, the numbers in a group of individuals undergo a stepwise decline. This may create statistical problems when average age at death is estimated. As physical conditions in the environment (e.g., pollution levels, weather, etc.) move away from the optima, subsequent mortality rates increase correspondingly. Immediate mortality can be caused by extremes of most physical conditions. Laboratory experiments on tolerance levels require consideration of cumulative effects of treatments, variability of response between individuals, and the problems caused by enclosing the insects. It is important to vary both the level of a factor and time of exposure to determine dose–time mortality relationships. Variability between individuals' responses means that techniques such as probit analysis (Finney 1971) may be needed to compare results. In analyzing responses to a variable in terms of developmental, survival, or some other rate function, attention to *means* is central, but the frequency distribution around *means* may also be important. In some cases, a small percentage of a population—that is, those with the most rapid developmental rates—produce the majority of progeny and contribute disproportionately to the intrinsic rate of increase.

One may observe the interaction between an insect and a natural enemy in the laboratory and discover who will "eat" whom, but quantitative data obtained from such studies are unreliable, since mortality rates in the field depend on a great many factors that affect the two species. Thus, to name only some of the influences discussed earlier, the number of prey killed by a predator can initially be affected by its hunger, number of prey available, size distribution of prey, distance between successive prey, and complexity of the relevant environment. Obviously, the experimental conditions provided will affect the probability of prey mortality. If a predator–prey relationship is a major component of the life system, the conditions affecting it should be worked out in detail, preferably in the field. Analogous comments could be made about relationships of pathogen (or parasitoid) and host.

4. *Behavior, polymorphism, diapause.* The remarks concerning quantifying natural enemy relationships could be made about laboratory studies of any sort of behavior. We reiterate that such studies generally form a

small part of an ecological program and thus are rarely carried out with the rigor of the ethological discipline.

The factors influencing facultative polymorphism usually can be studied by laboratory experiments. If the sex ratio varies little, it often can be ignored in population dynamics studies, but if it is substantially influenced by environmental factors, the relationships should be determined. Females produce the eggs directly resulting in population increase; the number of males is generally less directly related to such increase. It is often valid, therefore, to consider only the numbers of females in population studies.

The well-known effects of wing polymorphism on the population dynamics of many Homoptera have often been studied in the laboratory. Major relationships to variations of temperature, feeding conditions, and crowding have been found. Interactions occur between factors and the stimulus provided by crowding seems to vary between species. All the possible forms of crowding should be allowed for in experiments. Among aphids, other types of polymorphism also occur—for example, sexual, parthenogenetic, and transitional forms between the two. Again, several factors may influence the probability that an individual will take a certain form, but in this instance the species may (also) use a token stimulus from the environment (e.g., a critical photoperiod) to determine the morph produced.

The problems in diapause studies are too numerous to consider here but are dealt with in Chapter 6.

11.4.2 Analysis of Results

Life system data are analyzed (a) to allow accurate description of events and to check their internal consistency, (b) to describe the processes of which the events are the outcome, and (c) to suggest the mechanisms for these processes.

11.4.2.1 Analysis of Field Data.

Elucidation of events is usually achieved by comparing "states" seen in successive samples, but, occasionally, insect remains or artifacts can be analyzed at the end of a population cycle [e.g., bark beetles (Beaver 1966, 1967, 1974)]. Numerical change between one occasion and the next may be real or merely sampling "noise." It is not possible to account for changes smaller than the magnitude of observational errors, which should be determined by statistical procedures valid for the sampling situation. A significant increase must be accompanied by reproduction or immigration. The long-term pattern in numbers of eggs (or other first-stage individuals) should show peaks at times of sudden increase thought to be due to reproduction. Evidence of immigration is usually seen as an increase in numbers of a mobile stage that cannot be accounted for by estimates of abundance of the prior stage.

As noted earlier, unless the corpses are available for counting, as for certain nest building Hymenoptera (Freeman & Jayasingh 1975), losses due to mortality and those due to emigration cannot be differentiated. One of these causes may be suggested by the recognition of relics of individuals remaining after death, by finding individuals ("marked" in some way) beyond the boundaries of the study area, or by indications of a predisposition to emigrate. Relics include corpses or remains bearing signs characteristic of some natural enemy's attack. Predisposition for emigration could be the appearance of alatoid aphid nymphs in otherwise apterous colonies or an appropriate color phase change among locusts. Any ascribed losses should be consistent with the numerical possibilities of the confirmatory evidence. Thus, losses from an aphid population thought to be due to flight of adults should not exceed the number of alatoid individuals observed previously in the colonies. Similarly, if all the relics of a type of mortality are not available for counting, other evidence that the causal agent was operating at a level greater than the relics seen would be needed to support the speculation. For example, some recognizable relics of a predator's attack would need the support of evidence of predator occurrence in sufficient numbers to cause the speculated loss.

Comparison and Observed Trends. Speculation on causes of population increases and decreases is aided by comparisons of population trends with concurrent trends in environmental factors. Use of logarithmic scale and physiological time will often better demonstrate associations between rates of change and environmental factors. Any sudden change in numbers following an equally sudden change in an environmental factor provides circumstantial evidence; it should be followed up by further observation and experiment. Care is needed in comparing observed numbers killed by natural enemies and population changes. Relics of an earlier generation may remain or be adhered to the substrate a long time and occur together and be confused with relics of the generation being assessed. A natural enemy (e.g., a parasitoid) that remains dormant within the host relic may also be subject to inappropriate assignment of its generation. Estimation of the impact of natural enemies requires detailed consideration of a diversity of such problems.

Using techniques such as those described above, it is sometimes possible (if the generations are distinguishable) to follow the successive declines in numbers of a well-defined cohort of individuals and to label each loss with a likely cause ("budget" of Richards 1961). The greater the synchronization of development among individuals, the more accurate is the population budget. In cold temperate-zone insects, it is usually easier to construct accurate budgets following winter when individuals emerging from some form of cold quiescence are more often tightly synchronized. In arid-zone insects or those from other, warmer climes it may be the opening of the rainy season that provides synchrony (cf. Chapters 3, 6, 22). However, synchrony is gradually lost even within that generation, so that two or more adjacent stages coexist (Clark et al. 1967). Problems of estimating the numbers entering each stage can be overcome in several ways (review by Manly 1974). The best of these, that of Kiritani and Nagasuji (1967) and Manly's (1976, 1977) later extensions, should be used if possible. It can be applied even to situations where individuals from successive generations coexist, so long as the same *stages* of the two generations do not coexist.

If it is not possible to assign individuals in the same stage to particular generations, however, the population budget method becomes inapplicable. Most warm-temperate to tropical insect populations show rapid development and persistence of generation overlap, as in the scale insect described by Williams (1970). Because of generation overlap "classical" population budgets ("age-specific life tables") have less applicability than their prominence in texts suggests. Equally, methods based on population budgets (e.g., key factor analyses of Morris 1959 and Varley & Gradwell 1960) seem unsuitable for insects whose generations overlap. Some of the problems posed by overlapping generations can be overcome (e.g., Hughes 1962, 1963). If the overlap is complete enough (usually over three coexisting generations), temporarily consistent patterns of survivorship and reproduction give rise to a recognizably stable age distribution of individuals. If such age distributions occur, it is possible to compare potential increases of two or more populations from the numbers of individuals in successive age classes (Farrell 1976). The difference between potential and observed increase gives a measure of total population losses. Using this method, budget-type data collected on two occasions a short time apart can be analyzed to show relative roles of known subtractive factors in the population change observed. This approach was called a "time-specific life table" by Southwood (1966, 1978). It was used in the IBP study of the natural mortality of *Myzus persicae* (Blackman 1976).

Comparison of Derived Trends.   It is sometimes valid to combine two or more sets of observations to form other measures of population dynamics. A good example is the ratio of numbers of eggs or other first stage individuals to the number of reproducing females. This ratio is a crude measure of reproductive rate. A comparison of its trend with those of environmental variables will often give clues to relationships in the life system (e.g., to population density, food supply, etc.). The proportions of individuals in each possible state available to them provide other forms of combined data. The proportion of fourth-instar aphids forming wing pads often shows trends related to environmental variables, such as crowding, day length, temperature, and so on. Unusual increases in the proportion of gravid females (in species where egg development continues through adult life) may indicate temporary delays of oviposition relatable to ambient conditions, as in the bushfly (Hughes 1977a).

Preliminary Regression Analyses.   Occasionally one sees a correlation between population data and a single environmental variable. Because there is usually a lag between the operation of such a variable and its effect on the population, a correlation with *current* levels of a factor generally indicates that it is biasing the sampling results. Some correction of the data would then be required. With knowledge of the life history and phenology of an insect it is possible to interpret correlations with previous environmental conditions. The *Thrips imaginis* study of Davidson and Andrewartha (1948) showed that previous average temperatures and rainfall were each important to current abundance. Nonetheless, since photoperiod, radiation, average temperature, average humidity, rainfall (some regions), food supply, and abundance of associated organisms all vary together, the causal nature of a correlation to any one variable is open to doubt. Where several independent factors appear to be operative, some form of multiple regression may help to sort out causal relationships.

11.4.2.2 Analyses of Laboratory Data. Because laboratory observations and experiments, if well planned, can be done under rigorously defined conditions, the results should give unequivocal evidence of the existence and quantitative form of relationships. Rigorous significance tests should be applied and adequate descriptions of each relationship should be given. If the form of a relationship is clear, the constants in a simple equation can be determined by curve fitting using maximum likelihood techniques. Where the data suggest a simple but unknown form of relationship, curve fitting and testing may give some clue to its nature. Quantitative descriptions using complex polynomial expressions are rarely of more value than original data polygons for interpreting events. Existence of interactions between factors can be important to understanding. Their occurrence is best shown by analysis of factorial experiments, but demonstration of the quantitative form of the interaction requires fitting of a three-dimensional surface to the data.

Tabulation of the numbers of individuals in a sample cohort surviving to successive ages under specified conditions is termed a *life table*. Life tables allow the simple calculation of life expectancies (first applied to human populations for life insurance purposes). If during the *experimental* determination of a life table for a species under given conditions, the females are allowed to reproduce and the live progeny counted, a more useful form of tabulation can be assembled. A table of age-specific survival (l_x) and fertility (m_x) can be used to calculate the net replacement rate (R_0) intrinsic to the population and, once R_0 is known, the instantaneous rate of increase (r). Subsequently the innate capacity for increase (r_m) for a stable age distribution may be estimated (Andrewartha & Birth 1954). Harcourt (1969) reviews the construction and analysis of life tables. When r_m values are determined for each of a range of combinations of environmental conditions representative of the normal environment, the effects of such environmental changes on the potential increase of the population can be envisaged. The experimental conditions under which the highest value of r_m

(called r_{max}) is obtained are regarded as optimal for the population and, if fully representative, for the species. With due caution, r_{max} may be used as a benchmark in comparative studies. However, if it is planned to progress from the life system study to a *computer simulation* involving relationships to changing environmental variables, construction and analysis of life tables for particular combinations of constant conditions ("stationary life tables") and use of r statistics have little value, since the simulation itself is a "dynamic life table" (Hughes 1973).

11.4.3   Synthesis of Results

The final phase in a life system study involves showing that the set of identified and described components and relationships is complete enough to function as a system. Any such demonstration requires the components of a life system to be assembled in a form, taking due account of the order, magnitude, relative importance, and interactions of each relationship, which will adequately predict subsequent states of the system.

Clark et al. (1967) suggest that the life system determines "the existence, abundance and evolution of a particular population." The adequacy of a life system description may thus be judged by its ability to explain the observed spatial distributions (Jones 1974), but more critically, to predict changes in population abundance (Meats 1974a,b), and finally, to offer evolutionary explanations of fitness. Evolutionary tests may be possible. Fisher's (1930) theory asserts that natural selection maximizes "fitness" of any interbreeding population. From that basis it should be "...broadly true that any arbitrary, forced changes in population parameters away from the values actually observed should cause population fitness to decline" (Gilbert et al. 1976). It has sometimes proved possible to use evolutionary criteria. Maximization of a measure of fitness at the observed values of population parameters has been shown by Gilbert and Hughes (1971) for *Diaeretiella rapae*, a parasitoid of the cabbage aphid, for the thimbleberry aphid, *Masonaphis maxima*, by Gilbert and Gutierrez (1973), and for the cotton boll-weevil, *Anthonomus grandis*, by Gutierrez et al. (1979). These three kinds of tests require progressively more formal descriptions; the evolutionary test requires a full-scale quantitative model.

11.4.3.1   Life System Models

Verbal Models and Flowcharts.   Once a large proportion of the relationships are recognized, it is possible to develop a qualitative description of how the life system functions. Purely verbal models are difficult to communicate or evaluate as, for example, in some studies described in Clark et al. (1967), simply because the linear nature of language makes it inappropriate to describe networks of relationships. Consequently, various ways of visualizing verbal models have been devised. The best known of these are flowcharts and causal loop diagrams. These improve on verbal description, since the system can be explored in more than one dimension and any grouping of factors and relationships can be considered together. Ideas of time (e.g., lags due to development of individuals) can also be incorporated. Chains of relationships are made clear, so the directness of an influence can be judged. Causal loop diagrams emphasize *qualitative* the chain of responses of systems to any sort of change. Visualization of a life system may suffice for understanding the causes of gross spatial or temporal variations in abundance, but it is inadequate for predicting precise *quantitative* changes.

Key Factor Analyses. In the simplest circumstances of insects which have well-synchronized and separable generations, the life system components are naturally assembled in usable form, and the most significant components of intergeneration numerical change may be identified quasistatistically. This is done by comparing in each successive generation the magnitude of change caused by individual factors with the total change. Some authors partition the total change among the several life history stages and define *key factor* as the stage in which the part changes showed the greatest correlation with the total change. So-called key factor analyses were devised by Morris (1959) and Varley and Gradwell (1960). Luck (1971) confirmed the greater utility of the Varley and Gradwell analysis. This version was modified by Podoler and Rogers (1975). Since the basic methods are described in texts (from Southwood 1966, onwards), they are not detailed here. The combination of population budgets and key factor analysis has become the most common end point of life system studies. But it is not the *stage* in which an event (e.g., the mortality) occurs, but the factor(s) that *cause* the mortality that is critical ("key") to the total change. Moreover, it is important to identify those factors which tend to reduce change as well as those which cause change.

The method is easy to understand and apply, but these attractions have often resulted in its application to inappropriate data. Besides the limitations of separable, synchronous generations, only data from insects successively using the *same* habitat can be so treated. This is because insect numbers on temporary habitats (e.g., annual crops) can rarely be strictly related from one generation to the next, and unless the new immigrants which start the new generation represent all, and not more than, the survivors of the previous generation, the two sets of data are not comparable. In general, there are a number of technical difficulties with the method (cf. Southwood 1967, Smith 1973), but it is useful in drawing some conclusions from complex quantitative data of an appropriate type.

Computer Models. The difficulty in understanding and evaluating quantitative data from a complex life system lies in mentally combining knowledge of average levels of several population attributes, the ways that each responds to changes in environmental variables, and the effects of these responses on population change. With development of large computers in the 1960s, easy combination of large quantities of such information became possible and ecological synthesis in numerical terms became practicable. The life system approach guided the assembly of some early simulation models, and, indeed, might have been designed for the purpose. Clearly, the pieces of information required to construct a realistic simulation model of population dynamics are, in fact, the essential components of a life system. Quantitative methods of life system analysis have thus been improved. Each life system illustrated in this chapter (below) has been described at one level or another as a computer simulation.

Models of life systems can be implemented in a variety of ways on a computer, but because nonlinear relationships are invariably present, algorithmic models are usually favored. These serve particularly well as integrating devices, and the possibility of comparing their output directly with field population data has tremendous heuristic value. For the level of understanding of population dynamics normally required (e.g., for a heuristic purpose or to guide timing of a control measure), predictive accuracy at the level of comparable magnitude and pattern is quite satisfactory. This may be because the balance between additive and subtractive components of most life systems has a potential for wild fluctuation, and so a reasonable fit to independent field data is unlikely to occur by chance.

For a direct practical purpose (e.g., design of pest management strategies) more accurate predictions may be desired and attempts made to make the model more precise. However, due to the partly stochastic nature of population processes in the real world, *testing* the apparent goodness of fit of a simulation outcome to the observed population pattern has a built-in statistical difficulty. Each observed pattern is a unique stochastic outcome of relationships between a particular group of individuals and a particular set of environmental conditions. Unless more than one population can be initiated with the same circumstances (at the least, difficult in the field), no measure of the variability of a subsequent abundance trend is possible. Measures of sampling error for one population are not measures of the variation *between* populations. In a similar way no measure of variability exists for a deterministic model of the same situation. Only if the model includes realistic stochastic variability of its parameters may the *minimum* confidence limits of the simulation be elucidated from replicated runs. Few insect ecologists have attempted to make comprehensive models stochastic, recent exceptions being the grain weevil models of Hardman (1976a,b) who observed the levels of variability in some major parameters.

Analytical Models. The mathematical formulations of ecological relationships put together in computer models of life systems are not always sound. Even if the relationships are well formulated, numerical (algorithmic) methods are usually employed for comprehending their synthesis because the systems are too complex for simple analytic solutions. Cuff and Hardman (1979) note that while the numerical method initially allowed rightful subordination of mathematical structure to correct definition of the components, it had two serious flaws: (1) the difficulty in communicating a complete and comprehensive description of the system and (2) "desirable refinements and extensions" become increasingly difficult to incorporate in the algorithm because of the mass of detail.

Relatively simple population models which have analytical solutions have been described (e.g., Lotka 1925, Volterra 1926, and others) but were rarely able to describe a real population. Proponents of simple models claim that these allow the "essence" of interactions to be captured without the ambiguity and technical problems introduced by biological realism (Pim & Lawton 1977, Chapters 12, 15, 16). Such models lend themselves to mathematical treatments [e.g., from the extrapolations of predator–prey interaction of Nicholson & Bailey (1935) to the stability analyses of May (1973)]. However, many such results, and their usefulness, have been questioned as having more to do with the formulation of the models than with field population dynamics (e.g., Wang & Gutierrez 1980).

To be realistic, population models must at least include age structure and the principal relationships affecting birth and death rates. Leslie (1945) proposed a matrix model which includes age structure and constant age-specific birth and death rates:

$$N_{t+1} = A_t \cdot N_t$$

where N is the vector of age classes and A is the matrix of age-specific birth and death rates determining change in age structure at the next time step. The continuous form of the model is the balance equation proposed by von Foerster (cf. Sinko & Striefer 1967). If the birth and death rates involved are held constant, however, the population derived from such models grows without bound after reaching a stable age distribution. That is, of course, unrealistic as the current birth and death rates are partially functions of several restricting physical and biotic factors in the environment and because the assumption that changes in aging rates (Δa) are equal to changes in time (Δt) is not always met. Such factors can sometimes be built into extensions of Leslie matrix models, as was done by Wang and Gutierrez (1977), Gutierrez, Butler et al. (1977), Gutierrez, Leigh et al. (1977) for cotton systems, and Cuff and Hardman (1979) for *Sitophilus oryzae*. Such models are, in every respect, mathematical representations of a time-varying life table and they describe actual populations. Many simulations of insect populations (e.g., Hughes & Gilbert 1968, Gilbert & Gutierrez 1973, Gutierrez, Denton et al. 1974, Gutierrez, Nix et al. 1974, Gutierrez, Havenstein et al. 1974, Gutierrez et al. 1975, Gutierrez et al. 1976, Stinner et al. 1975, Tummala et al. 1975, Hughes & Sands 1979) could probably be cast in this form. The number of such models is now increasing.

Now that a greater richness of field relationships can be captured in a mathematically sound and rather more amenable form, the task remains to distill the "essence." Several approaches have been tried:

1. perturbation of model parameters with specific questions in mind (e.g., what happens if, say, fecundity is altered) (May 1973);

2. putting the model in a form suitable for optimization studies and asking various economic (Regev et al. 1976, Conway 1977, Shoemaker 1977) or ecological questions (Gutierrez et al. 1979);

3. using the model to generate the characteristics of the system for topological analyses (Clark et al. 1978a,b, Peterman et al. 1979) (cf. Chapter 24).

11.5   CASE STUDIES OF LIFE SYSTEMS

The need for a new and overall approach to study of the population dynamics of insect pests became increasingly obvious (e.g., cf. Auer 1968) as control of insects by chemicals became increasingly uncertain in 1955–65. Important sampling and analytical "tools" had been published in texts like Southwood's (1966); so a general new approach became possible. Ecologists such as Clark et al. (1967) redirected attention to the structure of the ecosystem in which insects live and, like many other entomologists, pointed to the need for more field population studies. Each year between 10 and 20 comprehensive population studies have been published from locales throughout the world, on species of esoteric interest [e.g., nycteribiid flies on bats (Marshall 1971, Funakoshi 1977)] to pests which in outbreak affect whole geographic regions [e.g., the red locust (Stortenbeker 1967)]. Studies on insects that are abundant [e.g., Colorado potato beetle (Harcourt 1971)] have been complemented by studies on rare ones [e.g., the milkweed beetle, *Labidomera* (Eickwort 1977)]. Sometimes more than one study has been made on a wide-ranging pest [e.g., the fall webworm, *Hyphantria cunea*, (Morris & Bennett 1967, Morris 1971, 1976 in Canada, and Ito et al. 1969 in Japan)]. Studies vary in length from little more than a year for an initial effort on the thimbleberry aphid (Gilbert & Gutierrez 1973) to, more commonly, 3–5 yr, or even very prolonged studies on some forest insects [e.g., 17 yr by Klomp (1968) on the pine looper].

The discussion of spatial variation in life-system-type studies has been added to that of time variation [e.g., in studies by Grimble & Knight (1970) on a cerambycid (*Saperda*) and in the retrospective analysis of nest relics of *Sceliphron* by Freeman (1977)] (cf. Chapter 14).

Only a small proportion of early studies fully exploited the life system approach—of which MacLellan's (1977, 1978) 12-yr studies of the codling moth and eyespotted bud moth are classic. The development of useful computer models from life system studies by Gilbert and Hughes (1971) and Southwood et al. (1972) helped stimulate studies to incorporate comprehensive approaches.

Three studies with which the authors are particularly familiar and which illustrate the scope of the life system approach will now be dealt with in some depth. The first, on the bushfly, was commenced in 1967 and is used primarily to illustrate some of the problems which had to be overcome to make the study comprehensive. The second, a study of the cabbage pest, *Pieris rapae*, illustrates how the life system provides a meaningful frame of reference for a specific study of movement and oviposition behavior. Finally, we discuss the study of plant–herbivore interfaces, showing how a plant population can be treated in a very similar way to the interacting insect population.

11.5.1 The Bushfly Life System

The bushfly, *Musca vetustissima*, is the Australian member of the *Musca sorbens* complex and thus very similar to the dog-fly of Hawaii and the market-fly of North and East Africa. The larval stages of the bushfly are innocuous but the adults seek protein-rich body fluids (tears, saliva, serous exudates, etc.) and in doing so is a serious nusaince to domestic animals and people. Although effective personal repellants are available (Waterhouse & Norris 1966), the need for more general bushfly control was a major stimulus for initiating the study. However, area suppression with chemicals proved impossible to maintain for more than several hours. In the mid-1960s a life system study of the bushfly was started, with the primary objective of *understanding* the problems of controlling this pest (R. D. Hughes & colleagues 1969–1978).

Very little research on the problem had been done, but information reported on its biology, phenology, and ecology was combined with knowledge available for other populations of the *Musca sorbens* complex to form a scenario for its life system. A picture emerged of a highly vagile species—moving between ephemeral larval food resources provided by fresh animal feces, with fly abundance favored by hot, but not too wet, conditions. A very significant feature was the absence of any diapausing or other stage tolerant of temperatures below 10°C, suggesting the bushfly could not exist in the southern third of Australia during the winter. Adult bushfly were certainly absent from the south in winter but could be found at that time further north where air temperatures exceeded 15°C during the day. Evidence for repopulation of the south each season was found by regular attempts to collect flies along a north–south transect in spring. When the first flies collected at each point on the transect were dissected, their development showed (Tyndale-Biscoe & Hughes 1969) they were too mature to have emerged locally and, in each case, flies collected further north on the same occasion were progressively younger (Hughes 1970). That the flies were being systematically displaced southwards on periodic warm-wind systems was suggested by data collected over 5 yr (Hughes & Nicholas 1974).

    Concurrent circumstantial evidence of rapid displacement on the wind
over hundreds of kilometers led to realization [confirming mark–release–
recapture data of Norris (1966)] that the full spatial limits of the bushfly
population were too large for a comprehensive study. Because the flies are
continuously on the move, the adult cohort in any local study area is numerically unrelated to the preadult stages present, except for eggs they
have just deposited. By the time larvae hatch out and develop, the associated
parent flies may be many kilometers away. The adult fly cohort of the
original area will *then* consist of other immigrant flies and any which have
emerged from eggs laid locally. The rate of development of the species
and the spread of oviposition over many days is such that there is a complete
overlap of generations. The local population dynamics depends on the
relative rates of (1) immigration of flies, (2) egg laying, (3) survival of
preadult stages to emergence, (4) survival of adult flies to oviposition, and
(5) emigration of flies.
    Numerical studies of the fly's population dynamics even in a local area
presented difficult sampling problems. While it would have been possible
to census the preadult stages in fecal material collected in the field, the
dispersion pattern of the population among the individual lots of feces
would have meant processing impracticably large quantities of material.
This was particularly true in summer when, for example, only about 1 in
40 cattle dung droppings contained bushfly stages. An alternative approach
was possible because adult flies which had just oviposited were easily recognizable on dissection and the number of eggs they had laid could be
estimated from the well-defined adult-size/number-of-eggs-in-batch relationship (Tyndale-Biscoe & Hughes 1969). So by sampling the local abundance of adult flies it proved possible to estimate the currently associated
egg numbers in feces. Similarly, the appearance of readily recognizable
newly emerged flies (Tyndale-Biscoe & Hughes 1969) among those adult
fly samples made possible estimates of their relative abundance. Measures
of egg-to-adult survival could then be derived (below). Unfortunately, there
is no completely satisfactory method of sampling abundance of bushfly
adults, and a form of trapping depending on their flight activity had to be
used (Hughes 1977b). Activity was obviously biased by temperature and
extremes of other conditions. The sample catches could be corrected for
temperature, and by avoiding other extreme conditions, reasonably satisfactory serial estimates of abundance were obtained (Hughes 1977a).
    Laboratory studies were used in working out the timing of the bushfly
life cycle. It has a development threshold temperature of approximately
12.5°C and preadult development required accumulation of 158 day-degrees above 12.5°C ($D°_{12.5}$) in the fecal and soil material occupied by the
eggs, larvae, and pupae. Thus, by comparing adult fly samples taken on
occasions estimated to be 158 $D°_{12.5}$ apart (dung and soil temperatures), the
estimated numbers of eggs laid and of flies newly emerged, respectively,
were used to calculate egg-to-adult survival (Hughes 1977a).

Measurements of adult survival in either laboratory or field cages suggested that females which had laid several batches of eggs should be common under field conditions in the summer. In fact, however, dissections of field-caught flies showed that older females were uncommon. The conditions in even very large cages apparently so ameliorated the effects of unfavorable weather and of movements between normally dispersed resources that such experimentally determined survival could not be applied to field situations. This problem was circumvented by Sands and Hughes (1977), who, by assuming that over most of the season flies *leaving* a locality are matched by ones *entering* it from similar nearby areas, concluded that the "physiological age" of females at the time of capture would be distributed with frequencies proportional to fly survival in the field population. By combining "age" distributions of all females caught in regular samples over four complete seasons, an average survivorship curve could be obtained. Analysis showed that mortality was not only age-dependent but also size-dependent. From the survival data it was possible to estimate the probability that female bushflies would survive to lay one, two, or more successive batches of eggs. Because both survival and number of eggs in a batch are size-dependent, so is the mean total of eggs laid per fly: from 95 for larger flies (2.5 mm headwidth) to 4 for small flies (1.5 mm).

Using the assumption concerning the balance of flies entering and leaving a study area, the population dynamics of the bushfly could thus be quantified, hypothetically, in terms of egg-to-adult survival levels, adult survival to oviposition, and mean total number of eggs (Hughes 1977a). Adult size had an obvious key role in bushfly population dynamics.

Studies of environmental influences which modify bushfly population dynamics were made in a series of laboratory experiments and observations. Of the weather influences, increased temperature not only speeded up development and phenology but also increased the ultimate size of larvae (and thus of adults and their fecundity) and the proportion surviving. Heavy rain was an important mortality factor in the egg stage and at times of larva–pupa and pupa–adult transitions. Otherwise the main environmental influences acted through food quantity and quality (Hughes & Walker 1970). The initial number of larvae/liter of fecal material affected both survival and ultimate larval size. However, the main determinant of size was the favorability of the fecal material as food for the larvae. In the case of herbivore dung (probably the major breeding resource) this favorability depended on the degree of maturity of herbage eaten by the herbivores (Greenham 1972). In any season the growth and phenology of herbage, and thus dung "quality," were determined by weather factors (Sands & Hughes 1976).

Figure 11.3 is a causal loop diagram for abundance changes of bushfly in a study area. The detailed properties of species and the relationships of the bushfly in a study area thus described have been assembled into an algorithmic model driven largely by weather data. Given initial abundances

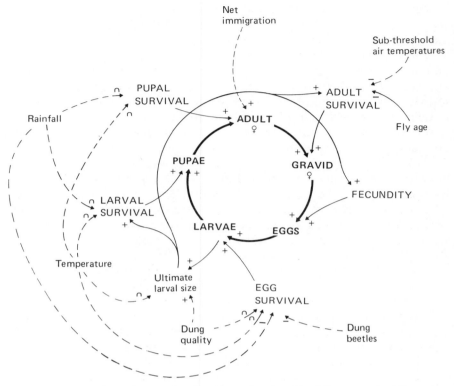

Fig. 11.3 A causal loop diagram for abundance changes of bushfly at one site. The arrows show the direction of causality and the symbols by the heads the nature of the effect: +, positive; −, negative; ∩, with a maximum in the normal range.

and "age" and size distributions of flies based on field samples of the spring immigrants, but otherwise assuming a balance between flies entering and leaving the study area, the model gave satisfactory simulations of trends in abundance and size of bushflies for five of nine sets of field data, collected at three widely separated areas (Hughes & Sands 1979). This, of course, leaves room for improvement. Two further sets can be satisfactorily simulated, allowing that the assumed normal balance between flies entering and leaving the area broke down occasionally. A concurrence between a net decrease in the one study area with a net increase in the other downwind suggests that a local abundance model might be used to study effects of displacement in a more comprehensive study. At present, however, we can only speculate upon the full extent of the bushfly life system. Flies in southeastern Australia during the summer seem to be part of a very extensive population which depends for its persistence on being able to shift the focus of its breeding from temperate Australia in spring and summer to subtropical Australia in autumn and winter. The favorability of fecal material as larval food is maximal in the temperate areas in spring and in the subtropical areas in autumn, corresponding to respective winter and summer rainfall patterns. Continuous breeding in one region is prevented by cold conditions in winter in the south and by drought conditions in spring in the north.

The nature of the bushfly life system clarifies the reasons for the difficulties found with attempts at local area suppression of the pest, that is, the high vagility of adults and the ephemeral, but constantly renewed, larval food resources in inhabited areas and range lands. Even where no animals are present, such as on drought-affected ranges, warm winds continue to carry the flies in, and once there, the lack of other food make them concentrate attention on any human visitors.

Knowledge of the bushfly life system has guided two attempts at biological control. In southern areas potential wasp parasitoids are hindered by the absence of the host insect throughout the winter. However, selection and introduction of parasitoids that attack all types of flies breeding in dung, including bushfly, but which could overwinter in the other hosts, present a possibility (Hughes et al. 1974, Hughes & Woolcock 1976, 1978). The other biological control has been a spin-off from introduction of cattle dung burying beetles from Africa and Europe. Such beetles have now established over large areas of tropical and subtropical Australia and demonstrably reduced local bushfly numbers through competition (Hughes et al. 1978).

11.5.2 Behavior and the Cabbage Butterfly Life System

The cabbage butterfly studies by R. Jones and colleagues is an example wherein a life system study provides the context for analysis of particular aspects of an insect's biology— in this case its searching behavior, movement patterns, and dispersal of progeny (Jones 1977a,b, Gossard & Jones 1977, Jones & Ives 1979). The life system study was an essential component of the behavioral study, since it allowed the investigators to evaluate the adaptive significance of the behavior patterns observed.

The behavior of caterpillars obliged to locate new food plants, and of adult females seeking hosts for oviposition, was studied and modeled in two places—Vancouver (Western Canada) and Canberra (Australia) (Jones 1977a,b, Jones & Ives 1979). The caterpillars behaved in much the same way in both populations, and *their* behavior is not discussed here. But the behavior of adult females of the two populations was quite different. The

Australian females flew in relatively straight lines, laid few eggs on any plants encountered, and were relatively unresponsive to differences in host plants, whereas the Canadian females flew in very convoluted paths, would visit and lay eggs on a single plant many times, and were highly responsive to differences in host plants. Simulation of these behavior patterns and experimental confirmation of the predicted outcomes showed that the Canadian behavior tended to produce a highly clumped distribution of eggs (with some plants receiving many eggs while many others received none) at a relatively low cost in flight time. The Australian butterflies, on the other hand, distributed their eggs relatively more randomly but at a considerably greater cost in flight time. To interpret the adaptive signifi- cance of these differences required an understanding of differences be- tween the life systems of this species in the two localities studied.

Considerable information was already available on the biology of this species and several of its natural enemies (cf. Richards 1940, Dempster 1967, 1969, Parker 1970, Harcourt 1963), although relatively little on its ecology in Australia or western Canada.

The basic developmental pattern of the two populations proved to be very similar. In each, eggs and larvae have a developmental threshold of about 10°C, with pupae somewhat lower at about 7°C. Developmental times are highly sensitive to food quality in the early instars, but on plants of high quality egg and larval development require approximately 220 $D°_{10}$, and pupal development about 130 $D°_7$ in the absence of diapause (which is induced by low temperatures and short photoperiods—Baker et al. 1963). When adults emerge, they require 60–80 $D°_{10}$ to reach peak egg laying (Gossard & Jones 1977) and live up to 400 $D°_{10}$. They may lay 50 eggs/10 $D°_{10}$ at the peak and up to 600 eggs during their lifetime. Although the basis of the development pattern is very similar in the two populations, their phenology, is in fact, very different. Canberra is generally much warmer than Vancouver. The temperature requirements for development are satisfied more quickly in Canberra, and there the butterfly may have as many as six generations in a year, in contrast to only two in Vancouver. The extent to which the Canadian population may increase during a single season is consequently restricted.

Survival rates of the juvenile stages, and the factors affecting them, also vary between the populations. Although survival as well as development of young larvae is sensitive to host plant quality, this is rarely a major factor in juvenile mortality in either population since the adult females discrim- inate strongly against poorer-quality hosts (Jones 1977a, Ives 1978, Jones & Ives 1979) and eggs are rarely laid on them. Parasites (especially *Apanteles glomeratus, A. rubecula,* and *Pteromalus puparum*) may kill a large proportion of older larvae later in the season in Australia. These parasites are rare or absent in Vancouver, but similar levels of late-season mortality on older larvae are inflicted by predatory wasps (*Vespula pennsylvanica*). Deaths from viral epizootics, which have been recorded as major influences on *P. rapae* mortality elsewhere, were of minor importance in these two populations. The major survival differences between the two populations occur early in the season and in very young larvae. In Australia survival of newly hatched larvae early in the season is generally very high, allowing a rapid buildup of population numbers. (High early-season survival has also been recorded in a number of studies elsewhere—for example, in the southern United States by Parker 1970). In Vancouver, however, survival of newly hatched larvae is uniformly poor but all the agents responsible for the high death rates are not known. As a consequence of the conjunction of low survival with a short growing season, populations of *P. rapae* in Vancouver almost never achieve high densities and the insect is a minor nuisance compared to its major pest status in Canberra and in most other places where it occurs.

As a result of the very different population densities achieved in the two areas, density-dependent influences on juvenile mortality, demonstrable by artificially increasing population numbers, seem almost never to be invoked naturally in Vancouver since individual plants rarely carry large numbers of larvae. But in Australia the density-dependent effects do occur frequently and indeed comprise a variety of different phenomena. The most important of these is egg cannibalism by larvae, especially by first instars (e.g., Jones & Ives 1979). Young larvae will eat eggs in preference to their host plant; as larval populations build up, therefore, egg survival is drastically reduced. Larvae at high density on a plant also develop more slowly than those at low density, and if densities are high enough, they may destroy the plant, resulting in starvation. There is also evidence that natural enemies—in particular the parasite *Apanteles glomeratus* and predatory ants (*Iridomyrmex* spp.)—attack larvae on densely infested plants more often than those on lightly infested plants (Jones, Gilbert & Nealis unpub.). Note that these particular effects operate on a per-plant rather than a per-unit-area basis.

Weather also exerts considerable effect on cabbage butterfly population dynamics, primarily via adult behavior. Females do not fly, hence do not lay eggs, in overcast or very windy conditions. Although there is then a compensatory spurt of egg laying when conditions improve, the compensation is not perfect; a succession of unfavorable days may strongly depress population growth (Gossard & Jones 1977, Dempster 1967, Biever et al. 1972). The frequency of overcast days during summer is much higher in Vancouver than in Canberra; thus population growth in Vancouver is much more likely to be restricted by this factor than in Canberra. Harcourt (1963) recorded a high frequency of larval deaths from drowning after rain in Ontario populations, but rain had little effect on survival in Vancouver or Canberra, except that those *few* eggs laid on the upper surfaces of leaves tended to be washed off. Dempster (1967) similarly found no effect of rain on mortality in English populations.

Thus, we may summarize the major differences between the life systems of these two populations as follows:

Vancouver	Canberra
Few generations per year	Many generations per year
Poor juvenile survival throughout the season	Early-season survival very good, declining as summer progresses
Flight and egg laying frequently restricted by bad weather	Flight and egg laying rarely restricted by bad weather
Per-plant density-dependent mortality of eggs and young larvae rare, since population densities very low	Density-dependent mortality higher, more frequent, since population densities much higher

A comparison suggests that the observed differences in searching behavior between the two populations may indeed be adaptive. In Vancouver the fact that bad weather may frequently restrict flying and egg laying implies that the time available to search for hosts may often be limiting. At the same time a high degree of aggregation of the eggs does not involve substantial disadvantage, since densities remain very low and local density-dependent mortality will seldom occur or will be light. In Australia, by contrast, search time is much less likely to be limited by bad weather and populations commonly reach high densities, but density-dependent effects often occur. If the butterflies aggregated their eggs to the same degree as Vancouver butterflies, the frequency and severity of these latter effects would be greater still. It is thus apparent that the relative "costs" of egg aggregation versus increased flight time to distribute them more widely differ between the populations in a manner consistent with the observed behavioral differences. It is also clear that the adaptive significance of the behavioral differences could only be elucidated by an understanding of the major differences between the life systems of each population.

11.5.3 The Plant–Herbivore Interaction

All population interactions involve the exchange of energy from one trophic level to another (e.g., Hodkinson 1973). In several studies of plant–herbivore interactions, Gutierrez and colleagues have used a surrogate of energy flow—change in dry weight—as a way of linking the dynamic changes within plants and between plants and herbivorous insect feeding on them and, as well, the other components in the life system of the plant (Gutierrez et al. 1980).

The Plant Population. The life system approach can equally be used to consider plant population dynamics. Plants do differ from insects, however, in that while their populations also consist of numbers of individuals of different ages, each individual plant has a varying number of plant *parts* (root and stem branches, leaves, flowers, and fruits) which also show age differences. Germination of new plants gives a birth rate, while the growth rates of the plant parts are expressed as assimilation of carbohydrate and other materials into their structure, that is, by increase of their dry weight. The quantity of carbohydrate utilizable for growth during a time interval is the amount of photosynthate produced plus any available reserves minus the amount needed for metabolism. The ratio of carbohydrate supply and demand determines the initiation, growth, and survival of the plant parts, that is, their number and the proportion of the dry-weight increment involved.

The supply of carbohydrate is modified by intrinsic factors such as the number of leaves and their age distribution and available reserves and extrinsically through features of the effective environment such as light intensity and quality, available water and nitrogen, and many other factors. The relative growth rates of the plant parts depends upon the species-specific (even biotype-specific) priority scheme under the ambient environmental conditions. The rates of development and other functions, and the "life span" of the plant parts are closely related to ambient temperature in the same way as with insects (Fig. 11.1), but temperature thresholds for development of plants are usually lower than those for herbivorous insects which feed on them—a natural adaptive feature for both organisms.

The plant's carbohydrate reserves clearly play an important role, maintaining metabolism during periods of little or no photosynthesis. Massive use of reserves takes place during regrowth after severe defoliation (by harvest or intense insect attack) or after a prolonged dormant period when carbohydrate demand exceeds rate of supply from the leaves. The proportion of dry weight constituting available reserves varies with the age and prior history and characteristically also with the type of plant and its genetic makeup. Perennial or biennial plants which normally have long periods of no photosynthesis build up large reserves. Annuals and continuously photosynthetic perennial plants rarely have high carbohydrate reserves. Plants die if all their reserves are used up before the carbohydrate supply exceeds metabolic demand. Thus, annuals and evergreen plants (nondormancy adapted) are more likely to die from severe defoliation than are the deciduous or other dormancy-adapted biennials or perennials.

Models of plant life systems have been developed in ways similar to those of insect life systems. As with the latter, simple plant models which ignore the age structure of the plants and their parts have been proposed (e.g., Fick 1975). Recently, comprehensive models including age structure have been developed (e.g., Gutierrez et al. 1975, Gutierrez & Wang 1977, Wang et al. 1977) (cf. Chapter 24).

11.5.3.1 The Herbivore Population. Gutierrez and Wang (1977) showed that plant–herbivore trophic interactions present most of the attributes of prey–predator trophic interactions, and that their population dynamics can be considered in similar ways. Although the plant life system is more complicated by the separate phases and dynamics of the plant parts, it does not have the complexities caused by movement. The herbivore population can, of course, be modeled in the way described above for insect life systems. However, net immigration into a new crop may have to be incorporated and the complexities of behavior built into the simulation of food relationships, as was done for a predator by Frazer and Gilbert (1976). Growth and survival of herbivore (and predator) populations may also be determined by supply and demand considerations (Gutierrez et al. 1981) like those described for plants. With mobile herbivorous insects—for example, plants or plant parts must be found and consumed (searching success) at a rate exceeding the herbivore's metabolic demand—if any excess is to be available for growth of individuals and reproduction. Unless they have a specially adapted nonfeeding stage, few insects have much food reserve and can rarely survive more than a short period without finding and consuming food. Hunger can greatly alter an insect's behavior and thus its food searching success (cf. Chapters 4, 16, 24).

11.5.3.2   The Plant–Herbivore Interaction.   Insectan herbivores generally exhibit varying degrees of host specificity and/or preference for certain plant parts. An insect may specialize as a leaf feeder or gall former on a closely related group of plant species (or a single species), whereas another may eat stems, leaves, flowers, and fruits from a range of host plant species. Some insects restrict attack to a certain age class of plant parts, for example, senescing leaves by *Myzus persicae* (Kennedy 1958). Some insects change their feeding preferences as they themselves age, develop, or increase in size (cf. Chapter 4).

To understand a plant–herbivore insect interaction, the population dynamics of the insect must be realistically coupled to the dynamics of the plant via the appropriate plant part(s). The insect will reduce the quantity of plant material and may alter the age structure of the plant population or the relative numbers of plant parts. Feeding on the plant may affect photosynthate supply (e.g., a sap feeder) or demand (e.g., a fruit consumer) or, of course, both (i.e., a defoliator). Because the ratio of photosynthate supply to demand is likely to be altered, a different pattern of plant growth may develop. As preferred plant parts are depleted or otherwise become relatively uncommon, the herbivore's success rate and/or its behavior may be greatly changed. It may then accept less preferred plant parts or it may emigrate. Hence its birth, death, and net immigration rates may also be greatly affected. Thus, we use the linkage of the two life systems at the energy flow level, the feature of major consequence for the dynamic interactions observed in the field. Essential to modeling a plant–herbivore interface are adequate models of the two life systems and realistic coupling of the two. For adequate understanding of the interaction, modeling the details of the life systems peripheral to their interface may not be necessary.

11.6 POTENTIAL APPLICATIONS

The value of the life system approach must ultimately be judged by its usefulness in either advancing our understanding of ecological principles and relationships or in its application to specific practical problems. With reference to theoretical contributions, first, and most importantly, the approach allows us to evaluate the frequency and significance of various ecological processes in real systems (including evolutionary adaptation) and hence to direct us to areas needing more, or justifying less, attention. We do not choose to expend a lot of effort on phenomena we consider of little consequence in the field. Second, life system studies, because each attempts a comprehensive overview of the entire ecology of an insect species provide, in time, a set of case studies against which the current interest in life history theory—for example, r and k selection—and in general the selective processes molding developmental and fecundity patterns can be considered.

With reference to crop or animal management, realistic life system models provide a sound basis for attempting to answer questions concerning specific situations (cf. Conway 1977, and Chapters 24, 25). In their simplest form, life system models may be used to predict population phenology or to provide insights concerning existent or probable future population densities or age structures. Coupled crop plant–pest models may provide realistic assessments of the damage potential of the pest. Caution must, however, be exercized because the model cannot predict all possible outcomes, and large model-generated disasters (e.g., loss of a valuable crop) may result. Large discrepancies between a model's predictions and observed trends commonly occur either because an important factor has not been included or the relationships have been formulated improperly.

The best use of life system models in management is as an aid to understanding (Regev et al. 1976, Shoemaker 1977), rather than as predictors of quantitative outcomes. For example, a detailed model of cotton growth and development provided critical insights for designing experiments to estimate the true rate of damage to cotton fruits caused by *Lygus* bugs (Gutierrez, Leigh et al. 1977). The horizon for living models developed from realistic, life system studies broadens daily.

REFERENCES

Andrewartha, G. G. and L. C. Birch. 1954. *The Distribution and Abundance of Animals*. University of Chicago Press, Chicago.

Auer, C. 1968. *Zeit. angew. Entomol.* **62:** 202–235.

Bailey, C. G. 1976. *Can. Entomol.* **108:** 1339–1344.

Baker, R. J., A. Mayer, and C. F. Cohen. 1963. *Ann. Entomol. Soc. Am.* **56:** 292–294.

Baker, R. R. 1968. *Phil. Trans. Roy. Soc. Lond. B,* **253:** 309–341.

Baker, R. R. 1970. *J. Zool.* **162:** 43–59.

Beaver, R. A. 1966. *J. Anim. Ecol.* **35:** 27–41.

Beaver, R. A. 1967. *J. Anim. Ecol.* **36:** 435–451.

Beaver, R. A. 1974. *J. Anim. Ecol.* **43:** 239–253.

Beddington, J. R. 1975. *J. Anim. Ecol.* **44:** 331–340.

Beddington, J. R., M. P. Hassell, and J. H. Lawton. 1976. *J. Anim. Ecol.* **45:** 165–185.

Benois, A. 1972. *Ann. Zool. Ecol. Anim.* **4:** 325–351.

Biever, K. D., D. I. Hostetter, and P. E. Boldt. 1972. *Environ. Entomol.* **1:** 440–443.

Bishop, J. L. and D. Crockett. 1961. *Va. Agric. Exp. Sta. Tech. Bull.* 153.

Blackman, R. L. 1976. *Phil. Trans. Roy. Soc. Lond. (B)* **274:** 437–488.

Bottrell, D. G. and C. B. Huffaker (Eds.). 1974. *Proceedings IPM Workshop on Evaluation of the Role of Predators in Crop Ecosystems.* Internat'l Center for Biological Control, University California, Berkeley. Mimeo.

Broadhead, E. and R. A. Cheke. 1975. *J. Anim. Ecol.* **44:** 767–793.

Campbell, A., B. D. Frazer, N. Gilbert, A. P. Gutierrez, and M. Mackauer. 1974. *J. Appl. Ecol.* **11:** 431–438.

Carne, P. B. 1969. *Aust. J. Zool.* **17:** 113–141.

Caughley, G. 1976. Pp. 94–113, *in* May (Ed.), 1976, referenced here.

Clark, L. R., P. W. Geier, R. D. Hughes, and R. F. Morris. 1967. *The Ecology of Insect Populations in Theory and Practice.* Methuen, London.

Clark, W. C., D. D. Jones, and C. S. Holling. 1978a. *Ecol. Modeling* **7:** 1–53.

Clark, W. C., D. D. Jones, and C. S. Holling. 1978b. Pp. 385–432 in J. S. Steel (Ed.), *Spatial Patterns in Plankton Communities.* Plenum, New York.

Conway, G. R. 1977. Pp. 177–200 in Norton and Holling (Eds.), 1977, referenced here.

Cuff, W. R. and J. M. Hardman. 1979. *Ecol. Modeling* **9:** 281–305.

Davidson, J. and H. G. Andrewartha. 1948. *J. Anim. Ecol.* **17:** 193–222.

Dempster, J. P. 1967. *J. Appl. Ecol.* **4:** 485–500.

Dempster, J. P. 1969. *J. Appl. Ecol.* **6:** 339–345.

Dempster, J. P. 1971. *Oecologia* **7:** 26–67.

Dingle, H. (Ed.). 1978. *Evolution of Insect Migration and Diapause.* Springer-Verlag, New York.

Doutt, R. L. 1964. Pp. 145–167 in DeBach (Ed.), 1964, referenced here.

Eickwort, K. R. 1977. *Ecology* **58:** 527–538.

Evans, G. C. 1976. *J. Anim. Ecol.* **45:** 1–39.

Farrell, J. A. K. 1976. *Bull. Entomol. Res.* **66:** 317–329.

Feeny, P. P. 1970. *Ecology* **51:** 565–581.

Fick, G. W. 1975. *ALSIM 1 (Level 1). User's Manual.* Cornell University, Department of Agronomy, Mimeo 75-20.

Finney, D. J. 1971. *Probit Analysis* (3rd. ed.). Cambridge University Press, Cambridge.

Fisher, R. A. 1930. *The Genetical Theory of Natural Selection.* Clarendon, Oxford.

Fleschner, C. A. 1950. *Hilgardia* **20:** 233–265.

Frazer, B. D. and N. Gilbert. 1976. *J. Entomol. Soc. Br. Columbia* **73:** 33–56.

Freeman, B. E. 1977. *J. Anim. Ecol.* **46:** 231–247.

Freeman, B. E. and D. B. Jayasingh. 1975. *Oikos* **26:** 86–91.

Funakoshi, K. 1977. *Jap. J. Ecol.* **27:** 125–148.

Gage, S. H., M. K. Mukerji, and R. L. Randell. 1976. *Can. Entomol.* **108:** 245–253.

Geier, P. W. 1964. *Aust. J. Zool.* **12:** 381–416.

Gilbert, L. E. and M. C. Singer. 1973. *Am. Nat.* **107:** 58–72.

Gilbert, N. and A. P. Gutierrez. 1973. *J. Anim. Ecol.* **42:** 323–340.

Gilbert, N., A. P. Gutierrez, B. D. Frazer, and R. E. Jones. 1976. *Ecological Relationships.* Freeman & Reading, San Francisco.

Gilbert, N. and R. D. Hughes. 1971. *J. Anim. Ecol.* **40:** 525–534.

Gossard, T. W. and R. E. Jones. 1977. *J. Appl. Ecol.* **14:** 65–71.

Greaves, T. and R. D. Hughes. 1974. *J. Aust. Entomol. Soc.* **13:** 329–351.

Greenham, P. M. 1972. *J. Anim. Ecol.* **41:** 153–165.

Greenslade, P. J. M. 1975. *Aust. J. Zool.* **23:** 511–522.

Griffiths, K. J. and C. S. Holling. 1969. *Can. Entomol.* **101:** 785–818.

Grimble, D. G. and F. B. Knight. 1970. *Ann. Entomol. Soc. Am.* **63:** 1309–1319.

Gutierrez, A. P. and J. U. Baumgaertner. In Press. *Can. Entomol.* **116:**

Gutierrez, A. P., J. U. Baumgaertner, and K. S. Hagen. 1981. *Can. Entomol.* **113:** 21–33.

Gutierrez, A. P., G. D. Butler, Y. Wang, and D. Westphal. 1977. *Can. Entomol.* **109:** 1475–1480.

Gutierrez, A. P., J. B. Christensen, C. M. Merritt, W. B. Loew, C. G. Summers, and W. R. Cochran. 1976. *Can. Entomol.* **108:** 635–648.

Gutierrez, A. P., D. W. DeMichele, Y. Wang, G. L. Cury, R. Skeith, and L. G. Brown. 1980. Pp. 155–186, *in* Huffaker (Ed.), 1980, referenced here.

Gutierrez, A. P., W. H. Denton, R. Shade, H. Maltby, T. Burger, and G. Moorhead. 1974. *J. Anim. Ecol.* **43:** 627–640.

Gutierrez, A. P., L. A. Falcon, W. Loew, P. A. Leipzig, and R. van den Bosch. 1975. *Environ. Entomol.* **4:** 125–136.

Gutierrez, A. P., D. E. Havenstein, N. A. Nix, and P. A. Moore. 1974. *J. Appl. Ecol.* **11:** 1–20.

Gutierrez, A. P., T. F. Leigh, Y. Wang, and R. D. Cave. 1977. *Can. Entomol.* **109:** 1375–1386.

Gutierrez, A. P., D. J. Morgan, and D. E. Havenstein. 1971. *J. Appl. Ecol.* **8:** 699–721.

Gutierrez, A. P., H. A. Nix, D. E. Havenstein, and P. A. Moore. 1974. *J. Appl. Ecol.* **11:** 21–35.

Gutierrez, A. P. and Y. Wang. 1977. Pp. 255–280, *in* Norton and Holling (Eds.), 1977, referenced here.

Gutierrez, A. P., Y. Wang, and R. Daxl. 1979. *Can. Entomol.* **111:** 357–366.

Hagan, K. S., E. F. Swall, Jr., and R. L. Tassan. 1970. *Proc. Tall Timbers Conf. on Ecol. Anim. Control by Habitat Manag.* **2:** 59–81.

Harcourt, D. G. 1961. *Can. Entomol.* **93:** 945–952.

Harcourt, D. G. 1963. *Proc. Entomol. Soc. Ont.* **93:** 61–75.

Harcourt, D. G. 1969. *Annu. Rev. Entomol.* **14:** 175–196.

Harcourt, D. G. 1971. *Can. Entomol.* **103:** 1049–1061.

Hardman, J. M. 1976a. *Can. Entomol.* **108:** 907–924.

Hardman, J. M. 1976b. *Can. Entomol.* **108:** 897–906.

Hassell, M. P. 1971. *J. Anim. Ecol.* **40:** 473–486.

Hassell, M. P. 1976. *The Dynamics of Competition and Predation.* Inst. of Biology, No. 72, Edward Arnold, London.

Hassell, M. P. 1978. *The Dynamica of Arthropod Predator–Prey Systems.* Princeton University Press, Princeton, New Jersey.

Hassell, M. P. and C. B. Huffaker. 1969. *Res. Popul. Ecol.* **11:** 186–210.

Hassell, M. P., J. H. Lawton, and J. R. Beddington. 1976. *J. Anim. Ecol.* **45:** 135–164.

Hassell, M. P., J. H. Lawton, and J. R. Beddington. 1977. *J. Anim. Ecol.* **46:** 249–262.

Hassell, M. P. and R. M. May. 1974. *J. Anim. Ecol.* **43:** 567–594.

Hassell, M. P. and G. C. Varley. 1969 *Nature (London)* **223:** 1133–1137.

Hodek, I., K. S. Hagen, and H. F. van Emden. 1972. Pp. 344, *in* H. F. van Emden (Ed.), *Aphid Technology.* Academic, New York.

Hodkinson, I. D. 1973. *J. Anim. Ecol.* **42:** 565–584.

Holdoway, F. G. 1932. *Ecol. Monogr.* **2:** 261–304.

Holling, C. S. 1959. *Can. Entomol.* **91:** 395–398.

Holling, C. S. 1961. *Annu. Rev. Entomol.* **6:** 163–182.

Holling, C. S. 1963. *Mem. Entomol. Soc. Can.* **32:** 22–32.

Holling, C. S. 1965. *Mem. Entomol. Soc. Can.* **45:** 1–60.

Holling, C. S. 1966. *Mem. Entomol. Soc. Can.* **48:** 1–86.

Howe, R. W. 1967. *Annu. Rev. Entomol.* **12:** 15–42.

Huffaker, C. B. 1944. *Ann. Entomol. Soc. Am.* **37:** 1–27.

Huffaker, C. B. (Ed.). 1971. *Biological Control.* Plenum, New York.

Huffaker, C. B. (Ed.). 1980. *New Technology of Pest Control.* Wiley, New York.

Hughes, R. D. 1962. *J. Anim. Ecol.* **31:** 389–396.

Hughes, R. D. 1963. *J. Anim. Ecol.* **32:** 393–424.

Hughes, R. D. 1970. *J. Anim. Ecol.* **39:** 691–706.

Hughes, R. D. 1973. *Entomol. Soc. New Zealand Bull.* **2:** 85–91.

Hughes, R. D. 1977a. *Aust. J. Ecol.* **3:** 43–54.

Hughes, R. D. 1977b. *J. Aust. Entomol. Soc.* **16:** 117–122.

Hughes, R. D. and N. Gilbert. 1968. *J. Anim. Ecol.* **37:** 553–563.

Hughes, R. D., P. M. Greenham, M. Tyndale-Biscoe, and J. M. Walker. 1972. *J. Aust. Entomol. Soc.* **11:** 311–331.

Hughes, R. D. and B. M. Mitchell. 1960. *J. Anim. Ecol.* **29:** 359–374.

Hughes, R. D. and W. L. Nicholas. 1974. *J. Anim. Ecol.* **43:** 411–428.

Hughes, R. D. and P. Sands. 1979. *J. Appl. Ecol.* **16:** 117–139.

Hughes, R. D., M. Tyndale-Biscoe, and J. M. Walker. 1978. *Bull. Entomol. Res.* **68:** 361–372.

Hughes, R. D. and J. M. Walker. 1970. Pp. 255–269, *in* Watson (Ed.), 1970, referenced here.

Hughes, R. D. and L. T. Woolcock. 1976. *J. Aust. Entomol. Soc.* **15:** 191–196.

Hughes, R. D. and L. T. Woolcock. 1978. *J. Appl. Ecol.* **15:** 145–154.

Hughes, R. D., L. T. Woolcock, and P. Ferrar. 1974. *J. Appl. Ecol.* **11:** 483–488.

Ito, Y., A. Shibazaki, and O. Iwahashi. 1969. *Res. Pop. Ecol.* **11:** 211–228.

Ives, P. M. 1978. *Aust. J. Ecol.* **3:** 261–276.

Ives, P. M. 1981. *Can. Entomol.* **113:** 981–997.

Jones, R. E. 1974. *Aust. J. Zool.* **22:** 71–89.

Jones, R. E. 1977a. *Behavior* **60:** 237–259.

Jones, R. E. 1977b. *J. Anim. Ecol.* **26:** 195–212.

Jones, R. E. and P. M. Ives. 1979. *Aust. J. Ecol.* **4:** 75–86.

Jones, R. E. and J. M. Walker, 1974. *Entomol. Expt. & Appl.* **17:** 117–125.

Kay, B. H. 1972. *J. Aust. Entomol. Soc.* **12:** 42–58.

Kennedy, J. S. 1958. *Entomol. Expt. & Appl.* **1:** 50–65.

Khare, B. P. and N. S. Agrawal. 1970. *Beitr. Entomol.* **20:** 183–188.

Kiritani, K. 1971. *Proc. Symp. Rice Insects.* 1971. Ministry of Agric. & Forestry, Tokyo, Japan.

Kiritani, K. and J. P. Dempster. 1973. *J. Appl. Ecol.* **10:** 323–330.

Kiritani, K. and F. Nagasuji. 1967. *Res. Popul. Ecol.* **9:** 143–152.

Kitching, R. L. 1977. *Aust. J. Ecol.* **2:** 31–42.

Klimetzek, D. 1970. *Zeit. ang. Entomol.* **66:** 84–95.

Klomp, H. 1968. Pp. 98–105, *in* Southwood (Ed.), 1968, referenced here.

Labeyrie, V. 1978. *Annu. Rev. Entomol.* **23:** 69–90.

Lack, D. 1966. *Population Studies of Birds.* Clarendon, Oxford.

Leslie, P. H. 1945. *Biometrika* **33:** 183–212.

Lotka, A. J. 1925. *Elements of Physical Biology.* Williams & Williams, Baltimore.

Luck, R. F. 1971. *Can. Entomol.* **103:** 1261–1271.

McClure, M. S. 1977. *Environ. Entomol.* **7:** 219–227.

MacKay, P. A. and W. G. Wellington, 1975. *Can. Entomol.* **107:** 1161–1166.

MacLellan, C. R. 1977. *Can. Entomol.* **109:** 1555–1562.

MacLellan, C. R. 1978. *Can. Entomol.* **110:** 91–100.

Madsen, B. L. and I. Butz. 1976a. *Oikos* **27:** 273–280.

Madsen, B. L. and I. Butz. 1976b. *Oikos* **27:** 281–287.

Manly, B. F. J. 1974. *Oecologia* **17:** 335–348.

Manly, B. F. J. 1976. *Res. Pop. Ecol.* **17:** 191–199.

Manly, B. F. J. 1977. *Res. Pop. Ecol.* **18:** 177–186.

Marshall, A. G. 1971. *J. Anim. Ecol.* **40:** 141–154.

May, R. M. 1973. *Stability and Complexity in Model Ecosystems.* Princeton University Press, Princeton, New Jersey.

May, R. M. (Ed.). 1976. *Theoretical Ecology, Principles and Application.* Blackwell, Oxford.

Meats, A. 1974a. *Oecologia* **16:** 119–138.

Meats, A. 1974b. *Oecologia* **16:** 139–147.

Meats, A. 1977. *Oecologia* **19:** 117–128.

Messenger, P. S. 1964. *Ecology* **45:** 119–131.

Miles, G. E., R. J. Bula, D. A. Holt, M. M. Schreiber, and R. M. Peart. 1973. *Simulation of Alfalfa Growth.* Am. Soc. Agric. Eng. Pap. 73-4547.

Morris, R. F. 1959. *Ecology* **40:** 580–588.

Morris, R. F. 1971. *Can. Entomol.* **103:** 893–906.

Morris, R. F. 1976. *Can. Entomol.* **108:** 1291–1294.

Morris, R. F. and C. W. Bennett. 1967. *Can. Entomol.* **99:** 9–17.

Murdoch, W. A. 1969. *Ecol. Monogr.* **39:** 335–354.

Nakamura, K. 1976a. *Jap. J. Ecol.* **26:** 49–59.

Nakamura, K. 1976b. *Jap. J. Ecol.* **26:** 125–134.

Nealis, V. G. 1977. *Can. J. Zool.* **55:** 138–147.

Neuenschwander, P., K. S. Hagen, and R. F. Smith. 1975. *Hilgardia* **43:** 53–78.

Nicholson, A. J. 1954. *Aust. J. Zool.* **2:** 9–65.

Nicholson, A. J. 1957. *Cold Spring Harb. Symp. Quant. Biol.* **22:** 153–172.

Nicholson, A. J. and V. A. Bailey. 1935. *Part I, Proc. Zool. Soc. Lond.,* pp. 551–598.

Norris, K. R. 1966. *Aust. J. Zool.* **14:** 1139–1156.

Norton, G. A. and C. S. Holling (Eds.). 1977. *Proceedings of a Conference on Pest Management,* Oct. 25–29, 1976. Laxenburg, Austria. International Institute of Applied Systems Analysis.

Obrycki, J. J. and M. J. Tauber. 1978. *Can. Entomol.* **110:** 402–412.

Parker, F. D. 1970. *Ann. Entomol. Soc. Am.* **63:** 985–994.

Peterman, R. M., W. C. Clark, and C. S. Holling. 1979. Pp. 321–341 in R. M. Anderson, B. D. Turner, and L. R. Taylor (Eds.), *Population Dynamics.* The 20th Symposium of the British Entomological Society, London.

Peters, T. M. and P. Barbosa. 1977. *Annu. Rev. Entomol.* **22:** 431–450.

Pim, S. L. and J. H. Lawton. 1977. *Nature* **268:** 329–331.

Podoler, H. and D. Rogers. 1975. *J. Anim. Ecol.* **44:** 85–114.

Price, P. W. 1971. *Ann. Entomol. Soc. Am.* **64:** 1399–1406.

Readshaw, J. L. 1975. *J. Appl. Ecol.* **12:** 473–495.

Regev. U., A. P. Gutierrez, and G. Feder. 1976. *Am. J. Agric. Econ.* **58:** 186–197.

Richards, O. W. 1940. *J. Anim. Ecol.* **9:** 243–288.

Richards, O. W. 1961. *Annu. Rev. Entomol.* **6:** 147–162.

Ricou, G. 1967. *Ann. Epiphytes* **18:** 451–481.

Rogers, D. J. 1972. *J. Anim. Ecol.* **41:** 369–383.

Rogers, D. J. and M. P. Hassell. 1974. *J. Anim. Ecol.* **43:** 239–253.

Royama, T. 1971. *Res. Popul. Ecol.* (Suppl.) **1:** 1–91.

Ruhm, W. 1970. *Zeit. ang. Entomol.* **63:** 212–227.

Sands, P. and R. D. Hughes. 1976. *Agric. Meterol.* **17:** 161–183.

Sands, P. and R. D. Hughes. 1977. *Bull. Entomol. Res.* **67:** 675–683.

Scopes, N. E. A. and S. B. Biggerstaff. 1977. *J. Appl. Ecol.* **14:** 799–802.

Sharpe, P. J. H., G. L. Curry, D. W. DeMichele, and C. L. Cole. 1977. *J. Theor. Biol.* **66:** 21–38.

Shiga, M. 1977. *Res. Popul. Ecol.* **18:** 284–301.

Shoemaker, C. A. 1977. Pp. 301–315, *in* Norton and Holling (Eds.), 1977, referenced here.

Sinko, J. W. and W. Streifer. 1967. *Ecology* **50:** 910–918.

Smith, R. H. 1973. *J. Anim. Ecol.* **42:** 611–622.

Solomon, M. E. 1949. *J. Anim. Ecol.* **18:** 1–35.

Southwood, T. R. E. 1966. *Ecological Methods.* Methuen, London.

Southwood, T. R. E. 1967. *J. Anim. Ecol.* **36:** 519–529.

Southwood, T. R. E. (Ed.). 1968. *Insect Abundance.* Blackwell, Oxford.

Southwood, T. R. E. 1978. *Ecological Methods* (2nd ed.). Methuen, London.

Southwood, T. R. E., G. Murdie, M. Yasuno, R. J. Tonn, and P. M. Reader. 1972. *Bull. World Health. Organ.* **46:** 211–226.

Springett, B. P. 1968. *J. Anim. Ecol.* **37:** 417–424.

Stinner, R. E., A. P. Gutierrez, and G. D. Butler. 1974. *Can. Entomol.* **106:** 519–524.

Stinner, R. E., G. D. Butler, Jr., J. S. Bacheler, and C. Tuttle. 1975. *Can. Entomol.* **107:** 1167.

Stoffolano, J. G. 1970. *J. Nematol.* **2:** 324–329.

Stortenbeker, C. W. 1967. *Agr. Res. Rept.* ITBON Med., Wageningen.

Tansley, A. G. 1935. *Ecology* **16:** 284–307.

Taylor, R. G. and H. G. Harcourt. 1978. *Can. Entomol.* **110:** 57–62.

Thompson, W. R. 1924. *Annals Fac. Sci. Marseilles* **2:** 69–89.

Tilman, D. 1978. *Ecology* **59:** 686–692.

Tummala, R. L., W. G. Ruesink, and D. L. Haynes. 1975. *Environ. Entomol.* **4:** 175–186.

Turnbull, A. L. 1962. *Can. Entomol.* **94:** 1233–1249.

Turnbull, A. L. 1964. *Can. Entomol.* **96:** 568–579.

Tyndall-Biscoe, M. and R. D. Hughes. 1969. *Bull. Entomol. Res.* **59:** 129–141.

Ullyett, G. C. 1950. *Phil. Trans. Roy. Soc. Lond. B,* **234:** 77–174.

Varley, G. C. and G. R. Gradwell. 1960. *J. Anim. Ecol.* **29:** 399–401.

Volterra, V. 1926. *Mem. Acad. Lincei* **2:** 31–113.

Waage, J. K. 1978. *Physiol. Entomol.* **3:** 135–146.

Wang, Y. and A. P. Gutierrez. 1980. *J. Anim. Ecol.* **49:** 435–452.

Wang, Y., A. P. Gutierrez, G. Oster, and R. Daxl. 1977. *Can. Entomol.* **109:** 1359–1374.

Waterhouse, D. F. and K. R. Norris. 1966. *Aust. J. Sci.* **28:** 351.

Watson, A. (Ed.). 1970. *Animal Populations in Relation to Their Food Resources.* Blackwell, Oxford.

Wellington, W. G. 1957. *Can. J. Zool.* **35:** 293–323.

Wellington, W. G. 1960. *Can. J. Zool.* **38:** 289–314.

Wellington, W. G. 1964. *Can. Entomol.* **96:** 436–451.

Wellington, W. G. 1965. *Can. Entomol.* **97:** 1–14.

Wellington, W. G., P. J. Cameron, W. A. Thompson, I. B. Vertinsky, and A. S. Landsberg. 1975. *Res. Pop. Ecol.* **17:** 1–28.

White, T. C. R. 1969. *Ecology* **50:** 905–909.

White, T. C. R. 1970. *Aust. J. Zool.* **18:** 105–117.

White, T. C. R. 1978. *Oecologia* **33:** 71–86.

Whittaker, J. B. 1965. *J. Anim. Ecol.* **34:** 277–297.

Whittaker, J. B. 1971. *J. Anim. Ecol.* **40:** 425–443.

Whittaker, J. B. 1973. *J. Anim. Ecol.* **42:** 163–172.

Williams, J. R. 1970. *Bull. Entomol. Res.* **60:** 61–95.

Woodburn, T. L., W. G. Vogt, and R. L. Kitching. 1978. *Bull. Entomol. Res.* **68:** 251–261.

Wylie, H. G. 1958. *Can. Entomol.* **90:** 957–608.

Chapter 12

Natural Control

of Insect

Populations

C. B. HUFFAKER, ALAN A.
BERRYMAN, and J. E. LAING

A *struggle for existence inevitably follows from the high rate at which all organic beings tend to increase. . . . There is no exception to the rule that every organic being naturally increases at so high a rate, that if not destroyed the earth would soon be covered by the progeny of a single pair. . . . Battle within battle must ever be recurring with varying success; and yet in the long run the forces are so nicely balanced, that the face of nature remains uniform for longer periods of time. . . .*

(CHARLES DARWIN 1859)

12.1 SCOPE AND PERSPECTIVE

This chapter is concerned with restrictive processes that act on populations to prevent unlimited growth as described in general terms by Darwin. However, the concept of natural control must accommodate all the changes, increases as well as decreases, in numbers and quality exhibited by natural populations. Populations may be regulated by single or multiple causes and the factors involved in limiting growth may change in time and space, particularly those buffeted by extreme fluctuations in the physical environment. For some populations density-dependent regulating factors may be engaged only for limited periods of time, while in others occupying more benign environments these factors may be engaged most of the time.

Recently there has been a tendency to look more closely at the role of natural selection acting on individual performance, and at the interplay between the resulting genetic changes and the functional dynamics and density-regulating mechanisms of natural control [e.g., den Boer & Gradwell 1971, Pimentel p. 366, Bakker 1980, Itô 1972, 1980]. Environments have been classified according to their stability and the kinds of species occupying them (e.g., Southwood 1977a). Nicholson (1957) and Pimentel showed that organisms may evolve under selection pressure from *environmental* or *competitive* interactions. Yet neither Nicholson nor Pimentel (in his later view) held that adaptive evolution can substitute, day by day and generation by generation, for conventional, *functional* regulation through density-dependent negative feedback.

Study of the natural regulation of populations has been a major focus of animal ecology for some decades. In the case of insects there was a period when emphasis was placed on the effects of temperature and moisture (physical factors). This was followed by interest in competition between members of a population and on parasitism and predation (biotic factors). In most cases the emphasis was on the day-by-day and generation-by-generation *functional* dynamics of the population. Little attention was given to evolutionary aspects (genetic dynamics).

In this chapter we stress the proximal functioning of the regulating process and its modification by the conditioning forces of the environment (the physical setting). Less emphasis is placed on the evolutionary aspects of populations, or of their associated allies, enemies, competitors, or food organisms. Nevertheless, we specifically recognize the importance of genetic feedback in the long-term stability of populations (Fig. 12.1). In this illustration the arrows not only indicate the interaction between functional elements of a population system, but also the pressures that selection exerts to cause changes in genetic properties of organisms—that is, changes in fecundity, escape behavior, food capture, and mutualistic or competitive interactions—which may give rise to coevolutionary processes. The right side of Fig. 12.1 illustrates significant evolutionary and/or geophysical changes that could, in time, completely alter the dynamics so that the equilibrium level and dynamic behavior of a population becomes quite different (e.g., it could become extinct).

"Natural control" implies limitation as a key feature and stresses the suppressive properties of population-regulating factors and *characteristic densities* that populations commonly exhibit. Thus, we emphasize population density and the stabilizing forces rather than population dynamics *per se*. Methods for assessing changes in density and detecting the immediate factors causing such changes are covered in Chapter 11. Various statistical or descriptive accounts of insect population dynamics are also covered in Chapters 1, 13, 14, 23–25, and so forth.

12.1.1 Perspectives in Viewing Populations

We define a *natural population* as an adapted and adaptable group of organisms of the same species occupying a natural area of sufficient size that reproduction and survival is sufficient to maintain the population for many generations and to permit normal dispersive and migratory behaviors.

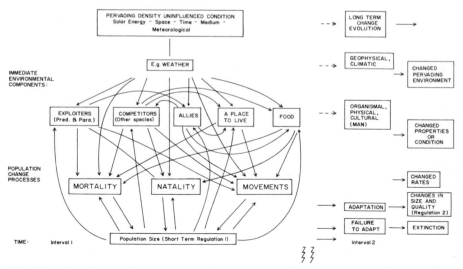

Fig. 12.1 Functional and evolutionary influences among factors acting on, and in interaction with, a population over time—arrows indicate paths of influence.

Natural control is the maintenance of numbers (or biomass) of a natural population over many generations within a restricted range (characteristic density) by the combined effects of all factors and processes of the environment, including the population's own members and, in the strict sense, necessarily including a density-dependent (negative-feedback) feature. To support this contention, we note the consistency in the places where species occur, in the size of their populations relative to other species, and in the range of variation in their densities (e.g., Doutt & DeBach 1964, MacFadyen 1963, Odum 1971). As Odum wrote, "A large area of forest might show an average of 10 birds per hectare and 20,000 soil arthropods per square meter, but there would never be as many as 20,000 birds per square meter or as few as 10 arthropods per hectare."

Much of the controversy in population ecology can be traced to vagueness in specifying the spatial and temporal dimensions of populations and to describing adequately the environments they occupy. Changes in quality of a population also become more important as the time dimension increases. In short-term studies (a season or so) changes in population size can be described without considering either changes in quality or regulating processes, but this view is obviously incomplete. The time span should be long enough to allow the population and the various factors which affect it to express their characteristic ranges of variation. Ideally, the time frame should be long enough to allow for short-term genetic responses, including coevolution with other important species (Chapter 9). Unfortunately, this is seldom practical.

12.1.2 The Life System View

By placing a population and its functional environment in a space–time frame (Chapter 11), we view it as a component of nature, with the realization that all components are linked through pathways of energy flow and material cycling (Chapter 2). Thus, a change in one component may reverberate through the system and eventually feed back to the original component. Another reality is that the components are not randomly mixed but are organized into a nested heirarchy (i.e., atoms, molecules, cells, tissues, organisms, populations, and communities). It is also implicit that each unit of each level has an environment from which essential matter and energy are obtained and in which there is space for the persuance of life's necessities and for disposal of waste products. *Time,* of course, is required for these events to occur.

12.1.3 The Evolutionary View

The evolutionary perspective requires criteria for "direction"; not to be confused with a teleologic or purposeful orientation. The criteria for direction commonly used by community ecologists are efficiency of energy transfer and stability of biomass production in the community, with secondary regard for the changing demographics or regulating processes of component species. In contrast, many population ecologists consider reproduction and survival of species (or fitness) to be the principal or ultimate criterion of evolutionary success and find demographics and regulation of stability to be of primary importance. In viewing adaptations, they look for survival strategies rather than strategies of energetics and productivity.

The focus on energetics and productivity is more holistic than that on survival, just as communities and ecosystems provide a more holistic perspective than do single-species life systems. However, the two views are complementary and should generate compatible and/or mutually heuristic concepts and theories. Survival of an organism is related to its stability in the community and this depends also on the stability of energy flow to, and from, it. Therefore, concentration on individuals and single population systems from a focus on fitness may lead to better understanding of community energetics, just as inclusion of community energetics may lead to better understanding of species fitness. Whether interest lies in the factors influencing survival of species or in the roles of the species in ecosystem energetics, it is useful to view a population and the main functional aspects of its environment as one "life system" of an hierarchically structured ecosystem (Chapters 2, 11).

12.2 CONCEPTS CONTRIBUTING TO UNDERSTANDING NATURAL CONTROL

Many authors (e.g., Solomon 1949, 1957, Cole 1957, Bodenheimer 1958, Huffaker & Messenger 1964a, Bakker 1964, Clark et al. 1967, McLaren 1971, Ricklefs 1979, Tamarin 1978) have reviewed the hypotheses advanced to explain both the changes in, and apparent long-term stability of, animal populations. In general, the basic views fall into two categories: (1) the hypothesis of regulation, according to which intrapopulation processes and relationships with natural enemies, food, shelter, and other factors intensify in their negative effects on the population as its density increases and relax in intensity as its density decreases—this gives rise to a negative feedback loop which, by definition, regulates the population at or around some characteristic equilibrium density (e.g., Malthus 1803, Verhulst 1838, Darwin 1859, Woodworth 1908, Howard & Fiske 1911, Pearl & Reed 1920, Nicholson 1933, 1954, Smith 1935, Utida 1941, Lack 1954, Ricker 1954, Solomon 1957, 1971, Huffaker & Messenger 1964b, Chitty 1960, 1967, Christian & Davis 1964, Wynne-Edwards 1962, Royama 1977, Berryman 1981); and (2) the hypothesis of nonregulation as dubbed by Solomon (1971), wherein at least some populations are said to simply vary in numbers, persisting without significant density-dependent influences, according to the shifting physical and biotic favorability and unfavorability of their environments, and shifts in their own properties (Bodenheimer 1928, Uvarov 1931, Thompson 1939, 1956, Andrewartha & Birch 1954, den Boer 1968, 1971, and Reddingius 1971). Whereas the hypothesis of regulation centers attention on factors which reduce fluctuations (i.e., tend to stabilize populations), the hypothesis of "nonregulation" focuses on factors which cause *changes* in density. The different viewpoints of the principal proponents and the historical development of the views are briefly summarized below.

1. *Malthus (1803), Nicholson (1933, 1954, 1958), Nicholson and Bailey (1935).* Malthus's concept of a "struggle for existence" forms the basis for the population regulation school of thought and, in addition, rests as a cornerstone of evolutionary biology (e.g., Darwin 1859). Nicholson, in particular, developed the concept further, proposing that intraspecific competition for resources (food and a secure place to live) is the underlying mechanism in the natural regulation of all populations. He recognized the presence of two groups of factors: (1) those which respond to population density (his density-governing factors or density-dependent factors of Smith 1935) which act to regulate density through negative feedback and (2) those which act independently of population size (density-independent factors) and determine the carrying capacity and rates of population growth or modify the level at which population regulation occurs.

2. *Thompson (e.g., 1929, 1939, 1956).* Thompson emphasized both physical and biotic factors and stressed the limitations in the adaptations of individual organisms and the heterogeneity of their environments in time and space. He developed the first generalized (algebraic) models of host–parasitoid interactions. However, these models showed that parasitoids either failed to control the host population or drove it to extinction. [It would take more complex models (e.g., Varley et al. 1973) to demonstrate stability of host–parasitoid interactions.]

3. *Andrewartha and Birch (1954).* Andrewartha and Birch considered that population changes in time and space (population dynamics) were affected, primarily, by environmental and individual variations, with the length of favorable periods being a crucial component. They recognized the presence of density-dependent factors but saw no practical reason to utilize the concept. They stressed the importance of chance in the synchronous occurrence of suitable qualities in the organisms to meet the challenge of changing conditions in their environments. They argued that, despite local extinctions, the movements of organisms lead to recolonization, so that populations persist in time and space without the action of density-dependent processes. This concept was developed further by den Boer (1968), Reddingius (1968), and Reddingius and den Boer (1970) as "spreading the risk". Andrewartha and Birch also emphasized the utility of the "rate of increase" r in describing population growth or growth potentials. They proposed the useful classification used in this chapter, di-

viding environmental factors into weather, food, a place to live, other organisms of the same kind, and organisms of other kinds.

4. *Chitty (1960, 1967).* Based largely on studies on voles and rabbits, Chitty proposed that animals have a capacity of self-regulation through their own intrinsic properties, and that changes in these properties are induced by population density. He proposed that certain types were more susceptible to weather factors, natural enemies, and other stresses, and that the direct percentage killed was a consequence of previous density-induced changes. Klomp (1966a,b) presented support for this view for an insect, the pine looper. He found that larval density affected the viability of larvae in the next generation.

5. *Milne (1957, 1962).* Milne argued that the only "perfect density-dependent factor" is intraspecific competition for resources, but that while many populations are regulated by these "perfect density-dependent factors," most are regulated by a combination of imperfect density-dependent factors (e.g., natural enemies) and density-independent factors.

6. *Huffaker 1958a, Huffaker and Messenger 1964b, Huffaker et al. 1976.* Huffaker and Messenger and their co-workers also argued that intraspecific competition is the only "completely reliable" regulating factor but recognized that it is not continuously in operation. They developed a classification of natural control factors and elaborated the differences in the roles of the density-dependent (= regulating) and density-independent (= conditioning) factors, insisting that neither, taken alone, forms a suitable basis for a theory of natural control. This argument is largely a restatement of Nicholson's views, but they acknowledged a greater role for density-disturbing factors than did Nicholson. They also emphasized the difficulty in specifying the "cause" of mortality. For example, cold may be a direct cause of death; however, the death may have been caused, indirectly, by competition for protective shelter. They emphasized that natural control could best be explained by considering *together*: (1) the environmental conditioning factors (temperature, moisture, currents, substrates, etc.) which furnish, support, or modify the necessary resources, influence rates of biological processes, and induce changes in density and biological properties independent of the population's members and the changes in these properties and (2) the regulating factors that tend to reduce fluctuations (i.e., density-dependent or negative feedback factors), which operate *in relation to* the **conditioning** factors and the population's intrinsic **properties**, the *third* factor.

7. *Wynne-Edwards (1962, 1965).* Wynne-Edwards suggested that all populations (amended later to "many") are self-regulated through intrinsic behavioral patterns which evolved by group selection. He argued that populations are not normally confronted by shortages of resources and are not normally restricted by natural enemies. Thus, behaviors such as territorial defense and social heirarchies (dominance) are likely to be the primary regulatory processes. However, other authors (e.g., Williams 1966, Ricklefs

1979) do not consider Wynne-Edwards's stipulation that the behavior must have arisen through group selection to be essential, arguing that it could arise through individual selection (cf. Chapter 17, and Wilson 1980).

8. *Pimentel 1961, 1968, Pimentel and Al-Hafidh 1965, Pimentel and Soans 1971.* Pimentel and his co-workers stressed the neglected fact that populations possess a capacity for even short-term, yet significant, genetic changes which can result in better adaptations (i.e., a better match between fecundity and the hazards faced in the environment) and that these adaptations affect the density and stability of the population. Studying host–parasitoid systems in the laboratory, they found that the system became more stable with time, suggesting that genetic feedback was taking place. They then proposed a theory of genetic feedback (coevolution) as an alternative to classical density-dependent population regulation. Later, however, they (Pimentel & Soans 1971) amended this theory, proposing that genetic feedback produces "coarse" regulation, with conventional density-dependence (density-dependent negative feedback) having the final role of "fine tuning" the regulatory process.

12.3 FACTORS INVOLVED IN NATURAL CONTROL

Population regulation may result from a single factor or a combination of factors acting concurrently or in sequence. The term *factor* is used in its dictionary sense to mean "an element that contributes to a result." We may thus speak of conditions, states, processes, energy, forces, and so on, as factors. A "process" is a special factor—sometimes a mechanism that arises from a single interaction, or a sequence of such, leading to a result.

12.3.1 Pervading Physical Conditions and Time

The pervading physical conditions of air, water, and substrate produce both direct and indirect influences on organisms. The sun, soil, basic climate, and meteorological conditions may be said to precondition or set the stage for success or failure of individuals of a population. These factors have influences that are pervading in the sense that they are necessary for, and influence the rates of change of all the other factors (Fig. 12.1). The pervading meteorological conditions determine the patterns of climate and weather (Chapter 13) over particular terrains, and this in turn affects the microweather and other conditions in the specific places where organisms live. These pervading conditions, therefore, determine the immediate factors of an organism's environment, both the essential ingredients of food and a place in which to live and the allies and enemies present. At the same time they influence the rates of population change, acting directly on natality, mortality, and migration and indirectly by modifying other environmental factors.

12.3.1.1 Time and Space. In the absolute sense, time and space are available to every population. Yet when taken in combination with weather, both the time *available* during favorable periods and *suitable* places in which to live may be very restricted (Andrewartha & Birch 1954). It is the intervals of favorability or unfavorability in the places where animals live, not time itself or absolute space, that limit or affect the rates of population growth.

Each organism that occupies a habitat provides opportunities for other organisms, adding still further to the potential diversity of the habitat (cf. niche and species packing, Chapters 2, 15). Ecosystem diversity can be partitioned (abstractly) into spatial, temporal, structural (trophic), and species diversity. The roots of ecosystem diversity lie in the pervading geophysical forces responsible for meteorological zonation, topography, differences in parent rock, soil, water, temperature, pressure, and air. Thus, one can associate broad patterns of distribution, movements, and abundance of different types of organisms, currently and over long geological periods, with broad geographic and climatic patterns (Chapters 6, 7, 10, 11, 13, 14). Yet, these patterns, as related solely to geophysical forces, are very indistinct for many organisms; they cannot be explained fully by "physical factors" alone.

12.3.1.2 Climate and Weather. Theories of climatic control of populations have been based upon the view of limiting and favorable levels of various climatic factors and upon correlations seen between the changes in weather components and the rise and fall of insect abundance. In environments exhibiting the most pronounced weather variation, insect population densities vary most spectacularly and workers in such environments have emphasized these *changes* in density. As Darwin wrote, "Climate plays an important part in determining the average numbers of a species, and periodical seasons of extreme drought or cold, I believe to be the most effective of all checks." But Darwin also accented the *indirect* actions of climate. He viewed climate, in part at least, as determining the food supply and the suitability of places to live, the objects of competition. He considered that climate acts, in the main, on the losers of competition—the less well-fed succumb to winter's cold.

However, weather may act directly in two distinct ways, each important in determining population size and stability. First, heat is the main variable which determines rates of growth, development, and reproduction of insects. Second, weather can have immediate effects on survival and thus can act either to upset or to promote stability (Chapter 11, and Section 12.3.6 and Figs. 12.1, 12.2, 12.5, 12.6, 12.13–12.16). Severe weather causes deaths, retards or arrests physiological and behavioral processes, and thereby reduces natality and changes dispersal patterns. If severe weather occurs at a time of relatively high density, it might contribute to stability. If it occurs at a period of low population, it might cause local extinction. These effects may or may not be associated in a density-dependent manner.

Density-dependent effects arise through competition for resources, but the abundance of these resources often depends upon weather and other physical factors. Weather, though varying independently of a population's density, frequently alters or modifies the favorability of the habitat and, consequently, helps determine the level of competition for favorable living places or food. Weather conditions, therefore, are intimate components of population regulation, setting the stage for competitive processes, disturbing population equilibria, and limiting the distribution and abundance of organisms in time and space.

12.3.1.3 A Place to Live. The environment supplies, or may fail to supply, suitable places for organisms to live in, grow, reproduce, escape from natural enemies, and where they are favored over competing species. Living places are often dispersed in space and may vary in quality, quantity, and distribution from time to time (Chapter 14). For instance, most insects require special sites for reproduction; for example, alkali banks for alkali bees, tree holes for certain mosquitoes, beds of dust for ant lions and weakened trees for bark beetles. Competition between individuals of a species (intraspecific) and between different species (interspecific) will often occur for these favorable sites. Competition is, therefore, the process by which these resources are allocated, and the intensity of competition is dependent upon the pervading factors, as well as the direct-acting, immediate environmental factors and the sizes of the competing populations (Fig. 12.1).

Southwood (1977a,b) has argued that the temporal stability or instability of habitats gives rise to selection pressures which result in the adaptation of populations to the particular habitat conditions. Thus, "r-selected species" are highly mobile and have high intrinsic rates of increase, being adapted as opportunists to unstable (unpredictable) habitats. On the other hand, "K-selected species" are superior competitors in more stable habitats and have lower rates of increase. Obviously, even the most opportunistic of species will, of necessity, require some competitive power against other opportunists, even if this is mainly a race to see which one can get to the habitat first (cf. Chapters 11, 13, 22).

12.3.1.4 Food. Food supply is also largely determined by the pervading conditions, and much of the foregoing discussion has parallels regarding suitable food supplies. Food is vital to every organism, and while some insects are able to delay need for it (e.g., by diapausing), food quality, quantity, and distribution in time and space, in relationship to other resources, often place limits on the density, dispersion, and variability of each species. Food may also become involved in mutual density-dependent relationships when interacting with its consumer population(s) (Chapters 4, 14–16). This subject is dealt with in regard to the influence of food on development (Chapter 3), reproduction (Chapter 5), diapause (Chapter 6), movement (Chapter 7), competition (Chapter 15), predation (Chapter 16), distributional dynamics and species composition (Chapter 14), in the context of an insect's "life system" (Chapter 11), and also in its broad ecophysiological context (Chapter 4) (cf. Fig. 12.1).

The way in which the food (or inhabitable units of the physical environment) is dispersed in the habitat can have marked consequences for the natural control process. Huffaker (1958b) and Huffaker et al. (1963), for example, demonstrated that increased heterogeneity in the spatial distribution of food and inhabitable units led to increased stability and persistence of an interaction between phytophagous and predaceous mites—for example, from approximately a week for the simplest system to 490 days for the most complex one; the interaction would clearly have continued well beyond this had not a disease pathogen appeared inadvertently "as a contaminant" and killed off the prey mite species. These experiments illustrate the principle that environmental heterogeneity "spreads the risk" (den Boer 1968) and reduces the chance of population extinction (cf. Chapter 14). Yet, it was apparent that competition for plant food by the prey and competition for prey between predatory mites provided the underlying regulatory mechanisms. Habitat heterogeneity, however, affected the stability of the regulatory processes.

12.3.1.5 Other Organisms of the Same Kind. In a specific environment there are three general types of relationships between members of a population, namely, cooperation, competition, and indifference. Cooperation, by definition, produces an expansive, positive feedback effect because increased density improves the chances for reproduction and survival (Berryman 1981); for example, massing together or mass activity may reduce the intensity of predation (sawfly aggregations), the effects of cold (hive warming in bees), increase the food supply (bark beetle aggregations), and increase the likelihood of mating.

Competition among individuals in a population, on the other hand, produces a regulating, negative feedback effect. Competition occurs when insufficient resources are available for all individuals (exploitation) or when individuals actively deter others from securing resources (interference).

Indifference (lack of interaction) is displayed when there is neither a positive nor negative influence between individuals. Indifference, and also cooperation and competition, may be displayed only at restricted times, places, and levels of a resource. The relative frequencies of these interactions vary among species and/or populations in space and time (cf. Chapters 11, 14, 15) and also with population density. They have profound effects on the dynamic behavior of populations, competition tending to stabilize and cooperation to destabilize population systems.

12.3.1.6 Organisms of Other Kinds. Other kinds of organisms may also affect individuals of a population in either a positive or negative way.

Organisms which benefit another species are allies or cooperators (Fig. 12.1), and the benefits may be one-sided or mutual. Mutually beneficial interaction between species produces a positive-feedback, destabilizing effect. Similarly, competition between species for common resources produces a positive feedback effect. The destabilizing effect of interspecific competition is the basic ingredient of the "competitive exclusion principle," which maintains that competition will result in the displacement of one species from the system (Gause 1934). Coexistence is only possible if the self-regulatory effect(s) of each competing species is eventually stronger than the interspecific competitive effect (Chapter 15). Interspecies competition, however, may play a role in regulating community structure and stability (Chapters 2, 22) or affect the density at which intraspecific competition operates and thus the degree of stability of each population. Coexistence of species utilizing the same resource usually occurs by sharing the resource in time and space, that is, partitioning the resource into different "niches" (the concept of species packing) (cf. Chapter 15).

Predation is an interaction between species where one benefits at the expense of the other, the prey or host; that is, the interactions between herbivores, predators, parasitoids, or pathogens and their food populations. These interactions involve different trophic levels where phytophagous insects are primary "predators" on plants, secondary "predators" (primary carnivores) feed on them, while secondary, tertiary, and even quaternary carnivores feed on corresponding higher trophic levels (Chapter 2, and Lawton 1978). Interaction between predators and their food produces a negative feedback effect, and hence these interactions are frequently important in natural control. Theoretically, secondary level carnivores (e.g., hyperparasites) may also stabilize interactions between primary carnivores and their prey (Chapter 16), and a phytophagous predator may be prevented from regulating its plant host population by the action of primary carnivores, the latter serving as allies of the plant (Fig. 12.1) (cf. Chapters 15, 16, 22).

12.3.2   The Regulating Process: Density-Dependence—Negative Feedback

In this section we concentrate on the regulating mechanisms, which by definition result from negative feedback in response to changes in population density (density-dependence). We define a *density-dependent factor* as any factor whose adverse effects on the population increase in *intensity* as population density increases and decrease in intensity as population density decreases, thus creating a negative feedback loop which tends to regulate population density.

The density-dependent regulation of population growth arises simply from the fact that as populations increase in size, at some point they automatically use up or fully occupy the things they need, or defile the places in which they live, or attract and generate, as a result of their own increase in density, elements in the environment or in the populations themselves that are detrimental to them (waste products, predators, parasites, disease, mutual interferences, or deterioration in quality).

Population regulation by density-dependent factors was expressed in general by Malthus (1803) and earlier students of human populations (cf. Cole 1957). Malthus observed that pressures against human population growth intensify as the limits of human resources are approached. Density-dependent factors, acting collectively or singly and in relation to the conditioning and genetic factors, tend to cause population density to return to a *characteristic* level. As a population is displaced farther and farther above or below its theoretical "equilibrium" density (i.e., down to the *undercrowding* position) (Allee 1947), the corrective or regulating tendency becomes stronger. This relationship is illustrated in Fig. 12.2. For an interacting prey and predator species it is illustrated in greater detail in Fig. 12.2B. Boiled down, the regulating process always arises as a result of intraspecific competition in one form or another, and this causes the population to tend toward its characteristic density following a disturbance if not too severe.

We can see that this general concept of population regulation also applies

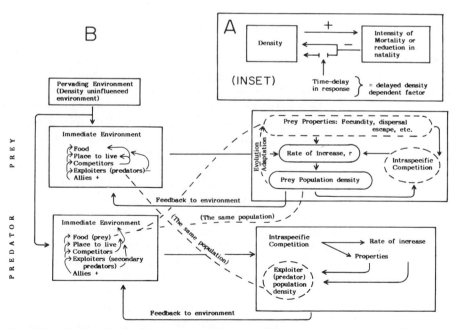

Fig. 12.2 (A) Negative feedback process of intraspecific competition and (B) the linkage of processes and events for one predator and one prey species. Note: The food of the predator's immediate environment is the prey species, and the predator of the prey's immediate environment is the exploiting predator species.

to the specific case where a population is controlled by its natural enemies (Fig. 12.2). As the prey population increases there is increased competition for hiding places or escape possibilities (Huffaker 1958a) and the risk of capture increases. The predator (parasitoid, pathogen), of course, finding a more abundant food supply which is more easily captured, reproduces more offspring, and this increased predator population then puts increasing pressure on the prey population. It is only in this linked, mutual (or reciprocal) density-dependent interaction that predators, parasites, and pathogens act as true and reliable regulating factors (Huffaker 1958a, Huffaker & Messenger 1964b, Berryman 1981). This general comment also applies to those cases termed *intrinsic regulation* which involve specialized or altered behavior, physiology, or vitality geared to population density. Thus, some populations exhibit restricted breeding behaviors (territoriality, social hierarchy, or genetic or psychophysiological deterioration) and this may limit population density below that where obvious shortages of food or places to live arise, and before predation, parasitism and disease become significant.

It is of interest that several authors (Gilbert & Singer 1975, Zwölfer 1975, Lawton 1978, Lawton & Schröder 1979) refer also to "competition for enemy-free space" as a means of limiting the species diversity of phytophagous insects on a particular plant species.

12.3.3 The Collective Nature of the Density-Dependent Process

From the previous discussion we can see that our view of population regulation is focused on intraspecific competition between members of the population and *not* on the stressing factors, *per se*, which this competition may evoke (e.g., predation, freezing for lack of shelter, etc.). The process as a whole includes any combination of stress factors that collectively intensify as the density of the population increases (density-dependent factors) (Huffaker 1958a, Huffaker & Messenger 1964b, Huffaker & Laing 1972, Berryman 1981). The total regulative power is, therefore, commonly greater than that indicated by the response of any single density-dependent factor alone. Few if any cases of regulation exist wherein a single density-dependent factor operates entirely alone without some contribution by other density-dependent factors. Commonly, two or more regulating factors operate together, as for example, when the Mediterranean flour moth is regulated by competition for food combined with contemporaneous attacks by a predatory mite and a parasitic wasp, but with each factor having characteristic periods of greater and lesser impact (White & Huffaker 1969).

Finally, we should mention evolutionary considerations. Two powerful natural forces are aligned in opposition. One is the high reproductive capacity of most insects and the other the density-dependent destruction and/or restriction of procreative capacity when density approaches environmental saturation or causes predators, parasitoids, and pathogen populations to increase. While evolution has adjusted reproductive ability in accordance with the density-independent hazards encountered, this can only achieve a rough balance. We accept Darwin's implication that evolution has necessarily provided every existing species with an excess in procreative ability above that needed to just meet the purely density-independent stresses encountered in its environment (Huffaker & Messenger 1964b). We cannot view density-dependent stress factors in the same way. These adjust population density through natality or mortality or dispersal to the limits imposed by time and space varying factors, hence attaining a finely tuned balance between the population and its environment. This was illustrated in Fig. 12.2, where we see that each population (predator or prey) is regulated by intraspecific competition. The pervading environment determines the immediate environmental factors [food, places to live, competitors, exploiters (predators, parasites), and allies] for both predator and prey, and these in turn affect the degree of competition for the required resources, the outcome being a change in the rate of increase of the population and, perhaps, in its genetic properties. The predator and prey populations are also linked, and this coupling results in regulation of the predator by intraspecific competition for prey, and the prey population by intraspecific competition for hiding places and escape "routes". These external features, then, strongly affect the level at which regulation occurs. Regulation, therefore, occurs *in relation to* the pervading and immediate conditioning factors and, of course, the properties of all the associated organisms. Variations in these features may alter the level at which the density-dependent processes regulate population density. Habitat suitability, in fact, embraces density-dependent as well as density-independent factors, and this suitability affects population replacement, characteristic density, and existence itself (Fretwell 1972).

12.3.4 Empirical Evidence for Density-Dependence

The theory of population regulation through density-dependent processes requires that the parameters of population survival and reproduction are inversely related to population density. Evidence supporting the theory is obtained from observing population dynamics in the field over long periods of time, by experimentally manipulating population densities, and by establishing relationships between density and various parameters which express population change.

12.3.4.1 Population Trends of Various Species in the Same Habitat. The results of Varley and Gradwell's (1958) censuses of insects on oaks recall Odum's classic comments illustrating the meaning of natural control. Certain oak-inhabiting species were always rare, others common, and still others abundant year after year despite the fact that certain species in one group (rare, common, or abundant) varied in a similar pattern to other species in a different density group. The authors concluded that, while density-independent environmental conditions might explain the *changes* in density (e.g., Chapters 11, 13, 14, 23), only some density-dependent factor could explain the consistent differences in mean densities. Klomp's (1966a,b) long-term studies of pine looper dynamics in the Netherlands illustrates this approach in a different way. Examining variations in mortality over a 12-yr period, he found year-to-year trends *toward* an equilibrium density for 8 of the 11 possible trend lines. Only 3 lines contradicted this assumption, and only 1 appreciably so.

12.3.4.2 Manipulation of Population Size.

Manipulations, either by destruction of a portion of a population or additions to the population and then comparing the subsequent dynamics with unperturbed controls, are often used to demonstrate density-dependence. There are innumerable examples of rapid resurgence after a high insect pest population has been suppressed by an insecticide (Ripper 1956, Huffaker 1971, DeBach 1974). Eisenberg (1966) altered adult densities of the snail *Lynmaea elodes* from a pond to approximately one-fifth and five times, respectively, the initial spring densities in snail-proof sections of a pond. Afterwards the populations had reached complete convergence of numbers of young snails. This implies that the natural control mechanism in effect in the unperturbed population was competition between snails because the only change made was in snail density.

The opposite situation is found when streams are overstocked with game fish, or a land area with pheasants (Hassell 1976). Even without increased fishing or hunting, one would expect such a population to subside to its environmental capacity. But if such populations were controlled solely by density-independent factors, as claimed or inferred to be a sufficient explanation of some cases of "natural control" (e.g., Thompson 1939, 1956, Andrewartha & Birch 1954), the altered populations should continue to vary about their new densities rather than returning to their former levels.

12.3.4.3 Parameters of Density-related Performance.

The per capita rate of population increase is determined by a number of properties, such as fertility, fecundity, oviposition, sex ratio, and development and survival rates. Demonstrating the effects of density on these parameters furnishes evidence for density-dependent regulation. The literature is replete with such examples, and we include several in Fig. 12.3. The common characteristic is that rates of performance of individuals decline with increase in density of the population. By inference, these declines are due to intraspecific competition for food, oviposition sites, security from predation, and so on.

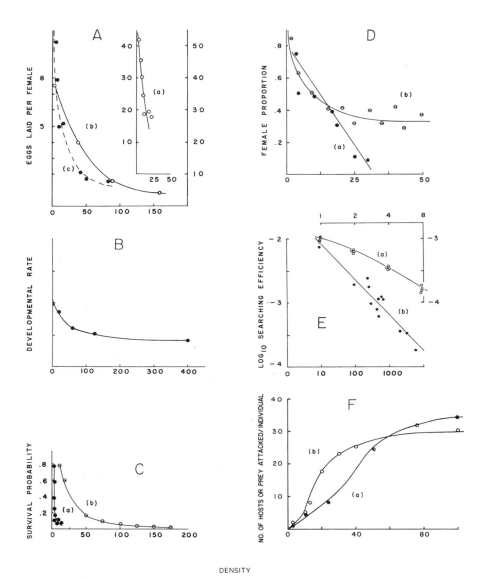

Fig. 12.3 Some density-related performance functions of various insects. Effect of population density (abcissa) on (A) the number of eggs laid by a bark beetle, (a) *Dendroctonus pseudotsugae* (McMullen & Atkins 1961), a flour beetle, (b) *Cryptolestes* (Varley et al. 1973), and a psyllid, (c) *Cardiaspina albitextura* (Clark 1963); (B) the developmental rate of *Endrosis sarcitrella* (Anderson 1956); (C) the probability of survival of a grain beetle, (a) *Rhizopertha dominica* (Crombie 1944) and the blowfly, (b) *Lucilia cuprina* (Nicholson 1954); (D) the proportion of females of (a) *Bracon hebetor* (Benson 1973) and (b) *Nasonia vitripennis* (Walker 1967); (E) the host-searching efficiency (log$_{10}$a) of (a) *Pseudeucoila bockei* parasitizing *Drosophila* larvae (Bakker et al. 1967), and (b) *Nemeritis canescens* parasitizing *Anagasta kuhniella* (Hassell & Huffaker 1969); and (F) the sigmoid relation in number of hosts attacked by (a) *Aphidius uzbekistanicus* parasitizing the aphid *Hylopteroides humilis* (Dransfield 1975), and (b) mosquito larvae attacked by the waterboatman, *Plea atomaria* (A. Reeve, unpubl., from Hassell et al. 1977). Note: These latter show the effect of increase in prey density in attracting a greater intensity of predation.

12.3.5 Cooperation and Positive Feedback

We have shown that intraspecific competition results in density-induced negative feedback which tends to lead to regulation of populations at characteristic densities. Competition, however, is not the only way organisms of the same species interact with one another; they may also cooperate during mating, hunting for food, or in defense against, or escape from predators. Cooperation occurs when individuals assist or benefit each other in some way. Thus an increase in density of cooperative individuals increases reproduction and/or survival of the population. This results in a positive feedback loop, commonly referred to as *inverse* density-dependence. The most obvious cooperation in sexually reproducing organisms involves mating. It has been argued that the probability of finding mates is directly related to population density and that when density becomes very sparse, extinction becomes a real possibility. This low-density, positive feedback (inverse density-dependent) relationship is often referred to as the "Allee" or "underpopulation" effect (Allee 1947, Allee et al. 1949). It illustrates a fundamental property of cooperative activities in that they tend to create unstable thresholds separating different dynamic patterns (Berryman 1981). For example, in sexually reproducing organisms there may exist a population threshold below which extinction occurs because of infrequent mating or reproduction, but above which population growth occurs.

The interaction between the two independent processes of cooperation and competition often produces unimodal or bimodal density-dependent functions such as that illustrated in Fig. 12.4. In this example, the cooperative activities of the bark beetle, *Scolytus ventralis,* in overcoming defense systems of their host trees result in higher reproduction and survival (productivity) as attack density rises (Berryman 1974). However, increasing attack density also gives rise to increasing competition for food and space (Berryman & Pienaar 1972). In the composite function we see that cooperative processes dominate at attack densities of less than 8/unit area while competition dominates at higher densities. While this figure may oversimplify the effects of interactions among individuals of the same species at different densities, it does illustrate the combined effects of competition and cooperation. It should also be noted that the relative dominance of cooperative or competitive processes may also shift again at higher population densities and give rise to more complex, multi-modal productivity curves (cf. Berryman 1981).

12.3.6 Relationships between Conditioning and Regulating Factors

As noted above, conditioning factors (density-independent factors) act directly or indirectly to cause population fluctuations (or perhaps little change), while others (density-dependent factors) reduce the scale of fluctuations, tending to regulate populations at a characteristic density which depends

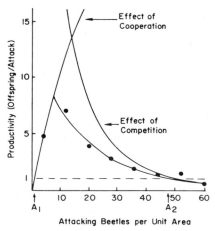

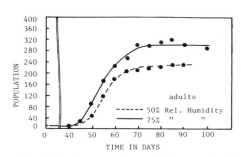

Fig. 12.4   Bark beetle survival as a composite function of attack density where the ascending function expresses the cooperative effect of attack density in reducing host resistance and the descending function expresses the effect of competition for food as density increases.

Fig. 12.5   A comparison of the growth of adult populations of *Tribolium* at 50 and 75% relative humidities—all at a constant temperature of 27°C (after Holdaway 1932).

on the properties of the environment and the species in question (Huffaker & Messenger 1964b, Huffaker et al. 1976). Long-term evolution of specific properties (Chapters 9, 10) and the shorter-term interplay of genetic feedback enter as factors that modify the regulation process. In reality, population movements, changes in specific properties, and both density-dependent and density-independent factors interact to produce the observed dynamics and the reality of natural control (e.g., Clark et al. 1979).

Because *conditioning* factors of the environment set the limits within which *regulating* factors operate, in the broad sense the environment rules (Milne 1957, Thompson 1939, Andrewartha & Birch 1954). Thus, the environment permits or denies existence and even dictates potential upper limits of abundance in places where existence is possible. Regulatory processes act to adjust population size in the "relative" sense, that is, relative to the *conditions* and within these broad potential limits and relative also to the *properties* of the organisms.

In a different way, Holdaway's (1932) experiments with *Tribolium* help explain the inseparably related but different *roles* of conditioning and regulating factors (Fig. 12.5). In this experiment populations leveled off at characteristic densities in both 50 and 75% relative humidities. However, the equilibrium levels at these different humidities were quite different. Humidity differences altered the conditions of the habitat, with higher rates of egg cannibalism in the drier habitat; this affected the *level at which the density-dependent mechanism (cannibalism) regulated population density.*

   General interrelationships between conditioning and regulating factors
in *determining* the distribution, abundance, and dynamics of a large population are illustrated in Fig. 12.6. The arbitrarily delimited habitats are
represented by (1) regions which have the most stable, continuously favorable conditions, (2) intermediate regions of favorability and permanent
occupancy, (3) very unstable conditions where permanent occupancy is
barely possible, and (4) regions of extreme changes in favorability where
permanent occupancy is impossible, but temporary occupancy derives from
immigrants from other areas (Huffaker & Messenger 1964b). The irregular
patches in each zone represent the minimum area of permanent occupancy;
the theoretical maximum occupancy would occur if the patches were to
expand to fill an entire zone. Thus, the spaces between the patches and
the zone limits represent the degree of variation possible in each habitat
due to waxing and waning physical conditions and associated biotic supporting factors such as food and beneficial species.
   In zone 1 density-dependent actions are hypothesized to dominate. There

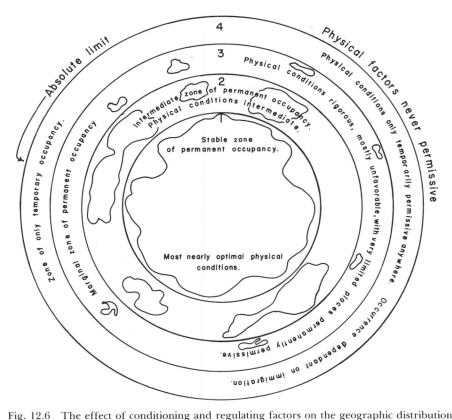

Fig. 12.6   The effect of conditioning and regulating factors on the geographic distribution
of a species population (After Huffaker & Messenger 1964).

is little change in physical favorability, and thus intraspecific competition for food and living places and action of natural enemies may largely account for both changes in density and characteristic population size. In zone 2 changes in conditioning factors are substantial and their effects produce variations in the areas of occupancy and also in population size and quality. Regulating and conditioning factors share the influence on changes in density. The effects of regulating (density-dependent) factors are never perfectly compensating, so they, too, can cause some fluctuation, even as they tend to reduce the scale of fluctuations. In zone 3 extreme variations in conditioning factors produce the dominant influence on *changes* in density except in the very limited places of permanent occupancy where density-dependent factors may share that role and regulate population *size* relative to the conditions. In zone 4 unfavorability is so extreme that only temporary occupancy is possible and changes in density are dominated by conditioning physical factors, or associated habitat conditions. Density-dependent processes may be engaged very infrequently, or only through their having operated in other zones that supply immigrants to this zone. One should note that even the cold arctic and hot deserts are favorable to well-adapted species and would be zone 1 habitats for them.

Figure 12.7 illustrates the temporal aspects of natural control operating in a variable environment. Following Solomon (1949), we refer to limitation of a population, meaning deceleration of population growth (*A*) as at (*B*). Subsequently, the population may be reduced or suppressed (*D*) or simply brought under extended limitation (*B* ⟶ *C*). With population decline, the density-dependent component(s) of repression relax in intensity, resulting in a conservation tendency (*E*), provided the population has not

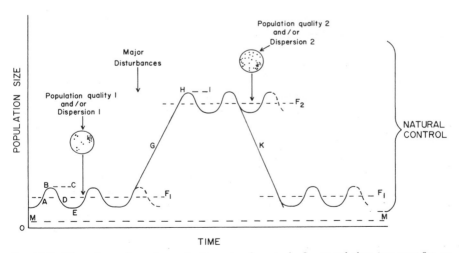

Fig. 12.7 Time-dependent phases in the natural control of a population (see text for explanation).

fallen below the extinction threshold (M). Below M an inverse density-dependent relationship (positive feedback) comes into play due to difficulties in finding mates or satisfying other social needs (Allee 1947). (The farther below M the population falls, the less likely it is that chance alone would save it from extinction.)

A population may proceed for an indefinite period of time under regulation about F_1 by a single density-dependent factor or a complex of such factors. If it is released from this control (G), it will grow until a new density-dependent factor(s) engage it at H producing a steady density ($H \longrightarrow I$) or oscillatory equilibrium around F_2. Furthermore, if the expanding population reaches such high densities as to induce numerical responses in natural enemy populations (e.g., a disease pathogen) or degradation of the habitat, a crash (K) would occur, bringing the population down to its original equilibrium at F_1, or overshooting such.

Different equilibrium levels may occur due to evolutionary changes in natural enemy efficiency or prey vulnerability (for laboratory examples, see Pimentel & Al-Hafidh 1965, Pimentel & Soans 1971, Pimentel et al. 1975). We emphasize, however, that we would not expect to see synchronized predator–prey interactions over extensive areas, so as to observe oscillations as implied in Fig. 12.7.

12.4   MODELS OF NATURAL CONTROL

In this section we outline some fundamental mathematical and synthetic graphical interpretations of population dynamics and natural control that illustrate the foregoing discussions.

12.4.1   The Competition Curve

Malthus (1803) and Verhulst (1838) were the first to consider the idea of a "law of diminishing returns" when individuals seek a common resource in short supply. In Nicholson's (1933) "competition curve," this law operates through the searching time required to find a unit of resource as the supply of the resource dwindles (Fig. 12.8A, B). Although the search for resources may be aided by environmental cues, the searching by nonsocial insects is generally unorganized and *random movement* (*in relationship to the resources*) *usually occurs* until the individual reaches a position to detect some cue. This idea of random search for a diminishing resource is captured by the Poisson distribution function, where the probability of exactly a occurrences (encounters with resource units) is given by

$$\frac{\bar{x}^a}{a!e^{\bar{x}}} \tag{12.1}$$

where $\bar{x}$ is the mean number of encounters and e is the base of natural logarithms. From this expression, the probability of not encountering a unit of resource (i.e., $a = 0$ and $a! = 1$) is

$$\frac{1}{e^{\bar{x}}} = e^{-\bar{x}} \tag{12.2}$$

so that the fraction of the resource encountered at least once is given by

$$1 - e^{-\bar{x}} \tag{12.3}$$

This expression forms the basis of Nicholson and Bailey's (1935) model of host–parasite interaction, Thompson's (1924) superparasitism models, and Utida's (1941) and Ricker's (1954) reproduction models (cf. Chapter 16). This more or less haphazard exploitation of a resource was called *scramble competition* by Nicholson (1933). In its pure form, this type of competition results in the resource being partitioned equally among all competitors, so that *all* die when the resource is insufficient to support them. In contrast, *contest competition* occurs when the resource is partitioned unequally among competitors, where the winners-take-all; for example, territorial behavior ensures that the winners of territorial conflict gain sufficient resources to survive and reproduce. Contest competition is exemplified by insect parasitoids which lay many eggs in a host but only one survives to fully utilize the resource. Scramble competition is exemplified by larval competition for liver in Nicholson's blowflies (e.g., 1954) (cf. Chapter 15).

12.4.2 Mathematical Descriptions of Population Regulation

A number of mathematical models have been proposed for describing effects of intraspecific competition on performance statistics of insect populations. Berryman (unpub.) has examined the fit of these models to a

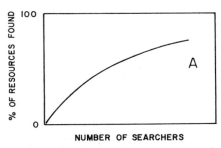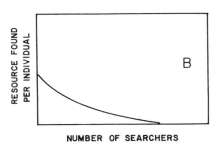

Fig. 12.8 (A) Relationship between number of searchers and the percentage of resource found (the "competition curve" of Nicholson 1933). (B) Relationship between number of searchers and the amount of resources found per individual searcher.

number of data sets and their ranked goodness of fit are presented in Table 12.1. In general, the Beverton and Holt (1957) and Berryman (1974) models best describe fecundity, survival, and parasite attack while sex ratio data, which are roughly linear, are best described by the Anderson (1956) model. Of course, better fits to data can often be obtained with three-parameter versions of these models (e.g., Hassell 1975, Berryman & Brown 1981).

12.4.2.1 Dynamic Single-Species Models. The "logistic" model is the best-known equation for population growth and regulation. It was deduced on the assumption of an instantaneous negative feedback response of population growth to population density (density-dependence). As advanced by Lotka (1925) it starts with the proposition that

$$\frac{dN}{dt} = f(N) \tag{12.4}$$

This equation must have a root at $N = 0$ since at least one individual is required to start growth. Expanding by Taylor's theorem we get

$$\frac{dN}{dt} = aN + bN^2 + cN^3 + \cdots \tag{12.5}$$

and the simplest expression which satisfies the condition of constrained growth (i.e., $dN/dt = 0$, $N > 0$) is

$$\frac{dN}{dt} = aN + bN^2 \tag{12.6}$$

with $a > 0$, $b < 0$. This equation embodies the assumption that the per capita rate of increase declines linearly with increase in population size.

Biologists are particularly interested in the saturation level, frequently called the environmental carrying capacity, K (i.e., $K = -a/b$). Carrying capacity refers to the density of the population which can be sustained indefinitely by the resources present in its environment. To incorporate the concept of carrying capacity, the "logistic" equation is commonly written as

$$\frac{dN}{dt} = r_m N \left(\frac{K - N}{K} \right) \tag{12.7}$$

where $r_m = a$ is the intrinsic or innate capacity for increase, and $K = -r_m/b$. One possible solution of the "logistic" is

$$N(t) = \frac{K}{1 + e^{-rt}} \tag{12.8}$$

Table 12.1 Some simple two-parameter models for intraspecific competition ranked by goodness of fit, χ^2, to four-survival, four-fecundity, three-sex-ratio, and four-parasite attack rate data sets[a]

Author	Model		Original use	Rank goodness of fit to data				
				Survival	Fecundity	Sex ratio	Attack	Overall
Anderson (1956)	$Y = a - bx$		Sex Ratio	8	7	1	8	7
Crombie (1944)	$Y = a + b/x$		Survival	7	5	6	6	7
Anderson (1960)	$Y = a - b\sqrt{x}$		Attack	6	6	2	7	6
?	$Y = a - b \ln x$		—	5	4	5	5	5
Beverton & Holt (1957)	$Y = (a + bx)^{-1}$		Survival	2	1	3	1	1
Watt (1959)	$Y = ax^{-b}$		Attack	3	3	5	3	4
Ricker (1954)	$Y = ae^{-bx}$		Survival	4	4	1	4	3
Berryman (1974)	$Y = ae^{-b\sqrt{x}}$		Survival	1	2	4	2	2

[a]Three-parameter generalizations of the Beverton and Holt model, $y = (a + bx)^{-c}$, and Ricker model, $y = ae^{-bx^c}$, have been proposed by Hassell (1975) and Gilpin and Ayala (1973), respectively. [After A. A. Berryman (class notes); Y = population statistic, x = population density, and a and b are parameters.]

383

where t is the elapsed time (Lotka 1925). This equation has been fit to some human populations (Pearl & Reed 1920) and to numerous populations grown in lab culture (e.g., Gause 1934, Fig. 12.9). Although some authors (e.g., Pielou 1969) argue that these remarkably good fits may be spurious, the model seems to remain firmly at the center of natural control theory.

Yet, although the logistic equation has provided reasonable descriptions for simple population systems, it is often inadequate in its classical form for describing many natural animal populations because of its following assumptions.

1. Density-dependent processes respond to population density without a time delay.
2. Density-dependent responses are linearly related to population density.
3. The parameters r_m and K are constant in time.
4. Generations overlap completely so that reproduction occurs continuously.
5. The distribution of age classes is stable.

Since one or more of these assumptions may not be met in natural situations, other formulations have been advanced, some of which are discussed below.

Models with Time Lag Effects. Hutchinson (1948) examined the theoretical consequences of relaxing assumption 1 by incorporating a time lag in the density-dependent response. Hence,

$$\frac{dN(t)}{dt} = r_m N(t) \left(1 - \frac{N(t - T)}{K} \right) \tag{12.9}$$

where T is the response time, or relaxation time, of the density-dependent process(es). The stability properties of the time-delayed logistic model have been evaluated in detail by May (1973b) (cf. Ch. 16). The main consequence is that, as T gets large, cycles of increasing amplitude and decreasing frequency are produced (stable limit cycles). May (1973b) has fit this model to Nicholson's cyclical blowfly data (Fig. 12.10).

Models with Nonlinear Feedback. In many real situations density-dependent processes do not respond to changes in population density in a linear fashion (e.g., Fig. 12.3). A number of authors have examined this problem, but perhaps the best general solution was proposed by Gilpin and Ayala (1973). They suggested that a kurtosis exponent, θ, be added to the negative feedback component; i.e.,

$$\frac{dN}{dt} = r_m N \left[1 - \left(\frac{N}{K} \right)^{\theta} \right] \tag{12.10}$$

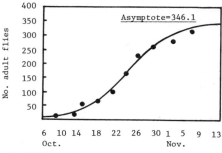

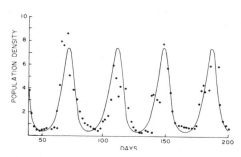

Fig. 12.9 Logistic growth of an experimental population of the fruit fly *Drosophila melanogaster*. The circles are observed census counts and the smooth curve is the fitted logistic (after Pearl 1927).

Fig. 12.10 Oscillations in density of Nicholson's blowflies *Lucilia cuprina* when regulated by competition for food (see May 1973b for analysis).

Gilpin and Ayala found that their non-linear response model provided a better description of intraspecific competition for several *Drosophila* species and for interspecific competition between two (or more) species. They speculated that invertebrates should have θ values less than unity, while vertebrates should have values greater than unity.

The logistic assumption 2 also implies that individual survival and reproduction do not decline at very low population densities (i.e. at the underpopulation or "Allee" effect). However, these complexities may be incorporated into the logistic framework by expanding the Taylor series (e.g., Berryman 1981).

Models with Parameters Varying in Time. It is often inadvisable to assume that the "logistic" parameters are constant in time or space. In particular, the maximum rate of increase of poikilotherms is dependent on environmental temperature and many other factors. Likewise, the potential carrying capacity of the environment is often dependent on the rate at which food is produced, and this is commonly determined by environmental conditions or affected by other external factors, including the exploiters.

A number of modifications to the logistic model have been proposed to handle these problems. Ricklefs (1979), for example, suggests

$$\frac{dN}{dt} = r_m N\left(1 - \frac{NM}{P}\right) \qquad (12.11)$$

where M is the quantity of resources required for maintenance of an individual, and P is the rate of production of the limiting resource.

Other authors have dealt with the problem by formulating r_m and K as functions of environmental parameters. For example, Long et al. (1974)

produced a model for intertidal gastropod populations by allowing r_m to change with water temperature and salinity and K to vary with incident solar radiation.

Models for Discrete Generations. Many insect species inhabiting temperate regions exhibit discrete generations, violating assumption 4 above. Of the numerous discrete time models that have been proposed, the one most seen in entomological literature is that of Haldane (1949):

$$N_{t+1} = N_t\lambda N_t^{-b} = \lambda N_t^{1-b} \qquad (12.12)$$

where λ is the finite rate of increase over the time increment t to $t+1$ (i.e., $\lambda = e^{r_m}$), and b is a coefficient of density dependence. This equation has received much attention as a basis for detecting density dependence, for which, however, it has considerable shortcomings (e.g., Hassell & Huffaker 1969, Itô 1972, Varley et al. 1973).

Cook (1965) and others (notably May 1974) suggest that the most appropriate discrete-time analogue of the "logistic" is an exponential density-dependent function often used by fisheries biologists (e.g., Ricker 1954). This equation is usually seen as

$$N_{t+1} = N_t\lambda e^{-\alpha N_t} \qquad (12.13)$$

where α is the density-dependent coefficient. If this equation is solved at equilibrium (i.e., when $N_{t+1} = N_t = K$), then we find $\alpha = (\ln \lambda)/K$ and, substituting in the above, we get

$$N_{t+1} = N_t\lambda^{1-N_t/K} \qquad (12.14)$$

This model can be generalized (Brown et al. 1978, Berryman & Brown 1981) to account for time delays and skewness in the environmental resistance expression as follows:

$$N_{t+1} = N_t\lambda^{1-(N_{t-T}/K)^\theta} \qquad (12.15)$$

and, as such, it has provided reasonable descriptions of both laboratory and natural insect populations (e.g., Fig. 12.11).

Models for Variation in Age Distribution. The assumption of a stable age distribution has been considered by Lotka (1925) for populations with continuous reproduction (the continuous case) and Leslie (1945, 1948) for those with discrete periods of reproduction. However, for many insect populations, a stable age distribution is attained rapidly or, for those with discrete generations, the problem is avoided if time increment t equals the generation span so that $\lambda = R_0$, the net reproductive rate per generation. (cf. Chapters 11, 24 for applications of the Leslie matrix).

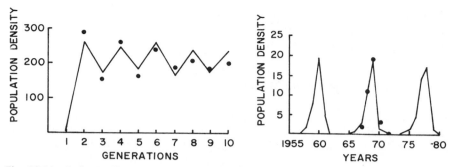

Fig. 12.11   Left: Oscillations in laboratory cultures of the southern cowpea weevil (Utida 1967) fitted in Eq. (12.15) with $N_0 = 16$, $K = 192$, $\lambda = 4$, and $\theta = 0.5$. (unpublished). Right: Population cycles of the Douglas-fir tussock moth, *Orgyia pseudotsugata* (data points from Mason 1974), fit to the discrete logistic Eq. (12.15) with a one-generation time delay (after Berryman 1978c).

12.4.2.2   Dynamic Two-Species Models.   The most elementary two-species population model is the so-called Lotka–Volterra equation (Lotka 1925, Volterra 1926) which can be formulated as

$$\frac{dN_i}{dt} = \alpha_i N_i + \beta_i \prod_{j=1}^{2} N_j \qquad i = 1, 2 \tag{12.16}$$

where α_i is the intrinsic rate of increase of the ith species in the absence of the other, and β_i is its coefficient of interaction with the other species. This system is unstable for cooperating or competing species and produces stable limit cycles for trophic interactions. However, stable solutions are possible in cooperating and competing populations and damped, stable predator–prey oscillations can be generated by including an intraspecific competition effect; that is, $\alpha_i = f_i(N_i)$.

A number of modifications have been proposed for the elementary "Lotka–Volterra" equation. For instance, for competing species one can extend the "logistic" model (e.g., Getz 1978) to

$$\frac{dN_i}{dt} = r_i N_i \left(1 - \frac{N_i}{K_i}\right) \qquad i = 1, 2 \tag{12.17}$$

where $K_i = f_i(N_{j \neq i})$. In this formulation each species is assumed to affect the saturation density or carrying capacity of the other.

Discrete-time predatory–prey models have also been built around the basic parasitoid searching model of Nicholson and Bailey (1935). Although the elementary model produces unstable oscillations it can be stabilized when predator or prey individuals compete among themselves, interfere with each other, occupy heterogeneous environments, or when predators aggregate (cf. Hassell 1978, and Chapter 16). However, all these models demonstrate that predator–prey interactions are basically self-regulating, and that stability is improved by additional intraspecific competition and spatial heterogeneity.

12.4.3 Graphic Population Models

From the classical theories of "logistic" growth and "density-dependence" has arisen what can best be termed "reproduction theory." Utida (e.g., 1941, 1957) and Fujita and Utida (1953) originally studied the relationship between population size from one generation to the next in laboratory cultures of the azuki bean weevil. They found a typical unimodal curve which was termed the *reproduction curve* by Ricker (1954). A reproduction curve is a plot of the density of organisms in one generation against the density in the previous generation (i.e., surviving progeny against their parents). Under density-dependent growth, the curves appear as illustrated in Fig. 12.12. The dynamics of population growth can be extracted from these curves by obtaining the density of the offspring from the starting parent population, carrying this value over to the 45° replacement line, which gives the new parent density, and so on. The 45° replacement line specifies the condition where one offspring is produced to replace each parent and it is on the intersection of this line with the reproduction curve that the population will reach equilibrium if the system is stable. The "logistic" population, of course, approaches equilibrium asymptotically. However, populations obeying Ricker's exponential survival model will frequently oscillate about the equilibrium.

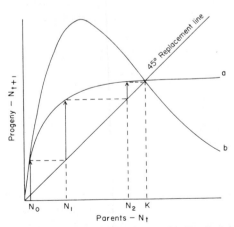

Fig. 12.12 Relationship between parent and progeny density for populations obeying Eq. (12.13) when $\lambda < 1$ (a) and $\lambda > 1$ (b) showing a growth trajectory $N_0 \to N_1 \to N_2 \to K$ in response to curve a; K = equilibrium density or carrying capacity.

12.4.3.1 Net Reproduction, Replacement Curves, and Stability. One may also plot the net reproductive rate ($\Delta N/\Delta t = N_{t+1} - N_t$) against population density. This provides a picture of the "surplus" reproduction for any given population size or the "sustained yield" which can be harvested at that population density. Maximum sustained yield, of course, is attained at the peak of the net reproduction curve.

When dealing with systems obeying the discrete "logistic" equation, it is often more convenient to examine the properties of the per capita replacement curve (Fig. 12.13a). The replacement rate is defined as

$$R = \frac{N_{t+1}}{N_t} \qquad (12.18)$$

where the time interval, t to $t+1$, spans one complete generation. Unlike reproduction or net reproduction curves, the dynamics of the system cannot be extracted directly from the replacement curve, although it can be inferred; that is, when $R > 1$ the population grows; when $R < 1$ it declines and when $R = 1$ it is in equilibrium. May et al. (1974) demonstrate the

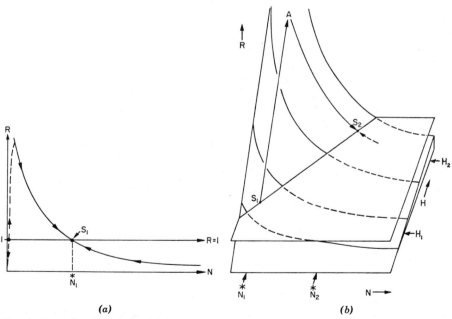

(a)   (b)

Fig. 12.13 Left: The relationship between the per capita replacement rate R and population density N in a density-dependent system with a single equilibrium point, S_1, occurring at density N_1^*. (from Berryman 1978a). Right: The relationship between the per capita replacement rate R, population density N, and basic habitat suitability H, showing the population dynamics following a shift in habitat suitability from H_1 to H_2 (from Berryman 1978a).

neighborhood stability of the equilibrium point (S_1 in Fig. 12.13*a*) depends on the slope *b* of the replacement curve at this point and the delay T in the density-dependent process. When $bT < 1$ the population grows, or declines, asymptotically ("logistically") to equilibrium; when $1 < bT < 2$ it reaches a stable state following damped oscillations; and when $bT > 2$ it is unstable.

Fretwell (1972) extended the theory of reproduction curves to consider the suitability, or "goodness," of the habitat, an idea which embraces both density-dependent and density-independent relationships. Habitat suitability can be thought of in terms of average success rate (and/or fitness) of the organisms in a given habitat, a good habitat contributing more offspring to the next generation than a poor one. The principal factors affecting suitability are food, weather and shelter, predators, allies, competitors, and the density of the subject population. Thus, Fretwell's theory includes density-dependent effects, expressed by the influence of population density on individual fitness, as well as density-independent effects, expressed by the *basic* suitability of the habitat when population density is very low and density-dependent effects are negligible. Fretwell's basic habitat suitability is represented by the "logistic" parameters r_m and λ.

12.4.3.2 Synoptic Models of Population Dynamics.

Berryman (1978a) developed a three-dimensional graphic model to explain the dynamics of populations occupying habitats of varying suitability. Extension of reproduction theory to include habitat variability allows us to hypothesize that insect epidemics can be caused by a rapid increase in favorability of a habitat in terms of physical and/or biotic factors. Using Fig. 12.13*b*, suppose a population is at equilibrium S_1 with density N_1^* in an unfavorable habitat H_1, but an improvement of the habitat to H_2 creates a new equilibrium at S_2. The realized replacement rate will respond by increasing to *A* and the population will grow and reach equilibrium at the new density, N_2^*. Should the environment revert to its original suitability (e.g., if the large population destroys its habitat or the weather returns to an unfavorable condition), the process will be reversed. This hypothesis may explain the dynamics of species occupying rather variable habitats. Population explosions of the bark beetle, *Scolytus ventralis*, for example, have followed outbreaks of insect defoliators which increase the food supply by weakening large numbers of trees (Wickman et al. 1973, Berryman 1973). The model may also describe many agricultural systems wherein insect habitats are drastically and repeatedly changed by production practices (cf. Chapter 23).

Some insect populations may have more complex replacement curves than those illustrated in Fig. 12.13. Morris (1963) first speculated that eastern spruce budworm populations may have more than one equilibrium position, and the computer model constructed by Holling et al. (1977) generates replacement curves which have equilibrium points at several population densities (Fig. 12.14). In this figure the three curves represent

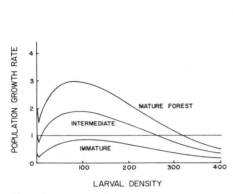

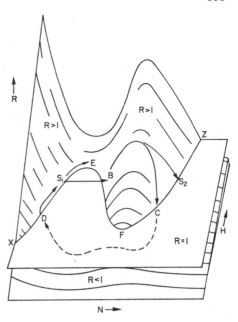

Fig. 12.14 Per capita rate of increase of the spruce budworm, *Choristoneura fumiferana*, at various budworm densities and three forest conditions. Potential equilibria occur whenever the growth rate intersects the horizontal replacement line (from Holling et al. 1977).

Fig. 12.15 Three-dimensional relationship between population density N, basic habitat suitability H, and per capita reproduction R when cooperative processes (positive feedback) operate at intermediate population densities (e.g., escape from predators, overcoming host resistance, etc.) (from Berryman 1978a).

stages of forest maturity, or differing degrees of basic habitat suitability. Berryman (1978a) then displayed the general effect of basic habitat suitability on this type of replacement curve in a three-dimensional graph (Fig. 12.15). Equilibrium populations are theoretically possible wherever the surface intercepts the $R = 1$ plane (i.e., along the shoreline $X \rightarrow Z$ of "lake equilibrium"). However, only the sections of shoreline $X \rightarrow E$ and $F \rightarrow Z$ represent potentially stable equilibria.

The dynamics of a population obeying this model can be interpreted as follows: Suppose we have a stable population at S_1; then it can only reach its potential size at S_2 if it is carried across "lake equilibrium" by an influx of individuals (e.g., an immigration from outside, $S_1 \rightarrow B$) or if the basic suitability of the habitat increases and carries the population around the "bay" (e.g., from $S_1 \rightarrow E$). Once it has exceeded the unstable shoreline, $E \rightarrow F$, replacement is greater than unity and it grows to the stable point S_2. However, if the large population destroys its habitat, as some epidemic populations like the spruce budworm do, then it may be carried around the cape, F, and crash back to the lower, stable shoreline, $X \rightarrow E$.

The biological explanation of this model usually assumes that the upper equilibrium shoreline is created by limiting resources, such as food and space and the lower one by parasites and predators (Takahashi 1964, Holling et al. 1977, Southwood & Comins 1976). However, although this explanation may apply to certain species, it is certainly not general. Berryman (1978b) argued, for example, that bark beetle populations are maintained at low densities by competition for a limited supply of severely weakened trees (cf. Chapter 22). However, once their populations have attained a level where they can overcome resistance of healthy hosts (i.e., the shoreline $E \rightarrow F$), population explosions erupt which eventually collapse as the forest is destroyed (the trajectory $C \rightarrow D$).

A more general explanation has been advanced by Isaev and Khlebopros (1977) who argued that the lower equilibrium is enforced by rapidly responding (low-inertial) density-dependent processes such as migration, starvation, interference, cannibalism, predator and parasite functional responses, and sometimes their numerical responses if they occur quickly relative to the prey's generation time (e.g., Huffaker et al. 1962, Huffaker & Kennett 1966). If the population escapes from these fast-acting and low-density responsive processes, then time-delayed, or inertial, processes may dominate and cause the typical boom and crash trajectory of an epidemic population. Inertial processes may include such things as the replacement rate of the basic food supply, nutritional effects on fecundity, physiology and behavior (Chapter 4), genetic deterioration, pollution, and numerical responses of some predators and disease pathogens (cf. Berryman 1978a). Berryman (1981) advanced this general interpretation by proposing that the unstable portion of the equilibrium isocline (E–F in Fig. 12.15) will be created by the dominance of cooperative intraspecific interactions, such as bark beetle aggregations overcoming host resistance, and other swarming or aggregative behavior which increases individual fitness at relatively high population densities.

Holling (1973) and Berryman (1978a, 1982) have argued that we should be less concerned with the potentially stable equilibria (i.e., the shoreline $X \rightarrow E$ and $F \rightarrow Z$) and more concerned with the unstable threshold $E \rightarrow F$, since low-level populations are usually inconsequential from an economic standpoint while outbreak behavior spells disaster. Holling terms the unstable shoreline a *resilience* threshold, meaning that the population can be "pushed around" and still regain its original or characteristic endemic state, provided this threshold is not transcended. Berryman (e.g. 1978a, 1982) suggests that if the boundary function separating endemic (sparse) from epidemic (dense) population behavior can be specified, a powerful tool for decision making is provided for pest managers. Boundary functions specifying the "epidemic threshold" can sometimes be defined if the critical variables are known and if the system has been observed operating under an array of these variables in endemic and epidemic modes. Berryman (1978b) defined the epidemic threshold for the mountain pine beetle in lodgepole pine stands as a function of stand resistance and phloem thickness.

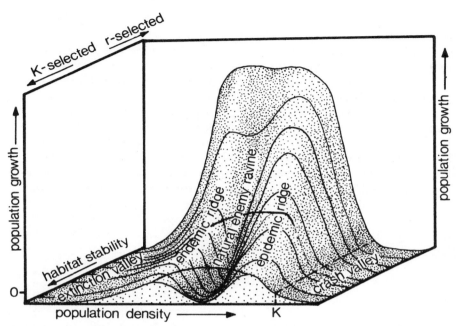

Fig. 12.16   Three-dimensional synoptic model describing population growth rate in relation to population density, habitat stability, and survival strategy—the r-K continuum. (Modified from Southwood, 1977a,b by Huffaker et al. 1977.)

12.4.3.3   Synoptic Models for *r* and *K* Strategies.   MacArthur and Wilson (1967) proposed the idea of an *r–K* continuum, with species adapted to stable habitats and having strong competitive abilities and low reproductive rates being termed *K* strategists, and those adapted to very unstable habitats and having high reproductive rates and strong dispersal abilities being termed *r* strategists. Based on these ideas, Southwood (1977a,b) and Southwood and Comins (1976) developed a three-dimensional synoptic model (modified as Fig. 12.16). In this model species are adapted to the durational stability, *D*, of the habitat. For example, $D = H/r$, where *r* is the generation time and *H* is the length of time that the habitat remains suitable for survival and breeding. Species occupying habitats with low durational stability tend to be small, mobile, and to have a short generation time. These are the *r*-selected species. Species occupying stable habitats tend to be larger, more likely to be territorial than migratory, have a long generation time, and often have fewer offspring, and practice parental care (Itô 1980, Itô & Iwasa 1981). Huffaker et al. (1977) warned that the size of individuals may be less important for insects than for certain other organisms, and we add or substitute broad physiological tolerance, as well as high rate of increase and high mobility, as being essential to the *r*-strategy makeup. Good searching capacity within the habitat locale, efficient utilization of resources, or competitive superiority in other ways characterize species of high K-strategy adaptation.

Huffaker et al. (1977) also discussed Southwood's model and reasoned that the natural enemy ravine should be deepened and extended to the most stable end of the habitat stability dimension. This extension of the model is intended to apply to the many organisms which are regulated by their natural enemies just as thoroughly (if not more so) in stable environments as in intermediate ones. They also note that cases where natural enemies do not regulate the population at low densities can be allowed for by *bridges* over the natural enemy ravine.

Lastly, the concepts and modeling of natural control features as presented in this chapter form only a part of the significant contributions to our understanding of natural control and the methods that can be used in its study. A rich assortment of data and examples are given throughout this book which in general exemplify and elaborate on the concepts and processes of natural control presented here.

ACKNOWLEDGMENTS

The authors wish especially to thank Dr. R. L. Rabb for material contributions during the development of this chapter, Dr. Wayne Getz for advice on some of the mathematics, Mrs. Nettie Mackey for innumerable typings and other assistance, and Mr. Richard Hom for work on the illustrations.

REFERENCES

Allee, W. C. 1947. *Encyclopedia Britannica, 1947*, pp. 971d–971r.

Allee, W. C., A. E. Emerson, O. Park, T. Park, and K. P. Schmidt. 1949. *Principles of Ecology.* W. B. Saunders, Philadelphia.

Anderson, F. S. 1956. *Oikos* **7:** 215–226.

Anderson, F. S. 1960. *Biometrics* **16:** 19–27.

Andrewartha, H. G. and L. C. Birch. 1954. *The Distribution and Abundance of Animals.* University of Chicago Press, Chicago.

Bakker, K. 1964. *Z. Ang. Entomol.* **53**(2): 187–208.

Bakker, K. 1980. *Netherlands J. Zool.* **30:** 151–160.

Berryman, A. A. 1973. *Can. Entomol.* **105:** 1465–1488.

Berryman, A. A. 1974. *Environ. Entomol.* **3:** 579–585.

Berryman, A. A. 1978a. *Res. Popul. Ecol.* **19:** 181–196.

Berryman, A. A. 1978b. Pp. 98–105 in A. A. Berryman, G. D. Amman, R. W. Stark, and D. L. Kibbee (Eds.), *Theory and Practice of Mountain Pine Beetle Management of Lodgepole Pine Forests.* College of Forest Resources, University of Idaho, Moscow.

Berryman, A. A. 1978c. *Can. Entomol.* **110:** 513–518.

Berryman, A. A. 1979. *Bull. Soc. Entomol. Suisse* **52:** 227–234

Berryman, A. A. 1981. *Population Systems: A General Introduction.* Plenum, New York.

Berryman, A. A. 1982. *Environ. Entomol.* **11:** 544–549.

Berryman, A. A. and G. C. Brown. 1981. Pp. 11–24 in D. G. Chapman and V. F. Gallucci (Eds.), *Quantitative Population Dynamics.* Int. Coop. Publ., Fairland, Maryland.

Beverton, R. J. H. and S. J. Holt. 1957. *Fish Invest.* **19:** 1–533.

Bodenheimer, F. S. 1928. *Biol. Zbl.* **48:** 714–739.

Bodenheimer, F. S. 1958. *Animal Ecology Today.* W. Junk, The Hague.

Brown, G. C., A. A. Berryman, and T. P. Bogyo. 1978. *Environ. Entomol.* **7:** 219–227.

Chitty, D. 1960. *Can. J. Zool.* **38:** 99–113.

Chitty, D. 1967. *Proc. Ecol. Soc. Aust.* **2:** 51–78.

Christian, J. J. and D. E. Davis. 1964. *Science* **146:** 1550–1560.

Clark, L. R., P. W. Geier, R. D. Hughes, and R. F. Morris. 1967. *The Ecology of Insect Populations in Theory and Practice.* Methuen, London.

Clark, W. C., D. D. Jones and C. S. Holling. 1979. *Ecol. Modelling* **7:** 1–53.

Cole, L. C. 1957. *Cold Spring Harbor Symp. Quant. Ecol.* **22:** 1–15.

Cook, L. M. 1965. *Nature (London)* **207:** 316.

Crombie, A. C. 1944. *Proc. Roy. Soc.* **131:** 135–151.

Darwin, Charles. 1859. *The Origin of Species by Means of Natural Selection.* Murray, London.

DeBach, P. (Ed.). 1964. *Biological Control of Insect Pests and Weeds.* Chapman & Hall, London.

DeBach, P. 1974. *Biological Control by Natural Enemies.* Cambridge University Press, Cambridge.

den Boer, P. J. 1968. *Acta. Biotheor.* **18:** 165–194.

den Boer, P. J. 1971. Pp. 77–97, in den Boer and Gradwell (Eds.), 1971, referenced here.

den Boer, P. J. and G. R. Gradwell (Eds.). 1971. *Dynamics of Populations.* Center Agric. Publ. & Doc., Wageningen.

Eisenberg, R. M. 1966. *Ecology* **47:** 889–906.

Fretwell, S. D. 1972. *Populations in a Seasonal Environment.* Princeton University Press, Princeton, N.J.

Fujita, H. and S. Utida. 1953. *Ecology* **34:** 488–498.

Gause, G. F. 1934. *The Struggle for Existence.* Williams & Williams, Baltimore.

Getz, W. M. 1978. *Ecol. Modeling* **5:** 237–257.

Gilbert, L. E. and M. C. Singer. 1975. *Annu. Rev. Ecol. & Syst.* **6:** 365–397.

Gilpin, M. E. and F. J. Ayala. 1973. *Proc. Natl. Acad. Sci.* **70:** 3090–3593.

Haldane, J. B. S. 1949. *Rice Sci.* **19** (Suppl.): 3–11.

Hassell, M. P. 1975. *Anim. Ecol.* **44:** 283–295.

Hassell, M. P. 1976. *The Dynamics of Competition and Predation.* Edward Arnold, London.

Hassell, M. P. 1978. *The Dynamics of Arthropod Predator-Prey Systems.* Princeton University Press, Princeton, N.J.

Hassell, M. P. and C. B. Huffaker. 1969. *Res. Pop. Ecol.* **11:** 186–210.

Holdaway, F. G. 1932. *Ecol. Monogr.* **2:** 261–304.

Holling, C. S. 1973. *Annu. Rev. Ecol. Syst.* **4:** 1–23.

Holling, C. S., D. D. Jones, and W. C. Clark. 1977. Pp. 13–90 in G. A. Norton and C. S. Holling (Eds.), *Proceedings of a Conference on Pest Management.* Int. Inst. Appl. Syst. Anal., A-2361. Laxenburg, Austria.

Howard, L. O. and W. F. Fiske. 1911. *USDA Agric. Bur. Entomol. Bull.* 91.

Huffaker, C. B. 1958a. Pp. 625–636 in *Proceedings of the 10th International Congress on Entomology,* Vol. 2, Montreal (1956).

Huffaker, C. B. 1958b. *Hilgardia* **27:** 343–383.

Huffaker, C. B. (Ed.). 1971. *Biological Control.* Plenum, New York.

Huffaker, C. B. and C. E. Kennett. 1966. *Hilgardia* **37:** 283–335.

Huffaker, C. B., C. E. Kennett and G. L. Finney. 1962. *Hilgardia* **32:** 541–636.

Huffaker, C. B. and J. E. Laing. 1972. *Res. Popul. Ecol.* **14:** 1–17.

Huffaker, C. B., R. F. Luck and P. S. Messenger. 1977. *Proc. 15th Int. Congr. Entomol.*, pp. 560–586.

Huffaker, C. B. and P. S. Messenger. 1964a. Pp. 45–73 in DeBach (Ed.), 1964, referenced here.

Huffaker, C. B. and P. S. Messenger. 1964b. Pp. 74–117 in DeBach (Ed.), 1964, referenced here.

Huffaker, C. B. and P. S. Messenger (Eds.). 1976. *Theory and Practice of Biological Control.* Academic, New York.

Huffaker, C. B., R. L. Rabb, and J. A. Logan. 1977. Pp. 3–38 in R. L. Ridgeway and S. B. Vinson (Eds.), *Biological Control by Augmentation of Natural Enemies.* Plenum, New York.

Huffaker, C. B., K. P. Shea, and S. G. Herman. 1963. *Hilgardia* **34:** 305–330.

Huffaker, C. B., F. J. Simmonds, and J. E. Laing. 1976. Pp. 41–78 in Huffaker and Messenger (Eds.), 1976, referenced here.

Hutchinson, G. E. 1948. *Annu. N.Y. Acad. Sci.* **50:** 221–246.

Isaev, A. S. and R. G. Khlebopros. 1977. Pp. 317–339 in G. A. Norton and C. S. Holling (Eds.), *Proceedings of a Conference on Pest Management.* Int. Inst. Appl. Systems Analysis, Laxenburg, Austria.

Itô, Y. 1972. *Oecologia (Berl.)* **10:** 347–372.

Itô, Y. 1980. *Comparative Ecology* (2nd ed.) (trans. by J. Kikkawa). Cambridge University Press, Cambridge.

Itô, Y. and Y. Iwasa. 1981. *Res. Popul. Ecol.* **23:** 344–359.

Klomp, H. 1966a. *Med. Landbouwhogeschool, Wageningen* **66**(3): 1–10.

Klomp, H. 1966b. Pp. 207–305 in J. B. Cragg (Ed.), *Advances in Ecological Research*, Vol. 3. Academic, New York.

Lack, David. 1954. *The Natural Regulation of Animal Numbers.* Oxford University Press.

Lawton, J. H. 1978. Pp. 105–125 in L. A. Mound and N. Waloff (Eds.), *Diversity of Insect Faunas.* Symp. Roy. Entomol. Soc. London No. 9, Blackwell, London.

Lawton, J. H. and D. Schröder. 1979. *Proceedings of the 4th International Symposium on Biological Control of Weeds, 1976,* Gainesville, Fla.

Leslie, P. H. 1945. *Biometrika* **33:** 183–212.

Leslie, P. H. 1948. *Biometrika* **35:** 213–245.

Long, G. E., P. H. Duran, R. O. Jeffords, and D. N. Weldon. 1974. *Theor. Popul. Biol.* **5:** 450–459.

Lotka, A. J. 1925. *Elements of Physical Biology.* Williams & Wilkins, Baltimore.

MacArthur, R. H. and E. O. Wilson. 1967. *The Theory of Island Biogeography.* Monogr. in Population Biology, 1, Princeton University Press, Princeton, N.J.

MacFadyen, A. 1963. *Animal Ecology, Aims and Methods* (2nd ed.). Pitman, London.

Malthus, T. R. 1803. *An Essay on the Principle of Population as It Affects the Future of Society* (2nd ed.). J. Johnson, London.

Mason, R. R. 1974. *Can. Entomol.* **106:** 1171–1174.

May, R. M. 1973a. *Am. Natur.* **107:** 46–57.

May, R. M. 1973b. *Stability and Complexity in Model Ecosystems.* Princeton University Press, Princeton, N.J.

May, R. M. 1974. *Science* **186:** 645–647.

May, R. M. 1976a. *Nature* **261:** 459–467.

May, R. M. 1976b. *Theoretical Ecology. Principles and Applications.* Saunders, Philadelphia.

McLaren, I. A. (Ed.). 1971. *Natural Regulation of Animal Populations.* Atherton, New York.

Milne, A. 1957. *Can. Entomol.* **89**: 193–213.

Milne, A. 1962. *J. Theoret. Biol.* **3**: 19–50.

Morris, R. F. 1963. *Mem. Entomol. Soc. Can.* **31**: 116–129.

Nicholson, A. J. 1933. *J. Anim. Ecol.* **2**: 132–178.

Nicholson, A. J. 1954. *Austr. J. Zool.* **2**(1): 9–65.

Nicholson, A. J. 1957. *Cold Spring Harbor Symposia on Quantitative Biology* **22**: 153–172.

Nicholson, A. J. 1958. *Annu. Rev. Entomol.* **3**: 107–136.

Nicholson, A. J. and V. A. Bailey. 1935. *Proc. Zool. Soc. Lond.* Pt. I, pp. 555–598.

Odum, E. P. 1971. *Fundamentals of Ecology* (3rd ed.). Saunders, Philadelphia.

Pearl, R. and L. J. Reed. 1920. *Proc. Natl. Acad. Sci., Washington D.C.* **6**: 275–288.

Pielou, E. C. 1969. *An Introduction to Mathematical Ecology.* Wiley, New York.

Pimentel, D. 1961. *Am. Nat.* **95**: 65–79.

Pimentel, D. 1968. *Science* **159**: 1432–1437.

Pimentel, D. and R. Al-Hafidh. 1965. *Annu. Entomol. Soc. Am.* **58**: 1–6.

Pimentel, D., S. A. Levin, and A. B. Soans. 1975. *Ecology* **56**: 381–390.

Pimentel, D. and A. B. Soans. 1971. Pp. 313–326 in den Boer and Gradwell (Eds.), 1971, referenced here.

Reddingius, J. 1968. *Gambling for Existence.* Ph.D. dissertation, Gröningen, Netherlands.

Reddingius, J. and P. J. den Boer. 1970. *Oecologia* **5**: 240–284.

Ricker, W. E. 1954. *J. Wildl. Manag.* **18**: 45–51.

Ricklefs, R. E. 1979. *Ecology* (2nd ed.). Chiron, New York.

Ripper, W. E. 1956. *Annu. Rev. Entomol.* **1**: 403–438.

Royama, T. 1977. *Ecol. Monogr.* **47**: 1–35.

Smith, H. S. 1935. *J. Econ. Entomol.* **28**: 873–898.

Solomon, M. E. 1949. *J. Anim. Ecol.* **18**: 1–35.

Solomon, M. E. 1957. *Annu. Rev. Entomol.* **2**: 121–142.

Solomon, M. E. 1971. Pp. 29–40 in den Boer and Gradwell (Eds.), 1971, referenced here.

Southwood, T. R. E. 1977a. *J. Anim. Ecol.* **46**: 337–365.

Southwood, T. R. E. 1977b. *Am. Scientist* **65**: 30–39.

Southwood, T. R. E. and H. N. Comins. 1976. *J. Anim. Ecol.* **45**: 949–965.

Takahashi, F. 1964. *Res. Popul. Ecol.* **6**: 28–36.

Tamarin, R. H. (Ed.) 1978. *Population Regulation.* Dowden, Hutchinson, Ross, Stroudsburg, Pa.

Thompson, W. R. 1924. *Annu. Fac. Sci. Marseille* **2**: 69–89.

Thompson, W. R. 1929. *Parasitology* **21**: 269–281.

Thompson, W. R. 1939. *Parasitology* **31**: 299–388.

Thompson, W. R. 1956. *Annu. Rev. Entomol.* **1**: 379–402.

Utida, S. 1941. *Mem. Coll. Agric. Kyoto Imp. Univ.* **49**: 1–20. Summary in Allee et al., 1949, referenced here.

Utida, S. 1957. *Cold Spring Harbor Symp. Quant. Biol.* **22**: 139–151.

Uvarov, B. P. 1931. *Trans. Entomol. Soc. Lond.* **79**: 1–247.

Varley, G. C. and G. R. Gradwell. 1958. *10th Int. Congr. Entomol.* **2**: 619–624.

Varley, G. C., G. R. Gradwell and M. P. Hassell. 1973. *Insect Population Ecology.* University of California Press.

Verhulst, P. F. 1838. *Corresp. Math. Phys.* **10:** 113–121.

Volterra, V. 1926. *Mem. Acad. Lincei* **2:** 31–113.

Watt, K. E. F. 1959. *Can. Entomol.* **91:** 129–144.

White, E. G. and C. B. Huffaker. 1969. *Res. Popul. Ecol.* **11:** 150–185.

Wickman, B. E., R. R. Mason, and C. G. Thompson. 1973. *USDA For. Serv. Tech. Rept.* PNW-5.

Williams, G. C. 1966. *Adaptation and Natural Selection. A Critique of Some Current Evolutionary Thought.* Princeton University Press, Princeton, N.J.

Wilson, D. S. 1980. *The Natural Selection of Populations and Communities.* Benjamin/Cummings, Menlo Park, Calif.

Woodworth, C. W. 1908. *Science* **28**(51): 227–230.

Wynne-Edwards, V. C. 1962. *Animal Dispersion in Relation to Social Behavior.* Oliver & Boyd, Edinburg.

Wynne-Edwards, V. C. 1965. *Science* **147:** 1543–1548.

Zwölfer, H. 1975. *Verh. dt. Zool. Ges.* **67:** 394–401.

Chapter 13

Weather

W. G. WELLINGTON and
R. M. TRIMBLE

13.1 INTRODUCTION

13.1.1 The First Half-Century

In 1929, there was a meteorological conference in London, at which methods of tackling bioclimatic problems were discussed. Two entomologists, J. J. de Gryse (1929) and B. P. Uvarov (1929), summarized North American and European literature on the effects of weather and climate on insects. The large body of work and the novel ideas outlined in these reviews prompted meteorologists in the audience to commend the entomological community for its modern outlook. Indeed, the ensuing discussion included predictions of continuing entomological leadership in applied bioclimatology.

Today the kindest thing that can be said about such forecasts is that they seemed reasonable at the time. By the end of the 1920s there was abundant evidence that economic entomologists were far ahead of other applied biologists in their treatment of meteorological factors. This impression seemed fully confirmed in 1931, when Uvarov published his monumental work, *Insects and Climate,* in which he reviewed more than 1150 articles written in 11 languages on every conceivable aspect of the subject.

Half a century later in many, if not most, pest problems, the entomological community seems to have missed the target— integration of biometeorological findings into practical control programs—that the pioneers of the 1920s had clearly in their sights. Furthermore, meteorologists nowadays would be hard put to find grounds for congratulating entomologists on their advanced treatment of biometeorological problems.

How—or why—did the methodology that showed such promise falter? Certainly not for lack of effort. Since the 1931 monograph appeared, there have been numerous shorter reviews that summarized advances in particular areas of the field. Many of these reviews have become classics in their own right. Buxton's (1932) treatment of humidity, for example, can still be read with profit, especially in conjunction with the more recent summaries by Edney (1957) and Cloudsley-Thompson (1962). Gunn's (1942) review of arthropod body temperatures also contains a great deal of relevant information, to which Flitters' contributions (1968a,b) should be added. And the papers by John Leighly (1937) and Warren Thornthwaite (1940) are still the clearest explanations of the process of evaporation that can be found in the ecological literature. Gates's (1967) work on energy exchange in the biosphere sets these summaries in their appropriate biophysical context.

There have been a number of discussions of the effects of weather on insect abundance that examine this complex topic from specialized points of view (Andrewartha & Birch 1954, Chauvin 1967, Haufe 1966, Johnson 1969, Varley et al. 1973, Wellington 1954a,b, 1957, Wellington et al. 1966, White 1974). In particular, the papers of Powers Messenger (e.g., 1959, 1971) should be consulted by everyone concerned with the biological control of pest populations in natural environments.

Lowry (1969) provided in-depth biometeorological treatment of a variety of organisms, including insects, during which he examined radiation, energy budgets, environmental temperature, moisture, and air movement in considerable detail. Geiger's (1965) classic treatment of boundary layer climates also provides a wealth of microclimatological information. Geiger's text should be read in conjunction with Oke's (1978) useful synthesis of more recent work on small-scale meteorological processes. [For larger-scale meteorological processes, the text by Barry and Chorley (1968) presents the information in ways that make sense to ecologists as well as meteorologists. Barry and Perry (1973) have also provided a comparable treatment of the problems and methodology of synoptic climatology.]

If space permitted, numerous other reviews and specialized surveys could be added to the few mentioned above. In addition, since 1931, many more thousands of papers have appeared which describe the effects of single variables, such as temperature, on one or another of the biological processes thought to be important to insects. There have been innumerable details of this sort added to the conceptual framework Uvarov provided in 1931, but woefully few major changes or additions to it. Since Uvarov himself said that his monograph should be regarded as a guide, not gospel, something other than its perfection must be responsible for the fitful progress since its appearance.

In this, as in any discipline, when there has been such a flood of information but so little real progress over such a long period, the practitioners should begin to suspect that their attitude towards their material may need revision. Accordingly, as we enter the 1980s, perhaps the most useful contribution a new chapter on insects and climate might make would be to identify reasons for this lost momentum and describe an alternative approach.

13.1.2 Required Changes in Attitude

Uvarov (1945) identified an important aspect of the problem; ". . . Most entomologists have only a superficial knowledge of meteorology. . . . It is a sad fact that much bioclimatic work on insects consists of painstaking studies of the insect, conducted against a background of semi-amateurish excursions into meteorology." That comment, written in 1945, for the most part still applies. Despite similar complaints by later reviewers, the vast majority of bioclimatic investigations since 1945 have continued to blend the most up-to-date biological information and methods with meteorological concepts that were already outmoded in the 1930s. It is therefore not surprising that the results of such investigations are not applicable in pest management, however intellectually satisfying they may be to physiologists.

Experience since 1945 has shown, however, that ". . . proper training in meteorology and climatology" (Uvarov 1945) will not invariably cure this ailment—or, in most instances, even alleviate it. The underlying problem seems to be a lack of motivation rather than a lack of knowledge. One need not become a full-fledged meteorologist to do useful biometeorological research. But to obtain meaningful results one *must* have a more sustained *interest* in weather, as such, than most people, including meteorologists, have nowadays.

Not long ago our forefathers were acutely aware of the problems weather created for plants and animals because they themselves frequently experienced the same problems. As our shelters improved, however, our instinctive understanding and sense of kinship faded. Nowadays, when even farmers and entomologists prefer to wait until the weather improves before venturing outdoors, people tend to equate such evasions with independence and therefore think of weather in terms of discomfort instead of danger. In this very real sense, mankind is out of touch with other life forms that must remain outdoors, day and night, in all seasons.

Although local catastrophes, such as floods or freezes, occasionally remind us how vulnerable to the weather our crops and livestock remain, even then we often fail to realize that we too remain vulnerable, through their frailty, if not our own. Consequently, in times when the hardiest outdoorsmen go swathed in heated sleeping bags, boots, and handwarmers, that missing sense of kinship can only be restored by some harsher personal experience. Before the looming shortage of petrochemicals provides that experience for everyone, it seems only reasonable to attempt to rejoin those other life forms, at least in spirit, before events force us to do so in reality.

The quickest and most memorable way to discover what plants and animals face in their habitats is to spend a winter night in an open field, without adequate clothing or any artificial shelter. Standing there unprotected even for an hour brings a rush of understanding. Readers who shy away from that much kinship with crops, livestock, or insects may prefer the alternative approach described here.

13.2 PROCESS, PATTERN, AND SCALE IN BIOMETEOROLOGY

13.2.1 Two Views of Scale

The subject matter of any discipline can be viewed in terms of processes, patterns, and scale. More often than not, biologists use scale as an artificial divider to group processes and patterns into separate packages; for example, at the cellular, organismal, population, and community levels. Therefore we are inclined to treat the physical environment in the same way during ecological studies, dividing its processes and patterns into microscale (where the things that concern us live), mesoscale (where we can see events taking shape as far away as the horizon), and macroscale (which covers those regional, subcontinental, and hemispheric events too vast to imagine and seemingly remote from the immediate concerns of the biosphere). A biologist concerned with meteorological events on the surface of a leaf is thus inclined to dismiss events in the hemispheric or regional circulations as irrelevant happenings of no consequence for his study.

In biology this tendency to compartmentalize events has probably impeded progress more than we care to admit, but clearly has not prevented it. In contrast, using scale as an artificial divider in biometeorology is a grave procedural error which allows the investigator to neglect the otherwise obvious connection between the smallest- and largest-scale meteorological processes and patterns. What is happening underfoot is always a direct result of circumstances aloft and may in its turn already be influencing events overhead. To ignore this interdependence is to invite failure in any attempt to apply biometeorological knowledge in entomological research. For without knowing *why, when,* and *where* particular temperatures may be expected to occur, one cannot apply one's knowledge of *how* such temperatures affect a biological system.

It was indifference to the reciprocal connections between large- and small-scale weather events that derailed earlier attempts to apply biometeorological information in pest management. Consequently, a new attitude, rather than new data, is required to restore the lost momentum. Instead of viewing scale as a convenient device for separating differently sized events, one should view it as the device that connects them. Used thus, the concept of a sliding, instead of a compartmented, discontinuous scale, emphasizes the direct links between the unimaginably large and the barely noticeable small. Continuity of scale then becomes a sort of biometeorological Rosetta Stone, allowing observers to translate mentally from one familiar to two alien environments, the vast and the very small, which they can never occupy physically. The following example shows one way to make such journeys.

13.2.2 Footprints and Valleys

Like wet cement, freshly rolled and raked soil invites footprints. Passersby who yield to this temptation can severely damage a newly seeded lawn in any season. But in the autumn trespassers with sufficiently large feet may also partially compensate for their damage by means of the biometeorologically interesting situation they leave in their tracks. On a south or west slope the steep edge of the footprint which faces the low autumn sun acts as a heat-reflecting wall, against which the adjacent seedlings grow much more quickly than their counterparts in the print's flat bottom. Because of our large size in relation to these events within the footprint, such small-scale differences might easily escape our notice. But to a thrips-sized gardener living on the steep "hillside" formed by the print's eroded edge, those thriving seedlings would seem to be flourishing like espaliered trees against a sun-warmed wall, in comparison with the free-standing vegetation down in the "valley."

After a clear autumn night that produced marginal frost (i.e., ice crystals forming only in the most favorable locations), we might notice that some of the grass sprouting in the bottom of the footprint had been frost nipped, whereas the grass on the aforementioned edge was still unharmed. Our thrips-sized resident, however, would see the valley below him as a large frost pocket, in which temperatures lowered by radiant cooling were further affected by cold air draining from his own hillside. Indeed, if the plot on which we have been observing this footprint were itself located on the sloping side of a real valley, we in our turn might be able to see frost damage in the bottom of our valley comparable with that which our diminutive colleague saw in his. In both instances the frost damage that occurred would be due to the same combination of radiant cooling and cold air drainage, though on vastly different scales.

Only one further change in scale is required to reinforce the point of this comparison. If the hillside on which we have been standing were not merely one side of a small valley but also lay well to the east of the Continental Divide (e.g., along the western edge of the Cypress Hills on the Alberta–Saskatchewan border), we would often hear complaints from neighbors occupying the land below us that their frost-free season was much shorter than ours or their friends' farther west. But in this instance only a space traveler would be in a position to notice that our hill was part of a natural barrier that hindered the downslope flow of surface air from the higher land to the west, creating a dam that penned cold air on our neighbors' low-lying farms, thereby increasing the possibility of damage during otherwise marginal conditions for frost.

On the planetary scale at which our passing astronaut would view the scene, this portion of Southern Alberta would be no larger than the footprint with which we began this example. If he were sufficiently alert, therefore, our astronaut would notice much the same local pattern that we saw in the footprint and correctly attribute its cause. Neither we nor our neighbors, however, from our positions on the scale, could immediately discover the reason for the increased incidence of frost on their land, any more than our thrips-sized observer could be aware of events outside his footprint valley.

Without further belaboring the point, we should note that the way we arranged these terrain features, from the footprint to the Eastern Slopes, exposed them simultaneously to the same regional weather. In other words, the regional weather pattern that covered Southern Alberta during that autumn night carried within its constituent air mass the conditions that triggered frost formation at every scale, from micro to macro. And, by and large, the same physical processes and comparable terrain configurations were present and were interacting at every scale. Finally, it is worth noting that observers in our size range are better placed on the scale than either the diminutive insect-sized observer or the astronaut because we can, if we wish, watch events unfolding up-scale as well as down-scale from our position. At a time when we earthbound observers can call on orbiting satellites to expand our horizons up-scale, all that is needed to improve our information on micro–macro interactions is more attentive observation at the small end of the scale and a bit more imagination at every point on it.

13.3 MESOSCALE INVERSIONS AND THE VERTICAL DISTRIBUTION OF OVERWINTERING MORTALITY

The abovementioned combination of nocturnal radiant cooling and its accompanying cold-air drainage often produces inversions in the air temperature gradient near the ground during still, clear weather. During such inversions, temperatures near the ground become colder than those a short distance above it, so that the temperatures in the affected layer of air actually increase, instead of decreasing, with height. Because such inversions can occur during a clear night in any part of the world, insects everywhere are frequently exposed to this phenomenon.

The weather that produces inversions can profoundly influence the distribution of mortality in populations exposed to it. An overwintering stage that cannot long endure intense cold is especially vulnerable. To date, such effects have been more noticeable in forest than in agricultural settings because the reduced defoliation that follows extensive insect mortality is usually easier to see in forests than in pastures. When a shallow pool of intensely cold air kills all the defoliators wintering in the bottom of a valley, but leaves unharmed those on the warmer slopes above, the resulting difference in abundance becomes dramatically evident soon after the survivors emerge in the following spring. The top of the wintertime pool of cold air can then be seen as a line on the hillside, separating the new defoliation above it from the undamaged trees below.

Although such inversion-induced patterns of mortality are common in the forested valleys of the Temperate Zones, most forest entomologists have not recognized their meteorological basis and so have incorrectly attributed them to other causes. Only a few observers (e.g., Edmunds 1973, Henson et al. 1954, Stark 1959a, Tenow 1975) have correctly identified their meteorological origin.

13.3.1 The Winter Moth

Tenow (1975) noticed that outbreaks of the winter moth, *Oporinia autumnata*, in the birch forests of northern Sweden were regularly killed back within cold-air pools that accumulated in low places in this hilly terrain during strong wintertime inversions. The overwintering eggs were killed in every low-lying place in which the cold air accumulated, so that the trees in those places were not defoliated during the following year. After the severe winter of 1954–55, in fact, this "green belt" extended upslope to the top of the overall inversion because temperatures in that whole layer fell sufficiently to kill all the eggs within it.

13.3.2 Black Pineleaf Scale

Edmunds (1973) observed comparable sequences of events in local populations of the black pineleaf scale, *Nuculaspis californica,* in the valleys of eastern Washington. These insects overwinter on the needles of their host trees. In normal circumstances gradually falling temperatures during the autumn slowly condition the scales to cold (Chapter 6) so that they can withstand the much lower temperatures of winter. Without such conditioning the insects cannot survive a sudden freeze.

In some valley systems cool air from the higher surroundings regularly drains into the lower valleys. Where the mouths of these valleys are sufficiently narrow, outflow is impeded, so the incoming air accumulates in the lower reaches. Depending on the prevailing weather, these accumulations of cooler air may have diametrically opposite effects on the insects.

When the regional weather is near freezing, the air accumulating in the low places eventually falls well below freezing. Such low temperatures kill all the scales in the valley bottom, so that only those higher on the slopes are spared. In some years, however, when the autumn weather is not so cold, the cool air draining into the lower reaches does not kill the scales there. Instead, it conditions them to lower temperatures. Thereafter, any sudden outbreak of very cold air that descends on the region harms fewer of these conditioned individuals than their unprepared siblings upslope. After such conditioning, therefore, the pattern of overwintering survival is altered in favor of the scales in the valley.

In fact, Edmunds noted that in such circumstances the zone where survival is best is near the top of the autumn cold-air pool. Temperatures in this hillside zone would normally be well above the drastically low temperatures produced on the valley floor by a wintertime inversion, but not so warm that scales living there would never experience the autumnal flows of cold air. Thus, the insects in this zone would be hardier than their counterparts higher up the slope and yet would escape the extreme temperatures to which their siblings on the valley floor would be exposed during winter inversions. Only a small difference in elevation, sometimes <40 m, may separate this refugium from the zones of higher mortality above and below it.

13.3.3 The Lodgepole Needle Miner

Radiant cooling, cold air drainage, and localized differences in topography also interact during particular kinds of regional weather to cause patterns of mortality in the lodgepole needle miner, *Coleotechnites starki,* that are very similar to those described for the winter moth and the black pineleaf scale (Henson et al. 1954; Stark 1959a,b). In the Canadian Rocky Mountains the lodgepole needle miner has a 2-yr life cycle, overwintering as a larva within a host needle during two consecutive winters. Its mortality during these winters is a major factor affecting its subsequent distribution and abundance within a valley. As with the pineleaf scale, the greatest amount of overwintering mortality occurs either in the valley bottom or on the upper slopes, so that the middle zone often becomes a refuge area for the needle miner. Unlike the scale insect's refugium, however, the needle miner's is not so much the outcome of conditioning versus the effects of suddenly severe weather as it is the result of the relative frequency with which particular kinds of weather invade the area.

Of the half-dozen types of air masses (Wellington 1954a) mainly responsible for North American weather, two, polar air from the North Pacific and arctic air from northern Canada and Alaska, alternately influence the winter weather along the British Columbia–Alberta border. When westerly, or zonal, air flow (Barry & Chorley 1968, Barry & Perry 1973, Wellington 1954a) dominates the region, storms from the Pacific cross the Great Divide in rapid succession and move eastward across the plains. Arctic air then can affect the region only during the very brief interludes between these Pacific storms. The associated changes in local temperatures, from cool to cold to cool, therefore take place in a matter of hours or, at most, in 1–2 days. Consequently, the air within the high valleys and mountain passes is repeatedly stirred and changed whenever Pacific and arctic air alternate so rapidly.

Since the time span of each temperature regime is so brief, temperature inversions have few chances to form, let alone intensify. Accordingly, the air temperature gradient in any valley usually decreases with height in the normal way (Barry & Chorley 1968, Geiger 1965, Oke 1978); that is, the upper slopes remain consistently colder than the more sheltered valley floor. In such circumstances weather-induced mortality among the overwintering needle miners is mainly confined to those cold upper slopes.

Sometimes, westerly flow over the region is temporarily blocked by an outbreak of arctic air flowing southward over the foothills. Occasionally, this air mass is so large that it straddles the whole Cordillera, extending westward to the Pacific and penetrating as far south as California, where it creates the dreaded "freezes" in the citrus groves that neither smudges nor wind machines can alleviate. Whenever this meridional flow (Barry & Chorley 1968, Wellington 1954a) reaches such subcontinental proportions, its central area of high barometric pressure becomes exceptionally well entrenched. In fact, it may remain in situ for several weeks.

During such prolonged stagnation continual radiant cooling from the surface layer of the air mass creates truly massive inversions in the high valleys. Surface temperatures may fall below $-50°C$, while those a few hundred meters upslope may be only -10 or $-20°C$. Since these differences may persist for nearly a month, needle miners near the bottom of the valley have no chance to survive unless they are completely buried in deep snow. The most likely survivors will be those situated on the slopes near the top of the inversion, where warmer temperatures prevail.

Winter mortality among needle miners occurs mainly in January, which is usually the coldest winter month in the Canadian Rockies (Henson et al. 1954, Stark 1959a). As noted, this mortality is distributed vertically in the same way as the zones of extreme cold, shifting from the valley bottom to the upper slopes as the frequency of winter storms increases and the duration and intensity of inversions decrease. A detailed survey of weather records back to 1920 and of yearly averages back to 1885 led Stark (1959b) to suggest that "climatic release" (cf. Thompson et al. 1979) of the needle miner populations of the 1940s took place during the final portion of a warming trend in the western climate. This trend was accompanied in western Canada by increased winter storminess (and therefore fewer prolonged "stays" of arctic air). Becoming more apparent in the late 1930s, this storminess reached its peak in the 1940s, and began to decrease in frequency again in the early 1950s. With the exception of those decades, therefore, since 1885 the winter climate in these high mountains has usually been too severe to sustain any lengthy outbreak by an insect that spends two consecutive winters inside a poorly insulated pine needle.

13.4 STORM TRACKS AND THE SPRUCE BUDWORM

Comparable long-term changes in weather patterns in other parts of the continent have also affected the population dynamics of the spruce budworm, *Choristoneura fumiferana,* and its western relatives (Greenbank 1956, Hard et al. 1980, Morris 1963, Otvos & Moody 1978, Pilon & Blais 1961, Shepherd 1959, Wellington 1952, 1954a,b, Wellington et al. 1950). These insects belong to a group that tie, roll, or web leaves in the larval stage. Thus they feed more or less concealed until they are ready to pupate.

Insects that occupy such individual shelters are, in effect, using sun-warmed greenhouses which offset the cool spells that frequently occur in the spring and early summer in the Temperate Zones. But this heavy dependence on solar radiation makes the whole group particularly vulnerable to the thick clouds and prolonged spells of rain that characterize the frontal storms of the North Temperate Zone. Such greenhouse builders function more effectively during the clearer, more stable weather that intervenes between frontal storms. Thus they are more likely to thrive in a locality when the seasonal frequency of frontal storms is much reduced and the duration of the intervening periods of sunny weather is correspondingly increased (Wellington 1952, 1954a,b, Wellington et al. 1950).

The spruce budworm is no exception. Even during winter dormancy early snow cover on the conifer branches kept there by continuing cool weather seems to provide the best conditions for hibernating larvae. Repeated, rapid thawing and freezing are detrimental. In the spring the emerging larvae need some direct sunshine to warm them sufficiently to allow them to reach the buds or needles in which they must establish themselves. And during their subsequent feeding period they require further solar warmth: (a) to raise their body and tunnel temperatures sufficiently above the cool surrounding air to permit feeding and (b) to increase the rate of evaporation inside their tunnels, to keep them from succumbing to the torpor that overcomes larvae kept in saturated air. In all periods of the larval stage, therefore, sunny, dry weather directly provides ideal conditions for rapid development.

Dry, sunny weather also is *indirectly* beneficial for the spruce budworm. Before the insect can develop large populations, the host trees must reach the appropriate stage of maturity (Morris 1963). In mature stands a dry spell when the staminate buds are forming will lead to a bountiful supply of staminate flowers during the next spring. The results of several studies (e.g., Greenbank 1956, Morris 1963) suggest that this enriched diet is more beneficial for the larvae than a diet consisting solely of developed or developing foliage. For such reasons, therefore, no substantial regional increase in numbers of spruce budworm will occur until seasonal climatic control of the indigenous, small populations is relaxed by a change from wet, cloudy weather to drier, clearer weather, particularly in the spring and summer months. If such weather recurs 3–4 yr in succession, the local populations respond extremely rapidly to this climatic release, reaching outbreak proportions wherever there is a suitable forest within the region.

Throughout most of the North American boreal forest (i.e., from northern Alberta to Newfoundland) such a prolonged change from very moist, cloudy weather to much drier, sunnier weather involves a major shift southward by the continental storm track and its associated jet stream (Barry & Chorley 1968, Barry & Perry 1973, Reiter 1967, Wellington 1954a). The track must shift sufficiently far south to prevent most of those humid air masses spawned near the Gulf of Mexico from entering the northern forests. In their absence the weather in these forests is mainly controlled by the much drier air masses from the continental and arctic source regions of northern Canada (Wellington 1954a,b, Wellington et al. 1950).

The western relatives of the spruce budworm respond similarly to comparable shifts in that portion of the continental storm track which controls their regional weather pattern. A change in the position of the jet stream that prevents moist Pacific air from overrunning the western region between Oregon and the Yukon automatically brings that region under the influence of much drier air from continental sources, and the western budworms thrive in the increased amount of sunshine.

In the much more rugged landscape of this western region, however, the local terrain may drastically affect even the most violent storms from the Pacific. For example, even in very cloudy weather, the views of southwestern British Columbia transmitted by meteorological satellites often show an X-shaped area of comparatively clear weather embedded in otherwise continuous cloud. This relatively clear zone is created by the intersection of two major valleys which dissipate the clouds. The area includes the

Lillooet refugium, within which endemic populations of western budworm persist through years of generally bad weather, spilling out from this climatic refuge only when the shifting regional circulation brings warmer, drier weather to the whole southwestern quarter of the Province. In rugged terrain, therefore, the topography itself has important effects on the large-scale weather systems passing overhead.

13.5 TERRAIN INFLUENCES

In rough terrain topographic variations can create remarkable differences in cloudiness over very short distances. Deep valleys that barely support clouds even when the sky is overcast lie next to ridges or peaks which are rarely without caps or scarves of cloud. In many parts of the world the cloud boundaries that mark the locations of these topographic extremes are very stable, reappearing in the same positions in the sky whenever a particular kind of weather recurs. In such places, therefore, the surrounding hills and valleys have become more than a handsome setting for the local biometeorological interactions; they play a major role in those interactions and may even control the outcome.

Clearly, these small- and medium-scale topographic features have not received the attention they deserve. Since our satellites have shown that there are many more folds than flat spots in our planetary landscape, we should seize the opportunities for improving ecological investigations which these terrain influences provide. In this and the following section we examine some of the possibilities for exploiting such opportunities.

13.5.1 Terrain Influences on Tent Caterpillar Population Dynamics

The western tent caterpillar, *Malacosoma californicum pluviale,* is another greenhouse-building insect. But the communal web which this colonial insect constructs during its larval stage is a much larger and more efficient apparatus for collecting and retaining heat than the small, individual tunnels or leaf rolls that tortricid larvae produce. In addition, although the dense sheets of silk that form the exterior walls of this communal tent are sufficiently waterproof to keep out the rain, they slow, but do not prevent, the passage of water vapor, thereby providing adequate control of interior moisture. The tent has allowed this insect to establish itself earlier in the season, and in colder surroundings, than its non-tent-building relative, the forest tent caterpillar.

But early emergence may have disadvantages as well as advantages. The moths lay their eggs in midsummer when there is seldom much difference between hill and valley climates. The larvae, however, emerge in March or early April of the following year, in a season when the west coast weather is notoriously capricious. Even in normal springs most of the hill climates are too cloudy or too wet for the young larvae. Only during exceptionally dry, sunny springs can the insects survive in the high as well as in the low parts of their habitat. Conversely, during very cloudy and wet springs, only the lowlands, with their thinner clouds and lighter rainfall, remain tolerable (Thompson et al. 1979, Wellington 1964, 1965, Wellington et al. 1975). The insect, therefore, must contend with this terrain-induced mosaic of tolerable and less tolerable local climates which may suddenly change between consecutive generations as the spring weather changes.

Air flowing from the northeast (i.e., from the land toward the sea) is usually so dry that clouds do not form even when the air is lifted as it crosses the coastal ridges. Therefore, northeasterly airflow brings comparatively cloudless weather to the coastal region. But a westerly flow from the ocean brings air which is so cool and moist that only slight lifting by the outlying coastal ridges is required to cap every hill with dense clouds and heavy showers. On the Saanich Peninsula of Vancouver Island, where most of the work on *M.c. pluviale* was done, a southwesterly airflow produces the most striking local variations in cloudiness and precipitation among the tent caterpillar habitats (Wellington 1965).

The differential dispersal required to cope with this climatic mosaic has been discussed at length elsewhere (Berryman & Safranyik 1980, Wellington 1977, 1980) and need not concern us here. The main purpose here has been to point out that topographic features only a few km² in extent can greatly influence events within large weather systems mainly, though not exclusively, by their effects on the growth and decay of clouds. Since the amount of cloudiness in turn affects the intensity of radiant heating or cooling and the amount and distribution of precipitation, the resulting patchwork of terrain climates affects the lives of resident insects at every stage, sessile or motile.

Just as terrain features measured in km² can produce mesoscale variations in cloudiness and rainfall, so much smaller features, encompassing at most a few hundred m², may have similar, more localized effects. The influence of these small-scale changes in relief on the intensity of rainfall and the thickness of low clouds that pass near them also may affect insect survival when the weather is sufficiently harsh. If the insects have already been subjected to nearly as much rain as they can tolerate, they may get the extra amount they cannot withstand if they are situated on a small knoll instead of in a hollow. The dividing line between barely tolerable and intolerable places may be remarkably sharp at such small scales. For example, in one part of the Saanich Peninsula, only 4 m separated 8 tent caterpillar colonies that were frequently drenched by passing showers from 10 others that remained dry. The first group was near the edge of a knoll that gave added lift to the passing clouds, whereas the others were in the adjacent hollow. None of the first group survived the larval stage. Five of the other 10 colonies reached the pupal stage (cf. Wellington 1965:627).

13.6 WEATHER AND INSECT DISPERSAL

13.6.1 The Biometeorological Framework of Dispersal

As might be expected, the different kinds of local weather created by topographic features can influence dispersing insects at least as much as they influence the development and survival of more sedentary stages. In view of the crucial importance of dispersal in population dynamics (Wellington 1980, and Chapter 7), the impact of local terrain on the processes involved deserves closer examination. Johnson (1969) has already carefully analyzed the influence of meteorological factors on the various stages of dispersal, and the movements of insects are dealt with in depth in Chapter 7. Consequently, this section of this chapter is concerned more with the biometeorological framework of dispersal than with its biological mechanisms.

We are constantly reminded that insects inhabit an alien world. But while they remain earthbound with us, we can at least visualize some of the problems they face. They become truly alien when they take to the air, entering a medium as remote from our everyday experience as the ocean deeps. That comparison is not farfetched. One of the main reasons we do not easily relate to aerial events is that we live at the bottom of the atmosphere, underneath it rather than in it. In conventional seas, a bottom-dwelling barnacle has a very different and far more limited view of the current patterns than a free-swimming salmon. As bottom dwellers in our atmospheric ocean, we have a comparably restricted view of the patterns overhead (Rabb & Kennedy 1979, Wellington 1979). This limited view has hindered many ecologically based investigations, but none more so than studies of aerial dispersal, an act that takes its participants out of one terrestrial habitat and into another by way of the atmosphere. Broadening our view of the atmospheric patterns and processes involved can only improve our understanding of this behavior and its population consequences.

We can achieve that greater breadth by learning to look down through the atmosphere instead of always looking up. Fortunately, we now have many kinds of remote sensors, including a variety of weather satellites, which can provide this essential overview. Before we can exploit the astronaut's vantage point, however, we must free ourselves from the narrow confines of the earth-bound point of view. And in this, as in the earlier biometeorological exercise in Section 13.2.2, scale again becomes the Rosetta Stone for translating the foreign into the familiar.

13.6.2 Midges and Microscale Convection

The earliest studies of airborne insects left the impression that the atmosphere at times was thick with aerial plankton. But airborne insects, whether drifting or flying, are no more uniformly distributed through their medium than marine plankters are in theirs. The greatest concentrations of airborne insects tend to form where there is some aerial support; that is, where there is upward motion in the atmosphere, rather than downdrafts.

The swarms of midges that frequently distress after-dinner strollers on the paved footpaths of lakeshore resorts are microscale examples of this relationship. On a still evening at the end of a sunny day these swarms may be sufficiently thick in lakeshore residential areas to clog the eyes and nose of anyone using the footpaths. Most strollers then either retreat or keep battling the swarms along the path until they reach their destination. They would fare better if they stepped off the pavement onto the bordering grass. For as long as the swarms hug the pavement, no midge will be found above the grass.

The main attraction of the pavement during the early evening is the gentle current of air that rises from the still-warm cement. But there is also some avoidance of the compensating downdrafts that form over the cooler grass on either side of the footpath. While the path remains the warmer place, no part of a swarm can be forced off it onto the surrounding grass. Even the most violent arm movements will not dislodge the insects from the path at such times. After sunset, however, the pavement cools more rapidly than the grass. Eventually, therefore, the distribution of vertical currents above the two substrates is reversed, so that the grass, now warmer than the pavement, creates the rising air currents while the downdrafts occur over the pavement. The swarms of midges then leave the footpath and reform above the grass. Strolling in comfort thus becomes a matter of staying on the cooler of the two substrates; the grass while outbound, and the pavement going home.

13.6.3 Convective and Turbulent Transport

The small-scale convective circulations supporting the lakeside swarms differ only in size and strength, not in origin, from the larger examples of local convection with which we are more familiar. The small circulation produced by the differing heat capacities of turf and pavement demonstrates in microcosm the physical basis for terrain-induced convection of any scale. In comparatively flat landscapes water and different kinds of soil or vegetation are the main causes of the differential radiant heating that produces thermals on sunny days. These thermals remain invisible to us until the rising air cools sufficiently to condense the accompanying water vapor. But whether or not we can see them, such updrafts provide one of the major transport systems that insects use during aerial dispersal. In fact, in comparatively flat terrain they and their low-level counterpart, turbulent wind, are the only aerial transport systems that may be regularly available if large-scale storms rarely enter the region (Wellington 1979, and Chapter 7).

Convection and turbulence each present dangers as well as opportunities for airborne insects. The cloudscapes they produce consist of clusters or networks of cumulus clouds separated by clear patches. From a drifting insect's point of view, each cloud marks the top of the thermal (or, in the case of turbulence, of the upward sweep of the wind) that had been providing the lift required to travel cross-country. Each clear zone between two clouds thus marks an area of subsidence, where descending currents soon reduce the altitude previously gained. As long as the total area covered by convective or turbulent clouds is far greater than the clear areas, drifting insects will encounter supporting currents more frequently than downdrafts and so may travel several kilometers before being returned to the ground. But when the clouds are few and scattered, there are rarely enough thermals to carry drifting insects very far before they are unceremoniously dumped by the intervening downdrafts.

Wherever there is more land than water beneath them, airborne insects that are returned to the surface by downdrafts are not likely to encounter any more hazards in their drop zone than they would meet in their original habitat. Where there are many lakes or marshes, however, both convective and turbulent transport systems become more hazardous. Marshes and lakes become obligatory drop zones for anything drifting over them, and many insects that are caught in their bordering downdrafts cannot avoid falling into the water (Wellington 1979).

13.6.4   Effects of Terrain-induced Circulation Patterns

In comparatively flat landscapes the patterns of turbulent or convective clouds produced by wind or solar heating may be very evenly spaced. In rougher terrain the patterns of air currents and their accompanying clouds are far less regular because the varying height of the terrain exerts important effects. We have already seen how stable the local patterns of clouds formed by major terrain features may be. That day-to-day stability offers a very useful transport system for small, flightless insects living in such places. Since, as noted earlier, a great deal of the earth's surface is uneven, one form or another of this stable transport system can be found across a great variety of habitats, where it remains available on a regular basis for any inhabitants already adapted for drifting.

The versions of this system which develop in narrow, steep-sided valleys are easier for novices to observe than the less obvious versions which develop over more rolling countryside. The cross-valley circulation system that develops in high mountain valleys is particularly interesting. At the same moment when this system is providing planktonlike drifters with a highly directional and very reliable means of traveling between similar habitats on opposite sides of the valley, it may also be hindering the directed travel of the more robust fliers dispersing or foraging over the valley floor.

The cross-valley transport system is part of the circulation pattern that begins to form in a narrow, steep-sided valley whenever sunlight warms only one side. For example, in the early morning, the low sun will warm the tops of the east-facing slopes, while the whole western face remains shaded. As the warmer air rises up the heated side, it is replaced below by air flowing across the valley floor from the opposite, shaded slope. A return cross flow soon develops near the ridgetops which compensates for this lower cross flow and completes the circulation pattern. The visible results of this differential heating are the clouds which form above the east-facing ridge soon after sunrise. Some parts of a ridge prove to be especially good cloud generators, producing lines of clouds that float out over the valley on the upper cross flow, dissolving slowly as they drift away from their parent ridge. These clouds show that the cross-valley transport system is functioning.

The various life zones in high mountain valleys are usually compressed into comparatively narrow bands by altitudinal variations in climate. The cross-valley circulation can carry inhabitants of the higher zones between equivalent habitats on the opposite sides of the valley, thereby minimizing the problems they might otherwise encounter if they landed in unsuitable habitats. This system is particularly valuable for sessile or wingless forms that would otherwise drift aimlessly on a less directional set of air currents. Its existence undoubtedly accounts for hitherto inexplicable appearances of wingless forms in out-of-the-way places. Since its indicator clouds make the cross-valley circulation easily visible from the ground (Wellington 1977), it is one of the most accessible systems for ecological work.

Some cloud lines may become sufficiently dense and continuous to cut off the polarized light from the overhead sky during most of the day, thereby affecting the movements of flying or crawling insects that use polarized light to maintain a straight course (Anderson & Kaya 1975, Wellington 1974a, 1975). Losing their sky compass while they are in unfamiliar territory either immobilizes such insects or turns them back into the clear areas between the cloud lines, where they can still use polarized skylight to direct their travel. Cloud lines thus may disrupt or channel dispersal (or the foraging activity that follows a period of dispersal) if they are sufficiently well developed to interfere with navigation.

13.6.5 Modeling Terrain Effects: The Gypsy Moth

In the narrow, steep-sided valleys just considered, the local terrain that produces the cross-valley circulation patterns can override the effects of winds and weather above the ridge tops. This overriding influence can be seen in the directional stability of the cross-valley cloud lines. The angle at which these lines (and thus, their supporting air currents) cross a narrow valley remains remarkably constant over a wide range of large-scale weather patterns. As might be expected, however, the direction of the overwind has more influence on the local circulations of broader, shallower valleys, so the paths of terrain-induced cloud lines are less constant in rolling

country. In this limited sense the directional stability of cross-valley circulations is inversely proportional to the breadth and depth of a valley.

This slight increase in uncertainty concerning the ultimate destination,
however, does not lessen the value of air currents as a vehicle for crossing
valleys in rolling terrain. Although the terminus may not be so precisely
pinpointed in broad as in narrow valleys, the expectations of a successful
crossing remain almost as high, even in large valleys. Recent field and
modeling experiments on the wind-borne dispersal of gypsy moth larvae
(Cameron et al. 1979, Mason & McManus 1980, 1981) dramatically emphasize this point. This work is a prime example of the explanatory power
that up-to-date meteorological concepts can bring to ecological investigations.

The atmospheric dispersion model that Mason and his colleagues developed and tested was a variant of those used in studies of atmospheric
pollution. Such models can be successfully applied to insects, such as young
gypsy moth larvae, that are passively transported once airborne. The model
incorporated the effects of terrain features that generate local wind fields
by interfering with the overwind or by creating local circulation patterns
similar to those already discussed. It allowed the investigators to predict
larval dispersal patterns from mathematical descriptions of these wind
fields and their associated terrain features. These predictions were tested
in field experiments that provided data for validating the model.

The model and the accompanying tests conclusively demonstrated that,
especially in relatively flat country, the majority of drifting larvae were
deposited much closer to the source than the folklore of airborne dispersal
would have us believe. (The occasional discoveries of small insects far at
sea have encouraged ecologists to ignore the fact that most things which
go up must soon come down again if they are not actively supporting
themselves while airborne.) Although initial distances traveled by the gypsy
moth larvae were often <60 m, even when there was some turbulence,
field trials suggested that some larvae could reenter the air stream after
their first landing and so cover a little more ground in a series of short
"hops." Nevertheless, the vast majority of those airborne did not travel
very far.

When airflow across parallel ridges was considered, the model predicted
that maximum deposition should be expected at points just below the top
of the next ridge downwind, rather than in the valley adjacent to the source.
According to the model, updrafts around the source ridge should carry
the larvae close to the facing summit of the downwind ridge, where they
would drop before reaching the next updraft. In fact, field observations
in Pennsylvania have shown that infestations spread from ridgetop to
ridgetop, leaving the valleys less heavily infested. But even in these ridge
and valley situations, distances achieved by aerial dispersal seem limited to
a few kilometers at most.

The model's predictions also explained larval distributions observed near
the New Jersey shore. In this region bands of heavy defoliation 2–3 km

wide appeared 8–10 km inland in successive years between 1970 and 1975. Coastal areas commonly have zones of convergence, at varying distances inland, where the onshore breeze meets the land breeze. Where these winds, in effect, meet head-on, their horizontal velocities are largely translated into vertical motion which provides some support for any object entering its sphere. When the settling velocity of newly hatched larvae was subtracted from the vertical velocity of the air currents in this convergence zone, the model predicted a band of increased population paralleling the shoreline some 11 km inland. This prediction closely agreed with the observed defoliation patterns of 1970–75.

13.6.6 Insect Aggregations in Convergent Wind Fields

Other recent examples of sea-breeze convergence zones (Greenbank et al. 1980, Rainey 1976) show that this mesoscale phenomenon is a very important mechanism for concentrating drifting and flying insects in narrow segments of the atmosphere. [Insectivorous birds had discovered the benefits of such concentrations long before entomologists even suspected their existence (Rainey 1976, Simpson 1967, Wellington 1979).] However, the sea breeze zone is only one of the several kinds of convergent wind fields that R. C. Rainey has discussed in a series of papers, beginning with his report on the relationship between locust swarms and the Inter-Tropical Convergence Zone (ITCZ) (Rainey 1951). The results of that brilliant application of synoptic meteorology to an intractable economic problem have been discussed in every ecological text since the 1950s (cf. Chapter 7), so they are not repeated here. As Rainey (1963) learned more about the interacting systems during further investigations, however, he found that other insects besides locusts (Rainey 1976, 1979) were associated with the ITCZ. In fact, the ITCZ carries with it its own complex of winged fauna of every sort, becoming a sort of movable ecosystem as it travels up and down Africa (Rainey 1978).

However, the ITCZ is merely the largest and most flamboyant of the convergent wind fields that concern us here. The abundant rainfall this vast system supplies as it passes over a region certainly is the essential trigger for the complex set of interacting biological processes that explodes behind it (Rainey 1978). But within it, as in every smaller scale example of a convergence zone (Greenbank et al. 1980, Mason & McManus 1980, Rainey 1976), it is the physical process of lifting associated with the converging winds that maintains the accompanying aggregations of airborne insects. Since these aggregations are the raw material on which most of the subsequent aerial and terrestrial interactions depend, the overriding importance that Rainey has attached to the *process* of convergence, at every scale, is fully warranted.

There is one other well-documented North American example of the concentrating effect that convergent wind fields have on airborne insects.

Greenbank et al. (1980) used radar and aircraft observations to expand earlier studies (Greenbank 1956, Morris 1963) of dispersing moths of the spruce budworm in New Brunswick. The radar observations provided data on the rates of climb and final altitudes reached by flying moths. The aircraft observations, among other things, provided additional data on the wind shift fronts and sea breeze fronts that were located by weather equipment on the ground. The results showed that the moths often disperse high above the ground and generally travel toward the northeast, in the direction of the prevailing wind. Decreasing temperatures at night in the north central highlands are probably partly responsible for concentrations of moths that collect there. In addition, however, sea breeze fronts moving inland at night from the coastal areas occasionally increase the concentrations in the highlands by virtue of the convergent wind fields they create. These studies are leading to better estimates of the actual amount of adult dispersal and its potential for influencing population trends. Although a directional component in the movement of moths within the region had been previously recognized, the magnitude of the nonrandom element that might be introduced by the various wind fields had not previously been suspected.

13.6.7 Commuting on Seasonal Winds

Other studies employing radar (Schaefer 1976) have shown that insects may themselves increase the concentrating effect of convergent wind fields by reacting to one another or to local air currents. Dragonflies and butterflies often congregate in narrow rows in the bordering updrafts of convective cells, thereby obtaining free lift and cross-country transport by restricting the height at which they travel (Schaefer 1976).

Variations on that behavioral theme probably occur more frequently than we suppose. Some insects display near-frenzied bursts of activity when there are small but rapid fluctuations in barometric pressure (Wellington 1946a, 1957, 1974b). Such bursts of activity may predispose many members of a local population to aerial transport within a rapidly moving summer cold front by causing them to fly just when the violent updrafts of the squall line appear overhead. Greenbank (1956) and Henson (1951) reported mass flights and depositions of spruce budworm moths that could be traced to such responses. In addition, Wellington (1974b) has described how resting blackflies took flight en masse whenever an approaching cumulus cloud was preceded by minute but extremely rapid fluctuations in pressure. Although there is no doubt that insects in such a state can be carried high into the clouds (Henson 1952, Knight & Knight 1978) and can often survive the subsequent journey (Greenbank 1956, Henson 1951), the reason for their initial, often frenzied reaction is still a matter for conjecture. In the Diptera the antennae and, in particular, Johnston's organ seem to be directly involved (Wellington 1946b). It is also possible that some indi-

viduals may be responding at least as much to changes in the electrical field
as to pressure changes (Edwards 1960, 1961, Maw 1960, 1961, 1964, Wellington 1957). Unfortunately, it is still impossible to separate electrical from
other influences in normal field situations, and attempts to measure electrical effects in the laboratory have often been disappointing (Perumpral
et al. 1978).

While studying the movements of coccinellid beetles between lowland
feeding areas and mountain hibernating sites, Hagen (1962) discovered a
particularly interesting way in which insects may exploit individual components of weather to assist their travels. The lowland beetles tower up-
wards from the fields on warm, calm mornings when they can exploit the
summertime convective activity. Because of the normal decrease in air
temperature with height, the climbing beetles eventually reach a level where
they become too cold to fly (11–13°C). From that level they apparently
drop down into a warmer layer and proceed in this oscillating manner in
the direction of the upper winds, which in summer commonly blow toward
the mountains that rim the San Joaquin Valley. On reaching the mountains,
they are then deposited on the upper slopes within the zone that intersects
with their temperature ceiling. In the following spring, when the temperature of the montane sites rises, the emerging beetles again fly upward and
encounter the seasonal winds that now blow back toward the valley. In
flight, the returning beetles once more oscillate up and down against their
temperature ceiling, but eventually they are forced down to the ground as
the height of this ceiling decreases at the end of the day. In carrying out
this investigation, Hagen made excellent use of current meteorological
concepts to show how the beetles exploit the seasonal changes in the regional weather patterns.

The timing, direction, and destinations of the southward migrations of
the monarch butterfly, *Danaus p. plexippus*, in North America and their
relation to autumn weather are now well known, thanks to the Urquharts
and their collaborators (Urquhart 1960, Urquhart & Urquhart 1977, 1978,
and cf. Chapters 6, 7). Gibo and Pallett (1979) have added an interesting
set of biometeorological observations on the late-summer exodus of the
butterflies from southern Ontario, which shows how the insects exploit the
winds during their long journey south. In laboratory trials Gibo and Pallett
compared the aerodynamic properties of the monarch butterfly with data
on one expert soarer, the white-backed vulture, and two man-made soaring
machines, a low-performance hang glider and a high-performance glider.
The sinking rates of gliding monarchs compared favorably with those for
white-backed vultures and were achieved at much lower air speeds. In the
field the butterflies were able to exploit tail winds and minimize hindrance
by strong head winds by varying their flight times and altitudes. When
thermals provided favorable conditions for soaring, the insects reduced
the duration and frequency of powered flight and soared instead. When
they flew with a tail wind, the insects were also able to use the rising air

currents and standing eddies around ridges and buildings to negotiate such obstacles with minimal effort. In short, they were experts in cross-country soaring, apparently able to detect thermal activity close to the ground, enter it, and stay within it to reduce their need for powered flight.

13.7 HARSH CLIMATES AND OTHER CONSIDERATIONS

In several preceding sections we have emphasized the sometimes dramatic differences in adjacent local climates that topographic differences can create. But larger scale climates also may vary dramatically over the range of a widely distributed insect. At higher latitudes and altitudes, and with increasing distance from the moderating effects of large bodies of water, the period favorable for insect activity shortens as the average daily temperature decreases.

Bursell (1974) has already summarized the extensive literature on the relationship between temperature and insect development. The developmental physiology of the local populations of a widely distributed species may be closely adapted to the peculiarities of local climates. For example, more rapid development at a given temperature has been observed in northern populations of lepidopterous species (Danilevskii 1965), aphids (Campbell et al. 1974), mites (Beck 1968), and mosquitoes (Trimble & Smith 1978). Additionally, in both aphids (Campbell et al. 1974) and lacewings (Tauber & Tauber 1976) the threshold for development is lower in northern than in southern populations of the same species. The more rapid rate of development and lower developmental threshold are adaptations to a cooler climate which serve to synchronize the growth, development, and reproduction of insects with local phenological events.

Although such adaptations allow an insect to cope with the kinds of spatial patchiness already discussed in this chapter, they also provide some of the mechanisms it requires to survive temporal patchiness as well. When the degree of temporal patchiness is expressed through seasonal, rather than shorter term variations in weather, insects respond physiologically as well as behaviorally to these longer term environmental changes. Thus, they may either move or become dormant while the environment is unsuitable.

The environmental regulation of migration is still poorly understood, despite the fact that a large number of species engage in seasonal movements (Dingle 1978). But the role of the physical and biological environment in the regulation of dormancy has received a great deal of attention (Beck 1968, Danilevskii 1965, Danks 1978, Dingle 1978, Masaki 1980, Saunders 1976, Tauber & Tauber 1976). Since the phenomena of diapause (Chapter 6) and insect movements (Chapter 7) have been treated in detail, neither topic requires further discussion here. But diapause and dispersal both lead to a biometeorological problem that has not been generally recognized, so that problem deserves additional comment.

The behavioral and physiological adaptations which insects have evolved for temporarily evading intolerable conditions are fascinating as well as impressive (Dingle 1978). But we must not allow our admiration for such dazzling evolutionary footwork to blind us to one other fact. Migration and diapause are merely temporary, and therefore very incomplete, solutions of a much larger, continuing problem—how to survive in a generally hostile environment (Wellington 1977, 1980). Running away or lapsing into a kind of coma to avoid seasonal deteriorations in climate are methods of side-stepping this problem which neither solve nor overcome it. The survivors of a period of diapause never again encounter exactly the same environment they knew before their change of state. Nor do the survivors of any diaspora ever find exact duplicates of the old environment they abandoned. So when the escape period has ended, both groups—the runners and the sleepers alike—face a similar problem: how to survive in a place that differs, perhaps radically, perhaps only subtly, but that differs, nonetheless, from the place they once knew. Their common problem is how to survive in these differently structured habitats. Whether or not they succeed depends on how well they can match all the elements of their life-style, including their capacities for dispersal and diapause, to these varying structures.

As biologists we are all aware of the complex linkages among the life forms that occupy any habitat (Chapters 2, 12). Often—perhaps too often— we become so preoccupied with those linkages that we neglect the physical part of the habitat—its weather. The biometeorological examples discussed in this chapter were selected to remind readers that weather, through the pervasive nature of its interactions with the inanimate as well as the animate components of a habitat, is more than a major structural element of habitat; it is also a major determinant of events and processes therein. Since this dual role leads inevitably to population consequences, weather deserves far more attention from ecologists than it has received since controversies over density-dependent and density-independent processes first appeared in the literature.

13.8 SUMMARY

We have endeavored to show how the weather patterns and processes which affect insects are linked along a scale that connects the smallest with the largest imaginable events. For example, despite their vastly different scales, there are similarities in the patterns of frost damage which may appear in a footprint, a valley, and a large depressed plain during the same clear night. These similarities arise because the same processes, radiant cooling and cold-air drainage, are involved in each situation, and these processes in turn are triggered within a single type of regional weather.

One can look down on a footprint and comprehend what transpires there. But without prior knowledge, no earthbound observer would be likely to realize that the depressed plain may sometimes be merely a very-large-scale version of a footprint. Consequently, if we are to understand fully the impact of weather on an insect, we must be able to project our imaginations up- and down-scale well beyond the limits our own size imposes. By doing so, we may begin to understand how to exploit the relationships between very large and very small events.

As a first example of pattern and process at work on a local scale, we discussed the interactions of terrain with invading arctic air masses which affected the vertical distribution of mortality in populations of winter moths, scale insects, and needle miners overwintering in high valleys. These examples provided further evidence that microscale events are directly affected by macroscale phenomena.

The effects of the changing frequencies and directions of migratory storms on the development of spruce budworm infestations were also used to introduce a large-scale phenomenon, the continental storm track, and its shifts in response to changes in the jet stream. Local differences in the development and survival of tent caterpillar populations on the west coast of North America showed how the area of suitable habitats may expand and contract as their surrounding topography produces different patterns of cloudiness in different kinds of spring weather.

The role of terrain in the production of various local circulation patterns was also examined to illustrate the different effects of convection and turbulence on aerial dispersal of insects. Air current patterns in valleys may often assist the dispersal of drifting, wingless forms, while the cloud patterns associated with these air currents simultaneously impede the dispersal of actively flying insects.

Sea breezes and other types of convergent wind fields, including the ITCZ, were discussed in relation to the Mason–McManus model of gypsy moth dispersal, radar and aircraft observations of dispersing spruce budworm, and R. C. Rainey's work on locusts. Dispersal and diapause were discussed only in relation to the meteorological framework within which they occur, since both topics were treated in considerable detail in Chapters 6 and 7.

ACKNOWLEDGMENTS

We are indebted to Linda Duncan, Daniela Bates, and Olga Schwartzkopf for helping with the bibliographic research and the many technical details involved in preparing this chapter.

REFERENCES

Anderson, J. F. and H. Y. Kaya (Eds.). 1975. *Perspectives in Forest Entomology*. Academic, New York.

Andrewartha, H. G. and L. C. Birch. 1954. *The Distribution and Abundance of Animals*. University of Chicago Press, Chicago.

Barry, R. G. and R. J. Chorley. 1968. *Atmosphere, Weather and Climate.* Methuen, London.

Barry, R. G. and A. H. Perry. 1973. *Synoptic Climatology: Methods and Applications.* Methuen, London.

Beck, S. D. 1968. *Insect Photoperiodism.* Academic, New York.

Berryman, A. A. and L. Safranyik (Eds.). 1980. *Dispersal of Forest Insects: Evaluation, Theory and Management Implications.* Cooperative Extension Service, Washington State University, Pullman.

Bursell, E. 1974. Pp. 1–41 in M. Rockstein (Ed.), *The Physiology of Insecta,* Vol. 2. Academic, New York.

Buxton, P. A. 1932. *Biol. Rev.* **7:** 275–320.

Cameron, E. A., M. L. McManus, and C. J. Mason. 1979. *Bull. Soc. Ent. Suisse* **52:** 169–179.

Campbell, A., B. D. Frazer, N. Gilbert, A. D. Gutierrez, and M. MacKauer. 1974. *J. Appl. Ecol.* **11:** 431–438.

Chauvin, R. 1967. *The World of an Insect.* World University Library, McGraw-Hill, New York.

Cloudsley-Thompson, J. L. 1962. *Annu. Rev. Entomol.* **7:** 199–222.

Danilevskii, A. S. 1965. *Photoperiodism and Seasonal Development of Insects.* Oliver and Boyd, London.

Danks, H. V. 1978. *Can. Entomol.* **110:** 1167–1205.

de Gryse, J. J. 1929. *Conf. Empire Meteorol., Agric. Sect.* **2:** 148–157.

Dingle, H. (Ed.). 1978. *Evolution of Insect Migration and Diapause.* Springer-Verlag, New York.

Edmunds, G. F. Jr. 1973. *Environ. Entomol.* **2:** 765–777.

Edney, E. B. 1957. *The Water Relations of Terrestrial Arthropods.* Cambridge Monographs in Experimental Biology, No. 5, Cambridge University Press, London.

Edwards, D. K. 1960. *Can. J. Zool.* **38:** 899–912.

Edwards, D. K. 1961. *Nature* **191:** 976, 993.

Flitters, N. E. 1968a. *Ann. Entomol. Soc. Am.* **61:** 36–38.

Flitters, N. E. 1968b. *Ann. Entomol. Soc. Am.* **61:** 923–926.

Gates, D. M. 1967. Pp. 1–29 in *Biometeorology: Proceedings of the Twenty-Eighth Annual Biology Colloquium,* Oregon State University Press, Corvallis.

Geiger, R. 1965. *The Climate Near the Ground* (4th ed.). Harvard University Press, Cambridge.

Gibo, D. L. and M. J. Pallett. 1979. *Can. J. Zool.* **57:** 1393–1401.

Greenbank, D. O. 1956. *Can. J. Zool.* **34:** 453–476.

Greenbank, D. O., G. W. Schaefer, and R. C. Rainey. 1980. *Spruce Budworm Moth Flight and Dispersal: New Understanding from Canopy Observations, Radar, and Aircraft.* Mem. Entomol. Soc. Can. No. 110, Entomol. Soc. Can., Ottawa.

Gunn, D. L. 1942. *Biol. Rev.* **17:** 293–314.

Hagen, K. S. 1962. *Annu. Rev. Entomol.* **7:** 289–326.

Hard, J., S. Tunnock, and R. Eder. 1980. USDA For. Serv., Nrn. Reg., Rept. No. 80-4.

Haufe, W. O. 1966. *Int. J. Biometeor.* **10:** 241–252.

Henson, W. R. 1951. *Can. Entomol.* **83:** 240.

Henson, W. R. 1952. *Nature* **169:** 40.

Henson, W. R., R. W. Stark, and W. G. Wellington. 1954. *Can. Entomol.* **86:** 13–19.

Johnson, C. G. 1969. *Migration and Dispersal of Insects by Flight.* Methuen, London.

Knight, N. C. and C. A. Knight. 1978. *Bull. Am. Meteorol. Soc.* **59:** 282–286.

Leighly, J. 1937. *Ecology* **18:** 180–198.

Lowry, W. P. 1969. *Weather and Life—An Introduction to Biometeorology* (2nd ed.). Academic, New York.

Masaki, S. 1980. *Annu. Rev. Entomol.* **25:** 1–25.

Mason, C. J. and M. L. McManus. 1980. Pp. 94–115, in Berryman and Safranyik (Eds.), 1980, referenced here.

Mason, C. J. and M. L. McManus. 1981. Pp. 161–202 in C. Doane (Ed.), *The Gypsy Moth: Research Toward Integrated Pest Management*. U.S. Dept. Agric. Tech. Bull. 1584, Washington, D.C.

Maw, M. G. 1960. *Can. Entomol.* **92:** 391–393.

Maw, M. G. 1961. *Can. Entomol.* **93:** 602–604.

Maw, M. G. 1964. *Can. Entomol.* **96:** 1482.

Messenger, P. S. 1959. *Annu. Rev. Entomol.* **4:** 183–206.

Messenger, P. S. 1971. *Proc. Tall Timbers Conf. Ecol. Anim. Control.* **3:** 97–114.

Morris, R. F. (Ed.). 1963. *The Dynamics of Epidemic Spruce Budworm Populations*. Mem. Entomol. Soc. Can. No. 31, Entomol. Soc. Can., Ottawa.

Oke, T. R. 1978. *Boundary Layer Climates*. Methuen, London.

Otvos, I. S. and B. H. Moody. 1978. Fish. Envt. Can., For. Serv. Info. Rept., N-X-150.

Perumpral, J. V., U. F. Earp, and J. M. Stanley. 1978. *Environ. Entomol.* **7:** 482–486.

Pilon, J. G. and J. R. Blais. 1961. *Can. Entomol.* **93:** 118–123.

Rabb, R. L. and G. G. Kennedy (Eds.). 1979. *Movement of Highly Mobile Insects: Concepts and Methodology in Research*. North Carolina State University Press, Raleigh.

Rainey, R. C. 1951. *Nature* **168:** 1057–1060.

Rainey, R. C. 1963. *Meteorology and the Migration of Desert Locusts. Applications of Synoptic Meteorology in Locust Control*. World Meteorol. Organ. Tech. Note No. 54, Anti-Locust Mem. 7, WMO-No. 138, TP64, Geneva.

Rainey, R. C. 1976. *Insect Flight*. Symp. No. 7, Royal Entomological Society of London, Blackwell, Oxford.

Rainey, R. C. 1976, Pp. 75–112 in Rainey (Ed.), 1976, referenced here.

Rainey, R. C. 1978. Pp. 33–48 in Dingle (Ed.), 1978, referenced here.

Rainey, R. C. 1979. Chapter 7 in Rabb and Kennedy (Eds.), 1979, referenced here.

Reiter, E. R. 1967. *Jet Streams*. Anchor Books, Doubleday, Garden City, New York.

Saunders, D. S. 1976. *Insect Clocks*. Pergamon, New York.

Schaefer, G. W. 1976. Pp. 157–197, in Rainey (Ed.), 1976, referenced here.

Shepherd, R. F. 1959. *Ecology* **40:** 608–620.

Simpson, J. E. 1967. *Br. Birds* **60:** 225–239.

Stark, R. W. 1959a. *Can. J. Zool.* **37:** 753–761.

Stark, R. W. 1959b. *Can. J. Zool.* **37:** 917–943.

Tauber, M. J. and C. A. Tauber. 1976. *Annu. Rev. Entomol.* **21:** 81–107.

Tenow, O. 1975. *Zoon* **3:** 85–110.

Thompson, W. A., I. B. Vertinsky, and W. G. Wellington. 1979. *Res. Popul. Ecol.* **20:** 188–200.

Thornthwaite, C. W. 1940. *Ecology* **21:** 17–28.

Trimble, R. M. and S. M. Smith. 1978. *Can. J. Zool.* **56:** 2156–2165.

Urquhart, F. A. 1960. *The Monarch Butterfly*. University of Toronto Press, Toronto.

Urquhart, F. A. and N. R. Urquhart. 1977. *Can. Entomol.* **109:** 1583–1589.

Urquhart, F. A. and N. R. Urquhart. 1978. *Can. J. Zool.* **56:** 1754–1764.

Uvarov, B. P. 1929. *Conf. Empire Meteorol., Agric. Sect.* **2:** 130–146.

Uvarov, B. P. 1931. *Trans. Ent. Soc. Lond.* **79:** 1–247.

Uvarov, B. P. 1945. *Quart. J. Roy. Meteorol. Soc.* **49:** 226–228.

Varley, G. C., G. R. Gradwell, and M. P. Hassell. 1973. *Insect Population Ecology; An Analytical Approach*. Blackwell, Oxford.

Wellington, W. G. 1946a. *Can. J. Res., D,* **24:** 51–70.

Wellington, W. G. 1946b. *Can. J. Res., D,* **24:** 105–117.

Wellington, W. G. 1952. *Can. J. Zool.* **30:** 114–127.

Wellington, W. G. 1954a. *Can. Entomol.* **86:** 312–333.

Wellington, W. G. 1954b. *Meteorol. Monogr.* **2:** 11–18.

Wellington, W. G. 1957. *Annu. Rev. Entomol.* **2:** 143–162.

Wellington, W. G. 1964. *Can. Entomol.* **96:** 436–451.

Wellington, W. G. 1965. *Can. Entomol.* **97:** 617–631.

Wellington, W. G. 1974a. *Can. Entomol.* **106:** 941–948.

Wellington, W. G. 1974b. *Environ. Entomol.* **3:** 351–353.

Wellington, W. G. 1975. Chapter 7 in Anderson and Kaya (Eds.), 1975, referenced here.

Wellington, W. G. 1977. *Environ. Entomol.* **6:** 1–8.

Wellington, W. G. 1979. Chapter 6 in Rabb and Kennedy (Eds.), 1979, referenced here.

Wellington, W. G. 1980. Pp. 11–24 in Berryman and Safranyik (Eds.), 1980, referenced here.

Wellington, W. G., P. J. Cameron, W. A. Thompson, I. B. Vertinsky, and A. S. Landsberg. 1975. *Res. Popul. Ecol.* **17:** 1–28.

Wellington, W. G., J. J. Fettes, K. B. Turner, and R. M. Belyea. 1950. *Can. J. Res., D,* **28:** 308–331.

Wellington, W. G., C. R. Sullivan, and G. W. Green. 1966. *Int. J. Biometeor.* **10:** 3–15.

White, T. C. R. 1974. *Oecologia* **16:** 279–301.

Chapter 14

Insects in Heterogeneous Habitats

FRED GOULD and
R. E. STINNER

14.1 INTRODUCTION

Most studies of insect ecology focus on the variation in an insect population's success that is due to differences in the average quality and/or quantity of its resources, including the effects of weather, natural enemies, and competitors. Much less attention has been devoted to the effect that spatial and temporal variation in these resources has on an insect population's survival and dynamics. Theoretical ecologists, as well as empirical ecologists, have historically shied away from tackling this aspect of insect interactions with their habitats, presumably due to the complexity of the problem, both conceptually and experimentally. Recently, however, there has been a surge of interest in this area, mostly from the theoretical perspective, but empirical study is also increasing. For a detailed review of the theoretical literature we refer the reader to a number of review articles and research papers (Chesson 1978, Cohen 1966, Ehrlich & Birch 1967, den Boer 1968, Reddingius & den Boer 1970, Levin 1974, Levin & Paine 1974, Roff 1974, 1980, Wiens 1976, Hassell 1980, Lomnicki 1980, Stenseth 1980). We attempt here to briefly summarize the theory while emphasizing the empirical work which deals with effects of spatial and temporal habitat heterogeneity on insect populations.

Although some insects may live in a habitat which is completely homogeneous spatially and temporally, it seems that the more one knows about an insect, the more heterogeneous its habitat appears. Take, for instance, a caterpillar feeding on an oak tree. A naive human observer may see the oak tree as a bountiful, homogeneous source of caterpillar food. From the caterpillar's perspective, life is not so simple. First of all, it may only survive if it feeds on young leaves which are tender and have not accumulated much tannin (Feeny 1970). Depending on when the caterpillar hatched, such leaves may be many or few. Since one young leaf does not offer sufficient nutrition to complete development, each caterpillar must, at some point, attempt to find another suitable leaf. Dependent upon the goodness of match between the color of the caterpillar and the twigs it must traverse to find the next leaf, the caterpillar may have more or less chance of being seen and attacked by an insectivorous bird. Additionally, the position of the leaf upon which the caterpillar feeds can dictate the caterpillar's body temperature and its rate of water loss. On some days, a leaf in the shade may be more suitable for growth than one in full sun. On other days, the relationship will be reversed. When we add to this the fact that, in terms of survival, the importance of each of these characteristics is different for each instar and may be influenced by the caterpillar's previous environment, we realize that the oak tree is a complex array of resource patches, varying in time and space.

There is a tendency among us to feel that we can deal with this problem by simply estimating the mean suitability of the oak leaves for the caterpillar and work with that value. It is the objective of this chapter to demonstrate that if we are to understand insect behavior, population dynamics, or evolution, it is not enough to know the "mean suitability" of the pieces of an insect's habitat. As difficult as it may be to study, the patchy nature or heterogeneity of an insect's habitat can dramatically influence the insect's ability to survive.

Before we discuss habitat heterogeneity in detail, we must first define our terms. We define *habitat* as the physical area encompassing the resources which support the existence of an individual insect or insect population for a specified period of time. *Heterogeneity* can be defined statistically as the state of being composed of significantly different parts. For our purpose, we must elaborate on this simple definition by dividing heterogeneity into two types. One type is that of presence–absence heterogeneity; the second, that of qualitative heterogeneity. For a habitat to exhibit presence–absence heterogeneity, there must be some areas, at some times, which have a given type of resource (e.g., food) while other areas lack it. This presence–absence heterogeneity is often synonymously referred to as resource patchiness. For a habitat to be qualitatively heterogeneous, a resource in some area at some time must be qualitatively different in suitability for the insect compared to the same type of resource in another area and time (e.g., oak leaves in shade or sun).

Obviously, qualitative habitat heterogeneity may occur without having presence–absence heterogeneity (resource patchiness). However, when we look at insect habitats, we usually see both types of heterogeneity. To help abstract and visualize the heterogeneity we will discuss, we present a series of topographic maps (Fig. 14.1) which illustrate a number of ways in which the heterogeneity of the insect's resources may be distributed within the habitat in time and space. In these maps any area which is not completely white contributes to survival of the insect in question and is considered a resource patch or part of a resource patch.

In Fig. 14.1a the resource patches are distributed contagiously (clumped) in space and remain constant over time in both space and quality. In Fig. 14.1b the patches are randomly (i.e., Poisson) distributed originally and become contagiously distributed at time period II and evenly distributed at time period III. (This change in resource patch distribution could occur due to various ecological processes, independent or dependent on the species population utilizing the resource.) In Fig. 14.1b the density of patches changes. In Fig. 14.1c the statistical distribution of the patches remain similar but their locations and their quality change. Some of the patches in all the figures are homogeneous in quality while others vary in the quality of the resource available in different parts of the patch. It is very important to realize that the temporal changes depicted in the figure may represent changes in the resources *per se* or changes in the abilities and needs of different developmental stages of the insect being considered. In these figures each patch is drawn approximately equal in size for illustrative purposes, although, of course, this is rarely the case in natural systems. For

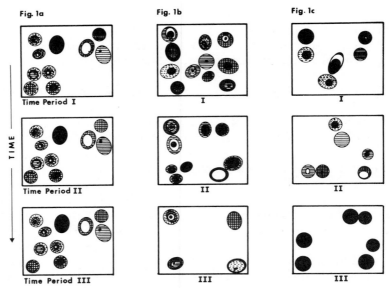

Fig. 14.1   A diagrammatic representation of habitat heterogeneity. The rectangles in *a*, *b*, and *c* circumscribe the area of the population's habitat. The quality and distribution of the resources in the three respective habitats are illustrated at three points in time. The quality of the resources within a resource patch range from 0 to 1 in terms of probability of survival of individuals using them. (⬤ = >0–0.25, ⬤ = >0.25–0.50, ⬤ = >0.50–0.75, ⬤ = >0.75–1.0). White areas denote the absence of any resource contributing to survival.

simplicity we have illustrated the habitat as a simple rectangular plot of the earth's surface. In reality precisely defining the area (or volume) of an individual's habitat is difficult, while doing this for an insect population is usually impossible. (The distinction between a population and a group of subpopulations becomes important in this regard and is dealt with in a later section of this chapter.) Not knowing a population's boundaries, we may have to predict the effects of resource patch distribution and quality on a population based solely on extrapolation from an understanding of the habitat defined at the individual insect level. Extrapolation from an understanding of how an ecological parameter affects an individual or small group of individuals to how it will affect a population is dangerous but, in empirical studies, is often necessary for practical reasons. Conceptual models, as in Fig. 14.1, often help guide our studies of real populations, and conversely, attempting to match real populations to our conceptual models help us add important complexities to the models which make them more realistic.

Earlier we examined some of the complexity of interactions encountered solely by the larval stage of a single caterpillar on a single oak tree. As we expand to the population level, we must include the entire life cycle of an insect and more than one individual. Accordingly, the picture becomes more complex. For example, let us examine some problems we would encounter in conceptualizing the interactions of another herbivorous lepidopteran, *Heliothis zea,* with its habitat. The first thing we note is that one cannot define resource patches based simply on the species of plant occupying a site (compare a and c in Fig. 14.2). A corn field at one phenological stage (silking) is much higher in quality for *H. zea* larval development than a soybean field at any phenological stage. However, a corn

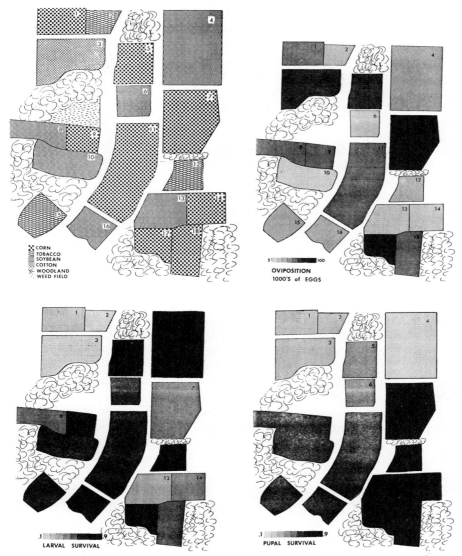

Fig. 14.2   Resource patches of *Heliothis zea* in eastern North Carolina. In *a* the patch types as viewed by a naive human observer are illustrated. In *b, c,* and *d* the gross patch distribution and quality as it may be perceived by the ovipositing adult, feeding larvae, and burrowing pupa, respectively, are illustrated. (From Stinner 1979.)

field at a later phenological stage may be equal in quality to a soybean field for the larval stage. Therefore, what we classify as a patch type based on our anthropocentric concept of a plant species requires tempering that is based on knowledge of the organisms zoocentric "viewpoint" (Wellington 1977). We must recognize that a high-quality resource patch for one life stage of this insect may be a low-quality resource patch for another life stage (compare b, c, and d in Fig. 14.2). In addition, we may notice in these figures that the spatial distribution of highly suitable patches may differ for the different life stages. For example, the variation in patch quality and distribution for the soil-inhabiting pupal stage, which is in a large part determined by moisture and tillage practices, seems to differ from the variation for the adult stage. There are differences in distances between good patches for the adult and the distances between good patches for the pupal stage. In some cases these differences, based on linear measurements, may accurately reflect what the insect experiences. However, in other cases, such measurements may be misleading. Adult *H. zea* are such strong fliers that a distance of 300 m between good patches is trivial, while for the pupa, which is virtually immobile, a distance of 300 cm is quite significant. These biological insights must be used carefully in conceptualizing and predicting the effects of habitat heterogeneity on the population dynamics of *H. zea.*

This problem of measuring distances "for" insects is an important one. We must firmly keep in mind the fact that the behavioral programs of many insects do not measure distances with a meter stick. Given two resources 5 ft from each other, a bee may "perceive" the downwind resource as very far away and the upwind resource as very close, while an airborne aphid may view them in the completely reverse order. On a terrain with 1-cm crevices such as the bark of a dogwood tree, a mite may have to crawl 10 times as far to get to the canopy as does a caterpillar, in terms of surface traversed, while on a beech tree with its smooth bark, the distance crawled would be almost equal for both arthropods.

We were purposefully vague at the outset to define a habitat in terms of "a specified period of time." The important, though somewhat arbitrary, decision of whether this period of time should be one generation, or 1 yr or more must depend in part on the characteristics of the species and the scope of predictions involved in the study. Taken to an extreme, a habitat may have to be defined large enough to make predictions about dynamics over an entire species range for 100 yr, or it may be defined to make predictions about the dynamics of a population in one pond for the larval stage of a semiaquatic insect. What determines the duration of the actual period of observation may in many cases reflect the point at which the temporal changes in the habitat become periodic in nature. For example, if the temporal changes are mostly seasonal, an understanding of the fluctuations over a 2-year period may account for 95% of the important changes in patch distribution and quality over 20 yr.

To add one last introductory note of complexity, we must recognize that whenever we deal with a population in a temporal context we must also contend with the possibility of genetic change. Given a habitat with a set distribution of resource patches, a population may evolve genetically to utilize the resources within certain patches more efficiently and/or may evolve to move more or less frequently between patches. The effect that this genetic change would have on the population's dynamics may be very significant.

Considering the complexity of this subject, it is advantageous to discuss the interactions between insects and their heterogeneous habitats at three somewhat arbitrary levels: (a) behavioral processes, (b) population processes, and (c) genetic processes. Many of the examples we have chosen could fit into more than one category. Our division has been more to provide a forum of discussion within the varying levels. At the behavioral level we look at the immediate local responses of individuals to habitat heterogeneity. At the population level we consider how physiological and behavioral processes of many individuals interact with each other and the resources in their habitat. At the genetic level we deal with the interactions emerging when both the habitat and the insects themselves contribute heterogeneity to the system.

14.2 BEHAVIORAL PROCESSES

Physiological and behavioral constraints dictate that no organism can efficiently utilize an entire spectrum of resource patch distributions. Some organisms are able to forage efficiently on contagiously distributed resources, while others forage more efficiently on evenly distributed resources. The behavior of some organisms enable them to find and utilize ephemeral resource patches; others have behaviors which enable them to maintain efficient utilization of temporally stable resource patches.

An interesting experimental study of this trade-off among insect species was conducted by Hubbell and Johnson (1978). They set out a grid of sugar baits in a tropical forest and observed the usage of these baits over time by six species of sympatric bees in the genus *Trigona*. The baits were set up every 20 m in an area 120 × 180 m², and the number of bees of each species at a bait was recorded at ½-hr intervals or less. Comparative data were obtained on a large number of foraging parameters, including time to initial bait discovery by a species, rate of discovery of additional baits, time until attainment of an approximate steady-state number of workers at baits and an equilibrium number of visited baits, frequency of number of workers per bait, and responses to presence of intraspecific and interspecific rivals at baits. They found unique aspects of the foraging patterns of each of the six species.

Two of the six species discovered a first bait with incredible swiftness (3 and 5 min), and the number of baits they discovered increased rapidly.

The equilibrium number of baits fed upon throughout the entire observation period was achieved after 0.5 and 1.5 hr. Two other species required 0.5 hr to find the first bait and one required 2.5 hr. Assuming that this difference in bait finding was not due to large differences in the total number of individuals of each species in the area, we might infer that the last species, *Trigona fuscipennis,* would be easily out-competed at the many highly dispersed flowers in that forest which offer nectar for a short time period. However, at food patches with more temporal stability, *T. fuscipennis* could do quite well, for once a worker of *T. fuscipennis* found a bait, it could efficiently recruit other workers to the bait. While the modal number of bees of *T. fuscipennis* at a bait was between 64 and 128, the modal number of bees per bait for all other species was between 1 and 2. Once *T. fuscipennis* found a concentrated resource it utilized it well and, being the most aggressive of the four common species, was able to defend it against other takers. So, although the study was not conclusive, *T. fuscipennis* would seem to do best on resources which are temporally constant and concentrated (clumped) spatially, while other species would do best on clumped, ephemeral patches or dispersed, ephemeral and/or constant patches. Considering the high spatial and temporal variability of some resources (nectar- and pollen-producing flowers) in this forest, Hubbell and Johnson felt that these six species may have been coexisting by partitioning the resources based on the dispersion of each resource in time and space.

Another group of insects which points out the contrasting interactions of spatial distribution of resource patches and behavior are the insects associated with collard crops. Root (1973) and Pimentel (1961) studied the effect of the spatial distribution of collard plants on the abundance of a number of herbivore species. In a 3-year study Root (1973) set out plots of collard plants in monoculture (large patches) and in polyculture (small rows of collards dispersed among other vegetation). The arthropod fauna of the collard plants was sampled periodically each season, and the data were analyzed to determine whether the collards in the monoculture patches or those in the polyculture rows had the higher herbivore load (i.e., biomass of herbivores/100 g of plant material). Additionally, data on a number of the common species were analyzed separately to determine under which conditions they were most abundant. In general, the herbivore load was higher in the large monoculture patches. However, examination of the data on individual species indicates that while some of the herbivore species which specialize on collards were most abundant in the monoculture patches, generalist herbivores and some of the specialized herbivores showed no higher density in the monocultures compared to the polycultures. In fact, one important specialist was more abundant in the polycultures. A number of different hypotheses related to insect behavior may explain these observations. Large patches of collards may have been more attractive to some herbivores due to concentration of the collard odor which they seek. Others, which may have searched in a more random fashion, would have had more of a chance of "bumping into" the small patches than the large patches since the ratio of the total patch circumference to the total collard leaf area is expected to be greater in small patches. Another factor which may have caused the observed differences was a difference in parasite and predator abundance due to their behavioral interaction with the two types of habitats. A follow-up study on the behavior of two of the collard specialists (Kareiva 1982) which had higher abundances in the monocultures demonstrated that for both of these species, although a similar number of individuals find both types of patches, a lower percentage leave the large monoculture patches than the smaller polyculture patches. Apparently the movement behavior of these specialists results in a positive correlation between the percent of individuals leaving a patch and the ratio of the patch circumference to the total volume of patch. (This relationship is somewhat analogous to water loss rates from organisms differing in surface area to volume ratios.)

   In both Root's (1973) and Hubbell and Johnson's (1978) experiments, the investigators were able to replace depleted resource patches. Such experiments mimic natural cases of renewable resources or resources which are not significantly diminished by the organisms using them. Figure 14.1a illustrates just this situation. However, in many cases the resources in a patch are eaten or otherwise depleted by the user and the abundance and distribution of the resources is changed. We must keep in mind the fact that not only does the resource distribution affect the insect, but the insect may affect the distribution of the resources. An example is that of *Cactoblastis cactorum*, a lepidopteran which was released in Australia to control the prickly pear cactus (Monro 1967). When this moth was first introduced, the cactus was widespread and common, but as the moth population increased, it destroyed most of the cacti. Eventually an equilibrium seemed to be achieved between the moth population and the cactus population at a point where cacti were so rare that the probability of a moth emerging at one site and finding a cactus at another site was very low. Monro (1967) also found that the moths tended to lay their egg masses in a clumped manner. Thus, instead of damaging all the cacti in an area, they would, because of their behavior, destroy a few cacti while others could reproduce. This would finally lead to an equilibrium like that in Fig. 14.1c where the patch distribution and abundance within the resource field remains the same but the actual location of patches shifts.

   Instead of contributing to a final long-term equilibrium, as in the case of *C. cactorum*, an insect's behavior may cause its resources to continually fluctuate in short-term distribution and quality. Heinrich (1979) found that upon encountering a new area, bumble bees initially visited a number of flowering plant species. They then began to specialize on what they perceived to be the most abundant single species which produced a suitable amount of nectar. For the first bees to forage in the area, this flower was usually jewelweed. In specializing on this single flower type, the bees became more efficient at foraging on it, but at the same time the abundance of nectar at these flowers was being depleted. When the suitability of this resource declined, the bees which had specialized on jewelweed began to forage in an unspecialized manner until they found another highly rewarding flower type. This allowed the jewelweed to rebuild its nectar supply. It is interesting that when the jewelweed nectar was depleted to the point where other flowers were more rewarding, the bees did not desert the jewelweed immediately. There was a lag time in which they foraged on it, even though higher quality flower types were available. Such examples illustrate the fact that the resource patches chosen by an insect because of its behavioral program will not always be the optimal choice in terms of energy accumulation or reproductive output.

Even the assumption that we can determine what an optimal choice should be is problematic (Pyke et al. 1977). In a series of theoretical and empirical studies, Real (1980, 1981) and Waddington et al. (1981) showed that insects may not only be assessing average gain from a specific choice but may also be assessing the variance around the average gain and may opt for a less variably rewarding choice, thus lowering their "risk." Given a choice between two types of artificial flowers with equal mean nectar reward, Real (1981) and Waddington et al. (1981) found that bumblebees preferred the flower type which varied less from flower to flower in the amount of nectar available. These results indicate that the mean quality of the patches in the habitat, although important, is not the only significant factor to these insects.

A number of recent studies indicate that ovipositional choice of females of polyphagous herbivores and parasites may be influenced by the first hosts they encounter upon emergence (Rausher 1978, Mark 1982, Prokopy 1982). This learning behavior could increase their egg-laying efficiency in environments where the identity of the first host encountered corresponded to its relative abundance in the female's habitat. In habitats where abundance of a host in the immediate vicinity of emergence did not correlate with average abundance throughout the female's habitat, this learning behavior could be maladaptive.

14.3 POPULATION PROCESSES

Patch size and distribution of resources in a habitat can greatly affect the size, quality, and stability of populations inhabiting it. One of the earliest and most heuristic demonstrations of the effect of patch distribution on population size comes from Leopold (1933). Although his interests did not directly involve insects, the general applicability of his results warrant inclusion here. Leopold was interested in the ecology and management of bobwhite quail, *Colinus virginianus*. He considered this bird's resources to be of four basic types. One is well-drained ground partially covered by grass for nesting, and partially barren so the young can dry off quickly after it rains. The second is food which consists of insects, seeds, and berries. The third is a place which offers protection from bad weather and predators. This may be a thicket or scrub or tangle of vines. Fourth, the quail requires an elevated area for sleeping which enables quick escape from predators.

The need for each of these resources varies during the year, but over an entire year the ideal area for survival of a covey of quail would be a spot where meadow, field, scrub, and woods meet. Leopold (1933) described two hypothetical areas differing in distribution of resources (Fig. 14.3*a,b*). Both areas had an equal quantity of each resource, but due to the difference in distribution of the resources, one area could support six times as many coveys of quail as the other. It is important to note from this example how the joint occurrence of a number of resources, not just the distribution of one resource, may factor into final population size. This is often the case with insects. In our earlier example of the corn earworm (Section 14.1), the joint occurrence of a suitable host plant for the larva and a suitable patch of soil for the pupa has a strong influence on population size since the larva cannot crawl far in search of a pupation site. A curious insect which we feel very clearly demonstrates the importance of joint occurrence of resources, is a Central American katydid, *Ancistrocercus inficitus,* which requires a host plant to feed on at night and a predator-safe shelter during the day (Downhower & Wilson 1973). A highly suitable habitat for it is one in which its host plant is found close to a wasp nest. Where this occurs, it feeds at night, then rests a few centimeters from the hanging wasp nest during the day. If a predator attacks, the katydid jumps

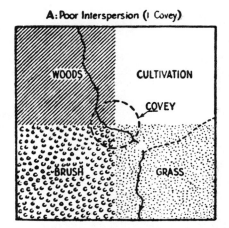 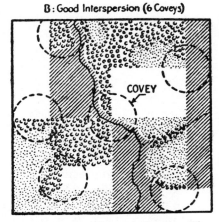

Fig. 14.3 The influence of interspersion of resource types on quail population size. The interspersion in A will only support one covey of quail since there is only one spot where the four resources meet. In B six coveys can be supported. Note that one covey (bottom left) lacks a wooded area and thus may be more vulnerable to predation. (From Leopold 1933.)

to the ground leaving the predator with agitated wasps for company. While the katydid and the corn earworm require both resources alternately, there is a difference between these two insects in the risk involved if one of the two resources disappears from the area. In the case of the adult katydid, if one resource disappears it can take to the air with some chance of finding another area with both resources. In the case of *H. zea*, once the egg is laid, long-range search has ended and disappearance of either resource means impending death of the individual and loss of its contribution to population size.

Not only is population size affected by resource distribution and the ability to move between resources, but the stability of the population may also be affected. A large body of theory has been developed on this topic, some of which we cover below. However, due to the complexity of some of the theory, we must refer the reader to recent papers for details (Crowley 1981, Levin 1974, Hastings 1978, Reddingius & den Boer 1970).

In terms of experimental demonstrations of the influence of resource distribution on population stability, the classic studies of Huffaker (1958) and Huffaker et al. (1963) are quite illustrative. These workers designed experimental resource arenas which involved three trophic levels. Oranges served as the producers, the six-spotted spider mite served as the primary consumer, and another mite, *Metaseiulus* (= *Typhlodromus*) *occidentalis,* served as the predator. When all three of these species were placed together on uniform trays (Fig. 14.4*a*), the six-spotted spider mite population increased for about 20 days, then began to decline due to predation by the *M. occidentalis* population which was increasing in size. The six-spotted spider mites were completely wiped out by the predators after 40 days. Ten days later, all of the predators had died of starvation.

In a number of further experiments Huffaker (1958) changed the distribution of the primary resource (oranges) by interspersing rubber balls with the oranges, by placing some oranges in remote areas of the tray, and by using a mazelike pattern of petroleum jelly to slow dispersal of the predators. The results of the most complex design of the first study (Huffaker 1958) are illustrated in Fig. 14.4*b* (cf. Chapter 12). The complexity of resource patch distribution within the experimental habitats allowed coexistence of the two species of mites for over 200 days. It is worth noting here again that not only does the resource distribution affect the consumer, but the consumer also affects the resource distribution.

In one of the rare field studies of effects of habitat heterogeneity on predator–prey interactions, Blau (1980) found that the population stability of the black swallowtail butterfly, *Papilio polyxenes,* in Costa Rica appeared to be influenced by the differential effect of the patchiness of its weedy host plant on it and its predators. The butterflies were able to stay one step ahead of their predators by finding and ovipositing on host plants soon after they began to flower. By the time predator density on the plants increased, many of the larvae had already pupated.

These findings regarding the interaction of habitat heterogeneity and insect population stability lead us back to a consideration of the theoretical ideas of Andrewartha and Birch (1954) and the extension of them by den Boer (1968), whose main premise was that the reason population numbers seem to be regulated is due to the fact that there is a "spreading of risk." At any point in time, some subpopulations would find themselves in good environments and others in poor environments. Due to density-dependent or density-independent interchange among the subpopulations, a buffered mean density within a subpopulation appears to be achieved (cf. Chapter 12).

This process can best be seen if we look at cases with extremely low interchange among subpopulations where we wind up dealing with a situation akin to "island biogeography" of single species (Benford 1979). What we envision is an archipelago acting as the entire population's habitat. The subpopulations on each island are relatively distinct in that very few migrants move from one island to the next within the generation time of the species. The subpopulations, therefore, fluctuate independently in response to conditions on each island; however, when the subpopulation on an island goes extinct due to random fluctuations, or other causes, there is a set probability that immigrants from other islands will reestablish a subpopulation in a given amount of time. What this probability is will depend on the subpopulation density on each of the other islands in the archipelago and their distances from the depopulated island. The closer a

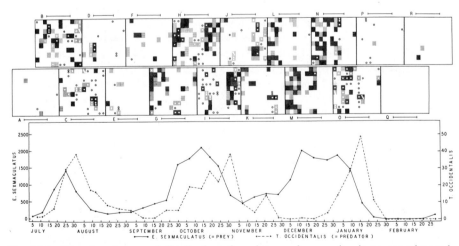

Fig. 14.4   The abundance and distribution of predator and prey mites in an experimental, heterogeneous habitat. A photograph of one of the more homogeneous habitats is shown in *a*. The graph in *b* depicts the population cycles of the two mite species in the most heterogeneous habitat. The checkered boxes represent the abundances of the oranges within the habitat over time.

populated island is to the depopulated island the more important its population density will be in setting the probability of reestablishment of the extinct subpopulation.

Very few studies have examined this situation in a natural setting. The best example we found is the long-term study by P. R. Ehrlich and colleagues (1965, 1975). Their work demonstrated the existence of extremely limited interchange between populations of the butterfly, *Euphydryas editha,* on Jasper Ridge in the San Francisco Bay region. They studied three groups of these butterflies, which were localized to small areas where their host plant, *Plantago erecta,* was abundant. These host plant patches were analogous to islands, for although the distance between the groups of butterflies were only a matter of hundreds of yards, they found no significant interchange of individuals. Their data indicated that the populations were fluctuating in numbers independently of each other. In the 4 yr one steadily increased in size from about 150 to 3200 adults, another fluctuated between 70 and 400, and a third dropped from 50 to extinction. Years later the third population was reestablished (Ehrlich et al. 1975). In the case of true "islands" an observer would probably identify the individuals on each island as a distinct population, but in other cases where population boundaries are not visually striking, Ehrlich and Birch (1967) point out that given a group of small, isolated populations, fluctuating independently, a casual observer may get the impression of one large population under tight "control." How common such cases of unidentified strong population structuring are is difficult to determine, but if such substructure is common, it has important implications for the way we interpret "population" regulation.

Earlier we mentioned problems that can arise in extrapolating from the individual level to the population level. In specific cases a habitat that is heterogeneous in quality for a single insect may be effectively homogeneous at the population level. This can be the case if there is heterogeneity among individuals in the population in the way they can utilize a resource. Lomnicki (1980) addressed this situation theoretically. As an empirical example, Wellington's (1964, 1965) work on tent caterpillars is useful even though it was not all technically rigorous (Papaj & Rausher 1983). He found more or less distinct physiological "classes" of larvae varying in their survival and development under specific environmental conditions. Under one set of environmental conditions in the habitat, one class of larvae were considerably more "fit" than the second, yet under another set of conditions, the relative "fitness" of each class may be reversed. This negative correlation in fitness of the two classes of larvae would tend to buffer the effects of habitat variation on the population's stability. For a given individual the habitat would "appear" to be very patchy, but for the population as a whole, the habitat suitability would be more constant. By producing at least some young which could survive well under either extreme of environmental conditions, a single female could be seen as ensuring her lineage survival by a "spreading of risk" strategy.

14.4 GENETIC PROCESSES

In the previous section we suggested that differences among individuals in a population can influence the population's response to habitat heterogeneity. Interestingly, differences among individuals are often caused by habitat heterogeneity, so we get a reinforcing effect which may further increase population stability (or perhaps finally lead to instability).

In the *Malacosoma* species, Wellington argued that variation among individuals was partially due to a cause intrinsic to the moth's physiology which dictates that not all eggs can be the same size. He felt that the mean and variance in egg size is, however, modulated by the environmental conditions which the female moth experienced during her development (see Papaj & Rausher 1983). A habitat with very variable environmental conditions (i.e., patch quality) could, therefore, lead to still more variation among eggs and the ensuing individuals in the population. In this system environmental variation may enhance individual variation through a physiological mechanism. In other cases environmental variation leads to more profound differences among individuals, that is, genetic differences.

The effects of habitat heterogeneity on genetic variation has recently received much theoretical attention. The surge of interest in this area was generated indirectly. In the 1920s through the 1940s many geneticists believed that populations of organisms for the most part consisted of wild-type individuals, that is, those individuals with the single genotype best suited for their environment (reviewed by Lewontin 1974). They felt that there were always a few mutations arising in the populations, but these were repressed unless they were more fit than the wild type. In the latter case these mutant genotypes would increase in frequency and would become the new wild type. This model assumed a single best genotype for a population.

In 1966, when Hubby and Lewontin used electrophoretic techniques to determine the genetic composition of populations, they found that there usually was no single wild type, "instead, populations were composed of a variety of genotypes, and individuals were often carrying more than one allelic form of single genes." This puzzled geneticists and required explanation. Some felt that this variation was neutral in terms of how the organism interacted with the environment (reviewed by Kimura 1979), but others felt that this was not possible (review by Lewontin 1974) and sought alternatives to the "neutralist" hypothesis. According to one view, if heterozygous individuals were more fit than homozygous individuals, genetic variation would be maintained. This hypothesis was difficult to support for a number of reasons, including the fact that it did not predict the enormous variation found in populations of haploid species that cannot produce heterozygous individuals. Detailed treatment of this topic is beyond the scope of this chapter, and the reader is referred to Futuyma (1979) and Lewontin et al. (1978) for further discussion.

Another mechanism which is thought to promote polymorphism within a population, and the one with which we are concerned here, is environmental heterogeneity in patch quality and distribution. One can envision a population inhabiting an area consisting of two qualitatively different patch types. Individuals in the population with genotype AA are better adapted to conditions in patch type I than individuals of genotype aa. The reverse is true in patch type II. Such conditions would give rise to balancing selection. Only under perfectly balanced selection would we expect the polymorphism to be maintained indefinitely, unless specific criteria were met in terms of the interaction between the genotypes and the patch types inhabited. These specific criteria were first outlined by Levene (1953). Since then, many refinements of the criteria in Levene's model have been proposed (e.g., Prout 1968, Maynard-Smith 1970, Gillespie 1977), and the properties of the subsequent models, as well as their match with biological realities, have been assessed (see Hedrick et al. 1976, Maynard-Smith & Hoekstra 1980, for reviews). As with many mathematical models, it is difficult to say which criteria or assumptions are most critical. However, one criterion which authors agree is very important in maintaining polymorphism is the degree to which genotypes can choose the patches in which they are most fit (Taylor 1975, 1976). That being the case, it behooves us to ask whether insects with specific genotypes can selectively choose environments in which they are most fit? Powell and Taylor (1979) and Jones (1980) point out that investigations of this phenomenon would greatly improve our understanding of how important patch heterogeneity may be to the maintenance of all the genetic polymorphism we find within natural populations, but to date few empirical studies have been conducted. Two studies in which genotype-specific habitat choice and survival have been examined are Kettlewell's (1955, 1973) demonstration of substrate choice by the melanic moth, *Biston betularia,* and studies by Powell and Taylor (1979) on *Drosophila* habitat choice.

Regarding the first study, Kettlewell (1955, 1973) noticed that pepper moths exhibited a curious pattern of movement when settling down on the bark of a tree. Moths would usually alight on the edge of a branch and walk a foot or more, then turn a circle on their own axis before settling down on the bark with wings and body appressed tightly to the substrate. Light-colored moths usually settled on the lightest area of bark they traversed and dark moths usually settled on similarly darkened bark. Kettlewell performed an experiment in which moths were placed in a barrel having both light and darkened areas. Dark moths chose darkened areas in a ratio of about 2 to 1 over light areas and the light-colored moths exhibited just about the opposite behavior. This substrate selection could obviously protect the genotypes from predation by maintaining an efficient camouflage system (Chapter 8). Without such substrate selection, not only would the population as a whole be more exposed to predation, it would also be more likely that one of the two color morphs would be lost from the population, dependent on the frequency of light and dark bark.

Of all the genera in the world, more studies of genetic polymorphisms have been carried out on *Drosophila* than any other genus, but the factors influencing the levels of polymorphism in *Drosophila* populations still remain elusive. Taylor and Powell (1977) found *Drosophila persimilis* in the area of Mather, California, quite heterogeneous with regard to frequencies of certain alleles and chromosome inversion. In an analysis of *Drosophila* from a biotically and abiotically heterogeneous study plot less than 1 km² in size, they detected genetic differences in the groups of *Drosophila* inhabiting subsections of the plot. To determine whether the *Drosophila* were capable of choosing among the available habitats in the area, Taylor and Powell (1978) conducted an interesting experiment in which *Drosophila* were captured and moved away from their original home site. They were then marked with fluorescent dust and released. They found that individuals collected from one type of environment (e.g., moist woods) usually returned to that environment in preference to another type of environment (e.g., open woods). Although the genetic basis for this behavior has not been elucidated, the evidence from this study is very suggestive. A similarly suggestive footnote in a study of Levins (1969) indicated that within one small area *Drosophila* of different sizes maintained a temporal difference in their habitat usage pattern. Flies that were caught at midday when desiccation rates were high were 5% larger than flies captured in the morning or evening.

In the two *Drosophila* examples habitat choice may have caused a disruption in random mating. That being the case, the patchiness of the environment could theoretically exert an even stronger force in maintaining polymorphism. Although the evidence presented in all of these studies sets the stage for genetic differentiation due to habitat patchiness, none of them proves this to be the case. Hedrick et al. (1976) point out that although such empirical proof is difficult to obtain, such data will be of great importance to our general understanding of population genetics.

In all of the above discussion on how variation in resource patch quality and distribution may maintain genetic polymorphism, we have been assuming that the variation in resource quality and distribution within the habitat remains constant. This is not always the case. Both resource patch distribution and quality may change over time. If the temporal variation in resource patch quality and distribution is independent of the genetic makeup of the population utilizing it, some contribution to maintenance of polymorphism in the population may be achieved, but it will be minimal (Hedrick et al. 1976). However, if the change in resource patch qualities and distribution is affected by the genetic makeup of the population utilizing it, a stable cycle of gene frequency fluctuations may be set up which preserves polymorphism. What this involves can be illustrated using two hypothetical parthenogenetic parasitoid genotypes, *AA* and *aa*, which are found in a habitat consisting of patches of host type I and host type II. Genotype *AA* does better on hosts of type I than does genotype *aa*. The reverse is true on hosts of type II. If there is originally a 1:1 ratio of the two parasitoid genotypes and a 1:1 ratio of the host types, we can only expect this to remain constant if the degree to which genotype *AA* is superior to genotype *aa* on hosts of type I is exactly equal to the degree of superiority of genotype *aa* over genotype *AA* on patches of host type II, and if the depletion of host types by individuals of the parasitoid population is equal. If one genotype is superior to the other, averaged over both patch types, and no significant depletion of the populations of either host type is caused by the parasitoids, one genotype may drive the other to extinction. However, if the consumers damage or deplete the host populations, the polymorphism may be maintained even if one parasitoid genotype is on average superior to the other in a habitat with a 1:1 ratio of the two host types, due to a process referred to as *frequency-dependent selection*. As an example of this process, let us assume that genotype *AA* has 100% survival on host type I and 50% survival on host type II. Genotype *aa* has 25% survival on host type I and 75% survival on host type II. (For simplicity, both genotypes will be assumed to be parthenogenetic and incapable of distinguishing the two host types.) Starting from a 1:1 ratio of genotypes and host types, we would expect an increase in genotype *AA* over *aa*; however, we would also expect more depletion of hosts of type I than hosts of type II due to the increasing abundance of genotype *AA*. If after a number of generations there were very few hosts of type I relative to hosts of type II, the average survival of genotype *aa* may now be highest. For example, if 90% of the hosts were of type II, the average survival of genotype *aa* would be 70% ($0.9 \times 75 + 0.1 \times 25$) and that of genotype *AA*, 55% ($0.9 \times 50 + 0.1 \times 100$). The direction of change in genotype frequency would be the opposite of that found when hosts were in a 1:1 ratio. This type of system, with host quality being affected by the population utilizing it, has been investigated experimentally by Pimentel (1968) with regard to insects and their parasitoids. Leonard and Czochor (1980) present a review of the theory and some experiments related to this.

Up to this point we have been discussing how variation in the qualities of patches within a single population's habitat may contribute to producing and maintaining genetic variability among individuals *within* that population. At another level we can examine the effects of habitat patchiness in generating genetic differences *among* populations. Before moving to a discussion of how patchiness may generate differences among populations, we must first understand that what we recognize as a single population or a group of populations is not defined simply, and that patchiness in a habitat may actually generate more populations of a specific organism in a defined area than would be generated in the same area with a homogeneous environment.

Populations are defined as groups of conspecific individuals that occupy the same area and interbreed and interact ecologically (Futuyma 1979, Ricklefs 1979). Depending upon the degree that interbreeding and ecological interactions between groups of individuals are curtailed, the population is considered a set of subpopulations or, at the limit, a set of distinct populations. As can be seen in the *Drosophila* studies described by Powell and Taylor (1979), environmental heterogeneity may cause division of a population into subpopulations of different phenotypes (and presumably different genotypes). Individuals in the subgroup of flies that prefer moist woods have a greater probability of having ecological interaction with each other than they do with individuals in the subgroup that prefers open woods. The same holds true for interbreeding since flies are more likely to mate with their nearest neighbors. In Levins's (1969) study the same effect may occur due to temporal distances between groups of flies.

If we focus more closely on the studies of Powell and Taylor (1979), we see that not only may population subdivision be influenced by the qualitative differences among patches, but this phenomenon of subdivision may also be influenced by the size and distribution of these patches and the characteristics of the species. Within the small area that Taylor and Powell studied, released flies usually returned to their original environment but some did not. In an undisturbed situation we might also expect some movement of flies from one type of environment to another. If we artificially made the patches of moist woods and open woods very small, we would expect more migration among environments due to stochastic movements at the boundaries of environments and, therefore, more interaction and mating among the different types of flies. On the other hand, if we consolidated all the small patches of each type so we had large plots of each environment, there would be a lower probability of mating and other interactions among the groups of flies and therefore more population subdivision.

Given the same patch size, different organisms may produce more or less population subdivision. Although an individual *Drosophila* may easily foray among patches of a given size, a mite species may rarely move among the same patches. At whatever point the patchiness causes the development of distinct populations, we must also realize that based on our original definition, the "habitat" has been subdivided into two or more qualitatively different habitats. With this achieved, it is easy to see how natural selection could cause genetic differentiation of the populations involved, for the lower the gene flow between the groups of individuals adapting to different conditions, the more extensive and rapid the genetic differentiation can be.

Let us further complicate the issue by assuming that patchiness itself is the quality that differs among two "habitats" and is the quality to which the populations within these habitats are adapting. We will address this problem with an empirical example in which patchiness differs among habitats. The case in point is that of the cinnabar moth, imported to the United States and Canada as a biological control agent of tansy ragwort. This case stands out not only as an example of how the structure of patches in a habitat may influence genetic differentiation of populations but also serves well as a summary of how behavioral processes, population processes, and genetic processes interact with patch structure. The biology of tansy ragwort and the cinnabar moth has been studied extensively in Europe (Dempster 1971, van der Meijden 1973, 1979) and in North America (Nagel & Isaacson 1974, Green 1974, Myers & Campbell 1976, Myers 1976, 1978, 1980). Two major types of patch structure have been studied. The first involves effects of plant size and interplant distance on the insect's biology. The second, in some respects functionally related to the first, is the temporal stability of the plant patches.

The cinnabar moth lays its eggs in masses of 10–150 eggs. Depending on rates of early larval mortality and numbers of eggs in a mass, a single plant may be completely devoured before the larvae are ready to pupate. When this occurs, and often before it occurs, larvae disperse to surrounding plants. Success in moving to another plant is related to interplant distance (Myers & Campbell 1976). Plant size and interplant distances vary from one population of ragwort to another. Myers (1976) simulated the population dynamics of the moth and demonstrated that high dispersal-related mortality associated with large interplant distances should theoretically stabilize the population dynamics of the moth, for even under outbreak conditions not all plants will be densely colonized and depleted, so many larvae will complete development. As interplant distance is decreased and successful dispersal is increased, most of the plants may be destroyed before the larvae are prepared to pupate, leading to mass starvation and population instability. Myers (1978) actually found that interplant distance in empirically studied habitats is correlated with moth population stability.

Although population stability is favored by clumping of eggs by females and low rates of dispersal by larvae to underexploited plants, natural selection at the individual level should favor those moths which do not clump eggs and those larvae which disperse to underexploited plants when chances of survival during dispersal are high. We would therefore expect selection for increased larval dispersal in a habitat with small interplant distances and selection for tenacity in habitats with large interplant distances. Myers and Campbell (1976a,b) found evidence somewhat suggestive of genetic differentiation of moth populations which agree with these predictions, for the tendency to migrate appeared higher in ragwort populations with small interplant distances. If we understand this system correctly, we must conclude that individual selection has not completely "adapted" the moths to the patchiness within the system or we would expect adult females to disperse their eggs more uniformly over the plants. If this had occurred, its effects on the population stability of the newly evolved lineage of moths may have been rather striking. As individuals, each moth would be highly adapted for utilizing the "patches" of tansy ragwort. In essence, the newly evolved lineage of moths would "view" what their ancestors viewed as a patchy habitat as one large, easily exploited homogeneous resource patch. Having eliminated the buffering effect of the patchiness, the new population would probably overexploit all of the tansy ragwort plants. As with the mites in Huffaker's (1958) study, after driving their resource to extinction, the moths might soon follow. Odd though it may seem, in this case adaptation at the level of single individuals would not necessarily be beneficial to the population as a whole.

14.5   CONCLUSION

We feel that the study of interactions of insects with their heterogeneous habitats is still an embryonic field of study. We hope that this chapter can serve to point out some of the conceptual and experimental difficulties inherent in the study of these interactions, while emphasizing the prime importance of such interactions in the ecology of insects. We suggest that there is an urgent need for researchers in ecological entomology to delve further into this area.

ACKNOWLEDGMENTS

We would like to thank W. Blau, A. Massey, L. Meinke, R. L. Rabb, M. Villani, M. Waldvogel, and K. G. Wilson for their insightful comments on earlier drafts of this manuscript.

REFERENCES

Andrewartha, H. G. and L. C. Birch. 1954. *The Distribution and Abundance of Animals*. University of Chicago Press, Chicago.

Benford, F. A. 1979. *A McArthur-Wilson Model of Archipelago Species Composition*. Ph.D. dissertation, Harvard University, Boston, Mass.

Blau, W. S. 1980. *Ecology* **61**: 1005–1012.

Chesson, P. 1978. *Annu. Rev. Ecol. Syst.* **9**: 323–347.

Cohen, D. 1966. *J. Theor. Biol.* **12**: 119–129.

Crowley, P. H. 1981. *Am. Nat.* **118**: 673–701.

Dempster, J. P. 1971. *Oecologia* **7**: 26–67.

den Boer, P. J. 1968. *Acta Biotheor.* **1**: 165–194.

Downhower, J. F. and D. E. Wilson. 1973. *Am. Midl. Nat.* **89**: 451–455.

Ehrlich, P. R. 1965. *Evolution* **19**: 327–336.

Ehrlich, P. R. and L. C. Birch. 1967. *Am. Nat.* **101**: 97–107.

Ehrlich, P. R., R. R. White, M. C. Singer, S. W. McKechnie, and L. E. Gilbert. 1975. *Science* **188**: 221–228.

Feeny, P. P. 1970. *Ecology* **51**: 656–681.

Futuyma, D. J. 1979. *Evolutionary Biology*. Sinauer, Sunderland, Mass.

Gillespie, J. 1977. *Evolution* **31**: 85–90.

Green, W. Q. 1974. *An Antagonistic Host/Plant System: The Problem of Resistance.* Ph.D. dissertation, Department of Zoology, University British Columbia, Vancouver, Canada.

Hassell, M. P. 1980. *Oikos* **35:** 150–160.

Hastings, A. 1978. *Theor. Popul. Biol.* **14:** 380–395.

Hedrick, P. W., M. E. Ginevan, and E. P. Ewing. 1976. *Annu. Rev. Ecol. Syst.* **7:** 1–32.

Heinrich, B. 1979. *Ecology* **60:** 245–255.

Hubbell, S. P. and L. K. Johnson. 1978. *Ecology* **59:** 1123–1136.

Hubby, J. L. and R. C. Lewontin. 1966. *Genetics* **54:** 577–594.

Huffaker, C. B. 1958. *Hilgardia* **27:** 343–383.

Huffaker, C. B., K. P. Shea, and S. G. Herman. 1963. *Hilgardia* **34:** 305–330.

Jones, J. S. 1980. *Nature* **286:** 757.

Kareiva, P. 1982. *Ecol. Monogr.* **52:** 261–282.

Kettlewell, H. B. D. 1955. *Nature* **175:** 943, 944.

Kettlewell, H. B. D. 1973. *The Evolution of Melanism.* Clarendon, Oxford.

Kimura, M. 1979. *Sci. Am.* **241:** 98–126.

Leonard, K. J. and R. J. Czochor. 1980. *Annu. Rev. Phytopathol.* **18:** 237–258.

Leopold, A. 1933. *Game Management.* Scribner's, New York.

Levene, H. 1953. *Am. Nat.* **87:** 331–333.

Levin, S. A. 1974. *Am. Nat.* **108:** 207–228.

Levin, S. A. and R. T. Paine. 1974. *Proc. Nat. Acad. Sci.* **71:** 2744–2747.

Levins, R. 1969. *Am. Nat.* **103:** 483–499.

Lewontin, R. C. 1974. *The Genetic Basis of Evolutionary Change.* Columbia University Press, New York.

Lewontin, R. C., L. R. Ginzburg, and S. D. Tuljapurkar. 1978. *Genetics* **88:** 149–170.

Lomnicki, A. 1980. *Oikos* **35:** 185–193.

Mark, G. A. 1982. *Entomol. Exp. & Appl.* **32:** 155–160.

Maynard-Smith, J. 1970. *Am. Nat.* **104:** 487–490.

Maynard-Smith, J. and R. Hoekstra. 1980. *Genet. Res.* **35:** 45–57.

Monro, J. 1967. *J. Anim. Ecol.* **36:** 531–547.

Myers, J. H. 1976. *Oecologia* **23:** 255–269.

Myers, J. H. 1978. Pp. 181–188, *in* Freeman (Ed.), *Proc. 4th International Symposium on Biological Control of Weeds, Gainesville, Fl., 1976.*

Myers, J. H. 1980. *Oecologia* **47:** 16–21.

Myers, J. H. and B. J. Campbell. 1976. *Can. Entomol.* **108:** 967–972.

Nagel, W. P. and D. L. Esaacson. 1974. *J. Econ. Entomol.* **67:** 494–496.

Papaj, D. R. and M. D. Rausher. 1983. Pp. 77–124 in S. Ahmad (Ed.). *Herbivorous Insects: Host-Seeking Behavior and Mechanisms.* Academic, New York.

Pimentel, D. 1961. *Annu. Entomol. Soc. Am.* **54:** 76–86.

Pimentel, D. 1968. *Science* **159:** 1432–1437.

Powell, J. R. and C. E. Taylor. 1979. *Am. Sci.* **67:** 590–596.

Prokopy, R. J., A. L. Averill, S. S. Cooley, and C. A. Roitberg. 1982. *Science* **218:** 76–77.

Prout, T. 1968. *Am. Nat.* **102:** 493–496.

Pyke, G. H., H. R. Pulliam, and E. L. Charnov. 1977. *Quart. Rev. Biol.* **52:** 137–154.

Rausher, M. D. 1978. *Science* **200:** 1071, 1072.

Real, L. A. 1980. *Am. Nat.* **115:** 623–638.

Real, L. A. 1981. *Ecology* **62:** 20–26.

Reddingius, J. and P. J. den Boer. 1970. *Oecologia* **5:** 240–284.

Ricklefs, R. E. 1979. *Ecology* (2nd ed.). Chiron, New York.

Roff, D. A. 1974. *Am. Nat.* **108:** 391–393.

Roff, D. A. 1980. *Oecologia* **45:** 202–208.

Root, R. B. 1973. *Ecol. Monogr.* **43:** 95–124.

Stenseth, N. C. 1980. *Oikos* **35:** 165–184.

Stinner, R. E. 1979. Pp. 199–211 in R. L. Rabb and G. G. Kennedy (Eds.), *Movement of Highly Mobile Insects: Concepts and Methodology in Research.* University Graphics, North Carolina State University, Raleigh.

Taylor, C. E. 1975. *Genetics* **80:** 621–635.

Taylor, C. E. 1976. *Genetics* **83:** 887–894.

Taylor, C. E. and J. R. Powell. 1977. *Genetics* **85:** 681–695.

Taylor, C. E. and J. R. Powell. 1978. *Oecologia* **37:** 69–75.

van der Meijden, E. 1973. *Netherl. J. Zool.* **23:** 430–445.

van der Meijden, E. 1979. *Oecologia* **42:** 307–323.

Waddington, K. D., T. Allen, and B. Heinrich. 1981. *Anim. Behav.* **29:** 779–784.

Wellington, W. G. 1964. *Can. Entomol.* **96:** 436–451.

Wellington, W. G. 1965. *Can. Entomol.* **97:** 1–14.

Wellington, W. G. 1977. *Environ. Entomol.* **6:** 1–8.

Wiens, J. A. 1976. *Annu. Rev. Ecol. Syst.* **7:** 81–120.

Chapter 15

Interspecific Competition in Insects

J. H. LAWTON and
M. P. HASSELL

In this chapter we review interspecific competition in insect populations; that is, situations where individuals of two or more species are striving against each other to secure some resource in limited supply (Varley et al. 1973). The chapter is divided into two major parts. In the first we examine the theoretical framework for two-species, or multispecies, competitive systems and discuss these largely in relation to laboratory systems. In the second part we provide a comprehensive, though not exhaustive, review of insect competition in the field and relate this where possible to theoretical predictions. The subject has been reviewed by DeBach (1966) and Miller (1967), which enables us to deal mainly with studies appearing since then. We focus almost entirely on contemporary interactions between species. The role of evolution in molding competitive interactions is dealt with in Chapters 8, 9, 20, and 22.

15.1 A BASIC MODEL

We commence with the logistic equation for population growth of a single species:

$$\frac{dN}{dt} = rN \left[\frac{K - N}{K} \right] \tag{15.1}$$

where r is the intrinsic rate of increase and K is the carrying capacity. This model is characterized by a monotonic approach of the population to a stable equilibrium at $N^* = K$ (cf. Chapter 12).

We now introduce a second species sharing the same resources as species 1, which on its own also shows logistic growth. When together, however, the growth rate of each species will in part be influenced by the abundance of the other, which leads to the familiar Lotka–Volterra competition equations (Lotka 1925, Volterra 1926):

$$\frac{dN_1}{dt} = r_1 N_1 \left[\frac{K_1 - N_1 - \alpha N_2}{K_1} \right] \tag{15.2a}$$

$$\frac{dN_2}{dt} = r_2 N_2 \left[\frac{K_2 - N_2 - \beta N_1}{K_2} \right] \tag{15.2b}$$

The parameters α and β are usually termed *competition coefficients* and express the effect of an individual of one species upon the other species. They are, in effect, merely a convenient means of rendering the negative feedback effects of N_1 and N_2 in each equation in the same units. Thus in Eq. (15.2a), for example, $dN_1/dt = 0$ when $K_1 = (N_1 + \alpha N_2)$.

The properties of Eqs. (15.2a) and (15.2b) are well displayed in phase space plots on which are drawn the zero growth isoclines of each species (i.e., the boundaries, where $dN/dt = 0$, that separate the regions of positive and negative growth). These equilibrium conditions are obtained when the respective carrying capacities are "filled" by some combination of the two species. Thus

$$\frac{dN_1}{dt} = 0 \quad \text{when } K_1 - N_1 - \alpha N_2 = 0 \quad\quad (15.3\text{a})$$

and

$$\frac{dN_2}{dt} = 0 \quad \text{when } K_2 - N_2 - \beta N_1 = 0 \quad\quad (15.3\text{b})$$

From Eqs. (15.3a) and (15.3b) it is clear that the zero isoclines must be linear with intercepts as shown in Fig. 15.1a.

This leads to the four general isocline configurations in Fig. 15.1, the relative position of which determines the outcome of competition. In Fig. 15.1a the isocline for species 1 lies above that for species 2, making the only stable solution at $N_1 = K_1$ and $N_2 = 0$. The situation is reversed in Fig. 15.1b: species 1 inevitably becomes extinct leaving species 2 to move to its carrying capacity. These cases conform to the notion of one species always being the superior competitor and so eliminating the other. Notice that the definition of a "superior" competitor in this model hinges upon the relative competition coefficients and carrying capacities but not the intrinsic rates of increase, which only affect how rapidly the final outcome is achieved (but see Vandermeer 1975). In Fig. 15.1, c and d differ from a and b in the isoclines intersecting, at which point a two-species equilibrium must occur. That in Fig. 15.1c, however, is unstable and one or other species becomes extinct, depending upon the initial ratio of numbers. Biologically, this corresponds to the situation where each species inhibits the growth rate of the other species more than itself. In other words, the destabilizing effects of interspecific competition outweigh the density dependence resulting from intraspecific competition. Finally, in Fig. 15.1d, all trajectories will inevitably move to the stable equilibrium where the isoclines intersect. Intraspecific effects now outweigh interspecific ones and so prevent extinction. Thus, the population size of species 1 required to eliminate species 2 is K_2/β, and since $K_2/\beta > K_1$ (Fig. 15.1d), species 1 can never become common enough to eliminate species 2. The same applies to the effects of species 2 on species 1.

The result in Fig. 15.1d provides the theoretical basis for the *competitive exclusion principle*, so succinctly defined by Hardin (1960) as "complete competitors cannot coexist." In order that intraspecific effects should outweigh interspecific ones for both species, there should be some refuge effect, or an unshared resource, that reduces the intensity of interspecific competition for at least part of the life cycle.

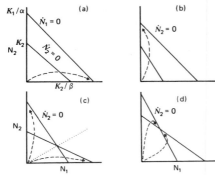

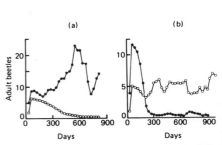

Fig. 15.1   The four possible outcomes from the competition equations [(15.2a) and (15.2b)] showing the linear zero growth isoclines of species 1 and 2. In each case the outcome is indicated by the broken arrow. (a) species 1 replaces 2; (b) species 2 replaces 1; (c) either species is replaced depending upon the initial ratio of numbers in relation to the dotted line; (d) both species coexist. The general configuration in (c) is only possible if $\alpha\beta > 1$, that in (d) if $\alpha\beta < 1$, and those in (a) and (b) if $\alpha\beta > 1$ or <1. The intercepts of the isoclines are given in (a).

Fig. 15.2   The effect of the parasite, *Adelina tribolii*, on the outcome of competition between *T. confusum* (○) and *T. castaneum* (●) at 29.5°C and 70% R.H. (a) *Adelina* present; *T. confusum* becomes extinct; (b) *Adelina* absent; *T. castaneum* becomes extinct.

15.2   COMPETITION IN THE LABORATORY

In seeking some correspondence between a general model such as Eqs. (15.2a) and (15.2b) and real systems, it is sensible to turn first to results of species interactions in the laboratory. Many vagaries of the outside world can here be eliminated and competition thus more easily analyzed. The first laboratory experiments on interspecific competition were those of Gause (1934) using two species of *Paramecium*. These were followed by many others, with insects being the favored subjects. Best known, are the experiments of Park and colleagues (e.g., Park 1948, 1954, Park et al. 1964) on competition between two tenebrionid flour beetles, *Tribolium confusum* and *T. castaneum*. Park found that replacement of one species by the other was the rule, but which species was to survive was critically dependent upon a variety of conditions. *T. castaneum* was always the superior competitor at 34°C and 70% RH and always inferior at 24°C and 30% RH. But at intermediate temperatures and humidities, the outcome would vary from replicate to replicate. Park also found an additional factor affecting the outcome—the presence or absence of a pathogenic microsporidian parasite, *Adelina tribolii*, which affects both species, but particularly *T. castaneum*. An example of this parasite's effect on the beetle populations is shown in Fig.

15.2. In terms of the Lotka–Volterra model, Park's experiments under conditions of extreme temperature and humidity all point to the situations in Fig. 15.1a,b with *Adelina* or the physical conditions determining which species' isocline lies above the other. It is tempting to assign the variable outcome observed at intermediate conditions to that predicted by Fig. 15.1c. Unfortunately, there is no evidence that the competitive outcome is at all dependent upon the initial numbers of the two species.

A further example, intermediate between laboratory and field, where temperature changes appear to alter the outcome is given by Chestnut and Douglas (1971). The grain moth, *Sitotroga cerealella*, is excluded from grain stores in southern Mississippi by the weevil *Sitophilus zeamais*, while in central Mississippi where temperatures are lower, the two species coexist. It is tempting to conclude that changes in temperature shift the competitive interaction from that shown in Fig. 15.1a,b to that in Fig. 15.1d.

In another classic series of experiments, Crombie (1945, 1946, 1947) also used stored product insects and was able to relate his results closely to the Lotka–Volterra model. This is well shown from experiments using *T. confusum* and the saw-toothed grain beetle, *Oryzaephilus surinamensis*, as competitors. With flour as food, *Tribolium* was always the superior species, probably because it would eat the eggs and pupae of *Oryzaephilus*, while *Oryzaephilus* would only eat *Tribolium* eggs. This result is shown in Fig. 15.3a

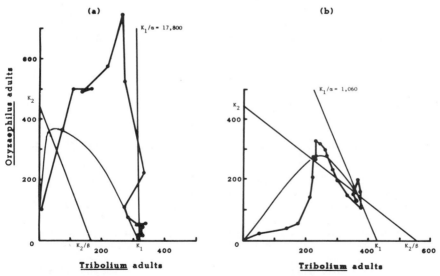

Fig. 15.3 Phase plane diagrams illustrating the outcome of competition between *Tribolium* (N_1) and *Oryzaephilus* (N_2) using either (a) fine flour or (b) wheat as food. The calculated trajectories come from Eqs. (15.2a) and (15.2b) using the following parameter values: (a) $K_1 = 320$, $K_2 = 440$; $\alpha = 0.02$, $\beta = 2.67$; (b) $K_1 = 425$, $K_2 = 445$, $\alpha = 0.4$, $\beta = 0.8$. Note that the equilibrium position in (b) is independent of the initial numbers of the two species. (After Crombie 1946.)

as a phase space plot similar to those in Fig. 15.1. Comparison of the observed and predicted outcomes shows considerable discrepancies in trajectories but good agreement in final outcome. Quite a different result occurred when cracked wheat instead of flour was used as food. *Oryzaephilus* could now pupate within the wheat grains, which provided a sufficient refuge for both species to coexist (Fig. 15.3b). Once again, the outcome is well predicted by the Lotka–Volterra model.

While such results are in general accord with the Lotka–Volterra model, other studies point forcibly to the model's inadequacies. Foremost among these are the experiments of Ayala (1969, 1970) on competition between various species of *Drosophila*. In several instances Ayala obtained coexistence of two species in situations where the conditions for coexistence in Fig. 15.1d did not apply. Instead of the isoclines being such that $\alpha\beta < 1$, the apparent opposite was found for most cases of coexistence; namely $\alpha\beta > 1$ with the isoclines appearing to intersect as in Fig. 15.1c. (The values for α and β were obtained by determining the K_1 and K_2 values from single-species experiments and then assuming linear isoclines passing through the observed equilibrium point as shown in Fig. 15.4a.) This anomaly, however, disappears when the notion of linear isoclines is abandoned in favor of curved ones, as shown in Fig. 15.4b (Gilpin & Justice 1972, Gilpin & Ayala 1973). The relative slopes of the isoclines close to the equilibrium

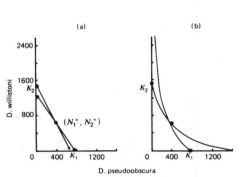

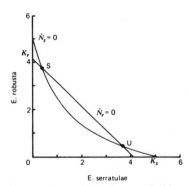

Fig. 15.4 Phase plane diagrams for the interaction of *Drosophila pseudoobscura* and *D. willistoni* in an experimental system. The points K_1 and K_2 were obtained from single species experiments and (N_1^*, N_2^*) represents the observed competitive equilibrium. (a) Linear zero growth isoclines drawn through the three equilibrium points, predicting (N_1^*, N_2^*) to be an unstable equilibrium; (b) the "true" isoclines separating the regions of positive and negative growth deduced from the vectors of population change from 19 experiments commencing at different points in the phase plane. (After Gilpin & Ayala 1973.)

Fig. 15.5 Phase plane diagram for the competitive interaction of *Eurytoma serratulae* and *E. robusta* larvae parasitizing the gallfly *Urophora cardui*. The curved isocline of *E. serratulae* is due to the inclusion of μ ($=0.6$) in Eq. (15.5a) describing the aggressive impact of *E. robusta* on *E. serratulae*. Points S and U indicate the stable and unstable equilibrium positions respectively. [After Zwölfer (1979), in which further details are given.]

are now once again appropriate for a stable equilibrium [cf. Borowsky's (1971) discussion of Ayala's results].

15.3 SOME ALTERNATIVE MODELS

In the face of results such as those in Fig. 15.4b, Ayala et al. (1973) and Gilpin and Ayala (1973) presented a number of competition models with nonlinear interaction terms, producing curved rather than linear isoclines. They may thus have multiple equilibria when the isoclines intersect more than once. Curved isoclines and multiple equilibria are also possible from rather different competition models. Schoener (1974a,b), for example, obtained curvilinear isoclines when such factors as the form of competition (e.g., interference or exploitation), the variety of resources, and the extent to which they are shared by competing species were explicitly considered. Zwölfer (1979), however, found that curved isoclines could be generated from the effects of direct aggression of one species on another. He studied two hymenopterous parasitoids, *Eurytoma serratulae* and *E. robusta*, competing for a common host, the knapweed gall fly, *Urophora cardui*. *Eurytoma serratulae* is the "extrinsically" superior species (i.e., the more efficient searcher) and *E. robusta* the "intrinsically" superior species in that its larvae often consume any *E. serratulae* larvae in the same host. The interaction was modeled by modifying the Lotka–Volterra equations to give

$$\frac{dN_s}{dt} = r_s N_s \frac{1}{K_s} (K_s - \mu N_s N_r - N_s - \alpha N_r) \qquad (15.4a)$$

$$\frac{dN_r}{dt} = r_r N \frac{1}{K_r} (K_r - N_r - \beta N_s) \qquad (15.4b)$$

where the subscript s refers to $E.$ *serratulae* and r to $E.$ *robusta*. The parameter μ is the amount that the carrying capacity of $E.$ *serratulae* is depressed by a single encounter between larvae of the two species, and α, β, and K are as used in Eqs. (15.2a) and (15.2b). (Zwölfer also introduced a variable difference in carrying capacities which is here omitted for simplicity.) The inclusion of $\mu N_s N_r$ in Eq. (15.4a) causes the isocline for $E.$ *serratulae* to become curvilinear, so permitting both a stable *and* unstable equilibrium, as shown in Fig. 15.5.

The mere inclusion of age structure effects can also generate curved isoclines as shown by Hassell & Comins (1976). They commenced with a general difference equation analogue of the differential Lotka–Volterra model:

$$X_{n+1} = X_n[f(X_n + \alpha Y_n)]^{-b} \qquad (15.5a)$$

$$Y_{n+1} = Y_n[g(Y_n + \beta X_n)]^{-b'} \qquad (15.5b)$$

where X and Y are the two competing species in successive generations n and $n + 1$, f and g are general functions, b and b' relate to the strengths of intraspecific competition, and α and β are again the competition coefficients. By being framed in difference equations in which the populations change over discrete time units, the model applies best to species with quite distinct generations. Equations (15.5a) and (15.5b) share one important property with the Lotka–Volterra model: the zero growth isoclines (e.g., $X_{n+1} = X_n$) are always linear, giving the same four configurations shown in Fig. 15.1. The one essential difference between the models lies in the local stability properties of the equilibrium populations. Instead of only a monotonic approach to the equilibrium being possible, models of the general form of Eqs. (15.5a) and (15.5b) permit monotonic or oscillatory damping, as well as stable limit cycles and chaotic behavior, simply because of the one-generation time delays in the feedback system (May 1974a, Hassell & Comins 1976).

When age structure is included, however, a major change in properties is apparent. Hassell and Comins (1976) illustrate this for a two-age-class extension of Eqs. (15.5a) and (15.5b); namely,

$$X_{t+1} = x_t \exp[- a(x_t + \alpha y_t)] \tag{15.6a}$$

$$Y_{t+1} = y_t \exp[- a'(y_t + \beta x_t)] \tag{15.6b}$$

$$x_{t+1} = \lambda X_t [\exp - c(X_t + \gamma Y_t)] \tag{15.6c}$$

$$y_{t+1} = \lambda' Y_t [\exp - c'(Y_t + \Delta X_t)] \tag{15.6d}$$

where x and y are the larval or juvenile stages of the two species, X and Y are the adults, α and β are the coefficients for interspecific competition between larvae, and γ and Δ are the corresponding coefficients for adults. The generations and stages are discrete and nonoverlapping (adults and larvae do not occur together), and the time interval from t to $t + 1$ represents the duration of adult or larval stages. The parameters λ and λ' are the finite net rates of increase of the two species, and a, a', c, and c' relate to the feedback term in the underlying single species model discussed by May and Oster (1976):

$$X_{n+1} = \lambda X_n \exp(- aX_n) \tag{15.7}$$

The zero isoclines may now be almost linear, concave, or convex and there can be up to three equilibrium points, as in Fig. 15.6c.

This work has been extended by Bellows (1979) in a laboratory study on competition between two species of bruchids (*Callosobruchus chinensis* and *C. maculatus*), whose larvae feed on legume seeds. A two-age-class model similar to Eqs. (15.6a)–(15.6d) was adopted for which all parameters were estimated from single and paired species experiments. The values of the competition coefficients so obtained suggest that among adults, *C. maculatus*

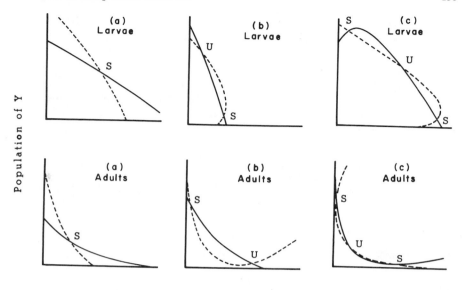

Fig. 15.6 Typical zero growth isoclines from Eqs. (15.6a)–(15.6d) yielding one or more equilibria. S = stable equilibrium; U = unstable equilibrium. Broken lines indicate the isoclines for species X; solid lines for species Y. (From Hassell & Comins 1976.)

has the competitive advantage, but that the larvae are more evenly matched, with *C. maculatus* having a slight edge, perhaps due to its larger size.

The zero growth isoclines predicted from Eqs. (15.6a)–(15.6d) are shown in Fig. 15.7 for both adults and larvae. They have a complex shape, arising solely from the age structure in the model, and they intersect once to produce an unstable equilibrium. The broken lines divide the phase space into regions in which *C. maculatus* or *C. chinensis* is the eventual survivor. The larger region in which the extinction of *C. chinensis* is predicted reflects the competitive superiority of *C. maculatus* mentioned above. In view of

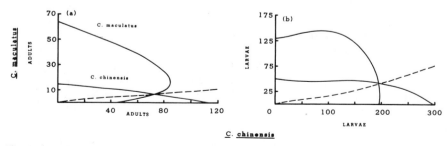

Fig. 15.7 The zero growth isoclines for (a) adult and (b) larval populations of *C. chinensis* and *C. maculatus* predicted from the two-age-class model of Bellows (1979). The equilibrium indicated by the intersection of the two isoclines is in each case unstable. (From Bellows 1979.)

this, it is at first surprising that *C. chinensis* is almost always the ultimate victor in long-term competition experiments. Out of 15 replicates with three different initial numbers of each species, *C. chinensis* replaced *C. maculatus* in 14 cases (T. S. Bellows & M. P. Hassell unpubl.). Development of a detailed simulation model has provided the solution to this anomaly (Bellows 1979). Whereas the models above have the implicit assumption of equal generation times, *C. chinensis* at 30°C and 70% RH completes development in 21 days and *C. maculatus* in 23 days. This difference is alone sufficient to counteract the competitive superiority that *C. maculatus* has within a single generation.

Clearly, the outcome of competition, even in the simplest of two-species systems, may depend critically on a variety of factors not included in the Lotka–Volterra equations [(15.2a) and (15.2b)].

15.4 MORE THAN TWO COMPETITORS

Natural competitive systems often involve more than two species showing varying degrees of resource overlap. A multispecies extension of the basic Lotka–Volterra equations [(15.2a) and (15.2b)] can be expressed as follows:

$$\frac{dN_i}{dt} = r_i N_i \left[\frac{K_i - N_i - \sum_{j \neq i} \alpha_{ij} N_j}{K_i} \right] \tag{15.8}$$

where there are now N competitors of the ith species and the competition coefficients α_{ij} denote the effect of an individual of species j on species i. [A discrete time analogue of this equation is given by Comins and Hassell (1976).] Strobeck (1973) shows how the necessary and sufficient conditions for coexistence of more than two species modeled by Eq. (15.8) depend not only upon the magnitude of the competition coefficients and the carrying capacities K_i (as in the two species case), but also upon the intrinsic rates of increase r_i. For n species there are $2(n - 1)$ inequalities which must be satisfied before stable coexistence is possible, increasing the number of inequalities by 2 for each new species added. In other words, the conditions become harder and harder to satisfy as n increases. An obvious inference to be drawn is either that natural species-rich assemblages are competitively very finely "tuned," or that most species do not compete effectively with most of their neighbors ($\alpha_{ij} \to 0$). The latter view is consistent with much of the data reviewed below (cf. Chapter 20).

By setting all $dN_i/dt = 0$ in Eq. (15.8), the equilibrium populations for n coexisting species is given by

$$N_i^* = K_i - \sum_{j \neq i} \alpha_{ij} N_j^* \tag{15.9}$$

for all i from 1 to n. From this it is clear that the equilibrium population of a given species (N_i^*) will be progressively lowered below K_i as more and more competing species enter the system. MacArthur (1972) called this *diffuse competition*. In theory, a species may be rare or excluded altogether from a habitat by the sums of the effects of several competitors that are individually relatively innocuous. Apparent examples of such diffuse competition are discussed in Sections 15.8.1 and 15.8.7.

15.4.1   Higher Order Interactions

An interesting extension to Eq. (15.8) is given by Vandermeer (1969):

$$\frac{dN_i}{dt} = \frac{r_i N_i}{K_i} \left(K_i - N_i - \sum_j \alpha_{ij} N_j - \sum_{jk} \beta_{ijk} N_j N_k \right) \qquad j \neq i, \; k \neq i \quad (15.10)$$

Imagine a community of three competitors. In addition to the possibility of one species competing with another (embodied in the first-order term $\sum_j \alpha_{ij} N_j$), there can now also be pairs of species acting in concert. This possibility is expressed by the second-order term ($\sum_{jk} \beta_{ijk} N_j N_k$). For a three-species system with $N_i = N_1$, the second-order term might represent the combined effects of species 2 and 3 on species 1 ($\beta_{123} N_2 N_3$) *over and above* their individual impacts on species 1 (α_{12} and α_{13}). When this is extended to larger systems of $n > 3$ competitors, the additional terms will now involve the multiplicative effects of not only pairs, but also triplets (third-order terms), or whole "gangs" of competitors.

The ease with which the behavior of multispecies competitive systems can be predicted rests upon the importance of higher order interactions in real communities. If rare, or absent, then measuring the effects of any one species upon another in simple pairwise experiments should predict well the outcome from mixtures of several competitors. This, however, would not be possible if higher order terms were important; knowing how any pair of species interact may then bear little relationship to their behavior in a larger community. Very few attempts have been made to assess the importance of higher order competitive effects, even in the laboratory, and some that have been reported have recently been questioned (Pomerantz 1981). More importantly, only a single example has been reported involving insects in the field (Seifert & Seifert 1976, 1979a). These authors studied the interaction between the species inhabiting water-filled *Heliconia* bracts in tropical forests. Most of the interaction coefficients (α_{ij}) between species in these systems were not significantly different from zero, but Seifert and Seifert did find 16 significant first-order terms (not all of them competitive) from 40 possible pairwise interactions. They also found one significant *third*-order interaction. Survival of an adult beetle, *Gillisius* "species 3," was significantly reduced by the joint presence of the fly larva, *Copestylum ro-raima*, adult beetles of a second species, *Cephaloleia neglecta*, and another fly larva, *Quichuana angustiventris*. None of these three species on their own affected *Gillisius*, and there were no significant second-order effects. Seifert and Seifert offer no explanation for these intriguing results.

15.4.2 Niche Overlap Along a Single Resource Axis

An interesting aspect of multispecies competitive interactions occurs when sets of species are orderly arranged along a single "niche dimension" or "resource gradient" (Fig. 15.8). Species then compete most strongly with immediate neighbors and progressively less so with species further along the spectrum. Several examples apparently conform to such an arrangement (see Sections 15.8.2, 15.8.4, 15.8.9). A theoretical basis for viewing this problem has been provided by MacArthur (1972), May and MacArthur (1972), and May (1975). Here we need simply note May and MacArthur's main conclusions. In a deterministic environment permitting no random fluctuations in availability of resources, there is no limit to niche overlap consistent with a stable equilibrium, short of complete overlap when coexistence is infeasible. As soon as any random fluctuations are permitted, however, a definite limit occurs to the degree of overlap that still allows all species to coexist. If the random fluctuations are severe, this limit depends primarily on the variance of the fluctuations. But for all more modest levels of stability the limit occurs when the separation of the species (d) is approximately equal to the spread of their resource utilization functions (w)—namely, when $d/w \simeq 1$ (see Fig. 15.8).

This work has inspired both developments and criticisms (Schoener 1974c, Abrams 1975, 1980, Feldman & Roughgarden 1975, Heck 1976, Mc-

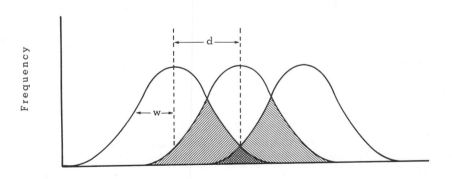

Fig. 15.8 Frequency distributions for the utilization of a resource (e.g., food of varying size) by three species; d represents separation of the curves and w their spread. (After May 1975.)

Murtrie 1976, May 1976, Nisbet et al. 1978, Turelli 1978, Angren & Fagerstrom 1980). A consensus from much of this literature is that the limit d/w will vary markedly and under some circumstances disappear altogether. This will depend upon the way niche overlap is related to the competition coefficients, when and how environmental disturbances are forced into the system, the degree of independence of the various points on the resource axis, and how similar or different the species are in their ability to harvest these resources.

We are left with three main conclusions. First, no one model of competition for resources distributed along a single niche dimension will be suitable for all circumstances, and it will thus be necessary to match model assumptions to data very carefully before predictions can be properly tested in the field. Second, both deterministic and stochastic models predict a limit to niche overlap under some, but not all, circumstances. Finally, it is extremely difficult to say what this limit will be a priori; but it is rather unlikely to be the pleasing answer, $d/w \simeq 1$.

15.5 SPATIAL HETEROGENEITY

Despite the seminal paper of Skellam (1951) and the intuition and experience of many biologists (e.g., Hutchinson 1951, Huffaker 1958), mathematical population theory has only in the last decade begun properly to face the problem of how heterogeneity and the potential for dispersal can together influence population dynamics (cf. Chapters 11, 12, 14, 23). Before this, most model populations were generally assumed to be uniformly distributed within homogeneous environments.

The influence of a patchy environment on the outcome of competition has attracted much attention from theoretical ecologists such as Skellam (1951), Levins and Culver (1971), Horn and MacArthur (1972), Levin (1974, 1977), Slatkin (1974), Fenchel (1975), and Whittaker (1975). In some cases the patches are assumed to be identical, but the species can differ according to dispersal and competitive ability (e.g., Skellam 1951, Levin 1974); in others (e.g., Horn & MacArthur 1972) the patches can also be of different kinds. The general conclusion is that spatial variability can act as a powerful stabilizing factor tending to oppose the destabilizing effects of interspecific competition. In addition, this literature points firmly to migration between patches helping "fugitive" species (Hutchinson 1951) to escape extinction at the hands of superior competitors within a single patch (May 1974b).

Simple and conceptually attractive models for competition in a patchy environment have been developed by Shorrocks et al. (1979) and Atkinson and Shorrocks (1981), specifically with insect populations in mind. They examine competition for a patchy resource in which the patches are temporary and regularly renewed (e.g., fruiting bodies of woodland fungi, which provide breeding sites for various species of *Drosophila*) and find that equilibria are easily obtained. Broadly speaking, coexistence becomes more likely if the patches are more finely subdivided and if the competitors exhibit an aggregated distribution between patches independently of one another. Particularly important for coexistence is the marked aggregation of the superior competitor, thus providing more patches in which it is absent and in which the inferior competitor can survive.

A rather different model, again developed with insects in mind, has been proposed by Lloyd and White (1980) for competition between three species of *Cicada*. The model focuses on the numbers of each competing species *j* in the immediate vicinity of any one individual of species *i*, as well as on the numbers of species *i* (i.e., on the "mean crowding" experienced by an individual due to the members of its own, and other, species). This model leads to much the same conclusions as those above. When each species tends to concentrate in different patches, coexistence is markedly enhanced, a conclusion that should apply to many, if not most, patchily distributed insect species.

15.6 PREDATOR-MEDIATED COEXISTENCE

Since the classic study of Paine (1966), in which he demonstrated that a "top predator" (the starfish, *Pisaster ochraceus*) can play a crucial role in maintaining prey species diversity, predator-mediated coexistence has received considerable theoretical attention (e.g., Cramer & May 1972, Steele 1974, van Valen 1974, Murdoch & Oaten 1975, Roughgarden & Feldman 1975, Comins & Hassell 1976, Fujii 1977, Hassell 1979). The central problem to which all these studies have been directed is how, if at all, polyphagous predators affect the coexistence of potentially competing prey species. Conventionally, the models have been in continuous time based on the Lotka–Volterra competition and predator–prey equations. Here we follow Comins and Hassell (1976) and describe a system framed in difference equations, which lends itself well to situations where the predator is an insect parasitoid (Hassell 1978).

We commence with two species competing on their own as given by Eqs. (15.5a) and (15.5b). The zero growth isoclines for this model are inevitably linear, giving the same general configurations as in Fig. 15.1. With these pictures in mind we pose the questions: (1) Can predation "neutralize" the dominant competitor (e.g., N_1 in Fig. 15.1a) in such a way that a stable equilibrium is created? and (2) Can predation reverse the isoclines in Fig. 15.1c and so convert the unstable equilibrium to a stable one as in Fig. 15.1d? In seeking answers to these questions we shall consider three types of predation: (i) "equivalent" predation, (ii) "preference," and (iii) "switching," in each instance using a particular case of Eqs. (15.5a) and (15.5b) to which a predator species attacking both prey is added, as follows:

$$X_{t+1} = X_t \exp[r - \frac{r}{K}(X_t + \alpha Y_t)]f(P_t)$$

$$Y_{t+1} = Y_t \exp[r' - \frac{r'}{K'}(Y_t + \beta X_t)]f'(P_t) \qquad (15.11)$$

$$P_{t+1} = X_t[1 - f(P_t)] + Y_t[1 - f'(P_t)]$$

where P_t and P_{t+1} are the predator populations in successive generations, r and r' are the prey rates of increase, and K and K' their carrying capacities. The functions f and f' are the fractional survival of X and Y from P_t searching predators and are given by

$$f(P_t) = \left[1 + \frac{aP_t}{k}\right]^{-k} \qquad f'(P_t) = \left[1 + \frac{a'P_t}{k'}\right]^{-k'} \qquad (15.12)$$

They represent the zero terms of a negative binomial distribution where a and a' are the searching efficiencies for the two predator species, and k and k' reflect the extent to which the distribution of attacks on prey are clumped. The provenance and properties of this model are discussed by May (1978). By varying k, the degree of clumping of predator attacks can be changed from random ($k \to \infty$) to very clumped ($k \to 0$), with the equilibria between the interacting populations becoming increasingly stable as the distribution of predator attacks becomes more and more clumped (i.e., as k gets smaller) (May 1978, May & Hassell 1981).

The customary use of the term *equivalent predation* implies not only that the predators show no distinction between the prey (i.e., $a = a'$ and $k = k'$) but also that the prey are equivalent in having the same growth rates ($r = r'$). Such predation, in continuous time models, can make no difference to the competitive coexistence of the prey species (van Valen 1974, May 1977). However, in systems with time delays, as in Eq. (15.11), an important difference is introduced. Instead of being inevitably stable as in the Lotka–Volterra models, the equilibrium in Fig. 15.1d may now be locally unstable, exhibiting limit cycles or higher order behavior. In such cases equivalent predation can have the major impact of creating a locally stable equilibrium where the prey alone showed cyclical behavior; or alternatively, of changing a locally stable competitive equilibrium to a locally unstable predator–prey system. In short, equivalent predation can affect local stability as long as a potentially stable equilibrium already exists.

More commonly, a predator will exhibit a degree of preference among its range of potential prey. Some prey may be easier to locate and/or provide a greater net energy return than others (Cock 1978, Hassell & Southwood

1978). Such preference may be introduced into Eq. (15.11) by assuming unequal searching parameters (i.e., $a \neq a'$ and/or $k \neq k'$). A situation impossible with "equivalent predation" can now arise, allowing the predators to create and maintain an equilibrium from a situation where the prey species alone exhibit no equilibrium whatsoever, as shown by the example in Fig. 15.9. In terms of Fig. 15.1, the predators are converting the isocline configurations of Fig. 15.1a,b to that in Fig. 15.1d, which they can do as long as there is a preference for the "dominant" competitor. The "upper" isocline is thus moved downwards more than the lower one, allowing the possibility for them to intersect and create a potentially stable equilibrium.

A predator population showing preference for the dominant competitor can therefore stabilize an interaction where no equilibrium existed in its absence, as long as some degree of niche separation ($\alpha\beta < 1$) exists. With complete niche overlap ($\alpha\beta \geq 1$), a stable equilibrium remains impossible. Unfortunately, the simplicity of this conclusion is a little misleading. The same effects can be observed where a predator shows no preference at all, but the prey have unequal rates of increase ($r \neq r'$), since this also influences the extent to which the isoclines are shifted (Hassell 1979). We thus have the interesting situation where an equilibrium is possible with *equal* predation as long as the less able competitor has the greater population rate of increase (May 1977).

Just as preference for some prey types can be widespread among polyphagous predators, so too might we expect certain predators to "switch"

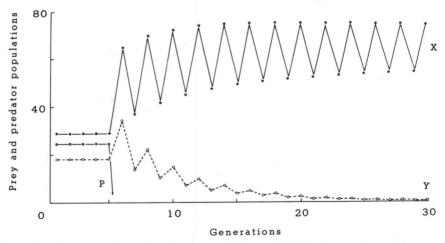

Fig. 15.9 Numerical simulation from Eqs. (15.11) and (15.12). The predator shows preference for prey species X (the dominant competitor) and is thus preventing species Y from becoming extinct. This is illustrated by removing the predator in generation 5. Parameter values: $a = 0.033$, $a' = 0.026$, $k = k' \to \infty$, $r = r' = 2.0$, $K = 65$, $K' = 90$, $\alpha = 0.54$, and $\beta = 1.5$. (From Hassell 1979.)

to whichever prey is the most abundant at the time (Murdoch 1969, 1977, Murdoch & Oaten 1975). This implies that the percent predation of a prey type changes from less than expected on the basis of the searching parameters a, a', k, and k' in Eq. (15.12) to more than expected as the relative abundance of that prey type increases. A good example is given by Lawton et al. (1974) from experiments using the aquatic predatory bug, *Notonecta glauca* (Fig. 15.10). Experiments of this type have tended to focus only on the relative abundances of the two prey species. Recently more extensive studies with *N. glauca* (R. Bonser & J. H. Lawton unpub.) show that the response in Fig. 15.10 is confined to a narrow range of *absolute* prey densities. With very low or very high overall densities of prey, *N. glauca* does not switch as the proportions of the two prey types change. Comparable studies with other predators have not been carried out, making it difficult to generalize on the importance of switching in natural populations. Intuitively we expect it to occur most often in a spatially heterogeneous environment with different prey types tending to occur in different patches. Switching would then result from the predator aggregating in the currently most profitable patches, in much the manner envisaged by Royama (1970).

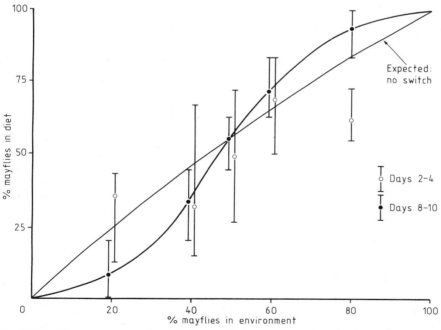

Fig. 15.10 The percentage of mayfly larvae (Ephemeroptera) in the diet of the aquatic predatory bug, *Notonecta glauca*, as a function of the percentage of mayfly larvae, *Cloen dipterum*, available in the environment. Alternative prey are *Asellus aquaticus* (Crustacea: Isopoda). Data are means and ranges of five replicates. The predators "switch" to whichever prey is proportionately the most abundant. (From Lawton et al. 1974.)

When a simple expression for switching is incorporated in Eq. (15.11), switching, like preference, can create a stable equilibrium where the competing prey alone showed no equilibrium, stable or otherwise (cf. Roughgarden & Feldman 1975). Unlike preference, however, a stable equilibrium is now possible even when the prey show complete niche overlap ($\alpha\beta \geq 1$). This is achieved by the switching causing the prey isoclines to *bend* which, if sufficient, can convert the necessarily unstable configuration of Fig. 15.1c to the potentially stable one of Fig. 15.1d.

Interestingly, a recent experiment on the influence of a switching natural enemy on the outcome of competition failed to show enhanced coexistence (Cornell & Pimentel 1978). Cornell and Pimentel found that the parasitoid *Nasonia vitripennis* switched markedly to whichever species of dipterous pupae was most abundant in paired tests: *Phormia regina* with *Phaenicia sericata*, *P. regina* with *Musca domestica* and *Phaenicia* with *Musca*. In the absence of *Nasonia*, *Phaenicia* was usually the competitively superior species in laboratory cultures, eliminating both *Musca* and *Phormia*. This outcome was not significantly changed by exposing the pupae to attack by *Nasonia*, either for 24 or 48 hr. Under slightly different conditions, however, with less space and more intense larval competition, parasitism by *Nasonia* completely reversed the outcome by exterminating *Phaenicia* and allowing *Phormia* to win. *Musca* was always eliminated, either by *Phaenicia* or by *Phormia*. Stable coexistence of the two host species as a result of parasitism by *Nasonia* was never achieved. These experiments with a switching natural enemy are unique, and in the absence of further examples, the model predictions on the effects of switching remain plausible but unproven.

The important conclusion to emerge from these models is simply that predation can have a marked influence on prey coexistence: preference and switching can create stable equilibria where none existed before. Even completely random predation across all prey types can do the same if the inferior competitors have the higher rates of increase. In addition, predation can have a profound influence on the local stability of interacting populations. Assuming that predation itself is stabilizing [i.e, k in Eq. (15.12) is small], the addition of polyphagous predators to a competitive interaction should enhance stability and thus the likelihood of prey coexistence.

15.7 APPARENT COMPETITION AND PREDATOR-MEDIATED EXCLUSION

Perhaps because the main focus of research on polyphagous predators has been to ask how they promote coexistence, the phenomenon which Holt (1977) calls "apparent competition" has been largely overlooked. Consider a single species of food-limited, potentially polyphagous predator in stable equilibrium with one species of prey, *A*. The introduction of a second prey species, *B*, to this system increases the predator's total food supply and hence may raise its average density, leading to a reduction in the density of prey A. That is, just as in true competition, experimental manipulation of prey species B (increasing or decreasing its density) will lead to changes in the density of species A of exactly the sort expected in classical resource-based competition. However, in this case, the interaction is via a shared natural enemy rather than a shared resource. While there are underlying similarities between models of these two types of systems (Williamson 1957), their detailed causes and the conditions under which they might be expected to occur are very different. Holt (1977) provides a detailed analysis of apparent competition from a set of differential equation models. He shows, for example, how a prey species may be completely excluded from a community by a polyphagous predator sustained by alternative prey. (Some possible field examples are discussed in Section 15.8.1).

This section really serves as a cautionary tale for the interpretation of field manipulation experiments. Just because one species increases as another declines, we cannot infer that they necessarily compete for limiting resources. They may, instead, share a polyphagous enemy. Even more worrying, as Holt shows, are effects transmitted through complex food webs via chains of prey and predators. When rabbits (*Oryctolagus cuniculus*) suffered catastrophic mortality in Britain due to myxomatosis, populations of the sun-loving ant *Lasius flavus* also declined as the vegetation thickened and recovered from rabbit grazing. In parts of their range, green woodpeckers (*Picus viridis*), which regularly fed on the ground on *Lasius flavus* mounds, then also declined and, although not documented, some of the alternative prey of the green woodpecker probably benefited from their demise. In this instance we know the complex chain of events, and nobody has suggested that *Lasius flavus* (or rabbits) compete with the alternative prey of green woodpeckers! But a similar mistake, with poorly known food-webs, would be much easier to make.

15.8 COMPETITION IN THE FIELD

In this section we review the evidence from various groups of insects for and against the importance of interspecific competition in the field. The guilds of species we examine are either all in the same taxonomic group (e.g., ants or dragonflies) or of mixed pedigree but share food resources (e.g., phytophagous insects). Without doubt some insect species are rarer in some habitats than they would otherwise be in the absence of competitors. Other species may be excluded completely. The important question for the field ecologist is no longer "Does interspecific competition occur?" but rather "On balance, how important is interspecific competition in structuring different ecological communities, in determining patterns of distribution, or in the moulding of species' morphologies and behavior by character displacement?"

We will undoubtedly leave the reader with the impression that competition is important, though much more so in some groups and habitats than in others. Unfortunately, for two reasons, the literature is inevitably weighted in favor of reporting examples where interspecific competition may be inferred. First, cases which on a priori grounds look interesting are more likely to be studied, and second, most ecologists are understandably reluctant to publish (and editors to accept) negative results. A notable exception is the exemplary study by Wise (1981) reporting an absence of competition between species of tenebrionid beetles. There is an urgent need for further experimental studies specifically designed to look for interspecific competition in the field, irrespective of whether or not it is expected in the first place. Only then will we have a clear idea of the general importance of interspecific competition in structuring natural insect assemblages.

The account which follows is not intended to be exhaustive, and if we have missed some favorite examples, we apologize. We have also been deliberately selective in one particular regard. The ecological literature, including the literature on insects, is full of studies on "ecological differences" between species, with a view to discovering the "reasons" for their coexistence. Moreover, it is not uncommon to argue that such differences have evolved to reduce interspecific competition. Certainly this is one possible hypothesis. Alternatively, differences between species might have nothing whatsoever to do with competition but represent the idiosyncratic adaptations of each species to some particular way of life (Wiens 1977, Huey 1979, Connell 1980). For this reason, we have chosen to ignore the many studies demonstrating that two or more species of closely related insects somehow "differ in their ecology." On their own, such observations tell us nothing about competition, and never did.

The examples we describe are organized roughly by trophic levels, starting with plant-feeding insects and other primary consumers, before moving on to insect parasitoids and then predators. Wherever possible, we refer back to the general insights provided by the laboratory experiments and models outlined in the first half of the chapter.

15.8.1   Phytophagous Insects

Insects feeding on green plants come from several major orders; principally the Lepidoptera, Hemiptera (Homoptera and Heteroptera), Coleoptera, Diptera, Hymenoptera, and Orthoptera. Since any one species of plant may provide food for guilds of insects assembled from any or all of these orders (Southwood 1961, Lawton 1978), it is sensible to look for evidence of interspecific competition among phytophagous insects as a whole, rather than deal with each order in turn. In marked contrast to some of the groups which follow, the most common case in phytophagous insects seems to be for species *not* to show evidence of interspecific competition. The available data are reviewed and detailed by Lawton and Strong (1981).

Different species of phytophagous insects often show considerable overlap in their feeding ecology, but this does not mean that they necessarily compete for food. In a pertinent study by Rathcke (1976), the larvae of 9 of 13 cooccurring species of stem-boring insects (8 Coleoptera, 2 Diptera, and 3 Lepidoptera) showed more than 70% overlap in resource exploitation, with at least 1 other member of the guild, but only 2 species had any perceptible negative influence on each other. A mordellid beetle larva, designated "Mordellidae 23," and a microlepidopteran caterpillar (*Epiblema* sp.) were found together in field stems less often than expected by chance. Laboratory observations indicated that the significant difference between observed and expected cooccurrences resulted from interference competition. Whenever larvae of these two species encountered one another within stems, the mordellid would attack, injure, and eventually kill *Epiblema*. Rathcke was even able to provide an estimate of the competition coefficient (α = 0.24) for the impact of the mordellid larvae on *Epiblema* caterpillar populations. The reciprocal interaction was negligible. The effect of the mordellid on *Epiblema* was therefore the only significant competitive interaction found within the guild.

Rathcke (1976) provides an excellent summary of papers which point to the infrequency of interspecific competition in phytophagous insect communities (cf. Rothschild 1971, Root 1973, Mamaev 1975, Yasumatsu 1976, Strong 1977, Strong & Wang 1977, Addicott 1978, Redfern & Cameron 1978, Simberloff 1978, Faeth & Simberloff 1981, Wise 1981). These studies cover a wide range of species and habitats and leave little doubt that interspecific competition among phytophagous insects is often absent or extremely weak. In a recent study of forest herbivores Futuyma and Gould (1979) conclude that "insect species are certainly not as equitably distributed over potential resources as a theory of species packing based on competition would lead one to expect."

Although normal population densities of many species of phytophagous insects often lie considerably below those necessary to produce interspecific competition (e.g., Harrison 1964, Pipkin et al. 1966, Le Quesne 1972, McClure 1974, Faeth & Simberloff 1981), it is possible that occasional outbreaks in one or more species may bring them into transient competition with others on the same plant (Varley 1949, Tilden 1951, Ross 1957, Hansen & Ueckert 1970, Edmunds 1973, McClure 1974). A sudden reduction in availability of the plant would have a similar effect (Halkka 1978). Surprisingly, actual examples of competition between such species during periods of maximal abundance are rare. McClure and Price (1975, 1976) demonstrated competition between tree-feeding species of leafhoppers (Cicadellidae) at high population densities; Darlington (1974), studying *Neuroterus* gall wasps (Cynipidae), also provides evidence pointing strongly in this direction. A further example that also shows how unexpected mechanisms of competition can be is given by Messenger (1975). The biological control of the walnut aphid (*Chromaphis juglandicola*), a pest on introduced walnuts in California, by the parasitoid *Trioxys pallidus* led to outbreaks of a second, formally rare aphid, *Panaphis juglandis*. A likely reason for this apparent suppression of *Panaphis* by *Chromaphis* was that honeydew excreted by *Chromaphis* fell onto the upper surface of the leaves where *Panaphis* fed and made the habitat unsuitable. *Chromaphis* feeds only on the sheltered undersurface and was thus not affected by its own excreta.

Various other introduced phytophagous insects, relieved of the impact of their own natural enemies, have so devastated their food plants that other species exploiting the same host have been excluded (e.g., Huffaker & Kennett 1969, Blakley & Dingle 1978). For example, the introduced chrysomelid beetle, *Chrysolina quadrigemina*, reduced St. John's Wort, *Hypericum perforatum*, to such low levels in California that three other species of insects that had also been released and established as control agents of the weed were eliminated from most areas (Huffaker & Kennett 1969). The closely related beetle, *C. hyperici* disappeared completely and a gall midge, *Zeuxidiplosis giardi*, and root borer, *Agrilus hyperici*, persisted for many years only in shady localities.

It is not at all clear how best to generalize from these examples. They clearly suggest competitive exclusion. On the other hand, the exceptional abundance and impact on host plants achieved by introduced insects relieved of their own natural enemies may not be typical of phytophagous insect communities in general.

Of course, we expect some species of plant-feeding insects to compete at normal population levels in natural communities. Good examples are provided by Gibson (1980) for grassland Miridae, by McClure (1980) for scale insects on hemlock (*Tsuga canadensis*), by Stiling (1980) for two species of leafhoppers on nettle (*Urtica dioica*), by Zwölfer (1979) for the guild of insects which mine and gall the flower heads of thistles, *Carduus nutans*, in Europe, and by Seifert and Seifert (1976, 1979a,b) for insects cohabiting in the flower bracts of tropical *Heliconia*.

Zwölfer's review is particularly elegant and suggests that cases of intense interspecific competition among phytophagous insects in the flower heads of *Carduus* (and related Compositae) may be common. In *C. nutans*, larvae of the tephritid fly *Urophora solstitialis* transform the heads into a compact gall, making the tissues quite unsuitable for larvae of two species of weevils, *Larinus sturnus* and *Rhinocyllus conicus*. A third type of inhabitant, solitary caterpillars of the microlepidopteran genera *Eucosma* and *Homoeososoma*, feeds on plant tissue and other insect larvae and is "intrinsically superior to all the other species in the system." They are aggressive, attacking and killing their cohabitants, and their underlying dynamics are presumably very close to those described in Eqs. (15.4a) and (15.4b). In other respects, this ecological microcosm also conforms closely to some of the general ideas outlined earlier. Competition manifests itself as a reduction in the proportion of *Rhinocyllus* eggs surviving to pupation, in the presence of either *Urophora* or *Larinus*. However, the intensity of this competition is ameliorated by some of the inhabitants, particularly *Urophora*, having highly clumped distributions which reduces the frequency of competitive interactions. The effect of *Larinus* may be one of simple exploitation competition. The effect of *Urophora* on *Rhinocyllus*, however, is likened by Zwölfer to an extreme form of interference competition because although "it is not capable of direct attacks," it "can eliminate competitors by destroying the surplus food." *Rhinocyllus* is simply denied access to its food supply. Finally, in the absence of competitors (and its enemies), *Rhinocyllus* introduced into Canada and the United States showed marked "competitive release," reaching larval densities 20 to 30 times higher than those in Europe.

As emphasized earlier, it is important not to generalize from these well-documented examples of interspecific competition and assume that the majority of phytophagous insects are so affected. On present evidence, at least, most such species, for most of the time, are not.

Finally, we should note an indirect, unfortunately disputed, demonstration of interspecific competition between phytophagous insects. Herbert et al. (1974) found that the most common adult macrolepidoptera (moths) caught at light traps were "taxonomically isolated" species, with few relatives in the same family. Rarer species tended to share their environment with many close relatives. Herbert and co-workers assumed, not unreasonably, that taxonomic affinity reflects, albeit roughly, ecological similarity, and they therefore attributed their results to the effects of "diffuse competition" [i.e., to the cumulative effects of ecologically similar species sharing limited resources]. Unfortunately, these patterns in the relative abundances of Lepidoptera at light traps have not been found in other studies (Taylor & Woiwood 1975), and it is debatable whether they are really due to competition.

The empirical evidence that competition between phytophagous insects is unusual, or at best not widespread, is in marked contrast to the view of Janzen (1973) who asserts that because "all parts of an individual plant are connected through the medium of its resource budget," insects feeding on that plant "automatically compete with *all* other species" on the plant. We would counter that the per capita effect of members of the same or different species on individual growth rates, fecundity rates, and survival rates are highly nonlinear (Hassell 1975, Hassell et al. 1976), making the effects of adding individuals on the dynamics of given species negligible, providing the overall abundances of the species in question do not rise above some critical threshold. On present evidence, the abundance of most (though certainly not all) phytophagous insect species seems to lie below such a threshold.

Why should competition for food between phytophagous insects be unusual? Our simple answer is that most phytophagous species are normally rare relative to their potential food supplies and maintained so by a potent combination of predators, parasitoids, diseases, and the vagaries of a harsh and unpredictable environment (Hairston et al. 1960). For some of these species, refuges are highly localized and unpredictable in time and space, or favorable environmental conditions rarely last long enough for the populations to reach their ceiling of food limitation (Andrewartha & Birch 1954, Andrewartha 1961, Thompson et al. 1976, Whittaker et al. 1979, and others). For others, and probably the majority (Lawton & McNeill 1979, Caughley & Lawton 1981), natural enemies, particularly insect predators and parasitoids, but sometimes also birds (e.g., Holmes et al. 1979), all play a key role in regulating individual populations well below the limits set by food and space.

It is also possible for polyphagous natural enemies to cause the extinction of one or more of their potentially competing prey populations. Some of the best-documented field examples of such apparent competitive exclusion caused by shared polyphagous enemies (Holt 1977) are provided by phytophagous insects. Goeden and Louda (1976) summarize several cases where insects introduced for biological control of weeds failed to establish because of the impact of polyphagous predators, and possibly parasitoids, maintained by the "resident" herbivores. The puzzling replacement of yellow scale, *Aonidiella citrina*, by red scale, *A. aurantii*, on citrus in California despite a superabundance of food may be an analogous phenomenon, although it is not clear whether or not the two species actually share a natural enemy(ies) in the field (DeBach et al. 1978). Less puzzling examples are provided by Zwölfer (1975).

As indicated earlier, we prefer to treat such cases of "apparent competition" via shared enemies as a process distinct from classical resource-based competition, although in practice the two might be very difficult to distinguish. Indeed, avoidance of the predators and parasitoids of neighboring species may be a major force driving niche diversification by phytophagous insects (Lawton 1978). This idea appears to have been discovered and rediscovered several times since Charles Darwin wrote, "from what we know of our own British Lepidoptera we may confidently believe that most of the different species of caterpillars would have different habits, or be exposed to different dangers from birds and hymenopterous insects" (from Stauffer 1975). Since then Brower (1958), Askew (1961), Huffaker (1971a), Gilbert and Singer (1975), Zwölfer (1975), and Heinrich (1979a) have all realized that differences in feeding ecology and appearance of phytophagous insects may have been molded by the impact of predators. Exactly the same arguments could also hold for other groups of insects but appear to have received no attention. We believe that "apparent competition" via shared natural enemies would repay detailed field study.

15.8.2   Ants (Hymenoptera: Formicidae)

Competition between species of ants is well documented both anecdotally and experimentally (Chapter 18). For example, artificially dispersed by man, *Pheidole megalocephala* has competitively excluded native ants on Bermuda and Madeira, only to suffer exclusion itself by the introduced *Iri-*

domyrmex humilis on both islands (Haskins & Haskins 1965, Crowell 1968). The complex history of competition between *Pheidole* and *Iridomyrmex* on Hawaii and other islands is reviewed by Fluker and Beardsley (1970). On a more local scale, Way (1953) observed the gradual replacement of one native ant species by another in a coconut plantation in Tanzania.

There have been a number of field experiments designed specifically to study behavioral interactions and interspecific competition between ant colonies (Chapter 18). The simple models of competition require for coexistence that the intraspecific effects be larger than the interspecific effects. Pontin (1969) thus surmised that the performance of a *Lasius flavus* colony would be depressed more by having another *L. flavus* nest dug up and moved adjacent to it than would a *L. niger* nest similarly treated. The magnitude of the difference between intra- and interspecific effects should depend upon the degree of ecological overlap shown by the species. Since *L. flavus* forages entirely below ground, while only one-third of the food of *L. niger* comes from subterranean foraging, changes in queen production (a measure of colony performance) should roughly mirror this difference. This is exactly what happened. The depression of queen production 1 yr after the manipulations were made was 3.7 times greater in *L. flavus* nests with another *L. flavus* nest moved next to them, than in *L. niger* nests receiving the same treatment. The performance of nests of both species adjacent to the removal sites, where competition was reduced, was also tested. As expected, queen production increased relative to the controls, and the response was greater in *L. flavus* nests relieved of intraspecific competition than in *L. niger* nests relieved of interspecific competition. However, the difference between the species was less marked than for nests around which introductions were made.

These experiments leave little doubt that *L. flavus* and *L. niger* colonies compete, the limiting resource being food. Competition does not exclude either species from the habitat. Instead, it leads to a stable equilibrium analogous to the case in Fig. 15.1d. Two years after Pontin's experiments queen production in the study colonies had returned to normal, although the manner of the readjustment (extinction, or movement of the introduced *L. flavus* colonies) is not specified.

Pontin's experiments were carried out on a scale sufficiently small to influence queen production but not the number of nests. Larger scale experiments demonstrating competition, in which most or all of the nests of one or more species were eliminated, have been carried out in Japanese grasslands by Yasuno (1965) and in Ghanaian cocoa plantations by Majer (1976). Majer describes the distribution of the ants as a "mosaic" maintained by both habitat preference and interspecific competition. His clearest result is shown in Fig. 15.11. The number of immigrating alate queens caught on the plots was directly proportional to the degree of removal of established colonies, either because workers in established colonies kill invading queens of their own or other species or prevent them from settling.

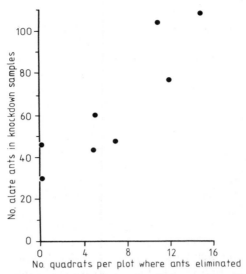

Fig. 15.11 Number of new queen (alate) ants invading experimental cocoa plots in Ghana as a function of the intensity of removal of established dominant ant colonies (Majer 1976).

Ants also allow us to examine multispecies competition in the field. One of the most interesting and detailed studies is that of Davidson (1977a,b, 1978) and Brown and Davidson (1977) on seed-eating ants in the southwestern deserts of the United States. A total of at least 15 granivorous ant species in five genera inhabit the region, although the number of species found at any one location varied between only two and eight, depending upon rainfall (Fig. 15.12).

Davidson suggests that resource partitioning on the basis of seed sizes collected by the workers plays a key role in organizing a local ant community, and closely mirrors the idealized niche-overlap model discussed in Section 15.4.2. The principal resource axis is seed size, with larger species of ants (i.e., those with larger workers) taking, on average, larger seeds (Fig. 15.13). Low rainfall deserts usually have two common and one rare middle-sized species. As rainfall increases, ants of all sizes, but particularly species with large (>9.0 mm long) and species with small (<2.0 mm long) workers, are added to the guilds. Some evidence of competition for a limiting supply of suitably sized seeds comes from cases of apparent competitive exclusion between ants of a similar body size (Chapter 18). An alternative explanation, however, cannot be lightly dismissed. From a "pool" of 15 species, there are a very large number of possible combinations of between two and eight species. If local ant communities were simply chance associations drawn at random from the pool, and if competition between the species played no part at all in structuring the assemblages, chance alone dictates that certain combinations of similar species will not be found

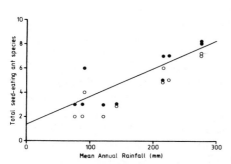

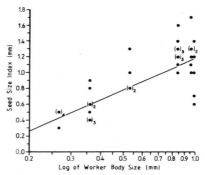

Fig. 15.12 The number of species of seed-eating ants in small local study plots as a function of mean annual rainfall at 10 sites in the southwestern deserts of the United States (from Davidson 1977a). Solid circles are total ant species; open circles include only common species. The equation for total ant species is $Y = 0.023X + 1.32$ (i.e., $F_{1,8} = 17.61$; $r = 083$; $P = 0.003$).

Fig. 15.13 Larger ants take larger seeds (Davidson 1977a). An index of the sizes of seeds exploited by six common desert ants is plotted against the log of their worker body size. Seeds are graded by sieving and the seed size index calculated from mean seed sizes, weighted by the abundance of the seeds in each size class.

together if only a limited number of "draws" are made (Connor & Simberloff 1979). Hence "missing combinations" are not strong circumstantial evidence for competition.

Although direct evidence for competition between the ant species in Davidson's study is lacking, competition between these ants and seed-eating rodents was successfully demonstrated by Brown and Davidson (1977). They found that the experimental removal of ants increased the biomass of seed-eating rodents by 24% and removal of rodents led to a 71% increase in the number of ant colonies. Hence, if ants and rodents compete in this way, it becomes very likely that the various species of ants also compete amongst themselves. Figure 15.13 may therefore provide a remarkably good description of resource division in this community.

A series of reasonable inferences are as follows. Seeds define the principal resource axis and are in limited supply. Different species of ants have different preferred seed sizes dependent upon their body size. Thus, to invade and remain in a community without displacing a resident, a species must have a preferred seed size sufficiently different from that of its neighbors; too much overlap in seed size leads to exclusion from the guild. Why then do higher rainfall deserts have more species of ants? Rainfall profoundly affects the productivity of the southwestern deserts, and although it appears not to have been measured directly, seeds are presumably more abundant in these higher rainfall areas. MacArthur (1972) suggested that if the diversity of food resources remains the same, consumer diversity should nevertheless increase as productivity increases, since previously scarce resources now become sufficiently abundant to support new species. In addition, in areas where the total abundance of seeds is greatest, species should tend to concentrate on those sizes of seeds which prove most profitable. Hence average niche widths (w in Fig. 15.8) should decrease with increasing seed availability and also, by inference, with increasing rainfall.

On drawing these arguments together (Fig. 15.14) we would expect that more ant species should be able to coexist in the higher rainfall deserts, exactly as has been observed. The chain of hypotheses and inferences summarized in Fig. 15.14 have not been tested in detail, but they have withstood a partial test centered on one species, *Veromessor pergandei* (Davidson 1978), described in Chapter 18. The crucial point is that in habitats with more species of ants, *V. pergandei* workers have a smaller coefficient of variation in mandible size (Fig. 15.15). Therefore, by inference, they are taking a narrower range of seed sizes and, by definition, have a narrower niche width w. Whether this variation in mandible size is phenotypic or genotypic is unknown. But on present evidence these results are entirely consistent with the hypothesis illustrated in Fig. 15.14 and with the general

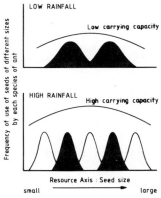

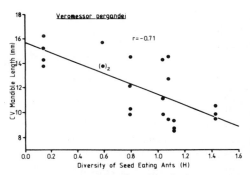

Fig. 15.14   Theoretical relationship between rainfall and niche width in desert ants. Low rainfall deserts (upper graph) support few species, each with a broad niche. As the carrying capacity rises with increasing rainfall and other ant species invade the community (lower graph), established species become more specialized in their choice of seeds (i.e., niche widths decrease).

Fig. 15.15   As the diversity of other ant species in the habitat increases, the workers of one common desert ant, *Veromessor pergandei* become less variable in size. The diversity of seed-eating ants is measured using a simple index of species diversity, H, which takes into account not only the number of ant species, but also their relative abundance. Variation in worker size is measured by the coefficient of variation (CV = standard deviation/mean) of mandible length within colonies (Davidson 1978). The inference from these data is that niche width (the range of seed sizes taken) should decrease as the number of potential competitors increases, in the manner illustrated in Fig. 15.14.

model outlined in Section 15.4.2. Apparently similar results with another ant assemblage have been obtained by Bernstein (1979), who showed that mean variation in head size within colonies decreased as the number of species of ants in the community increased.

Species morphologies provide at best only indirect tests of niche width and multispecies competition models, and it may well be that tests are generating the right results for entirely the wrong reasons. Such worries can only be resolved by further studies, but on present evidence it does seem that guilds of seed-eating (i.e., phytophagous) desert ants are strongly influenced by competition.

15.8.3   Termites (Isoptera)

Unlike ant populations where predation and parasitism are rare (Brian 1965), termite colonies often suffer heavy predation from ants (Wood & Lee 1971). What effect this has on competition between termite species is unknown, although in theory predation can reduce or completely eliminate competition. The limited, anecdotal data available suggest that termite populations behave rather like the ants discussed above (cf. Chapter 18). Wood and Lee (1971) summarize cases where *Armitermes evuncifer* enter mounds of *Cubitermes* and *Trinervitermes* in the Ivory Coast and displace the original owners, reminiscent of Way's (1953) results with Tanzanian ants. And just as shown by Pontin for *Lasius,* they always found their termite nests to be fairly regularly dispersed, with each species well separated from nests of other species.

15.8.4   Bees (Hymenoptera:Apidae)

In anthropomorphic terms, flowers produce nectar to ensure pollination. If a plant produces too little nectar it may not be profitable enough to attract pollinators; but if it produces too much, a single flower will satiate each visitor, and cross-pollination fails. Pollination systems therefore have rationing built into them, more or less guaranteeing competition between different species of bees for a limited resource, nectar.

Heinrich (1979b) provides an important review of competition in bees, set into the framework of their feeding ecology, physiology, and colony life. In this he points out that aggressive interference competition between species is mainly observed around highly profitable, easily defended food sources. Examples of interaction between foraging workers are summarized in Chapter 18. Translating such interactions into a competition coefficient between bee colonies is impossible without further information. However, individual behavioral interactions are almost certainly not trivial in population dynamic terms because colony performance reflects very closely the success of individual foragers (Heinrich 1979b). In Heinrich's view, however, examples of direct, aggressive interference competition between species of bees are rather rare. Usually, simple direct exploitation of scarce, scattered nectar resources plays a far more important role in structuring bee communities (cf. Chapter 20).

This view is consistent with an elegant model developed by Schaffer et al. (1979) to account for the apparent competition between three species of bees in Southern Arizona, all feeding on flowers of *Agave schottii*. The bees are the domestic honey bee (*Apis mellifera*) which has small workers but very large colonies; a bumble bee, *Bombus sonorus*, with intermediate-sized workers and colonies; and a larger carpenter bee, *Xylocopa arizonensis*, that provisions solitary nests. Schaffer et al.'s results are illustrated in Fig. 15.16. By excluding bees from the flowers, they showed that different patches of *A. schottii* differed in the total amounts of nectar produced, and it was therefore possible to rank the sites according to their nectar productivities. Honeybees predominated in the most productive habitats, *Xylocopa* in the least, with *Bombus* most abundant in patches of intermediate quality.

Zonation of species, or "habitat partitioning," in this way is very frequently attributed to interspecific competition, usually without much evidence. Schaffer et al., however, formulated a simple and convincing model to explain why these three particular species are distributed as they are. The key to understanding their results rests on the fact that individual foraging costs, measured in energy terms, scale in the opposite direction to the costs of colony maintenance. Small honey bee workers forage individually at low personal costs but need to fuel very large colonies; *Xylocopa* has no colony to support, but it is very large, making individual foraging costly; and *Bombus* is intermediate in both respects. Moreover, once bees start to harvest nectar, the average standing crop present in the flowers decreases. Indeed, and counterintuitively, in very productive patches large numbers of bees remove the nectar very quickly, and the average standing crop present at any one time becomes extremely low, too low in fact for a *Bombus* worker to "break even" on an individual foraging trip. An *Apis* worker, in contrast, with lower individual foraging costs can still make an energy profit. We therefore expect the following sequence along a gradient of increasing nectar production. First, no bees at all; then *Xylocopa* invades because it has no colony maintenance costs, but it is also driven out first by its high cost of individual foraging. It is followed by *Bombus* alone, and then by a zone with both *Bombus* and *Apis* present. Finally, as bee numbers build up to high levels and nectar is heavily harvested, the average standing crop of nectar falls below the individual break-even point for *Bombus* workers, and only *Apis* can survive. This model neatly explains the pattern shown in Fig. 15.16. It is formulated much more precisely by Schaffer et al. (1979) who examine models describing both nectar and bee colony dynamics. Our verbal description is, we hope, not too distorted a picture of their hypothesis (cf. Chapter 20).

Are resources really limiting for the bees? Schaffer et al. think so for

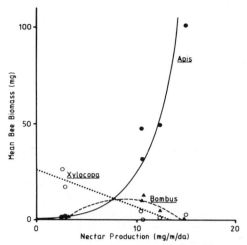

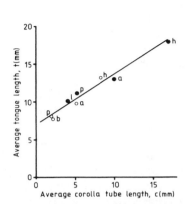

Fig. 15.16   Standing crop of nectar and bi-omass of three species of desert bees. *Xylocopa* is most abundant at sites with low nectar pro-duction; *Bombus* at intermediate sites; *Apis* is most common in the most productive sites. (From Schaffer et al. 1979.)

Fig. 15.17   The relationship between aver-age tongue length and average corolla tube length of flowers visited by queens (solid cir-cles) and workers (open circles) of *Bombus lu-corum* (*l*), *B. pratorum* (*p*), *B. agrorum* (*a*), and *B. hortorum* (*h*) (Brian 1957).

the good reason that by 4.00 p.m. on most days, the bees have exhausted all the nectar accumulated during the previous night. The crucial test of their hypothesis would be to remove *Apis* from the most productive habitats and see if *Bombus* can invade. Or, if both *Apis* and *Bombus* were removed, *Xylocopa* should occupy the whole habitat gradient. Such an experiment was actually carried out by Inouye (1978), who showed that after the re-moval of one species of bumble bee from its preferred species of flower, nectar levels rose sufficiently in the unexploited flowers to allow less effi-cient bumble bee species (with the "wrong" tongue length: Chapter 18) to make profitable use of them.

Although there are few such critical experimental studies, Heinrich (1979b) suggests that guilds of bees rather frequently divide up resources on the basis of their tongue length, coexisting species each choosing a different kind of flower. Figure 15.17 shows an example taken from the early work of Brian (1957) in Scotland. Other examples are given in Chapter 18. The parallel with the niche overlap model is again obvious. But just as with Davidson's ants, we should be wary of drawing too close an analogy between model and data until we know what the possibility is of seeing such patterns purely by chance.

Introductions of alien species offer a final demonstration of the impor-tance of competition in guilds of bees. Examples are again discussed in Chapter 18.

In summary, we find the evidence for competition between bee species convincing and believe that, in general, interspecific competition will be important in structuring local assemblages. But more work is required before this generalization can be made with confidence for a much wider range of species, habitats, and geographical areas.

15.8.5   Insects Associated with Dying or Dead Wood and Carrion

The miscellaneous assemblage of habitats centered on dying or dead wood, carrion, and dung, have a number of characteristics in common (Elton 1966). Each is small and distinct, consisting largely or solely of organic matter scattered as isolated, unpredictable "islands" in a "sea" of other environments. They also change quickly, largely from the activities of their inhabitants (Beaver 1977). Whatever competition there is between inhabitants (in particular larval Diptera and larval and adult Coleoptera) thus takes place for the kind of divided, ephemeral resources modeled by Shorrocks et al. (1979) and Atkinson and Shorrocks (1981). An important prediction of their models is that the less likely two species are to occur in the same patch of habitat by virtue of their independent, spatially clumped distributions, the more similar they can be in their ecology before competitive exclusion occurs (see Section 15.5).

Exactly this sort of situation has been reported for communities of fly larvae breeding in dead snails (*Cepaea nemoralis*) in North Wales (Beaver 1973, 1977) and for bark and ambrosia beetles (Scolytidae) tunneling inside fallen dead leaf stalks of Malaysian trees (Beaver 1979). These snail flies (10 species in 6 families) are little affected by predation or parasitism and might normally be expected to show strong interspecific competition and a smaller number of coexisting species. In practice, there was evidence for strong competitive interaction only between three of the species—all larvae in the genus *Sarcophaga*. The results show clearly that individuals of each fly species are almost always significantly aggregated when the numbers emerging per snail are determined, in accord with the model outlined in Section 15.5. Hence, although competition between larvae within any one snail may at times be intense, the possibility of any one of the species excluding another from the habitat as a whole is negligible.

Despite the apparently esoteric nature of dead snails or rotting tropical leaf stalks, neither is so very unusual. A vast range of habitats are patchy, ephemeral, and unpredictable for their insect inhabitants, and the general insights provided by Beaver's studies can probably be widely extended to many other situations.

15.8.6   Water Boatmen (Heteroptera:Corixidae)

Water boatmen are common aquatic bugs, feeding on algae, small animals and microorganisms in flocculent organic sediments. Several species usually occur together in the same body of water (Macan 1938), with some species apparently maintained by continuous immigration (Macan 1962).

Istock (1973) studied one such assemblage in a pond in Michigan, containing 14 species in two genera (*Hesperocorixa* and *Sigara*), of which at least 10 bred there. The two most abundant species, *S. macropala* and *H. lobata*, were selected for field manipulation experiments. Adding *S. macropala* to enclosures depressed the population size of *H. lobata* by roughly half, an important component of this being the failure of *H. lobata* to have a second summer generation in the presence of *S. macropala*. Reciprocal interactions were negligible. Istock (1977) then attempted to fit these data to the Lotka–Volterra equations [(15.2a) and (15.2b)], as shown in Fig. 15.18. The isoclines were fitted from independent estimates of K_1 and K_2, the supposed equilibrium densities being those in August 1967. Notice that α_{12}, the impact of *S. macropala* on *H. lobata* is, as expected, negative and large (-3.7), whilst α_{21}, the estimated impact of *H. lobata* on *S. macropala* is, if anything, beneficial ($+0.158$). However, as Istock noted, there are considerable errors in the estimation of the zero isoclines, and it could just as easily be that $\alpha_{21} = 0$. Notice that by assuming the existence of an equilibrium, Istock has ruled out the cases in Fig. 15.1a,b. The data can thus only predict either a stable equilibrium (which it does, and was observed) or an unstable one with either *S. macropala* or *H. lobata* as the victor.

Istock's application of the Lotka–Volterra model to his interaction can be criticized on several counts (Bergmans 1979), since populations of water boatmen conform to very few of the model assumptions. The population responses are not instantaneous, the age structure muddies the comparison, and seasonality makes it difficult to know if and when equilibria have been reached. Despite these problems, Istock's efforts to test population models with field data are valuable and provide a crucial step in developing better models and pointing to further experiments.

15.8.7 Parasitoids

We now move away from examples of insects as primary consumers to consider higher trophic levels. This section is concerned with competition between insect parasitoids, mainly Hymenoptera, but also parasitic Diptera.

That parasitoids sometimes compete in the field is strongly suggested by the example in Fig. 15.19 of three species of *Opius* (Braconidae) released sequentially in Hawaii for the biological control of the fruit fly, *Dacus dorsalis* (Bess et al. 1961). *Opius longicaudatus* was first introduced and established but caused low levels of parasitism. *Opius vandenboschi* then became established and *O. longicaudatus* virtually disappeared. With the final establishment of *O. oophilus*, this species increased rapidly, causing high levels of parasitism and the complete extinction of the other two species followed. Particularly interesting in this example is that the maximum percentage parasitism achieved is greater for each successive species established. This

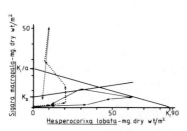

Fig. 15.18   Zero growth isoclines and population trajectories for two species of corixids in the field (Istock 1977). There is a predicted stable equilibrium at $N_1 = 41.8$ and $N_2 = 12.6$ mg dry wt. m^{-2}. The population trajectories are those followed at Istock's stations I and II (solid line) and stations III and IV (dashed line), respectively.

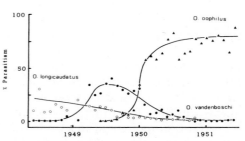

Fig. 15.19   Changes in percent parasitism of the fruit fly, *Dacus dorsalis*, by three species of *Opius: O. longicaudatus, O. oophilus,* and *O. vandenboschi.* (Data from Bess et al. 1961; after Varley et al. 1973.)

provides some support for the theoretical prediction that competitive displacement of one species of parasitoid by a second is more likely, other things being equal, if the searching efficiency of the second is greater than that of the first (Hassell 1978, 1979; May & Hassell 1981). Unfortunately, the relative searching efficiencies of the three species have not been determined.

A similar picture comes from the guild of natural enemies introduced into California for control of olive scale, *Parlatoria oleae* (Huffaker & Kennett 1966) and Californian red scale, *Aonidiella auranti* (DeBach et al. 1974). Coexistence, or competitive replacements in these parasitoid guilds assembled for biological control purposes constitute experiments on a grand scale. Unfortunately, few of the numerous cases of successful biological control have been sufficiently carefully studied or as well documented. [But see Annecke and Moran (1977) and Moran and Annecke (1979) for other exemplary post hoc analyses.] Of the many introductions that have failed (Clausen 1978), it remains unclear how many can be attributed to interspecific competition. Turnbull (1967) attributes many cases of failure to competition, although the evidence is very equivocal. Huffaker et al. (1971b) and C. B. Huffaker and M. Hoy (unpubl.) discuss some possible examples and conclude that most failures have probably been due to other causes.

Experimental manipulation of a parasitoid guild under controlled conditions has been carried out by Force (1970, 1974) for the species attacking the gall midge, *Rhopalomyia* (Fig. 15.20). In order to manipulate this system, Force carried out his experiments in a glasshouse and laboratory cages, using various combinations of the species shown in Fig. 15.20. Dramatically different patterns of abundance resulted, providing clear evidence for strong interaction between the species. The underlying mechanisms for these interactions are discussed in detail by Force (1970). Interestingly, laboratory

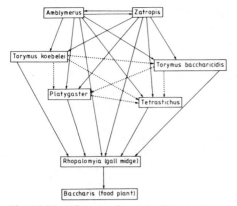

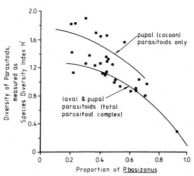

Fig. 15.20 The complex web of interactions shown by the parasitoids of the gall midge, *Rhopalomyia californica* (Force 1974). Solid lines denote direct attack and feeding by one organism upon another. Dotted lines indicate some other form of attack or inhibition. Arrows point from the exploiter to the victim.

Fig. 15.21 Apparent influence of the ichneumonid parasitoid, *Pleolophus basizonis*, on the diversity of other parasitoids attacking the larvae and pupae of the Swaine jack pine sawfly, *Neodiprion swainei* (Price 1970). As *P. basizonus* increases, so the diversity of other parasitoids falls, both for the total parasitoid complex (larval and pupal parasitoids) and the parasitoids attacking just the pupae.

coexistence of the three dominant species in the field (*Platygaster* and two species of *Torymus*), plus *Tetrastichus*, was only possible when the two facultative hyperparasitoids, *Amblymerus* and *Zatropis*, were also added to the system, suggesting that hyperparasitoids may sometimes be important in preventing competitive exclusion (cf. Chapter 22).

Gall insects appear particularly well suited to studies of this kind. Zwölfer (1979) studied a system based on the tephritid fly, *Urophora cardui*, that makes galls on creeping thistle (*Cirsium arvense*) flower heads, and two associated parasitoids, *Eurytoma serratulae* and *E. robusta*. A model for this system is described in Section 15.3 [Eqs. (15.5a) and (15.5b)]. If only small flower heads are present, the model *E. serratulae* are eliminated due to its preference for large flower heads. Otherwise, the two species coexist. Encouragingly, experimental field populations show either *E. robusta* dominating the system (as observed by Zwölfer in some populations) or the stable coexistence of both species.

Other work provides indirect evidence of competitive effects between guild members. Such is the case for Price's (1970, 1971) study of natural enemies of the Swaine jack pine sawfly, *Neodiprion swainei*, in Quebec (Fig. 15.21), and Dean and Ricklefs's (1979) analysis of Canadian Forest Insect Survey data for Southern Ontario between 1937 and 1948. Dean and Ricklefs thought that the average abundance of parasitoid populations should vary inversely with the number of competing parasitoid species (i.e., with the number of species sharing the same host)—essentially a "diffuse competition" argument. In contrast to Price's study, their correlation analysis failed to find any evidence of competition between parasitoid species. With one exception, the abundance of a particular parasitoid was either independent of, or positively correlated (rather than negatively so as demanded by the diffuse competition hypothesis) with, both the number and abundance of cooccurring parasitoid species. Dean and Ricklefs are sensibly cautious about their analysis. The insect collections on which it is based were made over a very wide area, ruling out any possibility of detecting interactions within local communities. Our impression of the basic data is also one of unease, since many species of parasitoids were represented by very few specimens. Similar reservations have been expressed by Bouton et al. (1980) and Force (1980).

15.8.8 Dragonflies (Odonata)

Most adult male dragonflies are territorial (Corbet 1962) and may exhibit conspicuous interspecific interactions (Moore 1964), which presumably restrict the total number of eggs laid by the female population. Subsequent survival of the aquatic, nymphal stages has been investigated by several workers on a variety of species (briefly reviewed by Lawton et al. 1980). Death rates are high and surprisingly constant throughout larval life, the available evidence pointing strongly to predation by fish and invertebrate predators as the major cause of mortality (Benke 1976, Benke & Benke 1975, Lawton et al. 1980). It would thus be surprising if interspecific competition for food between larvae sharing the same pond was at all common, a hypothesis elegantly confirmed by Benke (1978). Several species of Odonata coexisted in the littoral zone of Benke's study area, Dick's Pond in South Carolina. Some emerged and laid eggs in the spring ("early"), completing most of their oviposition before a second group of summer ("late") species emerged. Preventing the "late"-emerging species from laying in the pond had no significant impact on the densities or survival of the "early" nymphs. Competition between the two groups was negligible. There was, however, a suggestion from Benke's data of density-dependent survival within the "early" larval group. Small instars survived slightly better and grew slightly faster at low, compared with high, densities. Since the "early" group was made up of three species, there was clearly potential for some interspecific competition, but it must have been very small.

The most dramatic effect revealed by Benke's treatments was the impact of the "early" group of larvae upon the "late" group. Very simply, the "early" larvae tended to be larger and, by feeding on the smaller "late" larvae, greatly reduced the latter's abundance (Fig. 15.22). Whether or not such predation should be viewed as an extreme form of interference competition is debatable. Certainly there is only a fine distinction between the enhanced death rate of the caterpillar *Epiblema* in the presence of "Mor-

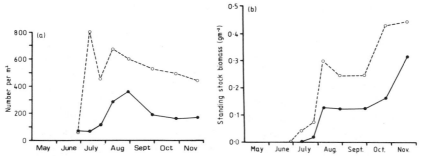

Fig. 15.22   The influence of "early" species of dragonfly larvae on (a) the abundance and (b) biomass of "late" species in a South Carolina pond (Benke 1978). The dotted line shows the abundance of "late" larvae in the absence of "early" species; the solid line their abundance in the presence of the "early" species. The difference between treatments is highly significant.

dellidae 23" and predation by one group of dragonflies upon another. On present evidence, interspecific competition for food between different species of larval dragonflies seems unlikely to be of much ecological significance, although predation on one group of species by another may be.

15.8.9   Predatory Coleoptera

Despite their importance in many terrestrial and aquatic ecosystems, very few studies have been made on interspecific competition between predatory beetles. The studies that have been done provide at best weak and indirect evidence for competition.

Species of *Pseudanophthalmus* are carnivorous, cave-dwelling carabid beetles, preying on a variety of smaller arthropods and are the top predators in the caverns where they occur (Zant et al. 1978). As such they may well be food limited. Zant et al. found adult *P. menetriesii* to be significantly smaller in three caves where the closely related, but larger, *P. pubescens* occurred, compared with the one cave where *P. menetriesii* was found alone. This result is consistent with the hypothesis that competition between the two species is reduced by character displacement in *P. menetriesii* when they occur together. However, other explanations are possible, and indeed more likely (Barr & Crowley 1981). A similar case, apparently involving differences in mandible size compatible with coexistence, was reported by Pearson and Mury (1979) for predatory tiger beetles on grassland sites in Arizona. They found that species with similar-sized mandibles taking similar-sized prey occurred together significantly less often than expected by chance.

The situation in even larger sets of predatory ground beetles can be extremely complicated. Southwood (1978) describes a guild of 28 common species from Silwood Park in England, drawing particularly on the work of Greenslade (1961). The species differ not only in adult body size but also in date of emergence, diurnal periodicity, and habitat utilization. The guild can thus plausibly be analyzed in terms of the niche overlap model (Fig. 15.8), although several resource dimensions rather than just one are now involved. Unfortunately, as in all studies of "ecological differences" between sets of close relatives, it is not at all clear in the absence of field experiments whether the differences have arisen from competition at some time in the past, are maintained by competition at the present time, or have nothing to do with competition past or present (e.g., Taylor 1980).

Predatory water beetles (Dytiscidae) display very similar patterns to those above, with guilds made up of sets of species of different sizes. They warn against simple explanations. In general the larger species prefer to eat larger prey, although there is considerable overlap in the range of prey sizes and species taken by each beetle species (J. H. Lawton & N. Webb unpubl.). Four kinds of water beetle occur widely in an old, long-abandoned canal near York in England, and another two, together with these four, occur in more restricted sections of the canal (Table 15.1). A total of 12 species live in the vicinity and form a pool of potential colonists. Other smaller bodies of water in the area have different combinations of species, usually four or five in total, with a tendency for one large, some intermediate, and a small species in each habitat. The similarity with other studies outlined previously on ants, bumblebees, and other beetles is striking.

Pairs and triplets of species from the canal were manipulated in experimental field enclosures. However, despite raising the natural densities of the adult beetle populations by an order of magnitude or more, no evidence of interspecific competition was found between any pair of investigated species. This was true even for those most similar in body size, and even when maintained at the higher densities for several weeks or even months. There was no evidence that the adult beetles competed for food, and the significance of the differences in body sizes is unknown. Larvae are never abundant, and the larval period is of very short duration relative to that of the adults. Hence, whilst it is possible that differences in adult body sizes reflect differences in larval ecology, data are lacking on this point, and the significance of body size differences in these beetles remains a mystery.

15.9 CONCLUSIONS

Many vertebrate ecologists have considered competition as a major force in structuring communities (e.g., MacArthur 1972, Lack 1971). On the other hand, some insect ecologists have been less impressed by the importance of competition, a view strongly expressed by Andrewartha and Birch (1954). Such a bald distinction is clearly an oversimplification and misleading. This review shows that the role of interspecific competition in insect communities varies widely from group to group.

Competition stands out clearly as most important in the social insects, particularly the ants and bees. This may be because predation on them is

Table 15.1 Predatory water beetles (Dytiscidae:Hydroporini) from water bodies near York in the North of England tend to form guilds of species in which the adults differ from each other in body size[a]

Species found in the vicinity of the canal (mean length in mm)	Pocklington Canal		Representative other habitats		
	Typical guild from weed beds	Guild from Phragmites reed beds	Hagg Bridge Pond	Melbourne Pond	Aughton Ditch
Graptodytes pictus (2.22)	*G. pictus*	*G. pictus*	✓		✓
Hygrotus inaequalis (2.29)		*H. inaequalis*		✓	✓
Porhydrus lineatus (3.20)			✓		
Hygrotus versicolor (3.33)	*H. versicolor*	*H. versicolor*		✓	✓
Hydroporus palustris (3.6)	*H. palustris*	*H. palustris*	✓	✓	
Hydroporus erythrocephalus (4.04)		*H. erythrocephalus*		✓	✓
Hydroporus planus (4.16)			✓	✓	
Potamonectes assimilis (4.23)					
Hydroporus dorsalis (4.59)					
Hyphydrus ovatus (4.60)	*H. ovatus*	*H. ovatus*	✓	✓	
Coelambus impressopunctatus (4.86)				✓	
Strictotarsus duodecimpustulatus (5.44)					

[a] A "pool" of 12 species (column one) is present in the area: the most intensively studied site is the Pocklington Canal (columns two and three) where 4–6 species coexist, depending upon the vegetation (J. H. Lawton & N. Webb unpubl.).

usually of minor significance and because the colonies are relatively long-lived, with a capacity to sustain numbers over unfavorable periods. In both respects there are parallels with many vertebrate populations. In the case of bees there is the additional factor that resource limitation is inherent in pollinator systems, without which pollination could not efficiently occur. At the other extreme are the phytophagous insects, in which competition is occasionally reported but will not commonly be a major determinant of community structure. In direct contrast to the social insects, natural enemies now frequently cause heavy mortalities, as also do several density-independent factors associated with environmental conditions. Between these two extremes lie groups for which the data are equivocal, such as the parasitic Hymenoptera and predatory beetles.

A particularly interesting insight to emerge from the examples where interspecific competition appears to be important is the frequency of strongly asymmetrical interactions; that is, where one species (or group of species) has a major impact upon its competitors, but the reverse relationships are effectively zero. Several examples have been summarized here, and a full summary is given in Lawton and Hassell (1981). Ecology textbooks sometimes distinguish such cases where $\alpha_{ij} > 0$ but $\alpha_{ji} = 0$ as "amensalism" (e.g., Odum 1971, Williamson 1972), treating them as a rather rare limiting case of a more usual competitive situation where both species affect each other. On present evidence from insect populations in the field, it seems that amensalism or something very close to it is the norm and symmetrical competition ($\alpha_{ij} \simeq \alpha_{ji}$) quite unusual.

Finally, we have focused almost entirely in this review on contemporary interactions between species, avoiding discussion of the role of competition in the evolution of morphological and ecological differences between species. The literature is full of examples of species which differ in their ecology, often interpreting such differences as being due to the "ghost of competition past" (Connell 1980) that is permitting coexistence at the present. We share with Connell, Huey (1979), and Wiens (1977) their misgivings about much of this work. "The notions of coevolutionary shaping of competitors' niches has little support at present" despite being widely held (Connell 1980). This does *not* mean that some, indeed many, contemporary differences between species have not been influenced by interspecific competition in the past (e.g., see especially Chapters 18, 20): it means only that hard evidence for such effects is mostly lacking. Herein lies a major challenge for future research.

REFERENCES

Anderson, R. M., B. D. Turner, and L. R. Taylor (Eds.) 1979. *Population Dynamics:* 20th Symp. Br. Ecol. Soc., Blackwell, Oxford.

Abrams, P. 1975. *Theor. Pop. Biol.* **8:** 356–375.

Abrams, P. 1980. *Theor. Pop. Biol.* **17:** 80–102.

Addicott, J. F. 1978. *Can. J. Zool.* **56:** 1837–1841.

Andrewartha, H. G. 1961. *Introduction to the Study of Animal Populations.* Chapman & Hall, London.

Andrewartha, H. G. and L. C. Birch. 1954. *The Distribution and Abundance of Animals.* Chicago University Press, Chicago.

Angren, G. I. and T. Fagerstrom. 1980. *J. Theor. Biol.* **82:** 401–404.

Annecke, D. P. and V. C. Moran. 1977. *J. Entomol. Soc. So. Afr.* **40:** 127–145.

Askew, R. R. 1961. *Trans. Soc. Br. Ent.* **14:** 237–268.

Atkinson, W. D. and B. Shorrocks. 1981. *J. Anim. Ecol.* **50:** 461–471.

Ayala, F. J. 1969. *Nature* **224:** 1076–1079.

Ayala, F. J. 1970. *Nature* **227:** 89–90.

Ayala, F. J., M. E. Gilpin, and J. G. Ehrenfeld. 1973. *Theor. Pop. Biol.* **4:** 331–356.

Barr, T. C. and P. H. Crowley. 1981. *Am. Nat.* **117:** 363–371.

Beaver, R. A. 1973. *J. Entomol.* (*A*) **48:** 1–9.

Beaver, R. A. 1977. *J. Anim. Ecol.* **46:** 783–798.

Beaver, R. A. 1979. *J. Anim. Ecol.* **48:** 987–1002.

Bellows, T. S. 1979. *The Modelling of Competition and Parasitism in Laboratory Insect Populations.* Ph.D. thesis, University of London.

Benke, A. C. 1976. *Ecology* **57:** 915–927.

Benke, A. C. 1978. *J. Anim. Ecol.* **47:** 335–350.

Benke, A. C. and S. Benke. 1975. *Ecology* **56:** 302–317.

Bergmans, M. 1979. *Am. Nat.* **113:** 759–761.

Bernstein, R. A. 1979. *Am. Nat.* **114:** 533–544.

Bess, H. A., R. van den Bosch, F. H. Haramoto. 1961. *Proc. Hawaii. Entomol. Soc.* **17:** 367–378.

Blakley, N. R. and H. Dingle. 1978. *Oecologia* **37:** 133–136.

Borowsky, R. 1971. *Nature* **230:** 409–410.

Bouton, C. E., B. A. McPheron, and A. E. Weis. 1980. *Am. Nat.* **116:** 876–881.

Brian, A. D. 1957. *J. Anim. Ecol.* **26:** 71–98.

Brian, M. V. 1965. *Social Insect Populations.* Academic, London.

Brower, L. P. 1958. *Am. Nat.* **92:** 183–187.

Brown, J. H. and D. W. Davidson. 1977. *Science* **196:** 880–882.

Caughley, G. and J. H. Lawton. 1981. Pp. 132–166 in May, 1981, referenced here.

Chestnut, T. L. and W. A. Douglas. 1971. *J. Econ. Entomol.* **64:** 864–868.

Clausen, C. P. (Ed.). 1978. *Introduced Parasites and Predators of Arthropod Pests and Weeds: a World Review.* USDA, Washington, D.C.

Cock, M. J. W. 1978. *J. Anim. Ecol.* **47:** 805–816.

Comins, H. N. and M. P. Hassell. 1976. *J. Theor. Biol.* **62:** 93–114.

Connell, J. H. 1980. *Oikos* **35:** 131–138.

Connor, E. F. and D. Simberloff. 1979. *Ecology* **60:** 1132–1140.

Corbet, P. S. 1962. *A Biology of Dragonflies.* Witherby, London.

Cornell, H. and D. Pimentel. 1978. *Ecology* **59:** 297–308.

Cramer, N. F. and R. M. May. 1972. *J. Theor. Biol.* **34:** 289–293.

Crombie, A. C. 1945. *Proc. R. Soc. B* **132:** 362–395.

Crombie, A. C. 1946. *Proc. R. Soc. B* **133:** 76–109.

Crombie, A. C. 1947. *J. Anim. Ecol.* **16:** 44–73.

Crowell, K. L. 1968. *Ecology* **49:** 551–555.

Darlington, A. 1974. Pp. 298–311 in M. G. Morris & R. H. Perring (Eds.), *The British Oak, its History and Natural History.* Classey, Faringdon.

Davidson, D. W. 1977a. *Ecology* **58:** 711–724.

Davidson, D. W. 1977b. *Ecology* **58:** 725–734.

Davidson, D. W. 1978. *Am. Nat.* **112:** 523–532.

Dean, J. M. and R. E. Ricklefs. 1979. *Am. Nat.* **113:** 302–306.

DeBach, P. 1966. *Annu. Rev. Entomol.* **11:** 182–212.

DeBach, P., R. M. Hendrickson, and M. Rose. 1978. *Hilgardia* **46:** 1–35.

DeBach, P., D. Rosen, and C. E. Kennett. 1971. Pp. 165–194 in Huffaker (Ed.), 1971a, referenced here.

Edmunds, G. F. Jr. 1973. *En. Entomol.* **2:** 765–777.

Elton, C. S. 1966. *The Pattern of Animal Communities.* Methuen, London.

Faeth, S. H. and D. Simberloff. 1981. *Ecology* **62:** 620–624.

Feldman, M. W. and J. Roughgarden. 1975. *Theor. Pop. Biol.* **7:** 197–207.

Fenchel, T. 1975. *Oecologia* **20:** 1–17.

Fluker, S. S. and J. W. Beardsley. 1970. *Annu. Ent. Soc. Am.* **63:** 1290–1296.

Force, D. C. 1970. *Annu. Ent. Soc. Am.* **63:** 1675–1688.

Force, D. C. 1974. *Science* **184:** 624–632.

Force, D. C. 1980. *Am. Nat.* **116:** 873–875.

Fujii, K. 1977. *J. Theor. Biol.* **69:** 613–623.

Futuyma, D. J. and F. Gould. 1979. *Ecol. Monogr.* **49:** 33–50.

Gause, G. F. 1934. *The Struggle for Existence.* Hafner, New York.

Gibson, C. W. D. 1980. *Oecologia* **47:** 352–364.

Gilbert, L. E. and M. C. Singer. 1975. *Annu. Rev. Ecol. Syst.* **6:** 365–397.

Gilpin, M. E. and F. J. Ayala. 1973. *Proc. Nat. Acad. Sci. USA* **70:** 3590–3593.

Gilpin, M. E. and K. E. Justice. 1972. *Nature* **236:** 273–301.

Goeden, R. D. and S. M. Louda. 1976. *Annu. Rev. Ent.* **21:** 325–342.

Greenslade, P. J. M. 1961. Studies in the Ecology of Carabidae (Coleoptera). Unpublished Ph.D. thesis, University of London.

Hairston, N. G., F. E. Smith, and L. B. Slobodkin. 1960. *Am. Nat.* **94:** 421–425.

Halkka, O. 1978. Pp. 41–55 in Mound and Waloff (Eds.), 1978, referenced here.

Hansen, R. M. and D. N. Ueckert. 1970. *Ecology* **51:** 640–648.

Hardin, G. 1960. *Science* **131:** 1292–1297.

Harrison, J. O. 1964. *Ecology* **45:** 508–519.

Haskins, C. P. and E. F. Haskins. 1965. *Ecology* **46:** 736–740.

Hassell, M. P. 1975. *J. Anim. Ecol.* **44:** 283–295.

Hassell, M. P. 1978. *The Dynamics of Arthropod Predator-Prey Systems.* Princeton University Press, Princeton.

Hassell, M. P. 1979. Pp. 283–306 in Anderson, Turner, and Taylor (Eds.), 1979, referenced here.

Hassell, M. P. and H. N. Comins. 1976. *Theor. Pop. Biol.* **9:** 202–221.

Hassell, M. P., J. H. Lawton, and R. M. May. 1976. *J. Anim. Ecol.* **45:** 471–486.

Hassell, M. P. and T. R. E. Southwood. 1978. *Annu. Rev. Ecol. Syst.* **9:** 75–98.

Heck, K. L. Jr. 1976. *Evol. Theory* **1:** 247–258.

Heinrich, B. 1979a. *Oecologia* **42:** 325–337.

Heinrich, B. 1979b. *Bumblebee Economics.* Harvard University Press. Cambridge, Mass.

Herbert, P. D. N., P. S. Ward, and K. Harmsen. 1974. *Nature* **252:** 389–391.

Holmes, R. T., J. C. Schultz, and P. Nothnagle. 1979. *Science* **206:** 462–463.

Holt, R. D. 1977. *Theor. Pop. Biol.* **12:** 197–229.

Horn, H. S. and R. H. Macarthur. 1972. *Ecology* **53:** 749–752.

Huey, R. B. 1979. *Oecologia* **38:** 249–259.

Huffaker, C. B. 1958. *Hilgardia* **27:** 343–383.

Huffaker, C. B. 1971a. Pp. 327–343 in P. J. den Boer and G. R. Gradwell (Eds.), *Dynamics Populations*, Centre Agric. Publ. & Doc., Wageningen.

Huffaker, C. B. (Ed.). 1971b. *Biological Control.* Plenum, New York.

Huffaker, C. B. and C. E. Kennett. 1966. *Hilgardia* **37:** 283–335.

Huffaker, C. B. and C. E. Kennett. 1969. *Can. Entomol.* **101:** 425–447.

Huffaker, C. B., P. S. Messenger, and P. DeBach. 1971. Pp. 16–67 in Huffaker (Ed.), 1971a, referenced here.

Hutchinson, G. E. 1951. *Ecology* **32:** 571–577.

Inouye, D. W. 1978. *Ecology* **59:** 672–678.

Istock, C. A. 1973. *Ecology* **54:** 535–544.

Istock, C. A. 1977. *Am. Nat.* **111:** 279–287.

Janzen, D. H. 1973. *Am. Nat.* **107:** 786–790.

Lack, D. 1971. *Ecological Isolation in Birds.* Blackwell, Oxford.

Lawton, J. H. 1978. Pp. 105–125, in Mound and Waloff (Eds.), 1978, referenced here.

Lawton, J. H., J. R. Beddington, and R. Bonser. 1974. Pp. 141–158 in M. B. Usher (Ed.), *Ecological Stability.* Chapman and Hall, London.

Lawton, J. H. and M. P. Hassell. 1981. *Nature* **289:** 793–795.

Lawton, J. H. and S. McNeill. 1979. Pp. 223–244, in Anderson, Turner, and Taylor (Eds.), 1979, referenced here.

Lawton, J. H. and D. R. Strong, Jr. 1981. *Am. Nat.* **118:** 317–338.

Lawton, J. H., B. A. Thompson, and D. J. Thompson. 1980. *Ecol. Entomol.* **5:** 39–51.

Le Quesne, W. J. 1972. *J. Entomol. A* **47:** 37–44.

Levin, S. A. 1974. *Am. Nat.* **108:** 207–228.

Levin, S. A. 1977. *Annu. Rev. Ecol. Syst.* **7:** 287–310.

Levins, R. and D. Culver. 1971. *Proc. Nat. Acad. Sci. USA* **68:** 1246–1248.

Lloyd, M. and J. White. 1980. *Am. Nat.* **115:** 29–44.

Lotka, A. J. 1925. *Elements of Physical Biology.* Williams & Wilkins, Baltimore.

Macan, T. T. 1938. *J. Anim. Ecol.* **7:** 1–19.

Macan, T. T. 1962. *Arch. Hydrobiol.* **58:** 224–232.

MacArthur, R. H. 1972. *Geographical Ecology: Patterns in the Distribution of Species.* Harper & Row, New York.

Majer, J. D. 1976. *J. Appl. Ecol.* **13:** 123–144.

Mamaev, B. M. 1975. *Evolution of Gall Forming Insects—Gall Midges* (trans. by L. Crozy). B.L.L.D., Boston Spa.

May, R. M. 1974a. *Science* **186:** 645–647.

May, R. M. 1974b. Pp. 1–50 in R. Rosen and R. Snell (Eds.), *Progress in Theoretical Biology.* Academic, New York.

May, R. M. 1975. *Stability and Complexity in Model Ecosystems* (2nd Ed.). Princeton University Press, Princeton, N.J.

May, R. M. 1976. Pp. 49–70, in May (Ed.), *Theoretical Ecology, Principles and Applications.* Blackwell, Oxford.

May, R. M. 1977. *Nature* **269:** 103, 104.

May, R. M. 1978. *J. Anim. Ecol.* **47:** 833–843.

May, R. M. (Ed.). 1981. *Theoretical Ecology.* Blackwell, Oxford.

May, R. M. and M. P. Hassell. (1981) The Dynamics of Multiparasitoid-Host Interactions. *The American Naturalist Vol.* **117,** 234–261.

May, R. M. and R. H. MacArthur. 1972. *Proc. Nat. Acad. Sci.* USA **69:** 1109–1113.

May, R. M. and G. F. Oster. 1976. *Am. Nat.* **110:** 573–599.

McClure, M. S. 1974. *Env. Entomol.* **3:** 59–68.

McClure, M. S. 1980. *Ecology* **61:** 1391–1401.

McClure, M. S. and P. W. Price. 1975. *Ecology* **56:** 1388–1397.

McClure, M. S. and P. W. Price. 1976. *Ecology* **57:** 928–940.

McMurtrie, R. 1976. *Theor. Pop. Biol.* **10:** 96–107.

Messenger, P. S. 1975. Pp. 201–223 in D. Pimentel (Ed.), *Insects, Science and Society.* Academic, New York.

Miller, R. S. 1967. *Adv. Ecol. Res.* **4:** 1–74.

Moore, N. W. 1964. *J. Anim. Ecol.* **33:** 49–71.

Moran, V. C. and D. P. Annecke. 1979. *J. Entomol. Soc. So. Afr.* **42:** 299–329.

Mound, L. A. and N. Waloff (Eds.). 1978. *Diversity of Insect Faunas: Symp. Roy. Entomol. Soc. London,* Blackwell, Oxford.

Murdoch, W. W. 1969. *Ecol. Monogr.* **39:** 335–354.

Murdoch, W. W. 1977. *Theor. Pop. Biol.* **11:** 252–272.

Murdoch, W. W. and A. Oaten. 1975. *Adv. Ecol. Res.* **9:** 2–131.

Nisbet, R. M., W. S. Gurney, and M. A. Pettipher. 1978. *J. Theor. Biol.* **75:** 223–237.

Odum, E. P. 1971. *Fundamentals of Ecology.* Saunders, Philadelphia, Pa.

Paine, R. T. 1966. *Am. Nat.* **100:** 65–75.

Park, T. 1948. *Ecol. Monogr.* **18:** 265–308.

Park, T. 1954. *Physiol. Zool.* **27:** 177–238.

Park, T., P. H. Leslie, and D. B. Mertz. 1964. *Physiol. Zool.* **38:** 97–162.

Pearson, D. J. and E. S. Mury. 1979. *Ecology* **60:** 557–566.

Pipkin, S. B., R. L. Rodriques, and J. Leon. 1966. *Am. Nat.* **100:** 135–155.

Pomerantz, M. J. 1981. *Am. Nat.* **117:** 583–591.

Pontin, A. J. 1969. *J. Anim. Ecol.* **38:** 747–754.

Price, P. W. 1970. *Ecology* **51:** 445–454.

Price, P. W. 1971. *Ecology* **52:** 587–596.

Rathcke, B. J. 1976. *Ecology* **57:** 76–87.

Redfern, M. and R. A. D. Cameron. 1978. *Ecol. Entomol.* **3:** 251–263.

Root, R. B. 1973. *Ecol. Monogr.* **43:** 95–124.

Ross, H. H. 1957. *Evolution* **11:** 113–129.

Rothschild, G. H. L. 1971. *J. Appl. Ecol.* **8:** 287–322.

Roughgarden, J. and M. Feldman. 1975. *Ecology* **56:** 489–492.

Royama, T. 1970. *J. Anim. Ecol.* **39:** 619–668.

Schaffer, W. M., D. B. Jensen, D. E. Hobbs, J. Gurevitch, J. R. Todd, and M. V. Schaffer. 1979. *Ecology* **60:** 976–987.

Schoener, T. W. 1974a. *Science* **185:** 27–39.

Schoener, T. W. 1974b. *Theor. Pop. Biol.* **6:** 265–307.

Schoener, T. W. 1974c. *Am. Nat.* **108:** 332–340.

Seifert, R. P. and F. H. Seifert. 1976. *Am. Nat.* **110:** 461–483.

Seifert, R. P. and F. H. Seifert. 1979a. *Ecology* **60:** 462–467.

Seifert, R. P. and F. H. Seifert. 1979b. *Biotropica* **11:** 51–59.

Shorrocks, B., W. Atkinson, and P. Charlesworth. 1979. *J. Anim. Ecol.* **48:** 899–908.

Simberloff, D. S. 1978. Pp. 139–153 in Mound and Waloff (Eds.), 1978, referenced here.

Skellam, J. G. 1951. *Biometrika* **38:** 196–218.

Slatkin, M. 1974. *Ecology* **55:** 128–134.

Southwood, T. R. E. 1961. *J. Anim. Ecol.* **30:** 1–8.

Southwood, T. R. E. 1978. Pp. 19–40 in Mound and Waloff (Eds.), 1978, referenced here.

Stauffer, R. C. (Ed.). 1975. *Charles Darwin's Natural Selection, Being the Second Part of His Big Species Book Written from 1856 to 1858*. Cambridge University Press, Cambridge.

Steele, J. H. 1974. *The Structure of Marine Ecosystems*. Harvard University Press, Cambridge, Mass.

Stiling, P. D. 1980. *J. Anim. Ecol.* **49:** 793–805.

Strobek, C. 1973. *Ecology* **54:** 650–654.

Strong, D. R. Jr. 1977. *Ecology* **58:** 573–582.

Strong, D. R. Jr. and M. D. Wang. 1977. *Evolution* **31:** 854–862.

Taylor, L. R. and I. P. Woiwod. 1975. *Nature* **257:** 160–161.

Taylor, V. A. 1980. *Ecol. Entomol.* **5:** 397–411.

Thompson, W. A., P. J. C. Cameron, N. G. Wellington, and I. B. Vertinsky. 1976. *Res. Pop. Ecol.* **18:** 1–13.

Tilden, J. W. 1951. *Microentomology* **16:** 149–185.

Turelli, M. 1978. *Proc. Natl. Acad. Sci. USA* **75:** 5085–5089.

Turnbull, A. L. 1967. *Bull. Entomol. Soc. Amer.* **13:** 333–337.

Vandermeer, J. H. 1969. *Ecology* **50:** 362–371.

Vandermeer, J. H. 1975. *Science* **188:** 253–255.

van Valen, L. 1974. *Theor. Biol.* **44:** 19–21.

Varley, G. C. 1949. *J. Anim. Ecol.* **18:** 117–122.

Varley, G. C., G. R. Gradwell, and M. P. Hassell. 1973. *Insect Population Ecology*. Blackwell, Oxford.

Volterra, V. 1926. Pp. 409–448 in R. N. Chapman, 1931, *Animal Ecology*. McGraw-Hill, New York.

Way, M. J. 1953. *Bull. Entomol. Res.* **44:** 669–691.

Whittaker, J. B., J. Ellistone, and C. K. Patrick. 1979. *J. Anim. Ecol.* **48:** 973–986.

Whittaker, R. H. 1975. Pp. 169–181 in W. H. van Dobben and R. H. Lowe (Eds.), *Unifying Concepts of Ecology*. Junk, The Hague.

Wiens, J. A. 1977. *Am. Sci.* **65:** 590–597.

Williamson, M. H. 1957. *Nature* **180:** 422–425.

Williamson, M. H. 1972. *The Analysis of Biological Populations*. Edward Arnold, London.

Wise, D. H. 1981. *Ecology* **62:** 727–735.

Wood, T. G. and K. E. Lee. 1971. *Pedobiologia* **11:** 341–366.

Yasumatsu, K. 1976. Pp. 121–137 in V. L. Delucchi (Ed.), *Studies in Biological Control. International Biological Programme*, Vol. 9, Cambridge University Press, Cambridge.

Yasuno, M. 1965. *Sci. Rep. Tohoku Univ. IV (Biology)* **31:** 181–194.

Zant, T. V., T. L. Poulson, and T. C. Kane. 1978. *Am. Nat.* **112:** 229–234.

Zwölfer, H. 1975. *Verh. dt. Zool. Ges.* **67:** 394–401.

Zwölfer, H. 1979. *Fortschr. Zool.* **25:** 331–353.

Chapter 16

Principles of
Arthropod Predation

R. F. LUCK

Predation is one of several factors that appear important in the structuring of biological communities (Paine 1966, Huffaker 1971a, MacArthur 1972, Connell 1975, cf. Chapters 2, 22). Moreover, arthropod predators and parasitoids have been shown to regulate a number of phytophagous arthropod populations (e.g., DeBach 1964, Huffaker & Messenger 1976, Huffaker 1971b). Many of these represent classical cases of biological control wherein these entomophages are employed to control arthropod pests. In several crop systems they are important elements in integrated pest management. Thus, both basic and applied rationale exists for seeking understanding of the predatory–prey (read also parasitoid–host) interaction. An extensive body of theory has grown up around this relationship (cf. Lotka 1925, May 1973a, Murdoch & Oaten 1975, Hassell 1978, Gutierrez & Getz in press, and references cited therein), too extensive to present here. Nonetheless, some current notions of arthropod predator–prey interactions as they apply in a field context can be gleaned from reviewing the active theoretical areas. Through the medium of models we will look at two of these areas as examples of predator–prey interactions and to foraging theory, which is intrinsic to such interactions, as an introduction.

16.1 PREDATOR–PREY AND HOST–PARASITE MODELS

Models are a means of describing systems of interacting components and processes. Those we discuss are deterministic in nature and attempt to describe predator–prey (host–parasite) interactions. They can best be thought of as mathematical analogues (representations) that attempt to mimic the processes involved. These models contrast with those used to mimic a life system in a field context with a practical purpose in mind, for example, the models used in integrated pest management (IPM) (cf. Chapter 24). By comparing simple mathematical models of predator–prey interactions we attempt to capture their essence and through this to identify principles and gain insights into the dynamics involved. Typically, such attempts take the form of stability analyses, the mechanics of which are presented by May (1973a), Maynard-Smith (1974), Murdoch and Oaten (1975), or Hassell (1978) and are not presented here. Some models, because of their complexity, defy analytical solution, so numerical solutions are obtained (e.g., computer simulations), but we do not deal with these either. Instead we review some of the basic predator–prey models with a view of their assumptions and predictions with respect to real-world experience.

However, before discussing these simple deterministic models, it is useful to define what is meant by a stability analysis since so many of the predictions of predator–prey models arise from such analyses. In our discussion the notions of May (1973a) and Maynard-Smith (1974) are followed. A stability analysis determines what happens to the population densities of a predator and prey (or host and parasite) if the density of one of the coupled populations is perturbed from its equilibrium density (state). If the coupled populations return to their equilibrium densities, stability is assumed. If they continue to increase, decrease, or oscillate at ever-increasing amplitudes, they are considered unstable. These two conditions are separated by one of neutral stability, wherein the population densities either remain at the perturbed level and/or exhibit oscillations of constant amplitude. These ideas are depicted graphically in Fig. 16.1.

To analyze these models we investigate their changes in density within the vicinity of their equilibrium points, that is, we characterize their neighborhood (local) stability. Since most such models are nonlinear, their global stability cannot usually be investigated. In addition to a mathematical analysis, these models can be analyzed graphically. Calculating the population densities for the interaction at the end of each generation, plotting these densities on a graph, one against another, and joining the points sequentially indicates whether the model is stable, given the specific values for the model's parameters based upon whether the points spiral inward, outward, or circle, as shown in Fig. 16.2. This process is repeated using different values for each parameter if we wish to determine under what conditions (parameter values) stability can be expected. Determining a model's stability using traditional mathematical (analytical) methods has generally been less time consuming, therefore more desirable, when this approach is possible. When it is not, numerical (computer simulation) techniques are employed. In Section 16.1.1.2 we illustrate a more general graphical analysis.

A nonlinear population model need not be characterized by a simple point equilibrium. It may exhibit a stable limit cycle, that is, the densities

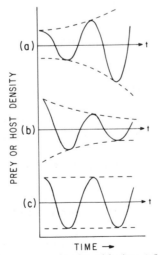

Fig. 16.1 Changes in predator or prey density with time (after May 1973). The responses depicted are (a) unstable, (b) stable, and (c) neutrally stable interactions.

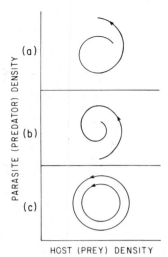

Fig. 16.2   Predator–prey interactions perturbed from equilibrium manifest themselves as spirals or circles when prey density is plotted against its corresponding predator density: (a) unstable, (b) stable, and (c) neutrally stable interactions.

change in a periodic but cyclical pattern. When these densities are perturbed, they return to those that characterized their cycle before perturbation. In contrast, the densities of populations that exhibit neutral stability continue to oscillate at an amplitude determined by the magnitude of the displacement. The difference between a stable limit cycle and neutral stability, however, cannot be detected without analyzing the interaction (model).

Thus, predator–prey interactions can be characterized by stable point equilibriums, unstable point equilibriums, stable limit cycles, or neutral stability. [Under certain conditions changes in population density may exhibit behaviors that range from periodic cycles, to apparently aperiodic fluctuations, to chaotic behaviors, but a discussion of these conditions is beyond our scope here (cf. May & Oster 1976).] It is the theoretical implications that arise from the assumptions implicit in these models, and their associated stability characteristics, that we wish to review since these assumptions bear on how we conceptualize and interpret the dynamics of field populations and how we might manage pest populations.

16.1.1   Models Mimicing Continuously Breeding Populations

Clearly, the logistics growth model is not widely used as a predator–prey or parasite–host model, but it has inspired and been used as a component in the development of predator–prey models. Moreover, it combines two general concepts that are important to all predator–prey models: the propensity for population densities to increase exponentially and for the environment to limit such increase. Thus, these two processes interact to maintain a population at a density characteristic of the organism and its environment (Pearl & Read 1920, Gompertz 1835, Verhulst 1838, cf. Chapter 12). This density is frequently referred to as the environment's *carrying capacity*. The interaction implies existence of a competitive density-dependent, or negative feedback, process (Smith 1935). This is accomplished in the model by incorporating a direct density-dependent component, the intensity of which is governed by the difference between the population density N at time t, and that sustainable by the environment K; the smaller the difference the more intense the negative feedback.

The model is commonly written as

$$\frac{dN}{dt} = rN \left(\frac{K - N}{K} \right) \tag{16.1}$$

where r is the intrinsic growth rate under a specific set of environmental conditions and N, t, and K are as previously defined. The model is based on a number of simplifying assumptions: (1) the population's density can be represented by a single variable, N (ignoring factors that cause heterogeneity such as age structure, sex ratios, genotypic and phenotypic variations, or dispersion patterns); (2) the effects of deaths and births are instantaneously felt throughout the population; and (3) changes in density are deterministic [implying that changes at $t + 1$ are unambiguously known and can be determined solely from the conditions (states) present at time t].

The effects of a predator or parasite are incorporated as a component of the negative feedback, that is, within the $(K - N)/K$ term of Eq. (16.1). In fact, Smith (1935) used the concepts embodied in this model to argue that prey (or host) populations were indeed limited by their natural enemies but he never formulated his concepts mathematically. May (1973a) presented the stability analysis for this model, showing that the model has two stability points, one at $N = K$, if and only if $r > 0$, and one at $N = 0$, if and only if $r < 0$.

16.1.1.1 Lotka–Volterra Predator–Prey Model.
Lotka (1925) and Volterra (1926) independently developed the first models that specifically dealt with a predator–prey interaction. They assumed that (1) the rate of change in the prey's population density (dN_1/dt) is a function of its intrinsic growth rate (r) and the rate at which the prey are removed from the population $(\gamma_1 N_1 N_2)$ and (2) the rate of change in a predator's population density (dN_2/dt) is a function of the predator's death rate in the absence of prey $(-d)$ and the growth rate of the predator population per unit of prey consumed $(\gamma_2 N_1 N_2)$. Thus,

$$\frac{dN_1}{dt} = rN_1 - \gamma_1 N_1 N_2 \tag{16.2}$$

and

$$\frac{dN_2}{dt} = -dN_2 + \gamma_2 N_1 N_2 \qquad (16.3)$$

where N_1 and N_2 are the prey and predator population densities, respectively, r is the intrinsic growth rate of the prey population, $-d$ is the intrinsic death rate of the predator population, γ_1 is the proportion of contacts between predator and prey which prove fatal to the prey, and γ_2 is the efficiency with which the predator converts consumed prey into offspring.

This model is, of course, also based on a number of simplifying assumptions in addition to those made by the logistic growth model: (1) In the absence of predation, the prey population increases exponentially and without limits; (2) the predator's rate of death ($-d$) is constant and thus independent of predator density or predator age; (3) the efficiency of capturing prey is independent of predator or prey age; (4) the efficiency of converting consumed prey to predator offspring is constant and independent of predator or prey age or density; (5) the movements and contacts between predator and prey are random; (6) the time involved in the prey's consumption is negligible (Maynard-Smith 1974); and (7) the predator's appetite is insatiable.

An analysis of this model shows it to be neutrally stable. The oscillations fail to either grow or dampen when the predator or prey's density is perturbed. This implies that the predator is unable to regulate the prey at a characteristic density (or set of densities in the case of a stable limit cycle). The model provides little insight into the mechanisms governing predator–prey interactions or the consequences of such interactions in terms that might explain changes in the respective population densities. Its assumptions are too naive. However, it does provide a base from which to explore potential mechanisms that might explain such interactions and that involve different assumptions about the predator's searching behavior that affects the rate of predation.

16.1.1.2 Modifications to the Lotka–Volterra Model. Rather than cataloguing and discussing all proposed modifications (e.g., May 1973a, Maynard-Smith 1974, Murdoch & Oaten 1975), we limit our discussion to one which provides a general analytical technique by which the effects of such modifications on stability can be judged. We then use this technique to summarize some of the results in the literature. This analytical technique, graphical in nature, can be used to analyze predator–prey models which possess an equilibrium. To illustrate, we compare the Lotka–Volterra model [Eqs. (16.2) and (16.3)] with a modification to the basic prey equation proposed by Volterra (1926). This discussion is based on work by Rosensweig and MacArthur (1963) and Maynard-Smith (1974).

The stability properties of the Lotka–Volterra model are analyzed by first finding the equations for the lines under the conditions that $dN_1/dt = 0$ and $dN_2/dt = 0$. From Eqs. (16.2) and (16.3), if

$$\frac{dN_1}{dt} = 0 \quad \text{then} \quad rN_1 - \gamma_1 N_1 N_2 = 0 \qquad (16.4)$$

and if

$$\frac{dN_2}{dt} = 0 \quad \text{then} \quad -dN_2 + \gamma_2 N_1 N_2 = 0 \qquad (16.5)$$

Solving Eqs. (16.2) and (16.3) we obtain

$$N_2 = \frac{r}{\gamma_1} \qquad (16.6)$$

$$N_1 = \frac{d}{\gamma_2} \qquad (16.7)$$

We can now plot these lines (Fig. 16.3a). We note that for values of $N_1 > d/\gamma_2$, dN_2/dt is positive and for values of $N_1 < d/\gamma_2$, dN_2/dt is negative. Similarly, for values of $N_2 > r/\gamma_1$, dN_1/dt is positive, and for values of $N_2 < r/\gamma_2$, dN_2/dt is negative. We can now insert arrows (vectors) on the graph. The arrows (resultant vectors) indicate the direction and the rate of change for the composite predator–prey interaction. For example, we can place arrows on the $dN_1/dt = 0$ line parallel to the predator (N_2) axis and on the $dN_2/dt = 0$ line parallel to the prey (N_1) axis (representing the resultant vector for the vectors showing the magnitude and direction in the rate of change in the prey and predator populations). These arrows can now be

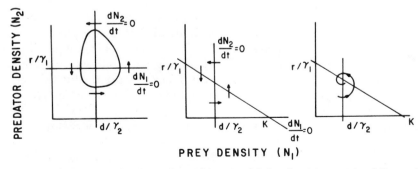

Fig. 16.3   Graphical analysis of a Lotka–Volterra model showing (a) neutral stability and (b) and (c) stability following Volterra's (1926) modification of intraspecific density-dependent interaction among members of the prey population for a resource in short supply (adapted from Maynard-Smith 1974).

joined to form a trajectory as in Fig. 16.3*a*. In most cases only the arrows associated with the $dN_1/dt = 0$ and $dN_2/dt = 0$ are needed to indicate the nature of the stability. For the Lotka–Volterra equations the analysis produces a closed trajectory, which indicates that the model is neutrally stable. Had the trajectory spiraled outward, this would suggest the model is unstable; had it spiraled inward, stability would have been suggested. However, stability analysis is necessary to confirm these suggestions.

Volterra's (1926) modification to the basic Lotka–Volterra model replaced the original assumption of unlimited growth by the prey population in the absence of predation to one of logistic population growth. In analyzing this model we only need concern ourselves with changing the equation for the prey population [i.e., Eq. (16.2)], since that for the predator population remains unchanged. Thus, the equation for the prey population is

$$\frac{dN_1}{dt} = rN_1 \left(\frac{K - N_1}{K} \right) - \gamma_1 N_1 N_2 \tag{16.8}$$

Again, solving for $dN_1/dt = 0$ yields

$$N_2 = \frac{r}{\gamma_1} - \frac{rN_1}{K\gamma_1}$$

Substituting for N_1 (i.e., Eq. 16.7), we obtain

$$N_2 = \frac{r}{\gamma_1} - \frac{rd}{K\gamma_1\gamma_2} \tag{16.9}$$

We note that Eq. (16.9) has a negative slope and that N_2 is positive when $r/\gamma_1 > rd/K\gamma_1\gamma_2$, that is, (by rearrangement) $K > d/\gamma_2$. We can now plot the lines for $dN_1/dt = 0$ and $dN_2/dt = 0$ and note that the line $dN_1/dt = 0$ has rotated clockwise from that in Fig. 16.3*a* (the Lotka-Volterra model). In plotting the arrows, one must keep in mind that they are placed parallel to the predator axis on the line $dN_1/dt = 0$ and parallel to the prey axis on the line $dN_2/dt = 0$ (Figs. 16.3*b,c*). By joining these arrows, we obtain a trajectory that spirals inwards toward the stationary point (the intersection of the two lines), indicating that the Volterra modification stabilizes the model.

We can generalize these results in the following manner, using the condition of neutral stability (Fig. 16.3*a*) as a frame of reference. Any factor added to the Lotka–Volterra model that rotates either or both lines (predator or prey isoclines) in a clockwise direction indicates increasing stability; rotation of either or both lines in a counterclockwise direction indicates decreasing stability. Thus, we have a straightforward graphical procedure for analyzing stability. Some of the modifications which have been found to improve the model's stability are (1) prey populations are resource limited (i.e., Volterra's modification discussed previously) (Maynard-Smith 1974, Murdoch & Oaten 1975), (2) a constant number of prey escape the predator (prey refugia) (Maynard-Smith 1974, Murdoch & Oaten 1975), (3) a prey's risk to predation is heterogeneous (a prey's risk to attack increases if it is in a high-density patch) (Murdoch & Oaten 1975), (4) spatial heterogeneity when different subpopulations have different parameter values for predators and prey and the predators must move between patches (Murdoch & Oaten 1975), and (5) a Holling type 3 functional response (May 1973a). A Holling type 2 functional response will yield an unstable model (May 1981). Detailed treatment of these assumptions can be found in May (1973a, 1981), Maynard-Smith (1974), and Murdoch and Oaten (1975). Ricklefs (1979) graphically summarizes the effects of several of these factors on the predator or prey isocline.

16.1.2 Models for Seasonally Breeding Populations

16.1.2.1 Nicholson–Bailey Model. Up to this point we have discussed models that theoretically relate to organisms whose populations exhibit continuous growth rates (e.g., aphids or spider mites) implicit in the use of differential equations. However, a parallel line of development has occurred with models describing organisms with discrete generations and that reproduce seasonally. These models utilize difference equations. The Nicholson–Bailey (1935) host–parasite model is a good example and is a homologue of the Lotka–Volterra predator–prey model (May 1973b). Nicholson (1933) and Nicholson and Bailey (1935) assumed that a parasitoid *population* randomly searches areas that contain hosts; thus, the searching of its members is "unorganized," and as the time spent searching an area increases, more time is spent searching areas previously searched. (N.B. that an individual parasitoid may search systematically but a parasitoid population searches at random.) He characterized a parasitoid's search path during its lifetime as the product of the path's length and width and denoted it as the area traversed. The area traversed was characteristic of a parasitoid species and specific to the context in which the parasitoid population was searching.

Nicholson also assumed that the number of hosts encountered, N_e, by a parasitoid population was directly proportional to (1) the total host area searched by an average parasitoid, (2) the density of the searching parasitoid population, P_t, and (3) the density of the host population, N_t; that is,

$$N_e = aN_tP_t \qquad (16.10)$$

where a is the proportion of hosts attacked per parasitoid (Nicholson & Bailey 1935). Furthermore, Nicholson initially considered a solitary parasitoid species; thus, a host parasitized one or more times issued only one adult wasp. Supernumerary larvae were eliminated by competition (Salt 1961). Since a parasitoid can encounter either an unparasitized or previously parasitized host (parasitoids, unlike predators, do not immediately remove their victims from the population), a proportion of these encounters, N_e, represents one or more contacts with previously parasitized hosts. Nicholson and Bailey (1935) reasoned, based on these assumptions, that the probability of a parasitoid contacting a host one or more times follows a Poisson distribution and, therefore, the probability of a host remaining unparasitized (i.e., P_0) is the zero class of this distribution; that is,

$$P_0 = \exp\left(-\frac{N_e}{N_t}\right) \tag{16.11}$$

where N_e/N_t is the mean number of hosts encountered. We can obtain N_e/N_t from Eq. (16.10) by manipulation:

$$\frac{N_e}{N_t} = aP_t$$

Therefore,

$$P_0 = \exp(-aP_t)$$

Thus, we have an estimate of the probability of a host escaping parasitization and can use this to calculate the number of hosts entering the next generation ($t + 1$); that is,

$$N_{t+1} = N_t\lambda \, (\exp - aP_t) \tag{16.12}$$

where N_t and N_{t+1} are the densities of the host populations in two successive generations, λ is the net number of offspring surviving to N_{t+1} (i.e., fecundity less the effects of migration and all mortalities except that due to parasitism), and $\exp(-aP_t)$ is the probability of a host remaining unparasitized. The number of hosts attacked one or more times can be estimated by summing the probabilities for all the frequency classes of the Poisson distribution other than that for the zero class or, more easily, by subtracting the zero class from 1.0; that is,

$$\text{Prob. \{no. of attacks} > 0\} = [1 - \exp(-aP)]$$

Hence, the number of hosts attacked (hosts that issue a parasitoid adult) is

$$N_a = N_t[1 - \exp(-aP)]$$

Since our parasitoid species is solitary, the number of emerging parasitoids (P_{t+1}) is equivalent to the number of hosts attacked one or more times:

$$P_{t+1} = N_t[1 - \exp(aP_t)] \tag{16.13}$$

Equations (16.12) and (16.13) constitute the Nicholson–Bailey model, and it follows the general format of a difference model for a host–parasitoid interaction (Hassell 1978); that is,

$$N_{t+1} = \lambda N_t f(N_t, P_t) \tag{16.14}$$
$$P_{t+t} = cN_t[1 - f(N_t, P_t)] \tag{16.15}$$

where N_t and N_{t+1} and P_t and P_{t+1} are the densities of the host and parasitoid populations in two successive generations, λ is the net number of offspring surviving mortality factors other than parasitism, and c is the average number of parasitoid adults issuing from a host, in our case $c = 1.0$. The $f(N_t, P_t)$ in the Nicholson–Bailey model is the $\exp(-aP_t)$ term and defines the probability of a host's escaping parasitization. It is this term that embodies the assumptions about the searching behavior of a parasitoid species.

The Nicholson–Bailey model makes a number of assumptions, some of which are the same as those made by the Lotka–Volterra model: (1) a parasitoid population searches at random, (2) the number of hosts encountered by the parasitoid population is a linear function of both the parasitoid's area of search and its population density and the host's population density, (3) for a given parasitoid density (P_t), percent parasitism remains constant with increasing host (N_t) density, (4) the parasitoid species is (at least initially) solitary, that is, only one adult issues from a host parasitized one or more times, (5) in the absence of parasitization the host population increases without limit, (6) each host is equally likely to be parasitized, (7) the probability of encountering a host is independent of whether it has been previously encountered one or more times (hosts encountered more than once have been parasitized on the first encounter), (8) the proportion of hosts attached per parasitoid is constant and independent of parasitoid density, and (9) the effect of parasitization on the host population is delayed until the host population reproduces. This latter assumption is implicit to difference equations when they are used to characterize a host–parasitoid model.

The equilibrium densities for the host (N^*) and parasitoid (P^*) populations can be calculated by rearranging the Nicholson–Bailey model and setting $N_t = N_{t+1}$ and $P_t = P_{t+1}$. The following equations can be obtained (see Hassell 1978):

$$P^* = \frac{1}{a} \ln \lambda \qquad (16.16)$$

$$N^* = \frac{\ln \lambda}{a(\lambda - 1)} \qquad (16.17)$$

The equilibrium density for the host population, N^*, depends on the parasitoid's area of discovery, a, and the net number of the host's offspring that survive, from factors other than parasitoids, to the next generation. This relationship is depicted in Fig. 16.4. The equilibrium density for the parasitoid population, P^*, also depends on the same two variables, but the relationship is more straightforward: the greater the area of discovery (a), or the lower the net number of host offspring that survive other factors, the lower the number of female parasitoids needed to maintain the host population at its equilibrium density. Although this model has an equilibrium, it is unstable. When either the host or parasitoid density is perturbed from its equilibrium value, the densities of both populations oscillate with ever-increasing amplitudes until they are both annihilated. Nicholson recognized this lack of stability and proposed that, in nature, interacting subpopulations of parasitoids and hosts may indeed become locally extinct. The populations of both persist, however, because the host–parasitoid in-

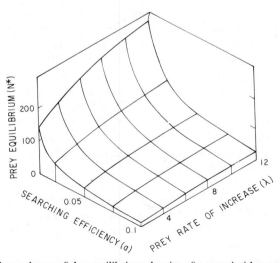

Fig. 16.4   The dependence of the equilibrium density of a parasitoid population, N^*, on the searching efficiency, a, and the host's rate of increase, λ (after Hassell 1978).

teractions are asynchronous in the many subpopulations that comprise the species populations. Immigrants from extant subpopulations reestablish subpopulations in localities where they are annihilated; hence both species exist in various stages of annihilation and reestablishment. Nonetheless, many of the assumptions are naive in light of findings concerning host selection behavior of some parasitoids (e.g., Salt 1934, Waage 1979, van Lenteren 1981, Galis & van Alphen 1981).

The unstable nature of the Nicholson–Bailey model arises from its difference equation format, with its implicit assumption of a time delay. Time delays arise because of the discrete periods of reproduction in the host population. This can be best seen by comparing the Lotka-Volterra [Eqs. (16.2) and (16.3)] and the Nicholson–Bailey models. Here we recast the Lotka–Volterra model into difference equations (changing May's 1973b notation) by using most of the assumptions of the Nicholson–Bailey model (cf. May 1973b for derivation of the model); we obtain the following set of equations:

$$N_{t+1} = \lambda N_t(1 - cP_t)$$
$$P_{t+1} = N_t cP_t$$

where c is equivalent to a, that is, the average proportion of hosts parasitized by a parasitoid.

We can compare the assumptions about a parasitoid's searching for hosts in the Nicholson–Bailey model and the recast Lotka–Volterra model by comparing their $f(N_t, P_t)$. For the Nicholson–Bailey model

$$f(N_t, P_t) = \exp(-aP_t)$$

and is thus an exponential function, while for the recast Lotka–Volterra model

$$f(N_t, P_t) = 1 - cP_t$$

is a linear function. Stability analysis of the recast Lotka–Volterra model shows that, in a difference equation format, it is just as unstable as the Nicholson–Bailey model (May 1973b). A stable difference model implies a stable differential homologue, but the converse is not true (May 1973a, Murdoch & Oaten 1975). Thus, time delays (e.g., seasonal births, etc.) implicit to difference models tend to be destabilizing (May et al. 1974).

16.1.2.2 Density-dependent Modifications to the Nicholson–Bailey Model.
Beddington et al. (1975) changed Nicholson's assumption of unlimited growth in the host population when parasitoids are absent to one of logistic growth. Their change is analogous to that made by Volterra

(1926) to the Lotka–Volterra model. The Beddington et al. model is written as

$$N_{t+1} = N_t \exp[r(1 - N_t/K) - aP_t] \tag{16.18}$$

$$P_{t+1} = N_t[1 - \exp(-aP_t)] \tag{16.19}$$

where $r = \ln \lambda$ and K is the carrying capacity of the environment. The term $(1 - N_t/K)$ is the same as that which appears in the logistic equation [Eq. (16.1)], that is, $(K - N)/K$. Note that the only change in the model occurs with Eq. (16.18) describing changes in host population density.

Addition of this density-dependent (competitive) limitation to growth adds a measure of stability to the interaction, but whether stability exists depends on the values of two parameters, $r(= \ln \lambda)$, the host's net reproductive rate, and q, a parameter that represents the ratio of the host's equilibrium density (N^*) to the density of the host population at the environment's carrying capacity (K), that is, $q = N^*/K$ (Fig. 16.5). When the ratio q is small (i.e., N^* much less than K) at low host reproductive rates (r), the model is unstable (unhatched area, lower left). As the host's reproductive rate r increases for any given q the model also becomes unstable (area to right of crosshatching). Thus, density-dependent limitation to growth involving certain values of N^*, K, and r causes the host and parasitoid populations to return to their equilibrium densities (crosshatched area). When these parameters $(N^*, K, \text{ and } r)$ have values that place the host and parasitoid densities in region B, the parasitoid population becomes extinct; when they have values that place the densities in region A, the interaction is characterized by everything from neutral stability and stable limit cycles to chaotic behavior (see May & Oster 1976 for a lucid presentation of such behavior).

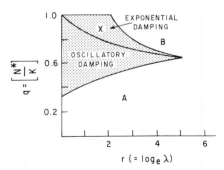

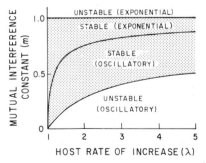

Fig. 16.5   Left: Stability boundaries for a Nicholson–Bailey model to which has been added an intraspecific, density-dependent interaction among members of the host population for a resource in short supply (after Beddington et al. 1975).

Fig. 16.6   Right: Stability boundaries for a Nicholson–Bailey model to which mutual interference has been added (after Hassell and May 1973).

16.1.2.3 Models Incorporating Mutual Interference. Hassell and Huf-faker (1969), while analyzing a *Nemeritis canescens* Mediterranean flour moth interaction, noted that the area of discovery of *N. canescens* varied with changes in its density. Ullyett (1950) had referred to such an effect as mutual interference. Hassell and Varley (1969) found this to be a common feature of the host–parasitoid interactions they reviewed and described the relationship with a regression equation:

$$\log a = \log Q - m \log P_t \tag{16.20}$$

the antilogarithm of which yields

$$a = QP_t^{-m} \tag{16.21}$$

We substitute QP_t^{-m} for a in Eq. (16.10) and, with some rearranging, obtain

$$\frac{N_e}{N_t} = QP_t^{1-m}$$

which can then be substituted into Eq. (16.11) to provide an estimate of the number of hosts that escape parasitism; that is,

$$P_0 = \exp(-QP_t^{1-m}) \tag{16.22}$$

We can now use Eq. (16.22) in our host–parasitoid model, that is, in Eqs. (16.14) and (16.15):

$$N_{t+1} = N_t \lambda \exp(-QP_t^{1-m})$$
$$P_{t+1} = N_t[1 - \exp(-QP_t^{1-m})]$$

where Q is the "quest constant" or the "area of discovery" when the parasitoid density is unity, m is the mutual interference constant (a measure of the frequency with which parasitoids encounter one another), and the remaining symbols (N_t, N_{t+1}, P_t, P_{t+1}, and λ) are defined as before. Mutual interference reduces the average efficiency of a parasitoid at high parasitoid-to-host ratios: A parasitoid is constantly being contacted or interrupted by other parasitoids while it is resting, searching, or ovipositing. This reduces its efficiency, which lowers its area of discovery. Hence, in contrast to Nicholson's assumption of a constant area of discovery, interference causes the area of discovery to decline with increasing parasitoid density at a given host density. Mutual interference of sufficient degree added to the Nicholson–Bailey model leads to a stable interaction (Fig. 16.6).

Royama (1971a) and Free et al. (1977) point out, however, that the relation between area of discovery and parasitoid density is unlikely to be linear since linearity implies that log a [Eq. (16.21)] increases without limit as the parasitoid density decreases. Free et al. (1977) showed that if parasitoids encountered one another at random and a constant amount of time was wasted per encounter, interference then produced curves. However, evidence is growing that mutual interference may be, in part, an artifact produced when the parasitoid-to-host ratios are small: the few parasitoids tend to aggregate on dense host patches. This leads to a larger area of discovery when compared to that produced under conditions of moderate-to-high parasitoid-to-host ratios, that is, when in addition to aggregating on dense host patches, the parasitoids are also forced to search host patches containing fewer hosts (Free et al. 1977). Thus, mutual interference arises from the parasitoid's response to denser host patches in host species that exhibit a patchy distribution.

16.1.3   Models Incorporating the Functional Response

16.1.3.1   The Basic Functional Response.
Solomon (1949) recognized two categories of predator responses to changes in prey density: (1) a functional response in which the number of prey consumed per predator (parasitoid) changes with changes in prey density and (2) a numerical response in which the predation density changes with changes in prey density. Holling (1959a) suggested that Solomon's functional response could be further subdivided into three (distinct) types: type 1, in which prey consumption increases linearly to a plateau with increasing prey density (Fig. 16.7a); type 2, in which prey consumption increases asymptotically to a plateau with increasing prey density (Fig. 16.7b); and type 3, in which prey consumption is S-shaped (sigmoid) with increasing prey density (Fig. 16.7c). [N.B. the lines for percentage kill are, in each case, quite different (Fig. 16.7d,e,f), and

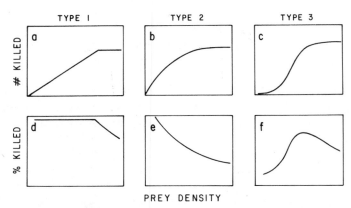

Fig. 16.7   The three types of functional responses (1, 2, 3) designated by Holling (1959a). For each type the number and percentage of prey killed per unit time by a single predator is graphed against prey density (after Huffaker et al. 1976).

only that in Fig. 16.7f possesses potential for population regulation through the functional response alone, and this only within a limited range of prey densities.]

Holling (1959b) considered that the number of encounters (N_e) by a single parasitoid with a host population is a simple function of two factors: host density (N_t) and the time a parasitoid spends searching (T_s); that is,

$$N_e = a'T_sN_t \tag{16.23}$$

where a' is the rate at which a parasitoid searches, that is, the amount of the area covered in a unit of time (T_s). However, for every host encountered, a certain amount of handling time (T_h) is required to both assess a host's suitability and to oviposit on or in it. If T_t is the total time during which the hosts are exposed to the searching parasitoid, then

$$T_s = T_t - T_hN_e \tag{16.24}$$

Substituting for T_s in Eq. (16.23) we get

$$N_e = a'(T_t - T_hN_e)N_t$$

which by rearrangement yields

$$N_e = \frac{a'T_tN_t}{1 + a'T_hN_t} \tag{16.25}$$

or Holling's (1959b) disc equation, which describes the type 2 functional response.

16.1.3.2 Nicholson–Bailey Model Incorporating a Type 2 Functional Response.
We now incorporate the basic functional response into a population interaction (numerical response) model appropriate for a parasitoid (i.e., the Nicholson–Bailey model). In order to obtain the number of hosts parasitized in a functional response model, we need to rearrange Eq. (16.25) so that it is in the form N_e/N_t and to scale it for the number of searching parasitoids (P_t). Thus Eq. (16.25) now becomes

$$\frac{N_e}{N_t} = \left(\frac{a'T_t}{1 + a'T_hN_t}\right)(P_t)$$

which we can now substitute into Eq. (16.11):

$$P_0 = \exp\left(\frac{-P_ta'T_t}{1 + a'T_hN_t}\right) \tag{16.26}$$

(N.B. that if $T_h = 0$ and T_t becomes the lifetime of the parasitoid, i.e., $T_t = 1.0$, then $\exp[-P_t a' T_t/(1 + a' T_h N_t)]$ reduces to the Nicholson–Bailey functional response model, i.e., $\exp(-a' P_t/1.0)$; thus the Nicholson–Bailey functional response model is a special case of the disc equation.) The number of hosts parasitized (N_a) is obtained by substituting Eq. (16.26) into Eqs. (16.14) and (16.15); that is,

$$N_{t+1} = \lambda N_t \exp[-a' T_t P_t/(1 + a' T_h N_t)] \tag{16.27}$$

$$P_{t+1} = N_t\{1 - \exp[-a' T_t P_t/(1 + a' T_h N_t)]\} \tag{16.28}$$

This is now a population interaction model which incorporates a type 2 functional response that describes a randomly searching parasitoid species; it makes many of the same assumptions about the parasitoid and host populations as the Nicholson–Bailey model. However, this model differs from that of Nicholson and Bailey in several important ways: (1) with each host encounter a specific amount of time, T_h, is required to assess the host's suitability and to deposit an egg(s) in or on it; thus the amount of time available for searching decreases with increasing host density (since more of the available time, T_t, will be taken up handling hosts as host density N_t increases) and (2) the assessment or handling time, T_h, is the same for all types of hosts encountered regardless of whether they have been encountered (or parasitized) one or more times.

Assumptions 1 and 2 point to an important distinction (assumption) between parasitoids and predators made by this model. As pointed out in a modeling context by Royama (1971a) and Rogers (1972), predators consume their prey and thereby remove them from an area; parasitoids do not. Consequently, a predator cannot rediscover and waste time assessing previously consumed hosts; therefore, more of a predator's total time is available for searching. This searching time is reduced only by the time required to pursue, subdue, and consume each prey (T_h) when it is initially encountered, that is,

$$T_s = T_t - T_h N_a \tag{16.29}$$

where T_s, T_t, and T_h are as previously defined and N_a is the number of prey attacked. Our modeling format, however, is in the form of the number of encounters that an average predator makes with prey. If we assume that the predatory population searches at random, then it will again search those areas in which it captured and consumed prey. We can treat this as though the predator recontacted those prey it consumed a second or subsequent time, except that at these subsequent (phantom) encounters the predator does not incur a loss in searching time due to rehandling of a prey. The number of encounters is again given by Eq. (16.23), that is,

$$N_e = a'T_s N_e$$

or

$$\frac{N_e}{N_t} = a'T_s \qquad (16.30)$$

We now substitute Eq. (16.29) for T_s in Eq. (16.30) and scale it for the number of searching predators, obtaining a model appropriate for predators; that is,

$$\frac{N_e}{N_t} = a'(T_t - T_h N_a)P_t$$

which is then substituted into Eq. (16.11); that is,

$$P_0 = \exp[-a'P_t(T_t - T_h N_a)] \qquad (16.31)$$

Equation (16.31) represents the probability of a prey escaping predation when the predators search randomly. However, we no longer can make the assumption that a predator offspring is produced for each prey consumed as we did in our parasitoid model, since a predator normally requires more than one prey to complete maturation or to produce offspring. This greatly complicates the predator equation [P_{t+1} of Eq. (16.15)], so comparatively little theoretical work has been pursued in this area. We do not consider predation beyond this point; however, these ideas can be pursued by consulting Murdoch and Oaten (1975), Hassell et al. (1976), Beddington et al. (1976), Frazer and Gilbert (1976), Hassell (1978), and Gutierrez et al. (1981), and the references cited therein.

The major difference between a randomly searching parasitoid and a randomly searching predator lies with the assumption that the parasitoid allocates equal handling time to hosts that are recontacted (which are assumed to have been parasitized during the first encounter). A predator does not, since it does not recontact consumed prey. To the degree that the parasitoid can detect previously parasitized hosts and reduce the amount of time it spends handling them, its searching behavior will approach that of a predator. Evidence is substantial that this is frequently the case (van Lenteren 1981).

The effects on stability of adding a type 2 functional response are straightforward. The Nicholson–Bailey model is obtained when the handling time is zero ($T_h = 0$). We also know the Nicholson–Bailey model is unstable. Since handling time decreases the amount of time a parasitoid can allocate to searching for hosts, a type 2 functional response added to the numerical response of a Nicholson–Bailey model increases the model's instability (annihilation will occur sooner) (Hassell & May 1973); the larger the value for the handling time the more rapid the annihilation.

16.1.3.3 Nicholson–Bailey Models Incorporating a Type 3 Functional Response.
It has been commonly assumed that the type 2 functional response characterizes invertebrate predators and parasitoids (Holling 1959a, Murdoch & Oaten 1975). However, Takahashi (1968) and van Lenteren and Bakker (1976, 1978) provided evidence that this is unrealistic. The type 2 functional response suggested for so many parasitoid species is probably a consequence of improper experimental design and not due to the parasitoid's behavior *per se* (van Lenteren & Bakker 1976, 1978, Luck et al. 1979). Consequently, when Hassell et al. (1977) and Hassell (1978) recognized the presence of an apparent sigmoid functional response in data from several invertebrate predator–prey laboratory experiments, they suggested that the attack rate a' is not a constant but increases with increasing host density; that is, the parasitoid searches more actively with increasing host density but at a decelerating rate. Based on this assumption they fitted an equation to the data of the form

$$a' = \frac{bN_t}{1 + cN_t} \tag{16.32}$$

The constant a' in Eq. (16.25) represents the rate at which the parasitoid searches an area for hosts during a given period of time. We can now substitute Eq. (16.32) into Eq. (16.25) for a' and obtain

$$N_c = \frac{[bN_t/(1 + cN_t)](T_tN_t)}{1 + [bN_t/(1 + cN_t)](T_hN_t)} \tag{16.33}$$

Manipulation and cancellation of Eq. (16.33) yields

$$N_e = \frac{bT_tN_t^2}{1 + cN_t + bT_hN_t^2} \tag{16.34}$$

With Eq. (16.34) we now obtain the number of hosts that are encountered by a single parasitoid. In order to obtain the number of hosts parasitized, we need to rearrange Eq. (16.34) so that it is in the form N_e/N_t and to scale it for the number of searching parasitoids (P_t). Thus, Eq. (16.34) becomes

$$\frac{N_e}{N_t} = \frac{bT_tN_t}{1 + cN_t + bT_hN_t^2}(P_t)$$

which can be substituted into Eq. (16.11); that is,

$$P_0 = \exp\left(\frac{-bT_tN_tP_t}{1 + cN_t + bT_hN_t^2}\right) \tag{16.35}$$

Equation (16.35) represents the probability of a host escaping parasitization. Thus, the equations for the host–parasitoid interactions become

$$N_{t+1} = \lambda N_t \exp\left(\frac{-bT_tN_tP_t}{1 + cN_t + bT_hN_t^2}\right) \tag{16.36}$$

$$P_{t+1} = N_t\left[1 - \exp\left(\frac{-bT_tN_tP_t}{1 + cN_t + bT_hN_t^2}\right)\right] \tag{16.37}$$

So we now have a model depicting a randomly searching parasitoid species that incorporates a type 3 functional response into the basic Nicholson–Bailey model. This model [eqs. (16.36) and (16.37)], again, makes many of the same assumptions about parasitoid behavior and about the growth of the parasitoid and host populations as the Nicholson–Bailey model. It also includes the same assumptions as the model incorporating a type 2 functional response (above). It differs from the type 2 case, however, by assuming that the rate of searching by the parasitoid increases with host density but at an ever-decreasing rate. We can recover the Nicholson–Bailey model by assuming, as before, that no handling time is required to assess or oviposit in a host and that a' is a constant; thus the Nicholson–Bailey model, and that obtained by adding the type 2 functional response, are special cases of one which incorporates the type 3 functional response.

Unlike the differential models, such difference models depicting a specific one-host–one-parasitoid interaction and containing a type 3 functional response [i.e., Eqs. (16.36) and (16.37)] cannot stabilize a host–parasite interaction (Hassell & Comins 1978, Hassell 1978). However, an analysis of this interaction assumes that the parasitoid spends a constant amount of time handling a host even if the host has been previously parasitized. But evidence indicates that this may not be the case (van Lenteren 1981, Galis & van Alphen 1981). Nunney (1980) has shown theoretically that for predators which remove their prey items from the population (and presumably for parasitoids that avoid previously parasitized hosts), the type 3 response incorporated into a difference model can stabilize a one-predator–one-prey interaction because of a declining host population due to exploitation, a consideration not included in the Hassell and Comins (1978) analysis. Thus, the presence of a type 3 response in invertebrate predators and parasitoids has important implications for stability of a host–parasitoid (predator–prey) interaction. (N.B. The conditions under which the difference models are stable are more restrictive than are those for their differential homologues).

16.1.4 Models Incorporating Switching Behavior

A number of parasitoid species and most predaceous arthropods are non-host-specific (i.e., they utilize several host species). In such a case they may limit the growth of a host's (prey's) population if they switch from one host species to another. Switching, as defined by Murdoch (1969), implies that the proportion of hosts parasitized changes from less than to more than expected as the density of the first host species changes from lesser than to greater than the density of the second host species. If two prey species are present at densities N_{t1} and N_{t2}, the simplest response one might expect is that the predator eats them in proportion to their abundances, or

$$\frac{N_{a1}}{N_{a2}} = \frac{N_{t1}}{N_{t2}}$$

where N_{a1} and N_{a2} are the numbers of prey species 1 and 2 attacked, respectively, and N_{t1} and N_{t2} are their respective densities. But the two prey species may not be eaten in that proportion. One prey species may be more easily seen or subdued, thus we need to adjust this relationship for prey preference. Experimentally, this can be accomplished by exposing the predator to equal densities of both prey species and determining the preference ratio c such that

$$\frac{N_{a1}}{N_{a2}} = c\left(\frac{N_{t1}}{N_{t2}}\right) \tag{16.38}$$

We can use Eq. (16.38) to measure switching. If switching exists, we expect more of prey species 1 and less of prey species 2 to be attacked as the density of prey species 1 increases relative to that for prey species 2. Thus, we expect c to be an increasing function of N_{t1}/N_{t2}. Graphically, this relationship will be evinced as an S-shaped curve which characterizes a type 3 functional response (Fig. 16.8). If switching does not occur, then c will remain constant and a linear plot results (Fig. 16.8).

Switching has been studied with several predator (e.g., Murdoch 1969, Murdoch & Marks 1973, Lawton et al. 1974, Murdoch et al. 1975) and parasitoid (Cornell & Pimentel 1978, J. J. M. van Alphen pers. comm.) species. Several generalizations are suggested by the results from these studies (Murdoch et al. 1975, Cornell & Pimentel 1978): Switching is unlikely to occur if the preference for one of the two prey species, whether weak or strong, is consistent among the individual predators when the prey species are equally abundant (Murdoch et al. 1975); however, switching is likely if the preference for one prey species, whether weak or strong, is inconsistent among the individual predators when the two prey species are equally abundant (Murdoch et al. 1975, Connell & Pimentel 1978). A pop-

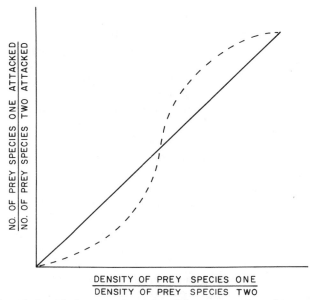

Fig. 16.8   The relationship between the ratios of the consumption of two prey species and their abundances. The dotted S-shaped line indicates frequency-dependent switching that results in a type 3 functional response.

ulation model which incorporates a type 3 functional response by a "switching" parasitoid—that is, one that is less dependent on a single host species—can stabilize an otherwise unstable host–parasitoid interaction (Royama 1971b, Murdoch et al. 1975).

16.1.5   Models Incorporating Special Searching Behaviors

The way in which prey or hosts are dispersed in the habitat may greatly affect the efficiency of a natural enemy's searching and the consequences for single prey species, or the members of a complex of species. Type 3 functional responses have been studied from the aspect of a predator's (1) exploiting patches containing single prey (host) species at different densities (e.g., Holling 1959b, Takahaski 1968, van Lenteren & Bakker 1976, 1978, Luck et al. 1979, Galis & van Alphen 1981); (2) exploiting patches containing two or more species of prey at different densities (e.g., Holling 1959a, Murdoch 1969, Murdoch & Marks 1973, Lawton et al. 1974, Cornell & Pimentel 1978, van Alphen & van Harsel in prep.); and (3) exploiting two species, each in a different habitat, and these species vary in density relative to one another (Royama 1971b, Murdoch & Oaten 1975). Theoretical studies have suggested such responses are capable of stabilizing a predator–prey interaction (Murdoch 1977, Oaten & Murdoch 1977a, 1977b, Nunney 1980).

Aggregative Response. The models so far examined assume that a predator or parasitoid searches randomly for hosts. However, prey are commonly distributed nonrandomly in their environment, frequently exhibiting a clumped distribution (Southwood 1978). If a predator is attracted to, or remains at a host patch because of a patch's larger size or because dense patches produce more of the cue(s) (e.g., kairomones), then the assumption that a predator, or a population of predators, searches randomly may be invalid. Moreover, spatial heterogeneity in the distribution of prey units, and thus the prey population, is known to be important in stabilizing predator–prey interactions in certain laboratory experiments (e.g., Huffaker 1958, Huffaker et al. 1963). Further, such heterogeneity has theoretically been shown to contribute to such stability (Maynard-Smith 1974, Hassell & May 1973, 1974, Murdoch & Oaten 1975, Murdoch 1977, May 1978, Hassell 1980, and cf. Chapters 12, 14). Since there are a number of examples wherein predators (parasitoids) spent more time searching denser patches (e.g., Hassell 1971, 1978, Galis & van Alphen 1981), we must look at the consequences of this behavior for population stability.

Hassell and May (1973) explored these consequences by distributing the parasitoid and host populations among n patches. They apportioned the hosts and parasitoids among the n patches by assigning α_i hosts and β_i parasitoids to the ith patch. The α_i and β_i do not have to equal one another but both the α_i and β_i for each patch must sum to unity; that is,

$$\sum_{i=1}^{n} \alpha_i = \sum_{i=1}^{n} \beta_i = 1.0$$

They assumed the parasitoid searched within each patch randomly, that is, according to the Nicholson–Bailey model. Thus, the probability of a host escaping parasitization in the ith patch is

$$P_{[0,i]} = [\alpha_i \exp(-a\beta_i P_t)]$$

To obtain the number of hosts escaping parasitism in all patches, we sum the $P_{[0]}$ probabilities over the n patches:

$$P_{[0]} = \sum_{i=1}^{n} [\alpha_i \exp(-a\beta_i P_i)]$$

Thus, the number of hosts surviving to the next generation is

$$n_{t+1} = N_t \lambda \sum_{i=1}^{n} [\alpha_i \exp(-a\beta_i P_t)]$$

The number of parasitoids present in the next generation is the sum of the probability classes of hosts parasitized one or more times:

$$P_{t+1} = N_t\{1 - \sum_{i=1}^{n} [\alpha_i \exp(-a\beta_i P_t)]\}$$

where N_t, N_{t+1}, P_t, P_{t+1}, λ, α_i, and β_i are as previously defined. Again, we assume that only one adult parasitoid emerges from a parasitized host. We also assume that a single egg is laid at each host encounter. If each patch received the same proportion of the parasitoid population ($\beta_i = 1/n$), then the model reverts to the Nicholson–Bailey. This model makes several assumptions in addition to those stated previously: (1) there are a constant number of patches available to the predator (parasitoid) and prey (host) ($=n$) throughout the duration of the interaction; (2) the same distribution of prey and predators per patch is generated each generation; that is, α_i and β_i are constants; (3) within each patch the parasitoid assumes the role of a Nicholson–Bailey parasite; (4) parasitoids do not search patches that lack hosts; and (5) an individual parasitoid spends all of its time searching only one patch (Murdoch & Oaten 1975).

When Hassell and May (1973) analyzed the model for stability they identified four trends: (1) Increasing parasitoid aggregation (β_i) increases stability and this factor alone can stabilize an otherwise unstable model; (2) the more even the parasitoid or host distribution the less stable the interaction; (3) a wider range of stability exists when approximately half the hosts occur in high-density patches; and (4) stability breaks down abruptly with increasing net reproductive rates (λ).

In a more elaborate analysis and modeling of predator aggregation on dense prey patches, Hassell and May (1974) generally confirmed their earlier findings. But in addition, they found that increased travel time between patches (cf. Murdoch & Oaten 1975) and marked differences between the minimum and maximum times spent searching sparse and dense patches, respectively, increased stability. However, because of their many parameters, these models presented problems when their analytical solutions were sought. Thus, May (1978) proposed a model in which the parasitoids' total aggregation behavior is subsumed in one parameter, k, the clumping parameter of the negative binomial (cf. May 1978 for elaboration).

16.2 OPTIMAL FORAGING

A second body of theory views predation from the perspective of an economic process consisting of benefits derived from capturing and consuming prey (energy and nutrients) versus the costs associated with obtaining energy (energy expended to pursue, capture, and subdue a prey) (e.g., Emlen 1966, MacArthur & Pianka 1966, Schoener 1971, MacArthur 1972, Pyke et al. 1977). The most successful predator maximizes its energy (and/or nutrient) intake per unit time or minimizes its time spent acquiring a given amount of energy. The latter strategy minimizes the risk of predation to the searcher (Schoener 1971, Pyke et al. 1977).

Prey or host items (species, age class, quality) are assumed to exist in patches—that is, areal units in which prey items can be found. It is also assumed that these patches differ in prey density or quality (Chapter 14). The forager exploiting patches chooses patches in a manner that maximizes its energy intake (or maximizes its host encounter rate, Cook & Hubbard 1977). As it is exploited, the value of a patch decreases—that is, with the decline in density or quality of the remaining prey; thus the likelihood of finding another, better patch elsewhere increases (MacArthur & Pianka 1966, Charnov 1976, Iwasa et al. 1981). Of principal interest are the rules of behavior used by a predator (or patasitoid) to "decide" how best to exploit patches. Four categories of behaviors are considered: (1) how to choose a patch in which to forage, (2) how long to forage in a patch, (3) what prey items to utilize in a patch, and (4) what pattern of search to use to find these patches. The theory assumes that these rules, evinced as behavior, are subject to natural selection and those predators which develop the most effective set of rules (e.g., those that lead to maximizing energy intake) are the most fit (i.e., they contribute most to the reproducing gene pool of the next generation).

In the following section we focus on optimal patch use, or how long a predator should forage in a patch. This question has received the most attention with insect predators (e.g. Cook & Hubbard 1977, Hassell & Southwood 1978, Hubbard & Cook 1978, Waage 1979).

16.2.1 Optimal Patch Use: Marginal Value Theorem

A preliminary model of optimal patch use might assume that a forager perceives the area in which it forages as composed of patches that differ in resource (prey) density or quality, and that the forager allocates its time among and within patches in such a way as to maximize acquisition of the resources, that is, maximizing energy gained (E) during the time spent foraging (T) or E/T. Foraging time consists of (1) that spent foraging within a patch (T_{si}) and (2) that spent traveling between patches (T_t). Thus, average exploitation time includes the time traveling to a patch from the last one:

$$T = T_t + \Sigma P_i T_{si}$$

where P_i is the proportion of patches of productivity i. The average gain in (net) energy is

$$E = \Sigma P_i E_i - T_t E_t$$

where E_i is the average net energy gained from a patch of productivity, i, and E_t is the (average) energy expended to travel between patches. Therefore, the average net energy gained per unit time (E/T) is

$$\frac{E}{T} = \frac{\Sigma P_i E_i - T_t E_t}{T_t + \Sigma P_i T_{si}}$$

The optimal time to spend in each patch type (T_{si}) is found by simultaneously setting $d(E/T)/dT_{si} = 0$ for all patch types and solving the equations. The results yield

$$\frac{dE_i}{dT_{si}} = \frac{E}{T}$$

for all patches regardless of quality. Thus, a forager should leave a patch when the instanteous rate of net energy intake (dE_i/dT_{si}) assumes the same value as that for the foraging area as a whole, that is, E/T.

Graphically this finding is depicted in Fig. 16.9. Curves OA and OB contrast the net energy gained by exploiting two patch types, one of high quality (curve OA) and one of low quality (curve OB). The shapes of these curves arise because as a patch is exploited it becomes more difficult to find hosts; the rate of return for effort expended diminishes with time. Two different average travel times between patches are considered, a long (distance OL) and a short (distance OS). The lines from points L and S tangent to OA and OB predict the time a predator should spend foraging in a patch of specific quality, given the travel time between patches. (This time is determined by the distance from point of tangency on curves OA and OB.)

Two predictions arise from this graphical model. First, a predator should

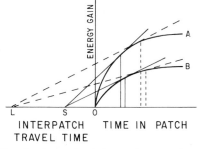

Fig. 16.9 Graphical solution for the optimal time to be spent in patches while foraging in a habitat containing patches of uniform quality, but comparing the times under two different travel times between patches; OS, short travel times; OL, long travel times. Habitats containing patches of two qualities are compared; OA, high-quality patches; OB, low-quality patches (after Krebs 1978).

spend more time exploiting a patch of given quality when more time is required to travel between patches. Second, for a given interpatch travel time, less time should be spent exploiting patches when the habitat is composed of high-quality patches. This latter prediction arises because when a predator leaves an exploited patch in a habitat composed of high-quality patches, it "expects" to encounter another high-quality patch.

In a habitat composed of patches of mixed quality, a predator should stay in a patch until its rate of capture (time interval between captures) equals that for the habitat as a whole, that is, E/T. Figure 16.10 depicts this prediction. Curve OB is net energy gain by exploiting an average patch in this type of habitat. The line SP from S tangent to OB indicates the optimal time for exploiting this average patch. To determine the optimal time to spend exploiting a high-quality patch (curve OA), a line is drawn parallel to SP and tangent to OA. A vertical line drawn from this tangent to the abscissa predicts the optimal time to spend exploiting the high-quality patch. This model predicts that more time should be allocated to a high-quality patch than one of average quality. The time that should be spent in a low-quality patch (OC) can be determined in a similar manner, again with the not too surprising prediction that less time should be spent foraging in a low-quality patch than one of average quality.

The marginal value theorem (cf. Charnov 1976) is based on several assumptions: (1) Prey items occur in patches and these patches are repeatable and predictable units; (2) the rate of (or time available for) energy intake is a limiting factor and this constitutes an important selection pressure on foraging animals; (3) the behavioral traits employed in "deciding" how to exploit a patch are heritable; (4) the predator acts as though it knows the average production of the habitat it is exploiting; and (5) the essential features of a foraging predator can be mimicked by a deterministic model.

Most of these assumptions are naive and are discussed relative to "rule 4" below. Patches are variable and unpredictable and are likely to exist

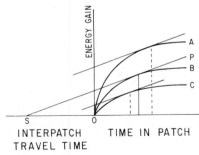

Fig. 16.10   Graphical solution for the optimal time to be spent in patches while foraging in a habitat containing patches of mixed quality (after Krebs 1978).

hierarchically, micropatches embedded in macropatches (Hassell & South-wood 1978, Waage, 1979, Tinbergen 1981). Some patch types are well defined, others are not (this may be more of a problem for the experimentalist than the forager). Moreover, as Oaten (1977) noted, chance is an important element in foraging decisions and a deterministic model such as one associated with the marginal value theorem cannot adequately mimic the circumstances.

Finally, the degree to which energy operates as the sole constraint on a forager's fitness is unclear. Cases in which energy intake is limiting can be envisioned (e.g., as when birds feed their young), but in many cases this constraint is likely to operate intermittently and in concert with other factors (e.g., access to mates). Nonetheless, Charnov's (1976) model is important in directing attention to important considerations in the foraging strategy. A number of "rules" governing how long an individual should exploit a patch have been proposed. They include: (1) a predator should exploit a patch until it utilizes a fixed number of prey (Gibb 1962); (2) a predator should spend a fixed amount of time exploiting a patch (Krebs 1973); (3) a predator should exploit a patch until the time interval between captures exceeds a fixed threshold value (Hassell & May 1974, Murdoch & Oaten 1975); (4) a predator exploits a patch until the time interval between captures equals that for the habitat as a whole, that is, the average time interval for the habitat (the marginal value theorem) (Charnov 1976); and (5) a parasitoid continues to exploit a patch so long as it successfully finds and oviposits in hosts prior to its habituation to the olfactory cue(s) associated with its host (Waage 1979).

"Rules" 1 and 2 have largely been rejected. Rule 4, the marginal value theorem, implies that the predator possesses complete knowledge of the size and value of all the patches comprising its habitat *prior* to exploiting them. This is an unreasonable assumption. Thus, we are left with rules 3 and 5.

Waage's (1979) "rule(s)" are based on a behavioral model wherein a parasitoid responds to a patch delineated by a chemical stimulus (cue) associated with its host. The cue(s) cause the parasitoid to arrest on the patch. Arrestment involves an orthokinetic response evinced as slowed walking and, perhaps, probing, and a klinotactic response evinced as turning back into the patch when it crosses the margin of the patch into noncue areas. Gradual habituation to the stimulus at the patch margin leads the parasitoid eventually to abandon the patch; higher concentrations of the cue(s) delay habituation, and the parasitoid remains longer. Ovipositions while in the patches also delay habituation. Eventually, though, the interval between ovipositions exceeds the threshold for response and the parasitoid leaves. These ideas are presented graphically in Fig. 16.11. Thus, the parasitoid spends a variable amount of time exploiting patches. The amount it spends depends upon the initial intensity of the chemical stimulus (CS_1, CS_2, CS_3, Fig. 16.11), the rate at which its responsiveness to the stimulus

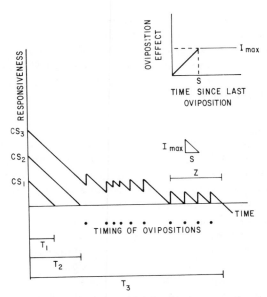

Fig. 16.11   Graphical representation of a model for the time spent by a parasitoid exploiting a patch, incorporating the effects of patch stimuli (CS_i) and ovipositional stimuli (I). The patch times, T_1, T_2, and T_3, for the hypothetical patch densities S_1, S_2, and S_3, respectively. The insert indicates how the time since the last oviposition influences the value of the increment (after Waage 1979).

wanes, and the time interval between its ovipositions. An oviposition causes an incremental increase in responsiveness to the chemical stimulus (height of the saw-toothed line in Fig. 16.11). If the responsiveness decreases to a threshold value (horizontal line in Fig. 16.11), then the parasitoid leaves. *Nemeritis* (Waage 1979) and *Asobara tabida* (van Alphen & Galis in prep.) behavior supports the essential features of this model—rule 5. Moreover, the time intervals between the last two ovipositions frequently exceeded that for the interval between the last oviposition and the parasitoid's abandonment of the patch. Thus, rule 3 (i.e., the fixed-threshold rule) seems an unlikely one governing the time a parasitoid spends exploiting a patch.

16.3   CONCLUSION

The models here reviewed attempt to explore predator–prey and parasitoid–host interaction by means of graphical and mathematical analogies. These models can be subdivided into three general groups: those that treat continuously growing predator–prey populations (overlapping generations), those that treat predator–prey populations with nonoverlapping generations, and those that view the predation process as one of resource exploitation (i.e., foraging theory). It is clear, however, that all of these models are, to one degree or another, inadequate. However, their utility does not lie in their accuracy but in the attention they focus on the various assumptions and processes associated with predation. This focus has identified several principles involved in the predator–prey interaction. For example, random encounters between predator and prey are of themselves insufficient to limit the growth of the prey populations. Moreover, predator aggregation on dense prey patches, habitat heterogeneity (both spatial and temporal), the presence of a sigmoid functional response (type 3) by a predator to single prey species occurring in patches that differ in density, or to multiple prey species (switching) as a frequency-dependent response, and competition in the prey species, all appear capable of stabilizing the predator–prey interaction. However, it is through the exploitation of prey patches that these processes are manifested. Thus, behavioral studies of a predator (parasitoid) exploiting a patch has assumed a central role in testing many of the notions implicit in these predation models. If this review has provided but a brief introduction into some of these notions, then its purpose has been well served.

REFERENCES

Beddington, J. R., C. A. Free, and J. H. Lawton. 1975. *Nature* **225:** 58–60.

Beddington, J. R., M. P. Hassell, and J. H. Lawton. 1976. *J. Anim. Ecol.* **45:** 165–185.

Charnov, E. L. 1976. *Theor. Popul. Biol.* **9:** 129–136.

Connell, J. H. 1975. Pp. 460–490 in M. L. Cody and J. M. Diamond (Eds.), *Ecology and Evolution of Communities.* Belknap, Cambridge, Mass.

Cook, R. M. and S. F. Hubbard. 1977. *J. Anim. Ecol.* **46:** 115–125.

Cornell, H. and D. Pimentel. 1978. *Ecology* **59:** 297–308.

DeBach, P. (Ed.). 1964. *Biological Control of Insect Pests and Weeds.* Chapman & Hall, London.

Emlen, J. M. 1966. *Am. Natur.* **100:** 611–617.

Frazer, B. D. and N. Gilbert. 1976. *J. Entomol. Soc. Br. Columbia* **73:** 33–56.

Free, C. A., J. R. Beddington, and J. H. Lawton. 1977. *J. Anim. Ecol.* **46:** 543–554.

Galis, F. and J. J. M. van Alphen. 1981. *Neth. J. Zool.* **31:** 596–611.

Gibb, J. A. 1962. *Ibis* **104:** 106–111.

Gompertz, B. 1835. *Phil. Trans.* **115:** 513–585.

Gutierrez, A. P., J. U. Baumgaertner, and K. S. Hagen. 1981. *Can. Entomol.* **113:** 21–33.

Gutierrez, A. P. and W. M. Getz. In press. In C. Shoemaker (Ed.), *Insect Pest Management Modelling.* Wiley, New York.

Hassell, M. P. 1978. *The Dynamics of Arthropod Predator–Prey Systems. Monogr. Popul. Biol.,* Vol. 13. Princeton University Press, Princeton, N.J.

Hassell, M. P. 1980. *J. Anim. Ecol.* **49:** 603–628.

Hassell, M. P. and H. N. Comins. *Theor. Popul. Biol.* **14:** 62–67.

Hassell, M. P. and C. B. Huffaker. 1969. *Res. Popul. Ecol.* **11:** 186–210.

Hassell, M. P. and R. M. May. 1973. *J. Anim. Ecol.* **42:** 693–726.

Hassell, M. P. and R. M. May. 1974. *J. Anim. Ecol.* **43:** 567–594.

Hassell, M. P. and T. R. E. Southwood. 1978. *Annu. Rev. Ecol. Syst.* **9:** 75–98.

Hassell, M. P. and G. C. Varley. 1969. *Nature* **223:** 1133–1137.

Hassell, M. P., J. H. Lawton, and J. R. Beddington. 1976. *J. Anim. Ecol.* **45:** 135–164.

Hassell, M. P., J. H. Lawton, and J. R. Beddington. 1977. *J. Anim. Ecol.* **46:** 249–262.

Holling, C. S. 1959a. *Can. Entomol.* **91:** 293–320.

Holling, C. S. 1959b. *Can. Entomol.* **91:** 385–398.

Hubbard, S. F. and R. M. Cook. 1978. *J. Anim. Ecol.* **47:** 593–604.

Huffaker, C. B. 1958. *Hilgardia* **27:** 343–383.

Huffaker, C. B. 1971a. Pp. 327–341 in P. J. den Boer and G. R. Gradwell (Eds.), *The Dynamics of Populations.* Centre Agric. Pubs. & Doc., Wageningen.

Huffaker, C. B. (Ed.) 1971b. *Biological Control.* Plenum, New York.

Huffaker, C. B. and P. S. Messenger (Eds.). 1976. *Theory and Practice of Biological Control.* Academic, New York.

Huffaker, C. B., R. F. Luck, and P. S. Messenger. 1976. *Proc. 15th Int. Congr. Entomol.* Pp. 560–586. Entomological Society of America, College Park, Maryland.

Huffaker, C. B., K. P. Shea, and S. G. Herman. 1963. *Hilgardia* **34:** 305–329.

Iwasa, Y., M. Hagashi, and N. Yamamura. 1981. *Am. Natur.* **117:** 710–723.

Krebs, J. R. 1973. Pp. 73–111 in P. P. G. Bateson and P. H. Klopfer (Eds.), *Perspectives in Ethology.* Plenum, New York.

Krebs, J. R. 1978. Pp. 23–63 in J. R. Krebs and N. B. Davies (Eds.), *Behavioural Ecology, an Evolutionary Approach.* Blackwell, London.

Lawton, J. H., J. R. Beddington, and R. Boneer. 1974. Pp. 141–158 in M. B. Usher and M. H. Williamson (Eds.), *Ecological Stability.* Wiley, New York.

Lotka, A. J. 1925. *Elements of Mathematical Biology.* Dover, New York.

Luck, R. F., J. C. van Lenteren, P. H. Twine, L. Kuenen, and T. Unruh. 1979. *Res. Popul. Ecol.* **20:** 257–264.

MacArthur, R. H. 1972. *Geographical Ecology Patterns in the Distribution of Species.* Harper & Row, New York.

MacArthur, R. H. and E. R. Pianka. 1966. *Am. Natur.* **100:** 603–609.

May, R. M. 1973a. *Stability and Complexity in Model Ecosystems.* Monogr. Popul. Biol., Vol. 6. Princeton University Press, Princeton, N.J.

May, R. M. 1973b. *Am. Natur.* **107:** 46–57.

May, R. M. 1978. *J. Anim. Ecol.* **47:** 833–843.

May, R. M. 1981. Pp. 78–104 in May (Ed.), *Theoretical Ecology: Principles and Applications.* Saunders, Philadelphia.

May, R. M. and G. F. Oster. 1976. *Am. Natur.* **110:** 573–599.

May, R. M., G. R. Conway, M. P. Hassell, and T. R. E. Southwood. 1974. *J. Anim. Ecol.* **43:** 747–770.

Maynard-Smith, J. 1974. *Models in Ecology.* Cambridge University Press, New York.

Murdoch, W. W. 1969. *Ecol. Monogr.* **39:** 335–354.

Murdoch, W. W. 1977. *Theor. Popul. Biol.* **11:** 252–273.

Murdoch, W. W. and J. R. Marks. 1973. *Ecology* **54:** 160–167.

Murdoch, W. W. and A. Oaten. 1975. *Advan. Ecol. Res.* **9:** 2–131.

Murdoch, W. W., S. Avery, and M. E. B. Smyth. 1975. *Ecology* **56:** 1094–1105.

Nicholson, A. J. 1933. *J. Anim. Ecol.* **2:** 132–178.

Nicholson, A. J. and V. A. Bailey, 1935. *Proc. Zool. Soc. Lond. 1935*, pp. 551–598.

Nunney, L. 1980. *Theor. Popul. Biol.* **18:** 257–278.

Oaten, A. 1977. *Theor. Popul. Biol.* **12**: 263–285.

Oaten, A. and W. W. Murdoch. 1977a. *Am. Natur.* **109**: 289–298.

Oaten, A. and W. W. Murdoch. 1977b. *Am. Natur.* **109**: 299–318.

Paine, R. T. 1966. *Am. Natur.* **100**: 65–76.

Pearl, R. and L. J. Reed. 1920. *Proc. Natl. Acad. Sci. USA* **6**: 275–288.

Pyke, G. H., H. R. Pulliam, and E. L. Charnov. 1977. *Quart. Rev. Biol.* **52**: 137–154.

Ricklefs, R. E. 1979. *Ecology.* Chiron, New York.

Rogers, D. J. 1972. *J. Anim. Ecol.* **41**: 369–383.

Rosenzweig, M. and R. MacArthur. 1963. *Am. Natur.* **97**: 209–223.

Royama, T. 1971a. *Res. Popul. Ecol.*, Suppl. 1, pp. 1–91. Kyoto University.

Royama, T. 1971b. Pp. 344–355 in P. J. den Boer and G. R. Gradwell (Eds.), *Dynamics of Populations.* Centre Agric. Pub. & Doc., Wageningen.

Salt, G. 1934. *Proc. Roy. Soc. Br.* **114**: 455–476.

Salt, G. 1961. *Symp. Soc. Exp. Biol. XV, Mechanisms in Biological Competition.* 96–119.

Schoener, H. S. 1971. *Annu. Rev. Ecol. Syst.* **2**: 369–404.

Smith, H. S. 1935. *J. Econ. Entomol.* **34**: 1–13.

Solomon, M. E. 1949. *J. Anim. Ecol.* **18**: 1–35.

Southwood, T. R. E. 1978. *Ecological Methods* (2nd ed.). Chapman & Hall, London.

Takahashi, R. 1968. *Res. Popul. Ecol.* **10**: 54–68.

Tinbergin, J. M. 1981. *Ardea* **69**: 1–67.

Ullyett, G. C. 1950. *Can. Entomol.* **82**: 1–11.

van Alphen, J. J. M. and F. Galis. In Prep.

van Alphen, J. J. M. and H. H. van Harsel. In Prep.

van Lenteren, J. C. 1981. Pp. 153–179 in Norlund, Jones, and Lewis (Eds.), 1981, referenced here.

van Lenteren, J. C. and K. Bakker. 1976. *Neth. J. Zool.* **26**: 567–572.

van Lenteren, J. C. and K. Bakker. 1978. *Neth. J. Zool.* **28**: 213–233.

Verhulst, M. 1838. *Mem. Acad. Roy. Belg.* **18**(1): 1–38.

Volterra, V. 1926. *Mem. Acad. Lincei* **2**: 31–113. (Transl. in R. N. Chapman, 1931. *Animal Ecology.* McGraw-Hill, New York.)

Waage, J. K. 1979. *J. Anim. Ecol.* **48**: 353–371.

Chapter 17

Do Populations Self-Regulate?

THOMAS J. WALKER

Voles probably exemplify a general law that all species are capable of limiting their own population densities without either destroying the food resources to which they are adapted, or depending upon enemies or climatic accidents to prevent them from doing so. If this is true, self-regulatory mechanisms have presumably been evolved through natural selection, and arguments in support of this view can certainly be advanced.

CHITTY 1960, p. 111

This sort of stabilization is often described in such terms as "the population adjusts itself to its food supply" or even "the population regulates its reproduction so as not to produce numbers in excess of what the environment can support." Such expressions imply that the density regulation is an evolved adaptation of the population as a whole, and that without such adaptations there would be no numerical stability.

These interpretations are utterly without justification. . . . the adjustment of fecundity to food supply is adaptive from the individual standpoint.

WILLIAMS 1966, p. 236

Whether insect populations "self-regulate" may well depend on the meaning assigned to self-regulation. In the broadest sense, *self-regulation* is a label for instances in which the actions of members of a population prevent it from exceeding its resources. Specifically individuals emigrate, die, or reduce the numbers or quality of their offspring at high densities, and their doing so is not the direct result of extrinsic factors other than conspecifics. For example, deaths from predators, pathogens, or starvation are not labeled self-regulating, but deaths from causes within the population (e.g., cannibalism, senescence, and density-induced hormonal imbalance) have been. Similarly, emigration upon exhaustion of food or cover at a site is not termed self-regulating, but emigration from a site having all needed resources has been.

Thus *self-regulation* (broad sense*) refers to density-regulating actions by members of the regulated population. The evolutionary origins of such actions are important to understanding them, and the reproductive consequences of the actions are the key to their evolution.

On the one hand the population regulating actions of individuals may be to their own reproductive advantage—for example, emigration from a habitat that is likely to become unsuitable to one that is likely to be bountiful; or death as a consequence of all-out reproductive effort at a time when later reproduction would be precluded. In these instances the evolutionary origin of the action need be no more complicated than differential reproductive success of genetically different individuals. The population-regu-

---

*A more restrictive definition of population self-regulation is that it involves *self-sacrificing* density-regulating actions by population members; such a definition may be of a null set.

lating consequence of the action may be an incidental effect of selection at the individual level; parsimony dictates that it be so considered in the absence of evidence to the contrary. To reiterate, an advantage or benefit to a population (or species, or ecosystem) does not require that the phenomenon evolved in that context; it need not be a population (or species, or ecosystem) adaptation. Adaptation at the population level or higher is a difficult concept and should not be invoked lightly. In the past it has been used as an easy way to avoid difficult questions (Williams 1966).

On the other hand the population-regulating actions of individuals may be detrimental to individual reproductive success—the individuals may be reducing their genetic contribution to future generations but benefitting the population. If this be so, natural selection at the level of the individual would tend to eliminate the behavior. Individuals that acted selfishly would leave more descendents than those that acted in the interests of the population. Within populations, genes that promoted selfishness would increase in frequency at the expense of their alleles, which promoted altruistic reproductive restraint, even if the population ultimately became extinct as a result. Extinction of populations with selfish individuals suggests that differential survival and perpetuation of groups of individuals (demes in this case) could effect adaptation at the population level (or higher). Such a process has been termed *group selection.*

The hypothesis of evolutionary change by group selection has a long history and many forms. V. C. Wynne-Edwards (1962) stimulated widespread discussion of its significance by proposing that much of the social behavior of animals concerned their regulating their own numbers and that the behaviors had evolved through group (i.e., interdemic) selection. Since 1962 group selection has been carefully analyzed, reviewed, and rereviewed (e.g., Maynard Smith 1964, 1976, 1978, Williams 1966, 1971, Ghiselin 1974, Gilpin 1975, E. O. Wilson 1975, D. S. Wilson 1975, 1980, Alexander & Borgia 1978, Wade 1978, Dawkins 1982).

This chapter is not the place for another detailed review of group selection, but some general ideas need to be stated before pursuing further the subject of self-regulation in insect populations. Group selection—in the sense of differential survival and perpetuation of genetically different groups—does occur. The debate over group selection concerns whether it has a significant role in determining the speed with which a trait changes and, in particular and of greatest importance, whether it can cause a trait to change in a direction opposite that produced by individual selection. The focal issue is whether traits detrimental to an individual and its close kin can become established because of their beneficial effect on the survival and propagation of the group. This issue has been met in two ways. First, modelers have postulated combinations of population phenomena that overcome the principal theoretical difficulties with selection among demes overpowering selection among individuals—namely, how to make extinction and reformation of demes common enough to act within the same time scale as individual selection, how to produce demes that differ significantly in the traits in question, and how to combine the isolation needed for demes to be different with the vagility needed for the successful groups to spread to the sites vacated by unsuccessful ones. The general nature of such models is that they work, mathematically, provided the population structure remains within the proposed narrow limits. Modelers have thus far had little success in identifying real populations that have the required characteristics, but so little is known of demographic structure under natural conditions and the particular requirements vary so much among models that there may be room for optimism. [cf. Maynard Smith (1976) and Wade (1978) for reviews of group selection models.]

The second approach to evaluating the importance of group selection in determining direction of evolutionary change is to study critically the traits that are supposed to owe their existence to it. To the extent that their existence accords with the simpler model of individual selection, the likelihood of group selection being important is diminished. With the possible exception of sexual reproduction (Williams 1966, 1975, Maynard Smith 1978) no trait has continued to receive strong support as being good for the population but detrimental to individual reproductive success.

Present evolutionary theory suggests that most, perhaps all, adaptation originated through selection at levels no higher than the individual and its close kin. If this is so, self-regulation in insect populations should be a by-product of individual adaptation and not itself an adaptation.

The remainder of this chapter will analyze seven phenomena that result in, or have been reputed to result in, population self-regulation. Emigration and sex ratio will be discussed at greater length than other phenomena because they are of general interest and importance and are especially likely to repay further study. The phenomena associated with population self-regulation are among the least understood in insect ecology; to some extent our ignorance results from their having been explained away as traits good for the population or species.

17.1 CANNIBALISM

Intraspecific predation is frequent among insects and usually has a density-dependent component that makes it population regulating (Fox 1975a, Polis 1981). In most cases older, larger individuals attack younger, smaller ones, but cannibalism within cohorts occurs (e.g., Fox 1975b, Duelli 1981), and pupation and molting sometimes reverse the vulnerabilities of age classes (e.g., Corbet & Griffiths 1963, Tschinkel 1981). When the cannibal is a predator and the conspecific prey is no kin, the origin and adaptiveness of the behavior are apparent—a predator that dines on a competitor has aided its own reproduction indirectly as well as directly. The double benefits of cannibalism reduce the surprise in finding that herbivorous insects are at times avid intraspecific predators. Hatchling butterflies and leafbeetles eat unhatched eggs or other hatchlings, and *Tribolium* larvae feed on eggs, pupae, and callow adults (Brower 1961, Eickwort 1973, King & Dawson 1972).

The population effects of cannibalism under natural conditions are difficult to isolate from other mortality factors, but studies by Fox (1975b) on the backswimmer *Notonecta hoffmanni* and by Polis (1980) on the scorpion *Pauroctonus mesaensis* establish that cannibalism can be a major mortality factor (more than 50% of first-instar *N. hoffmanni*) and a major source of nourishment (28% of diet biomass observed for *P. mesaensis*).

In some instances cannibalism is in defense of a specific space that has the food required for the cannibal's development (see Section 17.2 and Table 17.1). In grain-feeding insects, the first larva to occupy a grain may attack and kill any that follow (Crombie 1944). Similarly, the earliest parasitoid larva in a host may kill subsequent ones (Askew 1971). In the corn earworm (*Heliothis zea*) a larva occupying an ear tolerates no companions even though the food supply is sufficient for several, and cannibalism limits the population well below the limits of food resources (Barber 1936). Stinner et al. (1977) reported cannibal-caused earworm mortalities in excess of 75% as "not unusual" in North Carolina corn fields.

Cannibalism has self-evident population stabilizing effects and has never been mistaken for self-sacrifice. The remaining six phenomena are more problematical both in their effects and in their evolutionary origins.

17.2 TERRITORIAL BEHAVIOR

Some instances of territorial behavior in insects are population regulating but most are not. In either case, territoriality is understandable as an individual adaptation. Individuals defending an area—a *territory*—generally are maintaining exclusive use of certain resources within it. Territoriality has been reviewed by Wilson (1975) and Davies (1978b) and, relative to insects, by Otte and Joern (1975), Price (1975), and Burk (1979).

Territorial behavior in insects can be subdivided on the basis of the context in which it occurs, the most frequent being a male defending an area where he may mate with a female (Table 17.1).

Mating territories are of small consequence to total population and probably none at all to population growth. Nonterritorial males do not necessarily perish prematurely and females are not denied matings. Nonterritorial males may even secure some matings by adapting an alternative, "satellite" behavior (Alcock 1979, Cade 1979). The chief effect of mating territories is an increase in the variance of matings per male; when the area defended is merely a mating site, the result can be viewed as discrimination of male quality by females (Blum & Blum 1979, Bradbury 1981). Insect mating territories have been experimentally verified in the field in dragonflies (Moore 1964) (Fig. 17.1), grasshoppers (Otte & Joern 1975, Schowalter & Whitford 1979), and butterflies (Davies 1978a).

Table 17.1 Insect territoriality—three modes and some examples of each[a]

Principle function of territory—insect examples (References)

1. Mating—Male defends area from other males
   (a) Defended area includes one or more soon-to-be-receptive females:
   Solitary bees and wasps (Lin 1963, Alcock et al. 1978).
   Parasitoid wasps (Wilson 1961, King et al. 1969).
   (b) Defended area includes scarce resource required by female:
   Dragonflies—oviposition sites (Campanella & Wolfe 1974, Uéda 1979).
   Grasshopper—low-resin feeding site? (Schowalter & Whitford 1979).
   Big-legged bug—feeding site (?) (Mitchell 1980).
   Fruit fly—oviposition site (Bush 1969).
   Butterfly—oviposition sites (Baker 1972).
   Horned beetle—feeding site (Eberhard 1979).
   Bark beetle—feeding and oviposition site (Rundinsky et al. 1978).
   Solitary bee—feeding site (Alcock 1980).
   (c) Defended area merely a potential mating site:
   Damselfly (Bick & Bick 1965).
   Roaches (Bell et al. 1978, Gautier 1978).
   Meadow katydids (Feaver 1983).
   Field cricket (Cade 1979).
   Butterflies (Davies 1978a, Bitzer & Shaw 1980).
   Hawaiian *Drosophila* (Ringo 1976, Spieth 1978).
   Fruit flies (Tychsen 1977, Burk 1983).
   Solitary bees and wasps (Alcock et al. 1978, Gwynne 1980).
2. Progeny rearing—one or both parents defend area that has scarce resources required by progeny
   (a) Parent(s) defends site directly:
   Cricket (Walker 1980).
   Aphid (Whitham 1979).
   Burying beetles (Milne & Milne 1976).
   Dung beetle (Palmer 1978).
   Passallid beetle (Schuster 1975).
   Digger wasps (Steiner 1975, Brockman & Dawkins 1979).
   (b) Female marks site—other females avoid because resource is preempted.
   Bean weevil (Oshima et al. 1973).
   Butterflies (Rothschild & Schoonhoven 1977).
   Fruit flies (Prokopy 1981).
   Parasitoid wasps (Vinson 1976).
3. Feeding—juvenile or nonreproductive adult defends its food or feeding site
   (a) Rivals destroyed:
   Antlions (Wheeler 1930).
   Granary weevil (Assem 1971).
   Corn earworm (Barber 1936).
   Parasitoid wasps (Askew 1971).

Table 17.1 (*Continued*)

Principle function of territory—insect examples (References)

(b)	Nonlethal defense: Mantids (Edmunds 1976). Antlion (McClure 1976). Soldier beetle (Rausher & Fowler 1979). Mosquito (Gillett et al. 1977).

*a*Eusocial species excluded, see Chapter 18.

A second context of insect territoriality is one or both parents defending an area that has resources required for development of progeny (Table 17.1). Such defense prevents "scramble" or exploitation competition (Chapter 15) in which the developing juveniles may use up available resources and perish for lack of more. Blowflies, for example, may lay so many eggs on a piece of carrion that the maggots run out of food prior to obtaining enough to pupate (Nicholson 1954). On the other hand, a pair of burying beetles may not only defend a small carcass from other burying beetles

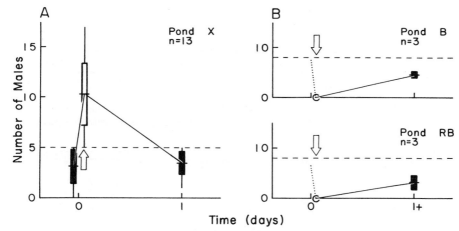

Fig. 17.1   Results of adding and removing territorial male odonates around bomb-hole ponds, Arne, England (Moore 1964). (A) Results of 13 addition experiments, *Cerigrion tenellum*, 1955 (from Table 9, Moore 1964; two cloudy days omitted); arrow indicates addition of 5–10 males to resident population. Horizontal dotted line is maximum number of males at pond under natural conditions. Vertical symbols show range (vertical line), mean (cross line), and ±1 standard deviation (vertical bar): left, prior to adding males; center, males added; right, 1 day later. By marking the males prior to release, Moore determined that they were only half as likely to remain as those already in residence, but this excludes the return to low level being a result of all introduced males leaving. (B) Results of six subtraction experiments, *Pyrrhosoma nymphula* (from Tables 4 and 8, Moore 1964); arrows indicate removal of all resident males. Other symbols as in A. Specific starting population not recorded; new "experimental" steady population (left symbol) noted after "a day or more."

(Milne & Milne 1976), but also introduce mites that consume blowfly eggs and small larvae (Springett 1968).

A more usual method of defense of resources for progeny is less direct and fails to satisfy some definitions of territoriality: the ovipositing female deposits a chemical that warns other females that added progeny will meet rigorous or overwhelming competition. This behavior is best known in parasitoids (Vinson 1976) (Fig. 17.2) and fruit flies (Prokopy 1981) and is diverse in detail: (1) females may leave chemical trails as they search for hosts (DeBach 1944, Price 1970, Greany & Oatman 1972); (2) females may chemically mark the host after ovipositing (Prokopy et al. 1977, Prokopy 1981); (3) females may chemically mark the host internally during oviposition (Greany & Oatman 1972, Guillot & Vinson 1972, van Lenteren 1976). In each case chemicals enable the depositing female and others to direct their searches or their eggs elsewhere. The advantage that a marking female gains by obeying her own marks is evident; the advantage to other females of refraining from ovipositing in a marked host are not as certain but two considerations probably contribute in varying proportions: (1) an offspring of a subsequent female will be younger than that of the marking female and will therefore have little chance of winning in competition with it; (2) even if later progeny are competitively equivalent to earlier ones, the offspring of a subsequent female will have no greater than 50% chance of

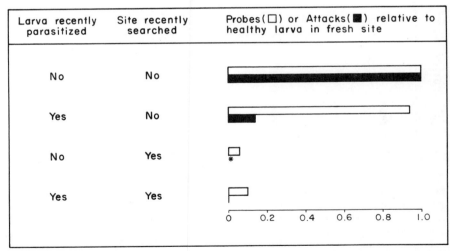

Fig. 17.2 Probe and attack rates of the braconid wasp *Orgilus lepidus* on larvae of potato tuberworm *Phthorimaea operculella* as influenced by site having been searched for the previous 2 hr (by other individuals) or by larvae having been parasitized less than 1 hr earlier. *Probes* were thrusts of the ovipositor toward or into a larva; *attacks* were ovipositor insertions of 2 sec or longer (time enough for oviposition). Rates are relative to those with healthy larvae at previously unvisited sites (uppermost bars) (data from Greany and Oatman 1972). Data on attack rate not available.

winning (case where no more than a single larva can mature) and may have none [all larvae in a multiply occupied host may have zero probability of acquiring enough food to mature (e.g., Jackson 1966)].

Whether marking chemicals are deposited continually prior to host finding, on the host, or in the host has important effects on how useful they are to the marking female and how likely it is that ovipositing females of hyperparasites (including parasitoids of fruit-feeding larvae), cleptoparasites, or competitors will use the chemicals to their advantage and to the detriment of the marking female (Vinson 1976, Prokopy & Webster 1978). External marks that last only a few hours may minimize the costs and perhaps not decrease the benefits; internal marks should perhaps last longer (Guillot & Vinson 1972). Previous oviposition may be detected by cues other than specially produced chemicals. For example, Cirio (1971) found that host juices smeared about by ovipositing olive flies, *Dacus oleae,* inhibit further egg laying, and Rothschild and Schoonhoven (1977) found that females of *Pieris brassicae* avoided ovipositing on cabbage leaves that had plastic models of eggs attached (though a volatile chemical deposited with the natural eggs was the greatest deterrent to continued oviposition).

In all of these cases parents act to insure their progeny access to needed resources. The effect may be prevention of excess population and crashes, but no function other than facilitation of individual reproduction need be postulated. In fact, at higher densities the behavior maximizes production of new adults and could then reduce the carrying capacity, causing a crash.

A final context of territorial behavior in insects is the defense of food or feeding sites by developing juveniles or nonreproductive adults (Table 17.1). Again there may be population-regulating effects, but the function is easily placed at the individual level—those maintaining territories have adequate food and probably reduced chances of predation (density-dependent predation, Chapter 12). In many cases attack is direct and the loser becomes food for the winner (Section 17.1). In other cases the defense is not lethal and may be so subtle as to make its function doubtful: The moist eggs of *Aedes aegypti* hatch in response to reduced oxygen, ordinarily a result of metabolism of microorganisms growing on their sculptured surfaces. The first larvae to hatch remain near the unhatched eggs and browse. Gillett et al. (1977) speculate that cropping of bacterial growth on unhatched eggs allows oxygen to increase and reduces the likelihood of additional hatch. Further hatch should occur only when uncropped bacterial growth again reduces dissolved oxygen. Eggs in a container thus hatch at a rate that is compatible with available food.

17.3 EMIGRATION

If a population produces higher proportions of emigrating individuals at higher densities, the effect is regulation; however, there is no need to consider emigration a population adaptation if it proves adaptive at a lower level (but see Lidicker 1962 and Taylor & Taylor 1979).

General features of insect migration and dispersal were discussed in Chapter 7. The concern here will be with *emigration*—leaving a place of abode or habitat. In the paragraphs below progressively more complicated circumstances will be considered; in each case selection at the individual or genic level can produce the described adaptations.

17.3.1 Transient Habitats

Many insects develop in habitats that remain suitable for no more than one or a few generations—for example, dung, carrion, fruit, logs, early stages of secondary succession. In such habitats permanent residence soon results in total failure. As old habitats vanish, new ones generally appear, and emigrants have unexploited habitats to colonize. Southwood (1962) showed a correlation between level of emigration and degree of impermanence of the habitat; he proposed (p. 172) that the "prime evolutionary advantage" of emigration is to enable "a species to keep pace with changes in the locations of its habitats." Its evolutionary origin is attributable to advantages at lower levels.

17.3.2 Fluctuating Habitats

Long-lasting habitats (e.g., lakes, grassland) may fluctuate in quality. If fluctuations in a region are not in phase, emigrants leaving a deteriorating habitat may find better circumstances nearby. If fluctuations are in phase, as in seasonal changes, only long-distance movement will be of advantage (e.g., monarch butterfly, Urquhart & Urquhart 1978). Insects generally escape seasonally synchronized fluctuations by diapausing (Chapter 6) rather than by emigrating.

17.3.3 Density-dependent Emigration

Emigration is often density dependent, with the proportion of emigrants increasing with population density (e.g., Naylor 1965, Shaw 1973, Hokyo & Kuno 1977, Tshinkel 1978).

The genetic basis for such behavior must be a coadapted set of genes that perpetuates itself more frequently than alternatives because it programs an individual to "emigrate if perceived density $>d$; stay if density $<d$." Density can be perceived directly [e.g., by tactile cues (Lees 1967)] or indirectly (e.g., through excretory products or food quality). Territorially can be considered a special case: "Emigrate if a territory can't be won; stay if one can." The program can be adaptively modified by incorporating previous experience—for example, "increase d if individual has emigrated once before"—or by restricting its use to adults of specific age or sex. The

existence of such specific genetic instructions is easily demonstrated in aphids because the parthenogenetic generations produced from a single overwintering egg or parthenogenetic female are genetically identical—the individuals constitute a clone (Lees 1961). Members of a clone show just the type of responses postulated, and the observed differences in emigration cannot be attributed to genetic differences, since none exist. [R. G. Harrison (1980) recently reviewed dispersal polymorphisms in aphids. Richard Dawkins (1980, 1982) reviewed the evolution of genetically determined behavioral alternatives in populations and included forceful statements of conditional strategies, as used above.]

17.3.4 Bet Hedging

In the previous paragraph the decision to emigrate or not is based on some circumstance or set of circumstances that correlates with reproductive payoff. For the responsible gene complex to increase in frequency, the correlation need not be perfect, but the average benefits must exceed those of competing complexes. Averages are only part of the story, however. What if a decision that is usually favorable is occasionally disastrous? For instance, what if staying home resulted in total reproductive failure once in every 5–20 generations—even though staying home, at low densities, is generally highly favorable? How may an individual (i.e., its relevant gene complex) avoid catastrophe relative to emigration? The most straightforward way for females is by widely spaced oviposition. A female may lay a portion p of her eggs at home and then emigrate, laying the remainder $(1 - p)$ in one or more other places. This is the strategy adopted by the spruce budworm; females lay approximately half of their eggs at their place of origin and the remainder at one or more sites that are generally 50 or more km apart (Fisher & Greenbank 1979). Such a strategy can be enriched, or compromised, by making p an inverse function of home habitat deterioration (e.g., spruce budworm, Blais 1953; a seedbug, Solbreck 1978).

A more tenuous method of spreading the risks relative to emigration decisions is for a female to cause a portion of her progeny to stay at home (p) and the remainder $(1 - p)$ to become emigrants. Such maternal influence is more easily accomplished and more likely to be adaptively correct the earlier the dispersal occurs in the ontogeny of the offspring; for example, the gypsy moth female controls the emigration tendencies of hatchlings by varying the amount of food in the eggs (Capinera & Barbosa 1976). Western tent caterpillar females likewise vary the nutrient content of their eggs, but here all hatchlings remain together and maintain an elaborate silk shelter. Nonetheless, the larvae from heavier eggs are more active foragers and eventually produce large adults that emigrate several kilometers. Larvae from lighter eggs produce light adults that remain close to the place of pupation (Wellington 1977).

The most convincing evidence of maternal influence extending to the emigration of adult progeny is from aphids. A female vetch or pea aphid (*Megoura viciae* and *Acyrthosiphon pisum*) can parthenogenetically produce varying proportions of winged and wingless daughters (Fig. 17.3). The determination is complete at birth and cannot be reversed by changing the crowding or food of the nymphs (Lees 1967, Sutherland 1969). Consequently, at birth some daughters are denied flight and others are programmed to develop wings and, to an unknown extent, to emigrate. In the bean aphid, *Aphis fabae,* maternal influence is more tentative. Nymphs that would otherwise develop into wingless daughters become winged ones if crowded (Shaw 1970a). Furthermore, at lower densities many of the winged morphs do not emigrate, and of those that do some first deposit progeny on the home plant (Shaw 1970b, 1973). Since the parthenogenetically produced daughter of an aphid is genetically identical to its mother, the reproductive interests of the two are the same, and maternal influence can operate without conflict between parent and offspring. Sexually produced progeny, being genetically different from the mother, have different genetic interests (Trivers 1974), and conflicts relative to maternal influence are expected. Nonetheless, some insects, in game theory terms, "play a mixed strategy" relative to emigration (Dawkins 1980). Stearns (1976) calls this type of risk spreading "bet-hedging."

17.3.5 Stable Habitats

When habitats are transient or erratically fluctuating in quality, the adaptive significance of emigration is clear. What if all sites suitable for colonization are permanent, stable, and already occupied, and migrant mortality is extremely high? Even so, substantial emigration should occur—according to a simple mathematical model developed by Hamilton and May (1977). The qualitative explanation is that genotypes that only stay home have no chance of taking over other sites; no matter how meager the chances of takeover,

Fig. 17.3   Wingless and winged parthenogenetic females ("apterous and alate virginoparae") of the vetch aphid, *Megoura viciae.* The two morphs are genetically identical; apterous virginoparae produce either morph, depending on crowding; alate virginoparae produce only the apterous morph. (Under some conditions virginoparae produce sexual morphs as well.) (From Lees 1961, *Roy. Entomol. Soc. Lond. Symp.* **1**:69.)

some emigration should win over no emigration, and substantial emigration is apt to win over minimal emigration.

The repertoire of evolutionary gambits operating on insect emigration at the individual or genic level is not exhausted in the paragraphs above but is at least revealed as adequate to accommodate the meager facts we have as to what are the actual patterns of insect emigration.

Models investigated by Roff (1975) and by Hamilton and May (1977) predict that levels of emigration that are evolutionarily stable (without group selection) differ in one direction or the other, from levels making for maximal occupancy of sites and minimal chances of extinction—that is, levels best for the species. In theory the models could be used to test for effects of group selection on emigration; in practice the assumptions of the models are too simple relative to reality to warrant such use.

17.4 SEX RATIO

Sex ratios vary within and between populations, sometimes in such directions and circumstances as to be advantageous to the population (e.g., Landahl & Root 1969, Mitchell 1970, Giesel 1972). Population regulation would result from changes in sex ratio if at low densities the proportion of males was just high enough to insure that all females were fertilized and at higher densities the proportion of males (i.e., nonegglayers) varied directly with density (Fig. 17.4, line a). With this hypothesis in mind Anderson (1961) reviewed the relations of density to sex ratio and found one case of a *decrease* in proportion of males with increased crowded (note line b, Fig. 17.4) and one case of an increase. In none of the remaining cases ($n \simeq 25$) was a significant change in sex ratio proved for the densities studied.

Although Anderson did not note it, most of the data he assembled fit deductions made by R. A. Fisher (1930, 1958) 30 years before. Fisher concluded that genic selection would result in equal parental expenditures on male and female progeny—that is, the most usual sex ratio (regardless of density) should be 0.50 males (Fig. 17.4, line c). He reasoned that in sexually reproducing organisms each individual receives half its genetic material from a male and half from a female; therefore, males and females *as classes* contribute equally to each generation. If one sex should be in the minority, the average individual of that sex will contribute more genetically to the next generation than the average individual of the majority sex. Provided individuals of the minority sex had not been more costly to produce, those parents that had produced an excess of the minority sex would, per parent, have more grandprogeny than those that had produced the two sexes equally or an excess of the majority sex. The consequence would be increased frequencies of the genes that had augmented the production of the minority sex and a shift in sex ratio toward equality. This would

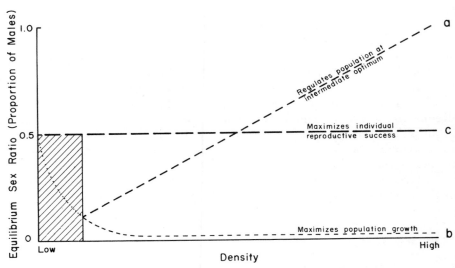

Fig. 17.4   Relation of sex ratio to population density under various assumptions: (a) sex ratio is a population adaptation that stabilizes the population at a favorable density, (b) sex ratio is a population adaptation that produces maximal population growth, (c) sex ratio is a result of selection of genes that maximize individual reproductive success. In shaded area, sex ratios for *a* and *b* may have to be high to insure that all or most females are mated. (It is assumed that males are capable of multiple matings, that costs of sexes are equal, and the mating is random.) (Modified from Williams 1966.)

continue, but as the sex ratio approached equality, the force of selection would weaken.

If sex ratios evolve entirely as Fisher proposed, they should have no role in regulating populations. Except perhaps at the lowest densities there should be more males than required to fertilize all females, and the sex ratio should remain constant rather than increasing with density.

Fisher's insights lay dormant nearly as long as Gregor Mendel's but eventually were seminal to the fuller development of the theoretical aspects of sex ratio evolution—germinating slowly (Shaw & Mohler 1953, Bodmer & Edwards 1960, Kolman 1960), blooming with a paper by W. D. Hamilton (1967), and maturing new seed in the recent book by E. L. Charnov (1982). Insects played prime roles in developing and testing sex ratio theory, and their potential is far from exhausted. Variations in insect sex ratios, within and between populations, will repay close examination. Situations in which sex ratios are expected to differ from equality (i.e., from 0.50 males) are discussed below.

17.4.1   Differential Mortality

Leigh (1970), Trivers (1972), Polis (1980), Marshall (1981), and Duelli (1981) have dealt with this topic. If one sex leads a more hazardous life than the other, the sex ratio will change with age. Fisher (1930, 1958) noted that if sex-biased mortality occurs during the time of parental investment (as in man), more of the sex at risk should be conceived and fewer should survive the period of dependency on parents. Sex-biased mortality after parental investment has ended should have no effect on earlier sex ratios. Adult males often take more risks than females, with the result that sex ratios that started at equality become heavily female-biased as the reproductive season progresses. For example, most females of the burrowing cricket *Anurogryllus arboreus* are still alive after all males have died (Weaver & Sommers 1969, Walker 1980).

Sex-biased emigration, immigration, or diapause can affect local sex ratios in ways analogous to mortality (Johnson 1969, Denlinger 1981, Shroyer & Craig 1981).

17.4.2 Sex Chromosomes with Meiotic Drive

The topic of sex chromosomes with meiotic drive has been dealt with by several workers (Hamilton 1967, Wood & Newton 1977, Dawkins 1982). If a gene on a sex chromosome subverts meiotic segregation in its favor— that is, causes its sex chromosome to be disproportionately represented among the functioning gametes—it will increase in frequency and distort the sex ratio. A simple case is a gene on a Y chromosome that causes all, or most, functional sperm to carry it. Such a gene will rapidly increase the proportion of males until the population becomes extinct (Hamilton 1967). A "driving-Y" gene (M^D) is known in *Aedes aegypti* (Hickey & Craig 1966a,b; Wood & Newton 1977); it apparently works by causing X chromosome breakage during spermatogenesis. Some X chromosomes are resistant to M^D (progeny of M^D males bearing these are approximately 0.50 male) and some are sensitive (progeny are as much as 0.99 males). Because resistant X chromosomes are widespread in natural populations, the value of M^D for control of this mosquito is limited. Wild populations with M^D have high frequencies of resistant X chromosomes and 0.50–0.60 males; populations lacking M^D have low frequencies of resistant X chromosomes and may be temporarily suppressed by introducing M^D (Wood & Newton 1977).

Driving-X genes are known in *Drosophila* (Novitski 1947, Bryant et al. 1982). They cause female-biased sex ratios and, in the absence of opposing forces, could lead to such a shortage of males that the population would become extinct.

For both driving-Y and driving-X genes, selection at loci on other chromosomes should reduce the distortion and return the sex ratio toward 0.50 males. (In the presence of excess investment in either sex those individuals with genes causing greater investment in the neglected sex have more grandprogeny—Fisher 1930). The facts that more than 0.50 males generally reduces population increase and that less than 0.50 males, in moderation, accelerates population increase are of little consequence in explaining the fates of these variations in sex ratio.

17.4.3   Local Mate Competition (LMC)

This topic has been dealt with by Hamilton (1967, 1979), Maynard Smith (1978), Waage (1982), and Charnov (1982). For the equilibrium sex ratio to be 0.50 males, competition for mates must be populationwide. If competition is local, sons may decrease the mating success of one another; then, even if males are a minority, producing more may not pay off in more grandprogeny. The effect is clearest at the extreme: When only brother–sister matings occur, producing more sons than required to fertilize all daughters decreases the expected number of grandprogeny. Hamilton (1967) listed 25 species of insects and mites, in 16 families, that usually sibmate; the proportions of males in typical sibships were 0.02–0.33, sibships of fewer than 16 generally had a single male (11 of 13 species). So far as known, males of the species listed by Hamilton are haploid and always come from unfertilized eggs (*arrhenotoky*). The females produce the strongly female-biased broods by releasing sperm from the spermatheca on all but one or a few eggs.

Hamilton (1967) and Maynard Smith (1978) calculated what evolutionarily stable sex ratios should occur for different degrees of local mate competition. Hamilton (1979) found that fig wasps of several genera had mean sex ratios roughly correlating with their probabilities of outbreeding (as estimated from other evidence), and Waage (1982) showed similar agreement with LMC theory in an analysis of sex ratios in 31 species of scelionid wasps (Fig. 17.5).

The ability of individual females to vary sex ratios of their broods in response to stimuli that predict the degree of LMC is considered in the next section.

17.4.4   Values of the Sexes Vary with Circumstances

Maynard Smith (1978), Hamilton (1979), Suzuki and Iwasa (1980), Werren (1980a,b) and Charnov et al. (1981) have dealt with this topic. Producing a male or a female may have different payoffs to a parent under different circumstances. A parent that can vary the sex to fit the circumstance has a reproductive advantage over one that cannot.

One well-studied example concerns females of certain parasitoid Hymenoptera that lay male-producing (unfertilized) eggs on smaller host individuals and female-producing eggs on larger ones (see Charnov et al. 1981 for references). Each host produces one parasitoid, the size depending on the size of the host. The smallest functional parasitoid males are generally smaller than the smallest females, and larger females apparently have a greater increase in fertility than do larger males. Consequently, by controlling egg laying in the manner described, a female produces more grandprogeny (and perhaps more progeny, since fertilized eggs may be wasted in minimal-sized hosts—Assem 1971). The overall sex ratio among a fe-

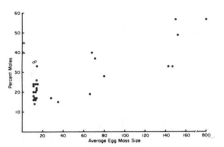

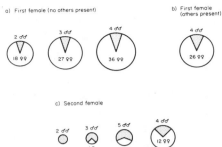

Fig. 17.5 (*Left*) Average sex ratios and average egg mass size of host for 31 species of scelionid wasps. When host egg masses are small, an ovipositing female can parasitize every egg, making it likely that her progeny will mate only with sibs (i.e., that there will be maximum local mate competition). When host egg masses are large (e.g., >50), an ovipositing wasp cannot attack every egg and leaves oviposition opportunities for other females—thereby reducing average LMC and selecting for less female-biased sex ratios. Reduced LMC is also characteristic of species that attack single eggs (△) and those that have unusually small daily fecundities (○). (Modified from Waage 1982.)

Fig. 17.6 (*Right*) Representative sibships of *Nasonia vitripennis*, a parasitoid wasp that attacks blowfly pupae and adaptively varies sex ratio (data from Werren 1980b). Each circle represents eggs laid by one female on one pupa. (See text for further explanation.)

male's progeny should still approximate 0.50 for an outbreeding population—provided an abundance of minimal-size hosts doesn't make sons cheaper to produce than daughters (see next section and Green 1982).

A second example of individual females adaptively varying their outputs of sons and daughters concerns circumstances that predict different degrees of local mate competition. The best-studied species is a pteromalid wasp that parasitizes blowfly pupae (Wylie 1966, Werren 1980a,b). If a female *Nasonia vitripennis* finds a host puparium that has not been previously parasitized, she lays 20–40 eggs on it. The females in the resulting sibship are generally mated by their flightless brothers. In keeping with LMC theory, males constitute only about 9% of each such sibship when the ovipositing female has no evidence of the presence of peers (Fig. 17.6a—Note that the sex ratio does not change with size of sibship.) However, if the female ovipositing on an unexploited host has come in contact with previously parasitized hosts or with other females (conditions predicting extra-sibship competition among males), she increases the proportion of males among her progeny ($\bar{x} = 14\%$) (Fig. 17.6b). When females oviposit on previously parasitized pupae, they lay fewer eggs ($\bar{x} = 9$) and the proportion of males ranges from 1.00 to approximately 0.25 depending on the number of eggs the second female adds and in keeping with the likely degree of LMC (Werren 1980a) (Fig. 17.6c). If the second female lays only a few eggs, she maximizes her reproduction (in terms of genes in grand-progeny) by making all eggs males and exploiting the scarcity of males in the first female's clutch; if the second female lays more eggs, she maximizes her reproduction by decreasing the proportion of sons to about 25% because LMC increasingly discounts the value of *her* sons as her progeny form an increasing proportion of the total. (The size of the second female's clutch is partly determined by the time since the first female's attack; if it is 24–48 hr or longer, the second female lays few eggs, in accord with the small share of host resources her brood is likely to garner.)

Although adjusting the proportion of males to the degree of local mate competition may have the effect of population regulation (I. Walker 1967), the details of the phenomenon do not support population regulation as a function: females increase their individual contributions to the overall gene pool by investing in higher proportions of males when populations are more dense.

Colwell (1981) and Wilson and Colwell (1981) emphasize that when two or more females contribute to a small mating group, female-biased sex ratios are selected because of the enhanced reproduction of the group (relative to groups with sex ratios closer to equality) and that within such a group, females having sex ratios closer to equality outreproduce those that make the group successful. In essence a female is confronted with winning within a group at the cost of losing populationwide or losing within a group but winning populationwide. Females that win populationwide are prevalent.

17.4.5   Sexes Unequal in Cost to Parent

Unequal sex-specific costs to the parent have also received attention (Fisher 1930, 1958, Trivers & Hare 1976, Maynard Smith 1978). Sons and daughters cost equally in most insects but not in sexually dimorphic Hymenoptera that progressively provision their larvae or provide sufficient stores at the time the egg is laid. In such cases the larger morph, most often female, receives correspondingly more provisions. As examples, cicada killers (*Sphecius speciosus*) usually give one cicada to their male progeny and two to female progeny (Lin 1979), while a solitary vespid wasp (*Euodynerus foraminatus*) provisions sons with about 130 mg of caterpillars and daughters with 204 mg (Cowan 1979). If such complications as LMC and uncertainties in assessing costs of the sexes are ignored, the predicted sex ratio should be the inverse of the cost ratio—a species in which females cost twice as much as males should manifest twice as many males as females. Trivers and Hare (1976) reported male-biased sex ratios (at emergence) for 11 of 13 species of solitary bees and wasps having females heavier than males. However, the degree of bias showed no sensitivity to the degree of weight discrepancy. Unequal costs are common among eusocial insects, but here sex ratios are further complicated by rearing being the function of workers with genetic interests different from their parents and one another (Trivers & Hare 1976, Alexander & Sherman 1977, Noonan 1978, Macevicz 1979, Metcalf 1980).

17.4.6   Other Eventualities

Other eventualities bear on evolution of sex ratio. For example, extra-chromosomal factors, including microorganisms in the cytoplasm, can cause females to produce all-female or all-male progenies (Poulsen & Sakaguchi 1961, Johnson 1977, Werren et al. 1981, Skinner 1982). Some "species" with female-biased sex ratios are actually two or more genetically isolated lineages: a bisexual line, presumably with a 50:50 sex ratio, and one or more all-female lines in which females mate with males of the bisexual line but produce only daughters genetically like themselves—the male genome is not incorporated in the zygote (e.g., bark beetle, Bakke 1968; geometrid moth, Mitter & Futuyma 1977). More in keeping with the subject of population regulation is *autoparasitism*, a phenomenon known in certain aphelinid wasps in which males develop as hyperparasites of conspecific females (Fig. 17.7) (Flanders 1967, 1969, Williams 1977). Data on autoparasitism are insufficient to guide speculation—for instance, it is unknown whether females, in nature, ever parasitize their own daughters. Colgan and Taylor (1981) proposed a model that presumes they may. R. I. Sailer (pers. comm.) showed that autoparasitism in *Encarsia lahorensis*, a citrus whitefly parasite recently introduced to Florida, has had the effect of population regulation. When parasitism of the whitefly hosts reached high levels, ovipositing females killed increasing proportions of the females of the next generation as they produced sons—the only, or cheapest, progeny they could produce under the circumstances. Consequently, the next generation of parasites had less potential for further reducing its host population. *Physcus seminotus* on Mauritius (Williams 1977) illustrates the complexities associated with autoparasitism (see Fig. 17.7g): Prior to 1969 (Fig. 17.7g, black arrows and bold numbers), the sugar cane scale insect (*Aulacapsis tegalensis*) was parasitized by an encyrtid wasp (*Adelencyrtus miyarai*) that was parasitized by a eulophid wasp (*Tetrastichus* sp.). Emergents from parasitized scales in 1968–69 were 33.5% *A. miyarai*, 65% *T.* sp., and 1.5% "other." The introduction of the aphelinid, *Physcus seminotus*, in which the female is a primary parasite of the scale insect and the male develops as a hyperparasite of any parasite in a scale (including a conspecific male!) complicated the trophic relations and changed the proportions of emergents, in 1975, to those in parentheses (Fig. 17.7g). Why are there so few males of *P. seminotus*? What determines population density of *P. seminotus*?

Much is yet to be learned of insect sex ratios and their effects on population growth and stability. Improved understanding will have important applications in parasite rearing and biological control.

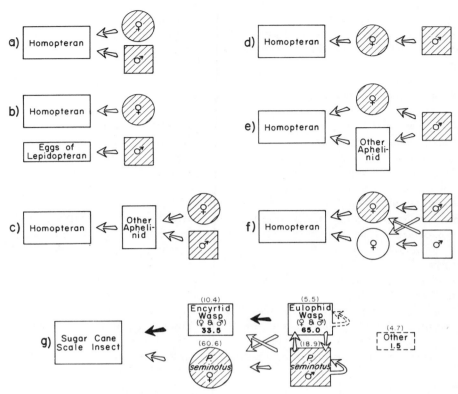

Fig. 17.7   Trophic relations of aphelinid wasps, including examples of autoparasitism (d-g). Cross-hatched squares and circles represent the males and females of the subject species. (a) primary parasites; female endoparasitic, male ectoparasitic; (b) primary endoparasites, sexes on different hosts; (c) hyperparasites; (d) obligatory autoparasitism (male develops as parasite of conspecific female); (e) facultative autoparasitism; (f) reciprocal facultative autoparasitism. Further complexities include females that develop parthenogenetically and females that lay male-producing eggs on healthy hosts with the eggs hatching only after another parasite has used the host to develop. (a–f from descriptions in Flanders 1967, 1969, no specific examples or data given for some categories); (g) *Physcus seminotus* on Mauritius (Williams 1977). (See text for explanation.)

17.5   DEATH AND SENESCENCE

Emerson (1960, p. 325) wrote that "among the attributes of the individual organism that would seem to be explicable only through the selection of whole population systems is the intrinsic limitation of the life-span and the incorporation of innate death mechanisms," and (p. 343) that "the innate regulation of numbers in unitary populations involves the evolution of death mechanisms in individuals." But Medawar (1952, 1957) and Williams (1957) had already shown that senescence should be expected as a side effect of individual selection. Williams's theory, later refined by Hamilton (1966) and Emlen (1970), depends on selection diminishing in force beginning with the age of first reproduction. Even without senescence, residual expected reproduction of each successive age is less and less because mortality from external causes continues.* Since the cumulative remaining reproduction diminishes with adult age, an allele that increases survival in young adults but reduces it in old adults will generally cause an overall increase in reproduction and therefore be increased by natural selection. Any modifying gene that postpones detrimental effects of other genes (i.e., to ages with less residual expected reproduction) will likewise increase in frequency over alleles that allow an earlier adult age of onset—but not over those that prevent the bad effect altogether: Selection indirectly causes senescence but acts directly against it, though less and less effectively at ages with lower and lower residual reproductive probability. Experiments using *Drosophila* and *Tribolium* provide support for Williams's theory (e.g., Rose & Charlesworth 1980).

Senescence need not be regarded as an adaptation at any level; it can instead be viewed as "a group of adaptively unfavorable morphogenetic changes that were brought in as side effects of otherwise favorable genes and which have only been partly expurgated by further selection" (Williams 1957, p. 402).

17.6 DENSITY-RELATED GENETIC CHANGES

Different alleles may be favored at high and low population densities, or in a population that is rapidly expanding versus one that is constant or declining. For example, optimal mate finding or predator-deterring behaviors may change with density, and different fertility schedules maximize reproduction in increasing and decreasing populations. To what extent such changes stabilize populations is by no means clear.

The common denominator of density-related genetic changes should be improved individual adaptation to the new density or rate or direction of population change. The specific adaptations need have nothing in common as to how they affect future density or change. For example, assume that a population has reached a higher density than it has had in the past. One adaptation that might result is aggressive defense of resources needed for reproduction (see Table 17.1; also Chitty 1967). This could reduce population density at the same time it increased natality. Another adaptation that might result is reduced minimal size for adults—permitting under-

---

*"Residual expected reproduction of each successive age" is from the vantage point of the zygote, *not* from the vantage point of an individual that has attained the specified age. Residual expected reproduction declines from the age of first reproduction and never increases [see Williams (1957) and Hamilton (1966) for fuller explanation]; it is not the same as Fisher's (1930) reproductive value.

nourished individuals some chance of reproduction. This could result in higher numbers of adults but might have no effect on natality or even decrease it. Further complexities could be introduced—such as time lags and frequency-dependent selection—but it should already be evident that density-related genetic changes do not reliably lead to population regulation.

Models relating density change and genetic change include those by Ford (1964), Chitty (1967), Carson (1968), Ayala (1968), and Tamarin (1978).

17.7 COEVOLUTION AND GENETIC FEEDBACK

Except in the most physically stressed, species-poor ecosystems evolution in any one species influences evolution in many others. This occurs because each species is important to the reproductive success of individuals of numerous other species—as food, enemies, allies, competitors, shelter, transportation, Batesian models, or any of a myriad other ways. With any two-species interaction, a change in one species will change the effective environment of the other. Numerous aspects of coevolution are discussed in Chapters 20–22. Here the discussion will center on David Pimentel's proposal (1961, 1968) that populations of herbivores, parasites, and predators are sometimes regulated by a "genetic feedback mechanism." Pimentel reasoned that a host (or prey) population would change genetically in response to the feeding pressure of a herbivore (or parasite, or predator) population and become more resistant to attack—thereby decreasing the reproduction and population level of the herbivore. With lower herbivore densities, the host population might lose some of its resistance and the herbivore population might regain some of its former numbers. "After many such cycles, the numbers of the herbivore populations are ultimately limited, and stability results" (Pimentel 1968:1433).

That there are reciprocal evolutionary changes between herbivore and host or predator and prey conforms to observations and to theory, but numerical stability is not a necessary consequence. Species that have evolved together exist at all population levels, and species that were at one level in the past are at other levels today. Genetic feedback has no "setpoint." It can result in abundance, rarity, or extinction. In fact Van Valen (1973, "Red Queen's Hypothesis"[3]) proposed that evolutionary interactions generate instabilities that are the principal cause of extinctions. He noted that the effective environment of every species is continually deteriorating because of evolution in other species—the herbivore that becomes better at finding its food reduces the fitness of its host; the plant that becomes more difficult to digest reduces the fitness of its herbivores, and so on. Only those species that change survive; yet changes evoke changes in other spe-

---

*"Now, here, you see, it takes all the running you can do, to keep in the same place." (L. Carroll, *Through the Looking Glass*.)

cies, diminishing or reversing the benefits. The ultimate result is extinction at a stochastically constant rate.

Genetic feedback may influence population numbers, but there is no evidence of it stabilizing them other than as an incidental effect of evolution in other contexts. [See Huffaker et al. (1976) for a critique of the role of genetic feedback in population regulation.]

17.8 FINAL REMARKS

As stated at the outset, whether self-regulation occurs in insect populations depends on how *self-regulation* is defined. Phenomena that regulate insect population density include many that are poorly understood, but none so far have been demonstrated to evolve through group selection opposing selection at lower levels. Since *self-regulation* has often been used with precisely this group selection connotation, the expression had best be avoided and unsullied ones substituted: for example, *regulation; population stabilizing effects.*

ACKNOWLEDGMENTS

I thank Jane Brockmann, Ted Burk, E. L. Charnov, Pat Greany, R. G. Harrison, J. E. Lloyd, and John Sivinsky for generous and important aid at one or more stages in the evolution of this chapter.

REFERENCES

Alcock, J. 1979. Pp. 381–402 in Blum and Blum (Eds.), 1979, referenced here.

Alcock, J. 1980. *Am. Sci.* **68:** 146–153.

Alcock, J., E. M. Barrows, G. Gordh, L. J. Hubbard, L. Kirkendall, D. W. Pyle, T. L. Ponder, and F. G. Zalom. 1978. *Zool. J. Linn. Soc., Lond.* **64:** 293–326.

Alexander, R. D. and G. Borgia. 1978. *Annu. Rev. Ecol. Syst.* **9:** 449–474.

Alexander, R. D. and P. W. Sherman. 1977. *Science* **196:** 494–500.

Andersen, F. S. 1961. *Oikos* **12:** 1–16.

Askew, R. R. 1971. *Parasitic Insects.* Elsevier, New York.

Assem, J. van den. 1971. *Neth. J. Zool.* **21:** 373–402.

Ayala, F. J. 1968. *Science* **162:** 1453–1459.

Baker, R. R. 1972. *J. Animi. Ecol.* **41:** 453–469.

Bakke, A. 1968. *Can. Entomol.* **100:** 640–648.

Barber, G. W. 1936. USDA Tech. Bull. 499.

Bell, W. J., S. Robinson, M. K. Tourtellot, and M. D. Breed. 1978. *Z. Tierpsychol.* **48:** 203–218.

Bick, G. H. and J. C. Bick. 1965. *Ecology* **46:** 461–472.

Bitzer, R. J. and K. C. Shaw. 1980. *J. Res. Lepidoptera* **18:** 36–49.

Blais, J. R. 1953. *Can. Entomol.* **93:** 446–448.

Blum, M. S. and N. A. Blum (Eds.). 1979. *Sexual Selection and Reproductive Competition in Insects.* Academic, New York.

Bodmer, W. F. and A. W. F. Edwards. 1960. *Hum. Genet.* **24:** 239–244.

Bradbury, J. W. 1981. Pp. 138–169 in R. D. Alexander and D. W. Tinkle (Eds.), *Natural Selection and Social Behavior.* Chiron, New York.

Brockmann, H. J. and R. Dawkins. 1979. *Behaviour* **71:** 203–245.

Brower, L. P. 1961. *Physiol. Zool.* **34:** 287–296.

Bryant, S. H., A. T. Beckenbach, and G. A. Cobbs. 1982. *Evolution* **36:** 27–34.

Burk, T. 1979. *An Analysis of Social Behavior in Crickets.* Ph.D. Thesis, Oxford University, Oxford. (Univ. Microfilms International, Ann Arbor, Mich.)

Burk, Theodore. 1983. *Fla. Entomol.* 66: 330–344.

Bush, G. L. 1969. *Evolution* **23:** 237–251.

Cade, W. H. 1979. Pp. 343–379 in Blum and Blum (Eds.), 1979, referenced here.

Campanella, P. J. and L. L. Wolf. 1974. *Behaviour* **51:** 49–87.

Capinera, J. L. and P. Barbosa. 1976. *Oecologia* **26:** 53–64.

Carson, H. L. 1968. Pp. 123–137 in R. C. Lewontin (Ed.), *Population Biology and Evolution.* Syracuse University Press, Syracuse, N.Y.

Charnov, E. L. 1982. *The Theory of Sex Allocation.* Princeton University Press, Princeton, N.J.

Charnov, E. L., R. L. Los-DenHartogh, W. T. Jones, and J. van den Assem. 1981. *Nature* **289:** 27–33.

Chitty, Dennis, 1960. *Can. J. Zool.* **38:** 99–113.

Chitty, Dennis. 1967. *Proc. Ecol. Soc. Aust.* **2:** 51–78.

Cirio, U. 1971. *Redia* **52:** 577–600.

Colgan, P. and P. Taylor. 1981. *Am. Natur.* **117:** 564–566.

Colwell, R. K. 1981. *Nature* **290:** 401–404.

Corbet, P. S. and A. Griffiths. 1963. *Proc. Roy. Entomol. Soc. A.* **38:** 125–135.

Cowan, D. P. 1979. *Science* **205:** 1403–1405.

Crombie, A. C. 1944. *J. Exp. Biol.* **20:** 135–151.

Davies, N. B. 1978a. *Anim. Behav.* **26:** 138–147.

Davies, N. B. 1978b. Pp. 317–350 in J. R. Krebs and N. B. Davies (Eds.), *Behavioural Ecology, an Evolutionary Approach.* Sinauer, Sunderland, Mass.

Dawkins, R. 1980. Pp. 331–367 in G. W. Barlow and J. Silverburg (Eds.), *Sociobiology: Beyond Nature/Nurture.* Westview, Boulder, Colo.

Dawkins, R. 1982. *The Extended Phenotype: The Gene as the Unit of Selection.* W. H. Freeman, San Francisco.

DeBach, P. 1944. *Ann. Entomol. Soc. Am.* **37:** 70–74.

Denlinger, D. L. 1981. *Evolution* **35:** 1247–1248.

Duelli, P. 1981. *Res. Popul. Ecol.* **23:** 193–209.

Eberhard, W. G. 1979. Pp. 231–258 in Blum and Blum (Eds.), 1979, referenced here.

Edmunds, M. 1976. *Zool. J. Linn. Soc.* **58:** 1–37.

Eickwort, K. R. 1973. *Am. Natur.* **107:** 452, 453.

Emerson, A. E. 1960. Pp. 307–348 in S. Tax (Ed.), *Evolution after Darwin,* Vol. 1. University of Chicago Press, Chicago.

Emlen, J. M. 1970. *Ecology* **51:** 588–601.

Feaver, M. 1983. Pp. 205–239 in D. T. Gwynne and G. K. Morris (Eds.), *Orthopteran Mating Systems: Sexual Competition in a Diverse Group of Insects.* Westview, Boulder, Colo.

Fisher, R. A. 1930. *The Genetical Theory of Natural Selection.* Oxford University Press, Oxford.

Fisher, R. A. 1958. *The Genetical Theory of Natural Selection* (rev. ed.). Dover, New York.

Fisher, R. A. and D. O. Greenbank. 1979. Pp. 220–229 in R. L. Raab and G. G. Kennedy (Eds.), *Movement of Highly Mobile Insects: Concepts and Methodology in Research.* N. C. State Univ., Raleigh.

Flanders, S. E. 1967. *Entomophaga* **12:** 415–427.

Flanders, S. E. 1969. *Entomophaga* **14:** 335–346.

Ford, E. B. 1964. *Ecological Genetics.* Methuen, London.

Fox, L. R. 1975a. *Annu. Rev. Ecol. & Syst.* **6:** 87–106.

Fox, L. R. 1975b. *Ecology* **56:** 933–941.

Gautier, J. Y. 1978. *Insectes Sociaux* **25:** 289–301.

Ghiselin, M. T. 1974. *The Economy of Nature and the Evolution of Sex.* University of California Press, Berkeley.

Giesel, J. T. 1972. *Am. Natur.* **106:** 380–387.

Gillett, J. D., E. A. Roman, and V. Phillips. 1977. *Proc. Roy. Soc. Lond. B.* **196:** 223–232.

Gilpin, M. E. 1975. *Group Selection in Predator-Prey Communities.* Princeton University Press, Princeton.

Greany, P. D. and E. R. Oatman. 1972. *Ann. Entomol. Soc. Am.* **65:** 377–383.

Green, R. F. 1982. *J. Theor. Biol.* **95:** 43–48.

Guillot, F. S. and S. B. Vinson. 1972. *Nature* **235:** 169, 170.

Gwynne, D. T. 1980. *Behav. Ecol. Sociobiol.* **7:** 213–225.

Hamilton, W. D. 1966. *J. Theoret. Biol.* **12:** 12–45.

Hamilton, W. D. 1967. *Science* **156:** 477–488.

Hamilton, W. D. 1979. Pp. 167–220 in Blum and Blum (Eds.), 1979, referenced here.

Hamilton, W. D. and R. M. May. 1977. *Nature* **269:** 578–581.

Harrison, R. G. 1980. *Annu. Rev. Ecol. Syst.* **11:** 95–118.

Hickey, W. A. and G. B. Craig, Jr. 1966a. *Genetics* **53:** 1177–1196.

Hickey, W. A. and G. B. Craig, Jr. 1966b. *Can. J. Genet. Cytol.* **8:** 260–278.

Hokyo, N. and E. Kuno. 1977. *Res. Popul. Ecol.* **19:** 107–124.

Huffaker, C. B., F. J. Simmonds, and J. E. Laing. 1976. Pp. 41–78 in C. B. Huffaker and P. S. Messenger (Eds.), *Theory and Practice of Biological Control.* Academic, New York.

Jackson, D. J. 1966. *Trans. Roy. Entomol. Soc. Lond.* **118:** 23–49.

Johnson, C. 1977. *Evolution* **31:** 603–610.

Johnson, C. G. 1969. *Migration and Dispersal of Insects by Flight.* Methuen, London.

King, C. E. and P. S. Dawson. 1972. *Evol. Biol.* **5:** 133–227.

King, P. E., R. R. Askew, and C. Sanger. 1969. *Proc. Roy. Entomol. Soc. Lond.* **44:** 85–90.

Kolman, W. A. 1960. *Am. Natur.* **94:** 373–377.

Landahl, J. T. and R. B. Root. 1969. *Ecology.* **50:** 734–737.

Lees, A. D. 1961. *Roy. Entomol. Soc. Lond. Symp.* **1:** 68–79.

Lees, A. D. 1967. *J. Insect Physiol.* **13:** 289–318.

Leigh. E. G. Jr. 1970. *Am. Natur.* **104:** 205–210.

Lidicker, W. Z. Jr. 1962. *Am. Natur.* **96:** 29–33.

Lin, N. 1963. *Behaviour* **20:** 115–133.

Lin, N. 1979. *Proc. Entomol. Soc. Wash.* **81:** 269–275.

McClure, M. S. 1976. *Biotropica* **8:** 179–183.

Macevicz, S. 1979. *Am. Natur.* **113:** 363–371.

Marshall, A. G. 1981. *Ecol. Entomol.* **6:** 155–174.

Maynard Smith, J. 1964. *Nature* **201:** 1145–1147.

Maynard Smith, J. 1976. *Q. Rev. Biol.* **51:** 277–283.

Maynard Smith, J. 1978. *The Evolution of Sex.* Cambridge University Press, London.

Medawar, P. B. 1952. *An Unsolved Problem of Biology.* H. K. Lewis, London.

Medawar, P. B. 1957. *The Uniqueness of the Individual.* Basic Books, New York.

Metcalf, R. A. 1980. *Am. Natur.* **116:** 642–654.

Milne, L. J. and M. Milne. 1976. *Sci. Am.* **235:** 84–89.

Mitchell, P. L. 1980. *Ann. Entomol. Soc. Am.* **73:** 404–408.

Mitchell, R. 1970. *Am. Natur.* **104:** 425–431.

Mitter, C. and D. Futuyma. 1977. *Entomol. Exp. & Appl.* **21:** 192–198.

Moore, N. W. 1964. *J. Anim. Ecol.* **33:** 49–71.

Naylor, A. J. 1965. *Ecology* **46:** 341–343.

Nicolson, A. J. 1954. *Aust. J. Zool.* **2:** 9–65.

Noonan, K. M. 1978. *Science* **199:** 1345–1356.

Novitski, E. 1947. *Genetics* **32:** 526–534.

Oshima, K, H. Honda, and I. Yamamoto. 1973. *Agric. Biol. Chem.* **37:** 2679, 2680.

Otte, D. and A. Joern. 1975. *J. Anim. Ecol.* **44**(1): 29–54.

Palmer, T. J. 1978. *Nature* **274:** 583, 584.

Pimentel, D. 1961. *Am. Natur.* **95:** 65–79.

Pimentel, D. 1968. *Science* **159:** 1432–1437.

Polis, G. A. 1980. *Behav. Ecol. Sociobiol.* **7:** 25–35.

Polis, G. A. 1981. *Annu. Rev. Ecol. Syst.* **12:** 225–251.

Poulson, D. F. and B. Sakaguchi. 1961. *Science* **133:** 1489, 1490.

Price, P. W. 1970. *Science* **170:** 546, 547.

Price, P. W. 1975. *Insect Ecology.* Wiley, New York.

Prokopy, R. J. 1981. Pp. 181–213 in D. Nordlund, R. L. Jones, and W. J. Lewis (Eds.), *Semiochemicals: Their Role in Pest Control.* Wiley, New York.

Prokopy, R. J., P. D. Greany, and D. L. Chambers. 1977. *Environ. Entomol.* **6:** 463–465.

Prokopy, R. J. and R. P. Webster. 1978. *J. Chem. Ecol.* **4:** 481–494.

Rausher, M. D. and N. L. Fowler. 1979. *Biotropica* **11:** 96–100.

Ringo, J. M. 1976. *Ann. Entomol. Soc. Am.* **69:** 209–214.

Roff, D. A. 1975. *Oecologia* **19:** 217–237.

Rose, M. and B. Charlesworth. 1980. *Nature* **287:** 141, 142.

Rothschild, M. and L. M. Schoonhoven. 1977. *Nature* **266:** 352–355.

Rudinsky, J. A., P. T. Oester, and L. C. Ryker. 1978. *Ann. Entomol. Soc. Am.* **71:** 317–321.

Schowalter, T. D. and W. G. Whitford. 1979. *Am. Midland Natur.* **102:** 182–184.

Schuster, J. C. 1975. *Comparative Behavior, Acoustical Signals and Ecology of New World Passalidae (Coleoptera).* Ph.D. thesis, Univ. Fla. Gainseville. (Univ. Microfilms International, Ann Arbor, Michigan.)

Shaw, M. J. P. 1970a. *Ann. Appl. Biol.* **65:** 191–196.

Shaw, M. J. P. 1970b. *Ann. Appl. Biol.* **65:** 197–203.

Shaw, M. J. P. 1973. *Ann. Appl. Biol.* **74:** 1–7.

Shaw, R. F. and J. D. Mohler. 1953. *Am. Natur.* **87:** 337–342.

Shroyer, D. A. and G. B. Craig, Jr. 1981. *Environ. Entomol.* **10:** 147–152.

Skinner, S. W. 1982. *Science* **215:** 1133, 1134.

Solbreck, C. 1978. Pp. 195–217 in H. Dingle (Ed.), *Evolution of Insect Migration and Diapause.* Springer-Verlag, New York.

Southwood, T. R. E. 1962. *Biol. Rev.* **37:** 171–214.

Spieth, H. T. 1978. *Evolution* **32:** 435–451.

Springett, B. P. 1968. *J. Anim. Ecol.* **37:** 417–424.

Stearns, S. C. 1976. *Q. Rev. Biol.* **51:** 3–47.

Steiner, A. L. 1975. *Quaest. Entomol.* **11:** 113–127.

Stinner, R. E., R. L. Rabb, and J. R. Bradley, Jr. 1977. *Proc. 15th Int. Congr. Entomol., 1976,* pp. 622–642.

Sutherland, O. R. W. 1969. *J. Insect Physiol.* **15:** 2179–2201.

Suzuki, Y. and Y. Iwasa. 1980. *Res. Popul. Ecol.* **22:** 366–382.

Tamarin, R. H. 1978. *Am. Natur.* **112:** 545, 555.

Taylor, R. A. J. and L. R. Taylor. 1979. Pp. 1–27 in R. M. Anderson, B. D. Turner, and L. R. Taylor (Eds.), *Population Dynamics.* Blackwell, London.

Trivers, R. L. 1972. Pp. 136–179 in B. Campbell (Ed.), *Sexual Selection and the Descent of Man, 1871–1971.* Aldine, Chicago.

Trivers, R. L. 1974. *Am. Zool.* **14:** 249–264.

Trivers, R. L. and H. Hare. 1976. *Science* **191:** 249–263.

Tschinkel, W. R. 1978. *Physiol. Zool.* **51:** 300–313.

Tschinkel, W. R. 1981. *Anim. Behav.* **29:** 990–996.

Tychsen, P. H. 1977. *J. Aust. Entomol. Soc.* **16:** 459–465.

Uéda, T. 1979. *Res. Popul. Ecol.* **21:** 135–152.

Urquhart, F. A. and N. R. Urquhart. 1978. *Can. J. Zool.* **56:** 1756–1764.

van Lenteren, J. C. 1976. *Neth. J. Zool.* **26:** 1–83.

Van Valen, L. 1973. *Evol. Theory* **1:** 1–30.

Vinson, S. B. 1976. *Annu. Rev. Entomol.* **21:** 109–133.

Waage, J. K. 1982. *Ecol. Entomol.* **7:** 103–112.

Wade. M. J. 1978. *Q. Rev. Biol.* **53:** 101–114.

Walker, I. 1967. *Ecology* **48:** 294–301.

Walker, T. J. 1980. *Evol. Biol.* **13:** 219–260.

Weaver, J. E. and R. A. Sommers. 1969. *Annu. Entomol. Soc. Am.* **62:** 337–342.

Wellington, W. G. 1977. *Environ. Entomol.* **6:** 1–8.

Werren, J. H. 1980a. *Science* **208:** 1157–1159.

Werren, J. H. 1980b. *Studies in the Evolution of Sex Ratios.* Ph.D. thesis, University of Utah, Salt Lake City. (Univ. Microfilms International, Ann Arbor, Mich.)

Werren, J. H., S. W. Skinner, and E. L. Charnov. 1981. *Nature* **293:** 467, 468.

Wheeler, W. M. 1930. *Demons of the Dust, a Study in Insect Behavior.* W. W. Norton, New York.

Whitham, T. G. 1979. *Nature* **279:** 324, 325.

Williams, G. C. 1957. *Evolution* **11:** 398–411.

Williams, G. C. 1966. *Adaptation and Natural Selection. A Critique of Some Current Evolutionary Thought.* Princeton University Press, Princeton, N.J.

Williams, G. C. (Ed.). 1971. *Group Selection.* Aldine-Atherton, Chicago.

Williams, G. C. 1975. *Sex and Evolution.* Princeton University Press, Princeton, N.J.

Williams, J. R. 1977. *Entomophaga* **22:** 345–350.

Wilson, D. S. 1975. *Proc. Nat. Acad. Sci.* **72:** 143–146.

Wilson, D. S. 1980. *The Natural Selection of Populations and Communities.* Benjamin Cummings, Menlo Park, Calif.

Wilson, D. S. and R. K. Colwell. 1981. *Evolution* **35:** 882–897.

Wilson, E. O. 1975. *Sociobiology, the New Synthesis.* Harvard University Press, Cambridge, Mass.

Wilson, F. 1961. *Aust. J. Zool.* **9:** 739–751.

Wood, R. J. and M. E. Newton. 1977. *Proc. 15th Int. Congr. Entomol., 1976,* pp. 97–105.

Wylie, H. G. 1966. *Can. Entomol.* **98:** 645–653.

Wynne-Edwards, V. C. 1962. *Animal Dispersion in Relation to Social Behaviour.* Oliver and Boyd, Edinburgh.

Chapter 18

Population Regulation in Social Insects

ROBERT L. JEANNE and
DIANE W. DAVIDSON

True social behavior, or eusociality, has evolved at least 13 times in the insect orders Isoptera and Hymenoptera. Each time the adaptation has enabled a lineage to exploit a share of resources that is large in proportion to the number of species represented. The ecological success of the more highly social species is truly remarkable and in no small way reflects convergence of the social insect colony on the biomass, life history, and homeostatic characteristics of vertebrate organisms. Yet by virtue of such typically insect traits as small body size and ectothermic metabolism and special adaptations related to sociality (e.g., caste and coordinated behavior), social insects are more energy-efficient consumers than vertebrates. Social insects thus combine some of the best features of both insects and terrestrial vertebrates. The evolution of social insects has also created new ecological niches, many of which have been filled by other social insects, including social parasites and such specialized predators as army ants, ponerine ants, and giant hornets.

In this chapter we discuss population regulation in these unique and important insects. We begin by documenting their ecological success and impact and then examine aspects of sociality that have promoted this success. In a third section we review the factors that impose limits to the density, distribution, and diversity of social insect populations. We conclude by drawing comparisons among the different groups of social species and noting some general ecological differences between social insects and their solitary relatives.

18.1 THE ECOLOGICAL SUCCESS OF THE SOCIAL INSECTS

Social insects exploit all major food resource categories. Termites, the oldest social group, are primitively decomposers, subsisting on dead and decaying wood, leaves, grasses, and organic matter in humus and dung. Many of the 2000–3000 described species (Howse 1970) are primary consumers of living woody tissue or live grasses and herbs. The ants are the largest group of social insects and are most diverse in their trophic habits. Extant species probably number more than 20,000, although only 11,000–12,000 species have so far been described (W. L. Brown, Jr. pers. comm.). Primitively, ants were predators of other arthropods, but most living species are omnivores, supplementing predation and/or scavenging with energy-rich plant materials, including sap, nectar, fruits, and seeds. The evolution of fungus cultivation in the attine ants in the New World (Weber 1966) and in the macrotermitine termites in Africa (Sands 1969) has enabled these advanced eusocial species to exploit leaves, the last major food resource not previously available to social insects. There are thought to be approximately 1600 species of social bees (C. D. Michener pers. comm.). (This estimate is very tentative because solitary or social status of the halictid bees cannot be determined morphologically.) Bees feed strictly as primary consumers, specializing on nectar and pollen. While the 800–900 known species of social wasps (Jeanne 1980) are primarily predators of other arthropods, they also consume nectar, honeydew, and sap, and some species opportunistically use vertebrate carrion. All four taxa of social insects also include social parasites of other social species. As a group, the social insects feed on energy-rich food sources; the exceptions enter into symbiotic relationships with flagellates (termites), bacteria (termites), or fungi (termites and ants), which break down cellulose into products the insects can digest.

The ecological importance of social insects is greater than their relatively low species richness might suggest. Although the 15,400–17,500 known species comprise only about 1.5% of insect species, in numerical abundance, biomass, and impact on their ecosystems, the social insects are the dominant arthropods of many terrestrial habitats.

Termites are conspicuous residents of tropical savannahs, and it is here that their numbers are best documented. Tree savannahs of northern Australia contain an average of 217 occupied mounds of *Amitermes laurensis*/ha (Wood & Lee 1971). In the Ivory Coast 1 hectare of typical Lamto savanna supports an estimated 7.5–10 million individual termites (Baroni-Urbani et al. 1978), and recent surveys place the density of humivorous termites in open savannah at 20 million/ha (D. Corveaule, cited by Baroni-Urbani et al. 1978). Wood and Sands (1978) conclude that the upper limit of termite population and biomass in any ecosystem is approximately 15,000 individuals/m^2. At these densities the biomass of termites (approximately 50 g/m^2) would equal or exceed that of grazing mammals (Lee & Wood 1971, Wood & Sands 1978).

High rates of food consumption and extraordinary assimilation efficiency make termites important agents of energy transfer in ecosystems. Johnson and Whitford (1975) estimated that subterranean termites consume fully 50% of the net production in Chihuahuan Desert ecosystems. These estimates are based on termite activity at baits, but they compare favorably with results of Bodine and Ueckert (1975), who measured increases in standing crop of grass and litter accumulation in response to termite removal experiments in a semiarid west Texas grassland. Relative to control plots, grass production increased by 22% and litter by 50% by the end of the second growing season after termite removal. The largest published estimate of consumption rate is for the humus-feeding *Cubitermes* and is a startling 11,300 g/$m^2 \cdot$ yr (Hébrant 1970). Among soil-dwelling organisms, termites are unique for the efficiency at energy processing (reviewed in Wood 1978). Symbiotic associations with flagellated protozoans (lower termites) and bacteria (higher termites) allow extremely efficient breakdown of polysaccharides, and assimilation efficiencies range from 54–93%.

Ants are also dominant members of many terrestrial ecosystems, especially in the tropics and subtropics. In Ivory Coast savannah densities of ant colonies may reach 7000/ha (Levieux 1967). Hocking (1970) calculated that the three most common species of arboreal ants in East Africa together have an average density of 29.2×10^6 individuals/ha. In the lowland seasonal tropics of Costa Rica ants comprised 5.7–24.4% of the individual arthropods in sweep samples of low vegetation (Janzen 1973). Ant population densities can also be impressive in temperate regions. Waloff and Blackith (1962) estimated $6–15 \times 10^3$ ants/m² or 15 g/m² for a single species (*Lasius flavus*) in English grassland.

By virtue of both the magnitude and selectivity of their foraging, ants have a considerable impact on the ecosystems they inhabit. In some habitats energy flow is greater through ant populations than through populations of endothermic vertebrates (Golly & Gentry 1964). *Atta sexdens* with a density of 10–18 colonies/ha may consume 52.5 kg/ha of plants daily, an amount sufficient to feed three calves (Amante 1967). *Atta* spp. probably reduce the reproductive fitness of preferred trees in neotropical forests (Rockwood 1973). Wood ants (*Formica*) significantly depress populations of herbivorous lepidopteran larvae and other forest pests (Stradling 1978). Ants are prominent seed consumers in arid environments, where they may harvest a significant fraction of seed production and alter the composition of the plant community by selective foraging (Whitford 1978, Inouye et al. 1980). By forming mutualistic associations with certain plants, ants may also favor plant species in some habitats by protecting their foliage (Bentley 1976) or dispersing their diaspores (Davidson and Morton 1981) in exchange for nest sites or food rewards provided by the plants.

The production ecology of bees and wasps has not been studied as intensively as that of ants and termites, but there is every indication that they too have achieved a measure of ecological dominance over their solitary and presocial relatives. As the primary pollinators of many groups of angiosperms, bees play a major role in most terrestrial ecosystems. Although the solitary bee species vastly outnumber the social species, the latter clearly dominate nectar- and pollen-foraging communities in many habitats. In one dry forest in Costa Rica, for example, stingless bees (Meliponinae) accounted for almost 30% of the flower visitors (Heithaus & Opler pers. comm. to Johnson & Hubbell 1974). In the same region bees in the family Apidae, which includes the most highly eusocial bees, constituted 8% of the species and 48% of the individual bees and wasps visiting flowers (Heithaus 1979).

Compared to the other three groups of social insects, the social wasps probably have the least ecological impact, yet they too are successful relative to solitary, flying hymenopterans. Social species comprised 26% of the wasp visitors to flowers in Costa Rican dry forest (Heithaus 1979). Colonies of certain species may be locally abundant. In one woodlot in Delaware the density of yellowjacket (*Vespula maculifrons*) colonies averaged 7.6/ha in 1977 (Lord & Roth 1979).

The biomass of social insects is higher than numbers of species would predict. This suggests that social insects have succeeded in diverting a disproportionately large share of resources to their use. Furthermore, since the mean production efficiency [net production ÷ (net production + metabolic heat loss)] of social insects is low and more comparable to that of fish and certain higher vertebrates than to nonsocial insects (Humphreys 1979), a relatively high proportion of food intake goes into maintenance. Thus, for a given level of insect productivity social insects have a greater impact on their food resources than do solitary insects of the same trophic level.

18.2 SPECIAL FEATURES OF SOCIAL INSECTS RESPONSIBLE FOR THEIR SUCCESS

In this section we attempt to account for the ecological success of social insects. We consider various adaptations that often parallel the evolution of sociality and examine their consequences for population regulation.

The three criteria that define eusociality are overlap of generations, cooperation in brood care, and reproductive division of labor (Wilson 1971). While these adaptations themselves may have conferred added fitness on the evolving eusocial species, other traits promoting the success of these insects arose only with relatively advanced levels of eusociality. Here we deal principally with these latter characteristics and refer the reader to other authorities (Wilson 1971, 1975, Lin & Michener 1972, Michener & Brothers 1974, Alexander 1974, West Eberhard 1975, Evans 1977, Bartz 1979) for discussion of the manner in which special features of genetic systems or various auxiliary factors may have facilitated evolution of eusociality. Suffice it to say that in the social insects natural selection may act on the colony as a whole as well as on its individual members. Although the interactions and relative importance of selection at these two levels remain to be elucidated, optimization models based on the maximization of colony fitness have proven to be excellent predictors of foraging behavior and worker caste ratios in some species (Taylor 1976, Oster & Wilson 1978, Wilson 1980a,b).

18.2.1 Life History Attributes and Population Demography

Protection and care of reproductives and immatures by workers has had the effect of transforming survivorship curves of established colonies from type III, typical of most solitary insects, to type I, characteristic of many higher vertebrates (Sakagami & Fukuda 1968, Wilson 1971). Although mortality during the nonsocial stage of the life cycle (dispersal and colony initiation by independent foundresses) is extremely high in many social species, this loss is compensated by production of prodigious numbers of sexuals in successful colonies. In army ants, apine and meliponine bees, and swarm-founding polistine wasps, which have socialized even the dispersal and colony founding stages, mortality and productivity are correspondingly reduced. Colonies of some swarming polistine wasps have a mean reproductive output of only 1–6 queens/queen per reproductive cycle (Richards & Richards 1951, Forsyth 1978). Established social insect colonies may have an expected survivorship of several to many years. A *Polybia scutellaris* colony in northern Argentina was observed by Bruch (1936) for 4 yr before it was abandoned, and the literature records colonies of other tropical wasps surviving for several years (Richards & Richards 1951). Among the ants, Forel (1928) cites a colony in the *Formica rufa* group that survived 56 yr. Certain more highly eusocial species have evolved ability to rear replacement reproductives, and colonies of these species are potentially immortal. Queen and king supersedure is common in termite colonies, and isopterans may hold the record for longevity. Grassé (1949), for example, reports that a mound of *Macrotermes* sp. was occupied for 80 yr, and Hill (1942) records a 60-yr-old colony of *Nasutitermes triodiae*.

As both competitive ability and reproductive potential increase with colony size, natural selection to enhance the colony's lifelong reproductive potential may produce delayed maturity and allocation of resources to nonreproductive castes early in colony establishment. In contrast, most adult nonsocial insects have relatively low expectations of survivorship and invest heavily in reproduction soon after metamorphosis. Delayed colony maturity and colony selection for competitive ability make the life histories of social insects more analogous to those of long-lived vertebrates than of nonsocial insects.

18.2.2 Homeostasis

A significant contributor to success of the social insects must be their remarkable capacity to maintain a steady state of physical, physiological, and social conditions in the colony, often despite harsh or unpredictable environments. The mechanisms by which homeostasis is maintained fall into three general categories. First, a measure of homeostasis is achieved simply through apportionment of colony resources among the many colony members. Worker and brood populations in effect act as buffers in times of stress. If individuals are lost to predators, aggressive encounters with competitors, or starvation during food shortages, the colony as a whole suffers a decrement in fitness, but the reproductive unit is cushioned against extinction. Production of smaller (nanitic) workers in the first brood of many ant and termite species is a particularly striking example of adaptation to distribute the risk of mortality over many individuals.

Secondly, many social insects exert a measure of homeostatic control over the supply of available resources. Established colonies with many individuals defend and harvest food intensively when it is abundant and store it for use during periods of resource depression. This, in effect, smoothes temporal fluctuations in food availability. Food storage is common in all four groups of social insects. Bumblebees and the highly eusocial bees store pollen and honey in quantity in the nest, and many wasps hold small droplets of honey in reserve for up to several days on the walls of nest cells (Richards 1978). In *Apis* spp., stingless bees (Meliponinae), and the "honey wasps" (*Brachygastra, Protonectarina,* and *Polybia*), honey storage is sufficient to maintain the adults through prolonged scarcity during tropical dry seasons or temperate winters (Michener 1974, Richards 1978). Protein sources (alate ants and termites) are stored briefly by some tropical polybiine wasps (Richards 1978). Among the ants, carbohydrate solutions are held in the repletes of the so-called honeypot ants (genera *Camponotus, Myrmecocystus, Proformica, Melophorus,* and others). Major castes of some ants can also serve as storage organs from which nestmates feed (Wilson 1974). Seed-harvesting ants in the genera *Messor, Veromessor, Pogonomyrmex, Pheidole, Chelaner,* and others commonly cache seeds in subterranean nest galleries. Food storage in termites is best developed in the grass harvesters [e.g., certain *Trinervitermes* in Africa and India, *Drepanotermes, Nasutitermes, Amitermes* and *Tumulitermes* in Australia, and *Cornitermes* in South America (reviewed in Wood 1978)]. Fungus cultivation in both leafcutter ants (tribe Attini) and termites (Macrotermitinae) represents a unique example of resource management for long-term use by colonies with high expectations of future survivorship. In addition to reserves of foraged food, many social insects use assimilated reserves in the form of oral larval secretion (wasps) or worker-produced trophic eggs (ants and stingless bees) (Maschwitz 1966, Wilson 1971, Michener 1974). In some termites and ants, when faced with extreme protein deprivation, the brood itself functions as a food supply of last resort to keep reproductives and workers alive (Wilson 1971).

A third form of homeostasis is especially developed in species where colonies have a high expectation of survivorship. As a mechanism for maximizing long-term fitness, these colonies may divert energy and resources to modification of the nesting and foraging environments. Nest architecture may aid in thermoregulation, enhance diffusion of gases, and afford protection from predators. In vespine wasps the multilayered involucrum dampens fluctuations in ambient temperature and helps retain metabolic heat (Himmer 1927). Large nest mounds of *Macrotermes natalensis* in Africa function as air conditioners to dissipate excess heat and CO_2 (Lüscher 1961). Many termites and some leafcutter ants forage under covered runways, and termites often build protective soil cases over vegetation they are dismantling. Ants (*Azteca* spp. and others) may also construct carton covers over feeding aphids and scales maintained by them as sources of honeydew (Sudd 1967). In social insects that nest in exposed places, nest architecture may be highly sophisticated and serve as a first defense against predators. Many social wasps surround the nest combs with an envelope having a small and easily guarded entrance that restricts access to the brood by ants and parasitoids (Jeanne 1975). Such wasps (Vespinae and many swarm-founding Polistinae) experience extremely low rates of parasitization (Jeanne 1979), while a variety of parasitoids heavily infest the open nest combs of *Polistes* (Nelson 1968, Jeanne 1979).

18.2.3 Caste and Polyethism

A caste is a subset of colony members that specialize on particular tasks for prolonged periods (Oster & Wilson 1978). Castes may be physical (morphologically specialized) or temporal (specialized within age classes). In either case division of labor (polyethism) on which caste is based contributes importantly to the ecological success of social insects. The most fundamental caste differentiation is the separation of reproductive and nonreproductive roles. Specialization of reproductives on egg laying and their relief from the risky and demanding tasks of foraging, defense, and nest building are undoubtedly of major importance in extending colony life expectancy. The expendable sterile workers not only construct and maintain the nest and gather the requisite resources but also act as a defending force to protect the colony's investment in reproductives.

Polyethism also occurs among the workers. Temporal division of labor is nearly universal in eusocial species, although more pronounced in ants, stingless bees, and honeybees than in halictine bees, bumblebees, wasps, and termites (Oster & Wilson 1978). Workers progress through a sequence of task specializations as they age. Typically they begin with intranidal activities, and later switch to foraging and defensive behavior (Wilson 1971). Older workers, which have the poorest expectation of survivorship, assume tasks with the greatest risk of mortality. Morphologically specialized workers are much less widespread. Physical worker subcastes are virtually absent from the bees and wasps and among the ants occur in only 44 of the 263 named extant genera, most typically in species that forage by recruitment and have relatively large mature colonies (Oster & Wilson 1978). Termites are distinctive for the near universality of worker polymorphism.

Long ago Wilson (1953) suggested that the evolutionary diversification of castes should reflect the number, distinctiveness, and predictability of the contingencies confronting the colony. Several investigations relate worker polymorphism to specialization on different food types (Wilson 1974, Davidson 1978, Bernstein 1979), foraging distances (Bernstein 1979), ambient temperatures (Bernstein 1976), and protection against predators (Eisner et al. 1976, Maschwitz & Maschwitz 1974). Work by Wilson (1980a,b) on the fungus ant *Atta sexdens* provides some of the best evidence that such specialization enhances worker efficiency. Actual foraging groups consist of those size classes found to be most efficient, as predicted by a model based on the assumption that evolution should maximize energy economy for the colony as a whole. Workers of a given age often perform spatially associated tasks (Wilson 1980a,b), and this adaptation may also enhance colony efficiency (Wilson 1976).

Aside from favoring energetic efficiency, division of labor within the colony may have contributed to the ecological success of social insects in one additional way. Oster and Wilson (1978) pointed out that the success of social insects is not based on innovation of behavior but rather on the fact that the major activities of the colony can be conducted concurrently, rather than sequentially as a solitary insect is obliged to do. An excellent example involves evolution of interference behavior in *Conomyrma bicolor* (Möglich & Alpert 1979). Colonies of *C. bicolor* deploy a small contingent of workers to inhibit foraging by neighboring colonies of competitors (*Myrmecocystus* spp.) by throwing stones into their nest entrances, while the main body of foragers remains free to harvest a contested resource.

18.2.4 Communication

Two categories of communication among nestmates are particularly relevant to our consideration of population regulation in the social insects: signals facilitating responses to predators and parasites and communication associated with acquisition of resources. The fundamental importance of effective defense is reflected in the observation that alarm communication is more widespread among the social insects than communication about any other environmental contingency. Alarm pheromones or substrate tapping (some termites) summon nestmates to repel predators or parasites, move brood to safer locations, or repair breaks in the nest. The importance of coordinated behavior in defensive responses can hardly be overestimated. With evolutionary increases in colony size, prodigious numbers of workers and soldiers can be recruited to repel attackers, so that large colonies are virtually immune to penetration by solitary arthropod predators. Nevertheless, larger size may increase a colony's susceptibility to attack by vertebrate predators, which are attracted to the high prey density and are likely to be deterred only by a rapid and massive attack. Defensive adaptations, including both coordinated responses and the distribution of risk over many individuals, are least effective against similarly coordinated and fractionated predators; thus, the worst enemies of social insects are other social insects (Wilson 1971).

In many termites, ants, and apine bees, communication plays a major role in acquisition of resources. Communication functions to enhance foraging efficiency by matching foraging behavior to the abundance, dispersion, and renewal rate of resources. Both evolved interspecific differences (Johnson & Hubbell 1975, Davidson 1977b, Hölldobler & Lumsden 1980) and behavioral plasticity in responses of individual species (Taylor 1976, Davidson 1977b, Hölldobler & Wilson 1978) provide evidence for the regulation of foraging responses by the pattern of resource availability. For example, in some harvester ants workers search for food independently of one another and in all directions surrounding the nest (Davidson 1977b). These species tend to take seeds from dispersed or low-density distributions. In contrast, species such as *Pogonomyrmex rugosus* of North American deserts forage from trunk trails that arise from recruitment to patches where seeds are concentrated (Hölldobler 1976). When resource levels are depressed seasonally, colonies of this species either curtail activity entirely

(Fig. 18.1), or a reduced population of foragers searches for food
independently.

Recruit foraging enhances a colony's ability to monopolize and defend
rich or concentrated resources. Bees and ants that recruit large numbers
of aggressive foragers often displace other species from a food source. In
Costa Rica *Trigona fuscipennis,* a group forager, specializes on dense clumps
of *Cassia biflora* and competitively displaces *T. fulviventris,* whose workers
forage solitarily, to widely spaced or isolated plants (Johnson & Hubbell
1975, Hubbell & Johnson 1978). Each of the several competing bee species
in the dry forests of this region may have specializations for exploiting

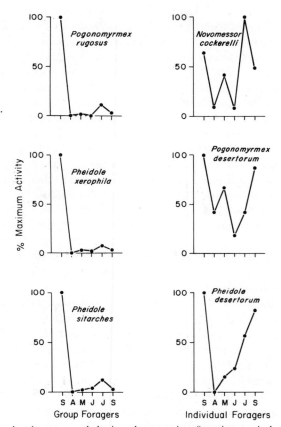

Fig. 18.1   Activity levels measured during the morning foraging periods for six species of
seed-eating ants at Rodeo, New Mexico. In the order recorded on the abscissa, the months
are: September 1974 and April, May, June, July, and September 1975. Maximum activity is
defined as the greatest activity level measured during any of these six observation periods.
Group foragers (left side) tend to have larger colony sizes and more energetically costly
foraging techniques, and to specialize seasonally on periods of resource abundance. (From
D. W. Davidson. 1977b. Foraging Ecology and Community Organization in Desert Seed-eating
Ants. *Ecology* **58:** 725–737. Copyright 1977, Ecological Society of America.)

resources that differ in density and dispersion. Thus, aggressive species that communicate food location specialize on high-density resources and exclude other species, while nonaggressive, individually foraging species have a higher probability of finding food sources and pursue a strategy of "insinuation" or visit low-density sources when foraging with aggressive competitors. Economic models generally predict a direct correlation between the costs and benefits of communication mechanisms. Species or colonies that exploit large or valuable resources are more likely to use energetically costly adaptations, such as chemical recruitment, to rapidly mobilize workers and monopolize prey. The costs of recruitment are difficult to measure, and this generalization has never been adequately tested, but circumstantial evidence supports the pattern, both within and among (Fig. 18.1) species (Johnson & Hubbell 1975, Taylor 1976, Davidson 1977b).

18.3   REGULATION OF ABUNDANCE AND SPECIES DIVERSITY

We have depicted the social insects as relatively long-lived and energy-efficient consumers highly resistant to solitary invertebrate predators and to environmental contingencies. What factors then limit the numbers and distributions of these insects? We will here consider several kinds of factors that are or may be instrumental in regulating the abundance and diversity of social insect species in ecosystems. We emphasize the effects of climate, productivity, colonizing ability, competition, and predation.

18.3.1   Climate

The impact of climate on the distribution and abundance of social insects is evident from latitudinal and altitudinal patterns in species diversity. Like many solitary types, social insects as a group reach their greatest abundance and diversity in the moist tropics. Low temperatures probably restrict the distributions of these ectotherms primarily by limiting the length of the season during which food is available and the insects can be active. Length of the foraging season is crucial to social insects since, by definition, these species have overlapping generations and must produce at least one brood of workers prior to reproduction.

Several adaptations enable some social insect species to persist in cold climates with abbreviated foraging seasons. Bumblebees (*Bombus*), which occur in colder climates than any other social insects, have evolved remarkable thermoregulatory abilities that permit them to forage at temperatures as low as $-3.6°C$ (Heinrich 1979). This genus ranges from Tierra del Fuego to the northern tip of Ellesmere Island at 81° N, only 880 km from the North Pole (Richards 1973, Michener 1979). Among the wasps, *Vespula* spp. forage at temperatures as low as 2°C, and species of *Vespula* and *Dolichovespula* occur well north of the Arctic Circle (Edwards 1980).

Rapid development and abbreviated periods of worker output prior to reproduction probably enable these wasps to adapt to the extremely short seasons at these latitudes. Shortness of season is probably also compensated by 24-hr foraging days around the summer solstice.

Ants and termites are even more disadvantaged and species-poor at high latitudes and elevations than social bees and wasps, which use more energy-rich food sources, on average, and generate considerable metabolic heat during flight. Ants that have been successful in cold climates include *Formica neorufibarbis gelida,* which ranges into Canada and Alaska and occurs at high elevations in lower latitudes. This ant exhibits a marked dimorphism in worker size and coloration and is adapted to a broad range of foraging temperatures (Bernstein 1976). In cool ambient temperatures of early morning, forager populations consist primarily of small workers whose dark coloration and high surface-to-volume ratio permit them to warm up rapidly. Large, red workers comprise an increasing proportion of the foraging population as temperatures climb throughout the day. Termites have the most restricted distributions of social insects, with respect to both altitude and latitude; they are absent entirely from alpine habitats and from latitudes higher than 45–52°, depending on continental region (Araujo 1970, Harris 1970, Weesner 1970).

Other abiotic factors also determine species distributions, but we cannot consider these in detail here. Most important are microclimate and soil type. To a considerable extent, microhabitat insolation determines the suitability of potential nesting and foraging areas, and thus species abundance and diversity may be influenced by vegetation structure (Sands 1965, Williams 1966, Chew & Chew 1980). Various properties of soils may also limit the distributions of ants and termites. Both very high and very low clay content, for example, can exclude termites from certain soils (Lee & Wood 1971).

18.3.2 Productivity

In part, latitudinal and altitudinal patterns in species diversity may be mediated by resource productivity, which increases with both temperature and moisture availability. Habitat productivity influences both the diversity and species composition of social insect faunas. In North American deserts of similar latitudes the abundance and diversity of seed harvester ants and the proportion of these ants having group-foraging colonies increase with mean annual precipitation, an index of productivity in arid regions (Davidson 1977a,b) (Fig. 18.2). Group-foraging species tend to have large colony sizes, more energetically costly foraging techniques, and to specialize on high-density seed distributions. Average levels of productivity may regulate species number and composition in communities of bees. Schaffer et al. (1979) argued that because colony biomass in *Apis mellifera* exceeds that in *Bombus sonorus,* the minimun environmental productivity required to main-

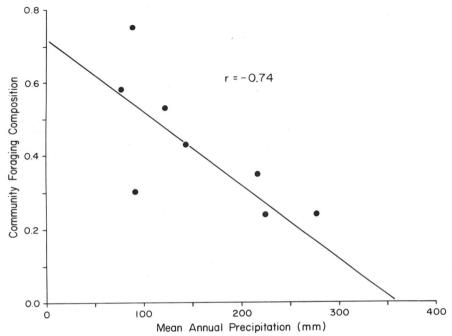

Fig. 18.2   Foraging score of granivorous ant communities as a function of the precipitation index of productivity. Scores are based on the relative numbers of individual versus group foraging species and weighted by their relative abundances. Higher scores indicate a greater proportion of individual foragers. (From D. W. Davidson. 1977b. Foraging Ecology and Community Organization in Desert Seed-eating Ants. *Ecology* **58:** 725–737. Copyright 1977, Ecological Society of America.)

tain a colony will be greater for *Apis* than for *Bombus*. *Bombus*, in turn, requires a greater threshold of productivity than the solitary bee, *Xylocopa arizonensis*. In support of this they find that where all three species feed primarily on *Agave schottii* in the mountains of southern Arizona, *Xylocopa* predominates in habitats where food production is lowest, with *Bombus*, then *Apis*, coming into areas of higher food productivity.

18.3.3   Colonizing Ability

The social insects are notoriously poor colonizers of isolated habitats and are absent from many oceanic islands. No social insects are native to the Hawaiian Islands (Wilson 1971, C. D. Michener pers. comm.). Only a few termite genera, largely members of the primitive Kalotermitidae and Rhinotermitidae, have colonized oceanic islands; the vast majority are restricted to continents (Emerson 1952). The high degree of endemism in insular faunas is also indicative of the poor dispersal abilities of these insects. On islands off the west coast of Africa in the Gulf of Guinea, the degree of endemism among the termites is positively correlated with distances separating islands from source pools, either on the continent or on other islands (Bouillon 1970). Although established colonies of social insects may be extraordinarily resistant to disturbance [ants were one of the few life forms to survive the recent eruption of Mount St. Helens volcano (J. MacMahon pers. comm.)], isolated and disturbed faunas may be slow to recover. Compared with other arthropods, ants were slow to colonize experimentally defaunated mangrove islands in the Florida Keys (Simberloff & Wilson 1969).

Dispersal of social insects normally occurs during reproductive episodes, and much of the variation in dispersal ability is explained by differences in reproductive biology. The best dispersers are those species in which the reproductive propagule is a single, inseminated queen (some of the ants, bees, and wasps). The flight of these females, especially if combined with passive wind dispersal, can carry them far from their natal colonies. In termites and many ants, in contrast, mating occurs at the end of the nuptial flight; for these species a reproductively viable propagule consists of a mating pair, and the probability is slight that a male and a female will find each other after being blown very far from the source population.

The poorest long-distance dispersers are species that reproduce by swarming or colony fragmentation. Distributional evidence suggests that the highly eusocial apine and meliponine bees, which initiate colonies by swarming, seldom cross broad geographic barriers. Stingless bees are absent from the Antilles, despite their abundance on the adjacent mainland (Michener 1979). Similarly, *Apis* is limited by broad water barriers, although *A. mellifera* swarms may disperse great distances across land, as evidenced by the rapid spread of the Africanized honeybee across much of South America during the past 25 yr. Among the wasps, swarm-founding polistines of the tropics appear to be the poorest dispersers. At least some of these maintain swarm cohesion during dispersal to a new nest site by scent marking vegetation along the route (Jeanne 1981). Ineffectiveness of this means of swarm orientation over even narrow water barriers probably accounts for the limited occurrence in the West Indies of swarming polistines, compared to species with solitary foundresses.

Colonization of remote islands is more likely to have occurred by rafting of entire colonies than by dispersal of reproductive propagules, especially for species that reproduce by swarming or budding. Species that nest above ground are much more likely to be spread by rafting than are subterranean nesters. In historical times, human activities have enlarged the distributions of many insect species regardless of reproductive biology. Man has transported the honeybee around the globe and inadvertently dispersed many of the tropicopolitan ants. Among wasps with solitary foundresses, *Vespula germanica* has become established in New Zealand, Australia, Tasmania, South Africa, Ascension Island, Chile, and the United States, all presumably by human transport (Edwards 1980).

18.3.4 Competition

An extensive, cosmopolitan literature documents the importance of resource limitation and competition in regulating social insect populations. The evidence, both direct and circumstantial, is stronger for ants and bees than for wasps and termites. Data bearing on competition come in a diversity of forms that are often idiosyncratic to particular social insect groups. We consider regulation by competition independently for each of the four major taxa.

18.3.4.1 Termites. Few studies have documented density-dependent population regulation in termites, probably because population measurements are difficult on large, diffuse, and often inaccessible colonies. Several kinds of evidence suggest, however, that competition influences termite abundance and species composition of termite faunas. First, a number of species are intraspecifically or interspecifically territorial (reviewed in Bouillon 1970). Territoriality in *Hodotermes mossambicus* has been tied to resource availability; territories defended against conspecifics contract or expand as food abundance increases or decreases (Coaton 1958). Overdispersion of nest sites is a frequent consequence of territoriality; Wood and Lee (1971) report significant overdispersion of mounds in 11 of 13 populations studied and provide evidence that the regular spacing of mounds is related to competition for food or space. In a dry sclerophyll forest near Adelaide, mounds of *Nasutitermes exitiosus* are overdispersed, and nearest-neighbor distance between mounds is positively correlated with sizes of neighboring mounds. In studies on *Amitermes laurensis* nest dispersion indices suggest less intense competition in an ungrazed area than a grazed one, where resources are probably more limiting. Colonies may compete and yet not show significant overdispersion, particularly where foraging territories coincide with food resources that are not regularly distributed, or where territories overlap but trails or galleries do not come into contact. For example, Josens (1972) found that although colonies of *Microtermes toumodiensis* and *Ancistrotermes cavithorax* were not spatially segregated, foraging groups of the two species overlapped significantly less on baits than predicted from their independent abundances on these baits. Over a larger geographic scale interspecific territoriality may produce a "checkerboard" distribution pattern of competing species. In Africa, for example, competing species of *Trinervitermes* occur as ecological replacements in disjunct patches of savannah separated by forest barriers (Sands 1965, 1967).

Niche differentiation may facilitate coexistence of termite species within a single habitat (reviewed in Wood 1978). Termite diets differ in relation to vegetation hardness, chemistry, stage of decomposition (especially important among decomposers), size, and fungal associates. Differences in seasonality of foraging and foraging depth (in humivores) may also separate species ecologically (Sands 1961, Williams 1966, reviewed in Wood 1978).

Unfortunately, we usually lack direct evidence that these forms of niche differentiation reduce the intensity of competition. One critical question is whether trophic or other differences enable coexisting species to exploit separately renewing populations of resource. Thus, termites feeding on the same plant species at different stages of decomposition may be competing intensively for that resource.

Because symbiotic associations have enabled termites to specialize on resources unavailable to other insects, their closest competitors outside the Isoptera may be unrelated organisms such as earthworms and mammalian grazers, which share with termites the ability to degrade cellulose (Coaton 1958, Lee & Wood 1971, Bodine & Ueckert 1975). In Texas grasslands termites consume 25% of the standing crop of grass when cattle are absent and only 17% in the presence of these grazers (Bodine & Ueckert 1975). Other studies (Wood & Lee 1971) also suggest that termites and large grazers compete for food.

18.3.4.2 Ants. Much evidence implicates competition in the regulation of ant populations. Overt aggression is a common and very obvious form of interference competition. Hölldobler and Lumsden (1980) described a unique form of interference competition involving highly ritualized intraspecific displays. In the southwestern deserts of the United States *Myrmecocystus mimicus* feeds on resources that are patchily distributed in both space and time, and foragers defend territories only where they happen to encounter intruders at close range. When patches of termites are contested by two ant colonies, workers from one colony swarm to the neighboring nest and engage alien workers in "tournaments," while their nestmates continue to forage. Hostile workers assume display postures by standing on "stilt legs" with upraised gasters and may kick and drum with the antennae on the opponent's gaster. Although individual encounters are usually resolved after 10–30 sec, contests between colonies may continue for several days and end in intraspecific colony raiding if the disparity in colony size is great. Highly ritualized contests such as those occurring between colonies of *M. mimicus* are the exception rather than the rule. Intraspecific and interspecific contests among workers frequently result in fights to the death (e.g., De Bruyn & Mabelis 1972, DeVita 1979). Mortality of founding queens may be very high as a result of interference competition from established colonies or other foundresses.

More subtle and indirect evidence for competition comes from patterns in the spacing of nests or feeding territories (reviewed in Wilson 1971). In populations of mature colonies territorial defense often leads to overdispersion of nest sites. Waloff and Blackith (1962) showed that although colonies of *Lasius flavus* are overdispersed in areas of high colony density, they are randomly distributed where population densities are low. A striking case of interspecific nest spacing is the mosaic of *Pheidole megacephala* and *Iridomyrmex humilis* colonies in Bermuda (Lieberburg et al. 1975). Indirect competitive interactions can result in positive as well as negative associations among colonies of different species. In a community of Chihuahuan Desert harvester ants, *Pogonomyrmex rugosus*, a species with large and aggressive workers, actively interferes with foraging and nesting by *Pogonomyrmex desertorum*. *Pheidole xerophila* and *Novomessor cockerelli*, two species that compete exploitatively with *P. desertorum* but are not subject to interference by *P. rugosus*, are spatially associated with colonies of the latter species (Davidson 1980). Examples of such indirect effects are well known in tropical forests (e.g., Room 1971). Here forest canopies are dominated by several species of ants that defend territories interspecifically. Distinctive assemblages of other (subdominant) ants occur in association with each dominant type and are competitively excluded from the territories of other dominants. The resulting patchiness in the distribution of ant communities has been termed *tropical ant mosaic*.

Observations and experiments of Pontin (1961) on *Lasius flavus* and *L. niger* provide direct evidence that competition reduces colony fitness. Where *L. flavus* occurs without *L. niger*, production of alate queens is directly proportional to the distance between colonies of conspecifics. In the presence of *L. niger*, which uses similar food resources, queen production is lower than predicted by the distance between *L. flavus* colonies. Experimental removal of one colony of *L. niger* caused a significant increase in queen production by neighboring conspecific colonies and an even more startling increase in reproductive output of *L. flavus* colonies in the vicinity. When 12 colonies of *L. flavus* were excavated and placed around each of 4 colonies of *L. niger*, alate production of these latter colonies dropped significantly.

Other less direct evidence suggests that competition regulates the abundance and diversity of ants on a larger geographic scale. Studies of Pacific island ant faunas (Wilson & Taylor 1967, Wilson & Hunt 1967) reveal checkerboard patterns of distribution of congeneric species that are likely close competitors. Although congeneric pairs may coexist on large islands or continental land masses, small islands typically contain one or the other species but not both. Similarly, in continental faunas there appear to be "forbidden combinations" of species with like ecological requirements. In harvester ant communities of southwestern United States species with similar worker body sizes and foraging behaviors are unlikely to coexist locally, and they defend territories interspecifically where their distributions meet (Davidson 1977a).

Coexisting ant species, like coexisting termites, often differ in niche requirements, and in some cases there is strong evidence relating niche differentiation to alleviation of resource competition. Shifts in morphology or resource utilization sometimes occur in response to changes in the competitive environment. Such a niche shift has been described for *Veromessor pergandei*, a harvester ant characterized by continuous worker size polymorphism (Davidson 1978). In the relatively unproductive deserts of southern California, *V. pergandei* typically occurs alone or with one other species, and worker polymorphism is pronounced. However, in the Sonoran Desert of Arizona, at the eastern boundary of this species' distribution, higher levels of productivity are associated with greater diversity of harvester ants. Here size polymorphism within colonies of *V. pergandei* is much reduced, specifically in the worker size classes that overlap with competitors. Worker size is directly proportional to prey size, both within colonies of *V. pergandei* and across species, and the divergence of *Veromessor* from competitors in zones of overlap strongly suggests that the degree of polymorphism is sensitive to competition for food resources.

At least one niche shift in response to competition has been documented experimentally (Stabaev & Reznikova 1972). In the steppes of southern Siberia small underground nests of *Formica subpilosa* coexist with relatively large, above-ground nests of *Formica pratensis*. While colonies of *F. subpilosa* not influenced by *F. pratensis* have their activity peaks in the afternoon, those either naturally or experimentally located near *F. pratensis* nests have activity peaks shifted toward evening (Fig. 18.3). Competition for nest sites is also suggested by niche shifts. Brian (1952) demonstrated that founding queens of *Myrmica ruginodis* are excluded from favored nesting sites in the presence of *Formica fusca* but move into these sites when *F. fusca* species is experimentally removed or absent.

18.3.4.3 Bees. As in ants and termites, there is much evidence that competition plays an important role in bee population dynamics. Some stingless bees use overt aggression to exclude nonaggressive species from concentrated, high-quality resources (Johnson & Hubbell 1974). These encounters are sometimes so intense as to result in death of the combatants, especially if they are from rival colonies of the same species. Intraspecific interference competition also occurs under certain circumstances in *Apis mellifera*. The first colony to discover a patch of flowers will defend it against intrusion by members of other colonies (Butler et al. 1943). When bumblebees are founding colonies, it is not uncommon to find one or more dead queens in the entrance of a queenright nest, suggesting that interference competition for nest site may be intense (Brian 1965).

Indirect evidence suggests that intraspecific competition for resources may limit populations of some meliponine bees. Of eight species of *Trigona* studied by Hubbell and Johnson (1977) in a Costa Rican tropical dry forest, four species compete aggressively at food sources. The intensity and duration of aggressive encounters at artificial baits rises sharply with increasing sugar concentration (Johnson & Hubbell 1974). The linear relationship between logarithm of home range area and logarithm of the weight of colony worker population implies that natural colony sizes are limited by food resources. Resource abundance may also regulate colony densities; intraspecifically, nests of the four species were regularly spaced (Hubbell & Johnson 1977). There was no evidence that availability of nest sites limited

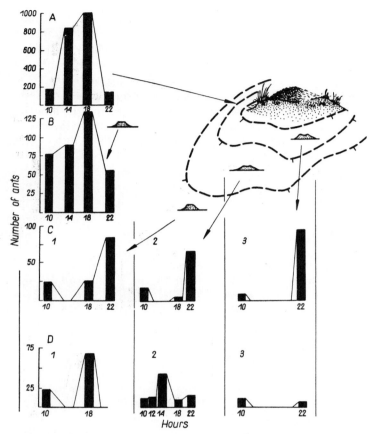

Fig. 18.3 Daily activity cycle of *Formica subpilosa* colonies situated in different parts of *F. pratensis* territory in comparison with *F. pratensis* daily cycle. The height of each column shows the number of ants leaving the nest during 5 min. (A) *F. pratensis* colony; (B) *F. subpilosa* colony situated in the neutral zone between two *F. pratensis* colonies. (C) *F. subpilosa* colonies situated in *F. pratensis* territory at (1) great, (2) intermediate, and (3) small distances from the *F. pratensis* nest; (D) *F. subpilosa* colonies dug far from *F. pratensis* nests and transplanted to *F. pratensis* territory at (1) great, (2) intermediate, and (3) small distances from the *F. pratensis* nest. (From Stebaev and Reznikova 1972.)

colony density. Four other *Trigona* species in the habitat are nonaggressive at food sources, and their nests were not uniformly dispersed, either with respect to conspecifics or to any of the aggressive species, suggesting that their densities were not limited by competition (Hubbell & Johnson 1977).

Another form of indirect evidence for resource competition is niche differentiation. Tongue length, for example, determines the efficiency with which a bee can manipulate flowers of a given corolla length (Ranta & Lundberg 1980); long-tongued bees spend less time extracting nectar from long-corolla flowers than do short-tongued bees, and vice versa (Heinrich

1979). The three or four species of bumblebees typically abundant in a given community differ in mean tongue length, usually by a factor of 1.2–1.4 (Inouye 1977). Specialization on flowers with different corolla lengths may facilitate coexistence of these species. For example, in the Colorado Rockies where *Bombus appositus* specializes on larkspur (*Delphinium barbeyi*) and *B. flavifrons* on monkshood (*Aconitum columbianum*), each species uses the plant whose corolla tube length best matches its tongue length (Inouye 1978). When Inouye removed either bee species from its flowers, the other soon increased its visits to the neglected flowers. While in North America some bumblebees have tongues as short as those of *Apis mellifera*, in Europe such short-tongued *Bombus* are lacking. There, the short-tongued niche is filled by the native honeybee (Inouye 1977). The introduction of honeybees to North America has undoubtedly had an adverse effect on populations of native bees. Heinrich (1979) calculated that the 200 kg of honey a healthy honeybee colony can collect is enough to produce 38,400 bumblebee reproductives.

Two species of bee may share the same flower species by partitioning the resource on the basis of differences in tongue length and diurnal activity schedules. *Bombus fervidus*, with a tongue length of 11 mm, is able to remove all the nectar from the deep flowers of jewelweed (*Impatiens biflora*), while *B. vagans*, with its 8-mm tongue, cannot. Heinrich (1976) found that in Maine *B. vagans* obtains rewards from this plant by foraging early and late in the day, when *B. fervidus* foragers are rare or absent. In years when *B. fervidus* is absent, *B. vagans* forages on jewelweed throughout the day. Morse (1977) showed that two species of *Bombus* partition goldenrod flowers on the basis of body size. *B. ternarius* forages mostly at the tips of goldenrod inflorescences, while the heavier *B. terricola* feeds more proximally, presumably because the flimsy tips do not bear its weight. If *B. terricola* is removed from an area, *B. ternarius* visits more proximal flowers.

Although we know of no studies that measure effects of competition on colony fitness in bees, Plowright et al. (1978) provide indirect evidence for the importance of competition. They compared foraging success of bumblebees in an undisturbed forest in southwestern New Brunswick with that of transplanted *B. terricola* colonies in nearby forest from which native bumblebees had been mostly eliminated by aerial spraying. Duration of pollen foraging trips averaged 1.8 times as long in unsprayed areas, and pollen input per unit of foraging effort was only one-fifth as great as in sprayed forest. These differences suggest that competition from other bees was reduced in sprayed areas and that competition for food should lead to reduced productivity in *Bombus* colonies of undisturbed forest.

18.3.4.4 Wasps. Overt aggression between queens at newly founded nests is common in yellowjackets and hornets and often results in death of one of the combatants (Akre et al. 1976, Matthews & Matthews 1979, Archer 1980). Archer (1980), in a study of *Vespula vulgaris*, interprets this behavior as manifestation of competition for nest sites and hypothesizes that it is the most important factor determining the number of colonies in a habitat. He believes that queens produced in years of abundance are "smaller and less well equipped for overwintering, leading to weak queens" (Archer 1980:206). Many of these queens are killed in aggressive contests over nest sites, and a year of low abundance results. Conversely, queens produced in years of low population density are stronger and more likely to produce successful colonies the following season. In England the alternation of good and bad years thus gives rise to the 2-yr cycle of abundance alternating with scarcity (Archer 1980).

Lord and Roth (Lord 1979, Lord & Roth 1979) postulated a similar mechanism for population regulation in *Vespula maculifrons*. In Delaware they found that in the second year, when the abundance of nests was 2.5 times that of the first year, the extra nests were accommodated primarily in previously unused areas. That these were areas of marginal habitat quality was suggested by the observation that productivity (nest biomass, number of queen cells, number of worker cells) of nests in these areas was significantly lower than that of nests in areas used in both years. The effect was that nest density in repeat-use areas increased only slightly in the second year. Lord and Roth hypothesize that queen–queen encounters occur more frequently in areas where queens are abundant and limit population density in prime habitats by the death of some queens and dispersal of others to marginal habitats. In marginal habitats the average number of queen-producing cells per colony dropped significantly in the second year, suggesting that increased interference competition among workers from different colonies later in the season reduced average colony productivity. Lord and Roth also found that in years of high abundance, queen cell size was significantly smaller, possibly due to increased competition for food. They suggest that these queens were less fit physiologically, resulting in fewer nests the next year and giving rise to a 2-yr population cycle.

Among foundresses of some species of *Polistes, Ropalidia,* and *Mischocyttarus* there is also conflict for newly founded nests, but rather than fight to the death, two or more related females may join forces on the same nest. If so, a dominance hierarchy is established and only one typically lays eggs, while the other's ovaries eventually regress (Pardi 1948). There is evidence that the higher the density of newly founded nests in an area, the greater the average number of founding females per nest (Gamboa 1978, Suzuki & Murai 1980). The result is a density-dependent reduction of effective population size in the area (Wilson 1971). Colonies with multiple foundresses have a lower productivity per foundress than singly founded colonies (Gibo 1978), but they experience a greater likelihood of surviving to reproduce, either by virtue of enhanced resistance to ant predation (Suzuki & Murai 1980), recovery from bird predation (Gibo 1978), or defense against colony usurpation by conspecifics (Gamboa 1978). Thus, foundress joining may be advantageous in dense populations where individual foundresses are faced with any of a variety of factors causing a high risk of early colony failure. Among Japanese *Polistes,* on the other hand, joining behavior is rare, but females of *P. chinensis antennalis* and *P. jadwigae* may turn to robbing brood from nearby nests if their normal prey is scarce (Sakagami & Fukushima 1957, Kasuya et al. 1980). Since this may result in extinction of the nest, it can reduce the population.

18.3.5 Predation

While evolution of eusociality undoubtedly provides protection against many small solitary predators (e.g., arthropods), large colony biomass has increased the attractiveness and susceptibility of social insects to predators with high energy requirements. The natural enemies of social insects include other social insects, relatively large solitary predators (e.g., vertebrates), and both social and infectious parasites. Exploitation of concentrated and continuously renewing sources of prey has enabled these predators to become specialized consumers of social insects. Among the vertebrates, many lizards and representatives of at least five mammalian orders feed almost exclusively on ants and/or termites. A number of species of army ants (e.g., genera *Dorylus, Anomma, Aenictus, Eciton, Neivamyrmex*), subsist partly or exclusively on the brood and adults of other social insects (ants, termites, and/or polistine wasps). Many *Vespa* (true hornets, native to the Old World) are specialized predators of other social wasps and bees (Matsuura & Sakagami 1973). The high benefit-to-cost ratio associated with foraging on such concentrated resources enables some of these predators (especially the migratory army ants) to maintain massive colonies that are themselves relatively immune to predation. More specialized social insect predators, such as the ponerine ant *Termitopone commutata,* which feeds only on termites of the genus *Syntermes,* have colonies of much lower biomass. Both slave-making ants and social parasites (occurring in all four groups of social insects) have capitalized on opportunities presented by social interactions within the colony to coerce alien species to labor in their behalf and to usurp the resources and reproductive potential of host colonies.

Despite the many records of predation and parasitism on social insects, little is known about the effect of these factors on population dynamics. Probably most predation on social insect colonies does not lead to death of the genetic unit. Various antipredator adaptations help to minimize the impact of such predation. Soil-nesting *Pogonomyrmex* ants reduce losses to horned lizards (*Phrynosoma*) by shutting down above-ground activity in response to removal of foragers (Gentry 1974, Whitford & Bryant 1979). Some polistine wasps [Richards (1978) lists 10 species] nest in close association with certain ants, gaining protection from army ants. Despite many such adaptations, it seems clear that predation and parasitism can constitute a severe energy drain on social insect populations and, in some exceptional cases, cause the death of the colony. Fuller (1915) and Hill (1922) document mortality in termite colonies preyed upon by ants. Colonies of the bumblebee *Bombus transversalis* in Amazonia are usually eliminated by predators and parasites (especially stratiomyiid flies) in less than 1 yr after founding (Michener 1974). High rates of parasitism are known to severely reduce the productivity of *Polistes* wasps (Nelson 1968). When attacked by army ants, adult populations of polistines abscond and renest elsewhere; although the colony is not killed, reproduction is delayed.

If energy drains from predation and parasitism fall disproportionately on certain species, these regulatory factors may influence the species composition of social insect communities (e.g., Fuller 1915, Bodot 1967, Rissing 1981). Because the defenses of social insect species are differentially effective, predation is often selective. Horned lizards, for example, harvest large numbers of ants in some deserts and feed selectively on species that forage in small groups that are incapable of mobbing the lizards (Rissing 1981). *Pogonomyrmex californicus* is preyed upon 60 times more than expected on the basis of its abundance in one Mojave Desert site, while *Pogonomyrmex rugosus* and *Veromessor pergandei*, species that forage from trunk trails, are underrepresented in the diet of *Phrynosoma platyrhinos*. This level of selective predation may have a marked effect on the structure of interactions in the community.

Influences of selective predation on community composition will depend on the interaction of predation with competition and other population processes. A particularly good example of the interaction between parasitism and competition comes from the work of Feener (1981). While studying the ant community of a Texas woodland, he showed that the outcome of competition between *Pheidole dentata* and *Solenopsis texana* was altered seasonally by the presence of parasitic phorid flies. An undescribed phorid (*Apocephalus*) appears to be specific to the soldier caste of *P. dentata* and oviposits on the heads of soldiers along recruitment trails. In the presence of phorids, major workers seek refuge in leaf litter or return to their nests. Feener showed that the alarm recruitment response of *P. dentata* to presence of a competitor, *S. texana*, was attenuated in the presence of *Apocephalus*. Phorids are inactive at the beginning and end of summer foraging of *Pheidole*, and this species wins a majority of confrontations with *Solenopsis* on tuna fish baits. In contrast, from May through October, when phorids are active, a majority of aggressive encounters are won by *Solenopsis*.

Among the bees, selective predation by giant hornets (*Vespa mandarinia*) may contribute to continued coexistence of *Apis cerana* and *A. mellifera* in Japan. Introduced in 1876, the large and aggressive colonies of *A. mellifera* exclude the native *A. cerana* from rich food sources, rob its hives of honey, and often exterminate its colonies (Sakagami 1959). However, *A. cerana* persists, especially in remote areas and at high elevations. In addition to foraging later in the fall and overwintering more successfully, *A. cerana* withstands predatory attacks of *V. mandarinia* by retreating into the nest or rapidly counterattacking en masse (Matsuura & Sakagami 1973). In contrast, *A. mellifera* workers are unsuccessful in defending against hornets individually, and thousands of their colonies are exterminated annually by the hornets (Sakagami 1959). Thus, differential vulnerability to predation and severe climate permit continued coexistence of the two species.

Social insects form numerous mutualistic associations with other organisms, and undoubtedly many of these interactions affect abundances and distributions. This should be particularly true of highly coevolved mutualisms. Even though individual mutualisms may not be highly coevolved or obligate, they are likely to be of consequence in aggregate because of their extraordinary frequency. Bees, for example, rely completely on floral nectar and pollen for their food, and a large proportion of all ant species tend sap-feeding homopterans. Unfortunately, we understand little about the effects of mutualism on population processes, but it is likely that mutualistic interactions affect the organization of social insect communities by, for example, enabling poor competitors or species heavily preyed upon to persist despite these sources of mortality. Mutualisms are of interest for their evolutionary as well as ecological consequences. Because mutualistic associates of social insects are often highly dependent on these insects and on occasion compete for their services (e.g., Pleasants 1980), the evolutionary interests of both participants are served by increasing the frequency of the interaction. Thus, natural selection on both predator and prey should increase the efficiency with which social insects locate and exploit their mutualists. Similarly, for wasps, ants, and termites that feed on nonliving food, evolutionary increases in feeding efficiency should be unopposed by evolution of resources.

18.4 CONCLUSIONS

Although social insect populations are limited by many of the same factors that regulate abundance of solitary insects, some attributes of social species have enabled them to achieve a significant degree of ecological success compared to solitary relatives. The most fundamental of these, and quite likely the one that served as the initial extrinsic stimulus for the evolution of sociality (at least in the Hymenoptera), is the enhanced defense provided against predators and parasites in an incipient social group (Michener 1974, Evans 1977). Similarly, increases in colony size among early social forms were probably also favored under predator and parasite pressure (Evans 1977). In support of the fundamental role of natural enemies is the widespread occurrence of some form of alarm communication among the social insects (Maschwitz 1964, Hölldobler 1977).

At least four adaptations have been important to supporting large colonies on foraging areas centered at the nest site. In conjunction with large colony size, efficient communication systems have enabled social insects to dominate and exploit rich or concentrated resources that might otherwise be exploited only by vertebrates. Secondly, social insect colonies actively forage for food over a long season; Heithaus's data on bees and wasps netted at flowers in Guanacaste, Costa Rica, clearly contrast the year-round activity of social species with the more seasonal occurrence of solitary species (Heithaus 1979). Thirdly, and in part correlated with long foraging seasons, most social insects are trophic generalists, rather than exhibiting the high degree of coevolution that characterizes interactions of many solitary insects with their food plants or prey groups. Together with food storage, the generalized diet has also promoted the success of social insects in harsh and unpredictable environments. A probable additional consequence of unspecialized diets is that the species composition of social insect communities is more sensitive to the spatial and temporal pattern of resource availability and to vegetation structure (mediated by its effect on insolation) than to the identities of resource species.

Finally, the culmination of social insect evolution with respect to population and community dynamics is the development of efficient resource acquisition and processing techniques. Among the ants and bees, efficiency is manifested largely in foraging economics. Communication enables species to match their foraging responses very precisely to abundance, dispersion, and renewal rates of resources. Competition from related social species may also be an important selective pressure favoring evolution of resource communication, for differences in social (communicative) aspects of foraging behavior directly influence the outcome of competition. Where they occur, caste and polyethism also contribute to colony efficiency. In termites, especially those species that live inside their food source and have negligible foraging and nest maintenance costs, efficiency is related primarily to food processing (Heinrich 1978). Mutualistic associations with protozoans and bacteria allow extraordinary assimilation efficiencies and permit termites to subsist on relatively low energy resources. Finally, it appears that for the disproportionately large numbers of social insects feeding as detritivores or obtaining food from mutualists, evolutionary increases in efficiency of food acquisition may be unopposed or even aided by the evolution of resources.

The ecologically dominant social insects, then, tend to have these traits in common: (1) reduction in risk of loss of the reproductive unit to predators and parasites, although these enemies may drain colony reserves and influence long-term fitness; (2) homeostatic control over fluctuations in resource levels or in the abiotic environment. To the extent to which mortality from predation and disturbance is reduced, populations of social insects may frequently approach the carrying capacities of their habitats. Consistent with this argument is the substantial body of evidence suggesting the importance of competition in limiting populations and regulating community structure. Two additional characteristics of social insects may have evolved in response to selection pressures favoring increased competitive ability: (3) delayed maturity and development of large and more or less perennial colonies, and (4) unusual efficiency in resource acquisition and processing, with this efficiency based on features such as communication and caste.

These characteristics have enabled the social insects to invade an adaptive zone available neither to solitary insects nor to vertebrates. They have converged on higher vertebrates with respect to the type I survivorship curves of established colonies, delayed reproductive maturity, and low replacement rates in populations, colony biomass measured in grams or kilograms, and ability to dominate and exploit rich or concentrated resources. Yet because they retain the small body size characteristic of insects, foraging costs of workers are low, and individuals economically exploit food types that would be unprofitable to even the smallest vertebrate. This special combination of characteristics sets the social insects apart from potential competitors among other insects and vertebrates and dictates that they compete most severely with each other. They are a truly unique category of consumers.

REFERENCES

Akre, R. D., W. B. Garnett, J. F. MacDonald, A. Greene, and P. Landolt. 1976. *J. Kansas Entomol. Soc.* **49:** 63–84.

Alexander, R. D. 1974. *Annu. Rev. Ecol. Syst.* **5:** 325–383.

Amante, E. 1967. *Biologico* **33:** 113–120.

Araujo, R. L. 1970. Pp. 527–576 in Krishna and Weesner (Eds.), 1970, referenced here.

Archer, M. E. 1980. Pp. 172–207 in R. Edwards (Ed.), *Social Wasps: Their Biology and Control.* Rentokil, E. Grinstead, England.

Baroni-Urbani, C., G. Josens, and G. J. Peakin. 1978. Pp. 5–44, in Brian (Ed.), 1978, referenced here.

Bartz, S. H. 1979, *Proc. Natl. Acad. Sci. USA* **76:** 5764–5768.

Bentley, B. L. 1976. *Ecology* **57:** 815–820.

Bernstein, R. A. 1976. *Psyche* **83:** 180–184.

Bernstein, R. A. 1979. *Am. Natur.* **114:** 533–544.

Bodine, M. C. and D. N. Ueckert. 1975. *J. Range Manag.* **27:** 353–358.

Bodot, P. 1967. *Insectes Soc.* **14:** 229–258.

Bouillon, A. 1970. Pp. 153–280 in Krishna and Weesner (Eds.), 1970, referenced here.

Brian, M. V. 1952. *Entomol. Mon. Mag.* **88:** 84–88.

Brian, M. V. 1965. *Social Insect Populations.* Academic, New York.

Brian, M. V. (Ed.). 1978. *Production Ecology of Ants and Termites.* Cambridge University Press, Cambridge, Mass.

Bruch, C. 1936. *Physis, Buenos Aires.* **12:** 125–135.

Butler, C. G., E. P. Jeffree, and H. Kalmus. 1943. *J. Exp. Biol.* **20:** 65–73.

Chew, A. E. and R. M. Chew. 1980. *Insectes Soc.* **27:** 189–202.

Coaton, W. G. H. 1958. *Dept. Agr. Sci. Bull. S. Afr., Entomol. Ser. 375,* **43:** 1.

Davidson, D. W. 1977a. *Ecology* **58:** 711–724.

Davidson, D. W. 1977b. *Ecology* **58:** 725–737.

Davidson, D. W. 1978. *Am. Natur.* **112:** 523–532.

Davidson, D. W. 1980. *Am. Natur.* **116:** 92–105.

Davidson, D. W. and S. R. Morton. 1981. *Oecologia* (Berlin) **50:** 357–366.

De Bruyn, G. J. and A. A. Mabelis. 1972. *Ekol. Pol.* **20:** 93–101.

DeVita, J. 1979. *Ecology* **60:** 729–737.

Edwards, R. 1980. *Social Wasps: Their Biology and Control.* Rentokil, E. Grinstead, England.

Eisner, T., I. Kriston and D. J. Aneshansley. 1976. *Behav. Ecol. Sociobiol.* **1:** 83–125.

Emerson, A. E. 1952. *Bull. Am. Mus. Nat. Hist.* **99:** 217–225.

Evans, H. E. 1977. *Bioscience* **27:** 613–617.

Feener, D. H. Jr. 1981. *Science* **214:** 815–817.

Forel, A. 1928. *The Social World of the Ants Compared with that of Man.* G. P. Putnam's Sons, New York.

Forsyth, A. B. 1978. *Studies on the Behavioral Ecology of Polygynous Social Wasps.* Ph.D. dissertation, Harvard University, Cambridge, Mass.

Fuller, C. 1915. *Ann. Natal Mus.* **3:** 329–505.

Gamboa, G. J. 1978. *Science* **199:** 1463–1465.

Gentry, J. B. 1974. *Ecology* **55:** 1328–1338.

Gibo, D. L. 1978. *Can. Entomol.* **110:** 519–540.

Golley, F. B. and J. B. Gentry. 1964. *Ecology* **45:** 217–225.

Grassé, P. P. 1949. Pp. 408–544 in P. P. Grassé (Ed.), *Trait de Zoologie,* Vol. 9. Masson, Paris.

Harris, W. V. 1970. Pp. 295–313, in Krishna and Weesner (Eds.), 1970, referenced here.

Hébrant, F. 1970. *Etude du Flux Énergétique Chez Deux Espèces du Genre Cubitermes Wasmann (Isoptera, Termitinae), Termites Humivores des Savanes Trópicales de la Région Ethiopienne.* Ph.D. thesis, Université Catholique de Louvain, Louvain.

Heinrich, B. 1976. *Ecology* **57:** 874–889.

Heinrich, B. 1978. Pp. 97–128 in J. R. Krebs and N. B. Davies (Eds.), *Behavioural Ecology.* Sinauer, Sunderland, Mass.

Heinrich, B. 1979. *Bumblebee Economics.* Harvard University Press, Cambridge, Mass.

Heithaus, E. R. 1979. *Ecology* **60:** 190–202.

Hill, G. F. 1922. *Bull. Entomol. Res.* **12:** 363–399.

Hill, G. F. 1942. *Termites (Isoptera) from the Australian Region.* Aust. Counc. Sci. Ind. Res., Melbourne.

Himmer, A. 1927. *Z. Vergl. Physiol.* **5:** 375–389.

Hocking, B. 1970. *Trans. Roy. Entomol. Soc., Lond.* **122:** 211–255.

Hölldobler, B. 1976. *Behav. Ecol. Sociobiol.* **1:** 3–44.

Hölldobler, B. 1977. Pp. 418–471 in T. A. Sebeok (Ed.), *How Animals Communicate.* Indiana University Press, Bloomington, Indiana.

Hölldobler, B. and C. J. Lumsden. 1980. *Science* **210:** 732–739.

Hölldobler, B. and E. O. Wilson. 1978. *Behav. Ecol. Sociobiol.* **3:** 19–60.

Howse, P. E. 1970. *Termites: A Study in Social Behavior.* Hutchinson, London.

Hubbell, S. P. and L. K. Johnson. 1977. *Ecology* **58:** 949–963.

Hubbell, S. P. and L. K. Johnson. 1978. *Ecology* **59:** 1123–1136.

Humphreys, W. F. 1979. *J. Anim. Ecol.* **48:** 427–453.

Inouye, D. W. 1977. Pp. 35–40 in W. J. Mattson (Ed.), *The Role of Arthropods in Forest Ecosystems.* Springer-Verlag, New York.

Inouye, D. W. 1978. *J. N.Y. Entomol. Soc.* **85:** 253–254.

Inouye, R. S., G. Byers, and J. H. Brown. 1980. *Ecology* **61:** 1344–1351.

Janzen, D. H. 1973. *Ecology* **54:** 687–708.

Jeanne, R. L. 1975. *Q. Rev. Biol.* **50:** 267–287.

Jeanne, R. L. 1979. *Behav. Ecol. Sociobiol.* **4:** 293–310.

Jeanne, R. L. 1980. *Annu. Rev. Entomol.* **25:** 371–396.

Jeanne, R. L. 1981. *Anim. Behav.* **29:** 102–113.

Johnson, K. A. and W. G. Whitford. 1975. *Environ. Entomol.* **4:** 66–70.

Johnson, L. K. and S. R. Hubbell. 1974. *Ecology* **55:** 120–127.

Johnson, L. K. and S. P. Hubbell. 1975. *Ecology* **56:** 1398–1406.

Josens, G. 1972. Doctoral thesis, University of Brussels, Brussels.

Kasuya, E., Y. Hibino, and Y. Ito. 1980. *Res. Popul. Ecol.* **22:** 255–262.

Krishna, K. and F. M. Weesner (Eds.). 1970. *Biology of Termites*, Vol. 2. Academic, New York.

Lee, K. E. and T. G. Wood. 1971. *Termites and Soils*. Academic, New York.

Levieux, J. 1967. *Insectes Soc.* **14:** 313–322.

Lieberburg, I., P. M. Kranz, and A. Seip. 1975. *Ecology* **56:** 473–478.

Lin, N. and C. D. Michener. 1972. *Q. Rev. Biol.* **47:** 131–159.

Lord, W. D. 1979. *Foraging, Colony Productivity, and Competition in Vespula maculifrons* (Buysson). M.S. thesis, University of Delaware, Newark.

Lord, W. D. and R. R. Roth. 1979. *J. N.Y. Entomol. Soc.* **86:** 304–305.

Lüscher, M. 1961. *Sci. Am.* **205:** 138–145.

Maschwitz, U. 1964. *Z. Vergl. Physiol.* **47:** 596–655.

Maschwitz, U. 1966. *Z. Vergl. Physiol.* **53:** 228–252.

Maschwitz, U. and E. Maschwitz. 1974. *Oecologia (Berlin)* **14:** 289–294.

Matsuura, M. and S. F. Sakagami. 1973. *J. Fac. Sci. Hokkaido Univ. Ser. VI. Zool.* **19:** 125–162.

Matthews, R. W. and J. R. Matthews. 1979. *Nat. Hist.* **88:** 56–65.

Michener, C. D. 1974. *The Social Behavior of the Bees: A Comparative Study*. Harvard University Press, Cambridge, Mass.

Michener, C. D. 1979. *Ann. Missouri Bot. Gard.* **66:** 277–347.

Michener, C. D. and D. J. Brothers. 1974. *Proc. Natl. Acad. Sci. USA* **71:** 671–674.

Möglich, M. H. J., and G. D. Alpert. 1979. *Behav. Ecol. Sociobiol.* **6:** 105–113.

Morse, D. H. 1977. *Science* **197:** 678–680.

Nelson, J. M. 1968. *Ann. Entomol. Soc. Am.* **61:** 1528–1539.

Oster, G. F. and E. O. Wilson. 1978. *Caste and Ecology in the Social Insects*. Princeton University Press, Princeton, N.J.

Pardi, L. 1948. *Physiol. Zool.* **21:** 1–13.

Pleasants, J. M. 1980. *Ecology* **61:** 1446–1459.

Plowright, R. C., B. A. Pendrel, and I. A. McLaren. 1978. *Can. Entomol.* **110:** 1145–1156.

Pontin, A. J. 1961. *J. Anim. Ecol.* **30:** 47–54.

Ranta, E. and H. Lundberg. 1980. *Oikos* **35:** 298–302.

Richards, K. W. 1973. *Quaest. Entomol.* **9:** 115–157.

Richards, O. W. 1978. *The Social Wasps of the Americas Excluding the Vespinae*. Br. Mus. (Nat. Hist.), London.

Richards, O. W. and M. J. Richards. 1951. *Trans. Roy. Entomol. Soc. Lond.* **102:** 1–170.

Rissing, S. W. 1981. *Ecology* **62:** 1031–1040.

Rockwood, L. L. 1973. *Ecology* **54:** 1363–1369.

Room, P. M. 1971. *J. Anim. Ecol.* **40:** 735–751.

Sakagami, S. F. 1959. *J. Anim. Ecol.* **28:** 51–68.

Sakagami, S. F. and H. Fukuda. 1968. *Res. Popul. Ecol.* **10:** 127–139.

Sakagami, S. F. and K. Fukushima. 1957. *J. Kansas Entomol. Soc.* **30:** 140.

Sands, W. A. 1961. *Entomol. Exp. Appl.* **4:** 277–288.

Sands, W. A. 1965. *Bull. Br. Mus. Entomol.* (Suppl.) **4:** 1–243.

Sands, W. A. 1967. *Proc. 5th Congr. Int. Union Study Soc. Insects, Toulouse.*, pp. 159–172.

Sands, W. A. 1969. Pp. 495–524 in K. Krishna and F. M. Weesner (Eds.), *Biology of Termites*, Vol. 1. Academic, New York.

Schaffer, W. M., D. B. Jensen, D. E. Hobbs, J. Gurevitch, J. R. Todd, and M. V. Schaffer. 1979. *Ecology* **60:** 976–987.

Simberloff, D. S. and E. O. Wilson. 1969. *Ecology* **50:** 278–296.

Stebaev, I. V. and J. I. Reznikova. 1972. *Ekol. Pol.* **20:** 103–109.

Stradling, D. J. 1978. Pp. 81–106 in Brian (Ed.), 1978, referenced here.

Sudd, J. H. 1967. *An Introduction to the Behaviour of Ants.* St. Martin's Press, New York.

Suzuki, H. and M. Murai. 1980. *Res. Popul. Ecol.* **22:** 184–195.

Taylor, F. 1976. *Behav. Ecol. Sociobiol.* **2:** 147–167.

Waloff, N. and R. E. Blackith. 1962. *J. Anim. Ecol.* **31:** 421–437.

Weber, N. A. 1966. *Science* **153:** 587–604.

Weesner, F. M. 1970. Pp. 477–525 in Krishna and Weesner (Eds.), 1970, referenced here.

West Eberhard, M. J. 1975. *Q. Rev. Biol.* **50:** 1–33.

Whitford, W. G. 1978. *Ecology* **59:** 185–189.

Whitford, W. G. and M. Bryant. 1979. *Ecology* **60:** 686–694.

Williams, R. M. C. 1966. *Trans. Roy. Entomol. Soc. Lond.* **118:** 73–216.

Wilson, E. O. 1953. *Q. Rev. Biol.* **28:** 136–156.

Wilson, E. O. 1971. *The Insect Societies.* Harvard University Press, Cambridge, Mass.

Wilson, E. O. 1974. *Psyche* **81:** 182–188.

Wilson, E. O. 1975. *Sociobiology: The New Synthesis.* Harvard University Press, Cambridge, Mass.

Wilson, E. O. 1976. *Behav. Ecol. Sociobiol.* **1:** 141–154.

Wilson, E. O. 1980a. *Behav. Ecol. Sociobiol.* **7:** 143–156.

Wilson, E. O. 1980b. *Behav. Ecol. Sociobiol.* **7:** 157–165.

Wilson, E. O. and G. L. Hunt. 1967. *Pac. Insects* **9:** 563–584.

Wilson, E. O. and R. W. Taylor. 1967. *Pac. Insects Monogr.* **14:** 1–109.

Wood, T. G. 1978. Pp. 55–80 in Brian (Ed.), 1978, referenced here.

Wood, T. G. and K. E. Lee. 1971. *Pedobiologia* **11:** 341–366.

Wood, T. G. and W. A. Sands. 1978. Pp. 245–292 in Brian (Ed.), 1978, referenced here.

Section IV
ROLES OR
EFFECTS OF
INSECTS IN
ECOSYSTEMS

There is so much interest in insects as pests that their diverse roles in nature receive relatively little attention, even in Entomology curricula. The next four chapters of this book are presented in an effort to partially correct this bias and to illustrate the functions of insects in communities and ecosystems, some of which are essential in maintaining an environment viable and aesthetically pleasing to mankind.

Insect activity, other than that which is annoying or otherwise undesirable, largely goes unnoticed because most species are so small and cryptic. Many species also are difficult to study because of their high mobility and nocturnal habits. Yet these characteristics, coupled with their high rates of reproduction and genetic adjustment, have allowed the Insecta to permeate terrestrial and freshwater ecosystems. The contribution of insects to the grazing and detritus food webs of these ecosystems are discussed in Chapter 19.

Honey production and pollination are among the most widely recognized benefits attributable to insects. On the other hand, the diversity of insect pollinators and the essentiality of their activities to the reproduction of plants are poorly appreciated by most people. Chapter 20 is offered to enhance this appreciation as well as to present a sample of the amazing array of morphological, physiological, and behavioral adaptions of plants and their insect flower visitors and pollinators.

The highly mobile Insecta also serve facultatively, obligatorily, and accidentally as disseminators of other organisms—seeds, other arthropods, nematodes, protozoans, fungi, bacteria, rickettsiae, mycoplasma, and viruses (Chapter 21). The variety of interrelationships and interdependencies which have evolved between insects and other organisms, insuring dispersal of propagules, comprises one of the most fascinating aspects of natural history.

Much of the energy flow and cycling of materials in nature is made possible by the myriad of insect consumers, which vary in feeding from monophagy to omnivory. The rich variety of herbivore–plant, parasite–host, and predator–prey interactions greatly influences the numerical performance and genetic composition of interacting lineages. From a broader view, these interactions influence the make-up, succession, and stability of plant–animal–microorganism associations. Such influences are discussed in Chapter 22.

Chapter 19

Insect Diversity and the Trophic Structure of Communities

JOHN N. THOMPSON

19.1 INTRODUCTION

"*Crawling at your feet,*" *said the Gnat* (*Alice drew her feet back in some alarm*), "*you may observe a Bread-and-butterfly. Its wings are thin slices of bread-and-butter, its body is a crust, and its head is a lump of sugar.*"

"*And what does it* live *on?*"

"*Weak tea with cream in it.*"

LEWIS CARROLL, *Through the Looking Glass*

Lewis Carroll time and again conjured up a bizarre array of creatures with diverse gustatory and other habits. The novelty of his creations rival those produced by natural selection. Like Lewis Carroll's creatures the real-world insects produced by natural selection exhibit a number of improbable habits, chew and suck on all forms of organic matter, living and dead, and some are even able to eat the leaves of tea without cream. These insects in turn become the food of a host of predators and parasites. The result at times is a bewildering complexity of trophic interactions within even small areas of habitat (Huffaker 1974). This chapter presents a synopsis of the importance of insect diversity in the trophic structure of communities.

In most communities insects occupy intermediate positions along food chains, acting as herbivores, saprophages, parasitoids, or predators of other animals, and they are subsequently eaten by other predators, parasitoids, and true parasites. In some habitats, however, insects may occupy the position of top predators as, for example, in some cave communities (Van Zant et al. 1978). Insects are most important as components of most terrestrial and freshwater communities. Herbivorous and parasitoid groups are the most species-rich, resulting partly from the increased specialization demanded in coevolving with their plant and animal hosts (Price 1975). Insects comprise a large percentage of the animal biomass in these communities. Golley et al. (1975) estimated that in a tropical, moist, Panamanian forest, arthropods accounted for 41% of the herbivore biomass, 10% of the frugivore biomass, 7% of the carnivore biomass, and virtually all of the decomposer biomass. Much of this arthropod component can be expected to be insect biomass.

In contrast to their abundance in most terrestrial and freshwater communities, insects play a relatively small role in marine communities. Most of the species found in marine environments occur in the supralittoral zone, rock pools, or marine algal wracks of intertidal and estuarine environments (Watt 1975, Neumann 1976). These environments have been independently colonized by members of the orders Coleoptera, Diptera, Hemiptera, and Trichoptera. Little is known about the trophic relationships of these species. The intertidal Chironomidae (Diptera) are either algal or algal detritus feeders (Neumann 1976). The larvae of some species are known to be preyed upon by young Pacific salmon (Annan 1958), mites, and several species of sculpin (Morley & Ring 1972).

19.2 INSECTS AS DECOMPOSERS

All living organisms eventually die, and locked within their dead bodies are nutrients essential for the future development of organisms within the community. Even while alive they shed parts that are nutrient-rich. Insects molt several times during development, mammals shed fur seasonally, and birds molt feathers. They all continually leave partly utilized food in the form of feces. Plants shed leaves and as much as 90% of net plant production in some temperate forests may fall to the forest floor as litter (Reiners 1973). Some of this litter is eventually deposited in streams and lakes, where it forms much of the basis for those food webs.

If this organic matter continues to accumulate within an ecosystem, primary production may conceivably be reduced. In most communities, however, a large and varied decomposer fauna utilizes these rich resources, and in terrestrial and freshwater systems a large percentage of the fauna is composed of insects. The habit of feeding on dead and decaying organic matter is a very old one and is observed in some species in most insect orders (Southwood 1972).

Much of the organic matter in decaying leaves, the most important component of detritus, is unavailable directly to the insect detritivores that ingest them. Few insects have the ability to degrade polysaccharides, for example, cellulose (Nielsen 1966, Bjarnov 1972). In most species of insect detritivores that have been studied, their selection of food is actually for the microorganisms on the decaying leaves (Berrie 1976). For example, Mackay and Kalff (1973) found that in two caddisfly species, *Pycnopsyche gentilis* and *P. luculenta,* individuals preferred leaf discs incubated with fungal culture to those incubated with bacterial culture, and both of these were preferred to sterile leaves. Hence, most insect detritivores actually constitute the second consumer level in detritus-based food webs, that is, after microorganisms. The major exception is termites, of which most species feed on woody litter. These species harbor in the gut either symbiotic protozoa or bacteria (depending on the family) capable of breaking down cellulose (Wood 1976); and at least one species apparently produces its own cellulose (Potts & Hewitt 1973). Therefore, termites differ from most other insect detritivores by eating much plant material before it has been attacked by saprophytic microorganisms (Wood 1976).

Generally, insect detritivores assimilate less than 10% of ingested material, most being returned to the soil or water as feces (Cummins 1973). However, by chewing through detritus and passing it through their bodies, they break it up into smaller pieces. This increases the surface area for future attack by microorganisms. Insect and other invertebrate feces become foci for microbial activity, thereby enhancing the rate of decomposition (Hargrave 1976). The major exception, again, is termites, which are highly efficient at assimilating food, partly because of their symbiotic microorganisms and partly because they reingest their own feces. This results in almost total utilization of ingested matter in some species (Wood 1976).

The role of detritivores in enhancing the rate of decomposition has been demonstrated experimentally, generally involving all invertebrate detritivores rather than simply insects. Weary and Merriam (1978) found that decay rates for litter during summer and fall in a red maple woodlot treated with a carbamate insecticide were significantly lower than in control plots, that is, 0.99–1.26 g/m²·day versus 1.48 g/m²·day. Other investigators have shown that rates of decomposition of litter placed in nylon mesh bags to exclude most insects and other invertebrates is slower than for litter exposed to these decomposers (e.g., Heath et al. 1966, Wiegert 1974, Strojan 1978). Confinement in mesh bags, however, apparently underestimates the normal rate of decomposition and results must be interpreted with caution (Witkamp & Olson 1963, Wiegert & Evans 1967).

The other major component of decaying plants is fallen logs. These logs are important colonization sites for forest plants (e.g., Thompson 1980) but are generally unavailable until decayed enough, through the action of microorganisms, mosses, and invertebrates, to have a humus layer on which a seed can germinate and become established. Insect communities in fallen logs can be quite complex (e.g., Graham 1925, Wallace 1953, Fager 1968) and may speed the process of decomposition.

Intermingled with plant litter throughout communities are the dead bodies and feces of animals. The insect communities built upon these resources consist primarily of Coleoptera, Diptera, parasitic Hymenoptera, ants (on carrion), and in warmer areas, termites. The number of species exploiting these transient resources is often quite large. Merritt (1976) catalogued the food habits of the 97 insect species known to be associated with cow dung in north central California. Among the immature insects alone were 24 species known or presumed to feed on dung, 8 species of predators, 16 parasitoid species, and 10 species of unknown feeding habits. The trophic structure and species composition change rapidly on these resources as they are broken down (e.g., Valiela 1974).

Insect activity often greatly enhances the rate of decomposition of carrion and dung. Payne (1965) compared the rates of decomposition of baby pigs placed on the forest floor during summer and attacked by insects with those for which insects were excluded and found decomposition to be significantly faster in the carrion exposed to insects. A more dramatic case of the role of insects in accelerating the rate of decomposition is reported from Australia. Prior to the arrival of Europeans the largest herbivorous animals in Australia were marsupials such as kangaroos, which produce small, dry pellets of dung (Waterhouse 1974). The native scarab beetles use this dung as food and as nest sites for their young. Therefore, the dung does not accumulate. Europeans, however, brought cows to Australia, and the larger, moister dung pads of cows were not utilized by the native dung-feeding species. These dung pads remained on the ground surface for months or years before finally disintegrating. As the number of cattle increased so did the magnitude of the problem; Waterhouse (1974) noted that every 30 min, 6 million cattle-dung pads are deposited on the surface of Australia. Bornemissza (1960) suggested importing dung beetles that would be attracted to bovine dung. Several species of dung beetle have since been imported from Africa and established, and rate of disappearance of dung has increased dramatically in areas where the beetles are established (Chapter 11). A similar problem occurred in the Intermountain West of North America, where cattle were introduced into areas that had not supported large herds of grazing mammals in the past (Mack & Thompson 1982). Although 34 species of the dung beetle genus *Onthophagus* occur east of the Rockies, none occurs in the Intermountain West (Howden 1966). One species, *O. nuchicornis*, has now been introduced into British Columbia (MacQueen & Beirne 1975).

The energy channeled through decomposer-based food webs may often be greater than that channeled through autotroph-herbivore food webs. Long-term studies at Hubbard Brook Experimental Station in New Hampshire have indicated that, except in years of caterpillar outbreaks, the food of insectivorous birds in that forest is linked predominantly to the decomposer food web (Gosz et al. 1978). Stream life is even more closely tied to the detritus food web, and such food webs in streams are derived primarily from detritus that falls into the streams. Thus leaves are colonized by microorganisms, primarily fungi, and then fed upon by a succession of invertebrates and vertebrates (Cummins 1973).

19.3 INSECTS AS HERBIVORES

Insects feed on a vast array of foodplants, ranging from algae to angiosperms. Each evolutionary development of a major new plant group (e.g. gymnosperms, angiosperms) has led to a radiation in insect species to exploit the plants. Because insects are small and able to subdivide resources much more finely than vertebrates, insect species generate a great deal of complexity in food webs at the herbivore level. Although insects have radiated onto most types of algae and plants (e.g., Gerson 1969, 1973, 1976, Balick et al. 1978, Hendrix 1980), the major radiation in herbivorous species has been in taxa that exploit angiosperms (Brues 1946). Angiosperms and insects have been coevolving since the mid to late Mesozoic and the interactions have resulted partly in the great proliferation in species of both groups (Ehrlich & Raven 1964). Much of the complexity in terrestrial food webs results from the large number of species of insect herbivores on angiosperms.

Insects are the major herbivores in many natural ecosystems and in some habitats may account for as much as 80% of the plant material ingested each year by herbivores (Gibbs 1976). Even so, overall estimates of consumption of annual plant production by herbivorous insects are generally less than 10%, except in Australia where estimates are higher (Bray 1961, Odum & Ruiz-Reyes 1970, Burdon & Chilvers 1974, Morrow 1977, Springett 1978) Most estimates are based on missing and damaged sections of leaves on the several dominant plant species in the community. Such damage to plants in a forest canopy results mostly from attack by insects and fungi. The estimates reflect actual total consumption only insofar as chewing insects are the most important herbivorous insects and the percentage of herbivory on the few most dominant species is reflective of the whole community. Sucking insects such as aphids and leafhoppers are important herbivores in some communities (e.g., Murdoch et al. 1972, Claridge & Wilson 1976), and estimates based on chewing insects alone would greatly underestimate total consumption by insects.

Estimates of total consumption are also deceiving in that consumption of plants is not random. Plant species differ in the amount of herbivore pressure they receive. White (1974, 1978) found that while New Zealand alpine grasshoppers rarely consume more than 1–2% of the foliage biomass, consumption of preferred species may reach 59%. In addition, predation on seeds by herbivorous insects may have important effects on the dynamics of plant species in communities that far outweigh the impression left by figures indicating low overall levels of herbivory (Janzen 1971, Thompson 1978, Louda 1982, and Chapter 22). Overall consumption rates, then, reveal little about the actual impact of herbivorous insects on the plant species in the community.

Insects appear to be less efficient than vertebrates in assimilating ingested plant food but more efficient than vertebrates at converting assimilated food into body weight, that is, secondary production (Wiegert & Evans 1967). Maximum gross conversion efficiencies (the ratio of secondary production to ingestion) for leaf-feeding Lepidoptera larvae range from 11–34%; net conversion efficiencies (the ratio of secondary production to assimilation) range from 29–69% (Calow 1977; cf. Scriber & Slansky 1981). This means that much of the food eaten by insect herbivores is returned rapidly to the decomposer food web via feces. The food that is assimilated—that is, absorbed across the gut wall—is often efficiently converted into secondary production, and this production is available to higher trophic levels.

Realization of the importance of insects as major herbivores has in recent years led to a body of literature suggesting that insect grazers may actually control or regulate primary production through their feeding by rapidly returning nutrients to plants (Golley 1972, Chew 1974, Lee & Inman 1975). Mattson and Addy (1975) suggested that the overall complex of insect grazers, by responding to changes in host plant conditions, tends to ensure optimal, long-term primary production within communities. Springett (1978) suggests an extreme view that the degree of regulation of primary production imposed by insects is related to the rates of consumption, high rates such as those found in eucalypt forests ensuring a high degree of control. He further suggests that the rate of regulation of primary production by these herbivores is a result of long periods of coevolution between plants and herbivorous insects. It is difficult to imagine, however, how this could result through natural selection.

The great species richness (diversity) and abundance of herbivorous insects also make them potential major competitors with vertebrates and other invertebrate selective agents on plants. Interactions between major taxa may affect the trophic structure of some communities and, over evolutionary time, the seasonal patterns of primary production. Pulliam and Brand (1975) suggest that the seasonal availability of different kinds of seeds in Arizona grasslands have evolved in response to the seasonal availability of different seed predators. Both ants and sparrows are major predators on seeds in the grasslands, but they differ in the time of year when they feed on seeds. Seeds difficult for ants to carry but easy for birds to handle are produced after the winter rains, when ants are most abundant and sparrows are concentrating on insects. Seeds difficult for birds to handle but easy for ants to carry and eat are produced after the summer rains when ants move underground and birds begin to concentrate on seeds. The seasonal pattern of plant–herbivore trophic structure appears to be an evolutionary result involving these three taxa at least. Thompson and Willson (1978, 1979) suggest that most fleshy fruited species in temperate regions ripen their fruits in fall because at that season a seed has a maximum probability of being dispersed by migrating birds before the fruit is destroyed by invertebrates, especially ants.

Brown and Davidson (1977) suggested that seed-eating rodents and ants are in competition in the Sonoran desert. Their conclusion was based on three lines of evidence: (1) the two groups overlap extensively in the sizes of seeds they eat, (2) exclusion of one group from a site leads to increases in the other group, and (3) the two groups are complementary in their responses to changes along geographic gradients of productivity. In contrast, Mares and Rosenzweig (1978) found no evidence for such complementarity in a larger geographic comparison. Experiments on seed removal rates by ants, birds, and mammals indicated that in the Sonoran Desert of North America rodents were the primary consumers, although ants also took large amounts of seed. In the Monte Desert of South America, similar in climate and flora to the Sonoran Desert, ants were the primary consumers of seeds and mammal consumption was unimportant. Consumption by birds was low in both deserts. Overall consumption, however, was more than an order of magnitude larger in the Sonoran Desert, indicating no evidence for complementarity between the groups.

The role of interactions between insects and other herbivores in affecting the seasonal patterns in primary productivity and the overall structure of communities is much in need of further study. Such studies provide an important level of investigation intermediate between studies of single taxa and studies of whole communities.

19.4 INSECTS AS PREDATORS AND PARASITOIDS

At higher trophic levels insects may affect the trophic structure of communities by modifying the absolute or relative abundance of herbivores and, therefore, indirectly, of plants. Much of the field of biological control emphasizes this point. Predatory and parasitic insects may also restrict the availability of herbivores to other taxa of predators, especially vertebrates. The relative contribution to food webs of insect predators and parasitoids, however, seems to vary considerably on a geographic basis, and patterns are only beginning to emerge.

Intensity of predation is thought by some ecologists to be greater in some natural tropical habitats than in temperate habitats, not only from insects but also from many predatory groups (e.g., Paine 1966, Elton 1973). If borne out by future study this could partly explain why many tropical habitats seem to have low abundance of insect individuals although high numbers of insect species (Bates 1863, Elton 1973, Janzen & Pond 1975). Elton (1973) suggested that the life histories of many tropical forest insects may be geared to low population levels as a result of a long evolutionary interplay between predators and prey. He cites especially the great swarms of army ants in the field layer as obvious candidates for maintaining current predation pressures.

Potentially higher predation by insects in some tropical habitats may also result in higher levels of competition between insects and other predators than in temperature regions. Pearson (1977) recorded only one, uncommon wood-creeping bird species in his study area in New Guinea and an absence or scarcity of wood-creeping birds other than woodpeckers on study areas in Gabon and Borneo. He attributed the paucity of these avian species at the New Guinea and Borneo sites partly to preemption of those resources by predatory tiger beetles. Predatory insects and spiders are also potentially major competitors in many temperate, as well as tropical, communities. Their relative contributions, however, to the trophic structure of communities have been largely unexplored. Janzen and Pond (1975) found that sweep net samples of secondary vegetation in Michigan and England yielded a higher proportion of spiders than did tropical samples. Much more quantitative comparison of habitats will be needed to test the generality of results of temperate–tropical comparisons than have been obtained thus far to be able to assess the relative influence of insects and other predators on the trophic structure of communities.

On the basis of available data, some parasitoid taxa appear to be equally or more abundant in temperate than in tropical food webs, in contrast to predatory insects. Owen and Owen (1974) found that Malaise trap samples of Ichneumonidae from sites in tropical Africa had no more species than did samples taken in temperate Europe. Using distribution data gathered from the literature, Janzen (1981) found that the richness of ichneumonid species in North America peaks between 37.5° and 42.5°. Janzen and Pond (1975) found that parasitic Hymenoptera accounted for a larger proportion of the arthropod fauna in midlatitude than in tropical sweep samples of secondary vegetation. They attributed the lower richness of parasitoids in the tropics to the scarcity of herbivore individuals to parasitize (cf. Jansen 1981). Finally, Rathcke and Price (1976) suggested an additional reason: if predation is more severe in the tropics, both parasitized hosts and their parasitoids will be selected against because parasitized individuals are particularly susceptible to predation. It is uncertain, however, whether or not the results for ichneumonids apply to other parasitoid taxa. Hespenheide (1979, cf. Morrison et al. 1979) argued that the few data that are presently available should not be accepted as exhibiting the general trend in latitudinal diversity of all parasitoid taxa, since the Chalcididae, another large group of parasitoids, seems to show the more usual trend toward higher diversity in the tropics.

Understanding local and geographic patterns in the contribution of insect parasitoids to trophic complexity is extremely important, because these insects may comprise 10–20% or more of the animal species on earth (Askew 1971, Price 1975, 1980). Parasitoid families tend to be much more species-rich than families of predatory insects in the same temperate region. Furthermore, contribution of parasitoids to trophic complexity may still be increasing: Price (1980) suggested that insect hosts are not yet nearly saturated with species of parasitoids.

19.5 INSECTS AS PREY

Insects form the mainstay of the diet for a large proportion of vertebrates as well as spiders and other arthropod taxa, including other insects. Most freshwater fish are apparently insectivorous for at least part of their lives. Some fish faunas are composed entirely of species that feed on invertebrates, with aquatic insects generally forming the major share of the diet, for example, New Zealand freshwater fishes (McDowell and Whitaker 1975). Lizards are mostly insectivorous and rather opportunistic, although some species have become specialists on ants or termites (Pianka 1975). The largest order of birds, Passeriformes, and the two largest orders of mammals, Insectivora and Chiroptera, are largely insectivorous. Clearly, the evolution of food web complexity in terrestrial ecosystems is due greatly to insect diversity.

Because of their diversity in habits and in the habitats they utilize, insects are exploited in a myriad of ways by larger carnivores. Guilds of insectivores are discernible in avian communities, each guild composed of one to several species that exploit insects in a similar manner. These guilds are divided generally on the basis of food habits, vegetation structure used most often for foraging, and foraging behavior (Root 1967). Snow (1971) argued that the great diversity of insects has favored the proliferation of avian species specialized in hunting techniques for different subsets of the insect fauna. Willson (1974) divided avian insectivores from grassland to shrub and woodland habitats in Illinois into 14 guilds ranging from low-foliage gleaners, to bark drillers, to high-canopy sallying species.

The role of insects in generating food web complexity in vertebrates is apparent along latitudinal and elevational gradients in vertebrate species diversity. Avian diversity is much higher in some tropical than in temperate habitats. Orians (1969) noted that this difference was a result mostly of the addition of species in the tropics with no temperate counterparts. He attributed the difference in diversity to the availability in the tropics of a greater range of resource types permanently above the minimum threshold level for exploitation. The most important resources available to birds at permanently exploitable levels in tropical but not in temperate habitats are large insects, fruit, and nectar (Orians 1969, Karr 1975). Similar arguments seem to apply to latitudinal gradients in mammalian diversity (Fleming 1973, Wilson 1974). Higher mammalian diversity in the tropics is due primarily to higher diversity of bats, which have radiated in species onto the greater diversity and year-round availability of insects, fruits, and flowers in many tropical habitats.

Elevational trends are also similar. Terborgh (1977) examined the patterns of avian species diversity along an elevational transect extending from 500 m to greater than 3500 m in the Cordillera Vilcabamba, Peru. Avian diversity decreased monotonically with increasing elevation. This decrease was due mostly to a dramatic decline in insectivores with increasing elevation, while species richness of frugivores and nectarivores declined much more slowly. Terborgh attributed the trend in diversity of avian insectivores to a decrease in the structural diversity of the vegetation at higher elevations, thereby limiting the opportunities for specialized ways of harvesting insects, and also to a general decline in the diversity and abundance of most insect taxa at higher elevations.

The other major species-rich taxon highly dependent on insects as prey and important in trophic complexity is the true spiders. Spiders exist in virtually all terrestrial and in some aquatic environments. Turnbull (1973) estimated that there may be 50,000 species of living spiders. They may be the dominant predators of insects in some terrestrial communities and may be important regulators of insect abundance in some communties, but quantitative data to support this supposition are few. Moulder and Reichle (1972) concluded that spiders are the dominant entomophagous predators, both in terms of biomass and abundance in the litter community of a *Liriodendron tulipifera* forest in Tennessee. By radioactively labeling members of the litter community, they estimated that spiders consumed 43.8% of the mean annual standing crop of litter animals.

Undoubtedly, then, insects as a group constitute the major prey type for most predators of animals in terrestrial and freshwater communities. In addition, even species that eat primarily foliage or fruits often rely on insects for part of their intake of protein. Few temperate species of bird are able to feed for long periods solely on a diet of fruit. Temperate avian species that are highly frugivorous—for example, some thrushes and mimids—need to feed also on insects (Berthold 1976a,b). A number of primate species that feed primarily on foliage or fruits also supplement their diet with insects and other invertebrates (Clutton-Brock 1977). In some species this may be necessary in order to obtain the proper balance of amino acids in the diet. Hladik (1977) found that chimpanzees in Gabon selected plant and insect species whose amino acids were complementary. Therefore, insects influence the pattern of trophic complexity not only through their effect on insectivores but also through this effect on species thought to be primarily herbivorous.

19.6 INSECTS AND THE DYNAMICS OF TROPHIC STRUCTURE

The trophic structure of insect-based communities is variable both in space and time. Some workers have suggested that insect trophic structure is a predictable result of the structure of the plant/detritus community. Heatwole and Levins (1972), using Simberloff and Wilson's (1969) data for recolonization of defaunated mangrove islands by insects, suggested that trophic structure rapidly returns to an equilibrium similar to that prior to defaunation. Simberloff (1976), however, argued that the data do not support this contention because the data from several islands were pooled, the numbers of species are too low for statistical analysis, and not all arthropods were included in the analysis, to list a few reasons. Others, however, have also suggested predictable and consistent trophic structure in communities. Samples of the invertebrate fauna of several broad-leaved tree species in South Africa and in Britain show a fairly consistent proportion of the phytophagous and carnivorous species in late summer: 24% phytophagous species and 20% carnivorous species (Moran & Southwood 1982). Proportions based on numbers of individuals or on biomass, however, differ greatly between tree species. Similarly, Evans and Murdoch (1968) noted a fairly constant ratio of phytophagous to predatory insects in a grassland.

There are two parts to this problem: (1) what determines the ratio of species in different trophic levels in the overall community and (2) how are the activity patterns of species at different trophic levels distributed throughout the season. In reanalyzing Evans and Murdoch's (1968) data, Cole (1980) showed that the consistent ratio of herbivores to predators throughout the season is explainable as a statistical result of the ratio of herbivores to predators in the overall pool of species. That is, the ratio throughout the season is simply a statistical subset of the overall ratio of 1.64 of herbivores to predators. The problem then collapses to the basic question of why the underlying ratio is 1.64 (Cole 1980).

As the species pool develops in a newly available habitat, the ratio of species at different trophic levels can change. The pattern of change in the ratio must depend certainly on the initial resource base in the habitat. A newly planted field of soybeans provides a different resource base for colonization by insects than does a fire-swept tall-grass prairie or a volcanic island. The soybean field has been plowed, leaving little detritus, unlike the prairie; the initial resource base on the volcanic island is likely to be detritus dumped on the shore by wave action. The ratio of phytophagous insects to predator and parasitoid insects on soybeans fluctuates widely over a season (Fig. 19.1). Moreover, the ratio differs between sites that are near the edge of the field (and near a woodland) as compared with sites at the center of the field. Herbivore species colonize the field before predators, skewing the ratio to herbivores early in the season (Price 1976, Mayse & Price 1978). The ratios are closest to unity at midseason.

The detritivore fauna was not included in Mayse and Price's (1978) analysis of soybean arthropods, but in some new habitats these may be the

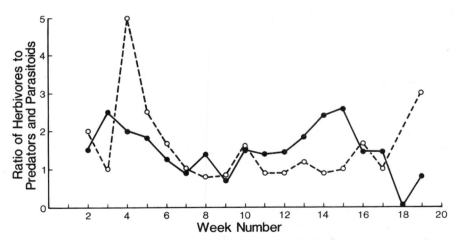

Fig. 19.1   Ratio of phytophagous insect species to predator and parasitoid insect species on soybeans through a growing season in east central Illinois. The ratios are based on the number of insect species found on 15 plants on each sampling date. Closed circles are for the site nearest the edge of the field and next to a woodland; open circles are for the site at the center of the field, 200 m from the edge. Number of herbivore species ranged from 0 to 18; number of predator and parasitoid species ranged from 0 to 10. During week 1, the edge site had no herbivores, predators, or parasitoids; the center site had one herbivore and no predators or parasitoids. (Data from Mayse & Price 1978.)

first colonizers. Dammerman (1948) characterized the order of establishment on Krakatau after its eruption as first scavangers and omnivores, then herbivores, and finally carnivores. The implication is that carnivores are able to become established only after herbivores and would be low in numbers and species in new habitats. On some older islands, however, there is an indication, at least within beetles, that predatory species are more likely to become established on islands than are herbivores. Sweep net samples of secondary vegetation on Caribbean islands yielded a greater relative species richness of predatory beetles than did samples on the mainland in Costa Rica (Janzen 1973). Similarly, Becker (1975), by comparing published faunal lists, noted a higher ratio of predatory to nonpredatory beetles on islands as compared to associated mainlands. Both Janzen and Becker suggested that predators are more likely to become established on islands because of their high level of polyphagy, while fewer herbivores will become established because they accept a narrower range of food species. However, analyses based solely on a single taxon must be viewed with caution if extended to conclusions for whole guilds or a whole trophic level, since patterns of species richness of one taxon may be complementary with other taxa.

Although we still know little about general patterns in arthropod trophic structure, we know even less in a quantitative way about the effects of shifts in arthropod trophic structure on vertebrate guilds or plant population structure. Some guilds of insectivores fluctuate more in local abundance and composition than other guilds (Karr 1980) and some of these variations may result from variation over time in insect trophic structure. What is clear is that among the four or five trophic levels common in many terrestrial and freshwater food chains (Cohen 1978), insects occupy the critical middle links. Their incredible diversity is a major determinant of how communities are structured at higher and at lower trophic levels.

ACKNOWLEDGMENTS

I thank C. B. Huffaker, R. L. Rabb, and G. P. Waldbauer for discussion or comments on an earlier draft of the manuscript.

REFERENCES

Anderson, J. M. and A. Macfadyen (Eds.). 1976. *The Role of Terrestrial and Aquatic Organisms in Decomposition Processes.* Blackwell, Oxford.

Annan, M. E. 1958. *Can. Fish. Culturalist* **23:** 1–3.

Askew, R. R. 1971. *Parasitic Insects.* Elsevier, New York.

Balick, M. J., D. G. Furth, and G. Cooper-Driver. 1978. *Oecologia* **35:** 55–89.

Bates, H. W. 1863. *The Naturalist on the River Amazon.* Reprinted by Dover, New York.

Becker, P. 1975. *J. Anim. Ecol.* **44:** 893–906.

Berrie, A. D. 1976. Pp. 328–338 in Anderson and Macfadyen (Eds.), 1976, referenced here.

Berthold, P. 1976a. *J. Ornithol.* **117:** 145–209.

Berthold, P. 1976b. *Ardea* **64:** 140–154.

Bjarnov, N. 1972. *Oikos* **23:** 261–263.

Bornemissza, G. F. 1960. *J. Aust. Inst. Agric. Sci.* **26:** 54–56.

Bray, J. R. 1961. *Oikos* **12:** 70–74.

Brown, J. H. and D. W. Davidson. 1977. *Science* **196:** 880–882.

Brues, C. T. 1946. *Insect Dietary.* Harvard University Press, Cambridge, Mass.

Burdon, J. J. and C. A. Chilvers. 1974. *Aust. J. Bot.* **22:** 103–114.

Calow, P. 1977. *Biol. Rev.* **52:** 385–409.

Chew, R. M. 1974. *Ohio J. Sci.* **74:** 359–370.

Claridge, M. F. and M. R. Wilson. 1976. *Ecol. Entomol.* **1:** 231–250.

Clutton-Brock, T. H. 1977. Pp. 557–584 in Clutton-Brock (Ed.), 1977, referenced here.

Clutton-Brock, T. H. (Ed.). 1977. *Primate Ecology: Studies of Feeding and Ranging Behaviour in Lemurs, Monkeys, and Apes.* Academic, New York.

Cohen, J. E. 1978. *Food Webs and Niche Space.* Princeton University Press, Princeton, N.J.

Cole, B. J. 1980. *Nature* **288:** 76, 77.

Cummins, K. W. 1973. *Annu. Rev. Entomol.* **18:** 183–206.

Dammerman, K. W. 1948. *Verh. K. Akad. Wet.* **4:** 1–594.

Ehrlich, P. R. and P. H. Raven. 1964. *Evolution* **18:** 586–608.

Elton, C. S. 1973. *J. Anim. Ecol.* **42:** 55–104.

Evans, F. C. and W. W. Murdoch. 1968. *J. Anim. Ecol.* **37:** 259–273.

Fager, W. 1968. *J. Anim. Ecol.* **37:** 121–142.

Fleming, T. H. 1973. *Ecology* **54:** 555–563.

Gerson, U. 1969. *Bryologist* **72:** 495–500.

Gerson, U. 1973. *Lichenologist* **5:** 434–443.

Gerson, U. 1976. *Revue Algologique.* **11:** 213–247.

Gibbs, G. W. 1976. *N. Zeal. Entomol.* **6:** 113–121.

Golley, F. B. 1972. Pp. 69–90 in J. A. Wiens (Ed.), *Ecosystem Structure and Function.* Oregon State University Press, Corvallis.

Golley, F. B., J. T. McGinnis, R. G. Clements, G. I. Child, and M. J. Denver. 1975. *Mineral Cycling in a Tropical Moist Forest Ecosystem.* University of Georgia Press, Athens.

Gosz, J. R., R. T. Holmes, G. E. Likens, and F. H. Bormann. 1978. *Sci. Am.* **238**(3): 92–102.

Graham, S. A. 1925. *Ecology* **6:** 397–411.

Hargrave, B. T. 1976. Pp. 301–321 in Anderson and Macfadyen (Eds.), 1976, referenced here.

Heath, G. W., M. K. Arnold, and C. A. Edwards. 1966. *Pedobiologia* **6:** 1–12.

Heatwole, H. and R. Levins. 1972. *Ecology* **53:** 531–534.

Hendrix, S. D. 1980. *Am. Natur.* **115:** 171–196.

Hespenheide, H. A. 1979. *Am. Natur.* **113:** 766–769.

Hladik, C. M. 1977. Pp. 481–501 in Clutton-Brock (Ed.), 1977, referenced here.

Howden, H. F. 1966. *Can. Entomol.* **98:** 1177–1190.

Huffaker, C. B. 1974. *Env. Entomol.* **3:** 1–19.

Janzen, D. H. 1971. *Annu. Rev. Ecol. Syst.* **2:** 465–492.

Janzen, D. H. 1973. *Ecology* **54:** 687–708.

Janzen, D. H. 1981. *Ecology* **62:** 532–537.

Janzen, D. H. and C. M. Pond. 1975. *Trans. Roy. Entomol. Soc. Lond.* **127:** 33–50.

Karr, J. R. 1975. Pp. 161–176 in F. B. Golley and E. Medina (Eds.), *Tropical Ecological Systems: Trends in Terrestrial and Aquatic Research.* Springer-Verlag, New York.

Karr, J. R. 1980. *Proc. 17th Int. Congr. Ornithol.,* Pp. 1192–1198.

Kuschel, G. (Ed.). 1975. *Biogeography and Ecology in New Zealand.* Junk, The Hague.

Lee, J. L. and D. L. Inman. 1975. *Ecology* **56:** 1455–1458.

Louda, S. M. 1982. *Ecol. Monogr.* **52:** 25–41.

Mack, R. M. and J. N. Thompson. in press. *Am. Natur.*

Mackay, R. J. and J. Kalff. 1973. *Ecology* **54:** 499–511.

MacQueen, A. and B. P. Beirne. 1975. *Can. Entomol.* **107:** 1215–1220.

Mares, M. A. and M. L. Rosenzweig. 1978. *Ecology* **59:** 235–241.

Mattson, W. J. and N. D. Addy. 1975. *Science* **190:** 515–522.

Mayse, M. A. and P. W. Price. 1978. *Agro-ecosystems* **4:** 387–405.

McDowell, R. M. and A. H. Whitaker. 1975. Pp. 277–299 in Kuschel (Ed.), 1975, referenced here.

Merritt, R. W. 1976. *Pan-Pacific Entomol.* **52:** 13–22.

Moran, V. C. and T. R. E. Southwood. 1982. *J. Anim. Ecol.* **51:** 289–306.

Morley, R. L. and R. A. Ring. 1972. *Can. Entomol.* **104:** 1099–1121.

Morrison, G., M. Auerbach, and E. D. McCoy. 1979. *Am. Natur.* **114:** 303–307.

Morrow, P. A. 1977. Pp. 19–29 in W. J. Mattson (Ed.), *The Role of Arthropods in Forest Ecosystems.* Springer-Verlag, New York.

Moulder, B. C. and D. E. Reichle. 1972. *Ecol. Monogr.* **42:** 473–498.

Murdoch, W. W., F. C. Evans, and C. H. Peterson. 1972. *Ecology* **53:** 819–829.

Neumann, D. 1976. *Annu. Rev. Entomol.* **21:** 387–414.

Nielsen, B. O. 1966. *Natura Jutl.* **12:** 191–194.

Odum, H. T. and J. Ruiz-Reyes. 1970. Pp. I69–I80 in H. T. Odum (Ed.), *A Tropical Rain Forest.* Div. Tech. Information, U.S. Atomic Energy Commission.

Orians, G. H. 1969. *Ecology* **50:** 783–801.

Owen, D. F. and J. Owen. 1974. *Nature* **249:** 583, 584.

Paine, R. T. 1966. *Am. Natur.* **100:** 65–75.

Payne, J. A. 1965. *Ecology* **46:** 592–602.

Pearson, D. L. 1977. *Condor* **79:** 232–244.

Pianka, E. R. 1975. Pp. 292–314 in M. L. Cody and J. M. Diamond (Eds.), *Ecology and Evolution of Communities.* Harvard University Press, Cambridge, Mass.

Potts, R. C. and P. H. Hewitt. 1973. *Insectes Soc.* **20:** 215–220.

Price, P. W. 1975. Pp. 1–13 in P. W. Price (Ed.), *Evolutionary Strategies of Parasitic Insects and Mites.* Plenum, New York.

Price, P. W. 1976. *Env. Entomol.* **5:** 605–611.

Price, P. W. 1980. *Evolutionary Biology of Parasites.* Princeton University Press, Princeton, N.J.

Pulliam. H. R. and M. R. Brand. 1975. *Ecology* **56:** 1158–1166.

Rathcke, B. J. and P. W. Price. 1976. *Am. Natur.* **110:** 889–893.

Reiners, W. A. 1973. Pp. 303–327 in G. M. Woodwell and E. V. Pecan (Eds.), *Carbon and the Biosphere.* Brookhaven Symp. Biol. 24, Nat. Tech. Info. Serv., Springfield, Va.

Root, R. B. 1967. *Ecol. Monogr.* **37:** 317–350.

Scriber, J. M. and F. Slansky, Jr. 1981. *Annu. Rev. Entomol.* **26:** 183–211.

Simberloff, D. S. 1976. *Ecology* **576:** 395–398.

Simberloff, D. S. and E. O. Wilson. 1969. *Ecology* **50:** 278–296.

Snow, D. W. 1971. *Ibis* **113:** 194–202.

Southwood, T. R. E. 1972. Pp. 3–30 in H. F. van Emden (Ed.), *Insect/Plant Relationships.* Symp. Roy. Entomol. Soc. Lond. No. 6, Blackwell, Oxford.

Springett, B. P. 1978. *Aust. J. Ecol.* **3:** 129–139.

Strojan, C. L. 1978. *Oecologia* **32:** 203–212.

Terborgh, J. 1977. *Ecology* **58:** 1007–1019.

Thompson, J. N. 1978. *Ecology* **59:** 443–448.

Thompson, J. N. 1980. *Am. Midl. Nat.* **104:** 176–184.

Thompson, J. N. and M. F. Willson. 1978. *Science* **200:** 1161–1163.

Thompson, J. M. and M. F. Willson. 1979. *Evolution* **33:** 973–982.

Turnbull, A. L. 1973. *Annu. Rev. Entomol.* **18:** 305–348.

Valiela, I. 1974. *Am. Midl. Natur.* **92:** 370–385.

Van Zant, T., T. L. Poulson, and T. C. Kane. 1978. *Am. Natur.* **112:** 229–234.

Wallace, H. R. 1953. *J. Anim. Ecol.* **22:** 154–171.

Waterhouse, F. D. 1974. *Sci. Am.* **230:**(4): 100–109.

Watt, J. C. 1975. Pp. 507–535 in Kuschel (Ed.), 1975, referenced here.

Weary, G. C. and H. G. Merriam. 1978. *Ecology* **59:** 180–184.

White, E. G. 1974. *N. Zeal. J. Agric. Res.* **17:** 357–372.

White, E. G. 1978. *Oecologia* **33:** 17–44.

Wiegart, R. G. 1974. *Ecology* **55:** 94–102.

Wiegart, R. G. and F. C. Evans. 1967. Pp. 499–518 in K. Petrusewicz (Ed.), *Secondary Productivity in Terrestrial Ecosystems.* Institute Ecol., Polish Acad. Sci., Warsaw.

Willson, M. F. 1974. *Ecology* **55:** 1017–1029.

Wilson, J. W., III. 1974. *Evolution* **28:** 124–140.

Witkamp, M. and J. S. Olson. 1963. *Oikos* **14:** 124–147.

Wood, T. G. 1976. Pp. 145–168 in Anderson and Macfadyen (Eds.), 1976, referenced here.

Chapter 20

Insects on Flowers

PETER G. KEVAN and
HERBERT G. BAKER

The harmonies between insectan and floral coadaptations are hard to separate. So interwoven are the structures, functions, behavior, physiology, and chemistry of both insects and flowers that mutually assure reproductive success that for their analysis artificial lines must be drawn. We will start with an entomological look at coadaptive structures and functions which will illustrate some behavioral interactions. Next we will address physiological features of insects and flowers which will lead to coadaptive chemistry and further behavioral phenomena. From there we will examine the ecology of communities in regard to pollination before ending with conservation issues. Figure 20.1 represents the facets of pollination biology in a general scheme. We have cited only major published works. Detailed bibliographies can be found (Baker & Hurd 1968, Proctor & Yeo 1973, Faegri & van der Pijl 1978, Richards 1978, Kevan & Baker 1983, Jones & Little 1983).

20.1 FLOWER-VISITING INSECTS

Proctor and Yeo (1973) present an extensive discussion of the most important orders of insects which visit flowers. The characteristics of the flowers they visit (i.e., color, shape, scent, and rewards) function as a unit and can be used to typify the flowers according to their pollinators. These arrays of floral characters, or floral syndromes, are discussed by Faegri and van der Pijl (1978).

20.1.1 Coleoptera

The most primitive pollinators are often thought to be, and to have been, beetles. By the time flowering plants became important in late-Jurassic and Cretaceous time, beetles were well differentiated. They were probably destructive in flowers, chewing on the ovaries, anthers, and other floral parts, as well as eating pollen and floral secretions. The flowers they visited are conceived as heavily constructed and bowl shaped, perhaps with the scent of aminoids or fermenting fruit. Pollination was accomplished in a "mess-and-soil" manner. *Magnolia* is often tendered as a model. Many flower-visiting beetles fit the above description. However, the syndrome of cantharophily is not found in many flowers that are also visited by beetles. Some specialized relationships exist as in the pollination of some orchids by beetles and, as in the neotropical beetle, *Nemognatha*, which has highly elongate maxillae used to reach deep into long tubular flowers, such as those of *Ipomoea*.

Most of the predatory Adephaga are not flower visitors, but among the Polyphaga there are many flower visitors. Some families (e.g., Mordellidae, Oedemeridae, and Melyridae) may be exclusively anthophilous as adults. Some show clear adaptations to floral feeding such as forward projection

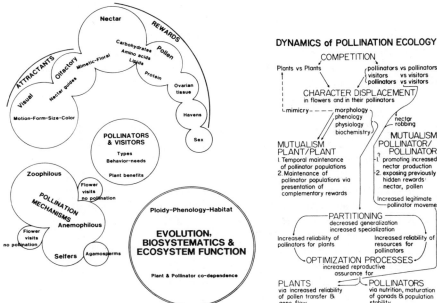

Figure 20.1   The dimensions of pollination. This entomocentric view shows the relationships of floral attractants and rewards on pollinators and insect visitors and the roles of the latter on plant reproduction. The processes involved (see Fig. 20.2) lead to understandings of evolution, ecology, and biosystematics.

Figure 20.2   The dynamics of pollination. This flowchart starts with aspects of competition in pollination systems and follows their consequences through main-stream processes of character displacement, partitioning, and optimization, all of which heighten the mutualism which is basic to evolution in pollination systems. Side issues of mutualisms between plants and mutualisms between anthophiles may heighten the effectiveness of the basic mutualism in assuring the reproductive success of both plants and pollinators.

and elongation of mouthparts, uptilting of the head, and elongation of the prothorax.

Curculionidae are known to be associated with many palm inflorescences, and recently *Elaeidobius kamerunicus* has been introduced from Africa to Malaysia to pollinate the oil palm *Elaeis guineensis,* also of African origin (Syed et al. 1982). The results have been startlingly successful and were evaluated at (USA) 115×10^6 per annum in increased oil crops in 1982.

20.1.2   Diptera

Diptera have also been suggested as early pollinators. The Nematocera are the most primitive. In most families the proboscis is short, although variable in form. Sciaridae have been recorded from the flowers of *Drimys* (Winteraceae), a primitive flowering tree of the tropics. Its flowers, like those visited by other Nematocera, have readily accessible nectar. Most flowers visited by flies have nectar which is exposed or partially exposed in short tubes (e.g., *Achillea, Senecio, Polygonum,* various Cruciferae, and Umbelliferae) or even hidden (e.g., *Salix*). Most Nematocera are small (Mycetophilidae, Cecidomyidae, Simuliidae, Chironomidae, Ceratopogonidae, etc.). For the most part, these insects seek nectar, although some feed on pollen [e.g., *Bibio, Scatopse, Sciara,* and *Atrichopogon* (Downes 1971)]. The larger Tipulidae are restricted to the same sorts of flowers, as they too have short mouth parts. The Nematocera with longer proboscides (e.g., Culicidae and Bibionidae) also visit such flowers, but included are some with deeper tubular corollas (e.g., Compositae and Scrophulariaceae). Hocking (1953) has detailed the nectar relations of northern biting flies, but in general the relations of Nematocera and flowers are poorly known. Fungus gnats have been shown to be important in the pollination of flowers in the Californian redwood forests (Mesler et al. 1980).

Although most flower visiting by Nematocera seems general, specialized relationships exist as in mosquito pollination of *Habenaria* orchids and various Nematocera in pollinating Araceae. In the latter the inflorescence and associated structures rely on heat, mimetic scents, and colors to attract the pollinators (below). Cocoa is pollinated by specialized Ceratopogonidae (cf. Winder 1978).

The Brachycera present a wider diversity of flower visitors. There are numerous records of Stratiomyidae, Dolichopodidae, Lonchopteridae, Phoridae, and especially Empididae and Bombyliidae as flower visitors. Probably, most adult Brachycera feed at flowers. Those with short mouthparts visit many kinds of flowers with easily accessible nectar. Those with longer mouthparts, especially the Bombyliidae, also visit more deeply formed and tubular flowers. The bombyliids are among the most specialized of dipteran flower visitors. Most have long sucking mouthparts. *Bombylius major* has a proboscis about 10 mm long; in *B. discolor* it is about 12 mm. This genus has been recorded on *Viola, Primula, Cardamine, Vaccinium,* and *Muscari.* The Empididae, although not as specialized for flower feeding as many bombyliids, are common flower visitors throughout their almost ubiquitous range. These flies use their tubular piercing mouthparts for killing and feeding on prey as well as for extracting nectar from open to short tubular flowers.

The Cyclorrhapha, the largest suborder of Diptera, is split into the Aschiza and Schizophora. In the Aschiza, Syrphidae is the most important family of anthophiles. These flies feed on nectar and pollen. Their mouthparts are variable, allowing different species to feed from the open-flowered umbelliferae to the deep flowers of Compositae, Labiatae, Scrophulariaceae, Violaceae, Primulaceae, Polemoniaceae, and so on. The hover flies with short mouthparts may be found on deep tubular flowers from which they feed on pollen; they are unable to reach the nectar. Pollen appears to be the protein staple of many Syrphidae (Gilbert 1981). Little is known of the anthophilous habits of other Aschiza; some families are flower visitors, others are not.

In the Schizophora the Acalypterae contain one noteworthy flower-visiting family, the Conopidae. These insects have long proboscides, about 6 mm in *Sicus* spp. and 4 mm in *Conops* spp. They seem to be restricted in their floral preferences, the Compositae being favored. They are recorded from other flowers. Their possible role in pollination is not well understood. Most families of the Acalypterae have been recorded on flowers. The Tephrididae oviposit on the heads of Compositae. Some Drosophilidae and Sphaeroceridae are sometimes found feeding and breeding in spathes of Araceae and in other flowers.

The Calypterae are an important group of anthophiles. The parasitic Tachinidae have elongate mouthparts and are frequently found on flowers, especially of Compositae. Calliphoridae have shorter mouthparts and may visit flowers when preferred food (e.g., dung and carrion) is not available. The Muscidae is a huge family with many well-known anthophiles. Many have short mouthparts and are not specialized anthophiles. Their importance in pollination is often discounted although they are abundant, frequently visit flowers, and effectively transport pollen. *Scatophaga* feeds at flowers for nectar and for prey. The Anthomyiidae are well-known flower visitors; some have quite long mouth-parts. They feed on nectar; some also feed on pollen which may be the protein staple for adult nutrition and ovarian maturation (cf. Chapter 4).

The sapromyophilous pollination syndrome is characterized by flowers which attract a variety of dung and carrion flies. The flowers are mimetically colored to resemble dung or carrion and release skatoles or aminoids to be appropriately smelly. *Rafflesia* (with flowers up to 1 m across), desert *Stapelias*, some Araceae, and orchids serve as examples. Some (e.g., *Ceropegia*) have light windows towards which the pollinators crawl, passing the sexual parts of the inflorescences in attempting escape.

20.1.3 Lepidoptera

Most adult Lepidoptera feed extensively on floral nectar. Some feed on a variety of other liquids, as in puddling, or on fruit juice, excrement, animal secretions, and even blood. A few do not feed at all, especially in the females. The Micropterigidae may feed directly on pollen. The characteristic long, sucking proboscis shows its epitome in *Xanthopan morgani* f. *praedicta* (Sphingidae) which feeds on nectar of, and pollinates, *Angraecum sequi pedale*, an orchid with a nectariferous spur 25–30 cm long. Generally, Lepidoptera imbibe only less viscous nectars, but some (e.g., *Plusia gamma*) may secrete saliva to dilute syrupy or crystallized nectars for imbibing, as do many Diptera.

The flowers visited by the diurnal butterflies are often colorful and may or may not be scented. Typically, they have long tubular corollas with extended petal lobes which form a platform on which the butterflies land to feed (e.g., *Phlox, Primula, Dianthus*). The head of many Compositae forms a similar platform. The flowers visited by nocturnal moths are typically pale and strongly scented. The scent acts as a long-distance attractant, and the color, contrasting with dark vegetation at night, may be enhanced by the floral parts being long and divided. Some flowers offer landing platforms, but those pollinated by hovering moths, such as Sphingidae, open more horizontally and are more trumpet shaped.

20.1.4 Hymenoptera

Bees are recognized as the most important pollinators, yet other groups of Hymenoptera are frequent visitors to flowers. The Symphyta have short mouthparts and no special adaptations to anthophily. Many visit the flowers of their larval host plants and flowers with easily accessible nectar. There they may feed on nectar, pollen, or floral parts. Their relations with flowers are nearly all unspecialized and little studied. In the Apocrita the Parasitica are also mostly unspecialized anthophiles. White flowers seem to be frequently visited. Nevertheless, some may be restricted in the range of flowers they will visit. Studies of the floral relationships of Parasitica are needed because of their potential as biocontrol organisms and the importance of nectar in their nutrition (cf. Chapter 4). Some highly specialized relationships are noteworthy, such as the pollination of *Cryptostylis,* an Australian orchid, by *Lissopimpla semipunctata* (Ichneumonidae) by pseudocopulation in which the males pollinate the flowers while attempting to copulate with them. The remarkable story of fig pollination by *Blastophaga* is described by Wiebes (1979). The Crysididae are the most specialized of the Parasitica for flower visiting, some having long proboscides, yet their role in anthecology is unknown.

Among the Aculeata, only the bees have elongated proboscides. The adults of many families of wasps visit flowers extensively (Spradbury 1973), mostly for their own nutrition. Some social Vespidae (e.g., *Vespula*) also gather sugary liquids to feed their larvae. "Wasp-flowers" have been characterized as dull, pinkish to brown in color, with easily accessible nectar and sometimes unpleasant smell (e.g., *Scrophularia, Cotoneaster, Epipactis*). Some Formicidae are frequent visitors to flowers, but their importance in pollination is not understood. The ant pollination syndrome is for low sprawling plants with flowers of different individuals being intermixed and with easily accessible nectar. The ants can walk between flowers of different plants carrying pollen. *Diamorpha smallii* (Crassulaceae) is a convincingly described example (Wyatt 1981). Other relations between ants and flowers involve the protective function of ants feeding at extrafloral nectaries and repelling other more destructive herbivores from developing flowers, as in *Helianthella* (Inouye & Taylor 1979).

The Apoidea are the most important and highly adapted anthophiles. Their mouthparts are especially adapted for imbibing nectar and their bodies for carrying pollen. They are highly diverse structurally, behaviorally, and taxonomically. Bumblebees and honeybees are quick at learning floral intracacies and behaviorally adept at manipulating complex flowers (von Frisch 1967, Laverty 1980). The syndrome of bee pollination is characterized by flowers which are zygomorphic (i.e., bilaterally symmetrical) with hidden rewards (usually nectar or pollen, but sometimes oils). Familiar examples are Labiatae, Violaceae, Leguminosae, Scrophulariaceae, Orchidaceae, *Aconitum,* and *Delphinium.* Broadly tubular flowers such as those of Ericaceae, Boraginaceae, and Campanulaceae and more narrowly tubular Caryophyllaceae, Primulaceae, and Cruciferae which are radially symmetrical are also bee flowers. Complex and highly specialized relationships exist in the brazil-nut (Lecythidaceae), in pseudocopulatory pollination in *Ophrys* (Kullenberg 1961), and in essential oil gathering and pollinating by male bees at orchid flowers as part of their mating behavior (Williams & Dodson 1972). The diversity of bees on flowers and their pollinating habits are given in broader studies on Onagraceae (cf. Linsley et al. 1963), creosote bush (Hurd & Linsley 1975), sunflower (Hurd et al. 1980), and squash (Hurd et al. 1974). Eickwort and Ginsburg (1980) review foraging behavior in Apoidea, as do Plowright and Laverty (1984) for bumblebees.

20.1.5 Minor Groups

Although most anthophilous insects are Holometabola, lower orders are also found on flowers (cf. Porsch 1958, Kevan & Baker 1983). Collembola ingest pollen, and sometimes nectar, from flowers all over the world (Kevan & Baker 1983). Plecoptera and orthopteroids also visit flowers (Porsch 1958, Kevan & Baker 1983). In the latter, floral mimicking predators (mantids) and herbivores (phasmids) are known. Blattids are known to visit flowers but are infrequently recorded. They may be pollinators. Tettigoniids, especially *Conocephalus,* may be frequent and destructive flower visitors. The peculiar Australian Zaprochilinae are adapted for anthophily, having narrowly prognathous heads (Rentz & Clyne 1983). Acrididae are frequently encountered on flowers, but discounted as incidental visitors. Earwigs hide in flowers where they are generally destructive.

Thrips are notorious flower visitors, and some have mouthparts especially adapted for piercing and sucking out pollen grains (Lewis 1973). Their role in pollination has been investigated in European Ericaceae (Hasterud 1974) and in Malaysian Dipterocarpaceae (Appanah & Chan 1981).

Heteroptera are conspicuous and common anthophiles. Nabidae, Miridae, Lygaeidae, Coreidae, and Pentatomidae are the most frequently found anthophilous families; they frequent flowers with easily accessible nectar (e.g., Compositae and Umbelliferae). Some Phymatidae use flowers as places to prey upon other insects (Balduf 1941). There is little information on the importance of the anthophilous habits of these insects to either the insects or the plants. There are occasional records of Neuroptera, Mecoptera, and Trichoptera as flower visitors feeding on nectar, or pollen, or both (Porsch 1958).

20.2   FLORAL ATTRACTANTS AND INSECT SENSES

The appreciation insects have of their environment can be understood through an ecological view of physiological adaptation. Anthophilous insects have finely attuned senses of vision, olfaction and taste, mechanoreception, and time, which extend especially to their appreciation of floral attractants.

20.2.1   Color and Color Vision

The visual spectrum of insects is shifted approximately 100 nm to the shorter wavelengths of the spectrum as compared with humans: their vision extends from about 300 (UV) to 650 nm (yellow–orange). Goldsmith and Bernard (1974) show that most insects so far tested have peaks of sensitivity in UV, blue–green, and yellow. In *Apis* and *Bombus*, color vision has been shown to be trichromatic, that is, using those three primary colors. Some flies appear to be deuteranopic (color blind, analogous to red–green color blindness in humans) and confuse blue through yellow but distinguish UV. Some insects may have only tonal, or black and white, vision. Kevan (1978, 1983) has placed insect color vision, as represented by *Apis*, into an ecological context, especially in anthecology, by considering the properties of daylight and the spectral reflectance of flowers across the insect visual spectrum. From this he has devised a method of colorimetry and color naming by adapting techniques used in the trichromatic color-naming scheme used for colorimetry in the human visual spectrum. Although this technique may have shortcomings, it provides a method whereby humans may start to have an appreciation of the diversity of color patterns in the insect world. These studies have stressed that UV is no more important to insects than their other primary colors and that all wavebands of concern to insects must be considered when attempting to understand floral colors as insects may see them.

Kevan (1983) examined whole floras of particular habitats—the Canadian high arctic and Canadian weeds—and showed that the colors of the flowers are more diverse and more discrete to insects than to humans. Furthermore, color patterns within flowers are more diverse and contrasting when looked at in the insect visual spectrum. These color patterns, or nectar guides, assist insects in obtaining rewards on complex or large flowers. Some of these are bulls-eye patterns as in *Myosotis* with its blue coloration and yellow center, or in many Compositae with yellow centers (insect-red) and yellow + UV peripheries (insect-purple). Others are patterns of stripes and spots such as can be seen on *Viola, Digitalis,* many lilies, and so on. In general, butterfly flowers show the highest incidence (83%) of nectar guides, followed by Zygomorphic flowers and then capitulate ones. Even about half of the bowl-shaped flowers examined have nectar guides. These patterns may change with age, telling the informed visitor the state of the flowers. In *Aesculus* the orange-spot nectar guides turn red as the flowers age and cease nectar production; they are then ignored by bumblebees. *Senecio* heads become brown in the center as they age, cease producing nectar and pollen, and are then ignored by hoverflies. Numerous legumes change the colors of their banner petals as they age (e.g., *Lupinus, Lotus, Oxytropis, Caesalpinia, Parkinsonia*). They often also change their shape, some by wilting, after pollination. Postpollination changes are often rapid in onset in orchids (cf. Gori 1983).

From the foregoing, it is obvious that flower color is important to anthophiles in their recognition of plant species and the potential for reward offered by the flowers. Other visual attractants also play a part in attraction. The size of flowers, inflorescences, or the corporate image of floral groups have been shown to be positively related to attractiveness over distance. Flicker fusion, that is, the speed at which flickering images blur together and appear to cease to flicker, is very much faster in insects than in humans. Thus, floral movement and the outlines of flowers where they contrast against the background, do not blur-out as the insect moves towards and about the flowers and the ommatidia of the compound eye are repeatedly and sequentially stimulated. Flowers with broken outlines or moving parts are generally more attractive, but these phenomena have been little studied.

Some generalizations on the color preferences of insect groups for flowers can be cautiously made. Flowers reflecting blue are frequented by bees, but these flowers are often structurally adapted to bee pollination (e.g., Leguminosae, Scrophulariaceae, Boraginaceae, Labiatae). Nocturnally pollinated flowers are pale, as are flowers of the deep forest, and contrast against dark or ill-lit backgrounds. Yellow flowers attract an almost unlimited diversity of visitors. Some unspecialized Coleoptera, Diptera, and Lepidoptera seem to show preference for yellow. Red flowers are mostly associated with bird pollination, but others have butterfly pollinators. Some butterflies have been shown to have red-sensitive vision. There are almost no UV flowers: *Papaver rhoeas* is one, being red (invisible to most insects) and UV. The UV reflective patterns on *Ophrys* flowers pollinated by pseudocopulation by male *Gorytes* wasps offer a "supernormal" visual image in mimicking the female wasp; the flowers have more UV insect reflectance than the model.

20.2.2 Odor

It is more difficult to generalize about floral odors. Odors are difficult to analyze and insects' powers of olfaction are more diverse than their powers of vision. Many floral odors have no counterparts outside blossoms: we associate the scents with flowers. In diurnal flowers, it seems that floral odors act as a close-in attractant to entice landing after long-distance attraction by general coloration and at intermediate distances by color patterns. However, in oranges and other plants, the corporate scent of large stands may act over long distances. In nocturnally blooming plants, the scents are often heavy and pervasive and the roles of color and scent reversed. The olfactory sense of bees seems quite similar to that of humans, although bees are more sensitive to floral scents and their own pheromones (von Frisch 1967). Butterfly-pollinated flowers are mostly weakly scented, possibly reflecting a poorly developed olfactory sense in these insects. We have already noted plants with mimetic scents that attract pollinators. These include skatoles and aminoids which are emitted from various Araceae, *Stapelia, Rafflesia,* and so on, and attract dung and carrion-seeking flies and beetles. Musky scent from *Arum conophalloides* attracts biting flies and fungal scents of some Aristolochiaceae and Araceae entice Mycetophilidae to enter and pollinate (Vogel 1978). The bee-orchids, *Ophrys,* are remarkable in producing a chemical mimic of the mating pheromones of bees and wasps; they resemble farnesol, hydroxycitronellal, and γ-cadinene (Priesner 1973) and induce copulatory behavior by the male insects on the flowers to bring about pollination. At *Catasetum* orchids, male euglossine bees collect chemically complex perfume droplets. The perfume is used by the bees in mating behavior, and the specific relations between species of euglossine bees and *Catasetum* act as isolating mechanisms in both the plants and insects (cf. Dodson 1975).

Recently, phenylacetaldehyde, which smells of lilac, has been used to bait black-light traps for moths. This compound is also emitted by the bladders of *Araujia sericifera* (Asclepiadaceae) flowers. The flowers are complex and only strong moths bring about pollination; the weaker ones are held by the tongue in the flower (Cantelo & Jacobson 1979). Different scent morphs in alfalfa have been shown to be preferentially pollinated. Galen and Kevan (1980) investigated this sort of phenomenon in the Rocky Mountain alpine plant, *Polemonium viscosum,* which may have either sweet or skunky flowers. The former are more visited by bumblebees and are more frequent at higher altitudes, as are bumblebees. At lower altitudes, but still above timberline, bumblebees are less common, skunky plants more common, and the flies which visit the flowers have fewer open bowl-shaped flowers from which to feed than they do on higher slopes.

20.3 THE REWARDS OF VISITING FLOWERS

As shown in Fig. 20.1 the insect rewards of visiting flowers are mostly nutritional, in nectar or pollen or both, or other floral tissue, but they also include havens from predation, for warmth, and sex. We will examine the chemical nature of nectar and pollen and place these in the context of the foraging energetics of insects. Baker and Baker (1975, 1983) and Kevan and Baker (1983) have reviewed the significance of nectar chemistry to pollinators, and Baker and Baker (1983) and Kevan and Baker (1983) have discussed pollen. Foraging energetics has been well reviewed by Heinrich (1975, 1979). Through the following discussion, the coadaptations of floral rewards and insect foraging and energetics will be seen to be as precise as the coadaptations for floral attractants and insect senses.

20.3.1 Nectar and Other Liquid Secretions

Nectar is a plant secretion derived from phloem sap. Its secretion is a complex physiological process of special glands (nectaries) that are not restricted to flowers. Autonomous rhythms in the plant, together with the plant's nutritional state, water balance, and responses to the physical environment, all affect secretion. After secretion, nectar may evaporate or absorb water, depending on atmospheric humidity, to become more or less concentrated (Corbet et al. 1979). Thus, generalizations about sugar concentrations of nectar must be made cautiously.

Nectar is a complex mixture of chemicals, of which sugars are the major constituents. Amino acids, proteins, lipids, antioxidants, alkaloids, vitamins, organic acids, allantoin and allantoic acids, dextrins, and inorganic materials such as minerals may be present and all probably have some role in pollination.

The three major sugars of nectar are glucose, fructose, and sucrose. These occur in different proportions in different plant families (e.g., Ranunculaceae tend to be sucrose-rich, whereas Compositae and Cruciferae tend to be hexose-rich). Open bowl-shaped flowers tend to be hexose-rich, have concentrated nectar because of evaporation, and are visited by generalist anthophiles with short lapping mouthparts. Flowers pollinated by long-tongued bees, butterflies, and moths tend to be sucrose-rich. Those pollinated by short-tongued bees may be either. The amount of sugar and nectar secreted also follows the same ranking of least in generalist flowers to more with increasing specialization. The least amounts are found in open bowl flowers and the most in the zygomorphic and stereomorphic flowers. Not all nectar sugars have the same nutritive value (Haydak 1970). Several are toxic to honeybees, but not to other insects (e.g., galactose, lactose, and raffinose). Although the various uses that insect groups put different sugars to may be different, the main outcomes of sugar ingestion are energy for locomotion (Hocking 1953), development, maintenance, and progeny production (Chapter 4), and sometimes in thermoregulation (Chapter 18). Sugars may be converted into fat, as in migratory insects (Johnson 1969, Chapter 7), and stored for long-distance flight and ovarian maturation. Sugar solutions fed to otherwise starved insects prolong their lives although they slowly lose nitrogen (cf. Chapter 4).

The presence of protein-building amino acids in nectar may be important in the nutrition of nectarivorous insects. However, there are few experimental studies which test this. Baker and Baker (1982, 1983) have surveyed floral nectars for the amounts of amino acids and found that those with the most amino acids are taken by insects which do not ingest pollen, for example, flowers pollinated by butterflies, settling moths, bees, and wasps. These nectars had 1.15–0.91 μmole/ml of amino acids, whereas those pollinated by insects which may also ingest pollen (flies and bees) had less than 0.56–0.62 μmole/ml. Exceptions occur on either end of the scale with dung and carrion-mimicking flowers at 12.5 μmole/ml and hawkmoth pollinated flowers at 0.54 μmole/ml. In the latter, huge amounts of nectar are imbibed, as much as 1 ml without satiation, and this would contain significant amounts of amino acids, which could be toxic if the amino acids were not weakly represented. All 20 protein amino acids can be found in nectars, but all are not equally available. Alanine, arginine, serine, proline, and glycine are the most commonly available, whereas tryptophan, histidine, and methionine are scarcest. Nonprotein amino acids may also be present. Apart from the nutritive function they probably have, amino acids may be feeding stimulants and taste modifiers.

Proteins in nectars probably have an enzymatic role and are present in small amounts. Lipids and oils are also present in many nectars [34% of those tested by Baker & Baker (1983)], most commonly in those imbibed by Hymenoptera and Diptera. Polyunsaturated fatty acids and sterols are two classes of lipids important to insects, as are fat-soluble vitamins. The nutritional role of these nectar constituents for insects is unknown, but the lipids must be regarded as beneficial. They may form a waterproofing monolayer on some nectars and thus retard evaporation of water (Corbet et al. 1979). Ascorbic acid and other antioxidants tend to be found together in nectars, elaiophores, and stigmatic exudates where they may prevent rancidity of the lipoidal compounds.

Oil droplets are secreted by some South American flowers in lieu of nectar [e.g., Scrophulariaceae and Melastomataceae (Vogel 1974)]. The secretory organs, elaiophores, may or may not be part of the nectary. Some Anthophoridae (*Centris*) have specialized tarsal brushes for handling this oil, which is mixed with pollen to form food pellets for progeny.

Potentially toxic or distasteful compounds such as nonprotein amino acids, glycosides, alkaloids, and phenolics are well known in some nectars and may find their way into honey. Minerals may be important nectar constituents but are not well studied. Potassium at 1500 ppm deters bees from taking onion nectar (Waller et al. 1972).

20.3.2 Pollen

Although pollen was probably the original reward sought by pollinating insects as they started their coevolution with flowering plants, it is now secondary to nectar. Nevertheless, it is a vital food for many insects such as springtails, orthopteroids, thrips, beetles, flies, and larval bees. Pollen is highly nutritive (cf. Stanley & Linskens 1974), being rich in protein, peptides, and amino acids. Free sugars are less important in pollen than in nectar. Starch may or may not be present and may be an alternative to lipids in other pollens as stored energy reserves for pollen development (Baker & Baker 1979). Lipid-rich pollen tends to be found in plant species which offer pollen as the only floral reward. Wind- and self-pollinated plants tend to produce starch-containing pollen. There is also a correlation with size of pollen grains: the smaller the grains the more the tendency to lipid storage. Lipids in pollen include sterols, which may provide consumer insects with the building blocks of hormones and pheromones. Vitamins are also present, along with pigments, enzymes, and occasional toxic substances.

There has been little experimentation on pollenivorous insects. Pollen is important in honeybee nutrition (Haydak 1970), and in some Diptera it is ingested more frequently by females than by males (by the former at the time of yolk deposition in ovarian maturation).

20.3.3 Other Rewards

Apart from nectar and pollen, some flowers offer special food bodies, sometimes staminodes (non-pollen-producing stamenlike structures), to pollinating insects. However, little is known about these. Flower-destructive insects, especially beetles, must gain considerable nutriment from eating floral parts such as ovaries and maturing anthers.

Perfume collecting is mentioned above for euglossine bees. It is not restricted to *Catasetum* orchids, being recorded for *Spathiphyllum, Anthurium,* and *Gloxinia.*

Insects may find protection in blossoms. Certainly thrips, anthocorids, small beetles, and flies may spend extended periods in flowers and may be pollinators (above). The use of flowers, and later fruits, as protected brood chambers is evident in the extraordinary mutualisms of *Tegiticula* moths and *Yucca* (Powell & Mackie 1966), agaonid wasps and *Ficus* (Wiebes 1979), and perhaps *Hadena bicruris* and *Silene alba* (Brantjes 1978). Some plants, such as Araceae, generate heat which may drive off volatile chemicals attractive to pollinators. Other flowers focus or absorb solar radiation and become warmed. Added heat may benefit the plant, the pollinators, or destructive insects in hastening their development (Kevan 1975a). *Serapias vomeracea,* a Mediterranean orchid, has flowers which entice solitary male bee pollinators to spend the night sleeping in them: the morning sun awakens them as temperatures rise to 3°C (Dafni et al. 1981).

Some bees and other insects mate in flowers, but the importance of this in pollination is not well known (cf. Kevan & Baker 1983). Insects using flowers as sites for ambushing prey are mentioned above (e.g., flower-mimicking mantids, *Scatophaga,* and Phymatidae). Balduf (1941) noted that the ambush bugs *Phymata pennsylvanica americana* were most abundant on flowers with the most pollen and nectar, which were well visited by prey species. Crab spiders (Thomisidae) are notorious and cryptic predators in flowers. Some mites also use flowers for finding prey or hosts (cf. Kevan & Baker 1983).

20.4 FORAGING, PHYSIOLOGY, AND BEHAVIOR

20.4.1 General

From the foregoing one can appreciate that the rewards provided by a flower are in accord with the nutritional needs of the pollinator. Most research has been centered on nectar rewards and energy needs of the pollinators. Thus, individual flowers must provide adequate reward to maintain the interest of the pollinators, but not so much as to satiate them before they make the required number of visits to neighboring flowers to bring about pollination (Heinrich 1975, 1979). The effectiveness of pollination is determined by floral structure, nectar characteristics (above), resource partitioning and competition between visitors (pollinating or not), and inter- and intraspecific competition between plants for pollinators. The pollinators interact in the same sorts of ways but must make optimal use of time and energy in foraging effectively. Figure 20.2 shows this dynamic interplay of mutualism and competition. One can see how selective pressures for excluding inefficient pollinators can promote a complex pattern of interrelationships in plant–pollinator communities. These pressures result in a frugal energy balance between the plant donor and its pollinators. Concomitantly, coadaptive floral characters increasingly heighten the precision of visitation and pollination through coevolution with pollinator anatomy, preferences, and behavior. At the same time, there is room for the breakdown of specialization and the favoring of less specialized relationships.

20.4.2 Physiology

The energy balance in pollination has botanical and zoological components. The amount of absorbed solar energy plants devote to nectar remains almost unstudied. Measuring energy consumption in insects is difficult. Respirometry is the most common method. As restrained insects are frequently used, the readings of O_2 consumed or CO_2 emitted are unnatural. Further, insects may change their metabolic rate 10-fold over a few minutes, with little outward sign. Heinrich (1975, 1979) reviewed these difficulties and found that the body temperature of bumblebees is a reliable indicator of energy expenditure. Most (80%) of their energy is released as heat.

Compared to other animals, flying insects consume vast amounts of energy relative to their weight. However, most are small and can obtain enough energy from floral nectar. During flight, honeybees and bumblebees consume about 10–11 mg of sugar/hr. Sphinx moths have slightly lower metabolic rates yet their weight range, from about 100 mg to 6 g, makes for sugar consumption between 8 and 840 mg/hr. Hovering flight of bumblebees, moths, and hoverflies is most expensive. Small flies (e.g., *Drosophila*) and butterflies (with large wings) have lower metabolic rates in flight.

Another energy component of foraging is mere preparation for flight: they must warm up. Some insects, such as flies, butterflies, and some bees, bask in the sun, sometimes in flowers. Nocturnal insects cannot do this. Bumblebees regulate their body temperature metabolically and may have metabolic rates of 1000–2000 times as great when thermoregulating than when torpid at 0°C. Sphinx moths, being larger, require more energy and time to warm up, which they do by thoracic muscular vibration (Heinrich 1981).

Once an insect has arrived at a patch of floral resources, it may forage by hovering in front of flowers (the most expensive way), flying between flowers and landing on them to feed, or walking between them. Sphinx moths and bombyliid flies do the first, butterflies and flies do the second, ants do the last, and bumblebees may do all three. On the patch, energy expenditure is determined by the spacing of plants with flowers, and of the flowers on each plant, and by the time it takes the visitor to extract the reward (handling time). Thus, one can see that walking, although energy-efficient, may become costly if the visitor spends too long in walking and could be feeding more frequently by flying.

With the ideas of foraging physiology, insect size, and anatomy in mind, one can understand the close correlation between the diversity of pollinators, their different energy needs, and the variety of flowers they visit and their nectars, according to composition, concentration, and caloric content. Some flowers pollinated by flies may have as little as 10 μg of sugar in them, which may be highly concentrated by evaporation to as much as 163 mg in *Acanthus mollis*. Hawkmoth pollinated flowers produce the most nectar and the most sugar; however, this sugar is relatively dilute and so not viscous and easily imbibed through the long tubular proboscis of the moths. Honeybee sucking rates decline markedly when syrup concentrations exceed 50–60% sugars. Flies and some butterflies spit on crystallized nectar to liquefy it. Honeybees have a taste threshold of about 10% sugars in syrup; below that they do not taste the sugar. This is a built-in safeguard against net energy loss as the bees will ignore too dilute nectars. Insects with lower metabolic requirement (e.g., Lepidoptera) show lower thresholds for tasting sugar solutions.

20.4.3 Behavior

Heinrich (1975, 1979) reviewed the energetics of foraging, especially by bees, and provided details on how nectar resources and foraging energetics are related. A pound of white clover honey represents the production of 8.7×10^6 flowers, which bees visit at a rate 500 per trip of 25 min. Thus, 17,330 bee-foraging bouts are required, taking 7221 bee hours of labor. *Bombus fervidus* forages at 40 flowers/min on red clover, which contains 0.05 mg sugar/flower; this is unavailable to shorter-tongued honeybees and bumblebees as it is out of reach. *B. terricola,* a short-tongued bumblebee, forages faster (110 flowers/min) on *Hieracium,* which has minute amounts of nectar in each floret. On the other hand, bumblebees foraging on *Chelone glabra,* with large complex zygomorphic flowers with an average of 3.3 mg of sugar each, visit for 2–8 min and spend up to 30 sec entering the flower. The time is well invested. In bumblebees, tongue length is an important anatomical feature, partitioning different species among flowers with corresponding corolla tube lengths (Inouye 1978). Learning is also important, as bees invest nonproductive time in learning how to forage from complex flowers, which in turn should offer greater rewards (Laverty 1980). The distance to patches of resources is important to insects which provision nests and must take food home. In honeybees, the waggle dance, signifying the direction and distance to food, is more intense the nearer the source (von Frisch 1967). If food for honeybees is 2–3 km distant, they can make 20 trips/hr whereas if it is 14 km away, only 1. For bumblebees on red clover, flying to a patch 1 km away costs 6.7 min or 267 blossoms.

The efficiency of pollinator foraging benefits both plant and insects. Theoretically, a flower visitor should move in such a way as to be optimally efficient in both foraging and dispersing pollen. Hence, the forager should travel short distances and not double back. Both features of optimal foraging have been demonstrated (Pyke et al. 1977, Pyke 1978). Interplant flight distances are generally short. Turning during foraging has also been quantified, and bees foraging in rich patches of plants with abundant nectar turn more frequently and with sharper angles than they do on less rewarding patches. This activity keeps the forager on the patch for a greater number of more frequent visits. Nevertheless, the mean angle of directional change is 0°, that is, the bees turn left or right but on average move ahead. Wind is an important factor because of the aerodynamics of upwind flight or downwind movement of floral odor or direction of wind over the flowers (Woodell 1978).

Part of this efficiency must also be in the forager's ability to recognize flowers and to learn how to manipulate them. As already mentioned, this requires an investment on the part of the forager so that, once having successfully probed a flower type and obtained a reward, it should continue to forage from those flowers as handling efficiency increases (Laverty 1980). So, floral constancy is developed. Honeybee foragers become recruited to patches of flowers through the waggle dance in the hive but will change to other flowers if resources decline. Bumblebees are less rigid and at any one time have specialty flowers (a "major"), secondary specialties ("minors"), and a testing behavior by which other flowers are investigated (Heinrich 1978). Real (1981) found that if rewards are variable, foragers will avoid the flowers, apparently eschewing uncertainty or averting risk.

As well as learning about flowers, foragers also learn their locations. Honeybees are unique in that they can communicate that information to each other. Particularly remarkable are trap-lining insects, such as euglossine bees, butterflies, and perhaps sphinx moths, which forage along extended routes linking isolated patches of resources. Clearly, these activities increase foraging efficiency, as the forager spends less time in site exploration.

So far we have examined the movements of foragers between plants. Most plants produce more than one flower at a time, some have many thousands in bloom at once. Foragers tend to advance from one flower to the next by moving more or less ahead to the nearest flower. On vertical inflorescences, they tend to start at the bottom and move up. Generally, the nectar is more dilute below, so gustatory saturation may be controlled. Also, the lowermost flowers are in the female stage and the uppermost in the male so that pollen flow is between plants. Examples are *Epilobium angustifolium, Delphinium, Aconitum,* many Scrophulariaceae, Leguminosae, and Labiatae. Highly linear arrangements of flowers, as in *Chamaedaphne calyculata, Polygonatum, Lotus,* and secund inflorescences, make forager movements even more predictable. Circular arrangements, as in Asteraceae, *Trifolium,* and *Allium,* require foragers to move around the inflorescence until they revisit the first flower visited. Plowright and Hartling (1982) have shown the precision of the relationship between optimal foraging by bumblebees and the optimal number of florets in blooms on inflorescences of red clover for optimal seed-set by outcrossed pollen.

The problems of pollination of massively flowering plants such as trees have been even less investigated from the viewpoint of pollinator movement. It is difficult to see why foragers should not sate themselves on one tree and thus not effect cross-pollination. Clearly, large pollinator forces are needed. In Costa Rica Frankie et al. (1983) found that bees follow the opening of flowers on individual trees and move between trees (Frankie & Haber 1983). Perhaps they move in an optimal pattern, yet eventually cross their own paths and then depart. Zimmerman (1979) argued for random movement and optimal foraging.

Not all insects which visit flowers are potentially pollinating. Some illegitimately remove resources, especially nectar. Inouye (1980) has categorized nectar robbing as involving the destruction of floral tissue and nectar theft as the removal of nectar without destruction or pollination. Some bees are frequent nectar robbers and chew holes in the corollas of flowers to get deeply hidden nectar which would be otherwise unavailable to them. A variety of small insects are commonly thieves, merely entering the flowers or feeding at the bases of the petals or sepals from outside. Pollen larceny also takes place in the same sorts of ways.

20.5 PHYSICAL ENVIRONMENT

The general effect of the physical environment on populations of anthophiles has not been well studied. Nevertheless, the effects of light, including day and night differences, and cloudiness, temperature, humidity, and wind on foraging insects have been examined for a few species. These factors also influence nectar secretion and characteristics (above). The effects of the physical environment on pollination have been reviewed by Kevan and Baker (1983).

It is well recognized that some insects are diurnal and others nocturnal. Moths are the most common nocturnal and crepuscular pollinators, but some are diurnal as in arctic (where there is no summer darkness) and alpine areas. Conversely, not all bees are diurnal: some desert species are nocturnal or matinal (cf. Linsley et al. 1963). Some insects are shade loving and pollinators of deep-forest flowers [e.g., Mycetophilidae (Mesler et al. 1980)]. Cloudiness may increase the activities of such insects outside deeply shaded habitats and stimulate crepuscular insects to activity. Many pollinators are sun-loving insects—syrphids and butterflies are conspicuously inactive during cloudy periods. The activities of other flies and bees are depressed by prolonged cloudiness. In some places, such as tropical cloud forests and cloudy mountains, and in rainy seasons some pollinator types may be absent (cf. Cruden 1972). Honeybees forage at the onset of a storm until light diminishes to below about 500 lux. Alkali bees, *Nomia melanderia*, stop foraging under heavy clouds regardless of temperature and may remain away from their burrows overnight if caught away from home.

Temperature changes are often closely linked to changes in light intensity. Honeybees generally start foraging when temperatures rise to more than 10°C, but this limit varies with the season; in spring it is 12–14°C, in early summer 14–16°C, and in summer 16–18°C. High temperatures, above about 40°C, inhibit honeybee foraging, and for bumblebees in the northeastern United States temperatures above 35°C inhibit activity. In especially hot environments (e.g., deserts) pollinating insects avoid the highest temperatures by becoming matinal or nocturnal (above) or by foraging well above the ground. In Timbuctu the crowns of trees are cool enough to permit continuous diurnal pollinator activity, even though the ground temperature is much higher (Hagerup 1932).

Wind, especially, influences insect activity. Honeybees cease foraging when wind speeds exceed 24–34 kph, or even less in exposed environments. On the coast of Israel honeybees are blown away from flowers of *Nigella arvensis* by winds over 14 kph. In the same area light sea breezes influence the activity of sphinx moths in pollinating *Pancratium maritimum*. Pollination activity takes place when wind speeds are minimum, at night when onshore–offshore breezes switch, and at the least windy time of year (cf. Eisikowitch 1978). Wind influences the directionality of pollinator movement through the aerodynamics of up-wind flight and down-wind movement of floral odors (Woodell 1978). In windy environments strong flight by pollinators and dwarfism and strong flower stalks are favored. In extreme conditions, such as mountain tops and oceanic islands, insects may become brachypterous or unwilling to fly and decumbent, spreading, or cushion plants are found. Nevertheless, flying pollinating insects, including flies, butterflies, moths, and bees, persist.

Humidity probably has its greatest effects on pollinators through its influence on nectar concentration (Corbet et al. 1979). It has been suggested that rain and humidity encourage pollen gathering by *Apis cerana* in India. Heavy rains inhibit pollinator activity and flowers may be used as shelters by some insects (e.g., Chironomidae and bees).

20.6 BEFORE ANGIOSPERMS

Before discussing the community ecology of pollination, it is useful to look at how the relationships between insects and flowers may have come about and how they are represented in the fossil record. Kevan et al. (1975) proposed that an evolutionary and ecological predisposition towards insect-mediated spore dispersal has existed since Devonian time, long before the explosive evolution of the angiosperms and insects in Cretaceous and later time. The spines and retrorse hooks of mid-Devonian spores suggest themselves as mechanical attachments to arthropod setae. At the same time heterospory becomes evident in the fossil record, along with retention of megaspores on the parent plants. By the end of the Devonian megagametophytes with attached tissues, micropyles, and possible "pollination" droplets are well known (Taylor 1981). The arthropods of the time may have been Collembola, various arachnids, or even the progenitors of winged insects for which there are no known fossils.

By mid-Mesozoic time, cycadoids were well established and some (e.g., *Williamsonia*) probably had showy inflorescences and attracted insects. Insects visit the inflorescences of modern cycads. Thus, we can infer that by Jurassic and Cretaceous time a firm trend toward insect pollination was established and became characteristic of the angiosperms. On the other hand, Crepet (1972) has suggested pollination in fossil cycadoids was by selfing and wind. Crepet (1979) has elucidated Cenozoic (Eocene) relationships of Coleoptera, Diptera, Hymenoptera, and Lepidoptera and their respective flower types represented in the fossil record.

20.7 COMMUNITY ECOLOGY

Pollination ecology at the community level is a new field, even though information about the plants and pollinators of discrete areas has been collected since the late 1800s. Particularly thorough are works reporting on the European alpine community, the arctic, and various geographic areas of Europe and North America. Kevan and Baker (1983) brought some of this information together, linking it to modern works.

In pollination in alpine and arctic regions Diptera seem to be most important, while bumblebees and Lepidoptera are also well represented. In the Chilean Andes butterflies are important at the highest altitudes. On islands flowers are often less showy and pollination syndromes less specialized than elsewhere. The same is true on remote and isolated mountain tops. Woodell (1979) points out that plants dispersed to isolated localities may leave their pollinators behind. Kevan and Baker (1983) refer to studies from New Zealand, the Galapágos, East African mountains, Australian alps, Aldabra in the Indian Ocean, and Norderney off the coast of northern Europe. Extreme environments may change pollination systems: sensitive pollinators may be replaced by less sensitive ones (cf. Cruden 1972); the periods of flowering and pollinator activity may be modified to coincide with relatively benign conditions; plants may adapt to abiotic pollination by wind or rain or may become adapted structurally to their environment, along with their pollinators (as discussed for wind and heat above); pollinators may change their habits from flying to walking in cold environments; and plants may avoid insect pollination by becoming self-pollinating or being apomictic.

Although it appears that more generalist pollination systems become more frequent as latitude or altitude increase, close inspection suggests this is not the case. But even in the arctic there is a significant proportion of specialization (e.g., *Bombus* and Syrphidae on Scrophulariaceae and Leguminosae) (Kevan 1972). Furthermore, Heithaus (1979) concluded that in tropical forests of Costa Rica there does not appear to be a high degree of specialized relationships. Studies from California, Colorado, Yorkshire, British Columbia, Canadian bogs, South American deserts, the Andes, and the Alps support this observation (see Kevan & Baker 1983). Different plant communities in a given area are likely to show differences, as Proctor (1978) showed in Ireland. There, in more stable and species-rich communities, the incidence of entomophily increases over anemophily, and entomophily is more specialized in the most stable and rich communities.

To maintain anthecological communities through the optimization processes of pollination energetics for both pollinators and plants, floral and pollinator population densities must be in some sort of equilibrium. When floral resources are heterogeneously dispersed, pollinators show less floral constancy, and as the rewards in a particular patch diminish to below a critical level, suggested to be that of the habitat in general, foragers should depart from the patch (Charnov et al. 1976). Thus, there are fluctuating pressures to select for specialization and generalization which require that genetic and behavioral flexibility be maintained. It is important to realize that factors which favor specialist or generalist pollinators are not necessarily the same as those which favor specialist or generalist flowers.

Pollination communities are highly dynamic, being based on the mutualism between plants and pollinators and on competitive relations of plants for pollinators and visitors for plants. Figure 20.2 shows the evolutionary and ecological pathways this mixed interplay of mutualism and competition may generate.

Competition between plants for pollinators is not well documented. Some weeds, such as dandelions and other "cornucopian" flowers, offer copious rewards and may draw pollinators from other plants and thus affect their reproductive success. Losers in such situations may respond by minimizing competition or its effects. They may become more specialized and reduce the spectrum of their pollinators but assure rewards to those they use. They may remove themselves temporarily from competition by shifting their flowering time. On the other hand, they may avoid competition by becoming self-pollinating, at least in part. These ideas are interesting but, as Kevan and Baker (1983) discuss, they are not established as fact. Phenological patterns probably have evolutionary significance, but this is not proven. Plant competition for pollinators may be only a weak force in the evolution and ecology of pollination systems (Brown & Kodric-Brown 1979). Mutualistic effects between temporally spaced blooming plants may then function: sequential blooms by different species maintain populations of long-lived pollinators such as hummingbirds or nests of social bees (Waser & Real 1979). These ideas lead to consideration of floral mimics of other flowers, another area wide open for research (Bierzychudek 1981).

Competition between flower visitors is better documented but still little studied. Interspecific dominance rankings place bumblebees as the dominant insects, followed by syrphids and butterflies (Kikuchi 1962–1964). Robustness seems to be an important factor, although small *Trigona* have been documented to drive off larger bees from artificial feeding dishes (Koeniger & Vorwohl 1979). The spread of the Africanized honey bee from Brazil is affecting populations of native pollinators as they come into competition (Roubik 1978). The outcomes of such competitive interactions are shown in Fig. 20.2 and are similar to those for interflower competition, that is, character displacement. Some examples of this may be represented by bumblebees in their tongue lengths and body sizes (above) and by similar features in species comprising pollinator assemblages of various plants (cf. Kevan & Baker 1983). However, these phenomena are difficult to study experimentally.

20.8 CONSERVATION

Conservation and pollination have come together centering mostly on honeybees, other managed bees (i.e., *Megachile rotundata* and *Nomia melanderia*) in agriculture, and the use of pesticides (Johansen 1977). There are few data available on other bees and nonagricultural systems, yet the implications of pollinator or plant reductions by large-scale man-made perturbations are clear (Kevan 1975b). Kevan and Baker (1983) briefly review available data.

One of the best documented cases of the demise of pollinators due to the wide-scale use of insecticides comes from New Brunswick, Canada. There, native pollinator populations on blueberry fields were severely reduced by applications of fenitrothion, which is highly toxic to bees, aimed at spruce budworm infestations in surrounding forest. Later, Dylox® was substituted as it is hardly toxic to bees and pollination forces have recovered (Kevan & LaBerge 1979). Plowright and co-workers have shown that forest pollination systems were also perturbed; bumblebee populations and seed set in entomophilous plants were reduced (NRCC 1981).

Habitat destruction has reduced native pollinator populations in Manitoba and England, where nesting sites of bees have been destroyed. The destruction of tropical forests may also destroy pollinator habitat and reduce floral density over wide areas so that pollinator populations may decline to ineffective levels. The plants then go increasingly unpollinated as the circle effect becomes more intense. The destruction of weeds in agricultural areas has been shown to reduce populations of parasitic insects, useful in biocontrol of pests, by depriving them of floral nutrient (e.g., Leius 1967). Lack of floral food may hinder the successful introduction of biocontrol agents from other parts of the world. Overcleanliness in cocoa plantations removes breeding sites for ceratopogonids and reduces populations of these pollinators (Winder 1978).

Introduced pollinators, especially *Apis mellifera,* may outcompete native pollinators, as does the Africanized honey-bee as it spreads through South and Central America (Roubik 1978, and above). On the other hand, the careful introduction of the weevil *Elaeidobius kamerunicus* from Africa into Malaysia has greatly improved pollination in oil palm plantations, which is aided by the beetle's great vagility and fecundity.

More attention must be paid to conserving and managing pollinator populations. The central role of pollination in maintaining the biomass and diversity of world ecosystems cannot be questioned. Hence, it is of paramount importance that pollination ecology be understood so that we may preserve the diversity of life as humans accelerate the destruction of natural areas.

REFERENCES

Appanah, S. and H. T. Chan. 1981. *Malays. For.* **44:** 234–252.

Baker, H. G. and I. Baker. 1975. Pp. 100–140 in Gilbert and Raven (Eds.), 1975, referenced here.

Baker, H. G. and I. Baker. 1982. Pp. 131–173 in M. H. Nitecki (Ed.), *Biochemical Aspects of Evolutionary Biology.* University of Chicago Press, Chicago.

Baker, H. G. and I. Baker. 1979. *Am. J. Bot.* **66:** 591–600.

Baker, H. G. and I. Baker. 1983. Pp. 117–141 in Jones and Little (Eds.), 1983, referenced here.

Baker, H. G. and P. D. Hurd. 1968. *Annu. Rev. Entomol.* **13:** 385–414.

Balduf, W. V. 1941. *Ann. Entomol. Soc. Am.* **43:** 204–214.

Bierzychudek, P. 1981. *Biotropica* (Suppl.) **13:** 54–58.

Brantjes, N. B. M. 1978. Pp. 13–19, *in* Richards (Ed.), 1978, referenced here.

Brown, J. H. and A. Kodric-Brown. 1979. *Ecology* **60:** 1022–1035.

Cantelo, W. W. and M. Jacobson. 1979. *Environ. Entomol.* **8:** 444–447.

Charnov, E. L., G. H. Orians, and K. Hyatt. 1976. *Am. Nat.* **110:** 247–259.

Corbet, S. A., P. G. Willmer, J. W. L. Beament, D. M. Unwin, and O. E. Prŷs-Jones. 1979. *Plant Cell Environ.* **2:** 193–208.

Crepet, W. L. 1972. *Am. J. Bot.* **59:** 1048–1056.

Crepet, W. L. 1979. *BioScience* **29:** 102–108.

Cruden, R. W. 1972. *Science* **176:** 1439–1440.

Dafni, A., Y. Ivri, and N. B. M. Brantjes. 1981. *Acta Bot. Neerl.* **30:** 69–73.

Dodson, C. H. 1975. Pp. 91–99 in Gilbert and Raven (Eds.), 1975, referenced here.

Downes, J. A. 1971. Pp. 232–258 in A. M. Fallis (Ed.), *Ecology and Physiology of Parasites.* Toronto University Press, Toronto.

Eickwort, G. C. and H. S. Ginsberg. 1980. *Annu. Rev. Entomol.* **25:** 421–446.

Eisikowitch, D. 1978. Pp. 125–131 in Richards (Ed.), 1978, referenced here.

Faegri, K. and L. van der Pijl. 1978. *The Principles of Pollination Ecology.* Pergamon, Oxford.

Frisch, K. von. 1967. *The Dance Language and Orientation of Bees.* Belknap, Cambridge, Mass.

Frankie, G. W. and W. A. Haber. 1983. Pp. 360–374 in Jones and Little (Eds.), referenced here.

Frankie, G. W., W. A. Haber, P. A. Opler, and K. S. Bawa. 1983. Pp. 411–447 in Jones and Little (Eds.), referenced here.

Galen, C. and P. G. Kevan. 1980. *Am. Midl. Nat.* **104:** 281–289.

Gilbert, F. S. 1981. *Ecol. Entomol.* **6:** 245–262.

Gilbert, L. E. and P. H. Raven (Eds.). 1975. *Animal and Plant Co-Evolution.* University of Texas Press, Austin.

Goldsmith T. and G. D. Bernard. 1974. Pp. 165–272 in M. Rockstein (Ed.) *The Physiology of Insecta 2* (2nd ed.). Academic, London.

Gori, D. F. 1983. Pp. 31–49 in Jones and Little (Eds.), referenced here.

Hagerup, O. 1932. *Dan. Bot. Ark.* **8**(1): 1–18.

Hasterud, H. D. 1974. *Norw. J. Bot.* **21:** 211–216.

Haydak, M. H. 1970. *Annu. Rev. Entomol.* **15:** 143–156.

Heinrich, B. 1975. *Annu. Rev. Ecol. Syst.* **6:** 139–170.

Heinrich, B. 1978. *Ecology* **60:** 245–255.

Heinrich, B. 1979. *Bumblebee Economics.* Harvard University Press, Cambridge, Mass.

Heinrich, B. (Ed.). 1981. *Insect Thermoregulation.* Wiley, New York.

Heithaus, E. R. 1979. *Ecology* **60:** 190–202.

Hocking, B. 1953. *Trans. Roy. Entomol. Soc. Lond.* **104:** 223–345.

Hurd, P. D. and E. G. Linsley. 1975. *Smithson. Contrib. Zool.* **193:** 1–74.

Hurd, P. D., W. E. LaBerge, and E. G. Linsley. 1980. *Smithson. Contrib. Zool.* **310:** 1–158.

Hurd, P. D., E. G. Linsley, and A. E. Michelbacher. 1974. *Smithson. Contrib. Zool.* **168:** 1–17.

Inouye, D. W. 1978. *Ecology* **59:** 672–678.

Inouye, D. W. 1980. *Ecology* **61:** 1251–1253.

Inouye, D. W. and O. R. Taylor. 1979. *Ecology* **60:** 1–7.

Johansen, C. A. 1977. *Annu. Rev. Entomol.* **22:** 177–192.

Johnson, C. G. 1969. *Migration and Dispersal of Insects by Flight.* Methuen, London.

Jones, C. E. and R. J. Little (Eds.). 1983. *Handbook of Experimental Pollination.* Van Nostrand, New York.

Kevan, P. G. 1972. *J. Ecol.* **60:** 831–867.

Kevan, P. G. 1975a. *Science* **189:** 723, 726.

Kevan, P. G. 1975b. *Environ. Conserv.* **2:** 293–298.

Kevan, P. G. 1978. Pp. 51–78 in Richards (Ed.), 1978, referenced here.

Kevan, P. G. 1983. Pp. 3–30 in Jones and Little (Eds.), 1983, referenced here.

Kevan, P. G. and H. G. Baker. 1983. *Annu. Rev. Entomol.* **28:** 407–453.

Kevan, P. G. and W. E. LaBerge. 1979. *Proc. 4th Int. Symp. Pollination. Md. Agric. Expt. Stn. Spec. Misc. Publ. No. 1,* pp. 489–508.

Kevan, P. G., W. G. Chaloner, and D. B. O. Savile. 1975. *Palaeontology* **18** (Part 2): 391–417.

Kikuchi, T. 1962–1964. *Sci. Rep. Tohoku Univ.,* Sec. 4, Biol. **28:** 17–22, 47–51; **29:** 1–8, 9–14, 107–115; **30:** 143–149.

Koeniger, N. and G. Vorwohl. 1979. *J. Apic. Res.* **18:** 95–109.

Kullenberg, B. 1961. *Zool. Bidr.* **34:** 1–340.

Laverty, T. M. 1980. *Can. J. Zool.* **58:** 1324–1335.

Leius, K. 1967. *Can. Entomol.* **99:** 444–446.

Lewis, T. 1973. *Thrips: Their Biology, Ecology, and Economic Importance.* Academic, New York.

Linsley, E. G., J. W. MacSwain, and P. H. Raven. 1963. *Univ. Calif. Publ. Entomol.* **33:** 1–24, 25–50, 59–98.

Mesler, M. R., J. D. Ackerman, and K. L. Lu. 1980. *Am. J. Bot.* **67:** 564–567.

NRCC. 1981. *Pesticide-Pollinator Interactions.* NRCC/CNRC Publ. No. 18471, Environ. Secret. Natl. Res. Council Canada, Ottawa.

Plowright, R. C. and L. K. Hartling. 1982. *J. Appl. Ecol.* **18:** 639–647.

Plowright, R. C. and T. M. Laverty. 1984. *Annu. Rev. Entomol.* **29:** 175–199.

Porsch, O. 1958. *Österr. Bot. Zeit.* **104:** 115–164.

Powell, J. A. and R. A. Mackie. 1966. *Univ. Calif. Publ. Entomol.* **42:** 1–46.

Priesner, E. 1973. *Zoon. Suppl.* **1:** 43–54.

Proctor, M. C. G. 1978. Pp. 105–116 in Richards (Ed.), 1978, referenced here.

Proctor, M. and P. Yeo. 1973. *The Pollination of Flowers.* Collins, London.

Pyke, G. H. 1978. *Oecologia* **36:** 281–293.

Pyke, G. H., H. R. Pulliam, and E. L. Charnov. 1977. *Q. Rev. Biol.* **52:** 137–154.

Real, L. A. 1981. *Ecology* **62:** 20–26.

Renty, D.C.F. and D. Clyne. 1983. *J. Aust. Entomol. Soc.* **22:** 155–160.

Richards, A. J. (Ed.). 1978. *The Pollination of Flowers by Insects.* Symp. Linn. Soc. Lond. No. 6. Academic, London.

Roubik, D. W. 1978. *Science* **201:** 1030–1032.

Spradbury, J. P. 1973. *Wasps: An Account of the Biology and Natural History of Social and Solitary Wasps.* University of Washington Press, Seattle.

Stanley, R. G. and H. F. Linskens. 1974. *Pollen: Biology, Biochemistry, Management.* Springer-Verlag, New York.

Syed, R. A., I. H. Law, and R. H. V. Corley. 1982. *Planter (Kuala Lampur)* **58:** 547–561.

Taylor, T. N. 1981. *Paleobotany: An Introduction to Fossil Plant Biology.* McGraw Hill, New York.

Vogel, S. 1974. *Akad. Wiss. Lit., Mainz Abh. Math. Naturwiss. K.* **7:** 1–267.

Vogel, S. 1978. *Flora* **167:** 329–366, 367–398.

Waller, G. D., E. W. Carpenter, and O. A. Ziehl. 1972. *J. Am. Soc. Hortic. Sci.* **97:** 535–539.

Waser, N. M. and L. A. Real. 1979. *Nature* **281:** 670–672.

Wiebes, J. T. 1979. *Annu. Rev. Ecol. Syst.* **10:** 1–12.

Williams, N. H. and C. H. Dodson. 1972. *Evolution* **26:** 84–95.

Winder, J. A. 1978. *PANS* **24**(1): 1–18.

Woodell, S. R. J. 1978. Pp. 31–39 in Richards (Ed.), 1978, referenced here.

Woodell, S. R. J. 1979. *Trans. Roy. Soc. Lond., B,* **286:** 99–108.

Wyatt, R. 1981. *Am. J. Bot.* **68:** 1212–1217.

Zimmerman, M. 1979. *Oecologia* **43:** 261–267.

Chapter 21

Insect Disseminators of Other Organisms, Especially as Vectors

EDWARD S. SYLVESTER

A variety of relationships exist where insects have either a facultative or obligatory role in the dissemination of other organisms. Some of these are discussed briefly.

21.1 DISSEMINATION OF SEEDS

Some ants actively disperse seeds, a phenomenon called *myrmecochory*, and the relationship may have coevolved. Elaiosomes (special oil-containing organs on seeds which attract ants) are indicative of ant dispersion. Ants feed on these bodies, leaving the embryo intact. The subject was monographed by Sernander (1906), and recent examples have been described by Berg (1966) and Handel (1976) (cf. Chapter 22).

21.2 DISSEMINATION OF ARTHROPODS

Phoresy, the transport of one insect or arthropod by another, is frequently observed. Whether phoresy is accidental or "intentional," in the sense of a special instinctive behavior that has evolved to become a requirement, is difficult to establish. Biting lice clasp hairs of culicid and hippoboscid flies. Fleas (Rothschild & Clay 1952) often are found infested with mites, some of which are riders, and transfer among hosts may be effected. Attachment and assumption of an ectoparasitic role by the scelionid wasp, *Mantibaria* (*Rielia*) *manticidus*, on adult *Mantis religiosus* is more obligatory. The wasps shed their wings and are positioned to directly parasitize mantid eggs (Chopard 1922, Coutrier 1941). Obligatory phoretic transport to essential food occurs with some Meloidae. First-instar meloid larvae attach to female *Anthophora* bees and, when transported to nests, have direct access to eggs, the essential food of these triungulins (Newport 1845). Nondiscriminatory dead-end attachment, while ending in starvation of triungulins, can cause losses among honey bees (Beljavsky 1933). Clausen (1976) reviewed phoresy among the entomophagous insects. Ants in the genera *Lasius, Oecophyllus, Solenopsis,* and *Crematogaster* can aid local transport and colony establishment of some aphids and mealybugs (Wheeler 1907, Hough 1922, Nixon 1951, Zwölfer 1958, Way 1963). Certain dung beetles serve as essential disseminators of specific phoretic mites, in which both species benefit (Krantz 1983).

21.3 DISSEMINATION OF MICROORGANISMS

A variety of relationships—accidental, facultative, obligatory, or mutualistic—of varying levels of complexity exist whereby insects disseminate nematodes, protozoans, fungi, and prokaryotes, as well as viruses and viroids (here considered organisms). Insects serve as hosts and disseminators of both symbiotic and pathogenic forms. Discussion will be limited to examples where insects are facultative or obligatory vectors.

21.3.1 Eukaryotes

21.3.1.1 Fungi. Part of the lore of entomology is the description of fungus-cultivating ants (New World) (Wheeler 1907, Weber 1966) and termites (Old World) (Sands 1969). Termites feed fungi to the young and royal castes (Sands 1969), and leaf-cutting ants grow fungi on substrates of gathered leaf material. Spawn is transported in intrabuccal pouches when new colonies are established (Ihering 1898).

Insects are not known to vector fungal pathogens of medical or veterinary importance but are disseminators of some plant pathogenic forms. Spores of *Fusarium moniliforme* var. *fici,* causing endosepsis of fig, are spread by ovipositing activities of the obligatory wasp pollinator of capri figs, *Blastophaga psenes.* Many insects pick up conidia of ergot, *Claviceps purpurea,* as they visit sticky surfaces of sclerotia among kernels of ripening, infected grain (Langdon & Champ 1954). Insects help hybridize some heterothallic fungi (Craigie 1927). Adaptations of rusts that may attract insects, including sugary exudate, fragrance, contrasting coloration, and so on, were discussed in 1889 by Plowright. Major attention has been given the associations of fungi and bark beetles (Francke-Grosman 1963), notably that of the ambrosia fungi that grow, sporulate, and serve as essential food for and within the communal or larval galleries of ambrosia beetles (Buchner 1928).

Economic losses are caused by xylem-invading *Ceratocystis* ascomycetes, namely, *ips, ulmi, fagacearum,* and *fimbriata,* causing blue stain, Dutch elm, oak wilt, and canker of stone fruits, respectively. Transport of all these fungi involves insects. Spores are casually moved by various wound-attracted insects, including fruit flies and nitidulid beetles in canker and in oak wilt, where insect activity among this heterothallic fungus insures production of resistant sexual spores (Hepting et al. 1952, Norris 1953, Boyce 1954, Morris et al. 1955). Blue stain, caused by a complex of fungal species and perhaps bacteria, is vectored specifically by bark beetles (Nelson & Beal 1929). Body structures are contaminated with spores, beetles attack weakened trees, and fungal invasion increases suitability of the host substrate for beetle development (Leach et al. 1934). In Dutch elm disease the principal vector, *Scolytus multistriatus,* may mate in a small niche cut into a crotch of a healthy twig (Svihra & Clark 1980) before moving to weakened elms for egg laying (Kaston & Riggs 1938). If inoculated, healthy trees later become a suitable brood site.

Sirex noctilio wasps and *Amylostereum* spp. fungal symbionts (Spradberg 1973) form a damaging complex for introduced *Pinus radiata* in New Zealand and Tasmania (Coutts 1969a,b). Larvae are believed to be nourished exclusively by the symbiotic fungus and wood cells. Eggs are laid in a protein-mucopolysaccharide sac contaminated with fungal spores. The mucus is toxic and rapid physiological changes in the tree follow, including premature senescence and foliar abscission. Both the toxin and the fungus may be needed for tree death.

21.3.1.2 Helminths.

Various arthropods are essential hosts of some helminths, both flatworms (Platyhelminthes) and roundworms (Nematoda). Among digenetic trematodes (including intestinal, hepatic, and pulmonary flukes) complex life cycles can include a definitive vertebrate host in which the sexual form (or forms) occurs and from which eggs pass, and one or more invertebrate hosts in which ciliated larvae (miracidia) invade and undergo a sequence of generations by polyembryony, the last of which escapes into the environment as cercariae. The cercariae may penetrate definitive hosts directly or, as with *Dicrocoelium dentiticum,* be carried to the nest by foraging ants and fed to larvae in which they encyst and form infective metacercariae (Krull & Mapes 1952). Ingestion of the ants infected with metacercariae by the definitive host completes the cycle.

Vectoring of plant pathogenic nemas by insects is rare. Adult palm weevils *Rhynochophorus palmarum* can be contaminated with *Rhadinaphelenchus cocophilus* nematodes (Fenwick 1962), the cause of red-ring disease of palms, but as the pathogen actively enters roots and leaf petioles, any vector role may be incidental (Hagley 1963).

Among vertebrates, however, nematodes in the subfamily Filarioidea, vectored by insects, cause several important diseases. Larvae produced in vertebrate hosts are ingested by, develop in, and are transmitted by arthropod hosts. The nematodes have a unique microfilarial stage, the circulation of which in the blood or lymph of vertebrate hosts can be periodic or subperiodic, to match the feeding habits of vectors (Hawking & Worms 1961). The developmental period in the vectors tends to coincide with intervals between blood meals. (Edeson & Wilson 1964). Damage to the arthropod host varies with the number of microfilaria ingested. Most of the 39 species in 16 genera of parasitic filarioid nematodes listed by Hawking & Worms (1961) are vectored by mosquitoes, but biting flies, black flies, fleas, ticks, and mites have been incriminated. Seven species infect man, each with a specific group of vectors. Microfilariae actively circulating in blood (microfilaremia) of vertebrates can be picked up by vectors: *Culex, Aedes, Mansonia,* or *Anopheles* mosquitoes in the case of *Wuchereria bancrofti* (found only in man); *Anopheles* and *Mansonia* in *Brugia malayi* (a disease of man and animals); and mango flies, *Chrysops* spp. in *Loa loa.* Mosquito-borne *Dirofilaria immitis,* the heartworm of dogs and wild canines, can infect humans (Beaver & Orihel 1965).

Microfilariae of *W. bancrofti,* causing elephantiasis, enter the blood from mother worms in lymphal tissue. Periodicity of peak numbers of circulating microfilariae varies with the feeding habits of major local vectors. Picked up by a vector, microfilariae move into thoracic muscles, transform into first-stage larvae, then into active filarial forms that migrate into the head and down the proboscis. The cycle takes about 2 weeks (Hawking & Worms 1961). Nemas are not injected by feeding moquitoes and usually enter wounds after mosquitoes cease feeding (Lavoipierre 1958). Development in vertebrate hosts is slow (Edeson & Wilson 1964).

Onchocercosis, or river blindness, caused by *Onchocera volvulus*, is characterized by subcutaneous fibrocytic nodules, opthalmitis, and blindness as microfilariae invade and persist in eye tissues. Microfilariae tend to remain in superficial lymphatic vessels and connective skin tissue and are ingested by *Simulium* vectors. Many vector species breed in fast-flowing streams, usually feed close to the breeding sites, and avoid entering well-shaded homes. Lacking a filaremia, no periodicity is involved (Duke 1971).

Loa loa causes a disease characterized by recurrent swelling of tissues invaded by the nemas. Damage to corneal conjunctivae as adult worms migrate across the eye is of particular importance. Mango flies, *Chrysops* spp., are obligatory intermediate hosts and active vectors. They breed in relatively clear, flowing streams in high-canopied rain forests. The adults come to clearings to feed during the day, and the microfilaremia has an accommodating diurnal periodicity. Other species of *Loa* infect monkeys and have a nocturnal periodicity (Duke 1959).

21.3.1.3 Protozoa.

Protozoans causing plant disease are rare. Flagellates of low pathogenicity (*Leptomonas, Herpetomonas,* and *Phytomonas*) occur in lactiferous plants (Lafont 1909, Bensaude 1925). Similar organisms occur in guts of hemipterans feeding on such plants (Lafont 1911, Franca 1922, Strong 1924), and cyclopropagative stages have been described. Seed transmission has been claimed (DuPorte 1925). Phloem necrosis, caused by *Phytomonas leptovasorum*, is of economic consequence in coffee in Surinam. The flagellates live in necrosing phloem and adjacent vascular tissues and can kill a tree within two months. An ant-attended hemipteran, *Lincus spathuliger*, is a suspected vector (Stahel 1933).

Many protozoans causing human disease are actively disseminated by insects. Casual transport of intestinal *Entamoeba histolytica* by filth flies is probably of minor importance. More significant are the diseases caused by vectored protozoans that have evolved using a vertebrate–invertebrate host system. These protozoans live in body tissues and blood. Malaria, the most important, continues to affect the lives of millions of humans and the economies of nations.

The many malarial (*Plasmodium*) parasites affect a myriad of vertebrate species, including man. All have a life cycle of alternating sexual (mosquito) and asexual (vertebrate) phases in a two-host system. Ubiquitous bird malarias are vectored by culicine mosquitoes, those of monkey and of man by anophelines. Four species (*vivax, falciparum, malariae,* and *ovale*) are essentially restricted to man. The first asexual stages are exoerythrocytic in cells of the reticuloendothelial system (birds) or in nonphagocytic liver cells (monkey and man). From here, daughter merozoites are released to invade erythrocytes. Generation time for one asexual multiplication cycle varies with the species. Each has a rather fixed rhythm of synchronization in humans. Gametocytes, ingested by a vector, transform in the midgut into macrogametes and flagellated microgametes. Motile zygotes (ookinetes) penetrate the stomach wall and encyst (oocysts) beneath the hemocoelic membrane. Maturation and bursting of the oocysts releases a large number of spindle-shaped sporozoites into the hemocoele; many reach the salivary glands, enter, and pass down the ducts into the hypopharyngeal tube of the proboscis to await injection into a vertebrate host with the next blood meal (Russell et al. 1963). All human malarial parasites can infect all species of *Anopheles* mosquitoes tested, although some species of mosquito are refractory. Malaria is limited by a mean summer isotherm of 15–16°C that prevents development of the parasite in mosquitoes (Harwood & James 1979). Transmission requires both stages of properly ripened gametocytes circulating in sufficient numbers. Man probably is the reservoir host. Good vectors are anthropophilic, benign biters, breed near human habitations, and readily enter domiciles to feed. They are highly susceptible to infection and produce abundant sporozoites and live long enough to take several blood meals. Various anophelines prefer different oviposition sites and require particular types of aquatic vegetation or plankton. Several species may occur in a locality, but one usually is the primary vector.

Anophelism without malaria occurs due to a variety of factors, namely, lack of (1) infected humans, (2) suitable vectors or vector density, or (3) a climate permitting development of the parasite in mosquitoes. In addition to chemotherapy, tactics used in malaria control aim to reduce vector populations to levels where effective transmission does not occur. Maintenance of such a level for 2 or 3 yr, can induce a state of anophelism without malaria. However, people with circulating gametocytes entering such a community can reinitiate the cycle.

Texas cattle, or redwater, fever caused by *Babesia bigeminia* is an important tick-borne tropical and subtropical disease. Asexual trophozoites in erythrocytes are acquired by feeding ticks (*Boophilus, Haemaphysalis,* and *Rhipicephalus* spp). The individual tick ingesting the trophozoites is not the vector. Rather the parasites penetrate developing ova infecting the next generation, and after multiplication in the salivary glands, transmission to the next mammalian host occurs (Faust et al. 1954). Other *Babesia* spp. infect sheep, goats, deer, equines, dogs, rodents, and more rarely humans (Spielman et al. 1977).

Several flagellate protozoan species affecting man need two hosts, a vertebrate and a hemophagic insect. Three *Leishmania* are vectored by *Phlebotomus* or *Lutzomyia* sand flies, two *Trypanosoma* by tsetse flies (*Glossina*), and one by triatomid bugs. Multiplication occurs in vector midguts, and except for *T. cruzi* (where the flagellates are voided with feces), all migrate forward to the insect's proboscis for release during feeding.

Leishmania donovani causes kala-azar in humans, a disease characterized by spleen and liver enlargement. The leishmania occur in cells of the vertebrate reticuloendothelial system. *L. tropica* causes a cutaneous leishmaniasis, beginning as a papule and eventually becoming craterlike. *L. braziliensis* tends to produce a mutilating leishmaniasis (Harwood & James 1979). Both forms infect dogs and rodents. The sandfly vectors pick up parasitized macrophages or temporarily freely circulating leishmania, which transform into flagellated leptomonads and, after replication, can be regurgitated from the foregut during subsequent feedings. Phlebotomines have the sterile gut required by human leishmanias which do not tolerate bacterial contamination (Adler & Theodor 1957). Sandflies, with a characteristic short hopping flight, rarely are found far from breeding sites. The females feed at night. Eggs are laid under stones, in masonry cracks, stables, poultry houses, and animal burrows, places that are dark and humid, with a supply of organic matter for larvae. Some species are distinctly anthropophilic. Leishmaniasis can be urban or rural, a disease of man alone, of man and domestic dog, or a zoonosis among arboreal rodents, depending upon local vector species (Adler & Theodor 1957).

Trypanosomes parasitize a wide range of vertebrates from fish to mammals. Those infecting land vertebrates in the central half of Africa have blood-sucking flies as vectors and may have evolved from blood-inhabiting commensals of antelope (Duggan 1970). Transmission usually involves a biological cycle. However, *Trypanosoma evansi*, causing surra in large animals in an area of the Old World from 15° W to 125° E, is mechanically transmitted by tabanid flies and has no obligatory insect cycle. It may have been derived from *T. brucei* (Hoare 1970). Nagana disease, endemic in game animals, is caused by a complex of species and strains (i.e., *T. brucei*, *T. congolense*, *T. vivax*, and *T. simiae*) that do not infect humans. Humoral tissue damage does not occur in native African hosts, but in domestic animals the severity of the disease has meant that only poultry can be raised in approximately one-quarter of the continent (Willett 1970).

Trypanosoma rhodesiense and *T. gambiense*, causal agents of African sleeping sickness, may be *T. brucei* mutants. Infection of man with *T. rhodesiense* is often fatal. The bushbuck is believed to be a reservoir host. *T. gambiense* causes a milder disease, and man is the important reservoir host (Hoare 1970). These trypanosomes, vectored by *Glossina* (tsetse flies), are acquired from an infected host, replicate in the midgut, and eventually move to the buccal cavity, up the hypopharynx, and into the salivary glands. Here a second replicative cycle results in infective trypanosomes. Inoculated during feeding, the trypanosomes multiply in the blood (parasitemia), invade the lymph nodes (lymphadenitis), and finally enter the central nervous system, producing the sleeping sickness syndrome in susceptible hosts.

Glossina spp. occur south of the Tropic of Cancer. Both sexes take blood in broad daylight and are attracted by moving objects. *Glossina palpalis* and *G. tachinoides*, major vectors of *T. gambiense* in west Africa, feed on most available hosts, including man. They are riverine species, breeding along forested edges of rivers and lakes and into dense savannah woodlands. *G. tachinoides,* capable of surviving in more zones than *G. palpalis,* carries the disease further northward (Nash 1970). *G. morsitans,* the most efficient vector of *T. rhodesiense,* ranges widely through grassy woodlands using shaded logs, treeholes, and rocks for breeding and open glades for hunting. It feeds mainly on suids and bovids (Weitz 1970). The closely related species *G. swynnertoni* uses drier thickets to breed, open spaces to hunt, and feeds mainly on suids.

Adult *Glossina* are restricted to high humidity zones, with woodlands and thickets for resting, seldom flying over extensive cleared areas. Strip areas cleared of vegetation can provide effective ecological control. In savannah conditions, natural isolation occurs where brush vegetation ends (Ford et al. 1970).

Trypanosoma cruzi, the cause of South American Chagas' disease, can naturally infect some 66 species of triatoma bugs, particularly *Panstrongylus megistus* and *Triatoma* spp., where it occurs as free flagellates or as intra-cellular leishmania within infected macrophages (Zeledon & Rabinovich 1981). The trypanosomes become flagellated in the midgut, multiply, but unlike other trypanosomes migrate to the hind gut to become infective. This adaptive modification suggests regurgitation from the gut is not a feature of hemipteran feeding. Infectivity can be retained for years by the bugs. Transmission to mammals occurs when insect excrement containing trypanosomes is rubbed into a feeding site. Active invasion of mucous membranes also can occur. Trypanosomal and leishmanial stages are sparse in mammalian blood, but infect reticuloendothelial cells of veins, cause spleno- and hepatomegala, adenopathy, and hyperplasia of bone marrow. Cardiac symptoms are common, and there can be extensive neuropatho-logical damage. The vectors are prevalent in dry areas of marginal agri-culture, and homes of adobe and palm thatch roofs provide an abundance of cracks and crevices in which they breed and hide between feedings. The bite is benign, and many domestic and wild mammals are reservoirs.

21.3.2 Prokaryotes

21.3.2.1 Bacteria. Transmission of bacterial pathogens by insects is of more than casual interest. Historically, fire blight, a progressive necrotic disease mainly of roseaceous hosts (e.g., pears, apples, and many orna-mentals) was the first association of a plant disease with a bacterium (*Erwinia amylovora*) (Burrill 1881). Some 10 yr later, evidence that the bacteria could be transferred on bodies of bees provided the first demonstration that insects could vector a plant disease (Waite 1891).

As Plant Pathogens. Plant bacteria need a port of entry to infect. Natural openings such as stomata, water pores, lenticels, and flower nectaries serve such a purpose, as do feeding and ovipositional wounds of insects. The relations between insects, bacteria, and disease may be casual, such as in fire blight, or more specialized as with some vascular-restricted bacterial parasites. Some phytopathogenic bacteria have the potential of hereditary transmission as surface contaminants of eggs [e.g., *E. carotovora* (Leach 1940)].

Erwinia tracheiphila causes bacterial wilt of cucurbits, a serious disease in the midwestern, north central, and northeastern United States. It also is known from Europe, South Africa, and Japan (Carter 1973). The bacteria are dependent on *Diabrotica* beetles for inoculation, dissemination, and perhaps overwintering (Rand 1915, Rand & Cash 1920), yet plant reservoirs may exist (Harrison et al. 1980).

Similarly, *Xanthomonas stewartii*, the cause of bacterial wilt (mainly on seed corn), and Stewart's leaf blight (principally of dent and hybrid corn) (Robert 1955) may overwinter in the gut of two *Chaetocnema* species of flea beetles (Elliott & Poos 1940), although other host plants may be reservoirs (Harrison et al. 1980). Spring infection is initiated by adult feeding. Seed infection in corn accounts only for a minor amount of infection, and even here soil root worms and maggots are believed to provide the wounds on seed coats for bacteria to gain access (Rand & Cash 1933). A high correlation exists between low winter temperatures and the number of overwintering beetles, and annual disease forecasts can be made as a guide to planting resistant varieties (Stevens 1937, Castor et al. 1975).

Olive knot disease in the Mediterranean area, caused by *Pseudomonas savastonoi*, is associated with the olive fly, *Dacus oleae*. While the disease spreads by means other than this fly (as it does in California), a symbiotic relationship exists between the pathogen and the fly (Petri 1910). The intestinal tract of *D. oleae* has a characteristic bacterial flora, with *P. savastonoi* usually present. Both feces and eggs are contaminated. Bacteria enter the micropyle, contaminate the embryo, and eventually become part of the gut flora. Pupation reduces the number of bacteria. The intestinal biota, including *P. savastonoi*, may be essential to larval nutrition (Hagen 1966).

As Animal Pathogens.   Mechanical transmission occurs with enteric bacteria responsible for bacilliary dysentery. *Shigella* infections often are associated with *Musca domestica* and other filth flies that can breed in animal excreta (Watt & Lindsay 1948). Cockroaches feeding on food, waste, or fecal material are potential casual vectors. However, most evidence of active transmission of bacteria by insects is correlative (Lindsay & Scudder 1956). Yaws, a tropical ulcerating disease caused by the spirochete *Treponema pertenue*, can be associated with wound feeding by *M. domestica*, *M. sorbens*, and *Hippelates* gnats (Satchell & Harrison 1953). The latter also are important in transmitting the bacterial complex causing an acute conjunctivitis known as "pink eye" (Bengston 1933). Tularemia, caused by *Francisella tularensis*, is mainly a zoonotic disease among rabbits and other small mammals. Ticks are important in maintaining this disease, but it can be mechanically transmitted by deer flies (Steinhaus 1946).

Transmission of bacteria that multiply in vectors occurs in several diseases, including endemic tick-borne relapsing fevers caused by *Borrelia* spp. The spirochetes are vector specific (Davis 1942) and multiply in extraintestinal tissues of ticks. Most are passed through the egg. Transmission occurs through bites, and coxal gland fluids are infectious if scratched into wounds (Burgdorfer 1951). The *Ornithodoros* vectors are interepidemic reservoirs (Philip & Burgdorfer 1961).

Louse-borne relapsing fever is caused by *Borrelia recurrentis* and is transmitted by the body louse, *Pediculus humanus humanus*. The organisms produce a lifelong, nonpathological, systemic infection in the hemolymph of lice, with transstadial but no transovarial passage. This febrile disease, with an exclusive man–louse cycle, is a disease of stress and poverty favored by crowding and poor personal hygiene. It has a high morbidity but low mortality rate and is most prevalent in the winter and spring. Lice usually live and oviposit on clothing, moving to the body only to feed. They are temperature-sensitive, and eggs will hatch only between 23 and 38°C. Human body temperature is optimal for development. Crushing infective lice, releasing infective spirochetes to invade excoriated skin, effects transmission (Harwood & James 1979).

Bartonella bacilliformis, the cause of Carrion's disease in man, is transmitted by, and multiplies in, tissues of sandflies, *Lutzomyia verrucarum.* Transmission is by bite (Hertig 1942). The two clinical forms, namely, a severe, often fatal anemia and a benign cutaneous nodular erruption, are found in high mountainous areas of Ecuador and Peru. The range appears to be determined by the climatic limits of the vectors. Above 3000 m night temperatures are too low, and below 800 m rainfall is insufficient. It is not considered a zoonosis (Harwood & James 1979).

Plague, a disease that was a molding force in the history of man, is caused by *Yersinia pestis,* a gram-negative bacillus. It is endemic among small rodents and transmitted by rodent fleas. Major pandemics during which plague has a pneumonic phase of droplet transmission belong to history, but the disease persists enzootically, with epidemic potential, in many foci in Asia, Africa, the Americas, and Hawaii. The most effective transmission is specific inoculation by "blocked" (Bacot & Martin 1914) fleas (proventricular valve obstructed by bacteria) (Bibikova 1977). The oriental rat flea, *Xenopsylla cheopis*, blocks readily and is the most common vector from domestic rats to man. The time between acquisition and transmission by *X. cheopis* averages 21 days. The bacteria multiply in the stomach of fleas and are not found in salivary glands or the body cavity, and rarely in the esophagus. Dark red or tarry fecal material contains virulent organisms. The rate of infection among fleas varies seasonally, and the human flea is not a competent vector. Urban plague involves commensal rodents, particularly *Rattus rattus,* the oriental rat flea, and man. Poor sanitation and large rat populations with intimate human contacts increase the danger of urban plague.

Zoonotic campestral or wild-rodent plague exists in permanent foci in desert and steppe zones of the world. Some 220 rodent species are susceptible. Lengthy bacteremias in wild rodents and the ability of some rodent fleas to survive long periods, even under starvation conditions, provide an effective reservoir maintenance. Man is threatened when transfer of plague by fleas occurs from wild rodents to commensal rats and when man intrudes into areas of active endemic activity.

21.3.2.2 Rickettsiae. Electron microscopy and antibiotic therapy has provided evidence that clover club-leaf, originally described as a virus disease, is caused by a phloem-inhabiting rickettsialike organism (Windsor & Black 1973). The disease was originally isolated when a transmission resulted using field-collected *Agalliopsis novella* leafhoppers. Both transstadial and transovarial passage occurs, and there is about a 3-week extrinsic (vector) incubation period. Infectivity persists for a long time and transovarial passage occurs. The disease has not been seen as a natural infection but has been an important laboratory model in transmission studies (Black 1950). Rickettsialike (as well as mycoplasmalike) etiology has been reported for psyllid-transmitted citrus greening (Garnier et al. 1975, Moll & Martin 1973), piesmid-transmitted beet latent rosette (Nienhaus & Schmutterer 1976) and for the etiology of Pierce's disease (Goheen et al. 1973, Hopkins & Mollenhauer 1973) and phoney peach (Purcell 1979).

The pathogens of the latter two diseases are xylem-restricted rickettsialike bacteria. Natural transmission is dependent upon xylem-feeding sucking insects. Pierce's disease of grape, alfalfa dwarf, and almond scorch have the same etiology (Purcell 1979). Pierce's disease in California is evidenced by leaf scorch, uneven cane ripening, and eventual death of vinifera grapes. Three sharpshooter species are important in vineyard spread. Two, namely *Draeculacephala minerva* and *Carneocephala fulgida,* are primarily grass feeders, feeding incidentally on grape. The other, *Graphocephala atropunctata,* feeds on both woody and herbaceous hosts and is an important vector in coastal areas. Bacteria have been isolated from vectors (Davis et al. 1978), and vector relations mimic those found with certain viruses. Acquisition or inoculation can occur in less than a 1 hr access to plants. The lack of a latent period and loss of infectivity during molting suggests regurgitation from the foregut as a mechanism of transmission (Purcell & Finlay 1979, Purcell et al. 1979). The pathogen has a wide plant host range. Most of the 75 susceptible species in 23 plant families are symptomless carriers (Freitag 1951). The number of reported vector species—24 sharpshooters and 4 cercopids (Severin 1950)—and the wide host range among plants makes the disease a serious threat when susceptible grapes border directly on, or encroach into, areas of high natural endemicity and vector activity (Purcell 1979).

Phony peach, also caused by a xylem-inhabiting organism, occurs primarily in the southern United States peach belt. The disease is characterized by reduced size of both trees and fruits and an intensification of green color. Purcell (1979) suggested that the northern limit of the disease is restricted by effects of low temperature on the causal agent, since the range of the xylem-feeding vectors is much greater than that of the disease.

In contrast to limited data on plant-infecting rickettsiae, those infecting animals, both vertebrate and invertebrate, are well known. The active vectors include lice, fleas, mites, and ticks.

Typhus fever, caused by *Rickettsia prowazekii,* is transmitted by the human body louse and like louse-borne relapsing fever is a poverty-associated disease of man in Europe, Asia, Africa, and at high elevations in Central and South America. It can become rampant whenever humans concentrate in close quarters, as in times of war or famine. Mortality, low in children, increases with age. The head louse, *Pediculus humanus capitis,* and the crab louse, *Pthirus pubis,* are hosts but not vectors. Acquisition occurs during feeding, and the rickettsia multiply in midgut epithelium. Infected gut cells swell, rupture, and release the rickettsiae into the lumen. Most infected lice die, but survivors remain infective for life. Contaminated louse feces or crushed lice are inoculum. Infective feces can persist at room temperature for more than 60 days as a source of respiratory infection. Symptomless human carriers are reservoirs. Recrudescences can introduce the disease into susceptible human groups. A zoonosis may exist among flying squirrels (*Glaucomys*) in northeastern United States (Bozeman et al. 1975). Reports of infection in domestic sheep, goats, and zebu in Ethiopia, and those of strains isolated from ticks and lice, need confirmation (Reiss-Gutfreund 1955).

Louse-borne trench fever is caused by *R. quintana* that multiply nonpathogenically in the lumen, not the cells of the midgut of infectious lice. Transmission again is by crushed bodies or contaminated feces. Man is the only known vertebrate host, and the disease may be asymptomatic (Harwood & James 1979). *Murine typhus,* caused by *R. typhi,* is a mild disease except in older people. Normally it is a zoonosis of rodents. Fleas, including *Xenopsylla cheopis, Nosopsyllus fasciatus,* and *Leptopsylla segnis,* have been incriminated as vectors, and possibly the rat louse, *Polyplax spinulosa,* and the tropical rat mite, *Ornithonyssus bacoti.* However, antibodies in *Rattus norvegicus* and *R. rattus* (commensal rats) have been correlated strongly only with the abundance of fleas. Human cases occur mainly in rural areas, with farm rats being the most important reservoir in the United States. Transmission is through infectious feces or perhaps mucous membrane contact (Traub et al. 1978).

In spite of the association of lice, mites, and fleas, with some rickettsial diseases of man, the primary natural vectors are ticks. Since most rickettsiae survive much longer in ticks than in vertebrate hosts, ticks are essential reservoirs in many cases.

Rocky mountain spotted fever, a febrilic condition characterized by a persistent rash and caused by *Rickettsia rickettsii,* is endemic in North America, Mexico, and parts of South America. Infection in man confers lasting immunity. The primary vectors are ixodid *Dermacentor* wood ticks, *D. variablis* in the eastern and *D. andersoni* in the western United States. A zoonotic cycle involves several species of ticks and a wide variety of vertebrate hosts, including 18 species of birds and 31 mammals, mostly rodents (Hoogstraal 1967). The short rickettsemia in mammals makes long-term maintenance in vertebrates doubtful. The transstadial and transovarial passage occurring in ticks indicates they are important reservoirs. Acquisition is by any stage. There is a 9- to 12-day extrinsic incubation period, and since ixodids tend to be three-host ticks (larvae, nymphs, and adults each on a different host), acquisition and transmission normally is not by the same stage. Transmission usually occurs by bite, less frequently by crushed ticks or excrement (Harwood & James 1979). Vectors winter as nymphs or adults. Strains, virulence, and possibly reactivation phenomena (Price 1954) add complexity to the epidemiology (Anacker et al. 1980). The organisms in infected ticks are found in cells of most tissue systems. Only 1–3% of ticks are naturally infected, and a minimum dose may be needed to establish transovarial passage (Burgdorfer 1963). In western United States spotted fever is a disease of occupational and recreational uses of forests; while in the east, incidence has increased with suburbanization (Hoogstraal 1981).

Siberian tick typhus, a low-mortality febrile disease occuring in Siberia, islands in the Sea of Japan, and in the Pacific Far East, resembles Rocky Mountain spotted fever. Disease incidence among humans increased in response to exploitation of virgin taiga forests (Hoogstraal 1981). *R. siberica* is the pathogen. Ticks, including *Dermacentor, Haemaphysalis, Hyalomma* (from rodents), and *Rhipicephalus* (from sheep and dogs) species, are vectors and reservoirs. Transstadial and transovarial passage occurs. Eighteen mammalian hosts are known. Rodents appear to be the most commonly involved, though birds may play a seasonal role (Hoogstraal 1967).

Boutonneuse fever is caused by tick-borne (*Amblyomma*) *Rickettsia conori* in Africa, the Middle East, and Southeast Asia. It causes a black buttonlike (boutonneuse) lesion at the site of tick attachment. Reservoirs appear to be a variety of tick species, with rodents and perhaps birds forming part of the biocoenosis. Dogs sustain only a short-lived infection but bring infected ticks in close contact with humans (Heisch et al. 1962). In Malaya the disease is primarily a forest zoonosis, with intruding dogs and humans acquiring infection (Hoogstraal et al. 1965).

Rickettsial pox, caused by *R. akari,* an urban zoonosis among house mice, is vectored by the mite *Allodermanyssus sanguineus.* It is a mild febrile disease lasting about 2 weeks with a generalized rash and an initial local lesion. Transmission to man is by incidental feeding by nymphs and adults of both sexes. Transovarial passage makes the vector an important reservoir.

Scrub typhus, caused by *R. tsutsugamushi,* is a chigger-transmitted (trombiculid mites) febrile disease of humans characterized by a rash and primary lesion at the point of attachment. The small initial lesion enlarges to 8–12 mm with a dark necrotic center. It may ulcerate and scar. Mortality rates vary, usually from 0.6–35%, but can reach 60% with secondary complications being contributive. The disease is endemic in Southeast Asia and adjoining islands. It typically occurs at low elevations (up to 600–900 m) in sandy bottomland overgrown with grasses and shrubs. Only larval stages are parasitic, and they receive the rickettsiae via the transstadial and transovarial routes (Miyajima & Okumura 1917). Foci need susceptible rodents, adequate ground moisture for mites, and the presence of rickettsiae. Infective mites may be further dispersed by larger mammals and birds.

21.3.2.3 Mycoplasma. No vertebrate-infecting mycoplasma are known to have an active vector, although the suckling mouse cataract agent, recently identified as a *Spiroplasma,* was isolated in 1961 from pools of rabbit ticks (Clark 1964). In the last 15–20 yr, using electron microscopy and tetracycline therapy, a series of presumptive virus diseases of plants was shown to be associated with phloem-inhabiting mycoplasmalike organisms (MLOs). Many of these plant mycoplasmas and those of a new taxonomic group, the *Spiroplasma,* have leafhopper vectors, but cixiids and psyllids also are involved (Davis & Whitcomb 1971). Confirmed culturing has been limited to spiroplasma forms related to citrus stubborn disease or to corn stunt. There are in excess of 40 recognized diseases presently associated with MLOs, so definitive research has just begun.

The MLOs are not seed borne, and although transmission can occur through root grafts and dodder, man is responsible for a good deal of spread through propagation of woody cultivars. Endemic survival of many species, however, is primarily dependent upon phloem-feeding suctorial vectors. All the leafhopper-transmitted mycoplasmas specifically tested multiply in their vectors, and although transstadial passage occurs, none are known to be transovarially passed. Most infections in vectors are asymptomatic, but one, namely, the western-X disease agent, reduces both longevity and reproduction of one of its vectors, *Colladonus montanus* (Jensen 1959, 1971a).

The *Western-X disease* agent has several experimental leafhopper vectors, none of which are primary feeders on peach and cherry where the disease is most significant economically (Jensen 1969). Experimentally, at least, it has a wide host range among both cultivated and native herbaceous species (Jensen 1971b). Whether any of these are natural reservoirs is unknown. Transmission from woody hosts is not easily accomplished, perhaps in part due to erratic seasonal or special distribution in infected hosts. The plant hosts may have an inherent resistance, or the attack rate by infective vectors is quite low, since orchards present a stationary, seasonally acceptable target for a number of years without being totally infected. X-disease has both an eastern and western form, each with a different set of local vectors. Choke cherry is an important reservoir host in eastern United States.

Leafhoppers are not long lived, and although adults may overwinter, a lack of transovarial passage means each generation of vectors must acquire the agent from feeding on infected plants. A systemic infection, including the salivary glands, must be established in the vector for transmission to occur. Temperature is probably the most important moderating element of the physical environment (Gold 1974). The spring cycle may be critical in developing a sufficient reservoir among biennial or perennial hosts, particularly so since infection, presumably in cases such as western-X disease, may reduce the probability of adult survival.

Aster yellows in the eastern United States and the Pacific region may overwinter locally where climatic conditions permit survival of *Macrosteles* vectors. In central United States there is an annually renewed cycle of infection. Infective insects migrate from populations overwintering in the southern Mississippi River Delta area. The stimulus for migration seems to be maturation of host grain crops. High summer temperatures may limit spread (Meade & Peterson 1964, Drake & Chapman 1965).

21.3.2.4 Viruses. Nowhere do insects and their relatives play a larger role in dissemination of pathogens than in the spread and maintenance of viral diseases. Viruses essentially are infectious nucleoproteins containing one kind of nucleic acid (RNA or DNA) that are completely dependent upon living cells for energy and replication. They are ubiquitous, affecting both prokaryotes and eukaryotes. They range in size from small polyhedrons of 18 nm to large complex structures $100 \times 230 \times 330$ nm (vaccinia) that contain a series of proteins, enzymes, and lipids. Given a suitable membrane to which they can attach, viruses can gain entrance to the essential replicating machinery of cells by processes varying from direct insertion of viral nucleic acid to active absorption of the entire virion with subsequent uncoating and release of the infectious nucleic acid within the cell. The protein shell, along with the bonding and folding of the nucleic acid, is believed to protect the infectious genome from inactivation by physical and chemical forces in the environment.

These obligate parasites continue to evolve relationships that provide for their perpetuation and adaptation to host cells in which they replicate. Their presence usually is indicated by a pathology induced in one or more of their potential hosts. The most generally accepted classification is based on their morphology and physiochemical structures (Fenner 1976). Our concern here is with viruses that are arthropod-transmitted to plants and animals.

Animal Viruses. Many major virus groups affecting animals, especially man, are not vectored. Rather, they are spread by contact or droplet infection: the double-stranded (ds) DNA adenoviruses (acute respiratory disease), herpes viruses (e.g., chicken pox), the papovaviruses (warts), as well as the single-stranded (ss) RNA orthomyxoviruses (influenza) and paramyxoviruses (e.g., mumps and measles).

Other viruses are mechanically transmitted by insects, particularly mosquitoes, fleas, and ticks. These include dsDNA myxoma (pox) viruses of rodents, particularly lagomorphs, bird pox viruses, Shope's rabbit papilloma virus transmitted by mosquitoes and bugs (Dalmat 1955), dsRNA nodamura reovirus (respiratory enteric orphan), as well as some ssRNA picornaviruses (i.e., enteroviruses, transmitted to an extent by filth flies). However, the fecal–oral route is dominant in the spread of this latter group.

More important are the arboviruses (arthropod-borne viruses) acquired by blood-sucking vectors from viremic hosts. These, by definition, multiply in an arthropod host, a process that typically results in a nonpathogenic, long-term, infective state during which vectors can transmit to susceptible hosts.

Over 200 arboviruses are recognized, most of which are considered distinct species. Classification is by physical and chemical characteristics, serology, epidemiology, circumstance of isolation, and effect on vertebrate hosts and tissue culture cells. The names of individual viruses may reflect (1) the geographical site of the first isolation,(2) the native name of the disease, (3) the distribution, or (4) the arthropod from which it was isolated. Included are different morphological and physiochemical types, namely, *orbiviruses* (enveloped dsRNA viruses with a segmented genome and transmitted by ticks (Colorado tick fever) or *Culicoides* biting midges (blue tongue of sheep and African horse sickness); *arenoviruses* [ssRNA viruses, of which one, lymphocytic choriomeningitis virus, has been associated with arthropods and perhaps nematodes (Lehman-Grube 1971)]; *rhabdoviruses* [ssRNA enveloped bacilliform viruses, e.g., vesicular stomatitis of animals, one strain reported as vectored by *Phlebotomus* sandflies (Tesh et al. 1971)]; *togaviruses* (ssRNA, enveloped, spherical viruses, some 40–60 nm in diameter); and *bunyaviruses,* a *group* of more than 80 viruses whose 100-nm virions contain a three-segmented ssRNA and an envelope with projections. Vectors are mosquitoes, ticks, and sandlflies.

Togaviruses are divided into an alpha group, many of which occur in temperate zones and are vectored by mosquitoes, and a flavivirus group, mostly tropical and associated with mosquitoes and ticks. Thus, major vectors of arboviruses are mosquitoes and ticks, with sandflies and biting gnats in a minor role.

Arboviruses alternate between invertebrate and vertebrate hosts, with the latter usually being the reservoir. All have a temperature-sensitive extrinsic (insect) incubation period. Transovarial passage may occur to a minor extent in mosquitoes (flavivirus group and several serotypes of the California encephalitis group of bunyaviruses—Watts et al. 1973, Coz et al. 1976, Beaty et al. 1980, Turell 1981) and in sandflies (rhabdoviruses—Tesh et al. 1971). Vertical transmission in sandflies of some viruses in the Phlebotomus fever serogroup of bunyaviruses is uncertain (Sabin et al. 1944). Vertical transmission is of greater consequence in some tick-borne viruses, but even here (except with Russian spring–summer encephalitis) is not considered a major factor in their ecology (Burgdorfer & Varma 1967).

Man is not a host for most arboviruses; rather, he is tangentially infected or is a dead-end host in the disease cycles. In spite of the severity of some arboviral disease syndromes, most have several vertebrate host species in which infection is inapparent, suggesting a prolonged adaptive evolution. Many human infections are subclinical with, at most, an undifferentiated fever (Theiler & Downs 1973). However, the ecology of many viruses is known only through isolation from a presumptive vector pool, and a cell culture or suckling mouse pathology and their human disease-producing potential has yet to be fully assessed.

Ecologically, arboviruses must circulate to survive. Normally the vertebrate reservoir host is infected and recovers with a lasting immunity. The viremia disappears and viral access to a vector is lost. The vectors, especially mosquitoes, are short-lived, lack (for the most part) an effective transovarial passage mechanism, and therefore transmission to a susceptible vertebrate must occur if the virus cycle is to be maintained. In the tropics the transmission cycles can be continuous, needing mainly a supply of nonimmune young produced by rapidly multiplying vertebrates with overlapping generations. In temperate zones the summer cycle can have a similar continuity, but winter poses a transmission hiatus that must be bridged. Overwintering in mosquitoes appears to be rare, and birds may be reservoir hosts for some. But during long winters there would have to be a latent infection, a loss of immunity, and a recrudescence in spring. This is not a proven survival mechanism. Another suggestion, also unproven, is a nonviremic infection in hibernating mammals with a renewed viremia in spring when hibernation is broken. Both mechanisms would mean temporary lapses in immunity.

Infection with arboviruses is often an occupational hazard, particularly when armies or work forces lacking any natural immunity enter a new ecological area of active endemic virus circulation. Changes in the environment, such as heavy timbering or development of vast irrigation and agricultural projects, also can bring the viruses to man and his animals.

Plant Viruses. Most insect-vectored plant-infecting viruses do not use insects as hosts. The viruses are separated into morphologic and physiochemical property groups (Fenner 1976), including groups dsDNA caulimoviruses, ssDNA gemini viruses, dsRNA reolike viruses, and a series of ssRNA viruses belonging to distinct groups with a variety of vectors. Symptoms induced by plant viruses vary from mild mosaics to lethal wilts and necroses. Most commonly, symptoms are described as mosaics, yellows, enation or tumor producing, stunts, mottles, and distortions. Many initially were recognized because of their impact on crop plants grown in areas where the disease was endemic in native plants. Van der Planck (1975) observed that native, unlike crop, trees rarely give evidence of viral infection and assumed native trees, being long-term stationary susceptible hosts, to have maximized selection of genetic factors for tolerance, resistance, or immunity to virus infection.

Plant viruses have a greater array of arthropod vectors than do the animal viruses, including aphids (Harris & Maramorosch 1977), beetles (Fulton et al. 1980), leafhoppers and planthoppers (Maramorosch & Harris 1979), mealybugs (Roivainen 1980), membracids (Simons 1980), piesmids (Proeseler 1980), thrips (Ananthakrishnan 1980), whiteflies (Muniyappa 1980), and eriophyid mites (Slykhuis 1980). Nematodes (Taylor 1980) and zoospores of fungi are also vectors (Teakle 1980). Aphids, leafhoppers, and planthoppers are the major groups. The ability of the latter to transmit plant-infecting rickettsialike and mycoplasmalike pathogens increases their importance.

Eriophyid mites are associated with numerous plant toxemias, as well as vectoring a few viruses affecting either Gramineae or woody hosts. Wind is important in dispersal of these vectors, and the diseases are especially serious where susceptible crops overlap. Transmission appears to be essentially mechanical, perhaps involving regurgitation, although the infective state does persist for varying lengths of time depending upon the virus. (Oldfield 1970, Slykhuis 1980). Agents causing virus diseases of woody plants that are mite transmitted, with the exception of fig mosaic, have bud-associated syndromes. The typical delay in bud burst, tissue distortion, and hypertrophy may be beneficial to mite vectors by prolonging a favorable environmental niche into the growing season (Oldfield 1970).

Mealybugs transmit the swollen shoot virus of cacao in Africa. Cacao was introduced into western Africa, and devastating losses followed virus infection and infestation by toxicogenic mirid bugs. Swollen shoot virus normally exists as an asymptomatic disease in native forest trees. Continuous canopy favors spread of both mealybugs and virus, and ants and wind also are involved. The toxic mirids are favored by an open canopy resulting from the decline of virus-infected trees (Todd 1951, Thresh 1960). Mealybugs also are involved in a wilt of pineapple, a complex disease that may result from a latent virus and a toxemia. It is highly dependent upon ants to spread the mealybugs into and within the fields (Carter 1963).

Mandibulate insects, particularly adult beetles of some species of Chrysomelidae, Curculionidae, and Apionidae transmit viruses. Many of these viruses induce bold mosaic patterns. Most affect herbaceous dicots, but a few are known from monocots and some are seed borne. Transmission by the vectors seems to be essentially mechanical, and the lack of persistence of vector infectivity suggests perennial plants as reservoirs (Fulton et al. 1980).

Four species of thrips transmit tomato spotted wilt virus (Best 1968, Ananthakrishnan 1980). This thermolabile, pleomorphic, lipid-containing virus infects a large number of plant species in temperate zones. Larvae, not adults, can acquire virus. Infective adults can transmit sporadically for life, although multiplication in the vector apparently does not occur. Susceptible crops near a source of both thrips and virus (e.g., a diseased flower garden) can become heavily infected. Tobacco in South Africa is economically damaged, but glandular hairs of tobacco trap thrips and effectively limit infective thrips to a single inoculation. Losses thus are a function of planting density (van der Plank & Anderson 1944).

Membracids are known vectors of but a single virus, namely, pseudo curly top (Simons 1962). The lack of extensive movement by infective insects limits the importance of the disease on tomato in Florida. Piesmids also are minor vectors, transmitting two viruses affecting sugarbeets, namely, beet leaf curl in Europe and savoy in midwestern United States. The former multiples in the insects, and infective adults are overwintering reservoirs (Proeseler 1980).

Whiteflies (*Bemisia tabaci* and two species of *Trialeurodes*) vector a series of viruses that induce disease, particularly in malvaceous (cotton), solanaceous (tobacco and tomato), cucurbit, and legume hosts in the tropics and subtropics. Adult whiteflies are active vectors and infectivity persists with transstadial, but not transovarial, passage. Perennial plants presumably are reservoir hosts (Muniyappa 1980).

Three types of relationships occur with leafhopper- and planthopper-transmitted viruses. The reolike—for example wound tumor (Black 1945, Black & Brakke 1954, Shikata et al. 1964, Whitcomb & Black 1959, Sinha & Black 1962, Chiu et al. 1966), rice dwarf (Fukushi 1933), and maize rough dwarf (Milne & Lovisolo 1977)—and rhabdoviruses (Martelli & Russo 1977) multiply in their vectors and some are transovarially passed. All rhabdoviruses have long extrinsic incubation periods. Transovarial passage, involving a minor percentage of offspring, occurs in potato yellow dwarf virus (Black 1953). Other viruses are circulative, that is, they pass through the gut wall into the hemolymph and are ejected in salivary secretions. Transstadial, but not transovarial, passage occurs but replication is doubtful. At least two viruses infecting graminaceous hosts have a semipersistent relationship with leafhopper vectors, that is, the vectors rarely remain infective for more than 3–6 days after acquisition. Neither transstadial nor transovarial passage occurs (Ling & Tiongco 1979, Nault & Bradfute 1979).

Reolike viruses transmissible to plants, while infecting vectors, exhibit little evidence of pathogenicity to insects, although both maize rough dwarf (delphacid transmitted) and rice dwarf (leafhopper-borne) are said to affect egg and larval viability (Milne & Loviøsolo 1977). Infection has been reported to increase the plant's suitability as a host for the vectors, a phenomenon also reported for the MLO-caused yellows diseases (Severin 1946, Maramorosch & Jensen 1963).

Maize rough dwarf is reported to overwinter in vectors, and other reoviruses use both diapausing planthoppers and infected perennial grasses as reservoirs. Hot weather, above 24°C, affects establishment of a maize rough dwarf infection in planthopper vectors, and in Israel there is a natural hiatus in the transmission cycle during the summer (Harpaz 1972). In Japan a synchronization of both leaf- and planthopper vectors with rice crops results in a persistent virus problem.

Aphids transmit more plant viruses than all other groups of arthropods combined. A single species, *Myzus persicae,* is a vector of more than 100 viruses (Kennedy et al. 1962). The viruses are sorted into groups using particle morphology and serological relationships. Most of these viruses have a nonpersistent relationship with their vectors, in which acquisition or inoculation occurs within a few seconds of probing. Feeding (i.e., ingestion of food) does not seem to be involved. It is unclear whether transmissible virus actually is carried on the aphid's stylet surfaces (Bradley & Ganong 1955, Day & Irzykiewicz 1954, Sylvester 1954) or on pharyngeal membranes, with transmission being a regurgitative process (Harris & Bath 1973). Various combinations of these options may be involved. However, once acquired the infective state is rapidly lost.

Some aphid-borne viruses have a semipersistent transmission pattern characterized by acquisition by feeding, rather than by probing. Retention of infectivity is inversely associated with temperature, but the half-life of infectivity is measurable in hours, not minutes as in the nonpersistent relationship (Sylvester 1955). Transstadial passage and multiplication do not occur in the nonpersistent and semipersistent pattern.

The persistent relationship, where the capacity to infect plants is retained for long periods of time, involves two mechanisms, one circulative, the other propagative. In the circulative type acquisition and inoculation efficiency are positively correlated with length of the feeding period. Acquisition is most efficient by the early larval instars. Inoculation usually can be accomplished more quickly than acquisition, and there is a short, temperature-sensitive latent period. Infectivity, while prolonged, is characterized by a gradual decline in the rate of transmission and may parallel decreasing feeding efficiency of aging vectors (Sylvester 1967).

Transstadial, but not transovarial (except perhaps potato leafroll virus, Miyamoto & Miyamoto 1971), passage occurs, and specificity is at a high level since the vectors must feed to acquire virus. Over 30 plant viruses have the circulative mechanism of transmission.

In the propagative relationship the aphid-borne viruses that multiply in the vectors are limited to the bacilliform rhabdoviruses with a large ssRNA genome.

Acquisition and inoculation are positively correlated with increasing host access time. There is a long temperature-sensitive, extrinsic incubation period and prolonged (but not necessarily for life) retention of infectivity. A low percentage (about 1% of the larvae) of transovarial passage occurs with the parthenogenetic viviparous morph in two of the viruses investigated (Sylvester 1969, Boakye & Randles 1974). Since neither of these viruses is known to be seed borne, a reoccurring transmission cycle among plants and insects is needed for virus survival.

Aphid-borne viruses reach their peak of variety and destructiveness in temperate zones. Normally the necessity of an overwintering diapausing egg furnishes a natural break in the transmission cycle. In some instances (e.g. beet yellows) the biennial nature of the major host plant (sugar beet) coupled with the fact that anholocyclic aphids (ones lacking an obligatory alternation of generations) can winter to a limited extent on infected beets in storage clamps assures an early start of the disease cycle in the new planted spring crops. Although drought and winter create natural breaks, perennial persistence of many plant species and a characteristic abundance of seed ensure a renewed supply of hosts for both vectors and the viruses in each growing season.

A final thought, in contrast to the situation for arboviruses, plants normally are not dead-end or short-term hosts, since they lack an immune system comparable to that found in vertebrates. The suitability of a plant, both as a host for viruses and vectors, may decline with maturity and age. Yet during periods of renewed growth perennial plants again become attractive to aphids and a source of high-titer virus. An analogous situation for arboviruses would be latency during hibernation and a renewed short-term viremia when hibernation is ended.

21.4 CONCLUSIONS

Insects are major disseminating agents, commonly casual, for a large variety of organisms and in many cases the intervention of insects is essential for the survival of the organisms that are moved. Examples vary from dispersal of seeds and phoresy, in which insects disperse other insects or mites with or without a parasitic or predaceous mode, to those in which adaptive mechanisms involving three systems—insects, microorganisms, and vertebrates or plants—are continually adjusted to fit the local ecological constraints that might endanger survival.

Instances occur (e.g., Nagana disease) where the presence of an insect-vectored trypanosome limits the type of animal husbandry and therefore a major source of animal protein, over a vast area of a continent, or where the potentially fertile river valleys have yet to be exploited because blackflies transmit a nematode (Onchocercosis). Malaria remains a potent force to be considered when assessing the potential for increasing the productivity of many third-world nations. Viruses and prokaryotes disseminated by insects pose a constant threat whenever new agricultural schemes, crops, or livestock husbandry are proposed to increase or diversify existing agriculture.

In their endemic state natural selection seems to have maximized most of the disseminating relationships for the survival of all elements involved. The intrusion of man and his need for food and fiber continually modify naturally existing ecologies. In many cases actions initiated to meet the increasing growth and progress requirement by humans have resulted in the appearance of significant or epidemic disease. What was once a functioning dynamic balance suddenly becomes a human problem to be assessed, investigated, and solved. On more than one occasion, insects have been the major disseminating agent of the pathogen involved.

REFERENCES

Adler, S. and O. Theodor. 1957. *Annu. Rev. Entomol.* **2:** 203–226.

Anacker, R. L., T. F. McCaul, W. Burgdorfer, R. K. Gerloff. 1980. *Infect. Immun.* **27:** 468–474.

Ananthakrishnan, T. N. 1980. Pp. 149–164, *in* Harris and Maramorosch (Eds.), 1980, referenced here.

Bacot, A. W. and C. J. Martin. 1914. *J. Hyg. Plague,* Suppl. III, Pp. 423–439.

Beaty, B. J., R. B. Tesh, and T. H. G. Aitken. 1980. *Am. J. Trop. Med. Hyg.* **29:** 125–132.

Beaver, P. C. and T. C. Orihel. 1965. *Am. J. Trop. Med. Hyg.* **14:** 1010–1029.

Beljavsky, A. G. 1933. *Bee World* **14:** 31–33.

Bengston, I. A. 1933. *Public Health Repts. (USA)* **48:** 917–926.

Bensaude, M. 1925. *Phytopathology* **15:** 273–281.

Berg, R. Y. 1966. *Am. J. Bot.* **534:** 61–73.

Best, R. J. 1968. *Adv. Virus Res.* **13:** 65–146.

Bibikova, V. A. 1977. *Annu. Rev. Entomol.* **22:** 23–32.

Black, L. M. 1945. *Am. J. Bot.* **32:** 408.

Black, L. M. 1950. *Nature (Lond.)* **166:** 852, 853.

Black, L. M. 1953. *Phytopathology* **43:** 9, 10.

Black, L. M. and M. K. Brakke. 1954. *Phytopathology* (Abs.). **44:** 82.

Boakye, D. B. and J. W. Randles. 1974. *Aust. J. Agric. Res.* **25:** 791–802.

Boyce, J. S. Jr. 1954. *Plant. Dis. Rep.* **38:** 212, 213.

Bozeman, F. M., S. A. Masiello, M. S. Willimans, and B. L. Elisberg. 1975. *Nature (London)* **255:** 545–547.

Bradley, R. H. E. and R. Y. Ganong. 1955. *Can. J. Microbiol.* **1:** 775–782.

Buchner, P. 1928. *Ergeb. Biol.* **4:** 1–129.

Burgdorfer, W. 1951. *Acta Trop.* **8:** 193–262.

Burgdorfer, W. 1963. *Exp. Parasitol.* **14:** 152–195.

Burgdorfer, W. and M. G. R. Varma. 1967. *Annu. Rev. Entomol.* **12:** 347–376.

Burrill, T. J. 1881. *Proc. Am. Assoc. Adv. Sci.* **29:** 583–597.

Carter, W. 1963. *Annu. N.Y. Acad. Sci.* **105:** 741–764.

Carter, W. 1973. *Insects in Relation to Plant Diseases.* Wiley, New York.

Castor, L. L., J. E. Ayers, A. A. McNab, and R. A. Krause. 1975. *Plant Dis. Rep.* **59:** 533–536.

Chiu, R. J., D. V. R. Reddy, and L. M. Black. 1966. *Virology* **30:** 562–566.

Chopard, L. 1922. *Ann. Soc. Entomol. Fr.* **91:** 249–272.

Clark, H. F. 1964. *J. Infect. Dis.* **114:** 476–487.

Clausen, C. P. 1976. *Annu. Rev. Entomol.* **21:** 343–368.

Coutrier, A. 1941. *Revu. Zool. Agric. Appl.* **40:** 49–62.

Coutts, M. P. 1969a. *Aust. J. Biol. Sci.* **22:** 915–924.

Coutts, M. P. 1969b. *Aust. J. Biol. Sci.* **22:** 1153–1161.

Coz, J. M. Valade, M. Cornet, and Y. Robin. 1976. *C.R. Acad. Sci. (Paris) Ser. D.* **283:** 109, 110.

Craigie, J. H. 1927. *Nature* **120:** 756–767.

Dalmat, H. T. 1955. *Smithson. Misc. Collect.* **125,** No. 1.

Davis, G. E. 1942. *AAAS Sel. Sym., Washington, D.C. 1941*, Pp. 41–47.

Davis, M. J., A. H. Purcell, and S. V. Thomson. 1978. *Science* **199:** 75–77.

Davis, R. E. and R. F. Whitcomb. 1971. *Annu. Rev. Phytopathol.* **9:** 119–154.

Day, M. F. and H. Irzykiewicz. 1954. *Aust. J. Biol. Sci.* **7:** 251–273.

Drake, D. C. and R. K. Chapman. 1965. *Univ. Wisconsin Res. bull. 261.*

Duggan, A. J. 1970. Pp. xlii–lxxxvii in Mulligan (Ed.), 1970, referenced here.

Duke, B. O. L. 1959. *Annu. Trop. Med. Parasitol.* **53:** 203–214.

Duke, B. O. L. 1971. Pp. 213–222 in A. M. Fallis (Ed.), 1971, *Ecology and Physiology of Parasites.* University of Toronto Press.

DuPorte, E. M. 1925. *J Parasitol.* **11:** 183–194.

Edeson, J. F. B. and T. Wilson. 1964. *Annu. Rev. Entomol.* **9:** 245–268.

Elliott, C. and F. W. Poos. 1940. *J. Agric. Res. (Washington, D.C.)* **60:** 645–686.

Faust, E. C., D. C. Beaver, and R. C. Jung. 1954. *Animal Agents and Vectors of Human Disease.* Lea & Febiger, Philadelphia.

Fenner, F. 1976. *Classification and Nomenclatures of Viruses.* S. Karger, Basel.

Fenwick, D. W. 1962. *J. Agric. Soc. Trinidad & Tobago* **57:** 265–281.

Ford, J., T. A. M. Nash, and J. R. Welch. 1970. Pp. 543–556 in Mulligan (Ed.), 1970, referenced here.

Franca, C. 1922. *Bull. Soc. Pathol. Exot.* **15:** 166–168.

Francke-Grosmann, H. 1963. *Annu. Rev. Entomol.* **8:** 415–438.

Freitag, T. 1951. *Phytopathology* **41:** 920–934.

Fukushi, T. 1933. *Proc. Imp. Acad. (Tokyo)* **9:** 457.

Fulton, J. P., H. A. Scott, and R. Gamez. 1980. Pp. 115–132 in Harris and Maramorosch (Eds.), 1980 referenced here.

Garnier, M., S. Latrille, and J. M. Bove. 1975. *7th Conf. Int. Org. Citrus Virol.* (Abs.), p. 18.

Goheen, A. C., G. Nyland, and S. K. Lowe. 1973. *Phytopathology* **63:** 341–345.

Gold, R. E. 1974. *Pathogen Strains and Leafhopper Species as Factors in the Transmission of Western X-mycoplasma under Varying Temperature and Light Conditions.* Ph.D. dissertation, University of California, Berkeley.

Hagen, K. 1966. *Nature* **209:** 423, 424.

Hagley, E. A. C. 1963. *J. Econ. Entomol.* **56:** 375–380.

Handel, S. N. 1976. *Am. J. Bot.* **63:** 1071–1079.

Harpaz, I. 1972. *Maize Rough Dwarf.* Israel University Press, Jerusalem.

Harris, K. F. and J. E. Bath. 1973. *Ann. Entomol. Soc. Am.* **66:** 793–796.

Harris, K. F. and K. Maramorosch. 1977. *Aphids as Virus Vectors.* Academic.

Harris, K. F. and K. Maramorosch. 1980. *Vectors of Plant Pathogens.* Academic, New York.

Harrison, M. D., J. W. Brewer, and L. D. Merrill. 1980. Pp. 201–292, *in* Harris and Maramorosch (Eds.), 1980, referenced here.

Harwood, R. F. and M. T. James. 1979. *Entomology in Human and Animal Health.* Macmillan, New York.

Hawking, F. and M. Worms. 1961. *Annu. Rev. Entomol.* **6:** 413–432.

Heisch, R. B., W. E. Grainger, A. E. C. Harvey, and G. Lister. 1962. *Trans. Roy. Soc. Trop. Med. Hyg.* **56:** 272–286.

Hepting, G. H., E. R. Toole and J. S. Boyce, Jr. 1952. *Phytopathology* **42:** 438–442.

Hertig, M. 1942. *Am. J. Trop. Med.* (Supp.) **22:** 1–81.

Hoare, C. A. 1970. Pp. 24–59 in Mulligan (Ed.), 1970, referenced here.

Hoogstraal, H. 1967. *Annu. Rev. Entomol.* **12:** 377–419.

Hoogstraal, H. 1981. *Annu. Rev. Entomol.* **26:** 75–99.

Hoogstraal, H., H. H. Trapido, and G. M. Kohls. 1965. *J. Parasitol.* **51:** 433–451.

Hopkins, D. L. and H. H. Mollenhauer. 1973. *Science* **179:** 298–300.

Hough, W. S. 1922. *Entomol. News* **33:** 171–176.

Hoy, M. A., L. Knutson, and G. L. Cunningham (Eds.). 1983. *Biological Control of Pests by Mites.* University of California, Division of Agricultural Sciences.

Ihering, H. von. 1898. *Zoolgischer Anz.* **21:** 238–245.

Jensen, D. D. 1959. *Virology* **8:** 164–175.

Jensen, D. D. 1969. *J. Econ. Entomol.* **62:** 1147–1150.

Jensen, D. D. 1971a. *J. Invertbr. Pathol.* **17:** 389–394.

Jensen, D. D. 1971b. *Phytopathology* **61:** 1465–1470.

Kaston, B. J. and D. S. Riggs. 1938. *J. Econ. Entomol.* **31:** 466–469.

Kennedy, J. S., M. F. Day, and F. Eastop. 1962. *A Conspectus of Aphids as Vectors of Plant Viruses.* Commonwealth Institute of Entomology, London.

Krantz, G. W. 1983. Pp. 91–98 in Hoy et al. (Eds.), 1983, referenced here.

Krull, W. H. and C. R. Mapes. 1952. *Cornell Vet.* **42:** 603, 604.

Lafont, A. 1909. *C. R. Soc. Biol.* **66:** 1011–1013.

Lafont, A. 1911. *C. R. Soc. Biol.* **70:** 58–59.

Langdon, R. F. N., and B. R. Champ. 1954. *J. Aust. Inst. Agric. Sci.* **20:** 115–118.

Lavoipierre, M. M. J. 1958. *Annu. Trop. Med. Parasitol.* (*Liverpool*) **52:** 326–345.

Leach, J. G. 1940. *Insect Transmission of Plant Diseases.* McGraw, New York.

Leach, J. G., L. W. Orr, and C. Christensen. 1934. *J. Agric. Res.* (*Washington, D.C.*) **49:** 315–341.

Lehman-Grube, F. 1971. *Virol. Monogr.* **10:** 1–173.

Lindsay, D. R. and H. I. Scudder. 1956. *Annu. Rev. Entomol.* **1:** 323–346.

Ling, K. C. and E. R. Tiongco. 1979. Pp. 349–366 in Maramorosch and Harris (Eds.), 1979, referenced here.

Maramorosch, K. (Ed.). 1977. *The Atlas of Insect and Plant Viruses.* Academic, New York.

Maramorosch, K. and K. F. Harris. 1979. *Leafhopper Vectors and Plant Disease Agents.* Academic, New York.

Maramorosch, K. and D. D. Jensen. 1963. *Annu. Rev. Microbiol.* **17:** 495–530.

Martelli, G. P. and M. Russo. 1977. Pp. 181–213 in Maramorosch (Ed.), 1977, referenced here.

Meade, A. B. and A. G. Peterson. 1964. *J. Econ. Entomol.* **57:** 885–888.

Milne, R. G. and O. Lovisolo. 1977. *Adv. Virus Res.* **21:** 267–341.

Miyajima, M. and T. Okumura. 1917. *Kitasato Arch. Exp. Med.* **1:** 1–14.

Miyamoto, S. and Y. Miyamoto. 1971. *Sci. Rep. Fac. Agric. Kobe Univ.* **9:** 59–70.

Moll, J. N. and M. M. Martin. 1973. *Phytophylactica* **5:** 41–44.

Morris, C. L., H. E. Thompson, B. L. Hadley, and J. M. Davis. 1955. *Plant Dis. Rep.* **39:** 61–63.

Mulligan, H. W. (Ed.). 1970. *The African Trypanosomaises.* Allen & Unwin, London.

Muniyappa, V. 1980. Pp. 39–85 in Harris and Maramorosch (Eds.), 1980, referenced here.

Nash, T. A. M. 1970, Pp. 602–613 in Mulligan (Ed.), 1970, referenced here.

Nault, L. R. and O. E. Bradfute. 1979. Pp. 561–586 in Maramorosch and Harris (Eds), 1979, referenced here.

Neinhaus, F. and H. Schmutterer. 1976. Z. Pflanzenkr. Pflanzenschutz **83:** 641–646.

Nelson, R. M. and J. A. Beal. 1929. Phytopathology **19:** 1101–1106.

Newport, G. 1845. Proc. Linn. Soc. **28:** 268–271.

Nixon, G. E. J. 1951. The Association of Ants with Aphids and Coccids. Commonwealth Institute of Entomology, London.

Norris, D. M. Jr. 1953. Plant Dis. Rep. **37:** 417, 418.

Oldfield, G. N. 1970. Annu. Rev. Entomol. **15:** 343–380.

Petri, L. 1910. Zentralb. Bakteriol. II. **26:** 357–367.

Philip, C. B. and W. Burgdorfer. 1961. Annu. Rev. Entomol. **6:** 391–412.

Plowright, C. B. 1889. A Monograph of the British Uredineae and Ustilagineae. Kegan Paul, Trench & Co., London.

Price, W. H. 1954. Pp. 164–183 in F. W. Hartman, F. L. Horsfall, Jr., and J. G. Kidd (Eds.), The Dynamics of Virus and Rickettsial Infections. Blakiston, New York.

Proeseler, G. 1980. Pp. 97–113 in Harris and Maramorosch (Eds.), 1980, referenced here.

Purcell, A. H. 1979. Pp. 603–625 in Maramorosch and Harris (Eds.), 1979, referenced here.

Purcell, A. H. and A. Finlay, 1979. Phytopathology **69:** 393–395.

Purcell, A. H., A. H. Finlay, and D. L. McLean. 1979. Science **206:** 839–841.

Rand, F. V. 1915. J. Agric. Res. (Washington, D. C.) **5:** 257–260.

Rand, F. V. and L. C. Cash. 1920. Phytopathology **10:** 133–140.

Rand, F. V. and L. C. Cash. 1933. U.S. Dep. Agric. Tech. Bull. 362.

Reiss-Gutfreund, R. J. 1955. Bull. Soc. Pathol. Exot. **48:** 602–607.

Robert, A. L. 1955. U.S. Dept. Agric. Farmers' Bull. 2092.

Roivainen, O. 1980. Pp. 15–38 in Harris and Maramorosch (Eds.), 1980, referenced here.

Rothschild, M. and T. Clay. 1952. Fleas, Flukes and Cuckoos—a Study of Bird Parasites. The Philosophical Library, New York.

Russell, P. F., L. S. West, R. D. Manwell, and B. MacDonald. 1963. Practical Malariology. Oxford University Press, Oxford.

Sabin, A. B., C. B. Philip, and J. R. Paul. 1944. J. Am. Med. Assoc. **125:** 603–606, 693–699.

Sands, W. A. 1969. Pp. 495–524 in K. Krishna and F. M. Weesner (Eds.), 1969, Biology of Termites, Vol. 1. Academic, New York.

Satchell, G. H. and R. A. Harrison. 1953. Trans. Roy. Soc. Trop. Med. Hyg. **47:** 148–153.

Sernander, R. 1906. Kgl. Svenska Vetenskapsakad. Handl. **41:** 1–410.

Severin, H. H. P. 1946. Hilgardia **17:** 121–133.

Severin, H. H. P. 1950. Hilgardia **19:** 357–382.

Shikata, E., S. W. Orenski, H. Hirumi, J. Mitsuhashi, and K. Maramorosch. 1964. Virology **23:** 441–444.

Simons, J. N. 1962. J. Econ. Entomol. **55:** 358–363.

Simons, J. N. 1980. Pp. 93–96 in Harris and Maramorosch (Eds.), 1980, referenced here.

Sinha, R. C. and L. M. Black. 1962. Virology **17:** 582–587.

Slykhuis, J. T. 1980 Pp. 325–356 in Harris and Maramorosch (Eds.), 1980, referenced here.

Speilman, A., J. Peisman, and P. Etkind. 1977. J. N.Y. Entomol. Soc. **85:** 214–216.

Spradberg, J. P. 1973. Ann. Appl. Biol. **75:** 309–320.

Stahel, G. 1933. Phytopathol. Z. **6:** 335–357.

Steinhaus, E. A. 1946. *Insect Microbiology.* Comstock, Ithaca, N.Y.

Stevens, N. E. 1937. *Plant Dis. Rep.,* 102–107.

Strong, R. P. 1924. *Am. J. Trop. Med.* **4:** 345–385.

Svihra, P. and J. C. Clark. 1980. *Calif. Agric.* **34:** 79–9.

Sylvester, E. S. 1954. *Hilgardia* **23:** 53–98.

Sylvester, E. S. 1955. *Am. Soc. Sugar Beet Technol.* **9:** 56–61.

Sylvester, E. S. 1967. *Virology* **32:** 524–531.

Sylvester, E. S. 1969. *Virology* **38:** 440–446.

Taylor, C. F. 1980. Pp. 375–416 in Harris and Maramorosch (Eds.), 1980, referenced here.

Teakle, D. S. 1980. Pp. 417–438 in Harris and Maramorosch (Eds.), 1980, referenced here.

Tesh, R. B., B. N. Chaniotis, and K. M. Johnson. 1971. *Am. J. Epidemiol.* **93:** 491–495.

Theiler, M. and W. G. Downs. 1973. *The Arthropod-Borne Viruses of Vertebrates.* Yale University Press, New Haven.

Thresh, J. M. 1960. *Emp. J. Exp. Agric.* **28:** 193–200.

Todd, J. McA. 1951. *Nature (Lond.)* **167:** 952–953.

Traub, R., C. L. Wisseman, Jr., and A. Farhang-Azad. 1978. *Trop. Dis. Bull.* **75:** 237–317.

Turell, M. J. 1981. *The Role of Transovarial Transmission in the Natural History of California Encephalitis Virus in California.* Ph.D. dissertation, University of California, Berkeley.

van der Plank, J. E. 1975. *Principles of Plant Infection.* Academic, N.Y.

van der Plank, J. E. and E. E. Anderson. 1944. *Dept. Agric. & For., Bot. & Pl. Pathol. Ser. No. 3, Sci. Bull.* 240.

Waite, M. B. 1891. *Bot. Gaz. (Chicago)* **16:** 259.

Watt, J. and D. R. Lindsay. 1948. *Public Health Rep.* **63:** 1319–1334.

Watts, D. M., S. Pantuwatana, G. R. DeFoliart, T. M. Yuill, and W. H. Thompson. 1973. *Science* **182:** 1140–1141.

Way, M. J. 1963. *Annu. Rev. Entomol.* **8:** 307–344.

Weber, N. A. 1966. *Science* **153:** 587–604.

Weitz, B. G. F. 1970. Pp. 317–326 in Mulligan (Ed.), 1970, referenced here.

Wheeler, W. M. 1907. *Bull. Am. Mus. Nat. Hist.* **23:** 669–807.

Whitcomb, R. F. and L. M. Black. 1959. *Phytopathology* (Abs.) **49:** 554.

Willett, K. C. 1970. Pp. 766–773 in Mulligan (Ed.), 1970, referenced here.

Windsor, I. M. and L. M. Black. 1973. *Phytopathology* **63:** 1139–1148.

Zeledon, R. and J. E. Rabinovich. 1981. *Annu. Rev. Entomol.* **26:** 101–133.

Zwölfer, H. 1958. *Z. angew. Entomol.* **43:** 1–52.

Chapter 22

Insect Influences

in the Regulation

of Plant Populations

and Communities

C. B. HUFFAKER, D. L.
DAHLSTEN, D. H. JANZEN,
and G. G. KENNEDY

This chapter presents an overview and some specific concepts and examples concerning the diverse roles of insects in regulating or influencing plant populations. Also considered are some consequences of these population-centered roles and of other, broader roles of insects that affect plant community composition, functioning, and maintenance. These roles are seen as effects on density, age and quality, patterns of seasonal expression, spatial distribution, and stability of specific populations and, to an extent, the communities of which they are a part (cf. Chapters 2, 12). Both short-term functioning and more distant evolutionary processes are considered. The spectrum of these influences embrace (1) local movement, dynamics, and natural control and (2) larger-scale movement, dynamics, and evolutionary change, all commonly affected by various physical and biotic influences, including their heterogeneity in time and space (cf. Chapters 6–18, 23). Insects can serve such roles as exploiters, allies, or competitors of other species. Their roles as pollinators are presented in Chapter 20, and omitted here. We here emphasize examples where insects are significant in a regulatory sense (Chapter 12). Some populations of plants (as of animals) may be regulated through the reciprocal predator–prey interaction involving insects, and the structure, functioning, and stability of some biotic communities may thus be strongly influenced by insects.

Most terrestrial plants, and also animals and microorganisms, are in some way closely associated with insects. Coevolution of insects and various associated organisms has produced some clearly mutual benefits; and for others roles mainly as food, one for the other. For some, a stable predator–prey (host–parasitoid) relationship at the population interaction level can be seen, despite the violence of individual to individual interaction. Some insects mediate interspecific competition between other organisms, including plants, affecting their dynamics, resource sharing, and succession (Chapters 2, 15). Through their exploitation of epidemic populations and mediating roles, some insects contribute to reestablishing a more natural (primeval) community composition (e.g., through reducing the density of one species of plant where an increased density has been caused by human or other disturbance of balance, or in mediating the species richness of the biotic community). Over long periods of time, it is presumed that individuals' adaptations associated with such activities have undergone a fine tuning, contributing to existing community integrity and relative stability.

Epidemics of herbivores resulting from disturbance may cause destruction of much of certain plant populations. Whether this occurs in patches or more generally will affect the rate of restoration of the plant stand. Rate of restoration will depend upon the type and age distribution of the stand and the characteristics of the individual plant species. For small patches of destruction, compensations (e.g., growth of neighboring surviving plants or branches) may achieve a rapid adjustment; for larger patches, restoration could only be achieved by establishment and maturing of recruit plants. This would be a long time for a redwood forest but much less for vegetation of annual plants.

We first discuss some general roles, processes, and concepts related to influences of insects, in either direct or indirect ways, in the regulation of plant populations and communities. Following this we review the modes of influence and some specific cases wherein insects have been seen to have significant roles in regulating or influencing plant populations, including their roles as vectors of plant diseases. In the last two sections we treat the roles of insects at a broader, more integrating level, embracing not only their influences on specific tree species populations in forests, but whole forest phenomena such as tree species richness and composition, plant succession, and community stability. In this, we deal first with temperate climate forests and lastly with tropical forests.

22.1 ROLES, PROCESSES, AND CONCEPTS RELATING TO INFLUENCES OF INSECTS ON PLANTS

22.1.1 Insects as Mediators of Interspecific Competition (cf. Chapter 15)

Harper (1977) showed the spectacular effect that a small, commonly observed animal may have in mediating the interspecific competition between two plant species. A nematode, *Heterodera avenae*, alters the interspecific competition between oat and barley plants because the barley but not the oat plants are resistant to the nematode (Fig. 22.1). The dotted lines in Fig. 22.1a show the result corresponding to equal competitive coefficients in the two species. In the absence of *Heterodera*, oats were strongly competitive

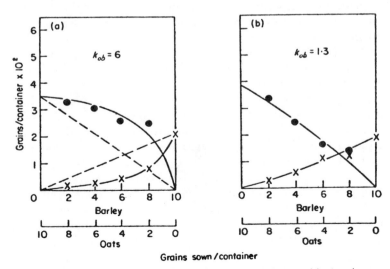

Figure 22.1   The growth of oats and barley in a replacement (competition) series experiment without (left) and with (right) an infestation of the nematode *Heterodera avenae*. x-----x = barley; •———• = oats. (After Sibma et al. 1964.)

(relative crowding coefficient k_{ob} = 6) against barley, not in its presence (coefficient reduced to k_{ob} = 1.3). Figure 22.1b shows that the competitive status of the two was then about equal. The biological control achieved by introduction of the thrips *Liothrips urichi* into Fiji for control of the weed *Clidemia hirta* is an example wherein an insect strongly affects plant competition. The plant and insect are native in Trinidad; by the 1920s the plant had invaded Fiji and become known as "Koster's Curse" in grazing lands. Search for enemies of the plant in Trinidad, and testing of their host specificity (for safety to introduce), resulted in introduction of the thrips even though in Trinidad it caused only minor effects on tip growth and did not retard its host populations. In Fiji, however, Simmonds (1933) reported that this herbivore effect was enough to tip the competitve advantage to valuable plant species, with the pest species greatly reduced (i.e., under substantial biological control). Utida (1953) also showed in a laboratory system with two competing bean seed weevil species that the hymenopterous parasite *Neocatolaccus mamezophagus* influenced the outcome of the seed weevil competiton. In the absence of the parasite, *Callosobruchus quadrimaculatus* always won over *C. chinensis*, but in its presence they coexisted. Janzen (1970) presented a theoretical model of observed mediation of competition between plant species by insects in certain tropical vegetation, with seed predation being the common route. (cf. Connell 1971)

22.1.2 Single Species of Insects Having a Direct or Suppressive or Regulative Effect

When an insect regulates a plant population's density, it commonly alters that population's competitive status, either through weakening or direct killing of individual plants. In some instances the competitive feature seems relatively minor, as in the biological control* of St. Johnswort in California by *Chrysolina quadrigemina* (or, e.g., of *Opuntia* in Australia by *Cactoblastis*, and the aquatic alligator weed in southern United States by *Agasicles*. St. Johnswort plants are killed outright by prolonged, severe defoliation and consequent retrenchment of the root system which becomes so small that the plants cannot obtain moisture during California's long dry season. In eastern Australia equally severe defoliation leads to much fewer deaths because of more rain in summer. There competition of the weakened plants with other species enters as a more definite feature in the degree of population reduction experienced (Huffaker 1967, cf. Section 22.2.5). The regulating process in a case like *Chrysolina* for St. Johnswort in California or *Cactoblastis* for *Opuntia* in Australia is the same as the regulation of an insect prey or host population by a host-specific enemy insect. There is a reciprocal density-dependent population interaction (cf. Chapters 12, 16).

---

*Biological control is the regulation or suppression of a host or prey population by one or more of its natural enemies, at a density lower than would be the case in the absence of the enemy(s)—that is, where they would reduce recruitment.

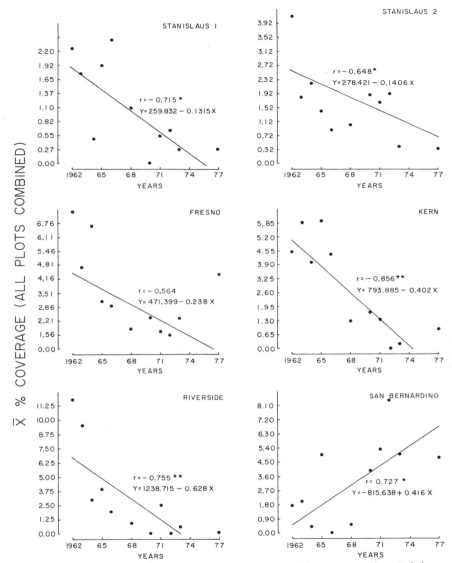

Figure 22.2 Regressions of percentage coverage of ground by puncture vine, *Tribulus ter-restris,* at six major areas in California as associated with introduced puncture vine weevils beginning in 1962 and through 1977 (declines in five of the six areas, but an increase in San Bernardino County despite the negative action of the weevils).

Even when *individual plants* (e.g., juveniles) are killed directly by an insect, competition of the *population* with other plant species may be significant because the reduction in individuals (e.g., seeds) lessens the population's competitive impact (Janzen 1970). Thus, while introduced *Microlarinus* weevils in California have their main effect on puncture vine (an annual) through direct seed limitation, the seeds that remain may face intense competition with other plant species and because of their reduced numbers, the weed's competitive potential is lessened. With approximately 46% seed prevention or destruction by *Microlarinus*, this pest plant has been substantially reduced over a 20-yr period (Huffaker et al. 1983). (Cf. DeBach 1964, MacArthur & Connell 1956)

22.1.3 Insects Sharing with Other Species or Factors in Complex Regulation

It is clear from the results of introductions of various exotic phytophagous insects for control of specific alien weeds, and from observations on the status of the same weedy plants in their native home areas, that some plant populations may not be subject to regulation or control by a single species of phytophagous insect (or even at all), whereas a complex of such enemies may serve such a role. This is suggested by the fact that the pattern of suppression of *Lantana camara* in Hawaii, beginning in the early years of this century with the first establishment of lantana insects (eight species) from Mexico, has been progressively improved through the accumulated effects from these species and a number of others introduced later, whose combined effects kill more plants at a given site and over a greater range of habitats (Andres & Goeden 1971, Andres et al. 1976). Hazeler (1981) elaborates beyond these basic reasons why a complex of species are required for *Lantana* control in Australia. He notes, too, that *L. camara* exists in Australia in many biotypes that differ in their susceptibility to damage by the various introduced insects (20 in Australia). With increase in the number of species established and with altered pasture management, the pest has gradually decreased in abundance. Other examples include the additive effects from introducing complexes of natural enemies for control of both prickly pear species and St. Johnswort in Australia, North and South America, Africa, and elsewhere (cf. Goeden 1978, Moran 1980) and of other complexes for control of alligator weed in southern United States (Maddox et al. 1971) and of *Tribulus* (e.g., puncture vine) in California (Huffaker et al. 1983) and Hawaii (Andres & Goeden 1971).

22.1.4 Insect Influences as Secondary Predators or Parasites

Exotic enemies of a natural enemy being introduced to control an exotic pest are commonly excluded. The assumption is, and the evidence is suggestive, that secondary enemies interfere with a primary enemy's effectiveness (e.g., DeBach 1974, Rosen 1981, Luck et al. 1981). However, in the native home areas the existing complex of secondary enemies of the primary enemies of insect pests usually do not seem to interfere significantly. Such species as have been introduced into other lands rather commonly appear to be equally effective in their native lands (despite their resident enemies) and in the new environments (Doutt & DeBach 1964). Yet, obviously, effective prevention of a depressing effect of a phytophagous insect on its plant host's population by a natural enemy of the phytophagous species is action of a secondary enemy, and it is precisely these sorts of enemies that have been so useful in classical biological control of insect pests of crops. Moreover, there are instances of secondary enemies apparently inhibiting effective action of primary weed-feeding enemies introduced for weed control [e.g., native Australian parasites shifted to *Zeuxidiplosis giardi* introduced for St. Johnswort control in parts of Australia (C. B. Huffaker unpubl.)], as may be so for secondary enemies attacking primary enemies of insect pests (Rosen 1981).

22.1.5 The Role of a Natural Enemy—Use of Manipulation

One can learn if a plant (or animal) population is being strongly limited (even regulated) by an exploiter species by use of manipulative methods: (1) by augmentation of the exploited population and (2) by exclusion, removal, or strong inhibition of an exploiter species suspected of having such an effect (DeBach & Bartlett 1964, DeBach et al. 1976, Harper 1977, and Chapter 12). The latter, but not the former, has been widely used in applied biological control. Little use was made of either method in field communities until recently, except for use of fences to exclude large ungulate grazers and rabbits in seminatural grazing lands and Connell's (1961) use in marine habitats. Procedures to inhibit or exclude the much more selective grazers among the insects have received little attention (Huffaker 1957, 1959, Harper 1977), but there are exceptions. Foster (1964) applied a complex of pesticides to quadrats in permanent grassland and reported that *Bellis perennis*, a constituent species, increased markedly. C. B. Huffaker and C. E. Kennett (unpub.) applied DDT to quadrats in a St. Johnswort-infested range under heavy attack by *C. quadrigemina* (cf. Section 22.2.5) and found that only in the treated quadrats did the wort survive beyond that summer, except as "trace" and depauperate individuals; other plant constituents flourished (cf. Huffaker & Kennett 1959, 1969). Cantlon (1969) applied insecticide to the ground in a Michigan woodland, and a rapid increase in the herb *Melampyrum* occurred. The cause was traced to the suppression of a polyphagous orthopteran which preferred *Melampyrum*.

22.1.6 Some Influences of Evolution

Evolution can modify characters associated in the regulation of a population. Genetic feedback may rather "permanently" alter properties (long-term evolution); this is seen in the deeply set characters of higher taxa and may be reflected in the density at which a population is regulated by density responsive factors. Or it may alter properties irregularly and during short periods of time and be reflected in altered population dynamics and regulation (cf. Nicholson 1957, 1960 and Chapters 9, 12). Such changes may result in response to any selective factor. In the interspecific competition experiments of Pimental and Al-Hafidh (1965), for example, shifts in competitive superiority of housefly and blowfly populations seemed to occur alternately, fostering coexistence of the two species.

A herbivore that has closely coevolved with a particular host species in some instances may be a more detrimental consumer to a relative of its natural host than to the natural host itself. This was true for the scales *Carulaspis visci* and *Lepidosaphes newsteadi* that invaded Bermuda. There they caused severe damage to *Juniperus bermudiana* which is much less resistant to these insects than their native hosts in the Orient (Thompson 1954). A spectacular case, not involving an insect, occurred with arrival in the United States of the fungus *Endothia parasitica,* to which its natural hosts were resistant. It quickly devastated our susceptible American chestnut. However, this type of occurrence is not typical of biological control of insects. Classical cases of highly effective biological control of insect pests have involved alien species which had colonized new areas of the world without their coevolved natural enemies being present (e.g., DeBach 1974, Wilson & Huffaker 1976). Usually, in the native home each species was relatively scarce and not a pest, suggesting possible effective biological control. Upon introduction of one or more effective enemies, the exotic pest population was then greatly reduced and so maintained subsequently [30–90 yr now for various cases (Huffaker & Messenger 1976)].

22.2   CASES OF INSECTS REGULATING OR INFLUENCING PLANT POPULATIONS

In Section 22.1 and in Chapters 2, 11, 12, 15, and 16, some general concepts concerning the roles of insects in the regulation of various populations were considered. In this section we detail specific instances of various influences of insects on plant populations, and in Sections 22.3 and 22.4, the roles that some insects may have in community structure and succession.

22.2.1   Modes by Which Insects May Affect Plant Population Regulation

Insects can affect plants by their feeding. The obvious and direct adverse effects can lead to a variety of significant indirect consequences. Certain species induce gall formations in their hosts. Feeding on foliage, for example, reduces photosynthesis and growth of the whole plant. It may alter moisture demand (e.g., mite and thrips feeding can increase moisture loss by the plant, thereby increasing the requirement) and affect flowering and seed production during the same or a subsequent year, and may lead to death of roots or other tissues, or the whole plant. Reduction in size or function of root systems reduces uptake of both water [sometimes beneficial

(Harper 1977) but lethal if a full root system is needed] and minerals. Extensive cambium destruction is often lethal. Borers in internal nonvital plant parts may introduce pathogens that destroy the plant or perhaps its mechanical support system. Beneficial influences of insects as pollinators are dealt with in Chapter 20, and in other ways in Chapter 18.

22.2.2 Seed and Flower Feeders

Direct flower feeding is often of less significance than feeding on seeds, juveniles, or adult plants. This is because there is a greater chance that such "deaths" of flowers may be dispensable rather than indispensable. Thus, flower feeding may indeed prevent seeds from developing, but the plants may compensate, to a degree, for the loss of flowers or young fruits, by producing replacements (e.g., Adkisson et al. 1964, Lloyd et al. 1962). For lack of space to consider both flower and seed feeders, we here emphasize the seed feeders.

Here we are interested in (1) the roles of insects as density-responsive regulators of the population densities and dynamics of the plants preyed upon and (2) their influences on the pattern of plant species occurrence and abundance in natural mixed communities. These two considerations involve not only the adaptive capabilities of individual plants and plant species populations to exist and compete with their siblings and other species in the absence of seed predation, but more importantly, the roles that seed predation and flower feeding (preempting seeding) might have in mediating such competition and its consequences for any of the aforestated considerations.

Seeds of many species may suffer heavy destruction by insects and sometimes insects transport and/or bury seeds, producing effects on patterns of growth and occurrence (e.g., Janzen 1971, 1977, 1980, Tevis 1958, Petal 1978, Carroll & Janzen 1973, Carroll & Risch 1981). Janzen (1971) reviewed the question of seed predation by animals, among which insects have some importance. The pattern of seed predation is highly structured and has involved evolutionary and coevolutionary relationships of chemical, morphological, and physiological nature. Pulliam and Brand (1975), for example, found that plants in an Arizona grassland appear to have adapted their seed morphology and reproductive phenology to minimize seed predation. A consequence is also an interclass (especially ants vs. sparrows) competition for seeds. Seeds produced after the winter rains are too smooth for ants to carry easily, and although these seeds are a type more readily eaten by sparrows, at this time sparrows are shifting their diet to insects. Seeds produced after the summer rains (ripening as cold weather arrives) have conspicuous awns and bristles and require husking, for which the sparrows' bills are poorly adapted, and although these seeds are readily carried by ants, this is a period when cold weather makes the ants inactive.

Seeds, flowers, and juvenile plants occupy a special role: supplying recruits for plant population maintenance and/or increase. The dispersal role of animals (mainly mammals, birds, and ants) is often closely associated with these animals' roles as predators (killing by eating), since even true seed predators drop viable seeds incidental to their feeding. This transport can generate a variable "seed shadow" pattern, with more spread and variation from birds and mammals than from ants. While some of the birds and mammals serve primarily as dispersers, others serve both roles; and the positive dispersal feedback can offset, or more than offset, the negative feedback from seed predation. The attributes of coevolved seed predators and their associated plant species have resulted from a mix of these classes of feedback (Gadgil & Bossert 1970). As Janzen (1971) noted, "The game is played by mobile predators in search of sessile prey plants; escape is through a single dispersal move, seed chemistry, parental morphology and evolutionary change." Janzen emphasized the ecological and evolutionary distinction between feeding on parent plant tissues (fruits) and killing of seeds and juveniles (recruits).

Predispersal seed predation may be light or heavy (1–100%). Such predation by prey-specific seed predators has a high potential of serving a direct density-dependent role in causing an adult plant population density to be lower in a mixed stand than would be the case if such seed predators were absent and interspecific competition were the only challenge. Thus, a biological control impact (by insects or other organisms, especially rodents, on seeds or other plant tissues) has been postulated to have a role in mediating and/or maintaining the high degree of plant species richness that exists in much of the tropics (e.g., Ridley 1930, Janzen 1970, Huffaker 1974; but cf. Johnson & Raven 1970); and to a degree, through removal of a strongly dominant plant species or more subtle effects on succession, in certain temperate forests (cf. Section 22.3, and, e.g., Moore 1942, Graham 1956, Smith 1970, Sartwell & Stevens 1975, Smith 1976) and in seminatural rangelands (e.g., Bond 1945, Dodd 1940, Huffaker 1957, 1959, 1974, Tevis 1958, Sharp & Barr 1960).

*Post*dispersal seed predation effects differ from those of predispersal predation effects in several ways. The pattern of the seed shadow (its heterogeneity, etc.) influences both the number of successful progeny and their placement positions. The character of the specific dispersers and of the propagules dispersed and influences of wind and streams produce in total a very complex aggregate seed shadow (Janzen 1970, 1971, Smythe 1970). Postdispersal seed predation also ranges from very light to very heavy for certain species and for the community as a whole (Janzen 1971, and references cited therein). Degree of litter or soil cover may be significant, including seed burying. The analysis of postdispersal predation has been difficult; assessment work is currently being intensified.

The chemical defenses of plants against herbivore feeding, including that on seeds and flowers, has become an extensive subject. Various workers (e.g., Janzen 1969, 1971, Jones 1966, Whittaker & Feeny 1971) consider that the abundance and diversity of "secondary compounds" in seeds that are toxic to seed predators are not likely to be waste products. It is suggested that seed chemistry has coevolved with the host specificity of the density-responsive seed predators, with the latter naturally influenced strongly by fluctuations in seed availability (cf. Janzen 1971). Interestingly, Janzen (1971) remarked that although insect predators of temperate zone seed crops characteristically support large secondary exploiter (parasite) populations as, for example, do bruchids of some common tropical hosts, there is "almost no evidence of entomophagous parasites of tropical *host-specific* seed predators." Janzen considers the latter to be due to the fact that these predators are serving, in the species-rich tropical vegetation, the role of secondary predators (parasitoids) superimposed on the lower trophic level, regulating predator–prey interaction, which he assumed to exclude the hyperparasites (parasites of the host-specific seed predators). He notes that even in the much less diverse northern temperate forests, parasites of seed crop predators usually attack an array of host species.

22.2.2.1 Satiation of Seed Predators and the "Mast Year" Phenomenon.
The question of satiation of predator requirements is important in the predator–prey ecology of both carnivores and herbivores. An extreme carnivore case is illustrated by the now extinct passenger pigeon. This bird nested in concentrations of hundreds of thousands, even millions, in a large primeval nesting habitat. It is reported that although enormous numbers of both avian and mammalian predators congreated at these locations and satiated themselves daily, they took but a small portion of the total recruitment. This bird's extinction in later years by humans, both as devastating predators and as destroyers of the prime breeding habitats, is of course not the point being made here. For plants the synchronous presence during a fruiting season of the seed crops of one tree species, and even of a group of species, may enable the seeds of a given plant or a whole species to escape predation through satiation (Janzen 1971, 1972). Seeds of one species may also be protected beyond the satiation effect by the greater predation on seeds of more preferred species. And, through selection, the timing of respective seed crops may be related to this potential of many predators to utilize several different species' seeds (Hurlbert 1970, Janzen 1969, 1970, Smith 1970).

The seed predator satiation effect has been significantly interrelated in the evolution in plants of the phenomenon known as "mast years," that is, years of heavy seed crops followed by one or more years without seeding. Thus, at the population and community level several workers (Lauckhart 1957, Smith 1970, Svardson 1957, Janzen 1971) consider that predator satiation is responsible for the spectacular coevolutionary and rather cyclical fluctuations in seed predators and their own enemies in northern forests of North America and Europe. It is conjectured that a weather event at some time resulted in failure of a seed crop one year, and due to the conserved nutrient reserves, a larger seed crop could be produced the next year, with satiation of seed predators resulting in more seed escapes. There is then a selection of plants hyper- or hyposensitive to the weather event and also for a physiology more responsive, in terms of greater seed production, the longer the plant has gone without producing a crop. For a tree species, there may have been a lengthening of this period without seed crops to 2–10 yr. This lengthening, however, is constrained by the competitive disadvantage of not putting seeds into the habitat, the costs of storing nutrients, and the probability of the tree suffering severe damage before it produces the delayed seed crop. Predators following a peak seed year are apt to be numerous and put heavy pressure on any seeds developed outside of the synchronized mast year pattern; hence predation generates selection against out-of-phase fruiting.

Janzen (1971) noted further, "The seed predator is confronted with the problem of waiting out the time between crops. A variable fraction of the insect population generally goes into diapause for 1 to 5 years, a behavior strongly reinforced by severe competition among the insects for the off years' seeds. Where diapause is highly synchronized with the plant population, percent seed predation by such predators, even in the peak years may, in fact, be higher than during the off years" (cf. Lyons 1957, Kraft 1968, Dalke 1953). Janzen (1971) also noted that in tropical communities with high species richness a shortage of dramatic weather-synchronizing cues, and so forth, such synchronization of extensive populations or whole communities as occurs in northern temperate zones is unlikely but does occur in one forest type that is dominated by Dipterocarpaceae in Malaysia.

22.2.3 Fruit Feeders

Animals that feed on fruits may have several distinct roles: namely, as seed dispersers or seed predators (previous section), as consumers of other parts of fruits, and as parasites or gall formers which often produce distinct malformations and may interfere with seed development (e.g., Varley 1937, Mellini 1952, Zwölfer 1967). Coevolution of attractively colored, tasty, and nutritious fruits and their specialist vertebrate consumers that coincidentally serve as dispersers has been commonplace (see Harper 1977). Insects, however, have had their main mutualistic coevolutionary role with plants at the pollinating (Chapter 20), rather than the seed dispersal level. But insect destruction of fruits (or seeds) reduces dispersal and thereby can change the species composition of communities.

There are innumerable insect feeders on fruits, and because of their economic importance, they have received extensive agricultural and silvicultural attention. The impact of *their* natural enemies on *them* has also been studied extensively. Bennett et al. (1976) summarized some of these relationships for tropical fruits and nuts, as did MacPhee et al. (1976) for temperate climate fruits and nuts. Many of these kinds of pests attack the rind or fleshy parts and only incidentally damage seeds. They are in effect, ordinary herbivores. Hence, Janzen's (1971) reference to the reasons for scarcity of host-specific parasites of seed predators in natural tropical vegetation would not apply. In fact, the insect pests of tropical, as well as temperate, fruits are frequently attacked by relatively host-specific, as well as more generalized parasitoid and predatory insects. Thus, introduction of exotic natural enemies of economically important fruit flies has had a prolonged history, with quite checkered results. In a few cases successes have been reported (Bennett et al. 1976), a major one being the biological control of *Dacus dorsalis* in Hawaii by a complex of *Opius* spp. (Chapter 15) and another that of *Dacus passiflorae* in Fiji, mainly by *Opius oophilus*. Rather little is known about the biological control of native fruit flies by native enemies.

22.2.4 Root and Stem Feeders

Root- and stem-feeding insects such as wireworms, rootworms, maggots, white grubs, wood borers, cambium feeders, bud feeders, gall formers, and so on, are often serious pests of crop plants, and as such have received much attention (cf. texts in economic and forest entomology).

The periodical cicadas in eastern United States are notorious for their great abundance and strictly synchronized periodicity. While we are not well informed as to their impact on host plants, the coevolution of the three species with their two life cycles is an intriguing subject. There are three species, *Magicicada septendecim*, *M. cassini*, and *M. septendecula*, and each appears as adults in natural woodlands and orchards after 13- or 17-yr periods of feeding on root xylem (White & Strehl 1978). The results of years of work on them have produced some hypotheses concerning competition, coexistence, predation, disease, and predator satiation as factors in cicada regulation, brood displacement, or synchronization of cycles (e.g., Lloyd & Dybas 1966a,b, Lloyd & White 1980, Dybas & Lloyd 1974, White 1980, White & Lloyd 1975). The three species occur together and are perfectly synchronized (linked) with one another over the 17-yr period in mid and northern areas (and 13 yr in valley and southern areas). The story and hypotheses are still evolving, but predator satiation on adults is thought to be the factor firmly linking the three species together; yet competition occurs and is thought to be a factor during larval development. The species have different preferences for the various tree species and specific sites in a general locality, and this contributes to their continued coexistence (e.g., White 1980). There is some evidence that the three species of 13-yr cicadas have been extending northward and driving out the three species of 17-yr cicadas, although 13- and 17-yr cicadas will hybridize every 221 yr, which slows down the process of displacement (Lloyd et al. 1983). Other stem and root feeders are dealt with below.

22.2.4.1   Regulation of a Shrubby Tree by a Stem Borer.   Species of prickly pears, *Opuntia*, became serious weeds in many dry areas of the world. By the early 1920s, some 60 million acres of range, timber, and arable lands in Queensland, Australia, were so heavily infested that the land could not be used. Some 50 species of insects were introduced to control it, many from southeastern United States where *Opuntia stricta,* a main pest in Australia is native. In the end, however, it was a phyticid moth, *Cactoblastic cactorum,* a native exploiter of cacti related to *O. stricta* in Argentina that proved effective. This illustrates that natural enemies used for biological control, though usually ones coevolved with the pest species in its native environment, need not necessarily have so coevolved (cf. Section 22.1.6). Within 3 yr of the moth's introduction the cactus had been reduced spectacularly (Fig. 22.3) (Dodd 1940). The pear now exists as a very sparsely scattered population; the insects are correspondingly scarce. Studies by Monro (1967) (cf. Birch 1971) reveal that *C. cactorum* clusters its eggs on *Opuntia* plants in such a way that while some plants are entirely destroyed others escape and this, it is argued, tends to maintain both *Opuntia* and *Cactoblastis* at rather constant, if low, levels over quite small areas. This testifies to its significance as a regulating biological control agent (cf. Laing & Hamai 1976, Goeden 1978). Haseler (1981) notes that for the *O. stricta–O. inermis* complex in Australia, completely satisfactory control is maintained by *Cactoblastis,* supplemented by *Dactylopius opuntiae,* over most of the infested environments, exceptions being in central and southern New South Wales and coastal areas of Queensland. This is a fine example of how host-specific herbivory can reduce a dominant, and therefore open up space for other species.

Interestingly, Moran (1980) studied the whole complex of *Opuntia* insects and concluded that "the co-evolution of Opuntia-feeding insects and their hosts has culminated in a community of specialist insects to the exclusion of nearly all generalist(s)"—with the immature stages possessing special adaptations that reduce the risk of attack by natural ememies. This Moran considers to be the consequence of the lack of hiding places on this structurally simple plant type.

22.2.4.2   Influences of Two Stem Feeders on Aquatic Weeds.   Two successes of note have been achieved in attempts to control aquatic weeds with insects. The case of partial success in control of alligatorweed, *Alternanthera philoxeroides,* by an introduced stem-feeding flea beetle, *Agasicles* sp. (cf. Maddox et al. 1971, Coulson 1977), stimulated much interest in the role of insects (and other natural enemies) on various other serious weeds in aquatic habitats throughout the world. One such weed is the fern *Salvinia molesta.*

Figure 22.3   Destruction of prickly pears (the *Opuntia stricta/O. inermis* complex) at a site in Queensland, Australia, within 3 yr from the introduction of the moth borer *Cactoblastis cactorum*. (After A. P. Dodd 1940 and DeBach et al. 1976.)

It develops dense floating mats on lakes and rivers. It and other *Salvinia* species have spread across Africa, India, and Australia, where it is reported as clogging pumps and irrigation channels. It also causes a depletion of other life. Herbicides are prohibitively expensive in some of these situations. Australian entomologists (see Room et al. 1981) obtained a bud-feeding beetle, *Cyrtobagous singularis*, in Brazil in the late 1970s. It multiplied so rapidly when released in Australia's Lake Moondara that a dense stand in 1978 was reduced to a placid scene of blue water in 1981 (Fig. 22.4). These authors suggested also that failures in other regions of the world may have been because of taxonomic misfits between the *Salvinia* present and the insect(s) introduced.

Figure 22.4   The clearing of Lake Moondara in Australia of the aquatic fern, *Salvinia molesta,* by introduction of a bud-feeding beetle, *Cyrtobagous singularis.* (After Room et al. 1981.)

22.2.5 Leaf Feeders

Leaf feeders produce the most obvious damage to vegetation. Apparently because of this and because the vegetation of the world remains "green," Hairston et al. (1960) proposed the proposition that plants are thus not limited by the herbivores that feed on them. They thus discerned a relative lack of foliage (or other debilitating) exploitation by herbivores. Their second conclusion was, therefore, that the natural enemies of these herbivores must be exerting a powerful depressing effect on the phytophagous species. Slobodkin et al. (1967) modified this concept to emphasize the "dominant components of the vegetation." Huffaker (1962) had pointed to the incompatibility of the view with respect to lower rank species of plants which may be under biologicial control by host-specific phytophagous insects. Ehrlich and Birch (1967) objected on this and other grounds. While being overstatements, these ideas do suggest the potential importance of leaf-feeding and other phytophagous insects as regulators of their plant hosts' populations, that is, if their own enemies were absent! Massive, contemporary defoliation (or killing) of their plant hosts over large areas of natural vegetation by insects is occasionally seen in temperate climate forests, mainly during insect outbreaks. In subtropical and tropical areas intense defoliation occurs on a much smaller spatial scale, primarily because the individual species of plants do not occur so *en masse* but are more spaced-out in the tropics, and this latter appears to be due in large part to effective regulating action by their own herbivorous seed predators.

One of the best documented cases of the role of a leaf-feeding insect in suppressing and regulating its host plant's populations in seminatural lands is that of the biological control of St. Johnswort, *Hypericum perforatum*. It is widespread in both natural and agricultural situations. Harper (1977) referred to the biological control of this plant (also known as Klamath weed) as ". . . perhaps the most exciting experiment in the whole of the science of plant-animal [population] relationships." The "experiment" has involved different species and ecological relationships in different parts of the world, and correspondingly different consequences. The plant was introduced into Australia in 1880 by a nostalgic German immigrant and inadvertently invaded California ca. 1900. Because it develops dense swards on favorable sites, crowds out grazing species, and causes skin phytosensitization and other toxic effects in livestock, the weed became a prime target for biological control attempts. In both Australia and California four principal insects were introduced: two leaf feeders, *Chrysolina quadrigemina* and *C. hyperici,* a root borer, *Agrilus hyperici,* and a leaf gall former, *Zeuxidiplosis giardi.*

Huffaker (1967) summarized the main factors accounting for the lesser degree of suppression in Australia, in contrast to California. He later observed in 1970–71 that the 1963 conditions (reported in 1967) were unusually favorable for biological control by the leaf feeder *C. quadrigemina* in eastern Australia, where in both 1970 and 1971 this beetle was scarce and therefore much less effective than in 1963. Many weed stands that were much reduced in 1963 had returned to their approximate densities prior to 1963. Thus, the contrast between this beetle's effectiveness in eastern Australia and California are even more pronounced than Huffaker (1967) contemplated; the reasons he gave for the differences existing in 1963 were even more relevant. In California a single complete defoliation continuing through the spring causes death in about 99.9% of the plants, with virtually no regeneration. Similar complete defoliation in eastern Australia in 1963 caused approximately 46% mortality of the plants, with extensive vegetative regeneration. In California C. quadrigemina is well adapted to the climate and is synchronized with the growth phases of the plant. It quickly builds up high populations and so reduces the foliage (and indirectly the root system) that this high mortality occurs because of and during the long rainless summers. In contrast, in eastern Australia the existence of some summer rain may seriously disrupt the cycle of the beetle and provides for recovery of fully defoliated plants. Western Australia, on the contrary, has a rainfall pattern like California and effectiveness of the beetle there is also comparable. Huffaker (1974) found that the action of C. quadrigemina converted northern California rangelands from a condition of very high dominance of H. perforatum, with few associated plant species of minor individual size and total mass, to a condition of considerably more species of much increased individual sizes and total biomass (Fig. 22.5, Table 22.1).

Table 22.1. Numbers of plant species dominant in microplot positions before removal of St. Johnswort by beetles (1952 at Loftus[a]; 1948, other sites[a]) and subsequently[a]

Year	Loftus	Loomis	Blocksburg	Ft. Seward
1946	—	20	24	28
1948	—	23[a]	29[a]	34[a]
1949	—	—	—	—
1950	—	23	37	41
1951	19	22	28	31
1952	24[a]	—	—	—
1953	28	30	36	37
1954	32	—	—	—
1955	33	26	34	30
1956	27	—	—	—
1957	29	25	32	34
$\bar{x}$ of post control years	29.8	25.2	33.4	34.6

[a]Year the beetles removed the weed; not yet sufficient time for much effect.
After Huffaker (1971a).

Figure 22.5   Removal of Klamath weed, *Hypericum perforatum*, by the introduced leaf beetle, *Chrysolina quadrigemina*, at Blocksburg, California. (A) 1948: Foreground shows weed in heavy flower while the rest of the field has just been killed. (B) 1950: Same location when the entire field had come under control and a heavy grass cover had developed. (After Huffaker and Kennett 1969 and DeBach et al. 1976.)

22.2.6 Insects as Vectors of Plant Diseases

Insects may also damage plants and suppress their populations through their vectoring of plant pathogens. The importance of pathogens in the dynamics of their plant hosts is related to the severity of the diseases caused and to their rates of spread. A number of factors influence the severity of disease expression. These include the host and pathogen genotypes, the age and physiological condition of the host, and the environmental conditions. Similarly, a number of factors influence the rate and pattern of a pathogen's spread through a plant population by an insect vector. These include the abundance and spatial and temporal activity patterns of the vector, the behavior of the vector, efficiency of pathogen transmission by the vector, abundance of inoculum and its location relative to other susceptible plants, and the abundance and spatial and temporal patterns of susceptible plants (cf. Chapter 21, and Carter 1973, Harris & Maramorosch 1977, Kennedy 1976, Maramorosch 1969, Harper 1977).

The effects of insect-borne plant pathogens on plant populations have been most studied in agroecosystems. In many cases effects have been devastating. Beet western yellow virus transmitted by at least nine aphids has reduced seed yields in lettuce by 44% in California (Ryder & Duffus 1966). Yield losses up to 95% were reported for rice infected with hoja blanca virus spread by the plant hoppers *Sogatodes oryzicola* and *S. cubanus* (Everett & Lamey 1969). In more natural ecosystems insect and other arthropod-borne plant pathogens can also have significant effects. Oak wilt, caused by *Ceratocystis fagacearum*, affects a number of oaks. The pathogenic fungus is spread by both insect vectors and natural root grafts. Long-distance spread is by insect vectors, the most important being nitidulids and scolytids (Rexrode & Jones 1971). From 1943 to 1965 it had spread to 20 states, causing much mortality of oaks (Carter 1973). In parts of Iowa up to 25% of the oaks were lost (True et al. 1960). Based on rates of spread from 1956 to 1965, it was estimated by Merrill (1967) that about 1% of the oaks in Pennsylvania would be infected 50 yr later, while in West Virginia 1% and 50% infection would occur about 25 and 40 yr later, respectively.

Dutch elm disease, caused by the fungus *Ceratocystis ulmi,* has caused extensive mortality of elms in Europe, where it is spread by several *Scolytus* beetles, and in North America, where it is spread primarily by *Scolytus multistriatus* and *Hylurgopinus rufipes*. By 1976 mortality of elms in southern England, following appearance of an aggressive fungus strain in the 1960's, reached 39% (Gibbs 1978). The pathogen was found in New York City in the 1920s and by 1940 had spread over 5500 mi^2. It has since spread to and widely killed elms in much of the United States. Species of elm differ in susceptibility, and thus the species composition of mixed stands of *Ulmus* are being changed by the disease. From 1951 to 1965, at one location in Illinois, 98% of the *U. americana* and 86% of the *U. rubra* died, while five other species, or hybrids, were virtually unaffected (Gibbs 1978).

In a diverse community with its plant hosts well separated, an insect-transmitted pathogen is spread rapidly and reliably only by a vector which has a good ability to search for its hosts over a wide area. Where the pathogen or vector characteristics are such that the probability of long-distance spread is low, while that for short-distance spread is high, healthy individuals are likely to persist in a plant population if they are widely separated. In Basutoland, Africa, where a number of potato viruses and their vector aphid, *Myzus persicae*, are common, "local" potato varieties planted in gardens quickly succumb to virus. Where volunteers grow as widely separated plants in fields or along roadsides, the disease is rare (Van der Plank 1948). In general, epidemic disease is rarely seen in natural vegetation except after major disturbance. Disease most often occurs where either the pathogen or the host has been introduced into a new area and coevolution has not occurred. A case in point concerns rough dwarf maize virus, which is endemic in Israel where it does not cause damage to its endemic weed hosts. Yet, it severely infects introduced maize, and is transmitted by leafhoppers (Harpaz 1972). It is possible that many natural invasions into new areas by plants have failed, at least in part, because of pathogens in the new area to which the plants were not preadapted (Harper 1977).

22.3 INSECT INFLUENCES ON PLANT POPULATIONS AND COMMUNITIES IN TEMPERATE CLIMATE FORESTS

Due in large part to their need of wood for shelter and warmth, humans have long been aware that insects damage trees and may influence forest composition. Continuing to present time, this need has dominated both attitudes towards, and research on, insects in forests. Thus, the principal attitude toward insects by forest managers and researchers has emphasized their negative effects, and research support has been mainly for work on those single species that do the most obvious damage to the most valuable tree species, and only at times of outbreaks (epidemics). Thus, our knowledge of the influences of insects on the less economically important species, and the longer term effects of these insects (and also those that cause epidemics) on the whole forest community has been accumulated very slowly. This applies not only to the insects that damage trees, but also to the large complex of natural enemies (parasites, predators, and pathogens) that prevent such damage and the varied complex of soil organisms in the forests. Thus, pest outbreaks were and still are treated as single pest phenomena, whereas in many cases a complex of pests, including plant pathogens, are actually involved. Most of the available information deals with the types of damage caused rather than the eventual *effects* at harvest and subsequently. Of course there has been a long-term interest, in that cost–benefit estimates must include the time from seedling establishment to harvest (30–100 yr) for a given stand. Over such a long period, the costs of extended chemical insect control, plus other costs, could easily exceed expected returns.

The major effects of pests in a forest stand are on stocking density and species and age composition. These features, in turn, affect both productivity and succession of the plant community as a whole. The broader and longer term effects are just beginning to be understood and for relatively few forest types. We emphasize in this section the specific influences of insects in these respects. Beginning largely with the works of Graham (cf. Graham 1956, Knight & Heikkenen 1980) there has been some emphasis on the need of a fuller understanding of the conditions that favor the various phytophagous insect populations and their longer term effects on forest dynamics. Two main concerns are the insects' effects on productivity of a stand and influences on succession, each of these being both ecologically and economically significant. We focus mainly on the ecological significance. Sacrifice of stands to save the forest has been used very little in forest management. Yet such an option could have great ecological as well as economic potential. Insects that have catastrophic effects on current productivity (e.g., bark beetles) produce different ecological (and economic) effects than those which cause growth losses, as with some defoliators, sucking insects, and shoot or tip insects. The latter can also have effects on final productivity as well as on tree species competition, and hence on species composition and succession.

Graham (1956) theorized that stands of mixed species and mixed ages, and ones not overstocked, are least subject to damage or heavy feeding by insects and other organisms. He considered stands low in species and age composition and/or overstocked to be unstable and conducive to outbreaks. In such stands the insects and other pests, by their thinning, in time alter the situation. This thinning can increase growth rates and productivity and foster or hinder succession. Thus, Graham considered that the *law of natural compensation* serves to return the forest toward a stable condition. Basically, this "law" states that in a natural biotic community compensating forces tend to keep each species in its coevolved proportion to others. Temperate forest ecosystems and their herbivores and carnivores have coevolved for thousands of years. Thus, Mattson and Addy (1975) argue that some low degree of insect feeding in forests is normal every year and that regardless of the low level, it should not be ignored, as has largely been the case. On the other hand, epidemic species like the spruce budworm (*Choristoneura fumiferana*), Douglas-fir tussock moth (*Orgyia pseudotsugata*), gypsy moth (*Lymantria dispar*), and several bark beetles (*Dendroctonus* spp.), among others, may not only cause drastic tree mortality but drastic changes in the plant community as well.

Major holocausts in forests (devastating winds, snow and ice storms, and fire) tend to favor single species forests. Outbreaks of insects, though they may cause heavy, intensive mortality, are somewhat different because insect feeding (and the resulting mortality) is more selective in terms of tree species and takes place over a longer period of time. In either case plant succession is influenced, but rather differently. A recycling of the whole successional process can occur with a major holocaust such as fire. In some cases it appears that insects tend to act like fire; in other cases they may only hasten or slow the on-going successional processes. With outbreaks, productivity may be reduced initially but increased in the long run.

Basically, there are two situations where insect outbreaks occur. In the first case the insects appear to be acting as scavengers and attack weakened trees. They may then build up large populations capable of successfully attacking healthy trees. These situations occur on poor sites (the species are not adapted to the site), or where the trees are crowded (overstocked) and/or declining in vigor from competition, or where they are simply overmature. The second situation is a density-related one in which vast amounts of favorable food, favorable conditions, and stand type (e.g., monocultures) are present. In some forests trees of the same species and age often occur together either naturally or by planting. At some point in development they rather uniformly become susceptible to a pest.

Examples of the influences of insects in specific temperate climate forests that illustrate these relationships include the following.

In eastern United States large areas of mixed white pine and hardwood forests were cleared for agriculture in the 1700s and 1800s. Many sites were later abandoned and were colonized by abnormally pure and commonly even-aged stands of eastern white pine, *Pinus strobus*. Thousands of acres of such stands developed in the early 1900s, presenting an ideal condition for the outbreaks of the white pine weevil, *Pissodes strobi*, which then occurred. This insect not only kills the tips (resulting in crooked stems) but eventually provides entrance courts for pathogens (Smith 1976). There are related situations from abandoned sites in southeastern United States involving loblolly pine forests (*Pinus taeda*). These sites, too, had formerly contained mixed hardwoods and pines. *Dendroctonus frontalis*, a tree-killing bark beetle, developed high and destructive populations, with the consequence of shifting these sites back toward the former condition. In Michigan pure aspen stands have resulted from early logging and repeated burning (Graham 1956). In 1910 these young aspen stands were abundant in areas that were once pine–aspen–birch forests. However, a complex of insects and pathogenic organisms were active at each growth stage of these stands, reducing the aspen and hastening a corrective succession. In some areas conifers are returning an in others a mixture of hardwoods.

The gypsy moth, *Lymantria dispar*, a defoliator, became established in eastern United States in the mid-1800s. The larvae feed on many species of trees but prefer oaks, particularly white oak and chestnut oak. The long-term effect of defoliation by gypsy moth has been to alter the species composition of some eastern hardwoods stands through their selective killing (Campbell & Sloan 1977). As a result, subsequent gypsy moth outbreaks have been less damaging on those sites—with the most severe damage being on weaker trees in the lower crown classes. It has also been reported that this insect is associated with disturbed and poor sites (Bess et al. 1947, Houston 1979). Disturbances due to fire and logging activity have hampered development toward more mixed hardwoods, thus favoring the oaks preferred by gypsy moths. The recent major outbreaks in New England have occurred in stands with low moisture availability, as on dry ridges or drained sandy soils. Trees on these sites are often the preferred hosts. Stands on moist sites are fast-growing and have more nonpreferred hosts in the mixtures. The straight-trunk trees on these sites also provide fewer refuges for the insect than those on poor sites. Houston (1979) considers that the gypsy moth generally hastens succession toward mixed hardwoods with fewer of the trees preferred by gypsy moths.

A close relative of the gypsy moth, the Douglas-fir tussock moth, *Orgyia pseudotsugata*, occurs in western United States and Canada. The larvae are defoliators of Douglas-fir, *Pseudotsuga menziesii*, and true firs, *Abies* spp., depending on location. In California the larvae feed almost exclusively on white fir, *Abies concolor*, except at high populations when they feed on almost any plant. Outbreaks occur every 9 to 10 yr. In northern Idaho, defoliation of grand fir, *Abies grandis*, was greatest on upper slopes and ridgetops and in the older stands of grand fir and Douglas-fir (Stoszek et al. 1981). The grand fir in the stand was also important, as was the stocking level. The higher the site occupancy, or stocking, the greater the damage. There appears to be a density-dependent relationship between the insect and grand fir in northern Idaho. Tussock moth activity would lead to changes in stocking level, species composition, and age composition on these sites. In California the results of another study were similar except that the greatest damage was in relatively open-grown white fir stands (Williams et al. 1979). Although the stocking levels were low, these stands were predominantly white fir. In another study it was found that white fir growth rate of the stand actually increased 10 yr following an outbreak (Wickman 1980).

The spruce budworm is probably the most important forest insect in North America. It periodically defoliates millions of acres of spruce-fir forests in eastern North America. The ecological implications are indeed interesting since it appears that the budworm is important in maintenance of the mixed spruce-fir type and perpetuation of balsam fir, *Abies balsamea*, even though *mature* fir is preferred over spruce. Those stands with a high content of mature balsam fir are most susceptible to budworm (cf. Chapters 24, 25 for details).

Bark beetles' (scolytids) relationships to trees or stands have been studied extensively. Most species are essentially secondary exploiters and act as decomposers or recyclers of nutrients in the community. Only a few tropical ambrosia beetles are considered primary exploiters; that is, they attack standing, healthy trees (Rudinsky 1962). Some of the most interesting bark beetles (e.g., *Dendroctonus, Ips, Scolytus,* and *Pseudohylesinus*) are intermediate, but basically secondaries that attack weakened and predisposed trees that are standing and still green. As a result, tree hazard rating systems have been developed to identify trees susceptible to attack (e.g., Salman & Bongberg 1942, Ferrell 1980, Schmid & Freye 1976). These insects can have a tremendous impact on forest communities in terms of stocking levels, species composition, age composition, and harvestable timber. Even on a longer term basis, insects may increase the species composition and stability of the forest. Since the beetles are dependent upon, and respond to, weakened or dying host trees, their population densities are closely related to availability of such host material. However, if the beetles have built up high populations (e.g., on slash), they can attack and kill apparently healthy green trees. And so, their epidemics have often been of great economic concern.

The effect of several bark beetle species that only attack following severe damage from messy logging, windstorms, or snow breakage is that of assisting breakdown of waste, recycling of nutrients, and renewal or revitalization of the forest (e.g., Nillson 1976). Other bark beetles attack trees that have been predisposed by such factors as drought, flooding, root diseases, smog injury, crowding or competition, and so on. These beetles often act mainly as thinning agents by removing or killing badly weakened trees, hastening deaths that would occur anyway due to the predisposing condition or other factors (insect or disease).

Spruce beetle, *Dendroctonus rufipennis,* infestations are commonly initiated in wind-thrown trees but then spread to and kill dominant and codominant standing green Engelmann spruce, *Picea engelmanii* (Schmid & Freye 1976). The result is a significant change in species composition to subalpine fir, *Abies lesiocarpa,* when this species is present as a component, or a change in age composition of Englemann spruce itself by release of suppressed trees of this species in the understory. Outbreaks of this beetle have occurred only at intervals of ca 50 yr, and it appears to be important in maintenance of Engelmann spruce stands (Miller 1970).

The role of the mountain pine beetle, *Dendroctonus ponderosae,* on any particular lodgepole pine site depends on whether lodgepole is seral or the persistant, climax type. In seral stands the continued role of this beetle depends upon fire, as lodgepole pine is perpetuated by periodic fires on these sites (Amman 1977); here the beetle has a beneficial economic as well as ecological role, killing the trees at an age whereby maximum timber productivity is maintained (Amman 1977, Safranyik et al. 1974).

In the mixed conifer forests of California's westside Sierra Nevada, both the mountain pine beetle and the western pine beetle, *D. brevicomis,* kill weakened ponderosa pines. In some situations trees are weakened by a root disease caused by *Verticicladiella wagneri,* which seems to occur in pockets. Beetle activity is common in these centers (Goheen & Cobb 1980). Susceptibility to beetle attack has also been associated with photochemical oxidant injury (Cobb et al. 1968). Other factors (e.g., *Ips*) may become involved. *Dendroctonus,* again in combination with root disease, seem to be important in succession to a mixed forest of very large trees. The key host, ponderosa pine, is shade intolerant and the pioneer species, whereas white fir, Douglas-fir, sugar pine, incense cedar, and black oak (also present) are all variously shade tolerant. Ponderosa invades open sites after fire or logging or both. The pine stands are nearly pure for several years until a canopy is formed and gradually the shade-tolerant species become established in the understory. As the crowded pines become older, they are thinned by root disease and *D. brevicomis* and the shade-tolerant species are "released." After a number of years, with the further influence of ground fires, an open, mixed stand of very large trees results.

What we have given here is only suggestive of the varied roles insects play in termperate forest succession, composition, and maintenance. (Cf. overviews by Mattson 1977, Rafes 1966.) From an understanding of these relationships one might have been able to predict some of the many adverse consequences that have ensued from ignoring these relationships and growing forest monocultures (cf. Knight & Heikkenen 1980, McFadden et al. 1981).

22.4 INFLUENCES OF INSECTS ON PLANT POPULATIONS AND COMMUNITIES IN TROPICAL FORESTS

Leaving aside the pollinators (Chapter 20), the herbivore load of a tropical plant population poses two major classes of influence on the members of the plant population. First, and mostly unappreciated, the long-standing existence of this herbivore load has selected for the plant to expend a substantial part of its resources on defensive traits such as resins, latex, polyphenols, toxic proteins, alkaloids, uncommon amino acids, cardiac glucosides, cyanogenic glucosides, indigestible storage polysaccharides, cyanolipids, phytoalexins, lignins, essential oils, urticating hairs, hard seed coats, heartwood, thick bark, and so on (e.g., Levin 1976, Rosenthal & Janzen 1979); that is to say that despite Hairston et al.'s (1960) opinion, the world is not always so green. Conceivably, cellulose may have been evolutionarily chosen over other structural polysaccharides because of its indigestibility to most herbivores. In any event, the plant pays a high price for its cellulose resistance in that it, likewise, cannot digest cellulose and thereby is deprived of recovering an energy-rich resource when an organ is discarded. This greatly restricts the plant's options for tissue repair and replacement. These observations should cause rejection of estimates of insect impact based only on amount of leaf surface eaten, estimates that usually range from 1 to 10% if averaged over many species in a habitat (e.g., Leigh 1975).

Secondly, herbivores cause two kinds of more direct damage to plants, and tropical insects are no exception: (1) There is the direct *parasitism* through sucking sap, eating leaves, and boring in stems, storage organs, and cambium. This damage lowers the fitness of the plant through reduction of progeny. This occurs because the damaged plant has directly lost resources and also loses competitive status, which in turn leads to lowered resource availability. (2) There is direct *predation* on seeds (and less commonly, seedlings) from the flower to already dispersed seeds (e.g., Janzen 1980). Just as we cannot know (anticipate) if introduction of yet another species of herbivore will result in a change in density or distribution of the host (prey) plant population, we cannot know the effect that a given species component of the herbivore load is having on a plant population unless we remove the species and see what happens, taking into account compensatory actions by other fractions of the herbivore load. We should add that herbivore loads illustrate very well the generality that herbivores compete *indirectly* through the medium of the resource and defense "budget" of the plant, as well as directly. Two herbivores may be in intense competition but never encounter each other or each other's direct damage (Janzen 1973). In short, a shoot tip eater affects the root feeder; the beetle that induces protease inhibitor formation in leaves by feeding on them may make the leaf inedible for a caterpillar.

Can we say anything unique about the impact of tropical phytophagous insects, parasites, and predators on their hosts and prey? First, there is much observational and circumstantial evidence and a bit of experimentation suggesting that, in tropical forests rich in tree species, insects are generally more effective at eliminating or severely reducing seed crops than in other forests. When the members of a population of large trees lose 50–70% of their seed crops to a single bruchid (e.g., *Pithecellobium saman* or *Merobruchus columbinus*—Janzen 1977), they have also had their tries at recruitment reduced by that much. Recruitment is a highly stochastic event. In the tropics every square meter has a finite but very small probablility of bearing a member of a tree's descendants in future years and the number of tries is very important in determining equilibrium density of adult trees present in the habitat (Janzen 1970). Why might insects be more destructive as seed predators in many tropical forests than in temperate forests? This may be because the climate is less effective at depressing their densities between major pulses of their food (seed crops), because their prey populations (the seeds) are not so thoroughly missing in between peaks of seed production in the tropics and because relatively fewer tree species are involved in supraannual synchronized seeding cycles than is so in extratropical regions (e.g., Janzen 1971, 1976, cf. Section 22.2.2.1).

Secondly, Janzen (pers. comm.) has considered that foliage removal for specific species of plants by tropical leaf eaters in *deciduous* tropical forests is as severe or more severe than that in extratropical deciduous forests. That it is severe in any sense may seem surprising. In any year there is substantial defoliation of certain species by their strongly to highly oligophagous herbivores, and the next year a different set of plant species have their turn being severely defoliated. The *forest* never appears to be heavily defoliated since there are likely to be 50–200 species of large woody plants within view at one time, but in species-rich tropical deciduous forests there is often a heavy shower of herbivore frass and 30–100% defoliation of some 5–30 species during the first 2 months of a rainy season (Janzen 1981). Gray (1972) has also viewed insect impacts in tropical forests to be much greater than commonly viewed by entomologists.

Thirdly, it is our impression that a given amount of defoliation of a tropical deciduous tree leads to larger losses in status, wood increment, future seed yields, height increment, and so on, than is the case with an average extratropical deciduous tree. We suspect this may be because extratropical trees are more accustomed to dealing with a variety of damages and physiological challenges (and therefore carry the reserves and have the physiological machinery for using them) than are their tropical counterparts. We suspect, too, that extratropical trees customarily have larger reserves because they are storing for mast crops at long intervals and because they are storing for heavy flushes at spring growth. Tropical plants, perhaps, may operate on a smaller margin of error than their extratropical counterparts.

Since the tropics are so rich in phytophagous insect species, it is tempting to assume that a tropical plant will be attacked by a larger array of insect species than a comparable extratropical plant of the same size and age. This does not appear to be so. Some studies in a Costa Rican deciduous forest with a flora of about 600 species of broad-leafed plants suggest that, if anything, an average tree there may have fewer species of insects to deal with than its extratropical counterpart.

If one thinks in units of years, everything happens faster in the tropics. Losses in status due to herbivore damage appear to be resolved more rapidly in tropical than in extratropical vegetation. Thus, if a noctuid larva eats off several terminal apices of a 2-m tall tree sapling in a tropical forest, the resultant loss in height may easily be adequate to result in that sapling losing out in competition with other saplings and being dead and gone in less than a year. The intense shading and continuously salubrious climate may cause a telescoping of the competitive process. However, we badly need more experiments in both temperate and tropical forests to substantiate such impressions as these from field observations.

Phytosociologists in northern latitudes have long gotten away with looking at only edaphic and climatic correlates with micro- and macro-habitat plant distributions. They rather ignored the insects and stressed the physical factors in defining the habitats in which plant recruitment can occur. However, in tropical habitats, and as we have seen, in temperate ones, it is certainly not safe to assume that the reason a tree species lives only in riparian bottomland vegetation, for example, is because that is a moist site during the dry season. We desperately need experiments on this subject in both tropical and extratropical forests.

The tropics are often thought of as extremely species-rich in trees ("diverse" in a sloppier terminology), but there are enormous areas of tropical vegetation where only one or a few species of trees (or other life forms) constitute the bulk of the plant matter. Even in some very species-rich forests there occasionally are single species that constitute up to 30% of the stand (e.g., Hartshorn 1975). Mangrove swamp forests are the best known examples; here virtually all of the self-supporting vegetation may be made up of 1–20 species of trees of tall to moderate height. This vegetation—leaves, bark, wood, seeds—is extremely rich in tannin, a powerful digestion inhibitor. There are herbivorous insects in mangrove forests, but it is clear that the bulk of the plant parts are simply inedible. Likewise, these plants grow very slowly compared to growth rates of trees on normal terrestrial sites. In short, they have traded their competitive ability for protection from herbivores and can get away with it because they live in a special edaphic habitat in which allospecific competitors have little chance of surviving. Another tropical habitat poor in tree species is a forest above about 2000–5000 m in elevation. These forests are continually cool and are commonly dominated by Fagaceae, Lauraceae, and Gymnosperms.

Perhaps the most interesting low-diversity tropical forests lie on white sand soils in Malasia. These "dipterocarp forests" have an upper canopy made up largely (50–90%) of 1–15 species of Dipterocarpaceae and an understory of many (up to some 400) species of smaller trees (e.g., Janzen 1974). Looking only at these dipterocarps, it is striking that they have foliar and reproductive biologies extremely similar to that of a hypothetical tree such as an oak in a mixed conifer–oak–ash stand in temperate United States; yet, these are evergreen rainforest trees by anybody's definition. They produce seeds in supraannual pulses that are highly synchronous at both the population and community levels. These seeds are heavily preyed upon by vertebrates and insects (several being weevils with roles like acorn curculios), but these animals are sufficiently satiated that dense lawns of dipterocarp seedlings appear after a seeding year, just as occurs in temperate climate oaks and pines after a seed year. The seeds are dispersed only a short distance from the parent. The adult trees appear to be site specialists and are rich in resins and tannins, and they appear to have similar relationships with forest insects as do evergreen oak and conifer trees in temperate climates. In short, in these habitats trees of this family have largely beaten the competitor and herbivore challenges and by and large come to dominate the system (Janzen 1974, 1978).

22.5 CONCLUSION

It would be sufficient if the scope and detail of concepts and roles of insects in the dynamics and regulation of other organisms as discussed in this chapter were to stimulate a deeper probing of the role of insects in the whole economy of nature.

ACKNOWLEDGMENTS

The authors wish especially to thank Dr. Charles Sartwell for considerable assistance in the review of concepts and events described concerning temperate forest ecosystems, and to Drs. R. L. Rabb and David Rosen for helpful suggestions on the organization of the chapter.

REFERENCES

Adkisson, P. L., R. L. Hanna, and C. F. Bailey. 1964. *J. Econ. Entomol.* **57:** 657–663.

Amman, G. D. 1977. Pp. 3–18 in Mattson (Ed.), 1977, referenced here.

Anderson, J. F. and H. K. Kaya. 1976. *Perspectives in Forest Entomology.* Academic, New York.

Andres, L. A., C. J. Davis, and R. D. Goeden. 1976. Pp. 481–499 in Huffaker and Messenger (Eds.), 1976, referenced here.

Andres, L. A. and R. D. Goeden. 1971. Pp. 143–164 in Huffaker (Ed.), 1971, referenced here.

Bennett, F. D., David Rosen, P. Cochereau, and B. J. Wood. 1976. Pp. 359–395, in Huffaker and Messenger (Eds.), 1976, referenced here.

Bess, H. A., S. H. Spurr, and E. W. Littlefield. 1947. *Harvard Forest Bull.* **22:** 1–56.

Birch, L. C. 1971. Pp. 109–126, in den Boer and Gradwell (Eds.), 1971, referenced here.

Bond, R. M. 1945. *N. Am. Wildl. Conf.* **10:** 229–234.

Brian, M. V. (Ed.). 1978. *Production Ecology of Ants and Termites.* Cambridge University Press, New York.

Bullock, J. A. 1967. *J. Indian Bot. Soc.* **46:** 323–330.

Campbell, R. W. and R. J. Sloan. 1977. *For. Sci. Monogr.* **19:** 1–34.

Cantlon, J. E. 1969. Pp. 197–203 in *Diversity and Stability in Ecological Systems.* Brookhaven Symp. Biology 22. Brookhaven National Laboratory, Upton, New York.

Carroll, C. R. and D. H. Janzen. 1973. *Annu. Rev. Ecol. & Syst.* **4:** 231–251.

Carroll, C. R. and S. J. Risch. In press. *J. Oecologia.*

Carter, W. 1973. *Insects in Relation to Plant Disease* (2nd ed.). Wiley, New York.

Clausen, C. P. (Ed.). 1978. *Introduced Parasites and Predators of Arthropod Pests and Weeds: A World Review.* USDA, Agriculture Handbook No. 480.

Cobb, F. W. Jr., D. L. Wood, R. W. Stark, and J. R. Parmeter, Jr. 1968. *Hilgardia* **39:** 141–152.

Connell, J. H. 1961. *Ecology* **42:** 710–723.

Connell, J. H. 1971. Pp. 298–323 in den Boer and Gradwell (Eds.), 1971, referenced here.

Coulson, J. R. 1977. *USDA Tech. Bull.* 1547.

Dalke, P. D. 1953. *J. Wildl. Manag.* **17:** 378–380.

DeBach, P. (Ed.). 1964. *Biological Control of Insect Pests and Weeds.* Chapman & Hall, London.

DeBach, P. 1974. *Biological Control by Natural Enemies.* Cambridge University Press, Cambridge.

DeBach, P. and B. R. Bartlett. 1964. Pp. 402–426 in DeBach (Ed.), 1964, referenced here.

DeBach, P., C. B. Huffaker, and A. W. MacPhee. Pp. 255–285 in Huffaker and Messenger (Eds.), 1976, referenced here.

den Boer, P. J. and G. R. Gradwell (Eds.). 1971. *Dynamics of Populations.* Centre Agric. Publ. & Doc., Wageningen, Netherlands.

Dodd, A. P. 1940. *The Biological Campaign against Prickly Pear*. Bull. Comm. Prickly Pear Bd., Brisbane, Australia.

Doutt, R. L. and P. DeBach. 1964. Pp. 118–142 in DeBach (Ed.), 1964, referenced here.

Dybas, H. S. and M. Lloyd. 1974. *Ecol. Monogr.* **44:** 279–324.

Ehrlich, P. R. and L. C. Birch. 1967. *Am. Natur.* **101:** 97–108.

Everett, T. R. and H. A. Lamey. 1969. Pp. 361–377 in Maramorosch (Ed.), 1969, referenced here.

Ferrell, G. T. 1980. USDA, Forest Service General Tech. Report PSW-39.

Foster, J. 1964. *Studies on the Population Dynamics of the Daisy, Bellis Perennis*. Ph.D. dissertation, University of Wales, U.K.

Gadgil, M. and W. H. Bossert. 1970. *Am. Natur.* **104:** 1–24.

Gibbs, J. N. 1978. *Annu. Rev. Phytopathol.* **16:** 287–307.

Goeden, R. D. 1978. Pp. 357–414 in Clausen (Ed.), 1978, referenced here.

Goheen, D. J. and R. W. Cobb, Jr. 1980. *Can. Entomol.* **112:** 725–730.

Golley, F. B. and E. Medina (Eds.). 1975. *Tropical Ecological Systems*. Springer-Verlag, New York.

Graham, S. A. 1956. *Can. Entomol.* **88:** 45–55.

Gray, Barry. 1972. *Annu. Rev. Entomol.* **17:** 313–354.

Hairston, H. G., F. E. Smith, and L. B. Slobodkin. 1960. *Am. Natur.* **94:** 421–425.

Harpaz, I. 1972. *Maize Rough Dwarf*. Israel University Press, Jerusalem.

Harper, J. L. 1977. *Population Biology of Plants*. Academic, New York.

Harris, K. F. and K. Maramorosch (Eds.). 1977. *Aphids as Virus Vectors*. Academic, New York.

Hartshorn, G. S. 1975. Pp. 41–59 in Golley and Medina (Eds.), 1975, referenced here.

Hazeler, W. H. 1977. Proc. 6th Asian-Pacific Weed Sci. Conf. **2:** 433–439.

Hazeler, W. H. 1981. Pp. 3–9 in E. S. Del Fosse (Ed.) *Fifth International Symposium in Biological Control of Weeds*. Commonwealth Sci. Ind. Res. Org., Melbourne, S. Australia.

Houston, D. R. 1979. *USDA, Agriculture Handbook* 542.

Huffaker, C. B. 1957. *Hilgardia* **27:** 101–157.

Huffaker, C. B. 1959. *Annu. Rev. Entomol.* **4:** 251–276.

Huffaker, C. B. 1962. *Can. Entomol.* **94:** 507–514.

Huffaker, C. B. 1967. *Mushi* (Suppl.) **39:** 51–73.

Huffaker, C. B. 1971a. Pp. 327–341 in den Boer and Gradwell (Eds.), 1971, referenced here.

Huffaker, C. B. (Ed.). 1971. *Biological Control*. Plenum, New York.

Huffaker, C. B. 1974. *Environ. Entomol.* **3:** 1–9.

Huffaker, C. B., Junji Hamai, and R. Nowierski. 1983. *Entomophaga*. **28:** 387–400.

Huffaker, C. B. and C. E. Kennett. 1959. *J. Range Manag.* **12:** 69–82.

Huffaker, C. B. and C. E. Kennett. 1969. *Can. Entomol.* **101:** 425–447.

Huffaker, C. B. and P. S. Messenger (Eds.). 1976. *Theory and Practice of Biological Control*. Academic, New York.

Hurlbert, S. H. 1970. *Bull. Torrey Bot. Club* **97:** 189–195.

Janzen, D. H. 1969. *Evolution* **23:** 1–27.

Janzen, D. H. 1970. *Am. Natur.* **104:** 501–528.

Janzen, D. H. 1971. *Annu. Rev. Ecol. & Syst.* **2:** 465–492.

Janzen, D. H. 1972. *Ecology* **53:** 258–261.

Janzen, D. H. 1973. *Am. Natur.* **107:** 786–790.

Janzen, D. H. 1974. *Biotropica* **6:** 69–103.

Janzen, D. H. 1976. *Annu. Rev. Ecol. & Syst.* **7:** 347–391.

Janzen, D. H. 1977. *Trop. Ecol.* **18:** 162–176.

Janzen, D. H. 1978. Pp. 83–128 in Tomlinson and Zimmerman (Eds.), 1978, referenced here.

Janzen, D. H. 1980. *J. Ecol.* **68:** 929–952.

Janzen, D. H. 1981. *Biotropica.* **13:** 271–282.

Johnson, M. P. and P. H. Raven. 1970. *Evol. Biol.* **4:** 127–162.

Jones, D. A. 1966. *Can. J. Genet. & Cytol.* **8:** 556–567.

Kennedy, G. G. 1976. *Environ. Entomol.* **5:** 827–832.

Knight, F. B. and M. J. Heikkenen. 1980. *Principles of Forest Entomology* (5th ed.). McGraw-Hill, New York.

Kraft, K. J. 1968. *Annu. Entomol. Soc. Am.* **61:** 1462–1465.

Laing, J. E. and Jungi Hamai. 1976. Pp. 685–743 in Huffaker and Messenger (Eds.), 1976, referenced here.

Lauckhart, J. B. 1957. *J. Wildl. Manag.* **21:** 230–234.

Leigh, E. G. 1975. *Annu. Rev. Ecol. & Syst.* **6:** 67–86.

Levin, D. A. 1976. *Annu. Rev. Ecol. & Syst.* **7:** 121–159.

Lloyd, E. P., M. E. Merkl, and G. B. Crowe. 1962. *J. Econ. Entomol.* **55:** 225–227.

Lloyd, M. and H. S. Dybas. 1966a. *Evolution* **20:** 133–149.

Lloyd, M. and H. S. Dybas. 1966b. *Evolution* **20:** 466–505.

Lloyd, M., G. Kritsky, and C. M. Simon. 1983. *Evolution,* **37:** 1162–1180.

Lloyd, M. and J. White. 1980. *Am. Natur.* **115:** 29–44.

Luck, R. F., P. S. Messenger, and J. F. Barbieri. 1981. Pp. 34–42 in Rosen (Ed.), 1981, referenced here.

Lyons, L. A. 1957. *Can. Entomol.* **89:** 264–271.

MacArthur, R. H. and J. H. Connell. 1956. *The Biology of Populations.* Wiley, New York.

MacPhee, A. W., L. E. Caltagirone, M. van de Vrie, and E. Collyer. 1976. Pp. 337–358 in Huffaker and Messenger (Eds.), 1976, referenced here.

Maddox, D. M., L. A. Andres, R. D. Hennessey, R. D. Blackburn, and N. R. Spencer. 1971. *Bioscience* **21:** 985–990.

Maramorosch, K. (Ed.). 1969. *Viruses, Vectors and Vegetation.* Wiley, New York.

Mattson, W. J. (Ed.). 1977. *The Role of Arthropods in Forest Ecosystems.* Springer-Verlag, New York.

Mattson, W. J. and N. D. Addy. 1975. *Science* **190:** 515–522.

McFadden, M. W., D. L. Dahlsten, C. W. Berisford, F. B. Knight, and W. W. Metterhouse. 1981. *J. Forestry* **79:** 723–726.

Mellini, E. 1952. *Boll. Inst. Entomol. Bologna* **19:** 97–119.

Merrill, W. 1967. *Phytopathology* **57:** 1206–1210.

Miller, P. C. 1970. *Am. Midland Nat.* **83:** 206–212.

Moore, A. W. 1942. *J. Mammal.* **23:** 37–41.

Monro, J. 1967. *J. Anim. Ecol.* **36:** 531–547.

Moran, V. C. 1980. *Ecol. Entomol.* **5:** 153–164.

Nicholson, A. J. 1957. *Cold Spring Harbor Symposia on Quantitative Biology* **22:** 153–172.

Nicholson, A. J. 1960. Pp. 477–521, Vol. 1, in Sol Tax (Ed.), *Evolution after Darwin.* University of Chicago Press, Chicago.

Nillson, Sten. 1976. *Ambio* **5:** 17–22.

Petal, J. 1978. Chapter 10 in Brian (Ed.), 1978, referenced here.

Pimentel, D. and R. Al-Hafidh. 1965. *Annu. Entomol. Soc. Am.* **58:** 1–6.

Pulliam, H. R. and M. R. Brand. 1975. *Ecology* **56:** 1158–1166.

Rafes, P. M. 1966. Pp. 5–68 in Academy of Sciences, USSR, Lab. For. Sci., Moscow. (Canada Dept. For. & Rural Devel. Translation No. 89).

Rexrode, C. O. and T. W. Jones. 1971. *Plant Disease Rept.* **55:** 108–111.

Ridley, H. N. 1930. *The Dispersal of Plants throughout the World.* L. Reeve & Co., Washford, England.

Room, P. M., K. L. S. Harley, I. W. Forno, and D. P. A. Sands. 1981. *Nature* **294:** 78–80.

Rosen, David (Ed.). 1981. *The Role of Hyperparasitism in Biological Control: A Symposium.* Agric. Sci. Pubs., University of California, Berkeley.

Rosenthal, G. A. and D. H. Janzen, (Eds.). 1979. *Herbivores—Their Interaction with Secondary Plant Metabolites.* Academic, New York.

Rudinsky, J. A. 1962. *Annu. Rev. Entomol.* **7:** 327–348.

Ryder, E. J. and J. E. Duffus. 1966. *Phytopathology* **56:** 842–844.

Safranyik, L., D. M. Shrimpton, and H. S. Whitney. 1974. *Management of Lodgepole Pine to Reduce Losses from the Mountain Pine Beetle.* Environment Canada, For. Serv., For. Tech. Report 1.

Salman, K. A. and J. W. Bongberg. 1942. *J. Forestry* **40:** 533–539.

Sartwell, C. and R. E. Stevens. 1975. *J. Forestry* **73:** 136–140.

Schmid, J. M. and R. H. Frye. 1976. *USDA, For. Serv. Res. Note RM-309.*

Sharp, L. A. and W. F. Barr. 1960. *J. Range Manag.* **13:** 131–134.

Simmonds, H. W. 1933. *Bull. Entomol. Res.* **24:** 345–348.

Slobodbin, L. B., F. E. Smith, and N. G. Hairston. 1967. *Am. Natur.* **101:** 109–124.

Smith, C. C. 1970. *Ecol. Monogr.* **40:** 349–371.

Smith, D. M. 1976. Pp. 3–20 in Anderson and Kaya (Eds.), 1976, referenced here.

Smythe, N. 1970. *Am. Natur.* **104:** 25–35.

Stoszek, K. J., P. G. Mika, J. A. Moore, and H. L. Osborne. 1981. *Forest Sci.* **27:** 431–442.

Svardson, G. 1957. *Br. Birds* **50:** 314–343.

Tevis, L. Jr. 1958. *Ecology* **39:** 695–704.

Thompson, W. R. 1954. Pp. 89–95 in *Rept. Comm. Entomol. Conf., 6th Meeting.* London.

Tomlinson, P. B. and M. H. Zimmermann (Eds.). 1978. *Tropical Trees as Living Systems.* Cambridge University Press, New York.

True, R. P., H. L. Barnett, C. K. Dorsey, and J. G. Leach. 1960. *W. Virg. Univ. Agric. Expt. Sta. Bull.* **448T.**

Utida, Syuno. 1953. *Ecology* **34:** 301–307.

Van der Plank, J. E. 1948. *Empire J. Expt. Agr.* **16:** 134–142.

Varley, G. C. 1937. *Proc. Roy. Entomol. Soc. Lond. A,* **11:** 109–122.

White, J. 1980. *Am. Natur.* **115:** 1–28.

White, J. and M. Lloyd. 1975. *Am. Midl. Natur.* **94:** 127–143.

White, J. and C. E. Strehl. 1978. *Ecol. Entomol.* **3:** 323–327.

Whittaker, R. H. and P. P. Feeny. 1971. *Science* **171:** 757–770.

Wickman, B. E. 1978. USDA, Forest Service Res. Paper PNW-244.

Williams, C. B. Jr., J. M. Wenz, D. L. Dahlsten, and N. X. Norick. 1979. *Bull. Soc. Entomol. Suisse* **52:** 297–307.

Wilson, F. and C. B. Huffaker. 1976. Pp. 3–15 in Huffaker and Messenger (Eds.), 1976, referenced here.

Zwölfer, H. 1967. *Urophora siruna-seva (HG) (Diptera, Trypetidae), a Potential Insect for the Biological Control of Centaurea Solstitialis L. in California* (Report 5), Comm. nw. Inst. Biological Control. Delemont, Switzerland.

Section V

APPLICATION

OF ECOLOGY

TO INSECT

POPULATION

MANAGEMENT

The biological resources, which mankind eclectically exploits, manages, and enjoys, continue to be available only as a function of the diverse but interdependent flora and fauna, including the Insecta, whose populations are essential to many of the ecological processes involved. From mankind's view the essentiality of insects to a viable environment is often overshadowed by concern with insects as competitors for resources and hazards to comfort and health, although a relatively small number of species can be objectively categorized as pests. Thus, the management of insect populations in a broader context than pest abatement is an integral part of resource management, the central objective of applied ecology.

In this section we discuss some of the potentials and problems of the ecological approach to pest situations. Our emphasis on insect pests is not to depreciate the view of the Insecta as a diverse and essential resource (Chapter 19). To the contrary, a chief motivation for applying ecological principles in managing pests is to protect the vast number of other organisms. Pest species are emphasized because they have constituted the target populations for the bulk of management attempts and there is more detailed information on their management.

The pest concept is based on anthropocentric values. The critical difference between a pest and a nonpest is that mankind has designated the former as undesirable (Clark et al. 1967). Some insect pests affect man directly by transmitting pathogens or causing pain or discomfort by their bites, stings, or other irritating attributes. Some depreciate the value of certain plants, animals, and products by direct destruction, disease transmission, or contamination. An insect pest, however, is such only in an environment conducive to its unwanted activities. In other situations the same species is not a pest and may be highly desirable. Thus, it is important to understand a species' role in nature (Chapters 12, 19–22) before planning its demise. Most pest species are pests either because man creates a situation where an innocuous population is elevated to pest status or else man moves into a new arena where he or his products become subject to attack. As a result of the increasing modification of the biosphere and the intrusion of man's expanding population into more and more of the world's limited space, the number and seriousness of pest species has increased, pollution has intensified, and perturbations of ecosystems throughout the world threaten the stability of the biosphere. Therefore, there is a rationale for concluding, as did Corbet (1970), that man is the chief earth pest.

The three chapters of this section serve to orient the reader to the complex, economically essential, and intellectually challenging field of pest management—a field that is on the leading edge of man's probing for a viable, productive, and stable relationship with other organisms. Man–insect confrontations have been one of the most consistently recurring themes of history, and in spite of the exponential growth of technology, these confrontations seem to loom more menacingly than ever before. However, perhaps one positive result of man's struggle to solve pest problems is a growing realization that man must in a sense cooperate as well as compete with other living creatures. To do so, he must act on the basis of available ecological knowledge and with respect for the probable importance of the biotic diversity which he so poorly understands. [But see Chief Seattle's (1854) eloquent statement of the native Americans' understanding].

All ecological principles (see previous chapters) are relevant to insect management, although relevance of particular principles varies widely among insect species. The number of such principles and their variability and generality necessitate a conceptual frame in which they may be organized and from which selections may be made to apply to specific pest situations. In the frame presented in Chapter 23, the authors relate some general ecological principles to be considered in manipulating specific components of pest life systems (Chapter 11). The manipulations in turn comprise candidate tactics for pest management. Successful pest management connotes the selection and use of one or more tactics giving a desirable level of protection in a manner acceptable economically, socially, and politically, with the least likelihood of reducing the natural order and aesthetic qualities of the environment. Ideally, each tactic, when used, should elicit no genetic or otherwise disruptive response intensifying future pest problems and should be innocuous to nonpest organisms, particularly biological control agents, the suppression of which might lead to elevation of potential pests to pest status. While these criteria have been met in some pest control efforts, they are rarely met in perennially serious, multiple-pest situations.

In response to the awesome complexity of the web of pest–crop interactions, scientists are now producing simulation and optimization models of crop–pest systems which promise a practical basis for assessing the economic impact of pest populations and for making management decisions (Chapter 24). These models, while simplistic and rudimentary, have already contributed appreciably to the efficiency of research, and, as discussed in Chapter 25, some are finding use, along with biological monitoring, in improved pest management programs characterized by the application of ecological principles.

REFERENCES

Clark, L. R., P. W. Geier, R. D. Hughes, and R. F. Morris. 1967. *The Ecology of Insect Populations in Theory and Practice.* Methuen, London.

Corbet, P. S. 1970. Pp. 191–204 in R. L. Rabb and F. E. Guthrie, (Eds.), *Concepts of Pest Management.* Proc. Conf. N. C. State Univ., Raleigh.

Seattle, (Chief), 1854. (Reply to U.S.A. President Franklin Pierce on tribal status)

Chapter 23

An Ecological Approach to Managing Insect Populations

R. L. RABB, G. K. DeFOLIART, and G. G. KENNEDY

23.1 INTRODUCTION

Approaching a pest situation ecologically connotes placing it in the context of what is known concerning the structure and function of nature at the population, community, and ecosystem levels. This has proven useful in discovering and improving control tactics (cultural, biological, and chemical) and in developing strategies for using them in relationship to ecological processes and specified human goals. Some of the latter have been set under heavy economic pressures without regard to ecological constraints and therefore have proven unrealistic and/or incompatible with stable, productive, environmentally sound, and aesthetically pleasing surroundings. Thus, the logic of integrating the expertise of natural and social scientists in goal setting is inescapable and forms a persuasive justification for integrated pest management (IPM). In this chapter our concerns are primarily ecological, and we give little attention to the economic, societal, and political constraints (cf. Chapters 24, 25). Our chief focus is on the dynamics of pests in managed areas and effects of manipulating life system components on their numbers, qualities, and distributional patterns.

Many, if not most, pest problems involve more than one pest species, but we shall first consider single-pest situations. Even so, the number of ecosystem components and processes involved are so numerous and complex that the conceptualization and modeling of population systems in complete detail and accuracy is impossible. Nonetheless, very simple models may provide useful answers to some questions (cf. Chapter 11).

Only those ecosystem features conceived to account for observed patterns of target populations in defined space–time arenas are included in population models. Because of the uniqueness of species (even biotypes of species), these features vary in detail and significance for different pests, and their selection requires information concerning biological attributes, resource requirements, and habitat–niche relationships of the species involved. However, very generalized models may serve to focus on major variables and processes. Figure 23.1, for example, is conceived as the life system of a generalized, herbivorous, agricultural insect pest.

The bulk of management inputs to most cropping systems are not directed toward insect pests but rather to provide a favorable physicochemical environment for crop growth. Nevertheless, these inputs greatly influence insect populations either through direct impact or, indirectly, through other life system components. Some of the major ecological processes through which the biotic factors interact with the target population are shown in Fig. 23.1; however, not shown are the processes through which solar energy drives the weather systems which determine in such large measure the suitability of the physicochemical media and microclimate of the entire biotic complex.

Individuals of the population respond physiologically and behaviorally

698

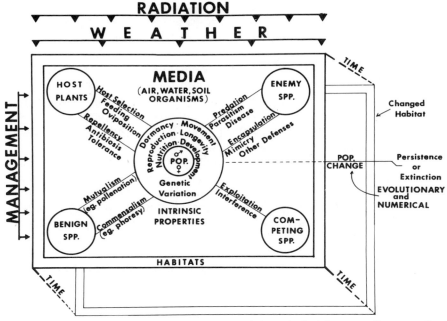

Figure 23.1   A much simplified representation of the principal features of a generalized insect life system.

to their life system components. These responses influence rates of growth, development, reproduction, mortality, and movement (previous chapters). These individual responses include intrapopulational adjustments to population density (Chapter 12), which vary in kind and significance among the insects, being most highly developed among social species. In addition to proximal adjustments of individuals within the constraints of existing genotypes, populations may change qualitatively (genetically) over time. Such changes may be expedited through heavy selective pressures of management.

Figure 23.1 fails to illustrate one very important aspect of every life system, that is, the arrangement within a defined space–time arena of the resources, refugia, and foci of mortality agents. Not only are quantitative and qualitative characteristics of these components of importance to populations, but distances between them and temporal synchrony among them (environmental heterogeneity) are crucial in determining population levels and distributional patterns (Chapter 14 especially). Another serious deficiency of Fig. 23.1 is that space–time dimensions of the target population and management arena are not defined (cf. Section 23.3.2.1).

If pest models are to be usefully realistic and predictive, the influence of mankind must be appropriately represented. This influence has increased dramatically since the industrial revolution and with the subsequent exponential growth of human populations and resource exploitation. As societies have evolved, high-energy subsidies have been used in exploiting the biosphere; and topography, soil, water cycles, and the species composition of biotic communities have been drastically altered. Indigenous biota has been stripped from large tracts and replaced with simplified crop communities, held in being through practices inserted to provide optimal physicochemical conditions for growth of largely exotic plant species with minimum competition from other species. Most inputs affecting major structural features of agroecosystems are not overtly directed toward insects, but influence them greatly. In many areas crop mixes and cultural practices have been adopted with little regard to their effects on actual and potential insect pests. This has led to heavy reliance on temporary suppression measures and collapse of certain agricultural systems (e.g., Adkisson 1973). Thus, before modifying management practices in efforts to control pests, it is logical to learn how current ones are influencing target populations. Agroecosystem structure is variable from farm to farm and year to year, and this variation must be reflected in the models of pest populations if they are to be useful in IPM.

Ecology might be conceived as an album of views of natural phenomena involving populations of organisms, accompanied by explanatory theories. By the same token, ecological pest management can be viewed as a set of guidelines for applying principles relating to control of pests. The organization of this album and presentation of these guidelines should be based on the holism of nature and should reflect the interdependencies and interactions involved.

Here, we can merely list the principles we consider of most general importance. Our few examples inadequately represent the rich array of information available [e.g., in Metcalf et al. (1962), NAS (1969), DeBach (1964), Huffaker & Messenger (1976), Metcalf & Luckmann (1975), Harwood & James (1979), Huffaker (1980), and Knight & Heikkenen (1980)]. Many texts are oriented to various categories of entities to be protected (e.g., humans, animals, crops, forests, household furnishings, and stored products) or to various major strategies of control (e.g., chemical, biological, cultural, and subclassifications of these). In contrast, we here attempt to view various management arenas as perceived by various insect pests (Wellington 1977, and Chapter 13) in an effort not only to examine the major cause-and-effect pathways through which management inputs may influence populations of pests and associated organisms, but also to illustrate the variations from species to species in how these pathways may be used.

Any classification of life system components or principles for their manipulation is potentially misleading, often subtly, because no component can be changed without affecting others (Chapter 2). Our choice is defensible only as an initial orientation to a complex and very important component of resource management.

23.2 SOME ECOLOGICAL PRINCIPLES FOR APPLICATION

23.2.1 Each Species Responds Uniquely to the Physical Factors and Condition of its Media

Pest populations and physical factors change in space and time according to general patterns (e.g., latitude, elevation, time of day, and season), and the ebb and flow of "patches" of pests are correlated, often closely, with patterns of climate and soil. Thus, knowledge of a pest's resource requirements and limits of tolerance to physical factor variation, coupled with maps depicting isopleths of temperature, humidity, rainfall, and soil types, is useful in predicting distributional patterns of pests and in developing management strategies. Information on the effects of physical factor variation on dormancy, development, and reproduction (Chapters 3, 5, 6) can be useful in selecting and timing control tactics if combined with local climatic data. There is also great potential in predicting pest behavior through studies of their responses to dynamic synoptic weather systems (Chapter 13).

Since early recorded history, man has used physical and mechanical means in responding to insect pests. Consequently, a diversity of such techniques have been proposed for insect control. Some are effective and in wide use, and others have potential for greater and/or future use, particularly in IPM systems (NAS 1969, and Chapter 25). These techniques may be categorized as (1) manipulating individual factors, (2) manipulating water and soil, and (3) mechanical destruction or exclusion.

23.2.1.1 Manipulating Individual Factors.
We speak of adjusting individual factors in relation to a pest's tolerance scale or response thresholds; however, a factor cannot be changed without altering other factors with which it is interlocked, and a change in a factor may affect an organism indirectly as well as directly. Nevertheless, manipulating temperature and/or moisture above or below tolerance limits has become standard practice in protecting some stored products from pests. Natural enemy action also may be enhanced through alteration of microclimate as exemplified by the high humidity requirements for fungal epizootics in many insect populations.

Insects of different species may be differentially attracted and/or repelled by various wavelengths of radiant energy, and this knowledge has stimulated development of a diversity of trapping techniques. Use of such techniques for pest detection and surveys of seasonal occurrence and abundance is of importance in many quarantine and pest management programs. There are also limited applications of light traps for insect control *per se* (NAS 1969).

23.2.1.2 Manipulating Soil and Water.
Many insects spend part or all of their life cycles in the soil and are affected positively or negatively by tillage

practices, according to the species, the type of tillage, and its timing (NAS 1969). Ovipositing females of *Agrotis orthogonia* are attracted to freshly tilled soil, whereas tillage in vineyards may bury overwintering cocoons and reduce spring emergence of *Paralobesia viteana,* and tillage to cover ("ridge") potato tubers, when properly timed, is effective in reducing tuber infestations by *Phthorimaea operculella.* Fall plowing kills many potentially overwintering insects, such as corn earworm pupae, directly through mechanical injury or indirectly through exposure to adverse weather factors or biotic control agents.

Water management may be used in controlling certain insects in soil and aquatic habitats. In dry land and irrigated areas withholding irrigation at critical times in the life history of certain insect pests may reduce their populations, and Genung (1976) had some success in controlling certain soil insects through flooding. Techniques for regulating water level and circulation in aquatic habitats are critical components of mosquito management systems, as exemplified by the open marsh water management (OMWM) method (Ferrigno & Jobbins 1968) in New Jersey for suppression of *Aedes sollicitans.*

23.2.1.3 Mechanical Destruction or Exclusion.

Handpicking hornworms from tomatoes and hand removal and destruction of borers in woody plants provide adequate protection from these pests, provided the labor is available. Such "primitive" methods may be practiced in home gardens and for small stands of shrubs and trees but are usually too labor intensive and expensive for most modern agriculture. Various mechanical devices have been invented for removing pests from crops, but their success has been very limited. Use of mechanical harvesters and equipment for pulverizing crop residues is important in suppressing certain insects such as the pink bollworm (NAS 1969).

Exclusion of insects is a major consideration in packaging food and certain manufactured articles and in screening homes and industrial buildings. This strategy also has limited application against agricultural pests (e.g., screening of vegetables in gardens to exclude certain Lepidoptera and use of banding to exclude cutworms).

23.2.2 The Quality, Quantity, and Distribution (Spatial and Temporal) of Food Influence Population Performance

23.2.2.1 Different Views of Food.

Insect species vary widely in food specificity: some require a particular mix of ingredients found only on or in a certain species of plant or animal; others utilize materials from many plant and animal sources, living and dead (Chapters 3, 4, 5); still others require a seasonal shift from one food source to another. Accurate knowledge of the quantity, quality, and position of food in a species' resource field (Chapter 14) can suggest powerful control tactics. In the narrow sense *food* is the ingested raw materials of metabolism. However, it is often used in a more general sense to represent food *sources* such as host or prey species or dead organic matter from which essential nutrients can be obtained. The ambiguity of equating food (essential nutrients) and food sources (matter or bodies from which obtained) may obstruct our view of the finer ramifications of trophic relationships. Nevertheless, the food sources are often the components of a pest's life system that are most amenable to manipulation. In manipulating them, it is significant that food sources (host plant, prey, dung, forest litter, etc.) comprise a conglomerate of entities and factors, widely scattered in the ecosystem, and often different for immature and adult insects. For many species, particularly immature stages, the food source is a course-grain resource (Chapter 14) which provides living space as well as nourishment. Therefore, manipulating such food also comprises microhabitat alteration.

From man's view, which dominates management decisions, insect food sources are of two types: (1) that which must be protected, that is, nonexpendable (humans, domestic animals, commercial timber, crops, stored products, etc.) and (2) that which is expendable (crop residues, alternate hosts of little value, animal wastes, etc.). Obviously, options differ in manipulating these two types; hence our rationale for discussing them separately below.

23.2.2.2 Food Man Wishes to Protect.

Methods that can be used to manipulate the food of insects of major economic importance include disruptions of synchrony and both nongenetic, and genetic alterations of food quality, as discussed below.

Disrupting Synchrony Between Food and Pest. Altering the spatial and/or temporal synchrony between a valued entity and its pests has long been used to reduce unwanted interactions. Very likely, this was a major benefit that accrued from the slash-and-burn agricultural systems used by primitive peoples. Even today, benefits in the form of reduced pest attack are reaped by growing crops on the periphery of, or outside the range of, potentially serious pest species. For example, soybean in the midwest has fewer serious insect problems than soybean in the southeastern United States; similarly, cotton in the western and southwestern United States is not plagued by boll weevil, which is restricted to the midsouth and southeast. Spatial isolation of seed potato production from the main potato crop production areas and other sources of aphid-borne potato viruses is one of the main tactics used in production of certified virus-free seed potatoes.

Synchrony between a crop and a pest population can be disrupted in some cases by not growing that crop in the same field year after year. Such rotation, while effective against species with limited powers of dispersal and few hosts, is essentially ineffective for highly mobile pests that have many hosts.

Certain insect species require a full-season association with a crop before even reaching a status of potentially serious pests. Larvae of grape colaspis, *Colaspis brunnea*, are of no known significance to soybean production during the first year of soybean culture in a given field. During this year, however, grape colaspis adults will move into these soybeans and produce larval populations which, after overwintering, may damage seedlings if the field is planted to soybean the second year. By avoiding continuous cropping of soybean in the same field, growers are, in essence, moving the crop away from potentially damaging populations produced in the previous crop. Effectiveness of rotation has been demonstrated for numerous species. But, even if effective against a single species, it may be impractical if it engenders problems with other pests or is incompatible with the grower's financial interests. For example, the corn rootworm was formerly widely controlled in midwestern United States by a 4-yr rotation involving corn, oats, and clover. Presently, rotation is not as profitable as continuous corn cropping. Thus, this pest is largely controlled with insecticides (Glass 1975).

Analogous to crop rotation is the "spelling of pastures" to reduce tick infestations of cattle by alternately moving cattle from one pasture to another and leaving each pasture unstocked for a long enough period that most of the tick larvae die. The method is discussed in Chapters 24 and 25 with respect to the management of the tick *Boophilus microplus* in Australia.

With many pests it is not possible or practical to avoid damage by creating discontinuities in *space*. Avoidance in time, however, may be a useful alternative. Many crops are susceptible to attack only during certain phenological stages of either the crop, or the pest, or both. Therefore, it is often possible to select planting times affecting the synchrony of crop and pest so that the most susceptible stage of the crop occurs when the pest population is at a low level or when it is associated with another host. A classic application of this principle was the avoidance of Hessian fly damage by delaying planting of winter wheat until after egg-laying adults were no longer active in the fall (Metcalf et al. 1962), although this practice is not as widely used as formerly because of widespread adoption of Hessian fly resistant wheat varieties. Another classic example, from animal rather than plant husbandry, involves the screwworm, *Cochliomyia hominovorax*. Before initiation of programs using genetic methods against this pest, livestock losses caused by it were reduced by restricting certain husbandry practices conducive to attack to months when screwworm adults are absent or in low numbers. Calving, dehorning, castration, and branding were scheduled before May when screwworm populations begin to increase. Relaxation of these tactics following introduction of genetic techniques contributed to the general increase in screwworms in the southwestern states during 1972 (Knipling 1979), although genetic change was also involved (Bush et al. 1976, and Chapter 9).

Establishing temporal discontinuity between a crop and sources of infectious pathogens or their vectors is an important method of reducing losses to certain insect-borne plant pathogens. In Israel, for example, heavy infections of aphid-borne potato virus Y and cucumber mosaic virus are avoided in bell peppers by delaying planting until the weed hosts of the virus have dried up (Loebenstein & Raccah 1980).

An understanding of the relationship between a crop's phenology and its susceptibility to attack by various pests, as well as the dynamics of host preference, is fundamental to ecologically sound pest management and has long been used in choosing minimum-risk planting dates, identifying fields with high risk of attack, and timing applications of temporary suppression measures. For example, in ovipositing, *Heliothis zea* is strongly preferential to fruiting stages of its many host species and to corn as compared to most other crops. In North Carolina, early-planted tomatoes, which ripen before corn matures, typically do not suffer significant losses to *H. zea;* however, late-planted tomatoes, which ripen after the corn is no longer attractive for oviposition, can be completely destroyed if no insecticides are applied. The *H. zea* situation is similar, although somewhat more complex, for soybean (Stinner et al. 1977).

In tropical and subtropical areas continuous or year-round production of a single crop in a given area can lead to large pest populations and losses. The brown plant hopper, *Nilaparvata lugens,* depends entirely on rice. Under continuous rice cropping, *N. lugens* migrates from fields of mature rice to fields of young rice. Since the injury resulting from subsequent generations developing within the field is closely related to the size of the immigrating population, reducing the initial density of migrants into a paddy is a major management component (Kiritani 1979). If practiced on an area-wide basis, rotating rice with annual crops or fallowing between two rice plantings for 6–8 weeks (eliminating continuous presence of rice) reduces the overall population of *N. lugens* and crop losses (Kiritani 1979, Kenmore 1980).

Similarly, where certain crops are continuously present because old and new plantings overlap in time, the old plantings often constitute major sources of aphids and aphid-borne pathogens which infect the new plantings. A mandatory crop-free period to break the aphid continuity and disease cycle can be effective. Aphid-borne celery mosaic virus which has few significant weed hosts was controlled in California and Florida through use of a celery-free period to eliminate overlapping presence of infected mature plantings and healthy new ones (Zitter 1977).

Another aspect of the role of spatiotemporal discontinuity in alleviating pest-induced crop losses involves the relationship between the size of a crop field and the invasion pattern of a pest population. Within-field position of initial invaders varies with many extrinsic factors such as wind, windbreaks, and sources of invaders. However, invasion patterns of highly mobile, actively flying species tend to show little if any "edge effects," whereas less mobile species and weak fliers tend initially to invade field margins nearest to overwintering sites or alternate hosts. For example, females of the stalk borer *Papaipema nebris* produced on weeds, oviposit largely in marginal rows of tillage corn. Thus, the proportion of plants damaged tends to be related to field size. However, in nontillage corn, where suitable weed hosts may be scattered throughout the field, plants may be damaged over the entire field (Gregory & Musick 1976).

Invasion patterns may also change seasonally. In northeastern North Carolina the two-spotted spider mite, *Tetranychus urticae,* invades corn in the spring primarily by crawling from weed hosts to corn plants near the field edge. As the season progresses, the infestation spreads deeper into the field. However, in all but the smallest fields, infestations remain heaviest near the field edges throughout the season. As the corn begins to senesce in late July and early August, tremendous numbers of mites are dispersed on the wind from the corn, and many are dropped on peanut crops. Because of this aerial invasion, the infestation in peanuts tends to be randomly distributed throughout the field. Steep gradients of spider mite abundance in peanut fields are typically observed only in fields adjacent to mite-infested corn fields (Brandenburg 1981).

Nongenetic Alteration of Food Quality. Insect attack and population growth varies among individual hosts (plants and animals), even of preferred host species in perfect spatiotemporal synchrony with the insects involved, as a response to variation in, or altered food quality brought about by, many factors such as availability of water and nutrients, temperatures, and the imposed chemical environment. Different species, however, cannot be expected to react similarly to a particular change in host quality. Thus, water stress imposed on a host plant may inhibit one pest problem but induce another. Severe water stress on seedlings of certain cucumber cultivars leads to an increase in resistance to two-spotted mites (Gould 1978). For certain cowpea cultivars, however, water stress increases susceptibility to attack and injury to *Sericothrips occipitalis* (Singh 1975).

Fertilization regimes have been shown to influence pest susceptibility in certain crops. Where added nitrogen has increased soluble N in plants, multiplication of aphids also increased (Van Emden et al. 1969). In rice heavy use of nitrogen favors dwarf virus disease by engendering increased populations of its vector, the green rice leafhopper, *Nephotettix cincticeps,* and also an increased susceptibility to infection (Kiritani 1979).

Herbicides and plant growth regulators can alter suitability of a crop as food for one or more pests (Tingey & Singh 1980) and can result in large changes in pest populations, as observed in tobacco following wide-scale and continuing use of maleic hydrazide (MH) to prevent production of suckers (axillary shoots). Though applied during preharvest, MH continues to retard sucker growth on live stalks left after harvesting. Prior to use of MH, thousands of acres of profuse sucker growth provided food for large populations of overwintering insects, including the tobacco hornworm, *Manduca sexta.* Consequently, winter carryover of this pest and its subsequent populations on commercial tobacco remained high year after year. Following widespread use of MH, decline in *M. sexta* populations was so great that this once serious pest became of secondary status in North Carolina (Rabb et al. 1976, Reagan et al. 1979). This tremendous effect of MH on hornworms could only have been achieved by its uniform, area-wide use (essentially all tobacco in southeastern United States), because of the high mobility and reproductive capacity of this insect.

Livestock condition also may be altered to reduce susceptibility to insect attack. For example, the "Mules operation" protects sheep from blow flies in Australia (Roberts 1952). Loose skin is removed from each buttock, altering growth of wool around the vulva, so that wool in this area is no longer wetted by urine, thus reducing attractiveness to flies.

Genetic Alteration of Food Quality. Suitability of individual plants or animals as food for insects is in part genetically controlled. Plant and animal breeders have been able to manipulate the plants' or animals' genetic composition to produce crop cultivars and breeds of livestock with varying degrees of resistance to particular pest species (Painter 1951, Maxwell & Jennings 1980, Harris 1980, Nelson et al. 1977).

Resistant plants may possess characteristics that reduce survival and fecundity of resisted insects and thereby reduce the pest populations in such plantings. Resistance of this type is termed antibiosis (Painter 1951). When resistant cultivars are planted on an area-wide basis, a dramatic effect on the size of an insect population may result. This was demonstrated by the tremendous reductions in the Hessian fly, *Mayetiola destructor,* following widespread use of resistant wheat cultivars (NAS 1969). Resistant plants may also possess characteristics that render them unacceptable or less acceptable than other hosts for either feeding or oviposition. Resistance of this type has been termed *nonpreference* (Painter 1951) or *antixenosis* (Kogan & Ortman 1978). It is illustrated by the high-level resistance of certain raspberry cultivars to the aphid, *Amphorophora agathonica* (Kennedy & Schaefers 1974). This latter type of resistance may have little or no effect on the size of the wide-area population of the insect but may force it on to other, more acceptable hosts. However, if no alternative acceptable hosts are available and the resistant plants are unacceptable (for feeding or oviposition) to a large proportion of the pest population, the wide-area population may be expected to decline.

Similar examples exist for livestock as well. Brahman cattle, *Bos indicus,* and some other breeds show a high-level resistance to the tick, *Boophilus microplus,* that contains elements of both of the preceding types of resistance. In work by Utech et al. (1978) in Australia, Brahman beef cattle were 99% resistant (measured by percent of larval ticks dying before maturity), followed by Brahman x *B. taurus,* 95–97%, and *B. taurus* British cattle, 85%. Wagland (1978) found that decreased mean weight of engorged female ticks paralleled the acquisition of resistance, suggesting that tick feeding is impaired by the resistance. Burns et al. (1977) concluded that nonlactating cattle carrying at least 50% Brahman blood are not greatly disadvantaged if they are not dipped to control ticks.

A third type of insect resistance, termed *tolerance* (Painter 1951), involves ability to withstand higher pest populations with less reduction in marketable yield than that of a standard crop cultivar or animal breed under a similar infestation level. Since tolerant cultivars in effect have higher economic treatment thresholds than conventional ones, higher populations of the given pest may develop if other factors permit. Where higher populations of species that are pests of several crops are allowed to develop in plantings of a tolerant cultivar, this may lead to subsequent invasion of, and unusually severe infestations in other, susceptible crops. This potential problem may or may not materialize in a given crop–pest situation, but the possibility of its occurrence should be considered in advance of widespread cultivation of a highly tolerant cultivar.

Tolerance is also a valuable attribute of some animal breeds. Brahman cattle are more tolerant to mosquito attacks than certain European breeds. As measured by numbers of blood-fed *Psorophora confinnis* resting on barn walls, populations that averaged 34 or more/ft^2 caused significant losses in Hereford weight gains/day, while 66 or more/ft^2 were required to cause significant gain losses in Brahman steers (Steelman et al. 1973). Heritability of the Brahman traits was evidenced in later tests showing that Hereford x Brahman steers were more tolerant than pure Herefords (Steelman et al. 1976). The data showed that mosquito density reached the injury threshold for crossbreds in only 2 of 5 test periods, but for pure Herefords, during 4 of the 5 tests.

In addition to the three types of resistance discussed above, plant breeders can contribute to pest management by altering the developmental rate of a crop in a way that the temporal synchrony between the susceptible phenological stage of the crop and peak activity of one or more insect pests is reduced. This is illustrated by some short-season cotton varieties, as detailed in Chapter 24, relative to bollworms and boll weevils.

23.2.2.3 Food of Little Value to Humans.

Insect pests rely on food sources other than the livestock and portions of crops people wish to protect. Manipulating or eliminating these expendable (from the human view) food resources often constitutes a valuable management strategy. Three ways to do this are: manipulating alternate hosts, trap cropping, and sanitation.

Manipulating Alternate Hosts. This approach is frequently used to inhibit spread of insect-borne plant viruses. Burning drainage ditches to destroy weed hosts of *Myzus persicae,* the vector of beet western yellow mosaic virus, reduced both aphid numbers and virus incidence in sugar beets and resulted in increases in yields of 1.5–2.3 tons/acre (Wallis & Turner 1969).

Trap Cropping. This method involves growing a relatively small planting in a way that is more attractive to target species than the main crop. Plots of soybean planted two weeks before the main crop "attract" the majority of overwintered bean leaf beetles, *Cerotoma trifurcata*, which then can be controlled chemically. This can greatly reduce total acreage needing treatment and also reduce the incidence of bean pod mottle virus which this beetle spreads. It is also effective against the southern green stink bug, *Nezara viridula* (Newsom et al. 1980).

Sanitation. Disposal of wastes that are breeding foci for many insects can suppress certain pests, as shown in control programs for various Diptera that are pests of humans and animals. Populations of many insect pests of crops also are influenced strongly by the manner of handling crop residues (e.g., with respect to pink bollworm control, cf. Chapter 25).

We have seen that performance of an insect population in a given area is dependent (among other things) upon the quantity, quality, and spatial and temporal distribution of its food (above, and Chapters 4, 14). An understanding of this principle and the specifics of these dependencies in real life systems of particular species can lead to development of effective management schemes. Yet, no matter how biologically sound these schemes may be, their widespread implementation will depend upon the economics of the manipulations required and their compatibility with other production components.

23.2.3 Competition for a Common Resource May Inhibit Population Growth and May Displace Weaker Competitors

Each species in a community exploits a certain set of resources. If an immigrant species comes into competition with a resident one for at least one limiting resource, there are three possible outcomes: the resident species can exclude the immigrant species, the immigrant can displace the resident, or the two can coexist (Chapter 15). Several factors can influence the outcome. In a given habitat one genetic strain of species *A* might win against species *B*, while another strain of *A* would lose; or one species may win under one set of environmental conditions, while the other will win under a different set. Or, if environmental conditions are cyclical, favoring each of the competitors part of the time, they may coexist. The habitat itself may be modified in some manner by the presence of one species that makes it unsuitable for another. The presence of active soldier fly, *Hermetia illucens*, larvae, for example, renders excrement physically unsuitable for development of house fly larvae (Furman et al. 1959, Tingle et al. 1975).

While many instances of competitive displacement in nature have been noted or assumed (Chapter 15), few have been deliberately initiated for vector or pest control. However, the possibility of using this approach is illustrated by the outcome of successive establishments of three *Aphytis* species, which are directly competing parasites of the California red scale, *Aonidiella aurantii* (DeBach 1966). Each successive *Aphytis* species displaced one of the others from a portion of the red-scale range, which varied environmentally from the coast inland, and added to the general suppression of the pest. For displacement by manipulation of its competitors to succeed, it is desirable that the pest requires, in addition to the resource to be protected, another resource that is neutral from our perspective (Moon 1980). Examples satisfying this feature are the nuisance flies that breed in cattle droppings, for example, the face fly, *Musca autumnalis*, and horn fly, *Haematobia irritans*, in North America, and the buffalo fly, *H. exigua*, and bush fly, *M. vetustissima*, in Australia. Populations of these flies have been reduced by importations of competing exotic dung beetles which bury and utilize bovine dung (Chapter 11).

Displacements of major vector species have occurred as indirect results of human activity. At least one has worked to man's disadvantage. With increasing urbanization in southeast Asia, the introduced domestic mosquito, *Aedes aegypti*, has progressively replaced the indigenous species, *Aedes albopictus* (Rudnick 1965). *A. aegypti* is the major vector of dengue virus; *A. albopictus* is a less efficient vector. In addition there have been a number of displacements involving vectors as a result of their differential tolerance to insecticides, for example, *Culex nebulosis* in Nigeria by *Culex quinquefasciatus* (Service 1966).

23.2.4 Predation, Parasitism, and Disease Contribute to Population Limitation and Regulation

Conservation and use of natural enemies, predators, parasites, and pathogens is fundamental in managing insects because, individually and collectively, they influence densities, and, to the extent that their actions are responsive to density, they constrain the degree of fluctuation above and below mean levels of abundance (Chapter 12). The history, principles, methodology, and various examples of the use of natural enemies are presented in DeBach (1964), Huffaker and Messenger (1976), Coppel and Mertins (1977), and other texts on this subject (cf. Chapters 16, 22). Three principal strategies for using natural enemies are: (1) conserving and enhancing indigenous natural enemies, (2) introducing exotic ones, and (3) releasing or applying them periodically for temporary suppression of pests.

23.2.4.1 Conservation and Enhancement of Indigenous Enemies. Many intensive agricultural practices suppress natural enemies more than their prey, thus improving conditions for pest outbreaks. The possibilities for modifying such practices to enhance natural enemy action are both numerous and diverse (Rabb et al. 1976). As with pests, manipulating their food is a powerful technique for managing natural enemies (Chapter 4); for example, provision of alternate hosts for parasites of a grape pest (Doutt

& Nakata 1965); use of food sprays and food supplements to focus natural enemy action in target areas (Hagen & Hale 1974); strip harvesting to retain natural enemies in alfalfa (van den Bosch et al. 1967); partial rather than complete removal of manure under caged poultry and use of fly adulticides rather than larvicides to preserve mite predators of the house fly, and the lesser house fly, *Fannia canicularis* (Axtell 1970). There also are many other opportunities to increase natural enemy action through habitat manipulation. Microclimatic favorability for entomogenous fungi may be enhanced through planting schedules and row spacing in soybeans (Sprenkel et al. 1979) and through irrigation schedules in alfalfa (Hall 1961). Provision of nesting sites for birds (Kroll et al. 1980) and predatory insects (Lawson et al. 1961) may serve desirably to concentrate foraging of these predators. Agricultural chemicals comprise an important component of the crop environment to which natural enemies react with great sensitivity. Thus, use of the economic threshold principle and selective use of toxicants for temporary pest suppression (Fig. 23.3) are critical in efforts to optimize the effects of natural enemies and pesticides (Chapters 24, 25).

23.2.4.2 Introduction of Exotic Natural Enemies. Perhaps no control technique has received more kudos, when successful, than classical biological control since it is environmentally innocuous, long-lasting, self-renewing, and relatively inexpensive (cf. Chapters 16, 22). Pests of foreign origin have been most amenable to this approach, which usually entails the introduction of one or more enemies from their native land. Lists of complete and partial successes are impressive and far outweigh the costs of failures (Laing & Hamai 1976). Pests of perennial crops and forests have seemed more susceptible to classical biological control than pests of annual crops in which frequent and heavy use of energy subsidies continually interrupt the pest–natural enemy feedback systems. The potential of this approach in intensive cropping systems, however, has not been fully explored and may prove of great value, particularly when combined with agroecosystem structural modifications conducive to more stable host–enemy relationships.

23.2.4.3 Natural Enemies for Temporary Suppression. Some natural enemies are not adapted for permanent residency where their hosts present problems, or, if so, are unable for various reasons to exert pest suppression. Some such enemies may be cultured and released at critical times and places to obtain desired results; however, the techniques and results are specific to the enemy species, the pest species, and the environment. In some cases small numbers of enemies are innoculated in a target area and subsequently reproduce and spread rapidly enough to exert control during the cropping period; in other cases large numbers are applied uniformly (inundatively) in the target zone (Stinner 1977, Ridgeway & Vinson 1977). More experience has been obtained from such "mass" releases with trichogrammatid parasites of a wide array of insect herbivores than any other group of entomophagous insects. Although results have been mixed and often inconclusive, large rearing and release programs are currently maintained in the USSR, China, Mexico, and a number of other countries.

Perhaps the most successful "biotic insecticide" is the bacterial pathogen *Bacillus thuringiensis,* currently available in several commercial formulations and effective for temporary suppression of a wide array of crop pests, including hornworms, cabbageworms, loopers, tobacco budworms, and gypsy moth larvae. More recently, another strain of this bacterium, *B. thuringiensis* var. *israelensis,* has proven effective for the control of mosquitoes (Garcia et al. 1980, Mulligan et al. 1980, Dame et al. 1981, and others) and blackflies (Gaugler & Finney 1982).

23.2.4.4 Qualitative Factors Influencing Natural Enemy Success. The merits of different kinds (i.e., predators, parasitoids, pathogens) and characteristics (i.e., ecological compatibility, synchronization, responsiveness to pest density, searching capacity, reproductive potential, host specificity, culturability, etc.) of natural enemies have been discussed by many authors (e.g., DeBach 1964, Coppel & Mertins 1977). While these discussions provide useful guidelines for selecting candidate agents, the performance of each can be established only through empirical tests. For a given pest situation, some authors (e.g., Turnbull & Chant 1961) recommended selecting only one species for release after careful studies have established its superiority, while many others (e.g., Doutt & DeBach 1964) find this unrealistic and support multiple-species releases. Regardless of the kind and number of species to be used, however, the genotype and condition of the introduced and/or cultured stock may influence results.

The genotype introduced may or may not be the best available for use in the target area; thus, various strategies have been proposed for selecting and combining samples from various locations within the enemy's geographic range. Strains differ in adaptability to environmental conditions and in capacity to inflict mortality. Turnock and Muldrew (1971), for example, found that a strain of *Mesoleius tenthredinis* lost its effectiveness against *Pristiphora erichsonii* because host strains developed ability to encapsulate the parasitoid's eggs. However, a Bavarian strain of *M. tenthredinis* was then found whose eggs the host has not been able to encapsulate. In addition to the effect of reciprocal or opposing genetic adjustments, a virus factor may prevent encapsulation in some host–parasitoid associations (Edson et al. 1981).

Genetic adjustment of enemy populations to pesticidal pressures also may be crucial to improved pest management, as exemplified by the role of resistant strains of predatory mites in the control of phytophagous mites in orchards (Croft & Brown 1975, Hoy 1982, and Chapter 25). The potentialities of breeding "improved" strain of natural enemies are as yet little explored (Mackauer 1976); however, some recent attention has been given this approach (Hoy & Knop 1981, Hoy 1982).

In summary, selecting enemies with genetic propensities for pest suppression (or improving them through breeding), producing vigorous stock free of disease and hyperparasites, and delivering it to target areas in proper condition at appropriate times must be accomplished for best results. To do this and at the same time avoid physiological and genetic deterioration is a continuing challenge.

23.2.5 Introduction of a Mortality Factor May Result in Pest Strains Resistant to the Factor

Qualitative changes are characteristic of populations because of interactions between environmental and genetic variability (Chapter 9). Through breeding programs, humans exert a measure of control on the direction of qualitative changes in crops and domestic animals. Unfortunately, in doing so and in imposing various management practices (e.g., use of chemicals) on them, they also influence the evolution of other organisms, including pest species, in ways counter to their desires.

23.2.5.1 Resistance to Insecticides. Since 1914, when resistance to lime sulfur spray in the San Jose scale was first reported, over 400 cases of insecticide resistance have been documented, a number of these resulting in crop disasters (Metcalf 1980). The rate at which this resistance has evolved is dependent upon a number of factors, including:

1. Genetic factors such as the frequency, number, and dominance of R alleles and the extent of their integration with other fitness factors;

2. Biological factors such as the species' reproductive rate and mating requirements (e.g., sex ratio, monogamy/polygamy or parthenogenesis), the isolation, mobility, and migratory tendencies of the population, and the fortuitous survival of individuals following treatment;

3. Operational factors such as the chemical nature of the pesticide and its relationship to previously used pesticides, the formulation used and the persistence of residues, the mode of application, and the application threshold and the life stages selected (Georghiou & Taylor 1977a,b).

The selection of insecticide-resistant pest biotypes can be delayed, if not avoided, by decreasing the frequency and intensity of genetic selection. This might be accomplished by applying insecticides only when and where absolutely necessary and in minimum acceptable amounts. Similarly, avoidance of insecticides with prolonged environmental persistence of treatments that select for resistance in both larval and adult stages will reduce selection pressures. Rate of change to pesticide-resistant genotypes may be reduced where cultural and biological tactics are included in management systems to reduce the frequency of damaging infestations and the number of pesticidal applications (cf. Chapters 8, 9).

23.2.5.2 Resistance to Other Mortality Factors.   Pest populations may change over time in response to any mortality factor. Natural enemies, for example, can produce selection pressure resulting in biotypes of hosts resistant to their attack (Turnock et al. 1976), and the control achieved by the enemy may become economically insignificant even though the ecological association between the insect and the enemy may persist in an altered form. This phenomenon, however, has rarely been observed, perhaps because in the coevolutionary process the enemy can also evolve to counter such resistance.

Developing crops resistant to pests is a popular pest management strategy. Yet, widespread cultivation of pest-resistant cultivars has in some cases resulted in selection of pest biotypes able to thrive on them (Gallun & Khush 1980). The occurrence of such biotypes is related to the genetic plasticity of the pest population, the genetics of the crop resistance, and the selection pressure the resistant cultivar exerts on the pest population. Resistance-breaking biotypes are most likely to develop when the plant resistance is of a high level and is controlled by a single gene, and where the resistant crop variety is widely grown and makes up a major portion of the resisted insects' potentially available food supply. Under these conditions the selection pressure for resistance-breaking biotypes is considerable.

Several strategies have been developed to minimize development of resistance-breaking insect biotypes where single-gene resistances are used. Where several different genes for resistance are known, varieties containing different genes can be intermixed in space and time by using multiline varieties (Jensen 1952, Borlaug 1958) or growing in rotation several varieties possessing different sources of resistance. In either case prolonged, intense selection for resistance-breaking biotypes is reduced. Combining several different major resistance genes into the same crop variety ("pyramiding") has also been advocated for reducing development of resistance-breaking biotypes (Nelson 1972). Gallun (1977), however, argued that this would lead to development of "super" biotypes able to overcome the combined resistances so that several valuable genes will be lost at once.

Where single-gene resistance is used, the probability of resistance-breaking biotypes developing is greatest in crops which constitute the principal or only food source for the insect, as in the case of wheat and the Hessian fly. If the resistant crop is but a minor component of that food source, as with tomatoes as a host for *H. zea* in a corn-dominated agroecosystem, selection pressure for resistance-breaking biotypes is likely to be much less and such biotypes unlikely to develop. Polygenic resistance is considered more stable with regard to potential for selecting resistance-breaking insect biotypes, but is generally of a low order and difficult to use in breeding programs (Gallun & Khush 1980). Yet, because of its potential value, improved procedures for utilizing such resistance continues to be sought.

Thus, the potentiality of resistance breakdown in crops, along with resistance development in pests to pesticides, comprise major justifications for using the IPM approach (below, and Chapters 24, 25).

23.2.6 Mutations and Other Genetic Alterations May Decrease Fitness of a Population

Introduction of altered genetic material is one of the most selective methods of manipulating pest populations. The goal is usually suppression or eradication, but some research has been directed toward replacement of less desirable genotypes with more desirable ones, for example, replacement of an efficient pathogen vector with a less efficient one (cf. Rogoff 1980, who proposes heritable behavior modification of stable flies, *Stomoxys calcitrans,* and face flies, *Musca autumnalis,* by competitive displacement).

To date, the only widely used genetic technique for insect suppression is the sterile male release (SMR). Competitive, sterilized males must be released in numbers large enough to ensure their involvement in a high proportion of matings. If such a release rate is maintained, and assuming complete sterility of released males, the efficiency of genetic techniques increases as the natural population declines because the ratio of sterile to fertile matings increases. This feature of maximum effectiveness when natural populations are low is in contrast with most conventional methods and is an especially attractive feature. Insecticides or other methods that are more efficient at high pest densities are often employed to reduce the target population prior to introduction of the genetically altered insects. In *some cases* it may also become desirable to apply insecticide after or between releases when the desired ratio of sterile to fertile insects is achieved in most, but not all, of a target area because of high pest densities in certain favorable parts of the habitat. In other instances an insecticide might be needed to reduce another pest species within a target area. Such applications are not antagonistic (Knipling 1979) as long as the insecticide kills equal proportions of sterile and fertile insects.

Genetic methods require a thorough understanding of the biology, ecology, and behavior of target pests and are more or less limited to consideration against major pest and vector species because of numerous technical requirements. Seasonal changes in density must be estimated with reasonable accuracy to determine if and when the natural population declines enough to permit its overflooding without supplementary suppressive measures. Influences of density-dependent factors must be known in order to anticipate changes in rate of increase at different population levels. Unless the target area is geographically or ecologically isolated, immigration of pest insects from adjoining areas and emigration of released insects may prove a problem, necessitating establishment of barrier zones. The width of such barrier zones will depend on the dispersal characteristics of the target species, including its long-range flight capabilities. Mass-rearing methods must be developed that produce enough insects to overflood the natural population, and the insects must be vigorous enough to compete successfully for mates within the natural population. It must be possible to induce a high level of sterility or other appropriate genetic defects in insects to be released without significantly reducing their competitiveness, and there must be suitable techniques for transporting them to the release sites, and widely dispersing them, without excessive mortality. In programs involving vectors whose females pose hazards or annoyances, it may be necessary to release only males, thus requiring efficient sexing. Although costly, as methods improve costs may decrease. Thus, development of a method that produced pupae that were 99.7% males (Seawright et al. 1978) reduced costs of releasing *Anopheles albimanus*, the major malaria vector in El Salvador, from $4000 to $500 per million males (Asman et al. 1981).

A major challenge in genetic alteration programs is to maintain quality control, especially competitiveness, of the mass-produced insects. The breeding structure of the target population must be monitored for evidence of assortative mating by cryptic species or subpopulations. If such occurs, it must be taken into account in selecting, maintaining, and releasing strains. In African *Aedes aegypti*, Trpis and Hausermann (1975) found that three overlapping races have different habitat preferences and are effectively isolated from each other. Even in the absence of such subpopulations initially, a population may become altered by selection pressures exerted by the control program itself.

The reared insects may also change. Bush et al. (1976) demonstrated that the 1972 outbreak of the screwworm, *Cochliomyia hominovorax*, largely resulted from genetic change in the mass-produced flies. The enzyme a-glycerol phosphate dehydrogenase (a-GDH) influences flight activity, and it was shown by gel electrophoresis to exist in two forms: a-GDH$_2$, attaining high frequencies in the mass-reared flies, while wild flies typically contained the alternate form, a-GDH$_1$. The two forms were shown to have quite different temperature ranges (Kitto et al. 1976), with the result that in the field the mass-produced flies did not become active until early afternoon, while wild flies were active throughout the day (Bush 1978).

Potentially useful genetic suppression mechanisms were placed in three categories by Knipling and Klassen (1976), based on whether there is infusion of genetic material into the target pest population and whether suppression is exerted beyond the generation of exposure. Although the chief strategy to date involves rearing and releasing sterile or other genetically altered insects, another approach is to use chemosterilants on the target insects themselves. The principles involved are quite different from those in the release of sterile insects and are more similar to those governing use of conventional pesticides. As Knipling (1979) noted, however, whereas 90% kill of a population (other factors remaining the same, which is unlikely where density-dependent factors operate) theoretically reduces reproduction by 90%, a 90% sterilization of a population reduces reproduction by 99%. Development of this approach to a practical level, however, awaits development of safer chemosterilants.

Utilization of the sterile male release tactic against the screwworm stands as an outstanding achievement. Through its use, the screwworm was eradicated in the southeastern and greatly suppressed in southwestern United States (Knipling 1959, 1979). It has been not only of great economic value to livestock producers but also has made significant scientific contributions applicable to other population management challenges. (For other applications of SMR and other genetic methods, cf. Davidson 1974, Pal and Whitten 1974, and Knipling 1979.)

23.2.7 Other Principles

We must omit discussion of a number of ecological principles of significance in IPM, for example, the use of hormones (or hormone mimics) and semiochemicals, that is, the internal and external chemical messengers which mediate physiological and behavioral repertoires, with their profound effects on population patterns. The selectivity and biological activity of hormone mimics make them attractive for use in IPM (e.g., Dunn & Strong 1973, Mulla & Darwazeh 1976), and efforts to remove technological and economic constraints on their use continue. Optimism, however, should be tempered with the caution that strains of insects resistant to conventional pesticides may show cross-resistance to these compounds. Among the semiochemicals, sex pheromones have received extensive research, both basic and applied (Roelofs 1981). Their greatest use in IPM thus far has been in highly sensitive monitoring of pest species (e.g., Rummel et al. 1980); however, they show promise for direct suppression (e.g., Beroza et al. 1975). Suggestions also have been made as to use of transpecific messengers (i.e., allomones, kairomones, and synomones) (e.g., Mitchell 1981, Nordlund et al. 1981); however, their practical use has been very limited to date.

23.3 INTEGRATING METHODS FOR PEST MANAGEMENT

23.3.1 Levels of Integration

The erroneous view that the mere presence of an insect pest species means that economic damage will occur often triggers human responses leading to wasted resources and undesirable effects. Thus, each suspected pest situation should be analyzed carefully before deciding that a response is justified. Even when a bona fide pest situation exists, there can be wide variations in its significance and complexity. At one extreme, use of a single tactic may be sufficient; whereas, at the other, several tactics may be required for a very diverse pest complex. Even in the simplest case, the tactic should have minimal side effects and blend harmoniously with other forces and processes within the system. As more tactics are added, their integration becomes more complicated.

Management systems may be conceived as sets of practices hierarchically arranged at different levels of integration. For illustration, we have chosen a generalized field crop pest situation and suggest (Fig. 23.2) that the process of IPM might proceed through seven levels. The plus and minus symbols represent positive and negative interactions among inputs and the inexorable necessity for some compromises. At level 1, attention is focused on a single pest representing a single discipline. Integration at this level involves inserting one or more tactics into the crop management system (CMS). The challenge at this level is to choose and use tactics that will augment rather than negate effects of natural control forces, for example, natural enemies, and which will have no other undesirable side effects. Preventive and/or remedial tactics (Fig. 23.3) may be chosen. In some cases a simple and agronomically acceptable adjustment in a cropping practice may prevent loss or at least reduce the need for remedial measures. Integration at level 2 occurs when tactics for two or more pests within a single disciplinary category are combined harmoniously. In many cases integration at this level presents problems, as in the increase in bollworms as a result of tactics used for boll weevil control (Chapter 25).

Interactions among the inputs of entomologists, plant pathologists, weed

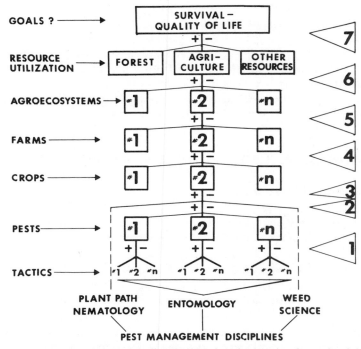

Figure 23.2   Management interactions in IPM. Numbers in the triangles to the right indicate seven levels of integration. The plus and minus symbols indicate that both positive and negative interactions among management practices may occur at all levels. (From Rabb 1978.)

scientists, and other disciplines are complex and require integration at level 3. While some of these interactions are positive [e.g., early crop residue disposal contributes to control of both plant pathogens and insect pests of tobacco (Rabb et al. 1976)], others are not [e.g., use of certain nematocides suppress natural enemies and exacerbate corn earworm problems on soybean (Morrison et al. 1979)].

Because of the interchanges of pests and natural enemies among crops and the significance of crop rotation to the intensity of local pest problems, successful IPM often requires a multicrop approach. Thus, as indicated at level 4, the total farm unit is a more natural pest management unit than a single crop. However, some pest problems cannot be managed satisfactorily on a single farm and may require community-wide action (level 5). In fact, some serious pests are so mobile that regional, national, or even international integration of efforts (level 6) may be desirable (Gunn & Rainey 1979, Rabb & Kennedy 1979).

Integration of practices influencing pest populations may be extended to the even more holistic level 7, where interactions among various cultural enterprises and activities may create and increase pest problems. For example, industrial air pollution may weaken trees or crops, increasing susceptibility to pests; or populations of organisms engendered by agricultural enterprises may comprise pest problems in residential, industrial, or recreational areas. Conflicts of interest involved in harmonizing enterprises may even create controversy as to the definition of the term *pest*. Weed

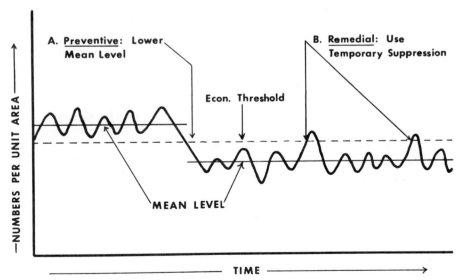

Figure 23.3   Differential effects of preventive and remedial control tactics on a hypothetical pest population from a population dynamics view. (From Rabb 1978.)

scientists often refer to weeds as plants out-of-place. In this context and in certain areas, crop and forest species are pests to real estate developers, mining executives, and engineers involved in highway and dam construction. While entomologists may have little influence on decisions at level 7, they must be aware of the interactions at this level if they are to use the ecological approach effectively in responding to ever-changing pest situations.

23.3.2 Constraints on Integration

23.3.2.1 Ecological Constraints. The holism of nature and the uniqueness of each species' habitat–niche characteristics (stressed repeatedly) pose the most inexorable limitations to scientifically feasible pest control. If a control tactic is to have minimum side effects, it must be highly species specific; however, due to the interdependencies and energy and matter pathways of ecosystems, any control tactic, no matter how selective, will have some degree of off-target effects. As noted earlier, many cases of negative interactions between control tactics for different pests have been discovered; however, our poor understanding of the interactions among management practices is a serious constraint to efficient management. The larger the pest complex and the more diverse the management inputs, the greater the likelihood of negative interactions and the more difficult the task of clearly perceiving them.

Arbitrary definitions of pest populations constrain the realization of IPM's full potential. For practical reasons the size of a pest population targeted for management is usually set by the size of the management unit, which in turn is determined by economic, crop-centered factors rather than ecological, pest-centered considerations. It is important, however, to relate to dimensions of the target population to other conspecific populations with which gene exchange may occur. For example, the spatial pattern of a hypothetical species population is depicted in Fig. 23.4. Theoretically, such a population may be studied or managed in toto (A), in deme units (B), or in intrademe units (C). Most target populations are intrademe units, managed with only vague notions of deme size or interactions with other demes. The misleading assumption is often made that entrances and exits of migrants are equal. The practical significance of these errors depends on many factors, particularly the location and size of the intrademe unit and the innate movement patterns of the species involved. Actual movement patterns of most pests are very poorly known (cf. Chapter 7), and lack of this information is a most serious barrier to improved understanding and management of agricultural pests. Since pest species vary greatly in mobility, the dimensions arbitrarily placed on their management arenas impose different constraints on the management of different species, as illustrated in Fig. 23.5. Range in size of possible management arenas is represented by rectangular areas. Circled numbers represent species with different *invasion ranges* (lines connecting circles to the centrally placed

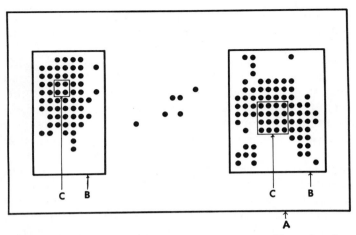

Figure 23.4 Diagrammatical representation of a species range, illustrating three types of populations targeted for study or management: (A) the entire species; (B) partially isolated demes; (C) intrademe units. Each dot represents a site occupied by a breeding population of the species. (From Rabb & Stinner 1978).

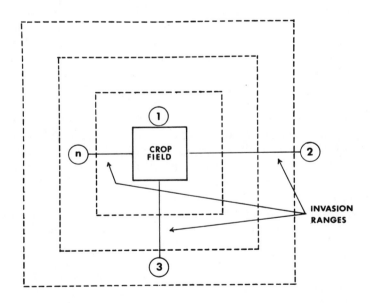

○ **Pest species 1,2,3,n**
⸢⸤ **Some Possible Management Units**
Invasion Range = Distance per Crop Season

Figure 23.5 The relationships between the dimensions of possible management arenas (indicated by broken lines) and invasion ranges of four hypothetical crop pest species. See text for discussion of the significance of the time dimension which is not illustrated. (From Rabb 1980.)

crop field). In this case we have defined the invasion range as that distance over which individuals will move to infest a given field during a given cropping year.

Our selection of the time constraint is just as arbitrary as the spatial constraints (above), and the space–time arena can be defined differently in any dimension. The longer the time dimension placed on the concept of invasion range, however, the less difference there will be between species in the spatial dimensions of these ranges because even very sedentary species become widely dispersed by both passive and active mechanisms, given enough time.

Determining appropriate management arenas in comparable terms for species differing in mobility and movement patterns is difficult. For example, assume that a tactic increasing mortality of a dormant stage is equally effective against pest species 1 and 2, provided it were applied to 95% of each species' overwintering refugia. For comparable results the tactic would have to be applied against species 2 in a much larger management arena— but how much larger? Our answers to such questions pertaining to real pest complexes are largely guesswork because we do not have adequate information on movement propensities or environmental patchiness as defined for each species. This problem is complicated by differences in types of seasonal movement and variations in the location of the management arena relative to the center and fringes of the species' ranges. Where a species overwinters outside the management arena, obviously, tactics directed to overwintering stages cannot be used unless cooperative arrangements are made to increase the management arena.

Geographic location and isolation greatly influence the feasibility of strategies and tactics. For example, tactics for reducing diapausing populations may be quite effective in temperate zones, as used for wide-area suppression of tobacco hornworms (Rabb et al. 1976) and boll weevil (Rummel et al. 1980), but are of little utility in tropical cropping systems, where within-season rotation of crops to disrupt sequential synchrony between pest generations and susceptible hosts has greater applicability. Isolated areas such as islands and irrigated desert valleys provide opportunities for a greater degree of local control over pest populations. And, as previously mentioned, manipulation of the physical factors of crop habitats is of greater potential in irrigated than in high rainfall areas.

The evolutionary process itself comprises a major constraint on the potentiality of pest management. Genetically based resistance to pesticides and genetic adjustments of pest species to pest-resistant crop varieties was discussed earlier, as were the opportunities of developing such varieties. However, the genetic systems involved remain open-ended and ever-changing, as do environmental systems (including management) which function as selective agents. Thus, management practices effective today may not remain so in the future. Pests comprise some of the most robust, adaptable species of our biosphere and thus are also among the most difficult to suppress permanently, or to eradicate. In this connection Rabb (1978) views consideration of species eradication as cause for serious concern scientifically and philosophically.

23.3.2.2 Economic, Societal, and Political Constraints.

The increase in complexity and size of management systems and arenas from the lowest to highest levels of integration (Fig. 23.2) is accompanied by an increase in economic, societal, and political impediments to harmonizing pest management with a stable, productive, and pleasing environment. Only a few of the many constraints can be mentioned (cf. Geier 1970, Corbet & Smith 1976).

Variation in the size of independently managed units is a basic societal constraint. Thus, privately controlled vegetable plantings have many pests in common with field crops, but in most cases are many orders of magnitude smaller. Ambient populations of many of the shared pests are highly mobile and largely produced on the much greater biomass of field crops. Thus, control practices in home vegetable plantings generally have minimal or negligible effects on the general populations of such pests. Vegetable growers, even collectively, can therefore make little practical use of tactics designed to lower the mean level of ambient populations unless comparable tactics are uniformly applied against populations on field crops. The tobacco hornworm, also a pest of tomatoes, illustrates this because its main production in the southeastern United States is on tobacco (Rabb et al. 1976).

A basic economic and psychological constraint on IPM is the variation in criteria used in decision making. These action thresholds may be set at such low pest levels that it is quite unfeasible to prevent pest populations from reaching them on a regular basis. The action thresholds are very low on fruits, vegetables, and other high-value commodities. There is, however, some evidence in many cases that because of cosmetic standards, thresholds are placed lower than necessary to prevent loss in yield and nutritional value of the produce. In addition, many growers opt for "insurance" treatments of pesticides rather than investing time and money in monitoring pest populations and using threshold guidelines. The risk factor in the absence of reliable predictive ability is also a partial reason for this choice.

Refining and generalizing the application of threshold principles in pest management decision making remains a central challenge. In entomology the economic threshold concept is used primarily in deciding if, and when, to apply insecticides for specific insect pests; however, when a situation involves a diverse complex of pests (weeds, nematodes, and pathogens as well as a complex of insects), the validity of thresholds for individual pest species is called to question, due to our poor understanding of the additive and synergistic effects of contemporaneous and sequential attacks. Among such complexes, innate differences in size, rates and patterns of reproduction and dispersion, and other biological attributes militate against uniformity in kind and accuracy of monitoring techniques. Thus, as emphasized by Newsom (1980), developing satisfactory action thresholds for pest complexes is a basic requirement for progress in IPM. The threshold concept also should be generalized to apply more effectively to decision making regarding preventive action (Fig. 23.3). To do so effectively, however, decision makers must have adequate information concerning pest ecology and be able to recognize high and low economic risk situations. With such knowledge, the selection of crop mixes, varieties, rotations, and practices reducing the necessity for remedial actions might allow more refined tailoring of management systems for different fields and agroecosystems. An extension of this threshold principle is deciding when a given pest situation justifies increasing levels of integration (Fig. 23.2), that is, to wide-area management or eradication programs. Needed would be scientifically sound information supplied in digestible form to the politicians who ultimately make such decisions.

Expanding levels of integration from single, privately owned units to communities or more comprehensive levels involves an increase in the number of people involved in management. The increase in number alone increases costs and complexity of organization, but perhaps an even greater problem is the accompanying increase in diversity of philosophies and personalities. Highly independent growers with little appreciation of the value of cooperative efforts in pest management present serious problems to those conducting wide-area programs, the success of which is dependent on uniform and efficient application of control tactics. Also, growers resist governmental regulations requiring compliance, and the general public is often hesitant to support programs implemented by government employees.

Many societal constraints vary widely among the political units of the world. Economic conditions in some countries preclude application of some potentially effective methods in common use in other areas. And too, highly centralized governments and closely regulated societies are conducive to mandatory pest control programs. From an ecological view, however, wide-area programs should be approached with caution. If sound, they may be of great benefit; if scientifically weak, the risks are great, as was the abortive attempt to eradicate fire ants with widespread applications of pesticide (Newsom 1978).

23.3.3 Current Trends and Challenges

Increasing awareness of environmental problems and many convincing demonstrations of the problems generated through sole reliance on pesticides are factors favorable for expansion of ecologically based methods. Support of IPM programs is increasing, perhaps prematurely. Those involved face difficult challenges. A fundamental problem is the need for a strong theoretical framework. Much of current ecological theory is of questionable application to highly managed ecosystems (Levins & Wilson 1980). Yet, managed areas provide excellent opportunities for developing and testing ecological theory. Thus, scientists face a challenge to transform current, or develop new, ecological concepts to strengthen the theoretical basis for IPM.

A companion challenge is to develop, maintain, and improve monitoring systems for insects and their major life system components. The unpredictability of the driving variables of life systems (weather and management inputs) and the open-endedness of population systems preclude development of useful models for decision making in pest management without continuous inputs of data on selected variables (Stinner 1979, and Chapter 25).

Experience with pesticide resistance and pests overcoming resistant factors in plants, as well as the ominous warnings of shortages of raw materials for synthetic toxicants, dictates that we should develop new types of temporary suppression tactics, as well as give more emphasis to preventive means. Meeting the food and fiber requirements of the current world population (without considering its disheartening rate of growth) will require agroecosystem structures and production levels engendering continuing use of ad hoc chemical and other remedial tactics. Thus, more effort in developing effective and safe chemical and biotic insecticides and methods of augmenting indigenous natural enemies, and search for new ones, is justified.

Perhaps the most profound challenge is to truly merge insect pest management with resource management, thus, giving appropriate consideration not only to insects as pests but also insects as resources essential to a viable environment. To meet this challenge will require a general philosophical adjustment among those involved in educational institutions, government, industry, and the general public.

REFERENCES

Adkisson, P. L. 1973. *Tall Timbers Conf. Ecol. Anim. Contr. by Habitat Manag.* **4:** 175–188.

Asman, S. M., P. T. McDonald, and T. Prout. 1981. *Annu. Rev. Entomol.* **26:** 289–318.

Axtell, R. C. 1970. *J. Econ. Entomol.* **63:** 1734–1737.

Beroza, M., C. S. Hood, D. Trefrey, D. E. Leonard, E. F. Knipling, and W. Klassen. 1975. *Environ. Entomol.* **4:** 705–711.

Borlaug, N. E. 1958. Pp. 12–27 in *Proc. First Intl. Wheat Genet. Symp.* The Public Press, Ltd., Winnepeg, Canada.

Brandenburg, R. L. 1981. *Aspects of the Ecology of the Two-spotted Spider Mite in Northeastern North Carolina.* Ph.D. dissertation, North Carolina State University, Raleigh.

Burns, M. A., J. F. Kearnan, J. Biggers, and K. B. W. Utech. 1977. *Queensl. Agric. J.* **103:** 521–524.

Bush, G. L. 1978. Pp. 37–47 in R. H. Richardson (Ed.), *The Screwworm Problem.* University of Texas Press, Austin.

Bush, G. L., R. W. Neck, and G. B. Kitto. 1976. *Science* **193:** 491–493.

Coppel, H. C. and J. W. Mertins. 1977. *Biological Insect Pest Suppression.* Springer-Verlag, New York.

Corbet, P. S. and R. F. Smith. 1976. Pp. 661–682 in Huffaker and Messenger (Eds.), 1976, referenced here.

Croft, B. A. and A. W. A. Brown. 1975. *Annu. Rev. Entomol.* **20:** 285–335.

Dame, D. A., K. E. Savage, M. V. Mesich, and S. L. Oldacre. 1981. *Mosq. News* **41:** 540–546.

Davidson, G. 1974. *Genetic Control of Insect Pests.* Academic, London.

DeBach, P. (Ed.) 1964. *Biological Control of Insect Pests and Weeds.* Reinhold, New York.

DeBach, P. 1964. Pp. 673–713 in DeBach (Ed.), 1964, referenced here.

DeBach, P. 1966. *Annu. Rev. Entomol.* **11:** 183–212.

Doutt, R. L. and P. DeBach. 1964. Pp. 118–142 in DeBach (Ed.), 1964, referenced here.

Doutt, R. L. and J. Nakata. 1965. *J. Econ. Entomol.* **58:** 586.

Dunn, R. L. and F. E. Strong. 1973. *Mosq. News* **33:** 110–111.

Edson, K. M., S. B. Vinson, D. B. Stoltz, and M. D. Summers. 1981. *Science* **2111:** 582, 583.

Ferrigno, F. and D. M. Jobbins. 1968. *Proc. N.J. Mosq. Exterm. Assoc.* **55:** 104–115.

Furman, D. P., R. D. Young, and E. P. Catts. 1959. *J. Econ. Entomol.* **52:** 917–921.

Gallun, R. L. 1977. *Annu. N.Y. Acad. Sci.* **287:** 223–229.

Gallun, R. L. and G. S. Khush. 1980. Pp. 63–85 in Maxwell and Jennings (Eds.), 1980, referenced here.

Garcia, R., B. A. Federici, I. M. Hull, M. S. Mulla, and C. H. Schaefer. 1980. *Calif. Agric.* **34:** 18, 19.

Gaugler, R. and J. Finney. 1982. *Misc. Publ. Entomol. Soc. Am.* **12**(4): 1–17.

Geier, P. W. 1970. Pp. 170–190 in R. L. Rabb and F. E. Guthrie (Eds.), *Concepts of Pest Management.* North Carolina State University Press, Raleigh.

Genung, W. G. 1976. *Tall Timbers Conf. Ecol. Anim. Contr. by Habitat Manag.* **6:** 165–172.

Georghiou, G. P. and C. E. Taylor. 1977a. *Proc. 15th Int. Cong. Entomol. (Washington, D.C.),* pp. 759–785.

Georghiou, G. P. and C. E. Taylor. 1977b. *J. Econ. Entomol.* **70:** 653–658.

Glass, E. H. 1975. *ESA Spec. Publ.* **75:** 1–141.

Gould, F. 1978. *J. Econ. Entomol.* **71:** 680–683.

Gregory, W. W. and G. J. Musick. 1976. *Bull. Entomol. Soc. Am.* **22:** 302–304.

Gunn, D. L. and R. C. Rainey. 1979. *Strategy and Tactics of Control of Migrant Pests.* The Royal Society, London.

Hagen, K. S. and R. Hale. 1974. Pp. 170–181 in F. G. Maxwell and R. A. Harris (Eds.), *Proceedings of the Summer Institute of Biological Control of Plant Insects and Diseases.* University Press, Jackson, Miss.

Hall, I. M. 1961. *Adv. Pest Contr. Res.* **4:** 1–32.

Harris, M. K. 1980. Pp. 23–51 in *Biology and Breeding for Resistance to Arthropods and Pathogens in Agricultural Plants.* Tex. Agr. Exp. Sta., Texas A&M Univ., College Station.

Harwood, R. F. and M. T. James. 1979. *Entomology in Human and Animal Health* (7th ed.). Macmillan, New York.

Hoy, M. A. 1982. *Entomol. Exp. Appl.* **32:** 205–212.

Hoy, M. A. and N. F. Knop. 1981. *Entomol. Exp. Appl.* **30:** 10–18.

Huffaker, C. B. (Ed.). 1980. *New Technology of Pest Control.* Wiley, New York.

Huffaker, C. B. and P. S. Messenger (Eds.). 1976. *Theory and Practice of Biological Control.* Academic, New York.

Jenson, N. F. 1952. *Agron. J.* **4:** 30–34.

Kenmore, P. E. 1980. *Ecology and Outbreaks of Tropical Insect Pests of the Green Revolution, the Rice Brown Planthopper, Nilaparvata lugens* (Stal). Ph.D. dissertation, University of California, Berkeley.

Kennedy, G. G. and G. A. Schaefers. 1974. *J. Econ. Entomol.* **67:** 311, 312.

Kiritani, K. 1979. *Annu. Rev. Entomol.* **24:** 279–312.

Kitto, G. B., R. Neck, and G. L. Bush. 1976. *Fed. Proc.* **35:** 1658.

Knight, F. B. and H. J. Heikkenen. 1980. *Principles of Forest Entomology.* McGraw-Hill, New York.

Knipling, E. F. 1959. *Science* **130:** 902–904.

Knipling, E. F. 1979. *USDA Agr. Handb.* No. 512, Washington, D.C.

Knipling, E. F. and W. Klassen. 1976. *USDA Tech. Bull.* No. 1533.

Kogan, M. and E. F. Ortman. 1978. *Bull. Entomol. Soc. Am.* **24:** 175, 176.

Kroll, J. C., R. N. Conner, and R. R. Fleet. 1980. *USDA Forest Serv. Agr. Handb.* No. 564, Pineville, La.

Laing, J. E. and J. Hamai. 1976. Pp. 685–753 in Huffaker and Messenger (Eds.), 1976, referenced here.

Lawson, F. R., R. L. Rabb, F. E. Guthrie, and T. G. Bowery. 1961. *J. Econ. Entomol.* **54:** 93–97.

Levins, R. and N. Wilson. 1980. *Annu. Rev. Entomol.* **25:** 287–308.

Loebenstein, G. and B. Raccah. 1980. *Phytoparasitica* **8:** 221–235.

Mackauer, M. 1976. *Annu. Rev. Entomol.* **21:** 369–385.

Maxwell, F. G. and P. R. Jennings (Eds.). 1980. *Breeding Plants Resistant to Insects.* Wiley, New York.

Metcalf, C. L., W. P. Flint, and R. L. Metcalf. 1962. *Destructive and Useful Insects, their Habits and Control* (4th ed.). McGraw-Hill, New York.

Metcalf, R. L. 1980. *Annu. Rev. Entomol.* **25:** 219–256.

Metcalf, R. L. and W. Luckman. 1975. *Introduction to Insect Pest Management.* Wiley, New York.

Mitchell, E. R. 1981. *Management of Insect Pests with Semiochemicals, Concepts and Practice.* Plenum, New York.

Moon, R. D. 1980. *Environ. Entomol.* **9:** 723–728.

Morrison, D. E., J. R. Bradley, Jr., and J. W. Van Duyn. 1979. *J. Econ. Entomol.* **72:** 97–100.

Mulla, M. S. and H. A. Darwazeh. 1976. *J. Econ. Entomol.* **69:** 309–312.

Mulligan, F. S., III, C. H. Schaefer, and W. H. Wilder. 1980. *J. Econ. Entomol.* **73:** 684–688.

National Academy of Sciences. 1969. *Insect Pest Management and Control. Principles of Plant and Animal Pest Control,* Vol. 3. U.S. Natl. Acad. Sci. Publ. 1695.

Nelson, R. R. 1972. *J. Environ. Qual.* **1:** 220–227.

Nelson, W. A., J. F. Bell, C. M. Clifford, and J. E. Keirans. 1977. *J. Med. Entomol.* **13:** 389–428.

Newsom, L. D. 1978. *Bull. Entomol. Soc. Am.* **24:** 35–40.

Newsom, L. D. 1980. *Bull. Entomol. Soc. Am.* **26:** 369–374.

Newsom, L. D., M. Kogan, F. D. Miner, R. L. Rabb, S. G. Turnipseed, and W. H. Whitcomb. 1980. Pp. 51–98 in Huffaker (Ed.), 1980, referenced here.

Nordlund, D. A., R. L. Jones, and W. J. Lewis. 1981. *Semiochemicals: Their Role in Pest Control.* Wiley, New York.

Painter, R. H. 1951. *Insect Resistance in Crop Plants.* MacMillan, New York.

Pal, R. and M. J. Whitten. 1974. *The Use of Genetics in Insect Control.* Elsevier, New York.

Rabb, R. L. 1978. *Bull. Entomol. Soc. Am.* **4:** 40–44.

Rabb, R. L. 1980. *Agr. Exp. Stn., Univ. Ark.* Spec. Rept. 99.

Rabb, R. L. and G. G. Kennedy (Eds.). 1979. *Movement of Highly Mobile Insects: Concepts and Methodology in Research.* University Graphics, North Carolina State University, Raleigh.

Rabb, R. L. and R. E. Stinner. 1978. Pp. 3–16 in C. R. Vaughn, W. Wolf, and W. Klassen (Eds.), *Radar, Insect Population Ecology and Pest Management.* NASA Conf. Publ. No. 2070, Wallops Island, Va.

Rabb, R. L., F. A. Todd, and H. C. Ellis. 1976. Pp. 71–106 in J. L. Apple and R. F. Smith (Eds.), *Integrated Pest Management.* Plenum, New York.

Reagan, T. E., R. E. Stinner, R. L. Rabb, and C. Tuttle. 1979. *Environ. Entomol.* **8:** 268–273.

Ridgeway, R. L. and S. B. Vinson (Eds.). 1977. *Biological Control of Insects by Augmentation of Natural Enemies.* Plenum, New York.

Roberts, F. H. S. 1952. *Insects Affecting Livestock.* Angus & Robertson, Sydney, Australia.

Roelofs, W. L. 1981. *Bull. Entomol. Soc. Am.* **27:** 3–8.

Rogoff, W. M. 1980. *Bull. Entomol. Soc. Am.* **26:** 121–125.

Rudnick, A. 1965. *J. Med. Entomol.* **2:** 203–208.

Rummel, D. R., J. R. White, S. C. Carroll, and G. R. Pruitt. 1980. *J. Econ. Entomol.* **73:** 806–810.

Seawright, J. A., P. E. Kaiser, D. A. Dame, and C. S. Lofgren. 1978. *Science* **300:** 1303, 1304.

Service, N. W. 1966. *Bull. Entomol. Res.* **56:** 407–415.

Singh, S. R. 1975. *Proc. IITA Collaborators Meet. for Grain Legume Improvement,* pp. 41–43.

Sprenkel, R. K., W. M. Brooks, J. W. Van Duyn, and L. L. Deitz. 1979. *Environ. Entomol.* **8:** 334–339.

Steelman, C. D., T. W. White, and P. E. Schilling. 1973. *J. Econ. Entomol.* **66:** 1081–1083.

Steelman, C. D., T. W. White, and P. E. Schilling. 1976. *J. Econ. Entomol.* **69:** 499–502.

Stinner, R. E. 1977. *Annu. Rev. Entomol.* **22:** 515–531.

Stinner, R. E. 1979. Pp. 199–211 in Rabb and Kennedy (Eds.), 1979, referenced here.

Stinner, R. E., R. L. Rabb, and J. R. Bradley, Jr. 1977. *Proc. 15th Int. Congr. Entomol., Washington, D.C.,* pp. 622–642.

Tingey, W. M. and S. R. Singh. 1980. Pp. 87–114 in Maxwell and Jennings (Eds.), 1980, referenced here.

Tingle, F. C., R. C. Mitchell, and W. W. Copeland. 1975. *J. Ga. Entomol. Soc.* **10:** 179–183.

Trpis, M. and W. Hausermann. 1975. *Bull. Entomol. Res.* **65:** 199–208.

Turnbull, A. L. and D. A. Chant. 1961. *Can. J. Zool.* **39:** 697–753.

Turnock, W. J. and J. A. Muldrew. 1971. Pp. 175–194 in *Biological Control Programs Against Insects and Weeds in Canada 1959–1968.* Commonw. Inst. Biol. Contr. Tech. Commun. 4.

Turnock, W. J., K. L. Taylor, D. Schroder, and D. L. Dahlsten. 1976. Pp. 289–311 in Huffaker and Messenger (Eds.), 1976, referenced here.

Utech, K. B. W., R. H. Wharton, and J. D. Kerr. 1978. *Aust. J. Agr. Res.* **29:** 885–895.

van den Bosch, R., C. F. Legace, and V. M. Stern. 1967. *Ecology* **48:** 993–1000.

van Emden, H. F., V. F. Eastop, R. D. Hughes, and M. J. Way. 1969. *Annu. Rev. Entomol.* **14:** 197–270.

Wagland, B. M. 1978. *Aust. J. Agr. Res.* **29:** 395–400.

Wallis, R. L. and J. E. Turner. 1969. *J. Econ. Entomol.* **62:** 307.

Wellington, W. G. 1977. *Environ. Entomol.* **6:** 1–8.

Zitter, T. A. 1977. Pp. 385–412 in K. F. Harris and K. Maramorosch (Eds.), *Aphids as Virus Vectors.* Academic, New York.

Chapter 24

Models for Managing the Economic Impact of Pest Populations in Agricultural Crops

A. P. GUTIERREZ and Y. H. WANG

24.1 INTRODUCTION

In Chapter 11 development and pursuit of life systems studies to provide basic understanding for formulating simulation models to depict the population ecology and dynamics of insect populations were dealt with. The requirements and methodology in securing the necessary data and a technology applicable to dealing with these data for a diversity of organisms and factors impinging on them were also presented in Chapter 11. Chapter 23 discusses in broad outline many of the more basic conceptions and principles of ecology that may be utilized to control insect pests in a more enlightened way than simple reliance on pesticides. Recently, the technology of modeling crop and pest interactions has received significant attention. There are many approaches and aspects of this modeling that may prove useful in gaining insight into the processes and interactions of consequence to farmers attempting to grow crops or produce livestock. This is the concern of this chapter. We use primarily examples from agriculture, but the concepts and methods discussed apply to veterinary and medical pest problems with only minor modification.

The interaction of a pest, the host crop or animal, and natural enemies of the pest as modified by the physical environment and man's management practices is a complex one, and man's attempts to manage pest impact for his own (or society's) benefit often meet with counterproductive or undesired side effects. We deal with this problem from the ecological perspective of energy flow in the ecosystem food web and with man's goal of redirecting this flow to maximize within-season yields of plants or animals, or in the long term to optimize his overall production returns through use of optimal strategy. Some of the current literature on simulation and optimization models that have been used to ascertain desired management strategies in various fields of resources or pest management is critically reviewed and some of the progress made is presented. A complete review is not attempted; we seek merely to find a path to the salient concepts in the application of modeling in assessing economic impact of pests on crops. Watt (1968) presented an early analysis and review of the problems and techniques for optimizing pest management strategies while Conway (1977), and Getz and Gutierrez (1982) present more recent reviews.

Here, we attempt also to integrate the various technical imputs of biologists and economists in a realistic systems analysis framework useful in evaluating and managing the impact of pest populations. There are clearly two phases. First, the analysis requires integration of mathematics, systems science, and economics into heretofore biological disciplines. Second, the analysis should be conducted with an aim of increasing our understanding of the system. It is the resultant understanding, not statistical prediction which often as not is impossible, that makes field implementation of integrated pest management (IPM) possible. In a pest management frame-

work any economic analysis of local area situations must of necessity be
kept in a physiological and ecological perspective: the local problem is first
and foremost a biological one, with economic constraints superimposed.
Economic analysis of consequences for producers, consumers, and the environment, when such results are to be adopted nationally, fall more in the
area of macroeconomics (e.g., Taylor & Lacewell 1977) and is not dealt
with here; our focus is at the farm level (microeconomics).

To assess and manage the biological and economic impact of pest populations on a resource, we must gain a reasonably detailed understanding
of the interplay between the interacting species (e.g., pests and crop plants)
as they change through time, as influenced by their own biologies, other
species, and agronomic practices (cf. Conway et al. 1975). Combining and
using this understanding to develop a sound strategy and to employ optimal
combinations of component tactics for economic management of pests is
what is meant here by integrated pest management. This is indeed an
awesome task, and only recently has the methodology for approaching the
whole problem become apparent (cf. Gutierrez et al. 1976, Conway 1977,
in press). Agricultural problems are much simpler than, say, grassland
ecosystems, and the questions, by comparison, may be easier to formulate.
This may be a technical advantage, but we must keep the implications of
crop production practices to larger ecosystem problems clearly in mind
(pollution, fish kills, etc.).

24.2 THE ECONOMIC THRESHOLD

The concept of economic threshold is very crucial to the problem of pest
management and its analysis. Figure 24.1A–C depicts the basic concept
relating to a single pest species from several points of view: (A) economic
entomologist (Stern et al. 1959), (B) economist (Headley 1972, Hall &
Norgaard 1973), and (C) ecologist (Gutierrez et al. 1979). Pestologists have
until recently thought of the economic threshold in Fig. 24.1A as static and
of pest damage as additive (i.e., losses would be partitioned, as e.g. 15%
insects, 10% weeds, 5% nematodes, 20% pathogens, etc.), when in fact
actual damage may have been compounded, or less than additive, and
damage in dollars was some function of pest numbers [$D = g(N)$]. Under
these assumptions, a control measure was warranted when the damage was
greater than or equal to the control costs. But we now know that the
economic threshold is not static and changes with crop phenology and
condition, pest numbers, age structure, and many other factors. The static
concept of the economic threshold also ignores the fact that primary pest
resurgence (as in Fig. 24.1A) and secondary pest outbreaks commonly occur
after chemical applications (e.g., Ehler et al. 1974).

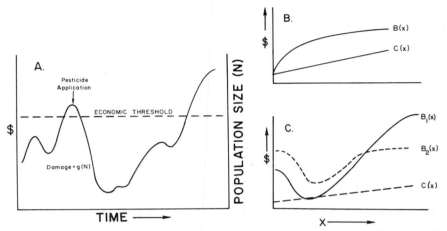

Figure 24.1   The concept of economic threshold: (A) economic entomologist, (B) agricultural economist, and (C) from an economic-ecological perspective: (A) population (N) fluctuation in relationship to a static economic threshold, (B) the assumed benefit [B(x)] and cost fluctuations [C(x)] for pesticide use (X), and (C) two benefit functions illustrating pesticide-induced losses from a key pest (B₁) and a minor secondary pest (B₂) (i.e., the dip in the benefit function).

The simplest loss model is a linear one and can be written as

$$C(x, N) = P_1 x + g(N)[1 - \mu(x)]P_2 \qquad (24.1)$$

Costs are the cost of pesticide + cost of unprevented damage, where C is costs; x is units of pesticide applied; P_1, P_2 is price per unit of pesticide and crop, respectively; N is pest population level; and $\mu(x)$ is mortality caused by x units of pesticide.

The benefits or gross revenue can be expressed by

$$B(x) = P_2 \cdot Y(x)$$

where $Y(x)$ is total yield, using x units of pesticides. Profit (Π) then is calculated as

$$\Pi = B(x, N) - C(x, N)$$

In fact, $C(x, N)$ should include other costs incurred by using pesticides.

Pestologists have generally focused on within-season profit maximization, but rarely have their methods achieved this because they have lacked both the understanding of the problem and the analytical methods to assess their need in a dynamic setting (i.e., $B(x)$ and $C(x)$ can rarely be predicted). Despite this weak biological framework, economists still seek comprehensive optimal solutions of such problems. Formally the problem solution can be written as

$$\max \Pi = B(x_1, x_2, \ldots, x_n) - C(x_1, x_2, \ldots, x_n) \qquad (24.2)$$

over x_i, $i = 1, \ldots, n$, such that $h_i(x_i, \ldots, x_n) \leq b_i$, and $x_i \geq 0$, where h_i represents the set of biological interactions of the system, b_i the set of imposed constraints, and the x_i's the set of management options (e.g., pesticide application(s), irrigation, fertilization, etc.).

Economists assume that management inputs (e.g., pesticides) impart some positive benefit, but that there is a point of diminishing returns, while costs are assumed to be proportional to the amount used, plus some fixed application cost (Fig. 24.1B). Restricting our focus to pesticides, the analysis implies that pesticides should be applied so long as the marginal benefit derived from using the quantity of pesticide $[B(x)]$ is greater than the cost $[C(x)]$. The derivatives (i.e., marginal changes) of these two functions are depicted in Fig. 24.2A (cf. Regev, Gutierrez, & Feder 1976) which shows that the optimal quantity x of pesticide that should be used is the point of intersection of marginal benefits and marginal costs. The model further implies that if unit cost is low, more pesticide should be used. However, the same model would predict that inclusion of environmental costs (E) associated with usage of the pesticide would reduce x (Fig. 24.2B) (cf. Regev, Gutierrez & Feder 1976). Development of pesticide resistance would lower the benefit function and possibly suggest lowered pesticide use (Gutierrez, Wang & Regev 1979, Sutherst et al. 1980).

The above formulations (Figs. 24.2A,B) are still unrealistic in that they ignore both adverse primary and secondary pest resurgences that are specifically induced by pesticide use (cf. van den Bosch 1978). Fig. 24.1C depicts a more plausible benefit function (Gutierrez, Wang & Regev, 1979). The model suggests a crop may be harvested even without use of pesticide $(x = 0)$, while inadequate amounts of pesticide may, as a result of upset

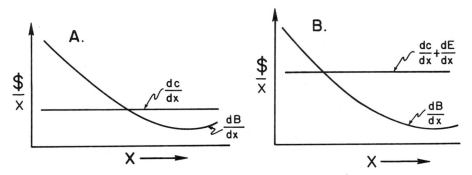

Figure 24.2   The optimal solution (A) for Fig. 24.1B and (B) the solution when environmental costs $[E(x)]$ are added to pesticide costs.

and resurgence problems, create adverse negative effects. Further increases in pesticide use may enhance yields up to some crop maximum (i.e., as modified by various factors). In practice, the benefit function for a very damaging pest (e.g., B_1-codling moth on apples) would be expected to differ radically from that of a crop with low pest damage (e.g., B_2-cotton in the San Joaquin Valley of California) where yield at $x = 0$ and $x = $ max are often equivalent. The economic threshold is then the quantity of x where

$$B(x) - C(x) \geq B(0) \tag{24.3}$$

while the optimal amount of x is x^*, such that

$$\Pi_{max} = B(x^*) - C(x^*) \tag{24.4}$$

While these notions are quite simple, three serious practical considerations must be overcome if models of this kind are to be of practical value: (a) the parameters of the biological interaction must be determined (Chapter 11), (b) the biological model must be formulated and validated, and (c) the economic optimization model must be formulated taking into account the relevant biology. This order of analysis is most important if the results are to be of practical value. We must realize, too, that finding optimal solutions first of all is dependent upon biological features and processes and is an economic problem only because the activity of the pests may be detrimental to humans. We seek to impose our will through chemical, cultural, biological, and other tactics available to us on the development, perhaps the evolution, of an ecological relationship. For this reason, (a) and (b) must be well understood before a realistic assessment of (c) can be made. Conway (1976), Regev, Gutierrez and Feder (1976), Shoemaker (1976a) and Sutherst et al. (1979) accomplished this transition for reasonably complex agricultural systems, while Yorque et al. (1979) attempted it for a more complex forest ecosystem. In these cases IPM decision models were distilled from detailed biological simulation models into a more compact mathematical form suitable for various optimization packages (cf. Appendix). The initial biological models were developed following methods outlined in Chapter 11 and reviewed in the next section. The IPM optimization–decision models were validated against the biological models, and only then used to formulate IPM decision rules. The results proposed by Regev, Shalit and Gutierrez (1976) for overwinter treatments of Egyptian alfalfa weevils, *Hypera brunneipennis,* are being field tested in California and can be reviewed from both the biological and economic perspectives in Gutierrez, Wang and Regev (1979).

To sum up the above, in this chapter we focus on the transition from the biological models of Chapter 11 to economic ones. At each level the focus must remain on the biological relationships, although the output variables may commonly be economic in nature.

24.3   ECOLOGICAL NATURE OF THE PEST CONTROL PROBLEM

The sun, via photosynthesis in plants, is the ultimate source of usable energy for nearly all life systems on earth. Figure 24.3 depicts the capture of energy in the metabolic pool via photosynthesis and from there to the various plant parts (e.g., fruit), fruit pests, and higher trophic levels. In each case we can visualize levels of interaction as tanks, with the resources of each tank allocated according to some priority scheme, and the fraction of the original resource to be allocated diminishing with increasing trophic level. In a stable ecosystem resources at each level are lost to natural consumption by decomposers or by higher trophic levels (e.g., predators) with fractionally less to reproductive rate. The literature on energy flow within and between ecosystems is vast and beyond our scope here (cf. Wiegert 1979, Odum 1971). The concept of a stable ecosystem assumes that populations fluctuate around some steady density or equilibrium position in nature (cf. May 1973) and the biological parameters at each trophic level are "optimized" with respect to the biology of other levels (cf. Gilbert & Gutierrez 1975). The flow rates of energy between trophic levels are represented by valves governed by various biological aspects of the relationships.

In crop systems man's objective is to minimize energy flow up this food chain or into noneconomic components of the food web and to maximize flow to plant parts (e.g., fruits in Fig. 24.3) which reach harvestable age. Agriculturists have selected plants for their growth form, profit potential, susceptibility to pest attack, and various other attributes. Plant protection

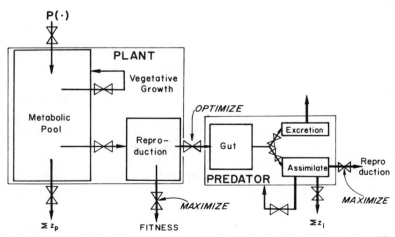

Figure 24.3   The production and flow of energy from the producer trophic level to a higher one in an agroecosystem. In evolved ecosystems, energy flows up the trophic chain, while in agricultural systems man seeks to minimize this flow using pesticides, resistant plants, and so on, to maximize the flow to some storage plant part which he wishes to harvest (e.g., fruits).

specialists have introduced exotic biological control agents (cf. Chapter 22) and used various chemical and physical methods for controlling pests (cf. Chapters 23, 25). The effects of plant breeding have in many cases not been well coordinated between pest disciplines, which adds to the complexity if not the instability of the system. Herein lies the crux of the problem: how do the costs of shunting the energy to a harvestable form by using inputs (x_i) compare with the cost of allowing it to be dissipated via its more natural course in the food web? In economic optimization the objective is to use the optimal amount of the various inputs (x_i's) to maximize profit (Π), as opposed to Nature's evolutionary "objective" of stabilizing relationships between trophic levels to maximize fitness for both host plant and herbivorous predators (Fig. 24.3). A realistic mathematical model is a convenient method for structuring the problem and analyzing the interaction.

24.4 MODELING

Agriculturalists have always been preoccupied with predicting pestilence. In biblical days it was the wrath of an angry God that punished his children. Recently, population ecologists have tried to use correlative methods to predict population outbreaks. They hoped the number of individuals N in generation t would provide a key to estimate the numbers in generation $t + 1$ (e.g., Morris 1959). Unfortunately, careful statistical analyses usually failed to give satisfactory predictions as the unexplained sources of variation (e.g., winter disappearance) were large and not likely to be predicted. The methods have limited applications because they lack the flexibility to incorporate in a dynamic way much of the relevant biology of the organisms involved, much less the vagaries of population survival through time as influenced by weather. The preoccupation was prediction, not understanding, although Morris (1959) and Varley and Gradwell (1963) sought understanding as a background for prediction.

Agriculturalists have also used statistical methods to assess crop losses. Simple regression models of field data may provide good predictors of yield loss for more determinate crops, and control decision rules may be easy to formulate (e.g., alfalfa—C. Summers, unpub. data, Fig. 24.4A), but such rules are more likely to fail for indeterminate crops such as cotton (Gutierrez 1976). The distinction between determinate growth forms (those having a basic, nearly deterministic phenological pattern) and indeterminate ones (those not having a fixed phenological pattern, but rather, a complex relation to their environment) is important, as it restricts the type of analysis most appropriate. In the latter case understanding of the biological relationship is often necessary before simple relationships can be distilled from them.

An example of this stems from research on bollworm, *Heliothis zea,* and boll weevil, *Anthonomus grandis,* damage on cotton in Nicaragua (Gutierrez

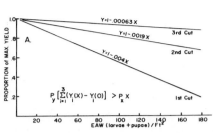

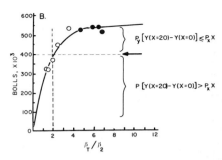

Figure 24.4   Decision rules for pest control in A) determinate (alfalfa) (C.G. Summers unpub.) and B) indeterminate (cotton) crops (from Gutierrez et al. 1981). Note that P_y is the price per unit yield (Y) and P_x is the price per unit pesticide (X).

et al. 1980). The relevant parameters driving the system where identified (i.e., fruit point production rates per unit area and pest damage rates) from intensive field and simulation studies of cotton growth and development (cf. Wang et al. 1977 for mathematical model). Several years of field data showed that yields were highly variable and related to intensity of pest attack, weather, and planting density (e.g., Fig. 24.5). However, despite the grossly different weather and pest levels, a simple relationship could be derived from the plots of observed yields (for all fields) on the index of the rate of fruit point production (β_T) to the rate of cumulative early and late season bollworms, plus boll weevil fruit damage (β_1 and β_2, respectively). From this, a very simple economic decision rule could be formulated [see Fig. 24.4B and the discussion for Eqs. (24.3) and (24.4)]. However, unlike the results in Fig. 24.4A for alfalfa weevil damage to alfalfa, the model for cotton could be formulated only after insights gained from construction of a more detailed mathematical simulation model (Gutierrez, Falcon et al. 1975, Wang et al. 1977) revealed the nature of the interaction (Fig. 24.4B).

24.4.1   Simulation Models

A good biological model of a crop–pest or animal–pest interaction and the subsequent economic analysis must be developed from sound field data. The model developed from such data may be of various forms—simulation models are but one class of model—but all have the limitation that they provide insights into the problem only to the extent that the various components of the problems are included accurately in their formulation. Among the various components or variables not often included in an IPM model are differences in the variety of a crop grown, for example, in degree of pest resistance, capacity to compensate for insect feeding, yield potential, and phenological growth and fruiting pattern. Walker et al. (1978) suggest that models developed to answer specific questions are more useful, but our experience to date does not support this view entirely. The model may

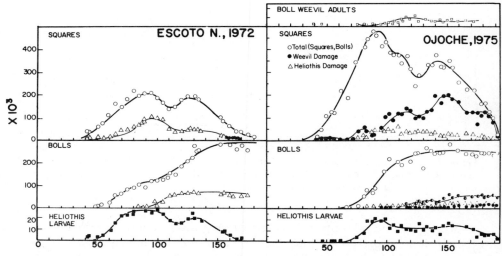

Figure 24.5 A comparison of fruiting and pest damage patterns in two cotton fields in Central America (cf. Gutierrez et al. 1981).

be specific or more general but it must be detailed enough to provide information in response to the question asked. A model should not attempt a one-for-one description of observed field events, as such models are likely to be as complex as nature itself.

The importance of sound biological data and correct mathematical form in constructing a simulation model of a biological process cannot be overemphasized. Some complex simulation models exist only as computer programs, but experience has shown that models are most useful if they have a good mathematical basis. There is little utility in biologically and/ or mathematically weak models. The construction of a simulation model of biological processes serves several purposes: (a) it forces the researcher to focus on the essential *ecological relationships* during the planning and execution of field and laboratory investigations and in formulating the mathematics; (b) once the model is formulated, it enables the *transfer of information* in a concise form to other workers within the same and other disciplines; and (c) if the model is an accurate mimic of reality it may have practical uses in IPM programs (see below, and Chapter 11).

24.4.1.1 Ecological Relationships.

Holling (1973) stressed that biological models should have realism, wholeness, precision, and generality. By "realism," it is meant that the model should mimic the real world, that is, nature; by "wholeness," that it should contain enough detail to represent observed real-world behavior; by "precision," that it should do so with a high degree of accuracy of detail; and by "generality," that it should have general applicability beyond being a description of events proceeding from the specific data used in its construction. These requisites apply whether one is formulating simulation models of crops and animal populations and their interactions under specific agronomic practices or any other model. These simulation models are in reality models of interacting populations, wherein the birth and death rate vary with time and age [i.e., time varying life tables (Hughes 1963, Southwood 1976)]. The approximate mathematical form of each model can be described as

$$\mathbf{N}_{t+\Delta t} = A\mathbf{N}_t \qquad (24.5)$$

where A is a matrix of age-specific time varying birth and death rates (Leslie 1945), wherein the elements contain the complicated biology, and N is the vector of age-distributed cohorts (mass, numbers, and energy units) in the population (Leslie 1945, 1948). The exact form of the model is determined by the biology of the organism, its interactions with other trophic levels, and the questions one seeks to answer. The model may specify a population in one field (e.g., alfalfa weevil, Ruesink et al. 1976) or be formulated to examine the population over a region (e.g., spruce budworm, Jones 1976; corn ear worm, Stinner et al. 1974; cowpea aphid, Gutierrez, Nix et al. 1974, Gutierrez, Havenstein et al. 1974).

During construction of the model, the modeler is forced to consider the animal and plant behavior and physiology, including temperature and other weather-dependent developmental relationships, and occurrences, if any, of time lags in development or other essential functions for each species. The researcher must recognize and account for the importance of population age structure, host or prey specificity and preference, and the dispersion of the various life stages within the habitat (Chapter 11). The simulation forces us to come to terms with and more explicitly specify our often vague notions about migrations, dispersal, and dispersion as these features may affect the simulation, but the major contribution is that it forces us to come to grips with the precision of quantitative ecology, not just verbal natural history. A model cannot describe all aspects of the biologies of the system, merely the dominant ones (Gilbert et al. 1976), and these are, of course, species specific. The relevance of highly simplistic, theoretical models to real-world problems is still in question (Wang & Gutierrez 1980), although Hassell (1980) has shown that some progress has been made (cf. Chapter 22).

Once we have described the important broad ecological relationships impinging upon the population in nature and structured these relationships in a realistic way, we can use simulations of the model to theoretically examine various aspects of the ecology of both the crop species and its consumers (at primary and higher levels) that are not possible to observe directly in nature. For example, Gutierrez, Butler et al. (1977) and Gutierrez, Leigh et al. (1977) found that their cotton model suggested simple field experiments that could be used to assess impact of lygus bugs and pink bollworm on cotton growth and development, problems that had eluded resolution for some time.

After the model has been thoroughly field tested, one can make additional checks of the model to see what happens when its parameters are modified (as if one of the species did something other than observed). This is commonly called *sensitivity analysis*, but it is also often useful in estimating optimal control strategies or examining basic ecological relationships. Before this can be done efficiently and in a meaningful way, we must formulate proper theoretical or applied questions to ask of the model. A theoretical question might relate to some aspect of species survival (fitness and adaptability). Gilbert and Gutierrez (1973), Jones (1977), Jones and Ives (1979), Gutierrez and Regev (1983), and Gilbert (1980) asked such questions and found that in the natural systems they studied, the populations characteristics seemed in general to have evolved so as to optimize at least some measure of fitness and adaptability (i.e., between-years reproductive or short-term survival "strategy"). We cannot at present predict how certain characters affecting processes or relationships might evolve or at what rate. An applied question might be the relative merits of various control strategy options (e.g., of the cattle tick, Sutherst et al. 1980). In this case, how clearly we perceive the question may determine whether simulation or optimization models should be used to answer the questions (see below).

Often the model brings to attention important bits of biology previously overlooked and may force us to go back to our data or, if necessary, to the field to obtain estimates of vital parameters (e.g., survival rate of pink bollworm infected squares, Gutierrez, Butler et al. 1977; damage rate by lygus bug on cotton squares, Gutierrez, Leigh et al. 1977). If the model has been constructed properly, it will simulate the real world (above, and Chapter 11). It is nearly impossible to achieve good comparison with independent data sets by artifice, especially if multiple validation criteria are used. Validation is an important aspect of developing IPM models. It is significant to remember that the goal should be a qualitative and quantitative understanding of the relationship, and hopefully, prediction when possible. Due to the large number of variables involved, and the imperfectness of the data used to formulate the model, we can never describe completely the biological system (even simple ones).

24.4.1.2 Transferability of Information.

One of the most serious criticisms of ecology for some years was that its information could rarely be transferred from one discipline to another because it lacked scientific rigor, i.e., it lacked precise mathematical or quantitative definition and formulation. While this criticism can still be made, considerable strides have been made in this area (e.g., in the mathematical formulation of a population model for cotton growth and development by Gutierrez, Falcon et al. 1975, Wang et al. 1977, Curry et al. 1978, 1980). The first two models were in part derived from modeling by McKinnion et al. (1974). Because it is mathematically formulated, the basic notions it employs of the complex biological interactions can be transferred to, and used by economists (Regev et al. 1976), engineers (Shoemaker 1976b), and others who have mathematical skills but may lack sophistication in field ecology. Examples of this are outlined in the section on optimization studies. Heretofore, this kind of disciplinary interaction and effort were greatly restricted (Gilbert & Gutierrez 1973; Hughes & Gilbert 1968). Because of this infusion, some of the rigor of the physical sciences is beginning to be felt in applied ecology (cf. Sharp & DeMichele 1977; Curry et al. 1981; Wang et al. 1977, Barr et al. 1973). This interdisciplinary activity has had a positive impact, and as an offshoot it is forcing pest control out of the "dark ages" criticized by Carson (1962) and van den Bosch (1978). It is forcing us to stop relying on half-guesses, intuition, and set patterns of management inputs (e.g., pesticides). It offers a more scientifically established basis for crop and animal production and protection. It hopefully will force theoretical ecologists to deal with real-world issues (cf., Gilbert et al. 1976, Wang & Gutierrez 1980).

24.4.1.3 Potential Uses in Agricultural Loss Assessment. For most crops of a given genotype, the weather, agronomic practices, and soil fertility set the upper limits to crop yields in the absence of pests. In many crops this potential may be reasonably stable (e.g., field corn) or may vary drastically (e.g., cotton). Pest injury must be superimposed upon this basic varying potentiality. Until recently, agriculturalists have used limited information to guess the extent of damage a pest may have caused and the extent of loss attributable to unfavorable weather. Furthermore, treated plots were often compared with a partially treated control, thus the experiments lacked a true check. This produced results specific to a given year and set of conditions, which could only be used with *great caution* in other years, seasons, and situations; such results usually do not provide the understanding considered essential to developing generalized pest control strategies.

Weather-driven simulation models which stress ecological relationships are not so limited in application. Such models can be used to evaluate development of the crop (i.e., up to a current point in the season) and with great care can be used to project plant and pest population dynamics and damage trends for short periods (for some crops, a few days) into the future. Thus, they can be used for short-term prediction. If weather could be predicted reliably, the models would have much greater utility with a longer view prediction potential. The accuracy of short-term predictions is enhanced by updating the model with each succeeding sample of data, such as the size and age structure of the pest population and the crop, and impact of natural enemies so that damage potential and crop compensations against damage can be evaluated relative to some economic injury level.

While the restriction to short-term prediction is a limitation, the models are useful in that they may be used to pinpoint or assess generation peaks in pest populations (Croft et al. 1979), emergence patterns of pests (Falcon et al. 1971, Gutierrez, Havenstein et al. 1974, Gutierrez 1976, Gutierrez, Butler et al. 1971, Hartstack et al. 1976, Welch et al. 1978), the susceptible periods of the plant (Gutierrez, Falcon et al. 1975), short-term population damage trends (Peart et al. 1974, Croft et al. 1979), and many other biological phenomena.

24.4.1.4   Potential Uses: Biological Control.   An obvious use for simulation modeling is in analysis of biological control. In the past biological control has been assessed on a rather simple empirical basis in that control has been credited if the pest insect population was reduced by a natural enemy to some acceptable level. It has merely been a case of bringing natural enemies into a target area, getting them established, and hoping that they do a good job. What has usually been lacking has been methods for evaluating the reasons for the apparent success or lack of success of specific introduced or indigenous enemies. Simulation models provide flexibility for performing these complex analyses. Hughes and Gilbert (1968), Gilbert and Gutierrez (1973), Fraser and Gilbert (1976), and Baumgaertner and Gutierrez (1982) have used simulation modeling to evaluate natural enemy impact. Frazer and Gilbert's (1976) study on predation efficiency of ladybird beetle adults, *Coccinella californica*, on pea aphid, *Acyrthosyphon pisum*, in alfalfa involved a mix of field and laboratory life table analyses and modeling. The results clearly showed the beetles are not the major regulating agent of the aphid at the densities observed in their area. Modeling and biological work by Gutierrez and Baumgaertner (In press.) on the interaction of alfalfa, two aphids, a host-specific parasite, ladybird beetles, and lacewings demonstrates the relative importance of each enemy alone and their interactions in various combinations. In general, these studies show that detailed simulation studies, while difficult, do not require large quantities of time or money—the budgets of both can be quite conservative if the approach outlined in Chapter 11 is taken.

24.4.1.5   Potential Uses in Pest Management Implementation.   Considerable effort is being made to use these models in the field to make pest control recommendations. Their utility was suggested by Watt (1968), but modeling systems for implementing pest control in the field have only recently been developed (Gutierrez, Havenstein et al. 1974, Peart et al. 1974, Croft et al. 1979, Croft 1981, and others). Moreover, until recently the computer hardware and software for implementing these systems has been beyond the resources of IPM practitioners. With the growing awareness of the utility of these models, state legislatures, agricultural college administrators, NSF, EPA, and USDA have begun to channel resources into such programs (at universities such as Purdue, Michigan State, Pennsylvania State, Texas A & M, University of California, and others); greater progress can be expected in the future. It is still too early to fairly assess the merits of delivery systems being implemented at these institutions, but in addition to providing model-generated information, these systems have the potential to provide weather summaries, information on pesticide registration and effectiveness, alternative methods for control of specific pests, and methods for data analysis and summarization.

24.4.2. Example of a Simulation Management Model: Control of Pests of Livestock

The methods discussed above have also been used with apparent success for examining pest management tactics and strategies for control of cattle ticks, *Boophilus microplus*, in Queensland, Australia (Sutherst et al. 1980). This study also provides a good transition between stimulation and true optimization models, and so is described in some detail here. As related in Chapter 25, in their study a modified Leslie matrix model of the tick population was employed to assess three control strategies for the pest: (1) use of acaricides; (2) pasture "spelling" in which cattle are removed from the pasture until the free-living tick larvae die, before restocking; and (3) use of tick-resistant cattle. The objective function was formulated as

$$\text{minimize} \quad PD + C$$

where

$$D = f_1(\theta) \qquad \theta = f_2(\theta_0, S) \qquad C = f_3(S) \tag{24.6}$$

and P is the price of beef per kilogram liveweight; D is the kilogram liveweight lost per beast caused by tick burden θ; θ is the tick burden per beast remaining after strategy; S is applied against burden θ_0, the untreated tick population; and C is the cost per beast of control strategy S.

The model was used to assess the impact of the control methods on the tick population. The results indicated that if acaricides are to be the main strategy, regular dipping of cattle at 3-week intervals five to six times annually, beginning either in spring or summer gave the best results. Yet such a strategy breaks down with the development of resistance (cf. Gutierrez, Regev et al. 1979). Use of single spelling periods each year was also compared with a rotational grazing system involving frequent movement of cattle between two pastures. Both strategies gave results similar to the optimal acaricide control program, but the fencing and other management costs might prove a deterrent to their adoption, even if pasture damage can be minimized. But in final analysis, use of resistant cattle, combined with use of a single spelling period or limited acaricide dipping, appeared to be the best strategy.

24.4.3 Optimization Models

Optimization models provide a more rapid method for finding solutions
to such complex problems (Watt 1968, Mann 1971) as finding economic
thresholds and developing economically and socially optimal IPM systems.
In seeking optimal IPM solutions using models, the most naive method is
to run the simulation model for all possible control strategies (i.e., the brute
force method). An example often used by C. Shoemaker to illustrate the
inefficiency of the method assumes the season is divided into 10 time
periods with the optimal control strategy being one of the 2^{10} possible
within-season combinations of spraying or not spraying. If there are 10
possible dosage levels (including zero), the possible combinations then number 10^{10}—hence, to simulate all possible combinations even for the first
example would be prohibitive in terms of time and money. (We affection-
ately term the 2^{10} simulations one "Shoemaker"). Of course, examining the
biology and selecting runs carefully may greatly restrict the number of runs
required by this method. The great advantage of true optimization methods
(cf. Appendix) is that the answer can be obtained in one or at the most a
few runs. The Appendix reviews the various common optimization methods in a simpler form than usually presented in most texts.

Optimization models may be mathematically difficult to formulate and
the problems difficult to solve, but, depending on the question asked and
the method chosen, the solution may even be obtained by relatively simple
optimization methods (cf. Walker et al. 1978). Realistic analyses of host–
pest population interactions should include age structure and various other
relevant biological details, but this often results in difficult if not intractable
optimization problems. Each new state variable (e.g., age structure variables) increases computer costs and memory required to solve the problem.
For example, the numbers of insects in several morphological stages (e.g.,
egg to adult) or several physiological age or activity categories (young, old,
in diapause or not, etc.) must be kept in account to all calculation of present
and future insect damage potential and population trends. These are important features in crop and pest management; for example, small worms
cause much less damage than large ones. The same relevance holds for
other features. It is not uncommon for such optimization problems to
involve a large number of variables and constraints (e.g., the 50 variables
and 40, mostly nonlinear, constraints in the problem solved by Regev et
al. 1976a).

The optimization concept is powerful, but it requires mathematical skills
not possessed by most biologists. Again, the ideal method for evolving such
programs is through close cooperation of biologists, systems analysts, and
economists. Such cooperative efforts are becoming more common, as in
the cooperative work of Yorque et al. (1979), Shoemaker (1976b), Ruesink
(1976), Fick (1975), Regev, Gutierrez and Feder (1976), Regev, Shalit and
Gutierrez (1976), Feder and Regev (1975), Gutierrez, Regev and Shalit

(1979), Gutierrez, Wang and Regev (1979), Talpaz et al. (1978), and Sutherst et al. (1980). The literature abounds with more theoretical works; review of these is quite beyond our scope here. We have sought to deal with more pragmatic examples—ones offering possibilities of being used in a practical context in the near future.

Due to the wide variety of biological phenomena, it is impossible to select any one optimization method which will work for all biological problems (cf. Appendix for discussion). This can easily lead to confusion in selecting the best optimization algorithm. A priori simplification of models is usually done for technical convenience to make the solution easier, but it may at the same time render the analysis useless. Considerable judgment must be exercised. In a simple case Walker et al. (1978) presented a preliminary analysis of the maximum stocking rates of cattle (C) and game, that is, impala (I), on a South African savanna (grass and browse) so as not to cause permanent damage to the vegetation. The usable amount of grass (G) and browse (B) have the constraints

$$G \geq g_c C + g_i I \qquad B \geq b_c C + b_i I \qquad (24.7)$$

and g_c, g_i is grass eaten per cow and impala, respectively, per unit time, and b_c, b_i are the respective values for browse. Figure 24.6 shows the optimal solution to the stocking problem (i.e., the corners 1–3), but any point in the shaded area satisfies the above constraints. The analysis is approximate and static, hence greater biological detail must be incorporated to evaluate the problem in a dynamic setting.

An early optimization effort in a practically based forest insect pest problem was conducted by Watt (1963), using dynamic programming. In that case the model of the spruce budworm, *Choristoneura fumiferana*, a pest

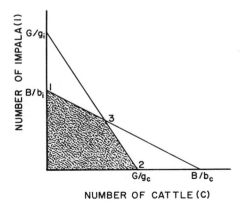

Figure 24.6   A linear programming model for optimal stocking rates of cattle (*c*) and impala (*i*) on browse (*B*) and grass (*G*). The optimal solutions are at points 1, 2, and 3 (see text, cf. Walker et al. 1978).

of spruce-fir forests, was used only to predict the density of pupae from one year to the next. In this simple model, using very few states variables, it was relatively easy to use dynamic programming as the optimization method. Shoemaker (1973) used dynamic programming in finding the optimal control policies applicable, theoretically, to an insect pest and its parasite, using the hypothetical Lotka–Volterra interaction model (Chapters 12, 16). It should be emphasized that in both the above models, dynamic programming could be used to find the best control strategy over time only because age structure was not included in the insect models. This lack of age structure implies some unrealism.

In a later study Shoemaker (1976a) used dynamic programming to analyze weevil (*Hypera postica*) control problems in alfalfa even with a greatly increased biological complexity in the model. She included three state variables: one for overwintering population density of weevil adults, one for parasite adults, and the other an index of alfalfa (a perennial) stand vigor. In this case the problem was simplified because the pest model, with the attendent problems of age structure, was included as part of each year's simulation, and hence the decisions made by the dynamic programming algorithm were not complicated by pest age structure. In general, dynamic programming is difficult to use in problems where pest control decisions significantly affect the age structure of the populations because changes in age structure significantly affect subsequent population trends.

Regev, Shalit, and Gutierrez (1976) used an optimization algorithm of the type commonly described as constrained nonlinear optimization techniques (see Appendix) in their study of Egyptian alfalfa weevil (*H. brunneipennis*). The model included age structure of the weevil and a simplified alfalfa stand model, which included phases of crop growth, harvesting, and stand deterioration. Dynamic programming would be impossible to use in this case because of the increased dimensionality created by age structure. The season was divided into several parts, and difference equations were written describing the various components' interactions during each time period. The equations were then solved numerically for the whole season (i.e., 50 variables and 40, most nonlinear constraints), using Abadie's generalized reduced gradient algorithm (Abadie & Guigou 1969). The model showed that the common practice of treating larval weevil populations during spring was suboptimal, while midwinter treatments against adults migrating into the field was economically and environmentally a better solution. These results are currently being field tested.

Another approach to optimization was taken by Talpaz et al. (1978) and Gutierrez et al. (1980) who investigated growth patterns of cotton as influenced by boll weevil feeding. The many state variables of the boll weevil and of cotton were included in a simulation model, and the entire model was then treated as a function to be minimized. To perform this task they used Powell's algorithm (see Appendix), a method of optimization which has few mathematical restrictions (Powell 1964). This method is reasonably efficient and enables one to ask a wide variety of economic and biological questions without extensive restructuring of the mathematical simulation model. The method suffers by lacking rigorous mathematical form (see Appendix) and might be used with disastrous results with mathematically weak simulation models.

24.4.4 Qualitative Analysis

It is difficult to evaluate the costs and benefits of alternative solutions to ecological problems having large time and/or space horizons using the optimization procedures described above. If cost–benefit estimates have been made at all, they have usually been *very crude* approximations. For this reason qualitative methods having the aim of understanding the problem, rather than finding optimal solutions to specific limited problems, have been used, chiefly in Canada by Yorque (1979), Jones (1976), Clark et al. (1977), and Peterman et al. (1978), to bridge the gap between the ecology of resource management (organism centered) and policy design (economics centered) to develop a science of ecological policy design. These studies are much larger in scope (they embrace broad social, industrial, and other endeavors in addition to agriculture) than the aforementioned case study analyses of crop systems. The resultant policies suggested by this technology tend to be couched in more qualitative than quantitative terms. The usefulness of this approach remains at this date to be shown.

The spruce budworm in conifer forests of New Brunswick and salmon fisheries in British Columbia have been used for case studies of this approach. Both systems, but especially the budworm–forest system (e.g., Morris 1963) have been intensively studied. The more traditional time-varying life tables (*sensu* Gilbert et al. 1976) for whole regions were developed and used as the starting point for these analyses (Jones 1976). But because of the large time–space dimensions of these problems, the more standard sensitivity analyses of the simulation model could not be used (Clark et al. 1977). However, the model of the spruce budworm was used to estimate between season (t to $t + 1$) replacement rates (R)

$$R = \frac{N_{t+1}(T)}{N_t(T)} \tag{24.8}$$

under different conditions of forest quality (age $= T$). The replacement rates were plotted on N_t (i.e., the initial number in the tth generation) (Fig. 24.7). Of special significance was the discovery that bird predators had a Type III functional response due to learning (cf. Holling 1965) and greatly affected spruce budworm survival, especially at low density (see inset, Fig. 24.7A). The equilibrium points (N^*) for each tree quality index occur at the value of N at the intersections of $R = 1$ and of the function relating R

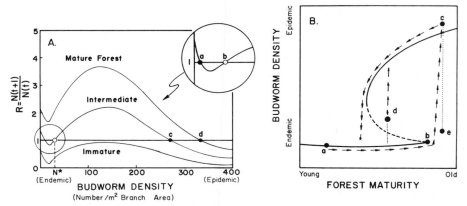

Figure 24.7   (A) The relationship between reproductive rate (R) and pest density (N) at a given tree quality (T_i). Note that N^* is an equilibrium point for tree quality (T_i). (B) A plot of $N^*(T_i)$ values from a, on their respective tree maturity values.

to N_t (see inset in Fig. 24.7A). The plot of the equilibrium points (N^*) (Fig. 24.7A) against forest quality categories (T_i) is shown in Fig. 24.7B, which is merely a slice from a three-dimensional model, intensity of predation being the third dimension.

Holling's (1973) concept of resilience* plays a dominant role in this analysis. In its specific application to spruce budworm (Jones 1977), the qualitative essence of the problem was captured, even though use of catastrophe theory, specifically in biological and social systems has been severly criticized by Zahler and Sussman (1977). Contrary to previous budworm studies, Holling and co-workers view the system as having multiple dynamic equilibrium points which change as a consequence of the short- or long-term history of the pest–plant–natural enemy interaction over time and space (Peterman et al. 1978; Clark et al. 1979) (cf. Chapter 12).

Simply, the model (Fig. 24.7B) shows the trajectory of stable equilibrium points as the system evolves (i.e., the upper and lower stable attractors of the system are the two solid lines, and the unstable boundary is the dotted line between them). The model suggests that so long as the forest remains of low quality, the budworm population tracks the lower endemic attractor $(a \rightarrow b)$, until some time when the system evolves to equilibrium point b, at which time the population jumps to c on the upper epidemic level. Biologically, this means that under the conditions at b, the reproductive potential of the host increases rapidly because of a highly favorable forest condition, hence an epidemic occurs.

Because of pest-induced tree mortality, forest quality, as measured by shifts in age–class diversity, is reduced (death of older trees, younger trees

---

*Resilience is defined as the ability of the system to absorb unexpected events and variability and to benefit from such change (Holling 1973, Peterman et al. 1978).

survive, and new trees become established) and the equilibrium shifts to the left until the population jumps from the upper attractor to the lower one, where the epidemic subsides. This has characteristically happened in New Brunswick forests where the intervals between outbreaks have been cyclic, with many years between outbreaks.

The model predicts that the effect of the introduction of large numbers of moths, say from the lower attractor to a point d across the stability boundary, would be sufficient to cause a spruce budworm outbreak. Furthermore, the model predicts how any management practice that does not alter forest quality (e.g., DDT applications) but would merely reduce the budworm population from the upper attractor to e, is doomed to failure because the population would resurge (i.e., because of natural enemy disruption as well as forest quality) and continually shift the forest back to a low forest quality condition and an endemic or low pest status. Both these qualitative notions predicted by the model have been observed in the field (Peterman et al. 1978). For example: (1) when two storm fronts meet, a large population of adult migrants may be deposited in a small area, causing local outbreaks and heavy, but local, mortality of unusually susceptible trees; and (2) repeated sprayings of pesticides have merely prolonged the epidemic, which now occurs yearly rather than cyclically. This result may be compared to that which occurs in agriculture when pesticide resistance develops and permits resurgence of pest populations.

24.5 CONCLUDING REMARKS

There is no set path that we must follow in assessing impact of pests in agricultural crops or other ecosystems. The methods may be quite varied and may depend as equally on the complexity of the biology of the species as on the mathematical tractability of the models that can be formulated. But it is clear that whatever the problem, the analysis usually spans an array of technical subjects which require interdisciplinary cooperation. Cooperative efforts must always imply a large measure of respect for the collaborator's area of expertise, as any attempt by biologists to assume that the biological information is central and the mathematical analysis is of secondary importance (and vice versa on the part of the mathematician) will doom such projects to failure. Analyses of biological systems are of greater complexity than most physical systems if for no other reason than that biological (as opposed to physical) laws are not well formulated, and biological populations have great genetic variability in both time and space and are greatly influenced by the prior and current physical environment. Characterization of complex biological systems for the purposes of making decisions (economic) as to their management involves extreme simplification and abstraction of the system and as such can be approached from numerous angles and can employ any or several of the various methodologies (determination of economic threshold, simulation modeling, optimization modeling, qualitative analysis) we have discussed above.

24.6 APPENDIX: AN OVERVIEW OF SOME OPTIMIZATION METHODS APPLICABLE TO INTEGRATED PEST MANAGEMENT

The advantages and shortcomings of some optimization procedures applicable to IPM are discussed in this appendix only in an introductory way. Readers are referred to references cited for details. Familiarity with models used in crop pest management and a working knowledge of elementary differential calculus are assumed.

The problem of minimizing the function $f(x_1, x_2, \ldots, x_n)$ is identical to that of maximizing $-f$ (i.e., its negation). Hence, without loss of generality, consider the following general optimization problem

$$\max_{x_1, \ldots, x_n} f(x_1, x_2, \ldots, x_n)$$

subject to the parameter constraints which dictate the biological interactions and restrictions (i.e., the parameter must not become unreasonable or violate assumptions)

$$g_i(x_1, x_2, \ldots, x_n) = 0 \qquad i = 1, 2, \ldots, p < n$$

$$g_i(x_1, x_2, \ldots, x_n) > 0 \qquad i = p + 1, \ldots, m$$

The general crop pest management problem as formulated in the foregoing sections is in the above framework. In this case

$$f(x_1, x_2, \ldots, x_n) = B(x_1, x_2, \ldots, x_n) - C(x_1, x_2, \ldots, x_n)$$

where $B(\cdot)$ and $C(\cdot)$ are the gross revenue and the cost functions, respectively.

Consider the following simple unrealistic example for the case $n = 1$. Suppose one wishes to know how much pesticide to apply at a specific time to maximize profit, with the assumptions that the number of pests killed is proportional to the amount of pesticides applied (γ), and the amount of damage is also proportional to the number of pests present (α). Then the problem can be formulated such that

$$B(x) = p(Y_0 - \alpha q(1 - \gamma x)) = \text{revenue} \quad \text{and} \quad C(x) = \beta x = \text{costs} \quad (24.9)$$

where Y_0 is the optimal amount of yield the crop can produce (pest free); q is the initial pest numbers; p is price per unit of yield; β is price per unit of pesticide; and α and γ are constants of proportionality (see above).

The optimization problem [max $B(x) - C(x)$] can thus be written as

$$\max_x f(x) = \max_x p(Y_0 - \alpha q) + (\alpha q p \gamma - \beta)x \qquad (24.10)$$

subject to the constraints, and $\gamma x \leq 1, x \geq 0$. The terms pY_0 and $p\alpha q$ indicate the pest-free yield and the max yield loss for pest level γ, respectively. The term $\alpha q p \gamma x$ denotes the benefit of pesticide use. The constraints indicate that one cannot kill more pests than are available, and the amount of pesticides to be applied is nonnegative ($x \geq 0$). The solution of Eq. (24.10) is simple and is shown in Fig. 24.8. Note that the maximum, depending on the value of the parameters, occurs either at 0 or at $1/\gamma$, which are the end points of the interval of x satisfying the constraints. This property is characteristic of linear optimization problems, for example, Eq. (24.10). In general, linear optimization problems take the form

$$\max_{y_1, \ldots, y_n} \sum_{j=1}^{n} \gamma_j y_j \qquad (24.11)$$

subject to $y_i \geq 0$, and

$$\sum_{j=1}^{n} \gamma_{ij} y_j = b_i \qquad \text{for } i = 1, 2, \ldots, m \qquad (24.12)$$

If there exists $(y_1^0, y_2^0, \ldots, y_n^0)$ satisfying Eq. (24.12), then the region determined by these constraints is convex; that is, it has no holes, and its boundary has no indentations, as in Fig. 24.9.

The solution of Eqs. (24.11) and (24.12) then occurs at an extreme point of this region (e.g., points A–F in the left–hand side of Fig. 24.9). It can usually be obtained in a finite number of iterations using a fundamentally simple procedure (the simplex method, Dantzig 1951). If the linear programming problem is solved by enumeration, the number of required iterations is the number of extreme points of the region, as dictated by the constraints. For large problems these points can be numerous.

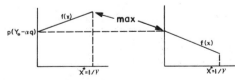

Figure 24.8   The solution (X) for a simple linear optimization model. Case 1: $\alpha q p \gamma - \beta > 0$; Case 2: $\alpha q p \gamma - \beta < 0$.

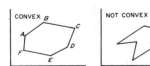

Figure 24.9   A linear programming model solution: (A) the convex solution; (B) no solution.

Unfortunately, the functions describing the biology of a crop ecosystem are rarely linear. In cases where the objective function or the constraints are nonlinear, methods of nonlinear and dynamic programming can be used. The formal development of the theory stems from the differential calculus. Consider a function, $f(x)$, with one independent variable. Then a point x^* is a maximum or minimum if $f'(x^*) = 0$ and $f''(x^*) < 0$ or > 0, respectively. There could be more than one point satisfying these conditions. In this case whether the optimum obtained is global or local has to be determined. In Fig. 24.10, x_1 and x_3 are the local and global maxima, respectively, and x_2 and x_4 are the global and local minima, respectively. For functions with more than one independent variable, similar conditions hold.*

Consider again a simple crop pest management problem similar to Eq. (24.10); only in this case the proportion of survived pests is a negative exponential of the amount of pesticides applied (see Fig. 24.11). In this case the optimization problem becomes

$$\max_x g(x) = \max_x pY_0 - \alpha q p e^{-\gamma x} - \beta x \qquad (24.13)$$

subject to the constraint $x \geqslant 0$.

To solve Eq. (24.13), one first takes the derivative of the objective function

$$g'(x) = \gamma \alpha q p e^{-\gamma x} - \beta$$

then point x^* is found such that $g'(x^*) = 0$.

$$\gamma \alpha q p e^{-\gamma x} - \beta = 0 \qquad e^{\gamma x} = \frac{\gamma \alpha q p}{\beta} \qquad x^* = \frac{1}{\gamma} \ln\left[\frac{\gamma \alpha q p}{\beta}\right]$$

Furthermore,

$$g''(x) = -\gamma^2 \alpha q p e^{-\gamma x} < 0$$

Hence the maximum of $g(x)$, in this case, is unique and occurs at x^*. However, Eq. (24.13) is not completely solved because of the constraint that $x \geqslant 0$.

Case 1: $x^* > 0$. The solution of Eq. (24.13) is then $\max[g(0), g(x^*)]$, where $g(0) = pY_0 - \alpha q p$ and

$$g(x^*) = pY_0 - \beta/\gamma \left[1 + \ln\left(\frac{\gamma \alpha q p}{\beta}\right)\right]$$

---

*The maximum (minimum) of $f(x_1, x_2, \ldots, x_n)$ occurs at the point where the gradient of f, vector of partial derivatives, equals zero and the matrix of second derivatives is negative (positive) definite.

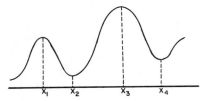

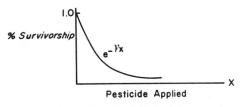

Figure 24.10   The maximum and minimum of a function.

Figure 24.11   Nonlinear pesticide survivorship model.

Case 2: $x^* < 0$. This implies that $\gamma \alpha q p < \beta$, which means that for all $x \geqslant 0$, $g'(x) < 0$. Hence, $g(x)$ is a decreasing function of x, for $x \geqslant 0$, and the maximum thus occurs at $x = 0$.

In the above example we have illustrated how differential calculus is used to solve nonlinear optimization problems. However, this method has its limitations. To locate the set of points such that $g'(x) = 0$ can be difficult. In fact, a well-known theorem in algebra states that the zeros of a polynomial of degree 5 and higher cannot be solved in a finite number of algebraic computations* (van der Waerden 1953).

For problems of great complexity such as a realistic crop–pest systems model, the techniques of differential calculus are usually inapplicable. In these cases the iterative methods of nonlinear programming can be used. These iterative methods can be classified into two main categories: those which use differential calculus as a theoretical basis and those which do not. The general idea behind all iterative procedures is to start at a point, searching in some systematic way, locating another point which is closer to the optimum than the existing one and repeating the process. There are various ways of locating the next point, and this gives rise to a spectrum of nonlinear programming algorithms. Each method has its advantages and disadvantages. A disadvantage common to all iterative methods in nonlinear programming is that no systematic procedures exist to determine when a global optimum (the optimum across all of the search space) has been found, hence each must be considered a local optimum (i.e., one of many optima that may be found). Some optimization procedures for nonlinear problems are briefly described below, as a complete coverage is quite beyond our scope here.

24.6.1   Methods Based on the Theory of Elementary Differential Calculus

These methods require information about the first and sometimes also the second derivatives of the objective function. Two of the best known classical optimization techniques of this category are the "steepest ascent" (Cauchy

---

*The algebraic operations are the additions, subtractions, multiplications, divisions, and the extractions of roots, $\sqrt{}$, $\sqrt[3]{}$, and so on. The theorem first proved by Abel is a consequence of Galois theory.

1847, Goldstein 1962) and the Newton–Raphson method. The "steepest ascent," as the name indicates, searches in the steepest direction, which is that of the first derivative of the objective function. Its greatest shortcoming is that its rate of convergence is unpredictable near the optimum. The Newton–Raphson method, on the other hand, requires both the first and second derivatives, and it searches on the local quadratic surface. This procedure, contrary to "steepest ascent," is very efficient if the initial estimate is close to the optimum, otherwise the method may fail to converge. There are many extensions and variations of the above two procedures, among which is the method of Goldstein et al. (1966).

While the property of rapid ultimate convergence, which we find in the Newton–Raphson method is desirable, the evaluation of second derivatives is often costly. Hence, another group of algorithms was developed which uses information of first derivatives to evaluate the second-degree terms. These methods are generally successful in practice, even though the reasons for success are as yet unknown. Some of these better known methods were those developed by Zoutendijk (1960), Fletcher and Reeves (1964), Powell (1964), and Davidon (1959); modifications are given by Fletcher and Powell (1963) and Stewart (1967). Stewart's procedure extends Davidon's variable metric method. In cases where even the computation of first derivatives is laborious, Powell's and Stewart's algorithms can be used, as these procedures only require function evaluation (i.e., they evaluate the entire algorithm simulation, etc.).

24.6.2 Direct Search Methods

Direct search methods have the following characteristics: (a) they use numerical and not analytical techniques, (b) they require only computations of function values, and (c) they can be applied to a wide class of optimization problems. These algorithms often compare favorably with the methods described just previously. The classical method of changing one variable at a time falls in this category (Spang 1962, Wilde 1964), as do those of Spendly et al. (1962) and Nelder & Mead (1965). The procedures discussed in the above papers are designed to solve optimization problems without constraints, but in any case there are ways to handle the constrained problem [Eq. (24.9)]. Kelly (1960) and Rosen (1960, 1961) extended methods of linear programming to handle constraints that are linear, or nonlinear constraints that can be appropriately linearized. Another approach is to convert the constrained problem into an unconstrained one, using Lagrange multipliers (Forsythe 1955). An application of the constrained nonlinear programming method to alfalfa–alfalfa weevil management can be found in Regev et al. (1976). [Reviews of nonlinear optimization techniques by Powell (1970) and Polak (1973) may be helpful, as also the more complete bibliography on optimization in Leon (1965)].

Problems of the form of Eq. (24.9) can sometimes be restructured so that the independent variables $X_1, X_2, \ldots, X_n$ form a sequence of inter-related decisions. The crop pest management problem as described in the text falls naturally in this category, where X_1 is the optimal amount of pesticide (or control) to be applied during the ith time interval.

For problems with this characteristic, a dynamic programming procedure can be used. This method, quite different from the methods discussed previously, was developed by Bellman (Bellman 1957, Bellman & Dreyfus 1962). To transform the optimization problem to the dynamic programming framework, the following procedure is usually followed: (a) determine the number of necessary state and decision variables, (b) define the optimal value function and its arguments (see example below), (c) write the appropriate recurrence relation, and (d) compute the appropriate boundary conditions. While the principle of dynamic programming is simple, the ability of good problem formulation requires experience and imagination (Dreyfus & Law 1977).

To illustrate the dynamic programming technique, consider an example similar to Eq. (24.13). Only in this case decisions of applying pesticides are made during two time periods. The optimization problem can thus be written as

$$\max_{x_1, x_2} pY_0 \underbrace{- \alpha pqe^{-\gamma x_1}}_{\text{stage 1}} \underbrace{- \alpha pqe^{-\gamma x_1}e^{-\gamma x_2} - \beta(x_1 + x_2)}_{\text{stage 2}}$$

subject to $x_1 \geq 0$, $x_2 \geq 0$. Note that the birth process is absent in this simple model. This problem can be reformulated so that it can be solved using dynamic programming. For simplicity, x_1 and x_2 take on values of 0 or 1 only, indicating to apply or not to apply. Following the process indicated above, we have

1. The decision variables are x_1 and x_2. There are three stages; at stages 1 and 2 a decision to spray or not to spray is made, and stage 3 denotes the end of season or time of harvest. The state variable is q, the current pest level.

2. The optimal value function is defined as $f_k(q) =$ the optimal return that can be obtained from stage k to stage 3, given that at stage k the pest level is q.

3. The recurrence relation can be written

$$f_k(q) = \max \begin{cases} \text{spray: } f_{k+1}(qe^{-\gamma}) - \alpha pqe^{-\gamma} - \beta & x = 1 \\ \text{not spray: } f_{k+1}(q) - \alpha pq & x = 0 \end{cases}$$

4. The boundary condition states that pY_0 is the maximum revenue

$$f_3(q) = f_3(qe^{-\gamma}) = f_3(qe^{-2\gamma}) = pY_0$$

Graphically, the problem can be portrayed graphically as in Fig. 24.12. To numerically solve the dynamic programming problem, we make the following arbitrary assignments: $Y_0 = 10$ lb, $p = \$2$, $\alpha = 0.05$, $q = 50$, $\beta = \$4$, and $\gamma = 1$. Solving backward, we have at stage 2

$$f_2(50) = \max \begin{cases} \text{spray (node } 2 \longrightarrow 5 \text{): } f_3(50e^{-1})\text{-}4\text{-}(0.05)(2)(50)e^{-1} \\ \text{not spray (node } 2 \longrightarrow 4 \text{): } f_3(50)\text{-}(0.05)(2)(50) \end{cases}$$

$$= \max \begin{cases} \text{spray: } 14.16 \\ \text{not spray: } 15.0 \end{cases} = 15.0 \text{ (not spray)}$$

$$f_2(50e^{-1}) = \max \begin{cases} \text{spray (node } 3 \longrightarrow 6 \text{): } f_3(50e^{-2})\text{-}4\text{-}(0.05)(2)(50)e^{-2} \\ \text{not spray (node } 3 \longrightarrow 5 \text{): } f_3(50e^{-1})\text{-}(0.05)(2)(50)e^{-1} \end{cases}$$

$$= \max \begin{cases} \text{spray: } 15.32 \\ \text{not spray: } 18.16 \end{cases} = 18.16 \text{ (not spray)}$$

and at stage 1

$$f_1(50) = \max \begin{cases} \text{spray (node } 1 \longrightarrow 3 \text{): } f_2(50e^{-1})\text{-}4\text{-}(0.05)(2)(50)e^{-1} \\ \text{not spray (node } 1 \longrightarrow 2 \text{): } f_2(50)\text{-}(0.05)(2)(50) \end{cases}$$

$$= \max \begin{cases} \text{spray: } 12.32 \\ \text{not spray: } 10.0 \end{cases} = 12.32 \text{ (spray)}$$

Hence, the optimum solution is to spray at stage 1 (i.e., $x_1 = 1$) and to not spray at stage 2 (i.e., $x_2 = 0$), and the return is 12.32. As a check, the brute-force enumeration gives the following results: $x_1 = 0$, $x_2 = 0$, return $= 10.0$; $x_1 = 1$, $x_2 = 0$, return $= 12.32$; $x_1 = 0$, $x_2 = 1$, return $= 9.16$; and $x_1 = 1, x_2 = 1$, return $= 9.48$. This agrees with the result using dynamic programming and can be visualized in Fig. 24.13.

The advantage of dynamic programming over brute-force enumeration is not evident from the above simple example. The power of the technique

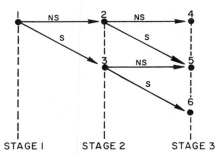

Figure 24.12 A dynamic programming model where the decisions are to spray (S) or not to spray (NS) at stages 1 and 2 in order to maximize yield at stage 3.

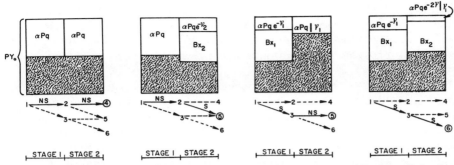

Figure 24.13   The graphical solution to the various possible spray combinations of the dynamic programming model of Fig. 24.12.

can be shown by considering an equipment replacement problem, where a decision of whether to replace or not to replace is made monthly, and the objective is to minimize cost over 20 yr. If we have available a digital computer which takes 10^{-5} sec to perform an addition or comparison, then the computer solution of the dynamic programming problem takes $1\frac{1}{2}$ sec. However, if we were to use brute-force enumeration, the same computer would require 10^{62} yr (Dreyfus & Law 1977). The machine replacement dynamic programming problem requires only one state variable and was efficiently solvable, but as the number of state variables increases relative to the number of possible decisions, the advantage of dynamic programming over brute-force enumeration disappears. Consider the following general crop pest management problem

$$\max_{x_1,\ x_2,\ \ldots,\ x_{20}} \quad B(\cdot, x_1, x_2, \ldots, x_{20}) - C(x_1, x_2, \ldots, x_{20})$$

where $B(\cdot)$ and $C(\cdot)$ are as defined in the text and the decision to apply or not to apply pesticides at the ith time interval is indicated by x_i. Furthermore, suppose that in the dynamic programming formulation of the problem, it is necessary to retain as state variables 10 age classes each of leaves and fruit, and each age class has 10 possible values. For this problem, then, the brute-force enumeration requires 2^{20} at 1.05×10^6 calculations of $(B - C)(\cdot)$, and the dynamic programming solution takes $2 \times 20 \times 10^{20}$ or 4×10^{21} function value computations. Hence, the applicability of the technique of dynamic programming to crop pest management depends among other things on the separability of the crop ecosystem simulation model from the optimization problem. In cases where the amount of damage can be determined from the number of pests alone, dynamic programming can probably be used to solve efficiently the optimization problem. If, however, the optimal value function depends also on the state of the plant, the dynamic programming procedure would most likely be impractical. Hence, the major restriction of dynamic programming to complex problems is the inability to handle even a moderate number of state variables. In cases where the dynamic programming procedure is not feasible, methods of nonlinear programming mentioned in previous sections might be more suitable.

From this brief introductory coverage of some existing optimization methods which might be applicable to crop or animal pest management, we conclude that the process of formulating optimization problems and selecting appropriate methods for solution is at this stage of development as much an art as it is a science. Nevertheless, the progress that has been made has led to practical benefits and offers greater possibilities for the future.

REFERENCES

Abadie, J. and J. Guigou. 1969. *Gradient Reduit Generalise*. HI 069/2 Unpub. Memor., Electricite du France.

Barr, R. O., P. C. Cota, S. H. Gage, D. L. Haynes, A. N. Kharker, K. Y. Lee, W. E. Ruesink, and R. L. Tummala. 1973. *Mem. Ecol. Soc. Aust.* **1**: 241–264.

Bellman, R. E. 1957. *Dynamic Programming*. Princeton University Press, Princeton, N.J.

Bellman, R. E. and S. E. Dreyfus. 1962. *Applied Dynamic Programming*. Princeton University Press, Princeton, N.J.

Carson, R. 1962. *Silent Spring*. Houghton-Mifflin, Boston.

Cauchy, A. L. 1847. *C. R. Read Sci., Paris* **25**: 536–538.

Clark, W. C., D. D. Jones, and C. S. Holling. 1977. *Lessons for Ecological Policy Design: A Case Study of Ecosystem Management*. IRE R-10-B.

Conway, G. R. (Ed.). In press. *Proc. Conf. Pest Management. ITASA, Laxenburg, Austria, November, 1979*. Wiley, New York.

Conway, G. R. 1976. Pp. 257–281 in R. M. May (Ed.), 1976, *Theoretical Ecology*. Blackwell, Oxford.

Conway, G. R. 1977. *Nature* **269**: 291–297.

Conway, G. R., G. A. Norton, A. B. S. King, and N. J. Small. 1975. Pp. 193–229 in G. E. Dalton, *Study of Agricultural Systems*. Applied Science Publishers, London.

Croft, B. A. 1981. *Pest Management for Deciduous Tree Fruits and Nuts*. Plenum, New York.

Croft, B. A., S. M. Welch, D. J. Miller, and M. L. Marino. 1979. Pp. 223–250 in Croft (Ed.), 1979, referenced here.

Curry, G. L., R. M. Feldman, and K. C. Smith. 1978. *Theor. Popul. Biol.* **13**: 197–213.

Curry, G. L., P. J. H. Sharpe, D. W. DeMichele, and J. R. Cate. 1980. *J. Environ. Manag.* **11**: 187–224.

Dantzig, G. B. 1951. *Maximation of a Linear Function of Variable subject to linear Inequalities, Activity Analysis of Production and Allocation*. Wiley, New York.

Davidon, W. C. 1959. *Variable Metric Method for Minimization*. A.E.C. Res. and Develop. Rep. ANL-5990.

Dreyfus, S. E. and A. M. Law. 1977. *The Art and Theory of Dynamic Programming*. Academic, New York.

Ehler, L. E., K. G. Eveleens, and R. van den Bosch. 1974. *Environ. Entomol.* **2:** 1009–1015.

Falcon, L. A., R. van den Bosch, J. Gallagher, and A. Davidson. 1971. *J. Econ. Entomol.* **64:** 56–61.

Feder, G. and U. Regev. 1975. *J. Environ. Econ. & Manag.* **2:** 75–91.

Fick, G. W. 1975. *ALSIM 1 (Level 1)—Users' Manual.* Agron. Mimeo 75-20. Dept. of Agron., Cornell University, Ithaca, N.Y.

Fletcher, R. and M. J. D. Powell. 1963. *Computer J.* **6:** 163–168.

Fletcher, R. and C. M. Reeves. 1964. *Computer J.* **7:** 149–154.

Forsyth, G. 1955. *J. Soc. Indus. Appl. Math.* **3:** 173–178.

Frazer, B. D. and N. E. Gilbert. 1976. *J. Entomol. Soc. Br. Columbia* **73:** 33–56.

Getz, W. M. and A. P. Gutierrez. 1982. *Annu. Rev. Entomol.* **27:** 447–466.

Gilbert, N. 1980. *J. Anim. Ecol.* **49:** 351–380.

Gilbert, N. and A. P. Gutierrez. 1973. *J. Anim. Ecol.* **42:** 323–340.

Gilbert, N. E., A. P. Gutierrez, B. D. Fraser, and R. E. Jones. 1976. *Ecological Relationships.* Freeman, San Francisco.

Goldstein, A. A. 1962. *Numer. Math.* **4:** 146–150.

Goldfeld, S. M., R. E. Quandt, and H. F. Trotter. *Econometrica* **34:** 541–551.

Gutierrez, A. P. 1976. *Management of Cotton Pests.* Proc. EPPO Congr., Paris, Oct. 1976.

Gutierrez, A. P. and J. U. Baumgaertner. In Press *Can. Entomol.*

Gutierrez, A. P., J. U. Baumgaertner, and K. S. Hagen. 1981. *Can. Entomol.* **113:** 21–33.

Gutierrez, A. P., G. D. Butler, Y. Wang, and D. Westphal. 1977. *Can. Entomol.* **109:** 1457–1463.

Gutierrez, A. P., R. Daxl, G. L. Quant, and L. A. Falcon. 1981. *Environ. Entomol.* **10:** 872–879.

Gutierrez, A. P., L. A. Falcon, W. B. Loew, P. A. Leipzig, and R. van den Bosch. 1975. *Environ. Entomol.* **11:** 1–20.

Gutierrez, A. P., D. E. Havenstein, H. A. Nix, and P. A. Moore. 1974. *J. Appl. Ecol.* **11:** 1–20.

Gutierrez, A. P., T. F. Leigh, Y. Wang, and R. D. Cave. 1977. *Can. Entomol.* **109:** 1375–1386.

Gutierrez, A. P., H. Á. Nix, D. E. Havenstein, and P. A. Moore. 1974. *J. Appl. Ecol.* **11:** 21–35.

Gutierrez, A. P. and U. Regev. 1983. *Acta Oecologica.* **4:** 271–287.

Gutierrez, A. P., U. Regev, and H. Shalit. 1979. *Environ. Entomol.* **8:** 101–107.

Gutierrez, A. P., C. G. Summers, and J. U. Baumgaertner. 1981. *Can. Entomol.* **113:** 21–33.

Gutierrez, A. P., Y. Wang, and U. Regev. 1979. *Can. Entomol.* **111:** 41–54.

Hall, D. C. and R. B. Norgaard. 1973. *Am. J. Agric. Econ.* **55:** 198.

Hartstack, A. W. Jr., J. A. Witz, J. P. Hollingworth, R. L. Ridgeway, and J. D. Lopez. 1976. *Mothzv-2: A Computer Simulation of Heliothis zea and Heliothis virescens Population Dynamics. Users' Manual.* ARS-5-127.

Hassell, M. P. 1980. *J. Anim. Ecol.* **49:** 603–628.

Headley, J. C. 1972. Pp. 100–108 in National Academy of Science, 1972, referenced here.

Holling, C. S. 1965. *Mem. Entomol. Soc. Can.* **45:** 1–60.

Holling, C. S. 1973. *Annu. Rev. Ecol. & Syst.* **4:** 1–23.

Hughes, R. D. 1963. *J. Anim. Ecol.* **32:** 393–426.

Hughes, R. D. and N. Gilbert. 1968. *J. Anim. Ecol.* **37:** 553–563.

Jones, D. D. 1976. Pp. 91–115 in Norton and Holling (Eds.), 1976, referenced here.

Jones, D. D. 1977. *Simulation* (July 1977): 1–15.

Jones, R. E. 1977. *J. Anim. Ecol.* **40:** 195–212.

Jones, R. E. and P. M. Ives. 1979. *Aust. J. Ecol.* **4:** 75–86.

Kelly, J. E. Jr. 1960. *SIAM J.* **8:** 703–712.

Leon, A. 1965. in T. Vogl and A. Lavi (Eds.), *Proc. Symp. on Recent Advances in Optimization Techniques.* Wiley, New York.

Leslie, P. H. 1945. *Biometrika* **33:** 183–212.

Leslie, P. H. 1948. *Biometrika* **35:** 213–245.

McKinnion, J. M., J. W. Jones, and J. D. Hesketh. 1974. *Proc. 1974 Beltwide Cotton Prod. Res. Conf., Nat. Cotton Council, Memphis, Tenn.*

Mann, S. H. 1971. *Biometrics* **27:** 357–368.

May, R. M. 1973. *Stability and Complexity in Model Ecosystems.* Princeton University Press, Princeton, N.J.

Morris, R. F. 1959. *Ecology* **40:** 580–588.

Morris, R. F. (Ed.). 1963. *Mem. Ent. Soc. Can.* **31:** 1–332.

Morris, R. F., C. A. Miller, D. O. Greenback, and D. G. Mott. 1958. *Proc. 10th Int. Congr. Entomol.* **4:** 137–149.

Nelder, J. A. and R. Mead. 1965. *Computer J.* **7:** 308–313.

Norton, G. A. and C. S. Holling. 1976. *Proc. IIASA Workshop on Pest Management Modeling.* Pergamon, Oxford.

Odum, E. P. 1971. *Fundamentals of Ecology.* Saunders, Philadelphia.

Peart, R. M., R. T. Huber, W. A. Blake, D. F. Wolf, and G. E. Miles. 1974. *Paper No. 74-5043,* Am. Soc. Agric. Eng., St. Joseph, Mich.

Peterman, R. M., W. C. Clark, and C. S. Holling. 1978. Pp. 321–341 in R. M. Anderson and B. D. Turner (Eds.), 1978, *Population Dynamics.* Proc. Symp. Brit. Ecol. Soc., April 1978. Blackwell, Oxford.

Polak, E. 1973. *SIAM Rev.* **15**(2): 553–584.

Powell, M. J. D. 1964. *Computer J.* **7:** 155–162.

Powell, M. J. D. 1970. *SIAM Rev.* **12**(1): 79–97.

Regev, U., A. P. Gutierrez, and G. Feder. 1976. *Am. J. Agric. Econ.* **58:** 187–197.

Regev, U., H. Shalit, and A. P. Gutierrez. 1976. Pp. 281–299 in Norton and Holling (Eds.), 1976, referenced here.

Rosen, J. B. 1960. *SIAM J.* **8:** 181–217.

Rosen, J. B. 1961. *SIAM J.* **9:** 514–532.

Ruesink, W. G. 1976. Pp. 80–89 in R. L. Tummala, D. L. Haynes, and B. A. Croft (Eds.), 1976, *Modeling for Pest Management.* Michigan State University Press, East Lansing.

Sharp, P. J. H. and D. W. de Michelle. 1977. *J. Theor. Biol.* **64:** 649–670.

Shoemaker, C. A. 1973. *Math. BioSci.* **18:** 1–22.

Shoemaker, C. A. 1976a. *Optimal Integrated Pest Management of Age Distributed Populations.* Department of Environ. Eng. Tech. Report 76.2, Cornell University, Ithaca, N.Y.

Shoemaker, C. A. 1976b. Pp. 301–315 in Norton and Holling (Eds.), 1976, referenced here.

Southwood, T. R. E. 1976. *Ecological Methods.* (second edition) Chapman and Hall, London.

Spang, H. A. 1962. *SIAM Rev.* **4:** 343–365.

Spendley, W., G. R. Hext, and F. R. Himsworth. 1962. *Technometrics* **4:** 41.

Stern, V. M., R. F. Smith, R. van den Bosch, and K. S. Hagen. 1959. *Hilgardia* **29:** 81–101.

Stewart, G. W., III. 1967. *J. Assoc. Comput. Mach.* **14:** 72–83.

Stinner, R. E., R. L. Rabb, and J. R. Bradley. 1974. *Environ. Entomol.* **3:** 163–168.

Sutherst, R. W., G. A. Norton, N. D. Barlow, G. R. Conway, M. Birley, and H. N. Comins. 1980. *J. Appl. Ecol.* **16:** 359–382.

Taylor, C. R. and R. D. Lacewell. 1977. *Southern J. Agric. Econ.* **9:** 129–135.

Talpaz, H., G. L. Curry, P. J. Sharpe, D. W. DeMichele, and R. E. Frisbie. 1978. *Am. J. Agric. Econ.* **60:** 469–475.

van den Bosch, R. 1978. *The Pesticide Conspiracy.* Doubleday, New York.

van der Waerden, B. L. 1953. *Modern Algebra.* Frederick Ungar, New York.

Varley, G. C. and G. R. Gradwell. 1963. *Proc. Ceylon Assoc. Adv. Sci.* **18:** 142–156.

Walker, B. H., G. A. Norton, G. R. Conway, H. N. Comins, and M. Birley. 1978. *J. Appl. Ecol.* **15:** 481–502.

Wang, Y., A. P. Gutierrez, and G. Oster. 1977. *Can. Entomol.* **109:** 1359–1374.

Wang, Y. and A. P. Gutierrez. 1980. *J. Anim. Ecol.* **49:** 435–452.

Watt, K. E. F. 1963. *Can. Entomol.* **95:** 525–536.

Watt, K. E. F. 1968. *Ecology and Resource Management: A Quantitative Approach.* McGraw Hill, New York.

Welch, S. M., B. A. Croft, J. F. Bruner, and M. F. Michels. 1978. *Environ. Entomol.* **7:** 487–494.

Wiegert, R. G. 1976. *Benchmark Papers in Ecology: Ecological Energetics.* Dowden, Hutchingson & Ross, Stroudsburg, Pa.

Wilde, D. J. 1964. *Optimum Seeking Methods.* Prentice-Hall, Englewood Cliffs, New Jersey.

Yorque, R., G. L. Baskerville, W. C. Clark, C. S. Holling, D. D. Jones, and C. S. Miller. 1979. *Ecological Policy Design: A Case Study of Forest, Insects, and Managers.* Wiley, New York.

Zahler, R. S. and H. J. Sussman. 1977. *Nature* **269:** 759–763.

Zoutendijk, G. 1960. *Methods of Feasible Directions: A Study in Linear and Nonlinear Programming.* Elsevier, Amsterdam.

Chapter 25

Applications of Ecology for Better Pest Control

B. A. CROFT, P. L. ADKISSON, R. W. SUTHERST, and G. A. SIMMONS

25.1 INTRODUCTION

Lack of a sound ecological foundation for insect pest control programs has been a critique of many applied entomologists and basic insect ecologists. This base was especially lacking during the pest control era of the 1940s–1960s when much information on the biology of insect pests and their natural, cultural, and other nonchemical controls was inadequately used.

Such critiques led to constructive modifications and wide acceptance of integrated pest management (IPM) as an improved philosophy of pest control (Stern et al. 1959, Huffaker & Croft 1976). Still, the application of basic ecological theory to insect control has been limited (Levins & Wilson 1980). In some cases this is with valid reason. Many agricultural crops are deliberately maintained as preclimax monocultures to secure maximum profit. They are thus in an unstable ecological state. The crop–insect relationships in such a simplified ecosystem do not attain a more natural or primeval state of balance (Chapter 22) but continue in a state of flux. Conditions often are too ephemeral and disruptive to permit long-term study of those ecological relationships that could provide a more ecologically based program. This research therefore was neglected and short-term solutions providing short-term benefits were developed. Long-term ecological perspectives as to the consequence of these actions often were lacking.

Despite these general failings, applied entomology has contributed much to basic insect ecology. For example, the roles of predation (predator or parasitoid) in regulating a prey (host) insect population has been clearly demonstrated by biological control specialists (e.g., Huffaker & Messenger 1976). DeBach and Sundby (1963), Huffaker and Kennett (1956), and others have shown that when competitive displacement of one host-specific natural enemy by another occurs, the one displaced has been the inferior control agent, and biological control has been improved. Parker et al. (1971) showed that for effective natural enemy action, continuity of prey (hosts) is essential and that this may be accomplished by small-scale releases of the host (pest itself) at times of great host scarcity. Other studies have shown that the habitat (van den Bosch & Telford 1964, Doutt & Nakata 1965) or the host plant (Rabb & Bradley 1968) can be managed to improve natural control through either increased natural enemy efficiency or host plant resistance (Chapter 23). Pest control systems based on such principles have been implemented in recent years and provide substantial economic, environmental, and social benefits.

Implementation of a successful IPM program for a crop based on strong ecolgial underpinnings requires integrating research of many disciplines, (e.g., ecology, entomology, agronomy, economics, and systems science). Usually the farm manager does not view them independently but as part of overall farm management. He must place the costs and benefits of pest control in context with other production elements. He must manage the system in an ecologically effective way so that returns exceed costs for management and investment. Because new insect pests and crop varieties are continually evolving, the pest–crop management system also must evolve. For example, rules for applying pest control measures may require change, with previously used economic thresholds being discarded and new ones established. In the end the basic principles of IPM apply; that is considering the ecosystem and its ecology must be paramount, but as the components vary the tactics must be varied.

The citing of several examples of IPM programs illustrating such approaches provides an appropriate final chapter of this book. Examples from an apple, a cotton, a livestock, and a forest production system are discussed.

25.2   INTEGRATED PEST MANAGEMENT FOR APPLES

Deciduous tree fruits, especially apples, have been fruitful systems for developing and implementing ecologically based IPM programs (Hoyt & Caltagirone 1972, Croft & Brown 1976, Croft 1978) for several reasons. Apples are perennials having ecological diversity and habitat continuity, including a complex of shelter (microhabitat) and food (including ground cover plants) features for a variety of arthropods. This diversity makes apple crops more similar to naturally occurring ecological systems, in contrast especially to annual crops.

Fortuitously, development of insecticide resistance among the major insect pest of apples (e.g., codling moth, *Cydia pomonella;* apple maggot, *Rhagoletis pomonella;* plum curculio, *Conotrachelus nenuphar*) has not occurred during the past 20 yr, even under intensive applications of organophosphate-based insecticides. Due to this long-term use of a single chemical group, a variety of useful natural enemies in apples have developed resistances to many of these materials. They include several phytoseiid mites (Croft 1977), a coccinellid beetle (Colburn & Asquith 1971), and a predatory cecidomiid (Adams & Prokopy 1977). Very likely, other beneficial species have developed resistance which has not been detected. Resistance allows some species to commonly persist and exert a continuing controlling action in commercial orchards where a variety of pesticides (insecticides, acaricides, fungicides, herbicides) are widely used. These organisms have responded to strong selective pressure and reassumed, to an extent, the roles they possess in a more natural ecological system. Study of such resistance features provides insights into the impact of pesticides at trophic levels above the primary consumer level. It has raised questions regarding the introduction of new pesticides (e.g., pyrethroids, carbamates) into an IPM system utilizing such insecticide-adapted species and one that is relatively stable with regard to natural control (Croft & Hoyt 1978). Furthermore, these developments suggest ways to better manage apple pests and natural

enemies relative to developed pesticide resistance based on a broader ecological perspective (Croft & Hoyt 1983).

25.2.1 The Ecology and Strategy of Apple Pest Control

In midwestern (e.g., Michigan, Ontario) and northeastern (e.g., Nova Scotia, New York, Pennsylvania) North America, apple harbors 40–50 potential insect and mite pests (e.g., Croft 1975, Hoyt & Gilpatrick 1976). In
western North America (e.g., California, Washington, British Columbia)
the number of pests attacking apple is fewer, although total pest damage
may be as great as in eastern regions (Hoyt & Gilpatrick 1976).

Pests in either the eastern or western regions can be classified into functional economic-ecological categories according to the severity, timing, and
frequency of attack, whether they attack the fruit directly or only other
parts of the tree, and according to their potential for control by predators,
parasites, or diseases (cf. Croft 1982).

Key pests of apple are species like the codling moth, apple maggot, and
plum curculio. They attack the fruit directly and are troublesome each year
and require insecticides, which can adversely affect other potential pest—
natural enemy relationships. Under current production conditions, they
usually have no, or limited, potential for biological control below a tolerable
economic injury level. This accounts for their being consistently troublesome.

Occasional pests differ in being more sporadic in occurrence; they may
be direct or indirect pests and may or may not have associated natural
enemies which can control them. Examples include such lepidoterans as
the eye-spotted bud moth, *Spilonota ocellana,* fruit tree leaf roller, *Archips
argyrospilus,* Oriental fruit moth, *Grapholitha molesta,* lesser apple worm,
Grapholitha prunivora, and green fruit worm, *Lithophane antenna.*

Secondary pests feed mostly on foliage and woody tissues (e.g., mites,
aphids, scales, leafhoppers) and can be tolerated at moderate densities in
apples. They have effective natural enemies which usually control them in
unsprayed orchards. When nonselective pesticides are applied for control
of key pests in commercial orchards, these natural enemies are reduced or
inhibited to a degree that outbreaks of secondary pests occur. Then these
pests also must be controlled by use of chemicals.

25.2.2 Apple Pest Control Tactics

In North American apples, key and occasional pests typically are suppressed
by relatively selective pesticides applied on the basis of improved surveil-
lance and more precise timing of treatments during management "windows" in the growing season (e.g., Welch et al. 1980). Practical commercial
programs based on intensive monitoring for entire pest complexes are now
being used in the major apple-producing states. For key pests and occasional
pests only a few nonchemical methods have been developed that are com-

petitive with pesticides, with respect to efficiency and cost (Croft 1975). The sterile male release method for codling moth control has been extensively evaluated in the Pacific Northwest (Croft & Hoyt 1983). While proven effective, its costs are greater than for insecticidal measures (Hoyt & Burts 1974). Use of pheromones to confuse males and prevent reproduction, suppresses the codling moth, red-banded leafroller, and oriental fruit moth (Mitchell 1977). However, a pheromone control system is not available for any whole pest complex, and it is unlikely that pheromones will replace conventional pesticide applications unless such a control system is developed. Efforts to develop pheromone control systems have stimulated much research on insect communication, dispersal, and mating behavior, which contributes to better use of other control measures.

The greatest advances in pest control on apples within the past two decades have been firmly based in ecological understanding and have focused on the prevention of outbreaks of secondary pests. Because of space limitations and to illustrate the ecological context for any life system component of larger programs, the remainder of this section on apples is confined to control and implementation programs for plant-feeding mites.

25.2.3 Integrated Control of Mite Pests

Integrated mite control was first implemented in Nova Scotia apple orchards during the 1940–50s (Pickett et al. 1958). Widespread interest in the United States gained momentum after the 1960s. Huffaker and Kennett's (1956) work established the value of a predatory mite in providing natural control of a pest mite on strawberries and stimulated studies on the Phytoseiidae, many of which have been shown to be important predators of spider mites (reviewed by McMurtry et al. 1970). Research by Hoyt (1969) did much to stimulate further interest among apple researchers. Today, integrated mite control programs have been developed and to some degree implemented in all major apple growing regions of the United States and Canada. These programs are based on effective control of spider mites by insecticide-resistant predators (see Table 25.1 for the pest mites and natural enemies involved). If unaffected by other nonselective pesticides to which they are not resistant, these predators often provide complete biological control of the pest mites. A variety of techniques using alternative prey, supplemental foods and releases, dispersal aids, and so forth have been developed for managing these species. In most programs possibilities for readjusting predator–prey ratios to favor the predators by using physiologically selective acaricides have been identified. By combining these tactics, effective control is achieved, with minimal pesticidal inputs.

The ecological research involved in development of the programs listed in Table 25.1 was extensive. Reviews by Huffaker et al. (1969, 1970) summarized much of the early work. An NSF/IPM research report on tree–fruit pest management contains a review of more recent ecological work

Table 25.1 Major integrated mite control programs on apple in North America

State or province	Principal pest controlled[a]	Principal natural enemies[b]	Research reference[c]	Implementation reference
Washington	Tm, Pu, As	To, Spp. Zm	Hoyt 1969	Hoyt et al. 1970
New York	Pu, Tu	Af, Tp	Watve & Lienk 1977	Tette 1977
Michigan	Pu, Tu, As	Af, Sp, Ag, Zm	Croft & McGroarty 1977	Croft 1975
California	Tp, Tu, Pu	To, Ss	Croft & Barnes 1971	
Pennsylvania	Pu, Tu	Sp, Af	Asquith 1971	Asquith 1972
North Carolina	Pu, Tu, As	Af, Sp	Rock & Yeargan 1971	Rock 1972
Virginia–W. Va.	Pu, Tu	Af	Clancy & McAllister 1968	
Illinois	Pu, Tu	Af	Meyer 1974	
Oregon	Pu	To, Tp	Zwick 1972	
Missouri	Pu, Tu	Af	Poe & Enn 1969	
Utah	Tu, Tm, Pu, As	To, Zm	Davis 1970	
Colorado	Tu, Pu	To	Qist 1974	
New Jersey	Pu, Tu, As	Af	Swift 1970	Christ 1971
Ohio	Pu, Tu, As	Af, Zm, Ag	Holdsworth 1968	Holdsworth 1974
British Columbia	Tm, Pu, As	To, Tp, Zm	Downing & Molliet 1971	Downing & Aarand 1968
Quebec	Pu, Tu	Af	Parent 1967	
Nova Scotia	Pu, As	several	Sanford & Herbert 1970	Anonymous 1970

[a]Pest of importance in each area listed sequentially: Tm = *Tetranychus mcdanieli*, Pu = *Panonychus ulmi*, As = *Aculus schlechtendali*, Tu = *Tetranychus urticae*, Tp = *Tetranychus pacificus*.

[b]Natural enemy of importance in each area listed sequentially: To = *Typhlodromus occidentalis*, Sp = *Stethorus punctum*, Zm = *Zetzellia mali*, Af = *Amblyseius fallacis*, Tp = *Typhlodromus pyri*, Ag = *Agistemus fleschneri*, Ss = *Scolothrips sexmaculata*, Spp = *Stethorus picipes*.

[c]References for this table are in Croft and Hoyt (1983).

768

(Croft and Hoyt 1983). This research covers a wide spectrum of ecological concepts associated with predator–prey relationships in apples: single species life tables (e.g., Tanigoshi et al. 1975a,b, Rabbinge 1976, Croft 1972); theoretical and applied predator–prey models (reviewed by Welch 1979); functional and numerical response relationships of mite predators to their spider mite prey. Franz 1974, Santos 1976, Blyth & Croft 1976, Rabbinge 1976); interspecific interference among predators (Hoyt 1970); intraspecific interference between prey and predators (Hoyt 1970, Franz 1974, Croft 1972); predator–prey switching features (Santos 1976); alternative prey and their effects on pest regulation (Hoyt 1969, Croft 1975, Croft & Hoyt 1983); distributional and searching coincidence of predators and prey (Hoyt 1969, Croft, Welch, et al. 1976, Franz 1974); relative contribution of predator life stages to total predation effects (Rabbinge 1976); influence of stage-specific feedings on predation (Franz 1974, Croft 1972); and competitive displacement among mite pests (Croft & Hoying 1977).

25.2.4 Implementation Features

Figures on the extent of commercial adoption of integrated mite management for the apple programs listed in Table 25.1 are difficult to obtain. However, the impact of these programs has been industry wide in Washington, Pennsylvania, and Michigan, where the greatest effort has gone into research and implementation. Where less effort has been made, the degree of use varies locally. Implementation in all states is affected by the status of pest control for other major orchard pests (e.g., apple scab, tentiform leafminer, white apple leafhopper).

Methods for implementing integrated mite management vary greatly. A simple approach is used in Nova Scotia, where growers are informed as to which pesticides are sufficiently selective to allow predators to exert appreciable biological control. No specific counting or other monitoring of predator or prey populations is recommended. In Michigan mite counting procedures are used for predators and prey, based upon grower request (Croft 1982, Tummala et al. 1976, Welch et al. 1979). An *index* based on predator–prey counts, or a population dynamics *model* incorporating current weather inputs, may be used to forecast probabilities for desired levels of biological control or whether selective sprays are necessary. A computerized extension delivery system (PMEX) integrates biological and environmental information with the mite management model and allows for telecommunication access by distributed computers and users from remote keyboard terminals located in strategic areas throughout the state's fruit belt (Croft et al. 1976, 1979).

In Michigan, as in most areas, several less extensive mite control implementation techniques are available which meet the needs of a wide spectrum of users. The goal is to increase the efficacy of the pest management operations using increased levels of implementation technology. Generally

speaking, the greatest benefits and most effective controls are achieved
when detailed monitoring and short-term forecasting are utilized. There
is also an increasing cost associated with these more technologically advanced programs. Therefore, the cost–benefits of these operations must
be carefully evaluated. Research to determine these trade-offs have been
made for mite control programs (Welch et al. 1980, Croft 1978, Edens &
Klonsky 1977).

While the economic and environmental benefits of integrated mite control are still being evaluated, some preliminary data are available. Washington scientists have measured the cost/acre of pesticides used for insect
and mite control for a 1000-acre block using a standard (1957–1966) versus
an IPM program (1967–1975) (Fig. 25.1). Data for 1970–1975 represents
the general trend for the 5-yr period (S. E. Hoyt, pers. comm.). Total control
costs have declined from about \$60 to \$20–30/acre despite inflation; of
the total reduction, the major savings have come from nearly complete
elimination of summer acaricide sprays. Similar success has been realized
in Pennsylvania and Michigan, although the present percentage reduction
in total pesticide use (30–40%) has been less than that in Washington due
to the greater complex of insect pests and diseases and smaller mite control
costs in relation to the total spray bill. Based on the *Stethorus punctum–
Panonychus ulmi* IPM system, acaricide use in Pennsylvania has been reduced
by 75% over the past 12 yr. In Michigan over a 6-yr period, use of *Amblyseius
fallacis* in an experimental plot gave about 90% reduction in mite control
costs, from about \$20–30 to \$2.50/acre/season. Over the same period, a

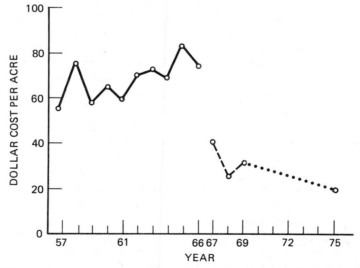

Figure 25.1   Cost/acre for mite control in a 1000-acre block of apples in Washington during
the period 1957–1975. (Adapted from Hoyt and Caltagirone 1971.)

60% decline in costs to about \$8/acre (in 1976) was attained in grower-implemented commercial orchards (Croft 1978).

25.3 INTEGRATED PEST MANAGEMENT FOR COTTON IN TEXAS*

Much of the success of IPM in row crops stems from its application in cotton (Reynolds et al. 1975, Bottrell & Adkisson 1977), following years of disasterous insecticide-oriented programs. The experience in cotton suggests possibilities for other annual or heavily disturbed crops. (cf. van den Bosch et al. 1976).

25.3.1 The Cotton Plant in Texas

In the important boll weevil–bollworm infested areas of Texas, the cottons grown are of the upland varieties, *Gossypium hirsutum*. Coventional *G. hirsutum* requires 180 days from planting to harvest. Recently developed short-season varieties can be matured in less than 140 days even under limited irrigation or rainfall (Namken & Heilman 1973).

Growth and development of cotton may be divided into several overlapping phases. The first is from plant establishment to first square, when the plant is in a strong vegetative stage and is establishing the root and branching pattern (25–45 days after planting). The next phase, fruit formation, begins with the first square (flower bud) and continues with some vegetative growth until the first boll is open. In Texas this phase requires about 105 days. The third phase is from boll opening until the crop is made. The new short-season varieties are capable of setting bolls amounting to 1–1.5 bales/acre during the first 40–50 days of flowering. This is the time of greatest vulnerability to insect attack, and strategies emphasize holding pest insect numbers at low levels during this period.

Under the best Texas conditions, and in the absence of pest insects, cotton sheds 50% of all squares and young bolls. Thus, the plant has tremendous capacity to compensate for insect-caused losses of squares (flower buds) and small bolls. Considerable loss of fruit may cause *no* loss of yield or delay in maturity, that is, if the loss occurs during early fruiting. Also, once a boll is 8–10 days old it will remain on the plant to maturity unless ravaged by an insect or disease organism. After it reaches 12–14 days of age it is safe from boll weevils. The plants essentially run a race to set more fruits than the insects remove. In most years the plant wins if grown under the best management program. Study of the ecology of the plant itself, relative to potential pest damage to it and its interaction with the pests, has been a major focus of IPM research.

*Others contributing to this section were W. L. Sterling, J. K. Walker, and R. E. Frisbie, Department of Entomology, Texas A&M University, College Station.

25.3.2   The Pest Complex

Texas cotton is attacked by many insects, the most important being the boll weevil, *Anthonomus grandis;* cotton fleahopper, *Pseudatomosceli seriatus;* pink bollworm, *Pectinophora gossypiella;* bollworm, *Heliothis zea;* and tobacco budworm, *H. virescens.*

The boll weevil feeds on squares and bolls and has three summer generations. The diapausing stage is the weak link in its seasonal history. Overwintering weevils are primarily controlled by prompt postharvest destruction of cotton stalks and insecticide treatments (Adkisson 1972). Cotton fleahoppers damage terminals of young, rapidly growing plants and cause squares to "blast" or shed. Their control with insecticides, however, is usually required only during a 10–14-day period just before and during early squaring (Talpaz & Frisbie 1975). The pink bollworm also attacks squares and bolls but prefers bolls. Feeding larvae destroy developing lint. Again, the overwintering, diapause stage is considered the weak link. A variety of cultural practices (e.g., early, uniform planting, early maturing varieties, use of defoliants and dessicants, and area-wide stalk destruction) reduce diapausing larvae in the fall and early emerging moths in the spring (Adkisson & Gaines 1960). The bollworm, *H. zea,* attacks cotton, corn, sorghum, soybeans, tomatoes, and peppers. On cotton, the larvae feed on squares and bolls in early season. It has many natural enemies (e.g., spiders, *Geocoris* sp., *Orius insidius*), which can keep numbers on cotton below economic levels unless upset by pesticides. The tobacco budworm, *H. virescens,* is almost identical to the bollworm in appearance, damage, and potential for control by natural enemies, but unlike the bollworm it is a pest in late season and is much more resistant to insecticides (Nemec & Adkisson 1969).

In central, southern, and eastern Texas, the boll weevil and fleahopper are key pests and attain damaging numbers almost every year unless controlled by insecticides. They are *key* to IPM because control of them may induce a complex of secondary pests, primarily *Heliothis* spp. Direct cultural control measures also must be applied against the pink bollworm to keep populations below damaging densities. The bollworm and budworm, although potentially major pests, are secondary, or occasional pests. They do not usually reach damaging numbers unless released from biological control through decimation of natural enemies by insecticides applied against key pests (above) or unless environmental conditions (e.g., weather, plant condition) become more favorable for their increase than for their natural enemies. The life histories and basic ecology of these pests are reviewed by Adkisson (1972), Walker and Niles (1971), Sterling and Adkisson (1971), and Gaylor and Sterling (1976).

25.3.3   The Rationale of Cotton Pest Control

Chemical control of boll weevil and cotton fleahopper always must be considered with regard to impact on biological control of budworm and bollworm, and especially the former which has such high insecticide resistance (Adkisson 1972). It matters little that the weevil and fleahopper can be controlled easily with insecticides (the dominant 1950s criterion) if the timing and/or the material used lead to an expensive multiapplication chemical program for the budworm that, even then, may not control the pest.

Texas farmers realize that *economic gain* is not necessarily synonymous with the *size* of the cotton yield, or how much money they invest in the crop, but is simply the amount of profit at the end of the season. Profit, within the required conformity to environmental and health constraints imposed by regulations, is the foundation on which insect pest management strategies are built. In simplest terms the problem is how to manage the crop and key pests (the boll weevil, cotton fleahopper, and pink bollworm) while avoiding natural enemy disturbances that induce outbreaks of the bollworm and budworm. This becomes the strategy of IPM, and a variety of tactics are required.

25.3.4 The Strategy and Tactics

As noted, research over many years has shown that boll weevils can be controlled with least disruption to the ecosystem by reducing diapausing adults. This is accomplished by promoting early crop maturity, use of desiccants and defoliants to terminate the vegetative and fruiting phases and causing early shedding of fruit, insecticidal treatment during the harvest, and stalk destruction, followed immediately by plowing crop residues. Of these, early maturity of the cotton and stalk destruction to remove late season food are the most important. Studies have shown that if the cotton can be matured, harvested, and the stalks destroyed area-wide in September and early October, boll weevil infestations the next year will be negligible (Brazzel 1961, Adkisson 1969). This combination of cutural practices is also effective against the pink bollworm, and insecticides are almost never needed for its control (Adkisson & Gaines 1960).

Other pest control measures based on manipulation of the cotton plant have evolved. Walker and Niles (1971) theorized that if a rapid fruiting, early maturing, short-season variety could be developed, it could make an average yield without late-season insecticide treatments, provided the boll weevil could be suppressed until its second generation. Scientists subsequently developed the Tamcot SP short-season cotton to fulfill this requirement (Bird 1978).

A cotton production system has been developed whereby 1–3 early season insecticide treatments are applied when overwintered weevils pose a problem. These treatments so reduce the F_1 weevils that plants can set a full yield of bolls before the F_2 weevils emerge. Even if the F_2 emerges in large numbers, control generally is not required even though weevil-infested squares and bolls of fruit formed in *late season* (not used for full yield) may be abundant. These fruit will normally be shed regardless of attack because early boll set already has matched the plant carrying capacity. This system allows development of large populations of the many arthropod enemies of the bollworm and tobacco budworm, which generally keep these two pests at low levels thereafter without insecticides (Parker 1978).

A second key problem is how to control cotton fleahopper without inducing bollworm or budworm outbreaks. Fortunately, early season insecticide treatments (methyl parathion at 0.1 kg/ha) for boll weevil also control fleahoppers. If boll weevil is not a problem, then lower dosages of selective insecticides (dimethoate at 0.1 kg/ha or trichlorfon at 0.28 kg/ha) will suppress the fleahopper, leaving many natural enemies of the bollworm and budworm (Sterling & Haney 1973). In addition, advantage is taken of the ability of cotton to compensate for square and early boll loss by doubling the old *economic threshold* once used to initiate treatments against fleahoppers, provided there is sufficient soil moisture to support plant compensation. If the cotton is fruiting well, producers avoid treatment irrespective of fleahopper numbers. In any case treatments are terminated 2–3 weeks before the cotton flowers to allow natural enemies of the bollworm and budworm to immigrate and establish in fields.

Cotton producers today do not react to bollworm–budworm infestations that would trigger insecticide treatment 20 yr ago. Yet, there are fields where, for various reasons, these pests reach damaging numbers despite the presence of their enemies and insecticide treatment is recommended. However, through research and experience, the old economic threshold for treatment of bollworms and budworms has been considerably raised, as for fleahoppers (above). On flowering cotton the economic threshold for *Heliothis* species has been changed, from "when 4 to 5 young larvae and eggs are present/100 plant terminals" to "avoid treatment unless 10–15% of the large squares or 5% of the small bolls are damaged." This strategy avoids unnecessary treatments that kill natural enemies and place the farmer on a treadmill of insecticide treatments.

Advent of the synthetic pyrethroid insecticides, which are effective against third- to fourth-instar larvae of resistant bollworms and budworms, has provided an additional tactic. These insecticides allow for use of microbials, *Bacillus thuringiensis* and *Baculovirus heliothis,* against the two pests when it appears that they may outstrip their natural enemies. Although these microbials are not as effective as the pyrethroids, their value lies in their ability to suppress the pests while conserving their natural enemies. Therefore, if the economic threshold for the bollworm and budworm is exceeded, one to two microbial treatments are applied. If these treatments (and the natural enemies) do not control the infestation, the producer may use the pyrethroids to kill the surviving third- to fourth-instar larvae. This is provided, of course, that the expected crop value gain is greater than the costs of application. The grower also may let the infestation go untreated, taking his chances that natural enemies will control the pests and the plants will produce additional cotton to compensate for the *Heliothis* damage. This choice may often be successful with short-season cottons.

Thus, profitable cotton production requires a high degree of management of the crop and the insects, both the pests and their enemies. The management system based on short-season varieties has improved profitability while reducing use of insecticides by more than 50%. Considerable improvement is anticipated as new and more pest-resistant varieties now being developed are incorporated into the total management package.

25.3.5 Implementing the Pest Management System

The change in views of cotton producers from an attitude that the only way to control their insect pests is by repeated season-long insecticide treatments to where chemicals are used only in a limited manner, has resulted from the realization that (1) the industry was being destroyed by the costs attendant to the continued development and spread of insecticide-resistant pests and (2) increased research has provided a more ecologically compatible, profitable, and less risky alternative.

Yet without effective technology transfer, the basic and applied research developed for IPM would find little practical adoption. To bridge the gap between research and application, a communications network was formed among researchers, extension specialists, private consultants, and cotton producers. The program was initiated in 1972 as a pilot project to provide intensive technical support to producers. In the program entomologists assist producers in organizing area-wide field scouting or monitoring. Data provided farmers through field reports allow for more intelligent control decisions, concerning need for treatment, materials and dosages to use, and so forth. Often the best decision is not to treat at all. Entomologists also provide an IPM education program including information on the profitability of field scouting and demonstrations on how best to manage the entire production system.

A computerized delivery systems (BUGNET) assists entomologists and farmers in predicting peak bollworm–budworm oviposition. The forecasting model, MOTH-ZV, permits 60-day predictions of oviposition by the two *Heliothis* species (Hartstack et al. 1977). This provides farmers with more precision for timing control measures. BUGNET also includes a boll weevil forecasting model and a dynamic economic threshold model and allows for data exchange between IPM programs in the state.

It was through the area-wide, county-based, educational delivery system that the short-season cotton production–insect management system has been extensively implemented by farmers (Frisbie 1977). The best estimate of use of these IPM procedures through 1978 is provided by the cotton acreage planted to the short-season varieties. This amounts to about 40% in the Lower Rio Grande Valley, 90% in the Texas Gulf Coast, southwestern Texas, and Pecos areas, 12% in the Blacklands, and 46% in the Rolling Plains. It is anticipated that within the next few years, most producers in these areas will have adopted the technology.

Table 25.2   A per acre comparison of cotton production costs with alternative cotton pest management practices: Frio County, Texas, 1974[a]

Item	Unit	Typical Systems[b]	Conventional high-input system	Short season (40")[c]	Short season (26")[d]
			Production System		
		Inputs			
Fertilizer	lb	80–40–0	116–62–0	24–24–24	24–24–24
Irrigation	ac in	20	18	12	12
Insecticides	lb	9.6	16.9	6.6	6.6
Total	1000 kcal	3264	3645	2445	2445
Cost	$/ac	278	326	281	279
Cost	¢/lb	47.60	42.56	33.84	26.90
		Production			
Yield	lb/ac	500	625	649	765
Gross[e]	$/ac	340	435	452	532
Net[e]	$/ac	62	109	170	252

[a]From Sprott et al. (1976).
[b]Based on enterprise budget published by the Texas Agricultural Extension Service.
[c]The cotton was planted on conventional rows spaced 40 in. apart.
[d]The cotton was planted with a high density, on rows spaced 26 in. apart.
[e]Based on a cotton price of $0.60/lb for lint and $120/ton for seed.

One factor responsible for the rapid change has been the sound documentation of the economic, environmental, and energy benefits that may be gained, as illustrated in Table 25.2 by data obtained in 1974. The data show that costs to a typical grower/acre of conventional, long-season, irrigated cotton in an area with boll weevils was $278, but the gross return on 1 bale/acre was $340, producing a net profit of $62/acre. The high-input system (used by the "best" cotton producers) used more fertilizer, irrigation, and insecticide than the typical system. The costs also were greater ($326/acre), but yields were increased so that the gross returns were $452/acre, producing a net profit of $109/acre.

The short-season cotton was grown with a greatly reduced fertilizer, irrigation, and insecticide input. The nitrogen was reduced by 70 and 80%, respectively, when compared with high-input systems. Irrigation water was reduced by 33–40% and inseticide use by 31–61%. Yields of the short-season cotton grown in 40-in. row spacing were 149 and 24 lb of lint/acre *greater* than for the high-input system, respectively, with corresponding increases in net profit of $108 and $61/acre. The short-season cotton planted in narrow rows (26-in. row spacings) produced even greater yields and profits (Table 25.2).

Table 25.3 Gross receipts, costs, and net returns per acre for 1973 and 1974 Texas cotton pest management program participants and nonparticipants[a]

	Rio Grande Valley		Blacklands		Trans-Pecos	
	Participants	Non participants	Participants	Non participants	Participants	Non participants
Total acres in sample						
1973	3781	6537	1270	1272	5069	3977
1974	3126	5192	1665	1341	541	2035
No. individuals sampled						
1973	22	22	22	22	14	14
1974	16	16	20	20	12	12
Avg gross yields receipts[b]						
1973	$312.40	$247.75	$211.60	$190.40	$397.13	$363.40
1974	461.66	448.56	124.52	99.04	369.65	354.39
Avg total costs						
1973	$207.91	$198.04	$118.56	$115.31	$271.70	$280.06
1974	293.00	295.63	132.83	126.24	365.54	412.12
Avg net returns[b]						
1973	$104.49	$ 49.18	$ 93.04	$ 75.09	$114.43	$ 83.84
1974	168.65	152.93	− 8.31	− 27.20	4.11	− 57.73
Difference in net return						
1973	+$ 55.31		+$ 17.95		+$ 30.59	
1974	15.73		18.89		61.84	

[a]From Frisbie et al. (1976).

[b]Averages were weighed by acre to account for acreage differences between samples.

Similar results with short-season cotton and narrow row spacing have been produced repeatedly since 1974 in other areas. In all cases above average yields have been made with reductions of 40–80% in fertilizer, insecticide, and irrigation water. This is why the number of farmers switching to this system has increased each year. However, the average cotton farmer has only gone part way in implementing the full IPM system and has not obtained the maximum economic returns possible as measured experimentally. On the other hand they have improved profits by increasing yield and decreasing costs. Table 25.3 shows a comparison of gross receipts, costs, net returns, and differences in net returns for several groups of farmers participating in IPM versus conventional production systems (Frisbie et al. 1976). Profits of IPM participants exceeded those of nonparticipants in each of three demonstration areas during each of the two seasons when compared. Budgets also showed a profit advantage for participants when the crop price was both high (1973) as well as when it was low (1974).

In conclusion, implementation of a successful IPM program requires close cooperation between the extension plant protection–production specialist and the farmer. To be successful, the IPM program must be practical for the farmer to manage and more profitable than the old system. The results presented here for cotton show that farmers will implement such practices when these requirements are met.

25.4 CATTLE TICK MANAGEMENT IN AUSTRALIA

The cattle tick, *Boophilus microplus,* is a blood-sucking parasite of cattle, deer, and buffalo in tropical regions of Africa, Central and South America, and Australia. It is a particular problem in the Americas and Australia where beef and dairy industries use European cattle, *Bos taurus.* On grazing lands of north and north-east Australia, the tick is the major pest, costing an estimated $40 million/yr (Anon. 1975). The ticks reduce liveweight gains and milk yields, downgrading of hides, and transmit tick fever (*Babesia* spp.). Failure to control ticks on European cattle may cause up to 100% mortality, by draining the animals' resources.

B. microplus has a simple life cycle involving only one host. Eggs laid in pastures hatch into larvae which transfer to cattle to feed, mature, and mate before fully fed females drop from their hosts to oviposit. There are 3–6 generations/yr, the number depending on geographic location. Fecundity is high, with each female producing about 2000 eggs. In the spring a small first generation is produced from overwintering eggs, and the population then increases progressively until winter. Weather affects the population densities directly and indirectly by affecting the resistance of the cattle hosts which causes density-dependent tick mortality (Sutherst, Utech, et al. 1979).

25.4.1 Control Methods

Several methods have been developed successively for cattle tick control. Prior to the 1970s, almost total reliance was on acaricides, applied by swimming cattle through dipping vats up to 15 times a year. Almost inevitably, such use led to high costs and to severe cases of developed resistance to acaricides (Wharton & Roulston 1970, Nolan & Roulson 1979). In 1966, and in 1970, new resistant strains emerged and no suitable alternative acaricides were immediately available.

Alternatives to dipping have long been sought. Wilkinson (1957), building on ecological work of researchers in the United States, showed that pasture "spelling" and dipping of cattle between transfers greatly reduced the dippings required each year. Implementation of spelling relied on detailed information on developmental rates of eggs and longevity of unfed larvae (Snowball 1957, Wilkinson 1962, Harley & Wilkinson 1964), on suitable subdivision of pasture, and on long-term planning for rotation of pastures. However, pasture spelling is rarely used on a whole property. More commonly, it has been applied to individual pastures when opportunities have arisen. Elder, Kearnar, et al. (1980) reported that 42% of cattlemen in Queensland practiced some kind of spelling; limited acceptance has been attributed primarily to its detrimental effects on pastures. Thus, extension recommendations to the industry, in favor of spelling, have been tentative until effects on cattle production are better understood.

A second innovation was developed by Norris (1957) who stressed controlling the small, often imperceptible spring generation of ticks in southeast Queensland to prevent buildup later in the year. Information on tick development combined with the knowledge that the parasitic phase of the tick has a 3-week duration produced a procedure termed *strategic dipping*. Six 3-week interval dippings in spring gave excellent control in south-east Queensland and New South Wales. Comparable dippings at similar intervals gave satisfactory results in north Queensland when geared to economic thresholds of 40 ticks engorging on 1 in 10 animals during 1 in 10 days (Harley & Wilkinson 1964, Wharton et al. 1969). Sutherst, Norton, et al. (1979, 1980), using a computer simulation model, showed that spring dipping in the south was marginally advantageous, despite the density-dependent mortality of the parasitic ticks. However, the main benefit of strategic dipping comes from matching dipping intervals to the 3-week duration of the parasitic phase of tick development. The 3-week interval dipping was reported by Elder, Emmerson, et al. (1980) to have been adopted by 20–30% of cattlemen in coastal areas of Queensland.

The major breakthrough in tick control came in the 1960s with revival of interest in host resistance. Resistant animals are those on which only a small proportion of ticks can feed successfully. Resistance is immunological and associated with Zebu cattle. Also, Riek (1962), Francis and Little (1964), and Wharton et al. (1969) showed that Zebu × European cattle were not

Table 25.4 Simulated costs and effectiveness—comparison of control strategies for cattle tick using single tactics and integrated control measures.[a]

	Dipping		Spelling			Integrated Control				
	5-dippings (6,9,12,15, 18)	3-dippings (15,18,21)	Single spelling (4,12)	Rotational spelling[b] (4,12,20, 26,38,46)	Resistant cattle	Single spelling & 2 dippings (5,13)	Rotational spelling & 1 dipping (4,21,20; 26,35,46)	Resistant cattle & 1 dipping (6)	Resistant cattle & single spelling (4,12)	Resistant cattle & rotations spelling (4,12,20, 26,38,46)
Combined Costs in Year 1 ($/head) — Average initial population (100,000 eggs)	4.58	5.09	3.25	4.35	1.85	1.79	4.27	1.63	1.20	3.70
High initial population (500,000 eggs)	5.54	9.81	6.74	5.85	5.01	2.15	5.27	3.05	1.50	3.91
Effect on population in subsequent seasons	+	–	0	–	0	+	+	+	+	+
Robustness	0	–	–	0	0	+	0	0	+	+
Acaricide resistance costs	–	0	+	+	+	0	0	0	+	+
Fencing/feeding costs	+	+	–	–	+	–	–	+	–	–
Handling costs	+	+	+	+	–	+	+	–	–	–

[a] Adapted from Sutherst, Norton et al. (1979). The figures in brackets denote the week(s) when control measures are applied. Unquantified performance criteria are assessed on a three-point scale, where + indicates a favorable performance, 0 a moderate performance, and – an unfavorable performance.
[b] Under the medium pasture constraint.
[c] Dipping carried out on the day 20.

severely affected by the smaller numbers of ticks they carried, even in areas highly favorable for tick propagation. The resistance associated with Zebu-type cattle has since become the basis for control of *B. microplus* in Australia (Wharton et al. 1973, Wharton & Norris 1979, Sutherst & Utech 1980).

Extensive research has been done on several features of tick resistant cattle. Techniques for measuring resistance were described (Wharton & Utech 1970, Wharton et al. 1973, Sutherst et al. 1978) and the association of resistance with different breeds defined (Seifert 1971a, Utech, Wharton, et al. 1978). Acquisition of resistance in response to tick infestation was demonstrated (Roberts 1968, Wagland 1975), together with the heritability of resistance (Wharton et al. 1970, Seifert 1971a, Utech, Seifert, et al. 1978) and its effects on tick populations in the field (Wharton et al. 1969, Sutherst, Wharton, et al. 1979). Also, the extent of damage caused by ticks to tick-resistant breeds was measured in different locations (e.g., Seifert 1971b, Turner & Short 1972, Holyroyd & Dunster 1978). As a result, it is now accepted that cross-bred Zebu × European cattle provide a satisfactory solution to the cattle tick problem in Australia.

The overall approach adopted by the beef industry has been to change to Zebu × European crosses, which grow rapidly and tolerate stress in the tropics and subtropics (Turner 1975, Frisch & Vercoe 1978) and to integrate control methods to produce strategies which are both near-optimal and robust. To define the role of each method, a thorough understanding of the ecology of *B. microplus* in different geographical regions of Australia and the effects of different control strategies (Sutherst & Wharton 1973) is necessary. To achieve this aim, a deterministic simulation model was built which is driven by meterological data and enables integrated effects of dipping, pasture spelling, and host resistance to be investigated (Sutherst & Dallwitz 1979, and Chapter 24).

Initially, the south-east Queensland region was chosen to develop a modeling approach to the design of control strategies. A simplified model was constructed in which parameter values describing the effects of weather were fixed to represent an "average" year in the region (Sutherst, Norton, et al. 1979). A systematic search was then made to compare single tactical versus integrated control approaches on cattle with different resistance levels (Sutherst et al. 1980), with emphasis on finding robust, rather than optimal, strategies. In Table 25.4 a comparison of control strategies using these approaches are summarized. It is envisaged that the weather-driven model will be used to simulate the local dynamics of *B. microplus* in other geographical regions. The parameter values will then be fixed to enable the same searching procedure to be used to define control strategies most appropriate for each region.

25.4.2 Implementation

In New South Wales, a compulsory strategic-dipping program to suppress ticks in a quarantined area is used. However, intensive tick control is not practical in most of northern Australia, which means that resistant cattle have a major contribution to make in the area. In Queensland a regional approach has been adopted, with the state divided into three geographically dissimilar regions—north, central, and southeast. Each region has its own multidisciplinary extension team and control strategies are being worked out within each region.

Adoption of tick-resistant cross-bred cattle is progressing at a high rate since these animals also have higher growth rates and greater heat and drought tolerance than European animals. Adoption rates of Zebu-type cattle range from 48% in the south-east to 90% in the northerly regions (Elder, Kearnan, et al. 1980). The first step in changing to an IPM program—the change of breeds—involves merely the purchase of Zebu bulls. The subsequent steps, aimed at minimizing costs of damage and control measures, are more demanding. Increasing further the phenotypic resistance of tick-resistant herds by culling susceptible animals, or the genetic resistance by selective breeding, require a much greater awareness and skill in cattlemen. Simplified techniques for assessing tick infestations are being developed to rank animals for culling. In practice, moderate dipping programs or simple pasture spelling may be the most appropriate means of recouping the small liveweight gain lost to ticks.

Cattlemen adopting tick-resistant cattle and their associated IPM programs benefit by removing the risk of severe acaricide resistance and by reduced requirements for labor and capital (Sutherst, Sutherland, et al. 1979). Additional benefits include increases in the useful life of acaricides and reduced acaricide residues in beef and milk. Adverse consequences are a slight decline in fertility in Zebu cattle and some deterioration of cattle temperament, even though this latter problem can be countered by appropriate handling. Also, control of the buffalo fly, *Haematobia irritans exigua*, which was enhanced by dipping for tick control, at least prior to the use of amidine acaricides, has declined.

25.5   INTEGRATED PEST MANAGEMENT FOR SPRUCE-FIR FORESTS*

The northern boreal forest extends across southern Canada to the Rocky Mountains, into Alaska and southward into Appalachia. The spruce budworm, (SBW), *Choristoneura fumiferana,* is a native insect, having coevolved with this forest for thousands of years. Unlike many agricultural pests, this insect is a regulator of forest primary production (Mattson & Addy 1975, and Chapter 22). Forests over 30 yr old experience "outbreaks" of budworms resulting in intense defoliation and mortality of balsam fir. Owing to inaccessibility and lack of demand, balsam fir was not heavily utilized by industry until after the 1900s. Thus, SBW outbreaks had minimal economic

*An extensive review by W. C. Clark, Institute of Animal Resource Ecology, U. British Columbia, Canada, was valuable to the completion of this section.

importance. Since 1930, pulp and paper production have become increasingly important in the eastern portions of the boreal forest. This exploitation of the forest eventually placed humans in more direct competition with spruce budworms.

Differing from most agricultural crops, the spruce-fir forest is a longer-term crop. Harvest begins only after about 50 yr of management. Unlike most agricultural insects, the population density for control intervention is high since only populations dense enough to kill trees are of concern. Also, the SBW is usually managed on a macroregional basis employing state, province, or federal cooperative programs. Generally, pest management decisions in agriculture affect production economics that year. By contrast, SBW-related actions may affect a regional economy for decades.

25.5.1 Ecology of The Spruce Budworm–Spruce-Fir Forest Ecosystem

The spruce budworm is univoltine. Eggs are laid in clusters of 5–50 on balsam fir and spruce needles during late July and August. After several days, nonfeeding larvae hatch and are wind dispersed among trees and between stands. Dispersed first instars hide among bark scales, staminate flower bracts, and lichens where they spin overwintering shelters. In May they emerge, are again passively dispersed, and begin feeding on single needles or staminate flowers. As buds burst and shoots expand, the larvae molt and feed on succulent shoots. Here they continue feeding and spin a sheltering web. There are six instars. Pupation takes place in late June, early July. After about 10–14 days, the adult emerges to mate and lay eggs for the next generation.

An average female will lay 170 eggs. Highest death rates are experienced by the small dispersing larvae and the large larvae and adults (Morris 1963). Small larvae dispersal leads to 60–80% mortality, while the mortality of the survivors in the large larval category is more variable (40–90%).

It is suspected that adults do not migrate at low endemic levels, whereas female moths readily disperse during outbreaks (Greenbank 1956, 1973). Most moths fly on the second night after emergence and after depositing about 50% of their eggs. A moth may travel 150 mi. downwind. The median dispersal distance is about 25 mi. Thus, a local population can lose many moths through dispersal or gain many from an influx. Such movements contribute to abrupt changes in population size. Ranges of total increase of 4–10,000-fold in 6 yr have been recorded in New Brunswick.

Budworm outbreaks can be triggered by at least three phenomena operating singly or in combination:

1. Local survival and recruitment can gradually increase as forests age.
2. Immigrating moths can release populations from natural controls.
3. Weather can favorably affect both dispersal and local survival (Clark et al. 1979).

The first recorded outbreaks of SBW began in 1909 but tree-ring analyses have disclosed epidemics occurring at irregular intervals of 20–90 yr since the 1700s (Brown 1970, Blais 1965, 1968, Weed 1977). Recent outbreaks occurred in the late 1940s and mid-1970s. Because this insect is native to North America, it is assumed that outbreaks have been recurring for centuries (cf. Chapter 22).

The effects of SBW damage are cumulative. Since feeding is confined to buds and foliage, several successive years of feeding are required to produce full defoliation and tree mortality. Sustained infestations produce top-kill in balsam fir after 2–3 yr and dead trees in 3–5 yr. Spruces can sustain 7–10 yr of intensive defoliation before dying. The threat of fire over large geographic areas increases after a SBW outbreak due to fuel buildup (Prebble 1975). In the absence of fire a new forest regenerates, favoring balsam fir, with some older spruces having survived. Thus, the new forest contains about the same percentage of balsam fir as before. The result is a forest that can sustain a new budworm outbreak within 30–40 yr. Outbreaks occur irregularly, depending on forest composition, growth, and maturity. Thus, the SBW is the key element governing forest dynamics.

25.5.2 Spruce Budworm Control Tactics

Although calcium arsenate dusts were used earlier, aerial spraying of organic pesticides for budworm control did not begin until 1952. By 1951, most northern New Brunswick and Quebec forests had already experienced three years of moderate to severe defoliation. Hoping to prevent large areas of tree mortality, aerial spraying was begun (Miller & Kettela 1975, Blais et al. 1975). New Brunswick has conducted an aerial spraying program every year since, except for 1959, over areas ranging from 0.2–10 million acres. Quebec adopted a similar tactic in 1952, as did Maine in 1954. Owing to a more diverse economy, Ontario adopted an "occasional spraying" strategy in small incipient infestations or small, high-value stands, and sacrificed lower-value stands (Howse & Sippell 1975). Over the years a number of different organic insecticides have been employed (Prebble 1975, Anonymous 1976), as well as experimental use of *Bacillus thuringiensis*, viruses, insect growth regulators, and combinations of microbial and chemical insecticides.

Compared to most agricultural programs, pesticide use for SBW control is minute. Single applications range from 1.5–8 oz AI/acre, and dosages have decreased with time and type of pesticide used. These materials are usually applied once a year by air, although some areas receive multiple treatments, with dosages reduced proportional to the number of applications. Such repeated, reduced dosages provide increased pesticide efficacy. The same areas rarely receive two successive years of treatment (Prebble 1975, Weed 1977). During the outbreak of the mid-1970s, 10 million acres (of 17 million in total) in New Brunswick, 9 million acres (of 20 million total) in Quebec, and 3.5 million acres (of 8 million total) in Maine were sprayed using these methods.

In the past there generally have been two control alternatives in the face of a budworm outbreak: (1) spray with insecticide to prevent tree mortality or (2) let nature take its course and sacrifice the area, hoping to reduce the extensiveness of the outbreak. With the first alternative forests are sprayed in an outbreak—generally where >1 yr of intense defoliation would result in tree mortality. This provides protection of spruce and fir, but also maintains the budworm food source, resulting in a prolonged and more extensive outbreak (Blais 1974, Clark et al. 1979). Some contend that if spraying were completely stopped (second alternative), even more severe outbreaks would encompass several provinces and states, resulting in millions of acres of dead trees, as is currently occurring in northern Quebec. Initially, some material could be salvaged-logged, but most would rot in the woods. In much of eastern North America, industry would wait 30 yr or more for new raw material! Thus, areas dependent on pulpwood would experience a "boom and bust" economy rather than a continuous economic flow permitted, in part, through spraying.

Much criticism has been leveled at spraying operations. Both the public and practicing professionals have questioned the long-term cost effectiveness, human hazard, and environmental sensitivity to current spraying tactics (Blais 1974, Prebble, 1975, Irland 1977). More recently, several groups, including specially chosen task forces, have examined the ramifications of improved tactics and of designing new management strategies (Baskerville 1976, Stedinger 1977, Clark et al. 1979). As discussed hereafter, these efforts have sought to identify a better array of integrated tactics for management of the boreal forests and ones which could provide for greater economic stability in the future.

25.5.3 Analysis of Current and Potential Budworm Management Policies

Fortunately, the forest–budworm system has been the subject of extensive scientific investigation for over 30 yr (e.g., Morris 1963, Jennings et al. 1979). Recognizing this fact and the ecological significance of the problem, a group working at the University of British Columbia and the International Institute for Applied Systems Analysis used the spruce budworm–boreal system as a case study to develop an approach to "ecological policy design" (Clark et al. 1979). Beginning with an overview of the socioeconomic political system in which the problem is embedded, an ecological descriptive computer simulation model of the system was formulated. Next, a spectrum of policy prescriptions, ranging from proven practical to highly theoretical ones, were identified. Each formulated policy was evaluated according to variables relevant to decision making, including employment in the industry, flow of products from the mills, aesthetic quality of the forest, number of acres sprayed/yr, and so on. Finally, the problems of implementing appropriate policies were considered.

In the case study the overview considered the historical record of SBW in North America and the current means of control. The New Brunswick control program was chosen for the analysis since it provided 25 yr of field data on SBW populations, weather, management actions, control results, and forest harvest records. Also, many people in the province, with insights gained from years of field experience, had strong commitments to developing better solutions. Lastly, New Brunswick contains 0.5 million people who depend upon forest industries for economic well-being.

The dynamic computer simulation model incorporated variables of forest branch density, foliage condition, budworm density, and weather (Jones 1979). A time horizon was selected to encompass at least two complete outbreaks—100–150 yr. A time resolution of 1 yr corresponded to the budworm life cycle and the management decision cycle. Budworm moth dispersal distances of 40 km set the spatial scale of the analysis. Initial efforts modeled a 4.5 million ha area subdivided into 265 spatial units of about 17,000 ha each (Clark et al. 1979). Forest–insect interactions per unit were assumed to be homogeneous and the units connected through moth dispersal and management activities. Policies examined were combinations of budworm killing at different population densities, times, and locations and forest cutting (management) at different forest ages, times, and locations.

Unlike tactical models used for detailed research simulations, the budworm–forest model was a strategic regional planning or application model used to project over large spatial areas and long time periods. Owing to its strategic nature, validation differed from those methods used for microtactical models. Rather than perform a detailed fitting to a specific time and space series of data, a more general, qualitative approach was utilized. A three-step procedure was followed (Holling 1978, Clark, et al. 1979):

1. Did the simulation model mimic the general outbreak pattern of historical unmanaged forest, that is, a 30–60 yr outbreak cycle, with outbreaks radiating outward at about 15 km/yr from their "epicenters"?

2. Did the model reproduce the radically different province-wide pattern of system behavior observed in New Brunswick?

3. Without "tuning" the model to each specific region, did the simulation produce behaviors exemplified in other regions to the extent that consulting scientists were comfortable with the results (Clark & Holling 1979)?

In all three cases the answer was yes. While differing from a quantitative validation, much more rigorous demands were placed on the model than those required for fitting to a specific time and space series. This established confidence in the model so that policies could be explored in new regions of systems behavior.

The high dimensionality, nonlinearity of relationships and stochastic nature of the budworm–forest interaction dictated development of many new techniques of computer modeling and optimization. Rather than trying to discover *the* optimal policy, an array of probes into policy space were attempted. Such probes, using a range of alternative objectives, were employed in an iterative process of policy evaluation, modification, and design. Alternative policies were evaluated by generating socioeconomic resources and environmental indicators meaningful to a policy user. Two classes of indicators were used: those of immediate concern and those related to policy resilience and robustness. Examples of immediate concern are: profits to the logging industry, cost of insecticide spraying, unemployment rates and continuity, volume of wood killed by the budworm, damage due to visible defoliation, age class diversity of the forest, proportion of an area sprayed, and so on. Visual inspection of selected indicators provides a decision maker with a time stream of each (Fig. 25.2). Often, such visual inspection is sufficient to suggest a useful policy without more vigorous analysis (Clark et al. 1979). Assessment of resilience and robustness requires an analysis of policy sensitivity to unknown events. Each indicator is reevaluated with questions of "what happens in policy failure" or "how different will it be to change objectives after policy initiation." Robust policies allow for such changes without creating economic, environmental, or sociological chaos.

Throughout the analysis open communications were maintained with scientists, resource managers, and other potential users. This required development of communications packages to simplify complex technologies. Slide–tape programs conveyed the features of the problem, the philosophy of the model, and the consequences of different policies (Bunnell & Tait 1974, Bunnell 1976). Workshops with scientists and forest resource managers were held.

Resource management slide rules allowed exploration of combinations of spraying and cutting policies without requiring computer interrogation (Fig. 25.3). The slide rules consisted of response surfaces (topographic maps) of management indicators. The axes of each response surface is the same: the *x* axis represents the threshold of budworm density and tree damage when spraying for budworm control is initiated (hazard index). The *y* axis is the age of the forest when harvesting takes place. To evaluate a combination of spraying and cutting relative to a particular indicator, one reads the response surface graph for that indicator at the point on the graph where the hazard index and cutting age intersect. Thus, the proportion of years when spraying is done at a hazard index of 6 and a cutting age of 40 yr is 0.08 from the graph in Fig. 25.3. Further, the relative sensitivity (or stability) of a particular combination of management acts is indicated by the response surface slope. A "steep" slope indicates a relatively unstable combination sensitive to minor changes in hazard index and cutting age; a "flat" slope indicates a comparatively stable combination insen-

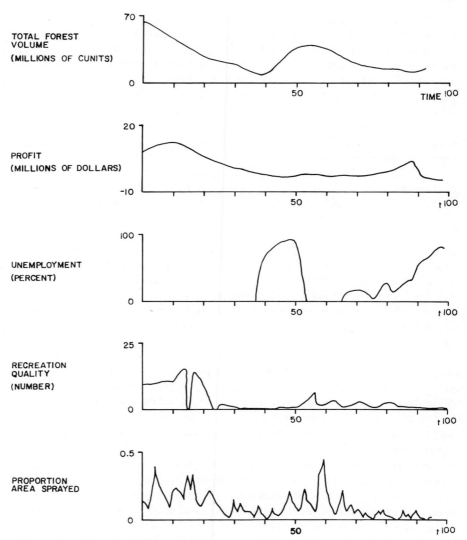

Figure 25.2   Evaluation indicators under historical budworm management. (From Holling et al. 1976.)

sitive to minor changes in management. The slide rules also allow for constructive dialogue and resolution of conflicts (Peterman 1977).

Responding to a request for a full analysis of budworm control options, a New Brunswick task force composed of private, public, and environmental personnel utilized these tools to explore policy alternatives (Baskerville 1976). They sought to: (1) maintain full employment in the forest industry, (2) improve the forest, and (3) reduce reliance on current spraying tactics. The analysis disclosed that there is no feasible policy alternative that

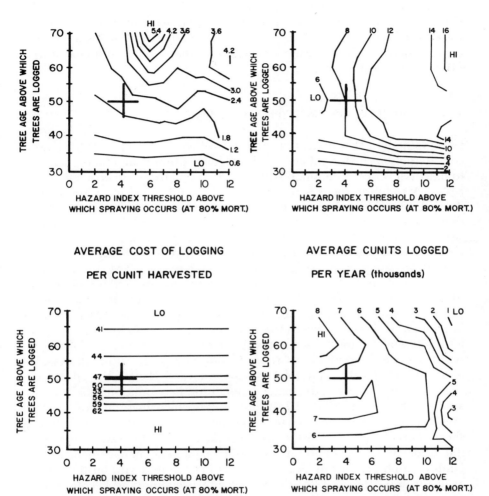

Figure 25.3 A resource management slide rule of the sort used in policy design dialogues with managers (modified from Peterman 1977). All axes are the same, representing two management acts which can be implemented at different levels. "Harvest Age" is the age at which a stand will be cut and "Hazard Index" is the threshold of insect abundance and tree damage when single annual sprays for budworm control are implemented. Each graph represents a single evaluation indicator of the sort shown. To evaluate a combination of spraying and cutting relative to a particular indicator, one reads the graph for that indicator at the point where the hazard index chosen and the cutting age chosen intersect. For example, the proportion of years when spraying is done at a hazard index of 6 and a cutting age of 40 yr is 0.08 from the graph. Further, the relative stability of a particular combination of management acts is seen by the slope of the graph for that combination. A "steep" slope indicates a relatively unstable combination sensitive to minor changes in hazard index and cutting age; a "flat" slope indicates a comparatively stable combination not dramatically effected by slight changes in management combinations.

does not violate at least one of these criteria. The task force, however, found the simulation and evaluation tools useful because "trial-and-error" decisions did not have to be made. Instead, an array of different policies were examined without risking real-world resources. Those tools provided development of policy-related conclusions that would influence future management actions.

25.6 SYNTHESIS OF PRINCIPLES FOR IPM

Development of IPM in the four pest control systems described above varies considerably in methodology and the degree of implementation achieved to date. All reflect a common strategic approach based on ecological principles. A common focus is on the ecosystem and the crop or animal production unit within which the pest management system is embedded. Rather than emphasizing incremental adjustments in tactics or minor technological issues, attention is on strategic design of programs and policies. Then tactical control methods to achieve the overall goals are sought.

With a strong emphasis on ecology in relation to IPM, the biological characteristics of the crop or animal production system and associated pest life systems become a focus for study. These attributes dictate which control strategy is most appropriate and how best to research the IPM system and implement it. Factors such as the weather, type of crop, mode of growth, longevity and fecundity of the pest, role of natural enemies, duration of pest life cycles, and modes of reproduction and potential for dispersal of both host plants (if applicable) and pests largely dictate the spatial and temporal dimensions of the management unit. Economic and social factors also determine many of the constraints under which a system must operate. We emphasize that the human organizational or institutional structures should complement the basic bioecological characteristics rather than being the primary determinants in the design of a pest management system. Undue emphasis on the latter forces artificial boundaries on the options available and often results in errors of interpretation and becomes counterproductive to cost-effective and environmentally sound pest control.

The biological characteristics, and especially the natural features, contributing to the control or regulation of a pest or pest complex are important in selecting which tactical options can be integrated for best continuous results. After detailed ecological study, including careful monitoring of physical factors, emphasis should be given maximum utilization of biological controls and host plant resistance features and to taking advantage of the catastrophic mortality (climatically or edaphically induced mortalities). This was well illustrated in the four examples presented here. Biological control of plant-feeding mites need only be supplemented occasionally with physiologically selective acaricides to readjust predator–prey ratios favoring the beneficial species. The majority of pests of cotton remain below economic levels due to natural control agencies, provided certain unfavorable environmental conditions do not occur or the natural enemies are not decimated by control factors such as broad spectrum pesticides. For those pests of apple and cotton which have no, or little, potential for natural control, detailed ecological research makes possible greatly improved timing and more effective and minimal use of chemical control agents. This point was also well illustrated in the cattle tick–pasture management. For the spruce budworm natural control or regulation is sufficient over long periods of time, during the endemic phases of population development. However, detailed ecological study and monitoring is necessary to understand and anticipate the outbreak phase and when the supplementing action of chemical treatments is essential to maintain the economy of the forest production system.

Lastly, for implementation the human organizational and operational components for IPM must be integrated with the biological complexities of the system to be managed. While the simplest form of system that will accomplish such synthesis is usually best, very often the demands for operations and communications require more complex systems and features as compared to more traditional means (e.g., unilateral chemical control) of pest control delivery. While these greater complexities are sometimes constraints to IPM implementation, they also reflect the greater ecological input into decision making and the substitution of information for pesticide use, which makes IPM such a desirable alternative to the more unilateral pest control approaches. In the future we must design more effective (and sometimes more complex) management systems based on ecological principles if we are to maintain our environment and a sustainable agriculture and forestry production. This book, *Ecological Entomology,* should help in achieving this.

REFERENCES

Adams, R. G. and R. Prokopy. 1977. *Mass. Fruit Note* **42**: 6–10.

Adkisson, P. L. 1969. *Conn. Agric. Exp. Sta. Bull.* **708**: 155–164.

Adkisson, P. L. 1972. *Proc. Tall Timbers Conf. Ecol. Anim. Control by Habitat Manag.* **4**: 175–188.

Adkisson, P. L. and J. C. Gaines. 1960. *Texas Agric. Exp. Sta., Misc. Publ. No. 444.*

Anonymous. 1975. *Cattle Tick in Australia.* Cattle Tick Control Commission Inquiry-Report 1973. Aust. Govern. Publ. Service, Canberra.

Anonymous. 1976. *Spruce Budworm Programs in Maine 1976–1981.* Maine Dept. Cons. For. Service.

Baskerville, G. (Task Force Leader). 1976. *Report of the Task Force for Evaluation of Budworm Control Alternatives.* Prepared for Cab. Comm. Econ. Dev. Prov. New Brunswick.

Bird, L. S. 1978. *Texas Agric. Exp. Sta., Leaflet, L-1672.*

Blais, J. R. 1965. *For. Sci.* **1**: 130–138.

Blais, J. R. 1968. *For. Chron.* **44**: 17–23.

Blais, J. R. 1974. *For. Chron.* **50**: 19–21.

Blais, J. R., P. Benoit, and R. Martineau. 1975. Pp. 113–125 in Prebble (Ed.), 1975, referenced here.

Blyth, E. J. and B. A. Croft. 1976. *Abstr. Proc. N.C. Branch, Entomol. Soc. Am.* **30**: 89.

Bottrell, D. G. and P. L. Adkisson. 1977. *Annu. Rev. Entomol.* **22**: 451–481.

Brazzel, J. R. 1961. *Texas Agric. Exp. Sta., Misc. Publ. 511.*

Brown, C. E. 1970. *Can. For. Ser., Dept. Fish. and For., Publ. No. 1263.*

Bunnel, P. 1976. *The Spruce Budworm: An Ecosystem Problem and a Modeling Approach to its Management.* Modular slide-tape presentation. U.B.C., Inst. Res. Ecol., Vancouver.

Bunnel, P. and D. Tait. 1974. *A Primer on Models: Why and How. A slide-tape presentation.* U.B.C., Inst. Res. Ecol., Vancouver.

Clark, W. C. and C. S. Holling. 1979. *Fortschr. Zool.* **25**: 29–52.

Clark, W. C., D. D. Jones, and C. S. Holling. 1979. *Ecol. Modelling* **7**: 1–53.

Colburn, R. and D. Asquith. 1971. *J. Econ. Entomol.* **64**: 1072–1074.

Croft, B. A. 1972. *Great Basin Natur.* **32**: 61–75.

Croft, B. A. 1982. Pp. 465–498 in R. L. Metcalf and W. H. Luckmann (Eds.), *Introduction to Pest Management.* 2nd Edition. Wiley, New York.

Croft, B. A. 1977. Pp. 377–393 in D. L. Watson and A. W. A. Brown (Eds.), *Pesticide Management and Insecticide Resistance.* Academic, New York.

Croft, B. A. 1983. Pp. 1–20 in Croft and Hoyt (Eds.), 1983, referenced here.

Croft, B. A. In *Proc. 4th USA/USSR Conf. on Integrated Pest Control, Yalta, USSR (1978).*

Croft, B. A. and A. W. A. Brown. 1976. *Annu. Rev. Entomol.* **20**: 285–335.

Croft, B. A., J. L. Howes, and S. M. Welch. 1976. *Environ. Entomol.* **5**: 20–34.

Croft, B. A. and S. A. Hoying. 1977. *Can. Entomol.* **109**: 1025–1034.

Croft, B. A. and S. C. Hoyt (Eds.). 1983. *Integrated Management of Insect Pests of Pome and Stone Fruits.* Wiley, New York.

Croft, B. A., D. J. Miller, S. M. Welch, and M. J. Marino. 1979. Pp. 223–250 in D. J. Boethel and R. E. Eikenbary (Eds.), *Pest Management of Tree Fruit and Nut Crops.* Plenum, N.Y.

Croft, B. A., R. L. Tummala, H. W. Riedl, and S. M. Welch. 1976. Pp. 97–119 in R. L. Tummala, D. L. Haynes, and B. A. Croft (Eds.), *Modeling for Pest Management.* Michigan State University Press, East Lansing.

Croft, B. A., S. M. Welch, and M. J. Dover. 1976. *Environ. Entomol.* **5**: 227–234.

DeBach, P. and R. A. Sundby. 1963. *Hilgardia* **34**: 105–166.

Doutt, R. L. and J. Nakata. 1965. *Calif. Agric.* **19**: 3.

Downing, R. S. and J. D. Aarand. 1968. *Integrated Control of Orchard Mites in British Columbia.* B.C. Agric. Entomol. Branch Publ.

Edens, T. C. and K. Klonsky. 1977. *The Pest Management Executive System (PMEX): Description and Economic Summary.* Michigan Agric. Exp. Sta. Memo.

Elder, J. K., F. R. Emmerson, J. F. Kearnan, K. S. Waters, G. H. Dunwell, R. S. Morris, and S. G. Knott. 1980. *Aust. Vet. J.* **56**: 212–218.

Elder, J. K., J. F. Kearnan, K. S. Waters, G. H. Dunwell, R. S. Morris, and S. G. Knott. 1980. *Aust. Vet. J.* **56**: 219–223.

Francis, J. and D. A. Little. 1964. *Aust. Vet. J.* **40**: 247–253.

Franz, H. G. 1974. *The Functional Response to Prey Density in an Acarine System.* Sim. Mono. Center for Agric. Publ. and Doc. Wageningen.

Frisbie, R. E. 1977. *The Texas Statewide Pest Management Program.* Beltwide Cotton Production—Mech. Conf. (January, 1977), Atlanta, Ga (unpublished paper).

Frisbie, R. E., J. M. Sprott, R. D. Lacewell, R. D. Parker, W. E. Buxkemper, W. E. Bagley, and J. W. Norman, Jr. 1976. *J. Econ. Entomol.* **69:** 211–214.

Frisch, J. E. and J. E. Vercoe. 1978. *World Anim. Rev.* **25:** 8–12.

Gaylor, M. J. and W. L. Sterling. 1976. *Environ. Entomol.* **5:** 55–58.

Greenbank, D. O. 1956. *Can. J. Zool.* **34:** 453–476.

Greenbank, D. O. 1973. *Can. Dept. Env. For. Serv., Rept. M-S-39.*

Harley, K. L. S. and P. R. Wilkinson. 1964. *Aust. J. Agric. Res.* **15:** 841–853.

Hartstack, A. W., J. L. Henson, J. A. Witz, J. A. Jackman, J. P. Hollingsworth, and R. E. Frisbie. 1977. *Proc. Beltwide Cotton Prod. Res. Conf. (1977)*, pp. 151–153.

Holling, C. S. (Ed.). 1978. *Adaptive Environmental Assessment and Management.* Wiley, London.

Holroyd, R. G. and P. J. Dunster. 1978. *Proc. Aust. Soc. Anim. Prod.* **12:** 277.

Howse, G. M. and W. L. Sippell 1975. Pp. 85–93 in Prebble (Ed.), 1975, referenced here.

Hoyt, S. C. 1969. *J. Econ. Entomol.* **62:** 74–86.

Hoyt, S. C. 1970. *Annu. Entomol. Soc. Am.* **63:** 1382–1384.

Hoyt, S. C. and E. C. Burts. 1974. *Annu. Rev. Entomol.* **19:** 231–252.

Hoyt, S. C. and L. E. Caltagirone. 1971. Pp. 394–421 in C. B. Huffaker (Ed.), *Biological Control.* Plenum, New York.

Hoyt, S. C. and J. O. Gilpatrick. 1976. Pp. 133–147 in J. L. Apple and R. F. Smith (Eds.), *Integrated Pest Management.* Plenum, New York.

Huffaker, C. B. and B. A. Croft. 1976. *Environ. Health Perspectives* **14:** 167–183.

Huffaker, C. B. and C. E. Kennett. 1956. *Hilgardia* **37:** 283–335.

Huffaker, C. B. and P. S. Messenger (Eds.). 1976. *Theory and Practice of Biological Control.* Academic, New York.

Huffaker, C. B., J. A. McMurtry, and M. van de Vrie. 1970. *Hilgardia* **40:** 391–458.

Huffaker, C. B., M. van de Vrie, and J. A. McMurtry. 1969. *Annu. Rev. Entomol.* **14:** 125–174.

Irland, L. C. 1977. *Univ. Maine School Forest Res.,* Tech. Note No. 67.

Jennings, D. T., F. B. Knight, S. C. Hacker, and M. E. McKnight. 1979. *Univ. Maine, Life Sci. Agric. Expt. Sta.,* Misc. Rept. 213.

Jones, D. D. 1979. Pp. 91–159 in G. A. Norton and C. S. Holling (Eds.), *Pest Management.* Pergamon, Oxford.

Levins, R. and M. Wilson. 1980. Ecological theory and pest management. *Annu. Rev. Entomol.* **25:** 287–308.

McMurtry, J. A., C. B. Huffaker, and M. van de Vrie. 1970. *Hilgardia* **40:** 331–390.

McKnight, M. E. 1968. *USDA For. Serv. Res. Pap. Rm-44.*

Mattson, W. J. and N. Addy. 1975. *Science* **190:** 522.

Miller, C. A. and E. G. Kettela. 1975. Pp. 94–112 in Prebble (Ed.), 1975, referenced here.

Mitchell, E. R. 1977. *Proc. Int. Controlled Release Pesticide Symp., Corvallis, Or.* Oregon State Univ. Agric. Exp. Sta. Publ., pp. 41–67.

Morris, R. F. (Ed.). 1963. *Mem. Entomol. Soc. Can.* **31:** 1–332.

Namken, L. N. and M. D. Heilman. 1973. *Agron. J.* **65:** 953–956.

Nemec, S. J. and P. L. Adkisson. 1969. *Tex. Agric. Exp. Sta.,* PR 2674.

Nolan, J. and W. J. Roulson. 1979. Pp. 3–13 in J. G. Rodriguez (Ed.), *Recent Advances in Acarology,* Vol. 1. Academic, New York.

Norris, K. R. 1957. *Aust. J. Agric. Res.* **8:** 768–787.

Parker, F. D., F. R. Lawson, and R. E. Pinnel. 1971. *J. Econ. Entomol.* **64:** 721–735.

Parker, R. D. 1978. *Interactions of Cotton Genotypes with Boll Weevil, Anthonomus grandis Boheman, Populations as Influenced by Planting Pattern, Irrigation Level and Insecticide Treatment.* Ph.D. dissertation, Texas A&M University, College Station.

Peterman, R. M. 1977. *Ecol. Modelling* **3:** 133–148.

Pickett, A. D., W. L. Putnam, and E. S. LeRoux. 1958. *Proc. 10th Int. Congr. Entomol. (Montreal, 1956).* **2:** 169–174.

Prebble, M. L. (Ed.). 1975. *Aerial Control of Forest Insects in Canada.* Env. Canada, Ottawa.

Rabb, R. L. and J. R. Bradley, Jr. 1968. *J. Econ. Entomol.* **50:** 778–784.

Rabbinge, R. 1976. *Biological Control of the Fruit-Tree Red Spider Mite.* Sim. Mono. Center for Agric. Publ. and Prot., Wageningen.

Reynolds, H. T., P. L. Adkisson, and R. F. Smith. 1975. Pp. 397–443 in R. L. Metcalf and W. Luckman (Eds.), *Introduction To Pest Management.* Wiley, New York.

Riek, R. F. 1962. *Aust. J. Agric. Res.* **13:** 532–550.

Roberts, J. A. 1968. *J. Parasitol.* **54:** 657–662.

Santos, M. A. 1976. *Ecology* **57:** 390–394.

Seifert, G. W. 1971a. *Aust. J. Agric. Res.* **22:** 159–168.

Seifert, G. W. 1971b. *Aust. J. Agric. Res.* **22:** 839–850.

Snowball, G. J. 1957. *Aust. J. Agric. Res.* **8:** 394–413.

Sprott, M. J., R. D. Lacewell, G. A. Niles, J. K. Walker, and J. R. Gannaway. 1976. *Texas Agric. Exp. Sta. MP 1250C.*

Stedinger, J. 1977. *Spruce Budworm Management Models.* Ph.D. dissertation, Harvard University, Cambridge, Mass.

Sterling, W. L. and P. L. Adkisson. 1971. *Texas Agric. Exp. Sta. MP 933.*

Sterling, W. L. and R. L. Haney. 1973. *Texas Agric. Exp. Sta. and Ext. Serv., Texas Agric. Prog.* **19:** 4–7.

Stern, V. M., R. F. Smith, R. van den Bosch, and K. S. Hagen. 1959. *Hilgardia* **29:** 81–101.

Sutherst, R. W. and M. J. Dallwitz. 1979. *Proc. 4th Int. Congr. Acarol., 1974,* pp. 442–446.

Sutherst, R. W., G. A. Norton, N. D. Barlow, G. R. Conway, M. Birley, and H. N. Comins. 1979. *J. Appl. Ecol.* **16:** 359–382.

Sutherst, R. W., G. A. Norton, and G. F. Maywald. 1980. Pp. 48–56 in M. G. Cooper and L. A. Y. Johnston (Eds.), *Ticks and Tick-Borne Diseases.* Aust. Vet. Assoc., Sydney, N.S. Wales, Australia.

Sutherst, R. W., I. D. Sutherland, and G. F. Maywald. 1979. *Proc. 2nd Int. Sym. Vet. Epidem. and Econ., 1979.* pp. 408–515.

Sutherst, R. W. and K. B. W. Utech. 1980. Pp. 385–407 in D. Pimental (Ed.), *Handbook of Pest Management in Agriculture,* Vol. 2. CRC Press, West Palm Beach, Fla.

Sutherst, R. W., K. B. W. Utech, J. D. Kerr, and R. H. Wharton. 1979. *J. Appl. Ecol.* **16:** 397–404.

Sutherst, R. W. and R. H. Wharton. 1973. *Proc. 3rd Int. Congr. Acarol., 1971.* pp. 797–801.

Sutherst, R. W., R. H. Wharton, I. M. Cook, I. D. Sutherland, and A. S. Bourne. 1979. *Aust. J. Agric. Res.* **30:** 353–368.

Sutherst, R. W., R. H. Wharton, and K. B. W. Utech. 1978. *Aust. CSIRO Div. Entomol., Tech. Paper No. 14.*

Talpaz, H. and R. E. Frisbie. 1975. *S. J. Agric. Econ.* (December): 19–25.

Tanigoshi, L. K., S. C. Hoyt, R. W. Browne, and J. A. Logan. 1975a. *Annu. Entomol. Soc. Am.* **68:** 972–978.

Tanigoshi, L. K., S. C. Hoyt, R. W. Browne, and J. A. Logan. 1975b. *Annu. Entomol. Soc. Am.* **68:** 979–986.

Tummala, R. L., D. L. Haynes, and B. A. Croft (Eds.). 1976. *Modeling for Pest Management.* Michigan State University, East Lansing.

Turner, H. G. 1975. *World Anim. Rev.* **13:** 16–21.

Turner, H. G. and A. J. Short. 1972. *Aust. J. Agric. Res.* **29:** 411–422.

Utech, K. B. W., G. W. Seifert, and R. H. Wharton. 1978. *Aust. J. Agric. Res.* **29:** 411–422.

Utech, K. B. W., R. H. Wharton, and J. D. Kerr. 1978. *Aust. J. Agric. Res.* **29:** 885–895.

van den Bosch, R. and A. D. Telford. 1964. Pp. 459–488 in P. DeBach (Ed.), *Biological Control of Insect Pests and Weeds.* Reinhold, New York.

van den Bosch, R., O. Beingolea G., M. Hafez, and L. A. Falcon. 1976. Pp. 443–456 in C. B. Huffaker and P. S. Messenger (Eds.), 1976, referenced here.

Wagland, B. M. 1975. *Aust. J. Agric. Res.* **26:** 1073–1080.

Walker, J. K. and A. Niles. 1971. *Texas Agric. Exp. Sta. Bull. 1109.*

Weed, D. 1977. *Spruce Budworm in Maine, 1910–1976, Infestations and Control.* Maine Dept. Cons., For. Serv.

Welch, S. M. 1979. Pp. 31–40 in *Recent Adv. Acarol. Vol. I,* Academic, New York.

Welch, S. M., B. A. Croft, J. F. Brunner, M. J. Dover, and A. L. Jones. 1980. *EPPO Plant Prot. Bull.* **10:** 259–268.

Wharton, R. H., K. L. S. Harley, P. R. Wilkinson, K. B. Utech, and B. M. Kelley. 1969. *Aust. J. Agric. Res.* **20:** 783–797.

Wharton, R. H. and K. R. Norris. 1979. *Vet. Parasitol.* **6:** 135–164.

Wharton, R. H. and W. J. Roulson. 1970. *Annu. Rev. Entomol.* **15:** 381–404.

Wharton, R. H. and K. B. W . Utech. 1970. *J. Aust. Entomol. Soc.* **9:** 171–182.

Wharton, R. H., K. B. W. Utech, and R. W. Sutherst. 1973. *Proc. 3rd Int. Congr. Acarol., 1971,* pp. 697–700.

Wharton, R. H., K. B. W. Utech, and H. G. Turner. 1970. *Aust. J. Agric. Res.* **21:** 163–181.

Wilkinson, P. R. 1957. *Aust. J. Agric. Res.* **8:** 414–423.

Wilkinson, P. R. 1962. *Aust. J. Agric. Res.* **13:** 974–983.

INDEX

The assistance of Dr. Linda S. Durston is greatly appreciated and acknowledged.